International Symposium on Vacuum Science & Technology and Its Application for Accelerators

(IVS 2012)

Journal of Physics: Conference Series Volume 390

Kolkata, West Bengal, India
15-17 February 2012

ISBN: 978-1-62276-853-0
ISSN: 1742-6588

Printed from e-media with permission by:

Curran Associates, Inc.
57 Morehouse Lane
Red Hook, NY 12571

Some format issues inherent in the e-media version may also appear in this print version.

Printed by Curran Associates, Inc. (2013)

For permission requests, please contact the Institute of Physics
at the address below.

Institute of Physics
Dirac House, Temple Back
Bristol BS1 6BE UK

Phone: 44 1 17 929 7481
Fax: 44 1 17 920 0979

techtracking@iop.org

Additional copies of this publication are available from:

Curran Associates, Inc.
57 Morehouse Lane
Red Hook, NY 12571 USA
Phone: 845-758-0400
Fax: 845-758-2634
Email: curran@proceedings.com
Web: www.proceedings.com

TABLE OF CONTENTS

PREFACE

The Indian Vacuum Society (IVS) was established in 1970 to promote vacuum science and technology in academic, industrial and R&D institutions in India. IVS is a member society of the International Union for Vacuum Science, Technique and Applications (IUVSTA). It has organized International and national symposia, short term courses and workshops on different aspects of Vacuum Science and Technology at regular intervals. So far 27 National symposia, 4 International Symposia and 47 courses have been organized at various locations in India. There has been an active participation from R&D establishments, universities and Indian industries during all these events.

In view of the current global situation and emerging trends in vacuum technology, the executive committee of the IVS suggested to us that we organize an International Symposium at the Variable Energy Cyclotron Centre, Kolkata from 15–17 February 2012. At the Variable Energy Cyclotron Centre we have a large number of high vacuum systems used in the K130 Cyclotron and K500 Superconducting Cyclotron. Also a large cryogenic system using LHe plant is in operation for cryopanels and a superconducting magnet for K-500 Cyclotron. The main areas covered at the symposium were the production and measurement of vacuums, leak detection, design and development of large vacuum systems, vacuum metallurgy, vacuum materials and the application of high vacuums in cyclotrons, LINACS and other accelerators. This symposium provided an opportunity for interaction between active researchers and technologists and allowed them to review the current situation, report recent experimental results, share the available expertise and consider the future R&D efforts needed in this area.

Keeping the industrial significance of vacuum technology in mind, an exhibition of the vacuum related equipment, accessories, products etc by various suppliers and manufactures was organized alongside the symposium. Participation by a large number of exhibitors clearly indicates that industry has advanced quite significantly. During the symposium, the Indian Vacuum Society honoured two distinguished personalities for their remarkable and significant contributions to the field of vacuum science and development of technology in the country. Awards were presented for both oral and poster papers during the symposium. A committee evaluated the scientific content and clarity of presentation of contributed papers. We believe that deliberations and discussions at the symposium will help gain a better understanding of the complicated and involved technology of vacuum science and be of benefit to scientists and technologists.

Subimal Saha Convener
Gautam Pal Co-Convener
V S Pandit Secretary
Surajit Pal Treasurer

International Advisory Committee		National Advisory Committee	
S Banerjee	*DAE/India*	R K Bhandari (Chairman)	*VECC*
Rockett Angus	*AVS/USA*	D L Bandyopadhyay	*IVS*
A V Dadve Cdr	*Pfeiffer Vac /India*	S B Bhatt	*IPR*
M Barma	*TIFR/India*	K G Bhushan	*BARC*
R K Bhandari	*VECC/India*	Alok Chakrabarti	*VECC*
R C Budhani	*NPL, India*	D P Chakravarty	*BARC*
Shekhar Chander	*CEERI/India*	Tushar Desai	*Mumbai Univ*
S C Chetal	*IGCAR/India*	R Dey	*VECC*
K L Chopra	*IIT Delhi/India*	S C Gadkari	*BARC*
Christian Day	*KIT/Germany*	S K Gupta	*IUVSTA/India*
Kraemer Dieter	*FAIR/Germany*	Shrikrishna Gupta	*BARC*
L M Gantayat	*BARC/India*	Rajendra Jathar	*Agilent Technologies*
R B Grover	*DAE, BARC/India*	S N Joshi	*CEERI*
P D Gupta	*RRCAT/India*	D Kanjilal	*IUAC*

Szajman Jakub	*VSA/Australia*	C Mallik	*VECC*
R N Jayaraj	*NFC/India*	S G Markandeya	*BRNS*
S Kailas	*BARC/India*	K C Mittal	*BARC*
P K Kaw	*IPR/India*	S Nagarjun	*HHV Bangalore*
Lalit Kumar	*MTRDC/India*	K G M Nair	*IGCAR*
Jean Larour	*Ecole/France*	Gautam Pal (Co-convener)	*VECC*
Marminga Lia	*TRIUMF/Canada*	Surajit Pal (Treasurer)	*VECC*
Shekhar Mishra	*Fermilab/USA*	V S Pandit (Secretary)	*VECC*
Ganapatirao Myneni	*Jlab/USa*	R G Pillay	*TIFR*
S V Narasaiah	*HHV/India*	Mohan Pradeep	*NPL*
K Radhakrishnan	*ISRO/India*	Y Ranga Rao	*Vac Techniques*
A S Raja Rao	*IVS/India*	R Ranganathan	*SINP*
T Ramasami	*DST/India*	Subimal Saha (Convener)	*VECC*
A K Ray	*BARC/India*	T K Saha	*BARC*
R Reid	*IUVSTA/UK*	Vikrant Sanglikar	*Edwards India*
Amit Roy	*IUAC/India*	D Sarkar	*VECC*
Milan Sanyal	*SINP/India*	Y C Saxena	*IPR*
V K Saraswant	*DRDO/India*	S K Shukla	*RRCAT*
E Schamiloglu	*UNM/USA*	Gurnam Singh	*RRCAT*
R K Sinha	*BARC/India*	P Singh	*BARC*
P Strubin	*CERN/Switzerland*	A K Sinha	*IUC-DAEF*
T Okano	*VSJ/Japan*	S K Thakur	*VECC*

Local Organizing Committee

Dr R K Bhandari	Shri Subimal Saha
Chairman	Co-chairman

Scientific Programme

1. Dr V S Pandit, Convener
2. Dr K C Mittal
3. Shri S K Gupta
4. Shri Gautam Pal
5. Dr Arup Bandyopadhyay
6. Shri Anjan Duttagupta

Reception & Registration

1. Shri C Mallik, Convener
2. Shri P Y Nabhiraj
3. Shri Manas Dutta
4. Kum Ranjini Menon
5. Shri Malay Kanti Dey
6. Shri Samit Bandyopadhyay

7. Shri Chinmay Nandi
8. Shri Anindya Roy
9. Shri R C Yadav

Transport & Accommodation

1. Shri N V S V Prasad, Convener
2. Shri S K Thakur
3. Shri Sumantra Bhattacharyya
4. Shri Debjit Gupta
5. Shri S R Gupta
6. Shri Jayanta Sur
7. Shri Sujit Saha

Auditorium

1. Shri Gautam Pal, Convener
2. Shri Tamal Bhattacharyya
3. Shri Tanmay Das
4. Smt. Seema Bhattacharyya, SINP
5. Shri Susanta Chakroborti, SINP
6. Shri S C Jena

Exhibition/Souvenir

1. Shri R. Dey, Convener
2. Smt Seema Bhattacharyya, SINP
3. Shri S K Thakur
4. Shri N Dutta
5. Shri S K Pati
6. Shri Yashwant Kumar

Website

1. Shri Tapas Samanta, Convener
2. Dr Surajit Pal
3. Shri Gaurav Saxena

7. Miss. Swantana Kumari
8. Smt Sudeshna Seth
9. Shri Anirban De

Publication

1. Shri Subimal Saha, Convener
2. Dr V S Pandit
3. Shri C D Dutta
4. Dr Tapas Bandyopadhyay
5. Dr Vaishali Naik
6. Shri Anirban De

Catering

1. Dr Arup Bandyopadhyay, Convener
2. Shri Asis Polley
3. Shri S. Chattopadhayay
4. Shri Debjit Gupta
5. Shri Tanmay Das
6. Shri R L Singh

Finance

1. Shri S Sambath, Convener
2. Dr Surajit Pal
3. Shri Asis Dey
4. Shri V K Khare

Cultural

1. Dr Alok Chakrabarti, Convener
2. Dr Vaishali Naik
3. Shri Dirtha Sanyal
4. Shri Suman Guha

Indian Vacuum Society

The Indian Vacuum Society (IVS) was established in 1970. It has over 900 members including many from Industry and R&D Institutions spread over the country. The society has an active chapter at Kolkata. The society was formed with the main aim to promote, encourage and develop the growth of Vacuum Science, Techniques and Applications in the country. In order to achieve this aim it has been conducting a number of short term courses at graduate and technician levels on vacuum science and technology on topics ranging from low vacuum to ultrahigh vacuum. So far it has conducted 47 such courses in different parts of the country and imparted training to more than 1500 people in the field. Some of these courses were in-plant training courses conducted on the premises of the establishment and designed to take care of the special needs of the establishment. Recently such a course was conducted at the Nuclear Fuel Complex, Hyderabad and Fundamentals of Vacuum Technology in general and Large Vacuum Furnaces, Vacuum Metallurgy in particular were the themes of the workshop. IVS also regularly conducts national and international seminars and symposia on vacuum science and technology with special emphasis on themes related to applications of vacuum. A large number of delegates from all over India take part in the deliberations of such seminars and symposia and present their work. IVS also arranges technical visits to different industries and research institutes. The society also helped in the UNESCO sponsored post-graduate level courses in vacuum science, technology and applications conducted by Mumbai University. IVS extended its support in standardizing many of the vacuum instruments and played a vital role in helping to set up a Regional Testing Centre along with BARC. As part of the development of vacuum education, the society arranges the participation of expert members on the subject to deliver lectures and take part in devising courses in the universities.

IVS has published the 'Bulletin of Indian Vacuum Society' quarterly since its inception, in which articles on Vacuum and related topics are published. NIRVAT, news, announcements, and reports are the other features of the Bulletin. The articles in the Bulletin are internationally abstracted. The Bulletin is distributed free to all members of the society. The society also publishes the proceedings of National, International Symposia/Seminars, manuals, lecture notes etc. It has published 'Vacuum Directory' containing very useful information on vacuum technology and it is in the process of updating this. The IVS has also set up its own website (http://www.ivsnet.org) in January 2002. The website contains information about IVS, lists of EC members, events and news, abstracts of articles published in the Bulletin of Indian Vacuum Society, utilities, technical data, announcements, reports, membership and other forms including advertisements. Our Society is a member society of the International Union of Vacuum Science, Techniques and Applications (IUVSTA) and many IVS members are nominated in its various committees since 1970. In 1983, IVS conducted an International Symposium on Vacuum Technology and Nuclear Applications in BARC, Mumbai under the sponsorship of IUVSTA. In 1987, IVS arranged, the Triennial International Conference on Thin Films in New Delhi, where more than 200 foreign delegates participated. IVS also hosted the IUVSTA Executive Council Meeting along with the conference. The society organized yet again an International Conference on Vacuum

Science and Technology (IVS-2007) during 28-30 November 2007 at TIFR, Colaba Camups, Mumbai.

IVS organizes the prestigious Professor Balakrishnan Memorial Lecture in memory of its founder vice-president. Leading scientists in the field from India and abroad are invited to deliver the talk. So far 24 lectures have been held in this series. IVS has instituted the 'IVS- Professor D Y Phadke Memorial Prize' in memory of our founder President the late Professor D Y Phadke in the University of Mumbai. The prize is given every year to the top ranker in the MSc (Physics) examination conducted by the Mumbai University. IVS Kolkata Chapter has established the Dr A S Divatia Memorial Trust with the objective to organize the Dr A S Divatia Memorial Lecture and a seminar once a year and to set up a vacuum testing and calibration facility. IVS has instituted an award in memory of late Shri C. Ambasankaran, its past president and pioneer of vacuum technology in India. This award is given to one of the best papers presented in the national symposium conducted by IVS. One more best Poster award 'Smt Shakuntalabai Vyawahare Memorial Prize' is established from the donation given by Shri. Mohan R. Vyawahare, a life member, in memory of his mother.

Recently, The IVS has decided to offer financial support to cover travel, registration fees and accommodation charges for a few selected students and participants (preferably IVS life members) for attending conferences in the country.

Shrikrishna Gupta
Gen. Secretary, IVS

Peer review statement

All papers published in this volume of *Journal of Physics: Conference Series* have been peer reviewed through processes administered by the proceedings Editors. Reviews were conducted by expert referees to the professional and scientific standards expected of a proceedings journal published by IOP Publishing.

International Symposium on Vacuum Science & Technology and its Application for Accelerators IOP Publishing
Journal of Physics: Conference Series **390** (2012) 012001 doi:10.1088/1742-6596/390/1/012001

Vacuum System of the Large Cyclotrons at VECC

G. Pal, C. Mallik, R.C. Yadav, J. Akhter, A. Datta Gupta, B. Mandal, A. Roy, A. Polley, M. Datta, C. Nandi, A. Sarkar, Srimantra Bhattacharyya, Sarbajit Pal, and R.K. Bhandari

Variable Energy Cyclotron Centre, I/AF Bidhannagar, Kolkata 700064, INDIA

Email: gautam.pal@vecc.gov.in

Abstract: Variable Energy Cyclotron Centre (VECC) has two large cyclotrons, K-130 cyclotron and K-500 cyclotron. The first beam in the room temperature K-130 cyclotron (RTC) was accelerated in June 1977. The cyclotron accelerated and delivered alpha and proton beams consistently to the cyclotron users for several years. Heavy ion beams were available in this cyclotron from 1997 to 2007. Presently, the cyclotron is working as a primary source for RIB production. The cyclotron has an acceleration chamber volume of about 28 m^3. The total length of beam line is about 65 m. Vacuum of the order of 1 x 10^{-6} mbar is presently maintained in the cyclotron and beam line using diffusion pumps. It is one of the largest vacuum systems operating in India. It is consistently being operated 24 x 7 round the year giving beam to the cyclotron users. A K-500 superconducting cyclotron (SCC) with K=520 has been constructed at Kolkata. SCC will be used to accelerate beams to 80 MeV/A for light heavy ions and about 10 MeV/A for medium mass heavy ions. Three turbo molecular pumps are connected to the acceleration chamber. Three cryopanels placed inside the lower dees in the valley gap of the superconducting magnet are available in the accelerating chamber for achieving high vacuum. The acceleration chamber having a volume of about 1.0 m^3 was operated using turbomolecular pumps, liquid nitrogen cooled panels and liquid helium cooled cryopanels at different stages during beam commissioning. Differential pumping is provided across the RF liner to avoid distortion. The first beam line of about 21 m has been installed in the cyclotron. The outer vacuum chamber of the cyclotron magnet cryostat has active pumping. The vacuum system of the superconducting cyclotron is also operating reliably round the clock throughout the year. The paper describes the details of the vacuum systems of the large cyclotrons at VEC Centre Kolkata India, its commissioning and operating experience.

1. Introduction

A K-130 cyclotron [1] and a K-500 cyclotron [2] are being operated at VEC Centre. RTC has a 224 cm water cooled magnet of about 262 Ton. The magnetic field is 2.12 T at the hill and 1.65 T at the valley. The radio frequency system operates at a frequency from 5.5 to 16.5 MHz with Dee voltage of about 70 KV. The beam chamber has a volume of about 28 m^3. It presently uses an internal PIG ion source and uses an electrostatic deflector for extraction of beam from the cyclotron. It is one of the largest vacuum systems operating in India. It is consistently being operated 24 x 7 round the year giving alpha and proton beams to the cyclotron users.

SCC has a superconducting magnet. The magnet uses NbTi wire immersed in a pool of liquid helium. The cold mass is about 7 Ton. The 80 Ton magnet iron is kept at room temperature. The radio frequency system operates at a frequency of 9 to 27 MHz. SCC will be used to accelerate heavy ion beams to energy of upto 80 MeV/A for light heavy ions and about 10 MeV/A for medium mass heavy ions. SCC is also operating reliably round the clock throughout the year with internal beam for commissioning trials

2. K-500 superconducting cyclotron

The major components in the vacuum system of SCC [3] are beam chamber, liner, injection line, external beam line, magnet cryostat and cryogenic system.

2.1 Beam chamber

The beam chamber is enclosed by the radio frequency liner at the top and bottom and the inner wall of the main magnet cryostat at the sides. An o-ring of diameter 1.3 m seals the interface of the liner and cryostat. The beam chamber has a volume of about 1.0 m^3. Figure 1 shows part of the beam chamber. The material for liner is copper CDA 101, the inner wall of the main magnet cryostat is made of iron, which has been elctroless nickel plated. The median plane of the cryostat forms part of the beam chamber. It is made of SS 316. The dee and the extraction elements are major components in the beam chamber. The dee is made of copper CDA 101, the magnetic channels from iron, the electrostatic deflectors from titanium. Table -1 gives the surface area of different components

Fig. 1. Main magnet cryostat assembled on the lower radio frequency liner

Table 1. Components of SCC Beam Chamber

Name of component	Material	Surface area m^2
Inner wall of main magnet cryostat	AISI 1010 Nickel plated	5.6 approx
Radio frequency liner	CDA 102 and CDA 110 copper	20.7 approx
Dee	CDA 101 copper	6.4 approx
Dee Stem	CDA 101 copper	6.0 approx

The median plane of the beam chamber (Fig. 1) has twenty penetrations having complex geometries. Two probes move in the beam chamber. They are equipped with maintenance chambers, isolation valves and independent pumping modules for connecting them to the high vacuum beam chamber. The radial probe has a drive of about 2.1 m and the bore probe has a drive of about 1.8 m [4]. The probes are connected to edge welded bellows to ensure leak tightness during movement and reliable operation of the probe. The maintenance chamber of the radial probe was modified to place a DN 63 pump below the maintenance chamber. It is equipped with a glass view port for viewing the probe.

There are three ports of 75 mm diameter in the magnet for connecting the beam chamber to positive displacement pumps. Three 550 litre/sec DN 160 turbo-molecular pump backed by scroll pumps have been connected to the beam chamber. The turbomolecular can tolerate an axial magnetic field of a 10 mT and a transverse magnetic field of 5 mT. During the magnetic field mapping of the superconducting cyclotron the axial magnetic field was also measured. Based on the results of the magnetic field measurement, it was decided to place the pumps at a elevation difference of 2.85 m from the magnet with piping of length 3.5 m so that the pump is placed in an acceptable magnetic field.

The cyclotron vacuum was initially operated with turbo molecular pumps. Considerable time of taken to achieve high vacuum for operation of the radio frequency system. The ultimate of 1 x 10^{-7} was achieved after about a month of operation and baking using the radio frequency system. The ceramic insulator of the radio frequency system had frequent failures. After achieving good vacuum in the beam chamber, internal beam was viewed in the cyclotron (Fig. 2) on the scintillation screen of the bore probe [4] (Fig. 3).

Fig. 2. Accelerated Neon beam spot on beam viewer

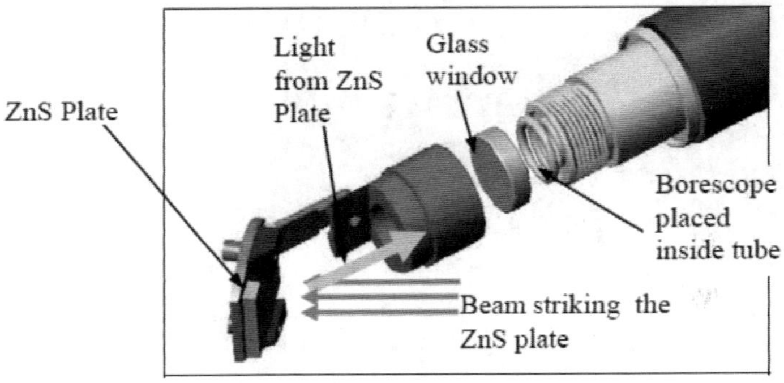

Fig. 3. Beam scintillation probe

Dee voltage, however, could not be raised because the beam chamber vacuum deteriorated as rf power to the cavity was increased. The chevron baffles (Fig. 4) were then cooled with liquid nitrogen. A pressure of 5 x 10^{-7} mbar could be achieved (Fig. 5) with the flow of liquid nitrogen in the chevron baffles [5] and the radiation shield having a projected surface area of about 0.27 m^2 per cryopanel. The operation of the cyclotron improved with the improvement in the vacuum of the cyclotron. The failures of the ceramic insulators of the cyclotron was reduced. The radio frequency system could be raised to higher voltages easily and beam was first extracted to the outer radius of the cyclotron.

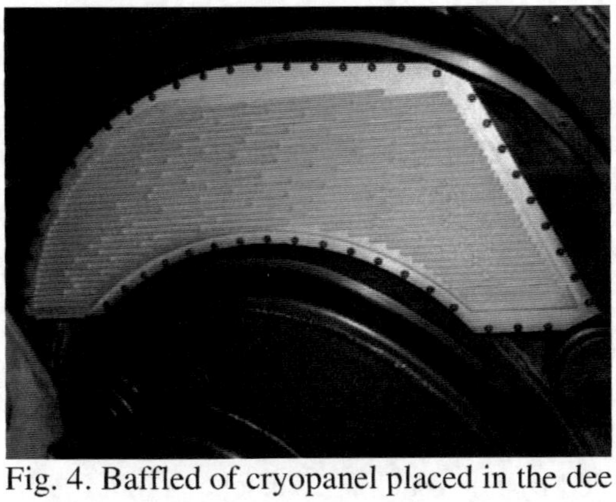

Fig. 4. Baffled of cryopanel placed in the dee

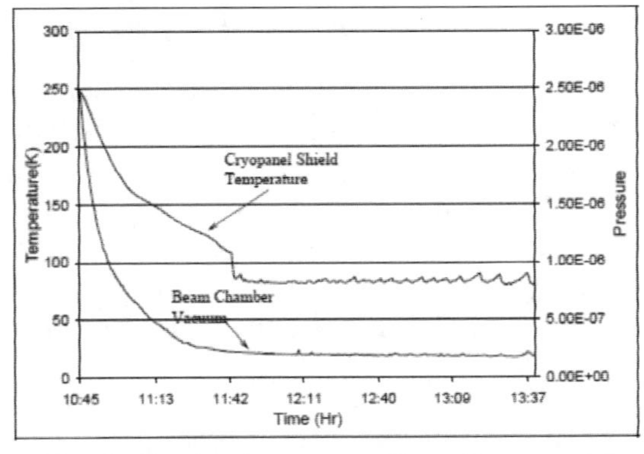

Fig. 5. Liquid nitrogen cooling of cryopanels

Commissioning trials were carried out in the cyclotron. But, it was observed that there was considerable beam loss in the outer radius. Due to conductance limitation, the effective pumping speed of the turbo-molecular pumps is about 135 l/sec. Although cooling the baffles and thermal shield of the cryopanel with liquid nitrogen gave a addition pumping speed of 30,000 l/sec, only moisture could be pumped using the surfaces cooled to about 90K. Liquid helium was then used to cool the cryopanels. There are three cryopanels of area 0.15 m^2 placed inside the lower dees (Fig. 6). Beam loss in the beam chamber of the cyclotron was considerably reduced (Fig. 7) [6]. The improvement of vacuum was considerable after cooling the cryopanels with liquid helium (Fig. 8). The vacuum in the cyclotron improved from about 5×10^{-7} to 5×10^{-8} mbar after initiating flow of liquid helium through the cryopanels.

Fig. 6. Cryopanel placed inside the radiation shield

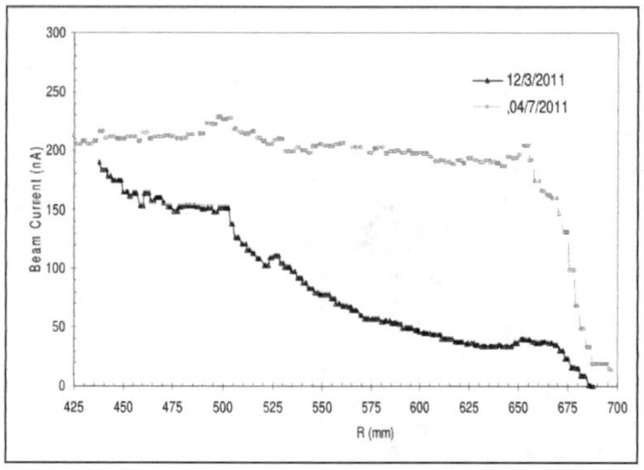

Fig. 7. Beam profile after operation of the cryopanels

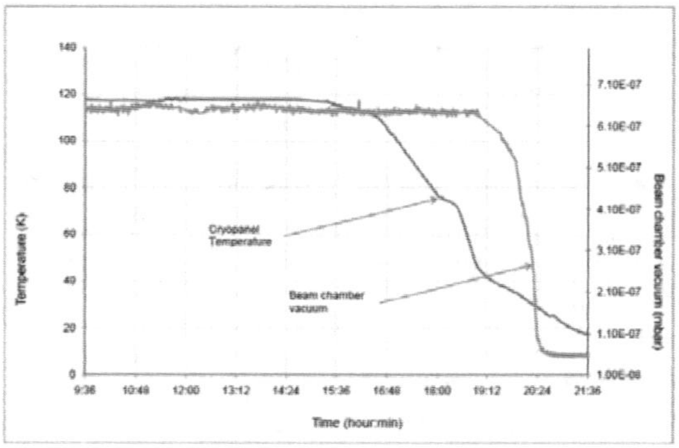

Fig. 8: Improvement of vacuum after liquid helium cooling of the cryopanels

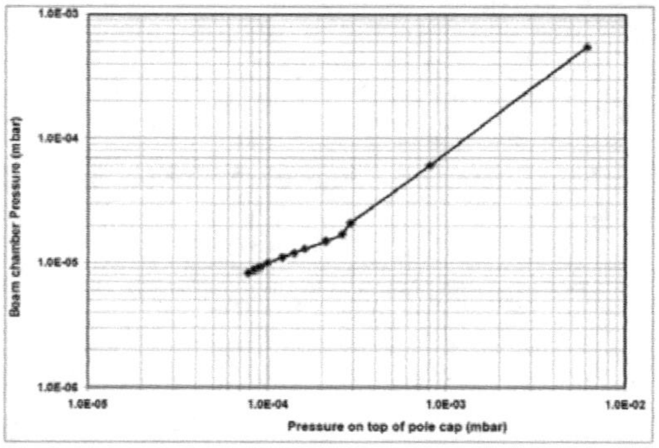

Fig. 9: Calibration of vacuum gauge

Measurement of vacuum in the cyclotron and its interpretation have some uncertainties. The vacuum gauge has to operate in the presence of high magnetic field. The gauge cannot also be mounted close to the beam chamber as the beam chamber is surrounded by the cryostat and the magnet yoke in the sides and the magnet yoke at the top. In order to eliminate the effect of magnetic field the gage was placed away from the magnet. In order to calibrate the gauge, a set-up was fabricated with a high vacuum pump and two gauges were placed away from each other with equivalent conductance between them. The calibration curve was generated by plotting the values observed by the gauges (Fig. 9)

2.2 Cryogenic system for cooling the cryopanels

The cryopanels are placed inside the dees. They are cooled by liquid helium. The chevron baffles and the thermal shield of the cryopanel is cooled by liquid nitrogen. The cold head for the liquid helium cooled cryopanel and the cold head for liquid nitrogen cooled baffle is placed at the end of a 6 m long 50 mm diameter compact transfer line. This transfer line is connected to co-axial bayonets for liquid helium and liquid nitrogen at the other end for easy dismantling. Due to the complex configuration of liquid helium flow, a liquid helium pump was used for circulating liquid helium in the circuit. Fig. 10 gives the pump characteristics of the liquid helium pump tested under ideal conditions and the calculated system characteristics. After complete installation and commissioning of the liquid helium pump in the cryogenic system, the desired head of the pump was not observed. Modifications were carried in the circuit to also use the head of the liquefier during operation of the system. The cooling power of the cold head for liquid helium cryopanel was tested and found to be about 10 watt at 10 K [7] (Fig. 11).

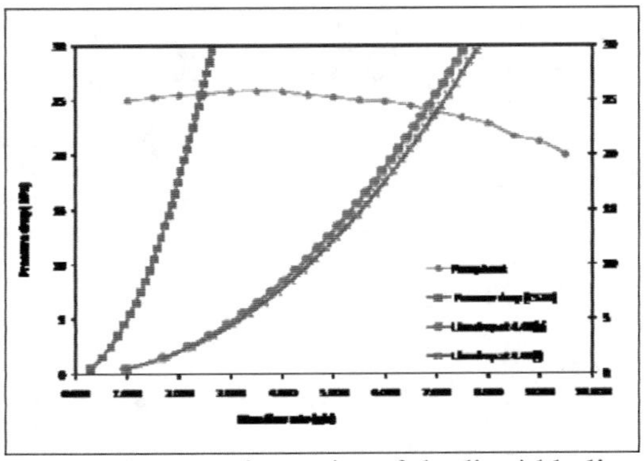

Fig. 10. operating point of the liquid helium pump

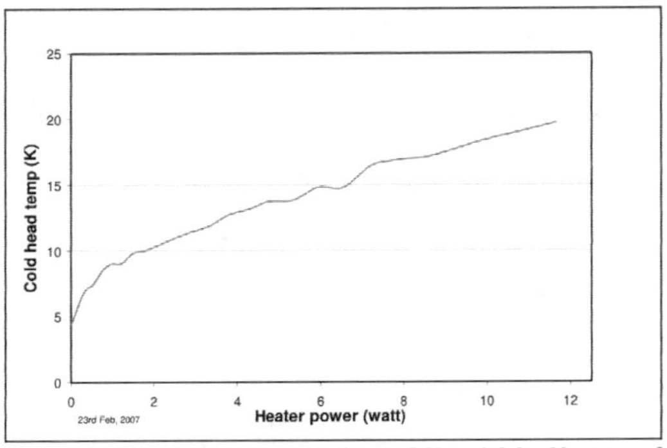

Fig. 11. Cooling capacity of liquid helium cold head

Different cyclotron laboratories have used different schemes for circulation of liquid helium through the cryopanels. The NSCL MSU cryogenic system [8] uses a sub-cooler and a pressurised liquid helium system. The cryogenic system at Texas A & M university, USA [9] has a reciprocating pump for liquid helium circulation through the cryopanels. The TRIUMF cyclotron used a cryogenerator initially and later used a helium refrigerator [10]. The AGOR cyclotron uses a cryocooler and heat pipes for cooling the cryopanels [11]. In VEC K-500 cyclotron, the cryogenic system was initially using a helial 50 for cooling the main cyclotron magnet. The cryogenic system for cryopanel was designed with the operating parameters of the helial 50 [12]. A liquid helium pump was added in the helium circuit to circulate liquid helium through the cryopanels [7]. In order to add to the redundancy of the system, later a helial 2000 was added [13]. The liquid helium system was also modified to operate the delivery dewar at a higher pressure. A sub-cooler was also added to the liquid helium system. These additions in the liquid helium system have made the liquid helium pump redundant.

Fig. 12. Copper rf liner

Fig. 13. Lower magnet iron

2.3 Liner vacuum

A thin copper sheet covers the contour of the hill and valley of the iron magnet. The space between the copper sheet (Fig. 12) and iron magnet (Fig. 13) is evacuated and a low quality vacuum is maintained in that region. It contains the iron pole tips, the copper liner, epoxy of the trim coils and insulation of the trim coils. It acts as a guard vacuum for the beam chamber and is required to ensure structural stability to the thin copper sheet forming the liner. At the beam chamber end, it is isolated using a 1.3 m o-ring placed below the cryostat. At the magnet end, there are 148 feed throughs which carry about 300 amperes to each trim coil. The upper and lower liner are connected together and evacuated using a 5 m^3/hr rotary pump.

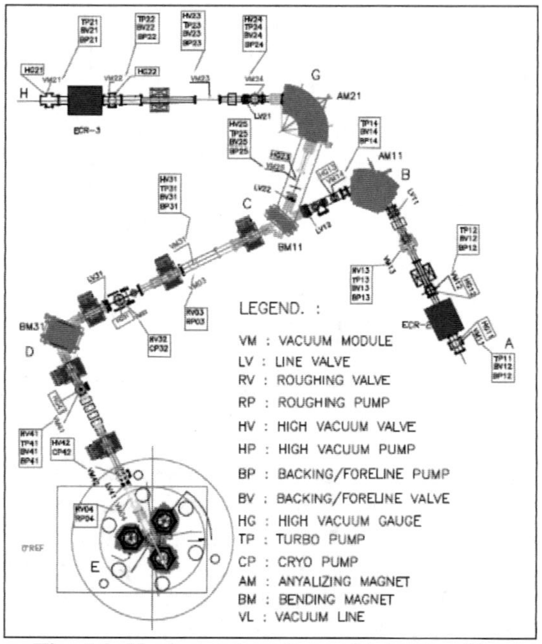

Fig. 14: Horizontal section of injection line

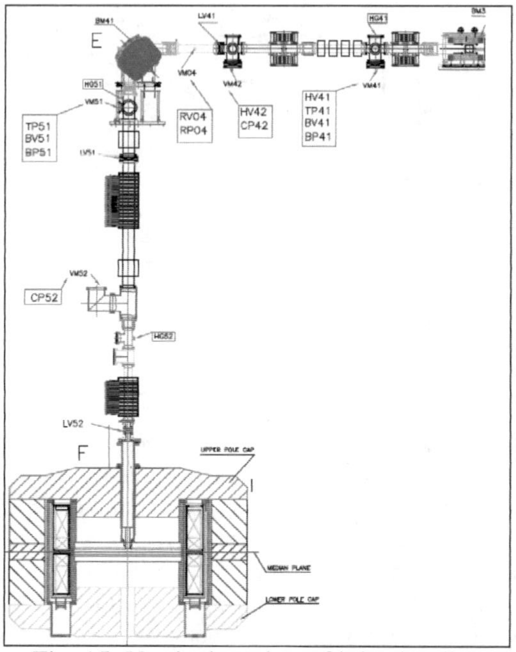

Fig. 15: Vertical section of injection line

2.4 Injection line vacuum

VEC K-500 superconducting cyclotron uses an ECR ion source located at Highbay above the cyclotron. The heavy ion beam from the ion source passes through an analysing magnet for charge state separation. Thereafter, the beam is transported though the horizontal section having a bending magnet (Fig. 14). At the end of the horizontal section, a vertical bending magnet bends it to the vertical section (Fig. 15) terminating at the inflector placed in the median plane of the cyclotron. Solenoid lenses are used in the horizontal and vertical beam line for focussing. Uncooled faraday cup, slits and collimators are placed in the beam line for diagnostics. The components have edge welded bellows for movement and are of welded construction. The length of the beam line is about 19 m. It is made of Stainless Steel SS 304, consists of vacuum chambers with copper gasket in all joints other than the ion source, where some viton o-rings have been used. It has a volume of about 0.2 m^3 and a surface area of about 8.4 m^2. A vacuum of 5 x 10^{-7} mbar is achieved by a combination of three DN 160 turbo molecular pumps and two DN 160 and one DN 200 cryopumps.

2.5 External beam line

Fig. 16 shows the beam line layout for the superconducting cyclotron, The quadrupole magnets for all the beam lines have been fabricated. The large switching magnets have been designed and fabrication is in progress. The vacuum chamber for the switching magnets have been designed and procurement is in progress [14]. The components for the first beam line have been assembled (Fig. 17). Turbomolecular pumps backed by scroll pumps have been used in the first beam line. The length of the first beam line is about 20 m with metal gasket joints. Faraday cup, beam veiwer and water cooled slits have been used. The movement of these diagnostic equipments use edge welded bellows to elimiate any air leak in the system during their movement.

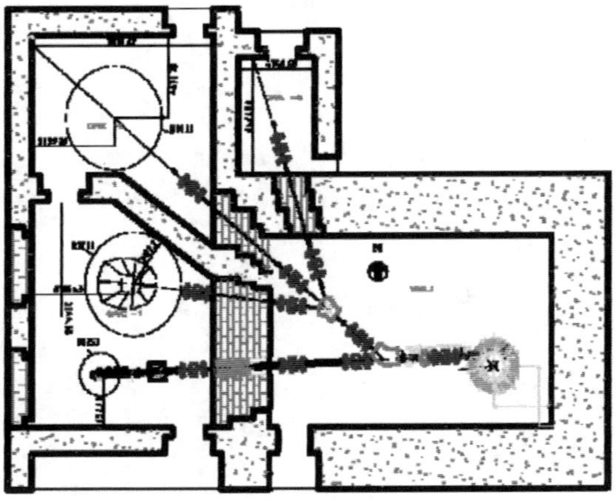

Fig. 16. Layout of beam line in SCC

Fig 17. Beam line in SCC vault

The beam line from the vault extends to the experimental cave (Fig. 18) with a large scattering chamber placed at the end. Shielding is placed in between the vault and the experimental cave and the beam line has a shield wall plug (Fig. 19) to bring the radiation at the experimental cave within allowable limits, when beam is accelerated in the vault and the experimental cave is occupied for preparation of experiments.

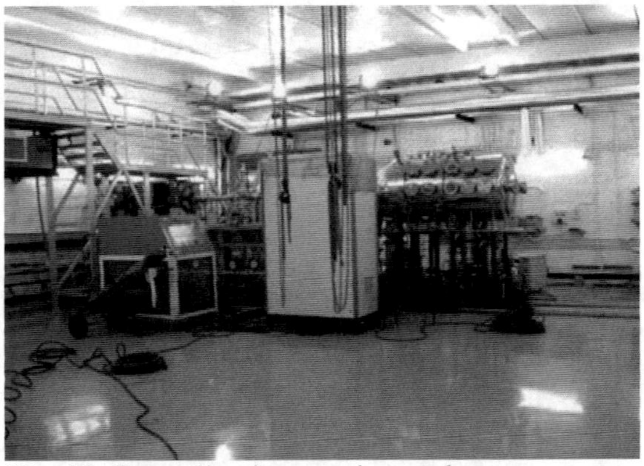

Fig. 18. Beam line in experimental cave

Fig. 19. Shield wall plug for first beam line

2.6 Magnet cryostat

The superconducting coil of the main cyclotron magnet is placed in an evacuated chamber. During the energisation of the magnet, it was observed that the insulation vacuum is deteriorating as the current flowing in the coil is increased to high levels (Fig. 20). At higher levels of magnet current, the level of liquid helium in the vessel was also reducing. It was initially felt that the degradation is because of the rise in magnetic field near the vacuum gauge. Subsequently, it was observed that a vacuum chamber placed in a similar location is not affected substantially by the magnetic field. Injection of helium in the outer vessel produced similar results [15]. Thereby confirming the conjection that a leak was present in hte helium chamber. Subsequently, additional pumping (Fig. 21) was provided to the cryostat and the capacity of the helium plant was increased [13] to operate the magnet to higher levels.

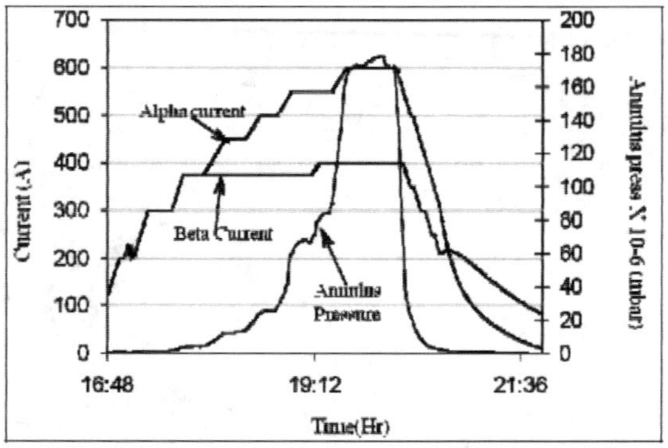

Fig. 20. Pressure variation during energisation

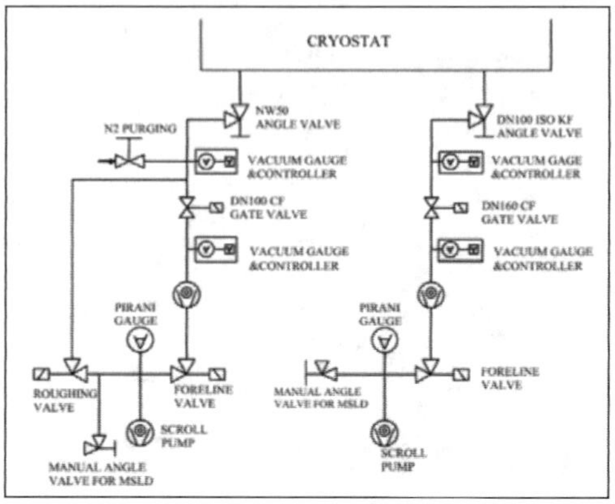

Fig. 21. Vacuum pumping for cryostat

2.7 Cryogenic system

The cold box for helium refrigerator (Fig. 22), cryogenic manifolds and the piping for transfer of liquid helium and liquid nitrogen to the magnet cryostat and the cryopanels form a large vacuum system. The cold box has a pump connected to the vacuum chamber of the cold box [16]. This pump is continuously operated. The dewars, cryogenic manifolds and the piping have about forty five independent volumes, a volume of 0.4 m^3, line length of 61.3 m and surface area of 3.5 x 10^5 m^2. The cold surfaces have activated charcoal for pumping the annular space during operation.

Fig. 22. Helial 50 cold box

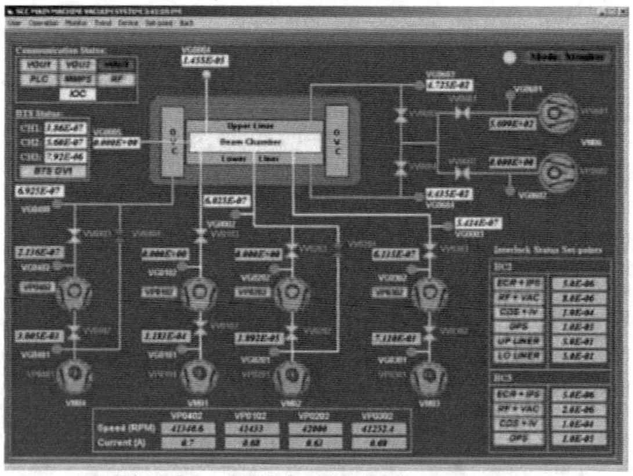

Fig. 23. OPI for K-520 vacuum

2.8 Supervisory control system :

The Experimental Physics & Industrial Control System (EPICS), a standard open-source dual layer software tool for designing distributed control system, was adopted to implement the supervisory control software for the vacuum system for superconducting cyclotron [17]. The indigenously developed Input Output Controller (IOC), which communicates with PLC, was in the lowest layer. The Operator Interface (OPI) sitting on top of IOC, communicates with the IOC for monitoring and supervisory control. The OPIs (Fig. 23) were developed in-house incorporating the features for system 'mimic' for ease of operation, on-line trending of selected parameters, audio-visual alarm, user authenticated secure mode for control of various components and modification of set-points, etc. These provide a diagnostic view of control hardware to the user.

3. K-130 cyclotron

The vacuum system of the K-130 cyclotron is operated by a fractionating oil diffusion pump having a speed of 42000 l/s for air. This is one of the largest vacuum pumps in India It has a 2.5 ton electro-pneumatically operated main gate valve and a refrigerated Chevron baffle at the top. A pair of above three assemblies is suspended from the two large opening in the resonator tank. Chevron baffles over the diffusion pumps are cooled to -60°C with the help of Freon refrigeration units. The cyclotron has been operating for over thirty years. The cooling coil of the diffusion pumps had lost contact with the body, cracks had developed in the backing line and the valves were not sealing properly. The vacuum system for K-130 cyclotron was refurbished [18] by changing these components. The vacuum chamber of analysing magnet formed by the pole pieces and side plates had leaks while energisation of the magnet. A new separate vacuum chamber was design, fabricated and replaced [19] to improve the beam line vacuum.

Fig. 24. K-130 cyclotron and beam line

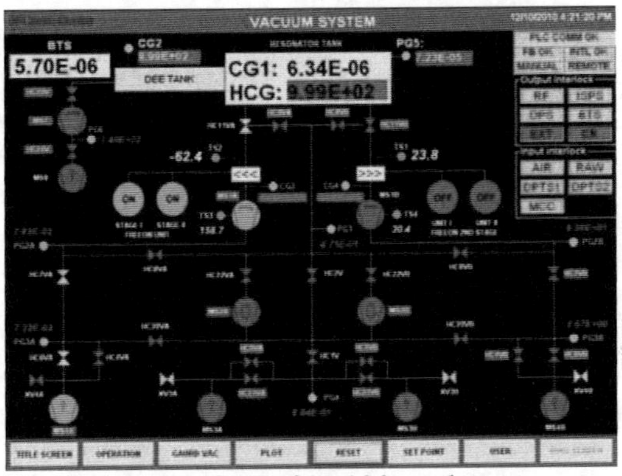

Fig. 25. OPI of K-130 cyclotron

3.1 Control system for K-130 cyclotron vacuum system
The control systems were implemented using Schneider make PLC with integrated Modbus-TCP connectivity [17]. The field components were distributed judiciously among the PLC input/output modules to overcome the single field component or PLC hardware failure. The system was designed to incorporate 'auto', 'manual' and 'maintenance' mode of operations. A hardware local panel was provided at each node for in-situ operation in manual mode and maintenance mode. The PC based remote terminal facilitates the user to monitor and operate the system in manual and auto mode. Fig. 25 shows the OPI of K-130 cyclotron vacuum system.

4. Conclusions

The vacuum systems for the large cyclotron at VEC Centre have been operating reliably. The vacuum in the beam chamber of K-520 cyclotron was considerably improved by cooling the cryopanels with liquid helium. The addition of charcoal to the helium cooled panels is likely to further improve the performance of the vacuum system of K-520 cyclotron. The vacuum system of the K-130 cyclotron was improved substantially with new components.

Reference

[1] Mallik C, Bhattacharyya S, Pal G, Nabhiraj PY, Taki GS, Som S, Bandopadhyay DL, Saha S, Kundu K, Banerjee AS, Mukhopadhyay B, Hazra DP, Mitra SK, Dutta M, Chakraborty PS, Polley A, Yadav RC, Chattopadhyay, Soor B, Pattanayak S, Prasad J, Dasgupta S, Das M, Pal Sarbajit, De TK, De R, Mukhopadhyay, Bhandari R K, Sinha B., 2006 Status of 224 cm Cyclotron at Kolkata *Proceedings of Indian Particle Accelerator Conference* (Mumbai India) p 30-31.

[2] Bhandari Rakesh Kumar, Sinha Bikash, 2007 Status of the K500 Superconducting Cyclotron Project at Kolkata *Proceedings of 18th International Conference on Cyclotrons and Their Applications* (Giardini Naxos Italy)

[3] Sur A., Bhandari R.K., 2008 Superconducting Cyclotron and its Vacuum System *Journal of Physics: Conference Series* **114** 012002

[4] Roy S, Bhattacharya S, Das T, Bhattacharyya T K, Pal G, Mallik C, Bhandari R K 2010 Beam Diagnostic Components for Superconducting cyclotron at Kolkata *Proceedings of 19th International Conference on Cyclotrons and their Applications CYCLOTRONS'10* (Lanzhou China) p 102-104

[5] Pal Gautam, Nandi Chinmay, Bajirao Sanjay, Mishra Santosh, Das Tanmoy, Bhattacharyya Tamal, Bhandari Rakesh Kumar 2010 Development of Cryogen Delivery System for VECC K500 Superconducting Cyclotron Cryopanels *Indian Journal of Cryogenics* **35 (1-4)** p 286 to 291

[6] Mallik C, Bhandari R K., 2010 First beam acceleration in kolkata superconducting cyclotron and its present status *Proceedings of 19th International Conference on Cyclotrons and their Applications CYCLOTRONS'10* (Lanzhou China) p 292-4

[7] Pal G, Nandi C, Bhattacharyya T K, Bhandari R K, 2008 Cryogenic System for VECC K500 Superconducting Cyclotron *Proceeding of International Cryogenic Engineering Conference 22, The Korean Institute of Applied Superconductivity and Cryogenics Publication* (Seoul South Korea) p 747 to 752

[8] Mallory M L 1978 Initial operation of the MSU superconducting cyclotron *IEEE transaction on Nuclear Science* **30(4)** p2061-2064

[9] Youngblood D H 1995 The Texas A&M K500 cyclotron facility *Nuclear Instruments & Methods in Physics Research: Section B* **56-57(2)** p991-995

[10] Sekachev I., 2008 TRIUMF Cyclotron Vacuum System Refurbishing *Journal of Physics: Conference Series* **100** 092022

[11] Horbowa A., Szott Ph. 2001 Vacuum cryopump systems supplied by cryogenerators and using heatpipes for heat transfer *Vacuum* **60** 207-212

[12] Panda U, Bhunia U, Mukherjee A, Maity T, Datta, N, Parate J, Dey R 2008 Modifications for optimum operation of helium liquefier *Indian Journal of Cryogenics* **33 (1)**

[13] Pal S, Dey R, Mukherjee A, Panda U, Maiti T, Datta N 2011 Integration of a new helium liquefier at VECC, Kolkata *Proceedings of Asian Conference on Applied Superconductivity and Cryogenics* (New Delhi India)

[14] Bhattacharyya S, Nandi C, Gayen S, Roy S, Mishra S K, Bajirao S R, Pal G, Mallik C 2012 Design of large vacuum chamber for VEC superconducting cyclotron beam line switching magnet *International Symposium on Vacuum Science & Technology and their Applications for Accelerators IVS-2012* (Kolkata India)

[15] Nandi C, Bhattacharyya T K, Das T, Bajirao S, Mishra S K, Pal G and Bhandari R K 2009 Operating Experience of Cryogen Distribution System for VECC K500 Superconducting Cyclotron *Indian Journal of Cryogenics* **34(1-4)** p101 to 105

[16] Pal S, Mukherjee A, Mandal A, Kar R, Nandy S 2012 Operational experiences in vacuum system of new helium liquefier and valve box in VECC *International Symposium on Vacuum Science & Technology and their Applications for Accelerators IVS-2012* (Kolkata India)

[17] Roy Anindya, Bhole R B, Akhter J, Yadav R C, Pal S, Sarkar D, Bhandari R K 2011 Vacuum System of the Cyclotrons in VECC, Kolkata *Proceedings of XIIth IEEE International Vacuum Electronics Conference IVEC* (Bangalore India) p 325-326.

[18] Yadav R C, Bhattacharya S, Bhole R B, Roy A, Pal Sarbajit, Mallik C, Bhandari R K, 2012 K-130 cyclotron vacuum system *International Symposium on Vacuum Science & Technology and their Applications for Accelerators IVS-2012* (Kolkata India)

[19] Mandal B, Saha S, Sarkar S C, Adak D, Viswanathan T, Hemram B, Chakraborty P S, Yadav R C, Mallik C, Bhandari R K 2012 Vacuum systems of the Large Cyclotron at VECC *International Symposium on Vacuum Science & Technology and their Applications for Accelerators IVS-2012* (Kolkata India)

Journal of Physics: Conference Series **390** (2012) 012002 doi:10.1088/1742-6596/390/1/012002

Vacuum System at IUAC

A. Mandal

Inter-University Accelerator Centre, Aruna Asaf Ali Marg, Post Box - 10502, New Delhi – 110067, Fax:011-26893666.

email: mandal@iuac.res.in

Abstract: Vacuum technology is an integral part of any accelerator system. At IUAC we have a 15UD PELLETRON, superconduting LINAC, Low Energy Ion beam Facility and a 1.7MV pelletron. Vacuum requirement in these accelerators is ~10^{-8} torr. Various types of Vacuum pump are used in different zones of the accelerators depending on load. Since the whole accelerator is quite long, distributed pumps are placed in different sections as per load. In ion sources displacement type pump viz turbo-pumps are usually used as the gas load is quite high. In other parts of the accelerator combination of getter and ion pumps are used. It is very much necessary to isolate different sections for maintenance purpose. Proper valves are used to isolate the sections and to avoid vacuum accidents proper interlock system is introduced. If air goes in some sections accidentally, valves will close automatically to protect other sections. The talk will cover different types of pumps and interlock used in accelerators at IUAC.

1.Introduction.

The vacuum system is an integral part of any particle accelerator and accelerator based research facilities. At IUAC we are having a 15UD Pelletron , superconducting LINAC and an ECR based low energy Ion beam facility. Vacuum system mainly consists of pumps, gauges for measurement of vacuum, valves for interlock and isolation of different zones and proper electronic interlock system for safe operation to avoid accidental vacuum leak. The design of vacuum system depends on the type of accelerator, high voltage used, type of ions to accelerate, volume of the system, gas load etc. For heavy ion accelerator the requirement of vacuum is ~ 10-8 torr as the particles while moving through beam line should suffer minimum collision with residual gases present in the system. Types of pumps needed depends on the ultimate level of vacuum required, gas loading in the system etc. While designing the vacuum system, the choice of material of the chamber is very important. Utmost care and clean handling of the components is necessary during installation of the vacuum system. After installation leak checking is must to ensure the quality of installation. To measure the level of vacuum achieved proper gauges at proper places should be attached. In a big accelerator, vacuum interlock is very important to save the important system in case of accidental leakage in any part. This paper describes the vacuum design in Pelletron, LINAC and Low energy ion beam facility.

2.Pelletron

15UD Pelletron is a vertical Tandem accelerator having terminal voltage 16 MV supplied by National Eletrostatic corporation (NEC), USA. The system is divided into several zones. A schematic diagram of the pelletron is shown in figure1. It has following zones: Ion source at the top where different negative ions are produced using SNICS ion source. The ions produced are pre-accelerated upto 300

Journal of Physics: Conference Series **390** (2012) 012002 doi:10.1088/1742-6596/390/1/012002

kV and mass analysed by a dipole magnet. The analysed ions are then injected into Tandem accelerator column which are kept in large tank filled with SF6 gas at high pressure. The details of system are described elsewhere [1]. The terminal voltage of the pelletron is 16MV. After passing through terminal the ions are passed through a gas stripper or foil stripper to convert into positive ions which are further accelerated downwards. The total energy achieved is (q+1)V where q is the charge state and V is terminal voltage. A big dipole magnet at the bottom analyses the ion of particular energy and finally transported to target in the experimental area in the beam hall downstream. The total system is ~ 100 meters.

2.1.Design of the vacuum system of Pelletron

In the 15UD PELLETRON Tandem Accelerator the ultra high vacuum (UHV) condition is maintained throughout the pre-accelerating, accelerating and post accelerating system and seven beam lines spanning over a length of more than 100 meters. The vacuum system is exposed to simple ambient pressure (1bar) as well as upto 8 bar SF_6 environment. The simplified view of the system is given in Fig 1. The system is designed keeping in mind the requirements of the distributed pumping stations taking care of conductance limitation of such a huge vacuum system. The pumps are controlled from the remote control using a computer and CAMAC highway.

All the pumps run in fully automated interlocked mode in a fail-safe condition. In case of any power failure no human interference is needed and the interlock system takes care of isolating the vacuum areas from venting and all the pumps are shut down. Within a few seconds our captive power is automatically put on to maintain the UHV vacuum level inside the accelerator by keeping the Ion Pump and Getter pumps on. It helps in restarting the accelerator immediately when the normal power is restored. In absence of the captive power the vacuum level would have deteriorated and it would have taken a couple of hours more to restore the UHV condition. With the interlocks, whole system is protected against any vacuum failure or accidental venting of the experimental chambers without closing valve before it.

In case of any vacuum accident the signal of rise in pressure above on pre-selected level is transmitted to various interlocked electro-pneumatically actuated vacuum valves and spring loaded fast acting valves to isolate the other vacuum regions. The individual sections of the vacuum system are isolated and the main accelerating region inside the high pressure (8 bar SF6) accelerator tank is totally isolated and maintained in UHV condition.

2.2 Types of vacuum valves and interlocks used in system

Manually actuated all metal right angle valve (NEC model VR1.5M) installed at all pumping ports for creating rough vacuum and leak checking of different beam line components. Two NEC manual all metal straight through valves (model VS2M) BLV T1 and BLV T2 are installed inside terminal for isolating terminal from accelerating tube during foil striper replacement. The NEC all metal pneumatic straight through valves(model VS2P) are installed all over beam line for isolating different sections . Closing time of valve is approximately 100 milliseconds.

BLV 01-1
BLV 02-2 ----- All metal Pneumatic Straight Through Valves.
BLV 04-1

Two high speed, spring actuated, all metal fast acting straight through valves are installed tank top and bottom side of accelerator for emergency closing in case of catastrophic failure of ultra high vacuum system. The fast valves are actuated by NEC coaxial pressure sensor (model CPS2.) and closing time of valve is approximately 35 milliseconds.

BLV 02-1 --- All metal Fast acting valves , closing by spring.
BLV 03-2

High pressure, pneumatically operated pendulum valves (BLV02-2 & BLV03-1) are installed on both sides of the accelerator for avoiding the loss of SF_6 gas in case of any major breakdown of the seal between high pressure SF_6 gas environment and UHV inside the accelerating tube.

The fast closing system protects only one vacuum line at a time. At IUAC, we have seven beam lines so one sensor was placed in each of the beam lines near experimental chamber and a multiplexer was incorporated into the fast closing system. At the time of experiment, the sensor of the active beam line is selected through the multiplexer and the Fast closing valve is interlocked with the selected sensor.

2.3 Types of vacuum pump used in system

Turbo Pump :Two turbo molecular pump (TP 01-1 & 02)of capacity 1500 l/s are installed before injector magnet for maintaining ultra high vacuum (~ 10^{-9} torr) through displacement of pressure load from ion source. Ion gauge controller IGC 01-1 and IGC 01-2 monitors vacuum near the ion source. Two turbo molecular pumps have been installed inside terminal (TP-T1 and TP-T2) for recirculating the gas to achieve better efficiency of gas stripper.

Getter Pump: The NEC air cooled titanium sublimation pump are used in different pumping location in combination with ion pump. Getter pump is effective upto pressure 1.0×10^{-6} torr to absorb active gases viz. oxygen and nitrogen. An array of six TS20 sublimator cartridge are used with special feed through flanged for long operation life time

Ion Pump : To maintain ultra high vacuum throughout pelletron accelerator and beam lines several D-I ion pump (Physical Electronics make) of different capacity are used. Ion pumps helps to remove the inert gases presents in the system. The combination of getter and ion pump helps to achieve vacuum of the order of 1.0×10^{-9} Torr . The pumping capacity of different ion pumps are given bellow.

IP 02-1 -- 80 l/s,,	IP 02-1a -- 220 l/s,	IP 04-1 -- 80 l/s ,
IP 03-2 -- 80 l/s,,	IP 02- -- 400 l/s,	IP 03-1 -- 400 l/s,

2.4 Types of vacuum gauge used in system

Conventional pirani gauges are used for measuring backing pressure of turbo molecular pump TP 01-1 and TP 01-2. Hot cathode ionization gauges are used for measuring high vacuum after ion source area using IGC 01-1 , IGC 01-2, pre and post accelerator section using IGC 02-1 and IGC 03-1. In rest of the area vacuum is measured calibrating ion pump current. Typical vacuum obtained at various locations are given in table1 .

Table 1. Ion gauge readings at different locations

Location	Pressure (Torr)	Location	Pressure (Torr)	Location	Pressure (Torr)
IGC 01-1	$< 1.0 \times 10^{-8}$	IP 02-2	$< 1.0 \times 10^{-9}$	IP T2	$< 2.0 \times 10^{-8}$
IGC 01-2	$< 1.0 \times 10^{-8}$	IGC 02-1	$< 1.0 \times 10^{-8}$	IP D2	$< 5.0 \times 10^{-9}$
IP 02-1	$< 1.0 \times 10^{-9}$	IP D1	$< 5.0 \times 10^{-9}$	IP 03-1	$< 1.0 \times 10^{-9}$
IP 02-1a	$< 2.0 \times 10^{-9}$	IP T1	$< 1.0 \times 10^{-8}$	IGC 03-1	$< 1.0 \times 10^{-8}$
		IP 03-2	$< 1.0 \times 10^{-9}$	IP 04-1	$< 1.0 \times 10^{-9}$

.

Journal of Physics: Conference Series **390** (2012) 012002 doi:10.1088/1742-6596/390/1/012002

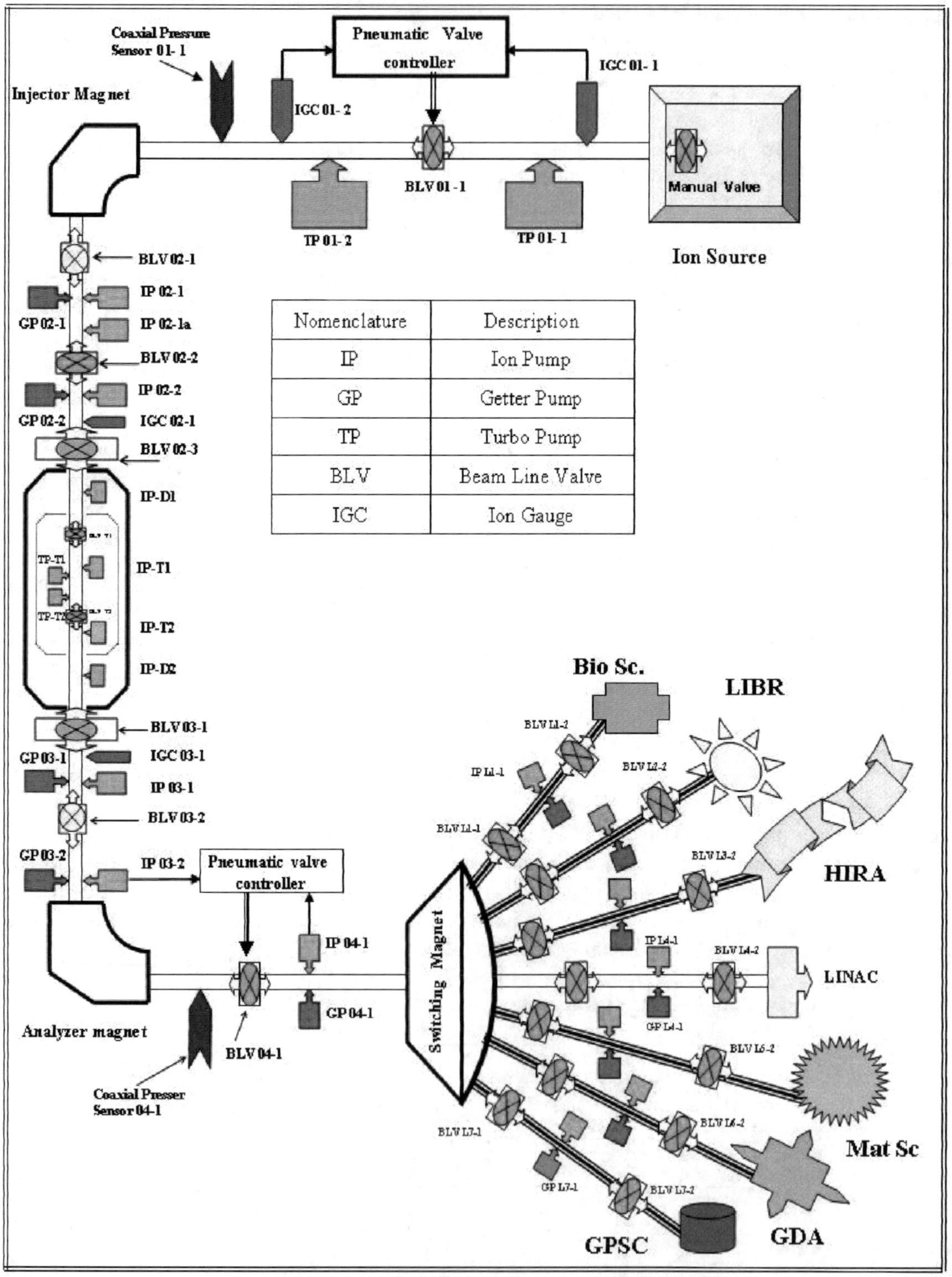

Nomenclature	Description
IP	Ion Pump
GP	Getter Pump
TP	Turbo Pump
BLV	Beam Line Valve
IGC	Ion Gauge

Fig.1. Schematic of vacuum system for pelletron.

3,Linear Accelerator (LINAC)

The zero degree beam line of Pelletron has been extended to LINAC. The bunched (~ns) beam from Pelletron is further bunched (~100ps) by a super conducting quarter wave resonator (QWR) and then injected into LINAC for further acceleration. The LINAC consists of three cryostats. Each cryostat has eight quarter wave resonators. The details are described elsewhere[2]. The accelerated ions from LINAC are re-bunched by two QWRS and then switched into four beam lines in BH-II. The whole LINAC system is shown in figure2.

3.1Vacuum design of LINAC

The distributed vacuum system for the LINAC is shown in figure [2]. The philosophy of vacuum system is similar to pelletron vacuum system. Each linac cryostat is evacuated by turbo pump of capacity 2300 lps backed by dry scroll pump (Edwards 30 m^3/hr). The beam lines are having distributed ION-getter pump combination similar to pelletron vacuum system.

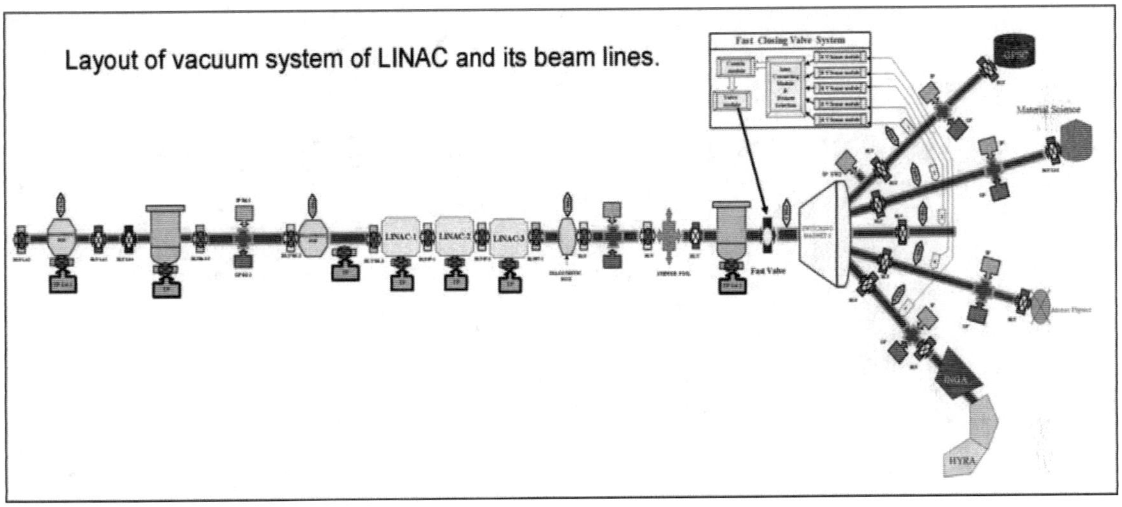

Fig. 2. Schematic of vacuum system for LINAC.

4.Low Energy Ion Beam Facility (LEIBF)

An ECR based LEIBF has been installed in new material science building for atomic -molecular physics and material sciences study. The accelerator consists of an ECR ion source, 400 kV accelerating column and an analyzing -cum switching magnet with three beam ports at 75, 90 and 105 degrees. It provides ions ranging from a few keV to a few MeV. Details of the system described elsewhere [3].

4.1 Vacuum system of LEIBF

In the ECR ion source region the gas load is quite high. A turbo pump having pumping speed of 500 lps on high voltage deck just after the ion source to maintain a vacuum ~10-6 torr. After the deck a 1000 lps turbo pump is used to maintain vacuum of the order of $< 1.0 \times 10^{-8}$ torr inside accelerating tube. In the beam line ion-getter combination is used to achieve a vacuum 1.0×10^{-9} torr. The layout of the whole system along with the vacuum system is shown in figure3. The whole system was leak tested using helium leak detectors before activating the pumps.

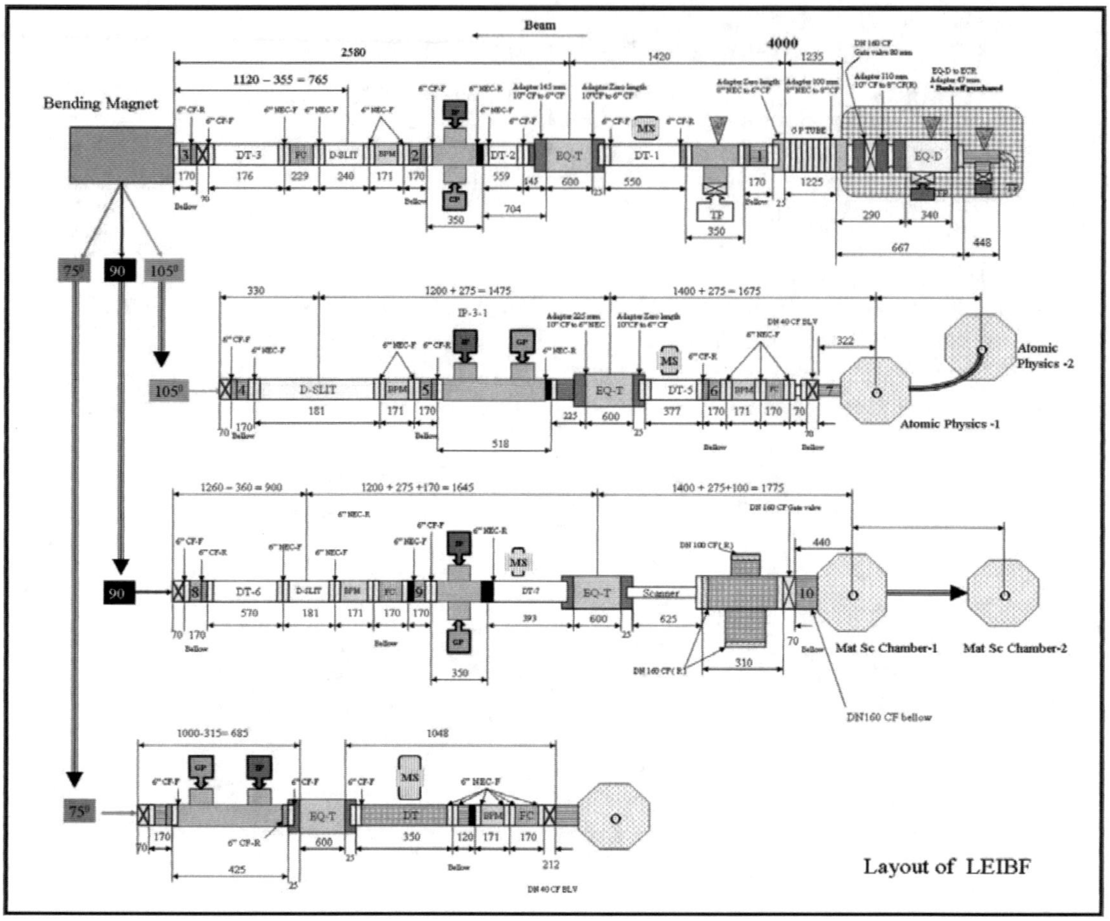

Fig.3. Layout of vacuum system for LEIBF.

Acknowledgement

The author acknowledges the contributions received from P. Barua and A. Kothari of vacuum group for design of vacuum system at IUAC..

References

[1] D.Kanjilal, S.Chopra, M.M.Narayanan, Indira S Iyer, Vandana Jha, R.Joshi and S.K.Dutta. Nucl. Instr and Meth, A 328(1993)97
[2] A.Roy, Curr. Sci.76(1999)149
[3] Abanimohan Mandal, Dinakar Kanjilal, Sarvesh Kumar, G.O. Rodrigues, IPAC2011, Spain.

Development of High Power Vacuum Tubes for Accelerators and Plasma Heating

Vishnu Srivastava

Microwave Tubes Division, CSIR-Central Electronics Engineering Research Institute, Pilani-333031, Rajasthan, India

E-mail: vs@ceeri.ernet.in, vsceeri@gmail.com

Abstract. High pulsed power magnetrons and klystrons for medical and industrial accelerators, and high CW power klystrons and gyrotrons for plasma heating in tokamak, are being developed at CEERI. S-band 2.0MW pulsed tunable magnetrons of centre frequency 2856MHz and 2998 MHz were developed, and S-band 2.6MW pulsed tunable magnetron is being developed for medical LINAC, and 3MW pulsed tunable magnetron is being developed for industrial accelerator. S-band (2856MHz), 5MW pulsed klystron was developed for particle accelerator, and S-band 6MW pulsed klystron is under development for 10MeV industrial accelerator. 350MHz, 100kW (CW) klystron is being developed for proton accelerator, and C-band 250kW (CW) klystron is being developed for plasma heating. 42GHz, 200kW (CW/Long pulse) gyrotron is under development for plasma heating. Plasma filled tubes are also being developed for switching. 25kV/1kA and 40kV/3kA thyratrons were developed for high voltage high current switching in pulse modulators for magnetrons and klystrons. 25kV/3kA Pseudospark switch of current rise time of 1kA/μ-sec and pulse repetition rate of 500Hz is being developed. Plasma assisted high power microwave device is also being investigated.

1. Introduction

High power vacuum tubes like magnetrons, klystrons, inductive output tubes (IOT) and gyrotrons are required both for the high energy particle accelerators, and for plasma heating in tokamaks [1-6]. High pulsed microwave power in MW range from 350MHz to over 10GHz is used for high energy linear accelerators. Lower frequencies, e.g., 350MHz and MW of power levels are used for proton accelerators. In an accelerator, propagating electron beam is energized by high microwave power that is injected into the RF propagating structure. Magnetron has the distinction of generating MW of pulsed microwave power in the frequency range varying from around 700MHz to well above 10GHz. It is the most compact, low cost, high efficient and the lowest operating voltage device but it has the disadvantage of phase and amplitude instabilities. Efforts are being made world over of improving magnetron coherency by injection locking techniques. Klystron has the distinction of producing pulsed power in MW range by amplifying low driving signal with high gain and low noise. Klystrons of MW of output power in the frequency range varying from around 350MHz to well above 10GHz are available. Single-beam klystrons of MW of output power are very large in size and have very high operating voltages compared to equivalent output power magnetrons. Multi-beam klystrons are new development for low voltage requirement and compact size. Klystrons of GW of pulsed output power are being investigated using sheet beam technology.

Microwave tubes of MW of CW power level are required for plasma heating in thermonuclear controlled fusion by generation of plasma and energizing ions suitable for energy production. In the operation principle of Electron Cyclotron Resonance (ECR) ion source, electrons confined by the magnetic field structure are heated to high energy by microwaves that have been injected into the plasma. These hot electrons then produce ions by successive electron impact ionization. Microwave frequency is decided by the electron cyclotron frequency. When electrons move in a magnetic field (B) they gyrate around the magnetic field lines due to the Lorentz force. The gyration frequency is called the cyclotron frequency ωc. If microwave radiation of the same frequency (ωrf) propagates into such a region, the electrons are resonantly accelerated or decelerated (depending on the phase of their transversal velocity component with respect to the electric field vector) when the following electron cyclotron resonance condition is fulfilled:

$$\omega rf = \omega c = (e/m) \times B$$

Here, e and m denote the charge and mass of the electron, respectively.

170GHz gyrotrons of MW of CW output power are used in European Tokamak. Optimum plasma heating requires high power vacuum tubes of generating low frequencies of 100's of MHz to 10's of GHz to 100'sGHz. Heavy ion particle like proton requires less frequency source around 350MHz. Tetrodes, klystrons and gyrotrons are therefore used in combination for plasma heating.

2. Development of high power vacuum tubes

CEERI, Pilani is engaged since long with the development of high pulsed power magnetrons and klystrons for medical, scientific and industrial accelerators, and with the development of CW klystrons and long pulse/CW gyrotrons for plasma heating. Plasma filled vacuum tubes like thyratrons and pseudospark switches are also being developed for high voltage/high current switching. The development of such high power microwave tubes are presented in this paper.

2.1 High Power Pulsed Magnetrons for Accelerators:
S-band 3 MW (pulsed) magnetron (as shown in Fig.1) of given specifications, was developed for particle accelerators.

Fig. 1. S-band 3MW (pulsed) Magnetron of following specifications:

Frequency	2998MHz,
Tunable:	2992-3002MHz,
Pulse duration:	4□s,
Duty cycle	0.001
Beam voltage:	50kV
Beam current:	120A

Two versions of S-band pulsed magnetrons of peak output power of 2.0MW with centre frequency 2998 MHz (for version-1), and 2856MHz (for version-2), tuning range 10MHz, pulse width 4.0μsec and duty cycle 0.001, were developed for industrial LINACS and MICROTRONS. S-band pulsed magnetron of peak output power of 2.6MW, centre frequency 2998 MHz, tuning range 10MHz, pulse width 4.5μsec and duty cycle 0.001, is presently being developed for medical LINAC. S-band tunable pulsed magnetron of 3MW pulsed power at centre frequency 2856 MHz, tuning range 10MHz, pulse width 5.0μsec and duty cycle 0.001, and beam voltage/current 50kV/120A is being developed for industrial accelerator. CEERI is working with Lancaster University, U.K., for injection locked, high efficiency, high power 704 MHz magnetron of CW output power of 800kW for proton accelerator.

2.2 High Power Klystrons for Accelerators and Plasma heating:
S-band, 5MW (pulsed) klystron, as shown in Fig.2, was developed for linear accelerator. Electromagnet is used for the electron beam focusing, and it is liquid cooled.

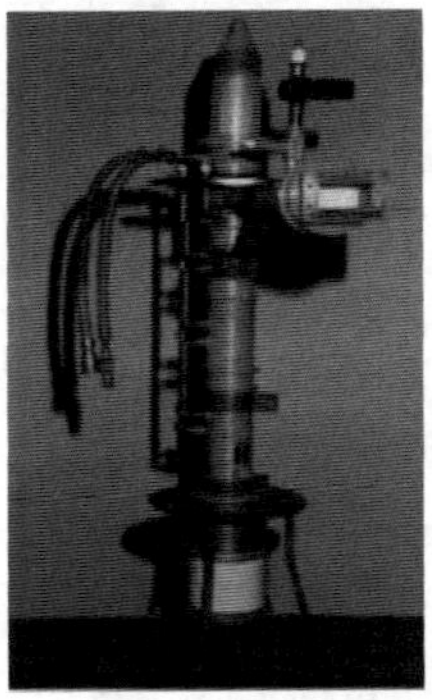

Fig. 2. S-band 5MW (pulsed) Klystron of following specifications:

Frequency:	2856MHz
Tuning:	10 MHz
Gain:	50dB
Efficiency:	40%
Pulse duration:	5ms
Beam voltage:	120kV
Beam current:	90A

Development of S-band pulsed klystron with rating of 6 MW (peak), 24 kW (average) and operating beam voltage 130kV, beam current 95A, pulse width 10μsec, PRF 400 is in progress. This klystron is used in 10MeV Industrial accelerator to accelerate an electron beam at high energy, to obtain an intense X-radiation. This radiation is used for many industrial applications related to food sterilization, decontamination, non destructive control, etc. 350MHz, 100kW (CW) klystron of operating beam voltage 30kV, and beam current 7A is being developed for proton accelerator. C-band 250kW (CW) klystron of beam voltage 60kV and beam current 10A is being developed for plasma heating.

2.3 High Power Long Pulse / CW Gyrotrons for Plasma heating:
42GHz, 200kW (CW/Long pulse) gyrotron, as shown in Fig.3, is under development for plasma heating. This is a multi-institutional project at CEERI, Pilani with the collaboration of IIT-Roorkee, BHU-Varanasi, IPR-Gandhinagar and SAMEER-Mumbai for the use of Indian TOKAMAK at IPR-Gandhinagar. Magnetron Injection Gun (MIG) of this gyrotron has been fabricated for operating beam voltage of 65kV and beam current 10A, and magnetic field 1.61 Tesla. The cavity for symmetric TE03 mode has been fabricated and tested. Other critical components like nonlinear taper, collector, and high power window have been fabricated. Full tube will be assembled within few months.

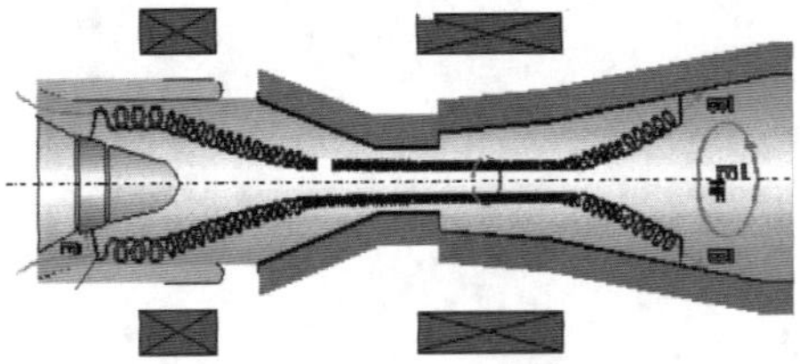

Fig. 3. Schematic diagram of 42GHz Gyrotron of output power 200kW

under development at CEERI-Pilani jointly with IIT-Roorkee, BHU-Varanasi, IPR-Gandhinagar & SAMEER.

120GHz, 1MW output power gyrotron is also being developed for plasma heating. The group is also working for design and development of gyrotrons at other frequencies.

2.4 High Power TWTs:

Traveling-wave tubes (TWTs) are also being explored for new generation ion sources because of their wide tuning range, lower operating voltage, and stable tube operation. TWTs are now available in wide frequency range from 1 to 100GHz of output power up to hundreds of kW (CW) and tens of MW (pulsed). High pulsed power C-band 75kW coupled-cavity TWT (Fig.4), was developed for a radar system. Bandwidth of this tube is 5.4 to 5.9GHz, and operating voltage and current are 34kV and 10Amps, respectively. This tube has gridded gun with both control grid and shadow grid, and coupled-cavity structure has cylindrical cavities coupled through double slots in their common wall.

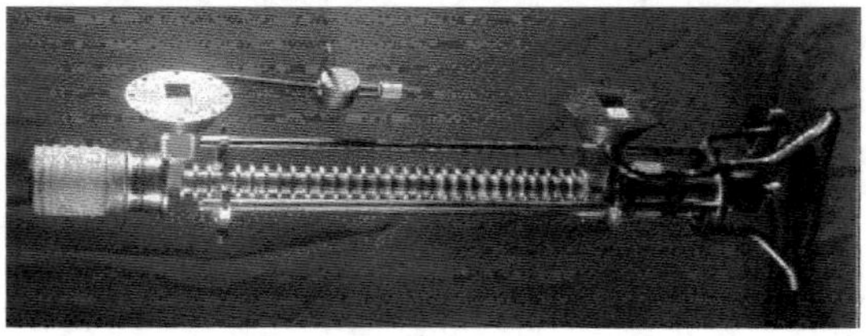

Fig. 4. C-band Coupled-cavity TWT of following specifications:

Pulsed output power:	75kW
Frequency range:	5.4 to 5.9GHz
Beam voltage/current:	34kV/10Amps

High efficiency C-band 60W (CW) TWT, Ku-band 140W (CW) TWT, and Ka-band 40W (CW) TWT are also being developed for their applications in satellite communication. THz TWT of frequency 94-100GHz and output power of 5-10W is also being developed for numerous applications.

2.5 Plasma filled tubes for switching:

25 kV/1kA and 40kV/3kA thyratrons (Fig.5) were developed and tested for high voltage high current switching in pulse modulators as required for magnetrons and klystrons.

Fig. 5(a). 25kV Thyratron

Peak Anode Voltage	25 kV
Peak Anode Current	1.0 kA
Avg. Anode Current	1.0 A
Time Jitter	3.0 nsec
PRF	60 kHZ
Di/dt	5000 A/μ sec
Hydrogen filled	

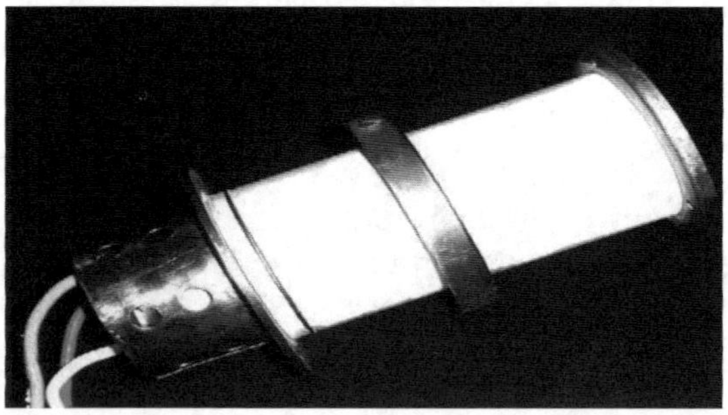

Fig. 5(b). 40kV Thyratron

Peak Anode Voltage	40 kV
Peak Anode Current	3.0 kA
Avg.Anode Current	3.0 A
Di/dt	5000 A/μsec
Anode delay	0.1 μsec
Time Jitter	1.0 nsec
Deuterium filled	

Plasma based 25 kV/5kA Pseudospark Switch and 40 kV/5kA Pseudospark Switch of current rise time of 1kA/μ-sec and pulse repetition rate of 500 Hz are being developed. Plasma based high power microwave devices like Pasotron are also being investigated using plasma based high current density electron gun.

4. Conclusion

In-house design and fabrication capabilities are developed at CEERI, Pilani, for the successful indigenous development of different types of high power microwave tubes and plasma filled high power devices for their applications in various accelerators and tokamaks. All efforts are being carried out for making these high power vacuum devices of low cost, compact and of high efficiency. Because of the many advantages of magnetrons over klystrons, CEERI is working with Lancaster University for injection locked 704MHz, 800kW magnetron for proton accelerators. Multi-beam and sheet beam technologies are also being developed for compact high power klystrons and for high power, very high frequency TWTs. Both multi-beam and sheet beam technologies have numerous advantages over single cylindrical beam because of significant reduction of space charge forces. CEERI is developing plasma filled devices for switching as well as for high microwave power generation. A plasma-assisted high power oscillator (*Pasotron*) is being investigated using plasma assisted electron gun.

Acknowledgement

Author is thankful to Dr Chandra Shekhar, Director, CEERI, for granting permission for this paper; to Dr SN Joshi, Dr RS Raju, Dr LM Joshi, Dr VVP Singh, Dr AK Sinha, Sh Udit Narain and all other colleagues of MWT Division for their full support in preparing of this paper.

References:

[1] Caryotakis G, High-Power Microwave Tubes, IEEE Trans. Plasma Sci. 22(5) 683-91, 1994
[2] Barker R J and Schamiloglu E, High Power Microwave Sources and Technologies (III series, Inc., New York: IEEE press) 2001
[3] Schamiloglu E, High Power Microwave Sources and Applications, IEEE MTTS, Fort Worth, Texas, 2004
[4] Shrivastava P., et.al., Research, Design and Development Activities on High Power Pulsed RF/Microwave Systems and Test Facilities for Particle Accelerators, IEEE-IVEC-2011, 53-54.
[5] Dexter, A.C., et.al., First demonstration and performance of an injection locked continuous wave magnetron to phase control a superconducting cavity, PHYSICAL REVIEW SPECIAL TOPICS - ACCELERATORS AND BEAMS 14, 032001-11, 2011.
[6] Choi, EM, et.al., Megawatt Power Level 120 GHz Gyrotrons for ITER Start-Up, Journal of Physics: Conference Series 25 (2005) 1–7.

Journal of Physics: Conference Series **390** (2012) 012004 doi:10.1088/1742-6596/390/1/012004

ECR Based Low Energy Ion Beam Facility at VECC, Kolkata

G S Taki[1]*, D K Chakraborty[1], Subhash Ghosh[1], S Majhi[1], Gautam Pal[1], C Mallik[1], R K Bhandari[1], J B M Krishna[2], K Dey[2], A K Sinha[2]

[1]Variable Energy Cyclotron Centre, Bidhanagar, Kolkata-700064, India
[2]UGC DAE CSR, Kolkata Centre, Bidhannagar, Kokata-700098, India

* E-mail: gstaki@vecc.gov.in

Abstract. A low energy heavy ion irradiation/implantation facility has been developed at VECC, Kolkata for materials science and atomic physics research, utilizing indigenously developed 6.4 GHz ECR ion source. The facility provides high charge state ion beams of N, O, Ne, Ar, S, Kr, Xe, Fe, Ti, Hf etc. up to a few micro amperes to an energy of 10 keV per charge state.The beam energy can be further enhanced by floating the target at a negative potential (up to 25 kV). The ion beam is focused to a spot of about 2 mm diameter on the target using a set of glaser lenses. A x-y scanner is used to scan the beam over a target area of 10 mm x 10 mm to obtain uniform implantation. The recently commissioned multi facility sample chamber has provision for mounting multiple samples on indigenously developed disposable beam viewers for insitu beam viewing during implantation. The ionization chamber of ECR source is mainly pumped by ECR plasma. An additional pumping speed has been provided through extraction hole and pumping slots to obtain low base pressure. In the ion source, base pressure of $1x10^{-7}$ Torr in injector stage and $\sim 5x10^{-8}$ Torr in extraction chamber have been routinely obtained. The ultra-high vacuum multi facility experimental chamber is generally kept at $\sim 1x10^{-7}$ Torr during implantation on the targets. This facility is a unique tool for studying fundamental and technologically important problems of materials science and atomic physics research. High ion flux available from this machine is suitable for generating high defect densities i.e. high value of displacement-per-atom (dpa). Recently this facility has been used for studies like "Tunability of dielectric constant of conducting polymer Polyaniline (PANI) by low energy Ar^{9+} irradiation" and "Fe^{10+} implantation in ZnO for synthesis of dilute magnetic semiconductor".

1. Introduction

In the present day, immense importance has been concentrated into the Materials Science Research with thin films, regarding basic physics as well as applications point of view. Exotic physics comes out from defects and disorders in two dimensional systems. Hence, charged particle irradiation brings forth enormous novel phenomena in thin films. Low energy (hundreds of keV) heavy ions from ECR ion source are ideal tools for generation of tailor made defects in thin films. The low energy nuclear projectile experiences significant energy loss through collisions in the target and its range may vary up to a few hundred Nano-meters. It causes an enormous generation of point defects through the so called nuclear energy loss in addition to the significant inelastic energy loss. Low energy ion beams also

[1] To whom any correspondence should be addressed.

provide a controlled means for modifying material surfaces for achieving desired properties. Recently, a low energy heavy ion irradiation/implantation facility has been developed at VECC, Kolkata for materials science and atomic physics research.

The facility comprises of a 6.4 GHz electron cyclotron resonance (ECR) ion source [1] which is a special types of hot plasma machine to produce multiply charged ions, a magnetic charge-state analyser, focussing magnets in the beam transport line and a multi facility implantation chamber. The total vacuum space of the facility is approximately 176 litres which is spread over a large length and area. Total length of the vacuum system is segmented and conductance limited by slits, collimators and diaphragms. The whole volume is pumped by several distributed pumping stations isolated by ultra-high vacuum gate valves. Besides magnetic confinement, operating frequency and microwave power of the ion source, the vacuum plays a major role in multiply charged ion beam production, analysis and transport. In ECRIS plasma discharge takes place at a vacuum level of ~10^{-7} Torr. The low-pressure condition is essentially needed to minimize charge exchange loss due to recombination of multiply charged ions with the lower charge state ions and the neutral atoms. A fixed ratio of neutral to electron density must be maintained for optimum production of a particular charge state in the steady state condition. As the electron density is proportional to square of the injected microwave frequency ($n_e \propto f^2$), it is understood that a particular operating pressure is essential for a specific charge state generation. The requirement of good vacuum in the beam extraction, analysis, transport and implantation area is again essential to minimize the space charge effect, beam scattering and implanted impurity caused due to residual environment.

2. The implantation facility

A low energy heavy ion implantation facility as shown in figure 1.has been developed at VECC using the indigenously built 6.4 GHz ECR ion source [VECECR IS] [1-3]. The ion source was basically designed to meet the injection parameters of the cyclotron and it operates with the maximum extraction potential of 10 kV, thus delivering ion beams at a maximum energy of 10 keV per charge state. The ion beam extracted from the ECR ion source is q/m analysed by a 90° double focussing analysing magnet with a resolution of 1 in 40 for 10 mm slit width. The analysed beam is optimized on a retractable type Faraday Cup through 10x10 mm slit. The multiply charged ion beam is then allowed to propagate through an ultra high vacuum beam line up to the multi facility experimental chamber for carrying out implantation and other experiments.

Fig. 1. Layout of ECR-based Low Energy Ion Beam Facility

2.1 VECECR Ion Source

VEC-ECRIS is a single stage high magnetic field (High-B) ion source with a biased electron repeller placed on the axis, near the injection mirror point [2]. The supply of cold electrons and use of low mass mixing gas improve the source performance and stability. High-B VEC-ECR has only two sets of coils with individual iron return yoke [Figure 2]. This topology facilitates with an easy tuning of the source and has negligible influence of the injection field on the extraction area due to an excellent isolation between the mirrors. B_{min} field is reduced but the peak fields are increased to 7.2 and 5.5 KGauss corresponding to mirror ratios to R_1=6.4 and R_2=5 respectively.

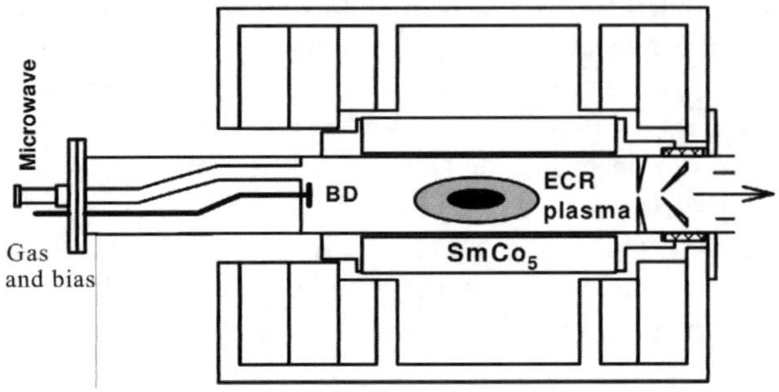

Fig. 2. ECR ion source schematic

2.2 Beam Transport and Experimental Facility
The mass analyzed ion beam is focussed by a pair of magnetic glaser lenses on the target film. The schematic of the beam line is shown in figure 3. Figure 4. shows the beam optical simulation result. The photograph shown in figure 5 is the beam spot seen on a beam viewer. The observed beam spot size (2mm dia.) shows a good co-relation with the simulation value.

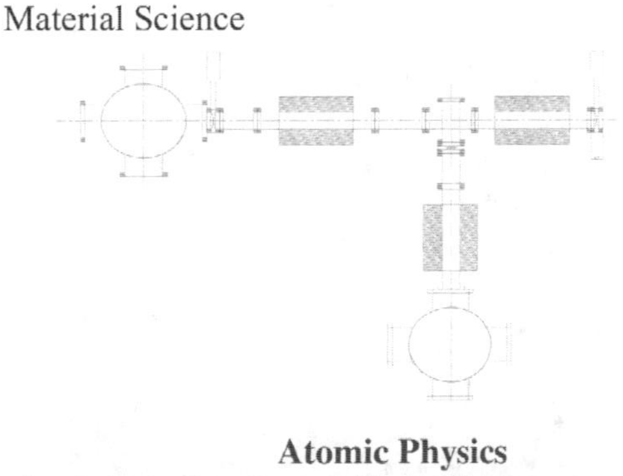

Fig. 3. Test chamber and beam lines

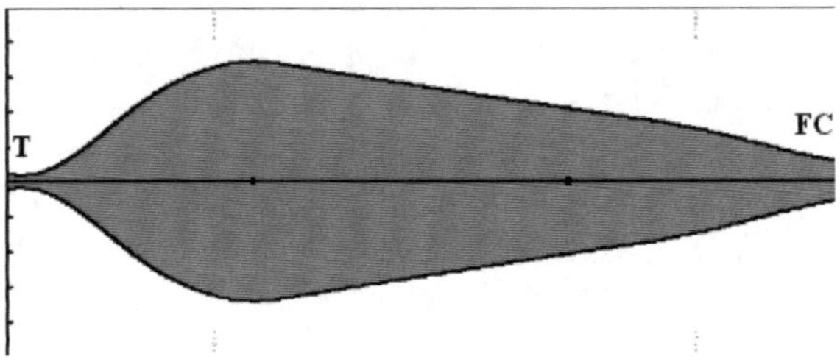

Fig. 4. Beam profile from faraday cup to target

Fig. 6. Multifacility experimental set up

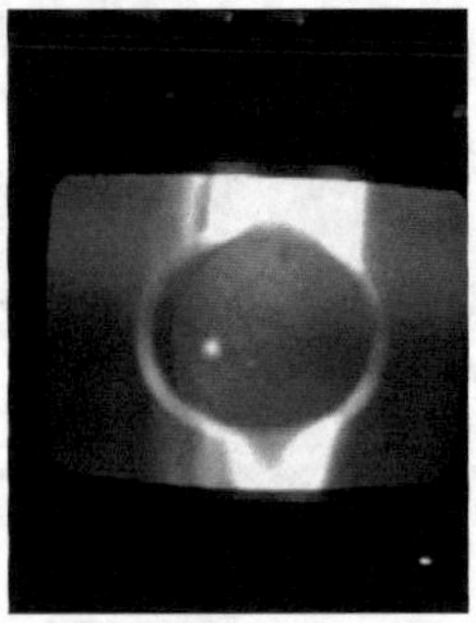

Fig. 5. Beam spot on a viewer at target ladder

The details of the experimental setup are shown in figure 6. At the top flange of the multi-facility experimental chamber, a four axes sample manipulator has been commissioned to accurately positioning an electrically isolated target ladder assembly for mounting multiple targets at a time. Each sample sits over a disposable type indigenously developed low energy beam viewer for viewing the beam during implantation.

With a view to deliver enhanced variable ion energy, the provision has also been made to float the target up to a variable potential of – 25kVolts. For most of the ground state experiments, total fluence is usually measured directly on a current integrator. But, in case of floating samples, to achieve higher projectile energy, the beam current is measured by an indirect method. Here also the total fluence is measured on the current integrator by monitoring the ion current captured by the collimator placed before the electrostatic scanner. For each such experiment, the collimator current, after scaling to a pre-calibrated factor, has been utilized as the total fluence measurement. During implantation on the metallic targets, the secondary electron produced from the target causes an appreciable error to the implantation doses. A suitable non-magnetic, 180 mm diameter semi cylindrical electron capture electrode has been kept isolated at 85 mm away from the target ladder assembly (figure 7a, b, c) to collect the produced secondary electrons. The secondary electron current thus produced is added algebraically to measured target current for actual fluence measurement. This arrangement facilitates us to measure the energy of the secondary electron in bulk or in discrete levels as required. This semi cylindrical electrode has two holes at a separation of 45^0 on the horizontal beam propagation plane. The 20 mm diameter hole which is concentric to the beam propagation centre allows the scanned beam for implantation on the target and another hole of 30 mm diameter enables us to online viewing the beam with the help of a CCD camera.

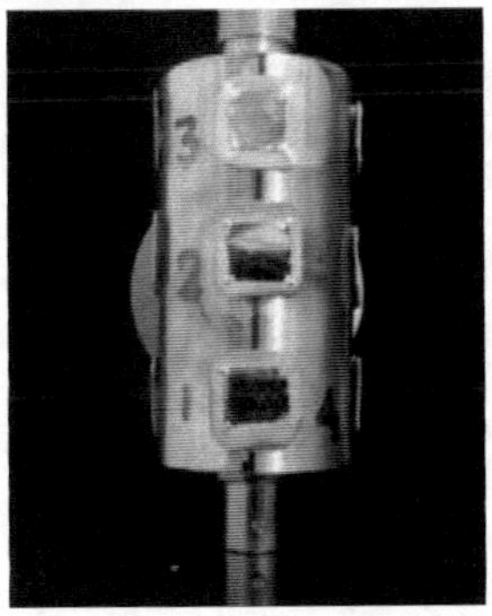

Fig. 7a. Target ladder assembly

Fig. 7b, c Integrated assembly of Collimator, beam scanner & electron capture electrode (180^0 views)

Since the irradiation/implantation is motivated for the modification of the properties of materials with proper understanding of the mechanism, it is essential to have uniform irradiation over the target area as well as the thickness. Uniform irradiation along the thickness is guaranteed by thin films. But to have an uniform irradiation over the whole area of the sample, it is necessary to scan the beam over the target surface, preferably with very low time periods compared to the irradiation time. Based on the beam transport calculations and also considering the ion optical properties of low energy beam to achieve small beam size on the target, the final focusing glaser solenoid has been placed nearest to the test chamber. An electrical scanner with a changeable collimator assembly has been designed and developed to scan the beam on the thin target films at higher scanning rate. The scanned area can be finely adjusted and shifted electronically in X-Y directions. To satisfy all the above mentioned requirements, an integrated collimator, electrical scanner assembly along with a well separated secondary electron capture electrode have been designed and developed in our laboratory (figure 7b, c). The integrated assembly is housed inside the test chamber concentrically to the floating target assembly. This set up has been used regularly for carrying out several experiments.

3. ECR source vacuum and charge state production

Ionization takes place when sufficiently energetic electrons collide with neutral molecules under certain environmental conditions. Single impact and successive ionization are the two basic ionization processes to be mentioned. But the probability of single impact ionization is small compared to that of successive ionization process regarding multiply charged ion production. In B_{min} field trap, the plasma is stable and the diffusion time of the ions is high enough to ionize the atoms to high charge state by successive ionization process. This process needs some time which is almost equal to the confinement time (τ_i) or exposure time of the ions to a cloud of hot electrons of density n_e with an impact energy T_e. The equilibrium charge state distribution is obtained by the balance between ion production rate and loss rate. In the steady state, the rate of change of ion charge state i is zero [4].

$$\frac{dn_i}{dt} = 0 = n_e(\sigma_{i-1,i} v_e)n_{i-1} + n_0(\sigma_{0,i+1} v_{i+1})n_{i+1} - n_e(\sigma_{i,i+1} v_e)n_i - n_0(\sigma_{0,i} v_i)n_i - \frac{n_i}{\tau_i} \quad (1)$$

where n_e, n_0, n_i are the density of electron, neutral particle and ion of charge state i respectively

τ_i =ion confinement time ≅ ionization time≅the life time of ion for ionization of i th state,

v_e = electron velocity,

v_i, v_{i+1} = ion velocities for charge state i and $i+1$ respectively,

σ= average ionization cross-section for respective cases (10^{-16} to 10^{-20} cm^2).

The probability of droping charge state by electron pick up is almost negligible compared to the probability of the charge exchange loss of produced ions colliding with low charge state ions and neutrals.

In the charge exchange process an ion and a neutral atom collide to form a molecular ion which then splits into two ions.

$$A^{z+} + B^0 \rightarrow (AB)^{z+} \rightarrow A^{(z-1)+} + B^+ \quad (2)$$

The cross- sections (σ_x) of the recombination process are very large and increase with z.

The charge exchange cross-section formula given Chibisov may be written as:

$$\sigma_x = 1.7 \times 10^{-15} z \text{ cm}^2 \quad (3)$$

As the charge transfer time limits the life time (τ_i) of a cold ion of charge state z, it may be written as:

$$\tau_i \approx \tau_{z \to z-1} \approx [n_0 v_i \sigma_x]^{-1} \approx 5 \times 10^8 \frac{\sqrt{A}}{z n_0} \quad (4)$$

where n_0 = neutral density (cm^{-3}),

A =the atomic mass number

$v_i \approx \frac{10^6}{\sqrt{A}} cm\ s^{-1}$ assuming ions are at thermal energy $T_i \approx 1 \rightarrow 5eV$

The well-known criterion for stepwise ionization may be written as:

$$\xi n_e \tau_i \geq 5 \times 10^4 (T_e^{opt})^{\frac{3}{2}} \quad (5)$$

where $\xi = \sum_{j=1}^{N} q_i$ = the total number of electrons in the outer shell

$T_e^{opt} \approx 5W_j^{max}$=the optimum electron temperature

The optimum plasma electron temperature is taken as five times the threshold energ W_j^{max}. It is easily understood from the above equation that totally stripped light ions need

$n_e \tau_i \sim 1.76 \times 10^{10} cm^{-3} s$ for electron temperature $T_e^{opt} \sim 5keV$.

To attain a charge state z the charge transfer time must be greater than the ionization time for the same z for a specified electron density n_e. When this condition is imposed to the stepwise ionization criterion, the upper limit of neutral density n_0 in the plasma is obtained for a specified electron density as:

$$\frac{n_0}{n_e} \leq 10^4 \xi \left[T_e^{opt}\right]^{-3/2} A^{1/2} z^{-1} \tag{6}$$

From this equation one can calculate the neutral density and the operating pressure of the discharge chamber for any charge state z at a particular electron density. For the ionization of hydrogen like ions $\xi = 1$. To produce Ne^{10+} in 6.4 GHz VEC-ECR a maximum typical value of electron density (n_e) can be taken as $n_{ce} \approx 5 \times 10^{11}\, cm^{-3}$.[5,6]

$$T_e^{opt} \approx 5W_j^{max} \approx 5keV \text{ , hence } \frac{n_0}{n_e} \leq 10^{-2} \text{ and } n_0 \leq 5 \times 10^9\, cm^{-3}$$

This corresponds to a pressure of 1.4×10^{-7} Torr. For Ar^{18+} the needed vacuum is about $5x10^{-8}$ Torr. So a higher frequency source operating at a higher magnetic field should give the same charge state production performance at a comparatively higher pressure. In a study [7], the vacuum requirements of ECR sources operating at different frequencies is shown in Figure 8.

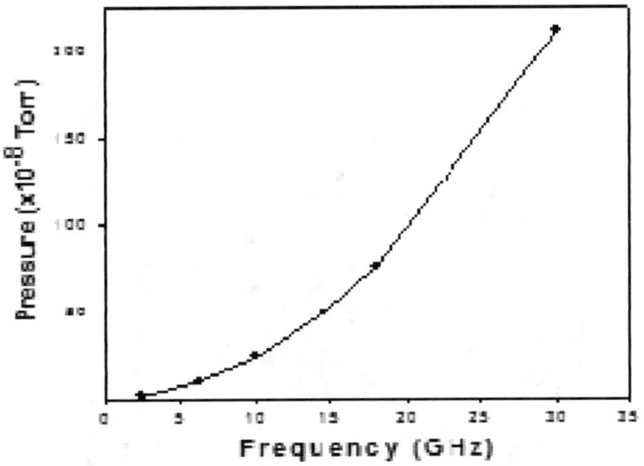

Fig. 8. Frequency vs operating pressure of ECR source [6]

Hence, the operating vacuum of the ionizer is extremely important parameter for high charge state production. The operating vacuum of the ionizer in the range of 10^{-7} Torr is essentially needed to reduce charge exchange loss and the pressure in the ion extraction chamber should be maintained lower than 10^{-7} Torr.

The main plasma cavity of the ECR ion source is 345 mm long and 108 mm in diameter which is made out of oxygen free high conductivity copper billet to meet the suitable surface condition essentially needed for good vacuum and also to satisfy the microwave properties. The chamber is water-cooled and carries 305 mm long rare earth sextupole magnet bars externally. The entire volume of the ion source is pumped by two 900 l/s special type oil diffusion pumps to achieve $5x10^{-8}$ Torr. Usually main plasma chamber is pumped by the plasma itself. Moreover an additional pumping speed (a few l/s) is provided through extraction hole and slanting pumping slots of the plasma orifice. The diffusion pumps are electrically isolated by ceramic tubes of low out-gassing rate and provided with a specially designed high voltage corona shield to avoid contamination by oil vapor fragmentation components. A pressure of about $1x10^{-7}$ Torr in the beam transport line is routinely obtained through distributed pumping by three turbo-molecular pumps to minimize scattering and charge exchange loss during beam transport.

4. Experimental vacuum chamber

As the facility has been developed for carrying out research related to atomic physics and material science with low energy beam, the sample/target essentially should be thin films and the typical target modification depth is few tens to hundreds of nano-meters. Usually the scattering loss will occur Depending upon the vacuum level there exists usual scattering loss to some extent. Beside this, there is the strong possibility that the chamber inner surface will be covered by the adsorbate which eventually will be buried into the film by the impact of the projectile ions. It introduces impurity in the implantation process producing erroneous result. The monolayer formation time basically can give us an idea about the desired vacuum. The monolayer time $T = \frac{1.94 \times 10^{-6}}{P}$ sec. Where p is the pressure in Torr. It is clearly understood that single monolayer is formed within two seconds time at a pressure of ~ 1×10^{-6} Torr. The impurity mixing rate will be reduced to one tenth due to an improvement of the vacuum by an order of magnitude. Basically the test chamber should be a bekable ultra high vacuum chamber. The multi-facility chamber used in the present facility is a similar one, which can produce ultimate vacuum of 3×10^{-9} Torr by continuous pumping.

5. Gas and Element Feeding System of ECR Source

Initially, the journey of the ion source was started to produce ions of the gaseous elements. However, due to the pressing demand of the materials scientists to use this facility, the ion beams of S, Fe, Ti and Hf have been developed using the MIVOC technique. In this method the source feed material is usually an organo-metallic volatile compound containing the element of the desired ions.

Fig. 9. MIVOC setup

We have observed that for obtaining good yield the vapour pressure of the compound should lie between 10 to 0.1 Torr at room temperature. A dedicated set up has been developed for this purpose which has the pumping and vacuum measuring facility of its own. This has been developed with a view to reduce the residual constituents and also to estimate approximately the vapour pressure of the compounds. The entire MIVOC setup is electrically isolated from the ion source, which enables its operation at ground potential. The vapour of the MIVOC compound is injected into the ion source using an electro-mechanically operated sapphire seal valve. With this arrangement it is possible to precisely control the elemental feed and one can optimize the ECR parameters for the desired ion species [figure 9]. We have observed that for certain ion species a suitable mixing gas is essential to improve the required beam yield. The facility exists to inject simultaneously three sample elements which is needed for very special experimental purposes.

6. Recent Experiments

The ion currents for some species are given in Table 1 below. Using this facility, several ion implantation experiments have been carried out. Some of these experiments were aimed to simulate the modifications and damages created due to very high dpa in nuclear structural materials. Because of the availability of very high currents from this machine (~10μA) such experiments could be carried out successfully. In addition to this, implantation experiments on semiconductors, organic semiconductors, organic single crystals for non linear optics etc. have been carried out. To mention a few are given below:

1. Experiments were carried out for the synthesis of β-FeSi2 by implantation of Fe-ions in single crystalline silicon samples. β-FeSi2 is a direct band gap semiconductor having a large number of applications in optoelectronics[8].
2. In an experiment, the Fe^{10+} ion were implanted in ZnO for synthesis of p-type ZnO [9].
3. Several charge states of Oxygen ion were implanted in PMMA samples to study the effect of charge states on the nature of material modification [10].
4. In another experiment, the tunability of dielectric constant of conducting polymer Polyaniline(PANI – Polyvinyl Alcohol (PVA) has been observed by low energy Ar^{9+} irradiation. The modified material can be used as the e-m shielding materials [11].

Users from several universities [12,13], BIT Mesra [9], IIT Rorkee[10], UGC DAE CSR [8], BARC and VECC [11,14,15] have utilized this facility for carrying out several successful experiments

Table 1. Ion Current of different species

Species	Charge States	Typical Currents (μA)
Ne	6^+ to 9^+	45 -1.5
Ar	8^+ to 13^+	100-1.5
Kr	12^+ to 19^+	45-6.5
Xe	20^+ to 31+	4 -1
C	2^+ to 5^+	100-20
O	5^+ to 7^+	120-12
N	4^+ to 6^+	130-31
Fe	10^+ & 11^+	10
Ti	11^+	7
S	10^+	2.5
Hf	12^+-29^+	5 -0.3

7. Conclusions

Low energy ion beam implantation has been proved to be an extremely valuable tool for atomic physics, surface physics and material science research. Ion loses energy by colliding with the atom of the material and displacing them from their lattice position generating enormous defects creating drastically new properties. Nuclear energy loss principally dominate at this energy regime. ECR based low energy facility can deliver very high ion currents for various elements at variable energies and very high value of DPA can be achieved easily for carrying out material damage experiments. Due to the low ion energy, the vacuum requirement for the whole facility is equally stringent as the production of multiply charged ions and delivering it on the target. A sufficiently adequate volume of the multi-facility chamber prompts us to incorporate many more on line experimental facility in the set- up. Facility for on line IR temperature profiling and the heating/cooling of the samples at desired temperature will be commissioned very soon.

References

[1] Taki G S, Chakraborty D K and Bhandari R K 2002 *PRAMANA- J. Phys.* **59** 775
[2] Taki G S, Sarma P R, Chakraborty D K, Bhandari R K, Ray P K and Drentje A G 2006 *Rev. Sci. Instr.* **77** 03A310
[3] Taki G S, Sarma P R, Drentje A G, Nakagawa T, Ray P K and Bhandari R K 2007 *Chinese Journal of High Energy Physics and Nuclear Physics*, **31**,170
[4] Lyneis C M 1986 *Report on Summer School on Accelerator Technology, Osaka, Japan* p.4-4
[5] Geller R 1996 *"ECR Ion Sources and ECR Plasmas" (IOP Publishing Bristol and Philadelphia)*
[6] Taki G S and Bhandari R K 2007 *Proc. DAE-BRNS-PSI Symposium on Ion Beam Technology and Applications SIBTA, Mumbai* p-194
[7] Ghosh Subhash, Taki G S, Mallick C and Bhandari R K 2008 *Journal of Physics, Conference Series, IOP publishing,* **114** 012066
[8] Rajesh P V, Pati S P, Krishna J B M, Ghosh B, Das D 2011 *Proceedings of DAE Solid State Physics Symp.*
[9] Kumar Ashutosh, Krishna J B M, Das D, Keshri Sunita *2012 Applied Surface Science* **258(7)** *2237-2245*
[10] Bind U C, Krishna J B M, Dey Kamal, Taki G S, Yadav K L and Dutta R K 2011 *Proc Nuclear andRadiochemistry Symposium (NUCAR- 2011), GITAM University, Vishakapatnam.*
[11] Himanshu A K, Bandyopadhayay S K, Sen P, Mondal N N, Talpatra A, Taki G S. Sinha T P 2011 *Radiation Physics and Chemistry* **80** 414.
[12] Bakiaraj G, Krishna J B M Taki GS, Dhanasekaran R 2011 *Proc. ICMAT11-A-2438 SUNTEC Singapore*
[13] Kannapan P, Krishna J B M, Taki G S, Dhanasekaran R 2011 *Proc. Symposium EE: CGCT-5-Fuctional Materials Crystallization, Characterization and Devices ICMAT11-A-2182*
[14] Bandhyopadhyay S K, Himanshu A K, Taki G S, Krishna J B M 2010 *Proc. Symposium on Current Trends in Condensed Matter Physics, NISER*
[15] Bandhyopadhyay S K, Himanshu A K, Taki G S 2010 *Proc. Symposium on Current Trends in Condensed Matter Physics, NISER,*

Journal of Physics: Conference Series **390** (2012) 012005 doi:10.1088/1742-6596/390/1/012005

Sustained performance of 8 MeV Microtron

Ganesh Sanjeev

Microtron Centre, Department of Studies in Physics, Mangalore University, Mangalagangotri – 574199, DK, Karnataka, INDIA

E-mail: ganeshsanjeev@rediffmail.com

Abstract. Energetic electrons and intense bremsstrahlung radiation from 8 MeV Microtron are being utilized in variety of collaborative research programs in radiation physics and allied sciences involving premier institutions of the country and sister universities of the region. The first of its kind electron accelerator in the country, set up at Mangalore University in collaboration with RRCAT Indore and BARC Mumbai, has been facilitating researchers since its inception with its inherent simplicity, ease of construction, low cost and excellent beam quality. A bird's eye view on the reliable aspects of the machine, efforts behind the continuous operation of the accelerator and important applications of the accelerator in physical and biological sciences are presented in this paper.

1. Introduction

An 8 MeV Microtron facility at Mangalore University, established in the year 1995, has become a major tool to the researchers fromnational laboratories and sister universities of the region for their systematic investigations in the area of radiation dosimetry, radiation effects on semiconductor materials and devices, radiation damage studies in biological systems, radiation processing of food items, cross linking of polymers, angular distribution of photo-fission and a number of other research studies in basic and interdisciplinary areas of science and technology. The results obtained have helped to understand the basic mechanism involved in the fundamental processes and have also been useful for application oriented studies. The salient features of the facility are discussed elsewhere [1].

The accelerator with its inherent simplicity, excellent beam quality, easy operation and maintenance aspects has become a focal point for variety of research programs in collaboration with premier laboratories of the country. The first of its kind versatile electron accelerator in the country has been delivering 8 MeV electrons, intense bremsstrahlung radiation of peak energy 8 MeV and neutrons of reasonably high flux successfully in the last one and half decade for conducting advanced research in several frontline areas of radiation physics and allied sciences. All these years, the indigenously developed facility has been maintained with a vacuum level around $7 x 10^{-7}$ mbar with the help of two sputter ion pumps (270 l/s & 35 l/s) and a turbo-molecular pump (300 l/s). Varieties of smaller experimental (vacuum) chambers were used in varieties of experiments at different vacuum levels.

2. Collaborative Research Programme

A systematic collaborative research program is being conducted through a triangular MoU signed by Mangalore University, RRCAT Indore and BRNS, DAE, Govt. of India. Several research groups from different institutions are involved in this program. A brief account of all these activities along with sample results is given below.

2.1 Dosimetry Studies

Radiation dosimetry plays an important role in any radiation installation. Dosimetry studies have been carried out for high energy photons and electrons using wide varieties of chemical dosimeters along with thermo luminescent dosimeters to measure radiation doses and dose distributions from the Microtron. Fricke dosimeter, FBX dosimeter, Glutamine, Alanine and Thermoluminescent dosimeters were used in the study for different dose ranges. An effort is also being made to use different materials as dosimeters based on their behavioural variation in the influence of radiation.

Radiation dosimetry studies have shown the electron beam distribution and its dependence on the beam line magnet fields. Dosimeters are used to obtain the dose details of the machine along with beam distribution along the axis through which it propagates. Several trials of this experiment at different distances from the titanium window showed that the machine has uniform electron beam distribution of 40 mm in either side of the beam axis in vertical and horizontal directions at distances of 30 cm from Titanium window. The distribution was observed without any kind of beam flattener or scatterer.

Thermoluminescent dosimeters were used in radiation mapping of the microtron room and the experiment gave an idea of the radiation environment of the microtron facility inside the microtron room.

2.2 Photo-fission experiments

The angular distribution of ^{237}Np fission fragments was measured over the energy range 7.4 - 9.2 MeV in steps of 0.4 MeV of bremsstrahlung radiation. A vacuum chamber and a target needed for these experiments have been designed and got fabricated in collaboration with Nuclear physics Division and Radiochemistry Division, BARC, Mumbai. Lexan polycarbonate films (SSNTD) were used to detect the fission fragments as the technique is a simple and advantageous over other methods.

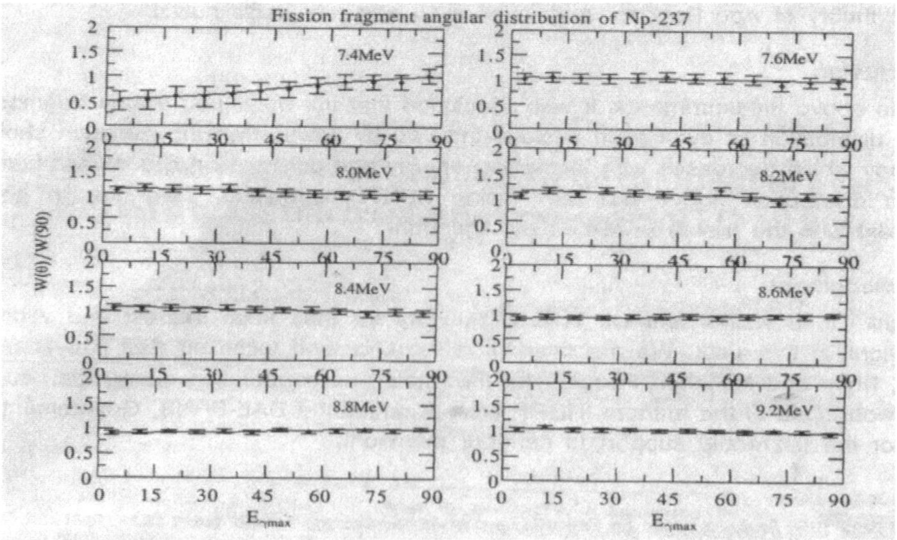

Fig.1. Fission fragments angular distributions of ^{237}Np induced by bremsstrahlung radiation

The photon intensity from the Microtron accelerator at a distance of 15 cm from the bremsstrahlung converter (tantalum target) facility was estimated to be 10^{10} photons/s using the code EGS-4. The photofission cross-section was calculated using the angular distribution of fission fragments and the results were compared with those obtained using the code EMPIRE-II and various barrier parameters of the RIPL-1, RIPL-2 libraries and with the new analytical fission barrier formula based on the HugenholtzeVan Hove theorem. The results are in good agreement with the prediction of the RIPL-1 barrier parameter and analytical fission barrier formula [2].

2.3 Studies on Relative Biological Effectiveness

Radiation biophysics study has been initiated to work out all possible physical and biological parameters for bremsstrahlung and electron beam from microtron. Systematic investigations of various parameters influencing biological response of the systems are in progress.

The extent of repairable and irreparable damage in aliving cell produced by ionizing radiation depends on thequality of the radiation. In the case of sparsely ionizingradiation, the dose rate and the pattern of energydeposition of the radiation are the important physicalfactors which can affect the amount of damage in livingcells.

Radio-sensitive and radio-resistive bacteria cells were exposed to 8 MeV pulsedelectron beam and the efficiency of cell-killing was investigated to evaluate the mean lethal dose. The dose to the cell was delivered in micro-second pulses at an instantaneous dose rate of 2.6 x 10^5Gy/s. The results were compared with those of gamma rays. The survival curve of radio-resistive Deinococcus-radiodurans (*DR*) is found to be sigmoidaland the survival response for radio-sensitive Escherichiacoli (*E-coli*) is found to be exponential without anyshoulder. Comparison of Do values indicate that irradiation with pulsed electron beam resulted in more cell-killing than was observed for gamma irradiation [3].

The differences in the cell survival efficiency and dose rate effect in diploid yeast strains Saccharomyces cerevisiae X2180 and Saccharomyces cerevisiaeD7 under euoxic and hypoxic condition have been quantified. Irradiation was carried out using a dose per pulse of 0.6 Gy and pulse width of 2.3 μs, which correspond to an instantaneousdose rate of 2.6 x10^5Gy/s. A significant difference in the dose response has been observed under euoxic and hypoxic condition. Dose rate effect has been studied by changing the pulse repetition rate of the electrons beam. A significant dose rate effect was observed under euoxic condition for Saccharomyces cerevisiaeX2180 but the same was absent under hypoxic condition. The dose rate effect was absent for Saccharomyces cerevisiae D7 under irradiation condition. The survival curves are found to be sigmoidal in shape under both conditions but with a wider shoulder under hypoxic condition. Extensive work in this field of research is in progress.

2.4 Irradiation Effects on Semiconductor Devices

The study on irradiation effects on semiconductor materials and devices was carried out with an aim to investigate modifications in basic characteristics on inducement of damages in semiconductor materials and devices due to high-energy photon/electron irradiation. Semiconductor devices such as diodes, transistors, solar cells, photo-detectors etc., that need to be operated in space or in radiation environment are often subjected to various types of radiation. These components therefore need to be tested for radiation exposure to determine their tolerance and suitability in different applications. The study of the effect of ionizing radiation on semiconductor materials / devices is not only important to have broader understanding of the damage process leading to various modifications, but is essential to assess the device performance when it needs to be operated in radiation environment.

A comparative study on the effect of 8MeV electron irradiation on electrical characteristics of CdTe and CIGS thin film solar cells was carried out after having an in-depth study on influence of radiation on different types of solar cells. I–V characteristics of the cells under AM 1.5 illumination condition and capacitance-frequency measurements before and after irradiation were carried out. Solar cell parameters like short circuit current (I_{sc}), open circuit voltage (V_{oc}), fill factor (FF) and efficiency (η) were calculated from I-V characteristics. It is found that the efficiency of the CIGS and CdTe solar cells decreases as electron dose is increased to 75 kGy. The other cell parameters such as FF and V_{oc} do not seem to be affected much by irradiation.

p-CdTe/n-CdS thin film solar cells were exposed to electrons and they were characterised in dark and illumination conditions in order to test the device stability in radiation environment. Solar cell parameters like short circuit current, open circuit voltage, fill factor, conversion efficiency, saturation current and ideality factor have been considered. The study reveals that the thin film solar cells exhibit good stability against electron dose up to 100 kGy.

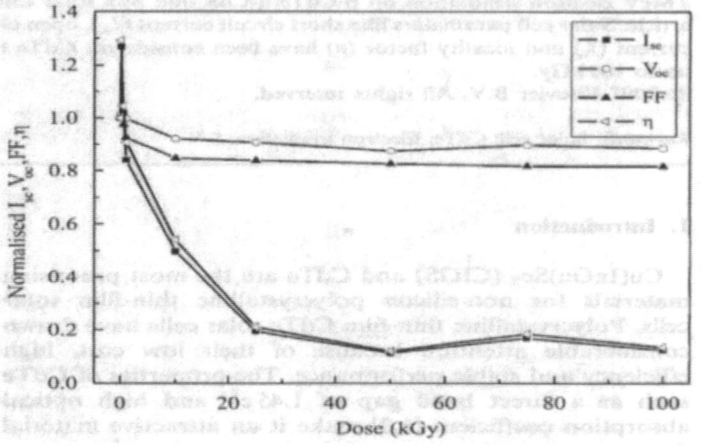

Fig.2. Normalised solar cell parameter as a function of dose

Irradiation studies on p-Si and SS/CdTe/Au Schottky diodes explain that the effect of irradiation is more pronounced at higher voltages than at lower voltages and the main effect of irradiation is reduction in forward current with increasing dose. The degradation in the diode properties may be due to the introduction of radiation induced interfacial defects via displacement damage.

Effect of 8 MeV electrons on Silicon Photo detector fabricated by diffusion of phosphorous into the p-type mono-crystalline silicon wafers of <100> orientation, was carried out. A p+ back surface field layer was created on the rear surface of the silicon wafer by depositing aluminium. As front and back ohmic contacts, metallic coating consisting of titanium, palladium and silver deposited using ion beam sputtering were used. The silicon photo detectors were characterized and the solar cell parameters were determined under dark and AM0 conditions. From the study it is observed that both forward and reverse currents increased systematically with electron dose [4].

A systematic study on solar cells in radiation environment is being carried out using different varieties of solar cells with different conversion efficiencies. Electrical characterizations at different conditions are being studied extensively in collaboration with UNAM Mexico and NREL, USA.

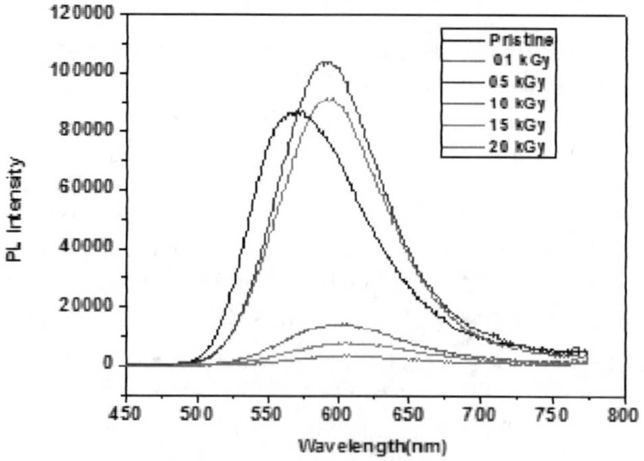

Fig. 3. Photolumenescent Spectra of CdTe quantum dots at various electron doses

2.5 Radiation effects on Quantum dots

A novel one pot synthesis of CdTe Quantum dots (QD) using hydrothermal technique has been developed using Na_2Teo_3 as Te source and $CdCl_2$ as cadmium source. CdTe QD was exposed to various electron doses, ranging from 1kGy to 20 kGy. The characterisations of the samples have shown interesting changes in photoluminescence properties as shown in Figure 3. An increase in emission intensity for doses up to 5 kGy and decrease in emission intensity for doses more than 5 kGy were observed. A Red shift in emission spectra was also observed. [5]

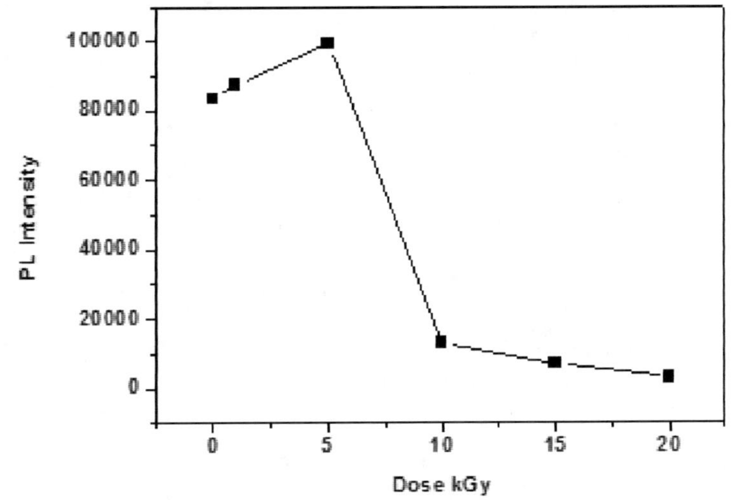

Fig. 4. Photoluminescent yield (peak) at different electron doses

The results were ascertained using time resolved emission spectroscopy and X -ray photo electron spectroscopy of the QDs. The CdTe QDs can be used in Bio tagging as CdS layer formed (upon irradiation) on CdTe Core is Bio friendly and formation of CdS can enhance photo emission, giving rise to bright emission for easy detection of tagged DNA (any bio material). Detailed studies in this area are in progress.

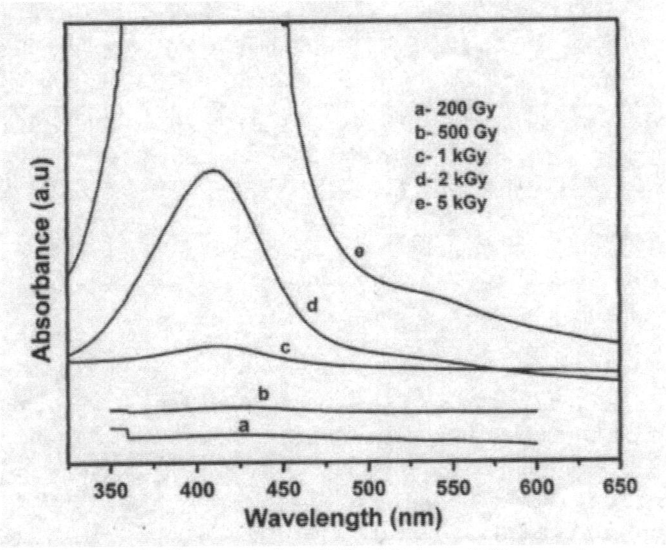

Fig. 5. UV-Vis Spectra of $AgNO_3$:PVA at different electron doses

2.6 Nanoparticle Synthesis

Stable, non toxic Ag nano-particles have been synthesized by exposing an aqueous solution of AgNO3 and PVA having two different degree of hydrolysis acting as the stabilizer, to 8 MeV electrons. The rate of formation of nanoparticles was controlled by changing either the electron dose or the relative concentration of the precursors. The size, shape, and the rate of formation of the nanoparticles depend on the final dosage, as well as the weight ratio of AgNO3 and PVA. The formation of Ag nanoparticles was confirmed by UV–Vis spectroscopy and the nanoparticle size was established through TEM result analysis. Increasing the irradiation dosage seems to favour the formation of polygonal nanostructures. DSC measurements show that there exist a strong interaction between the PVA matrix and the Ag nanoparticles. The method of synthesis of nanoparticle is a clean method may find these useful for medical applications [6].

2.7 Radiation effects on thin films

From the investigations carried out on the electrical behavior of silver particulate films deposited on electron beam irradiated polystyrene (PS) coated substrates held at a temperature of 455K in a vacuum of 8 x10^{-6} torr at a constant deposition rate of 0.4 nm/s, it is observed the polymer-metal interaction can be brought about in inert polymers like PS by electron irradiation. Further, it was shown that the morphology is dependent on polymer-metal interaction. The results indicate that the films deposited on PS irradiated to a dose of 20kGy and 25 kGy give rise to smaller clusters with smaller inter-cluster separation, better suited for sensor applications. The induced polymer-metal interaction is attributed to the creation of free radicals due to electron beam irradiation [7,8].

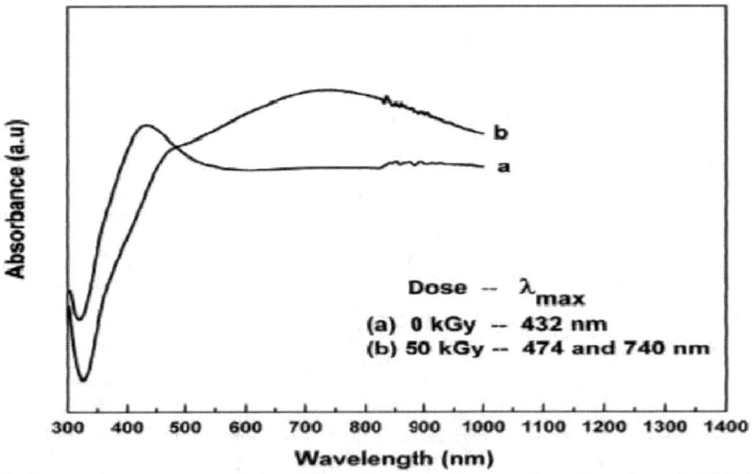

Fig. 6. Optical absorption spectra of Ag films deposited on unirradiated and 50 kGy irradiated PS

2.8 Radiation Processing

Varieties of polymers and materials have been studied by exposing them to energetic electrons at graded doses. A few of them have been detailed below:

2.8.1 Poly Vinyl Alcohol

PVA is a widely used, non-toxic, water soluble, biocompatible, and biodegradable synthetic polymer with excellent mechanical properties such as flexibility and good film forming quality with excellent gas barrier ability [9]. Irradiation effects on the microstructural parameters in PVA films have been studied using Wide Angle X-ray Scattering (WAXS) data. The crystal imperfection parameters like crystal size <N>, lattice strain (g in %) and enthalpy ($\square^{*}$) have been determined by Line Profile Analysis (LPA) using Fourier method of Warren.

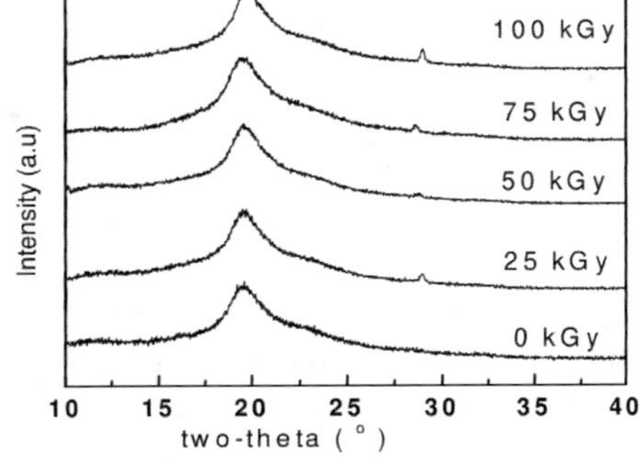

Fig.7. XRD scans

From WAXS study, significant change in the values of micro structural parameters was observed after irradiation even though there was not much change in the position of the X-ray reflection. The crystallinity of the polymer also found to be decreasing with increase of radiation dose.

2.8.2 Polyaniline (PAni)

Electron beam irradiation of PAni has yielded crystallisation in interfacially polymerised polyaniline. This was ascertained from UV-Visible spectra, XRD and impedance analysis. PAni crystallises in orthorhombic system at an optimum electron dose 20 kGy, which is the best observed feature. Though there is an improvement in crystallinity in low dose electron irradiated PAni, the conductivity decreases due to the electron induced chain scission, whereas in 40 kGy dose, crosslinking of the polymer chain leads to amorphization. The impedance analysis of PAni after irradiating it to different doses, gives Nyquist plots.

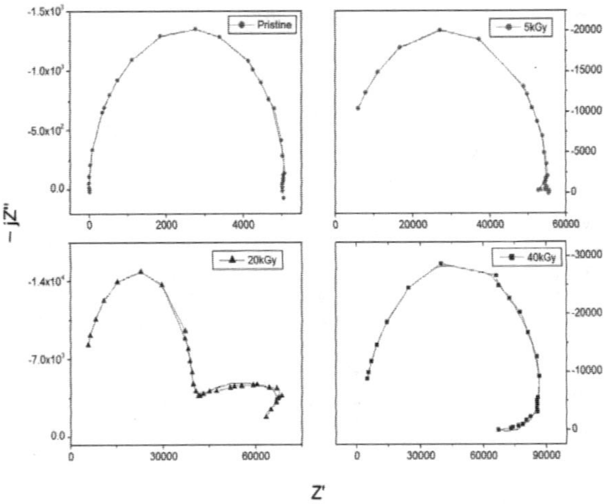

Fig.8. Nyquist plots for pristine and electron induced polyaniline nanomaterial

The above figure shows the Nyquist plots of pristine, 5, 20 and 40 kGy dose electron irradiated PAninanomaterial. The pristine PAni shows a single semi circle indicating a single relaxation process. In the 5 kGy dose irradiated PAni, the Nyquist plot shows a single semi circle with a pseudocapacitance loop at low frequency region.

Interfacially polymerised PAni irradiated with 8 MeV electron beam shows electron induced chain scission atlow doses and crosslinking. The structural change is also reflected in the electrical properties of PAni. This can be seen in the 20 kGy dose, in which the formation of subgrains results in two semicircles–grain boundary and grain interior arcs. It is concluded that the crystalline polyaniline can be obtained by 8 MeV electron beam irradiation for which 20 kGy is the optimal dose [10].

2.8.3 PMMA

Free volume related fluorescence properties of electron irradiated chalcone doped PMMA films have been studied using positron annihilation and fluorescence spectroscopic techniques. In this polymer composite, enhancement of fluorescence at lower doses and reduction at higher doses has been observed under electron irradiation. From Positron annihilation studies suggests that at lower doses of irradiation induced crosslinking which affect the free volume properties and inturn hinders the chalcone molecular rotation. At higher doses chain scission process affect matrix relaxation. Under the restricted condition the chromophore molecules likely to emit enhanced fluorescence and its mobility is directly related to the free volume around it [11].

2.8.4 Lexan Polycarbonate films

Modifications in the properties of Lexan polycarbonate (Lexan) films exposed to 8 MeV electrons have been studied. The UV-Visible spectroscopy results showed that there is decrease in optical energy gap, optical activation energy and increase in number of carbon atoms per cluster with increase in electron dose. The chemical changes in electron irradiated polymers due to chain-scission and reconstruction have been observed from FTIR spectroscopy results. The correlation of positron lifetime study with optical measurement is obtained and an attempt is made to understand the electron irradiation induced microstructural modifications within the polymer. The XRD pattern shows that the crystallite size and percentage of crystallinity decreases due to the breakage of bonds resulting change of semicrystalline structure of polymer to amorphous. The surface morphology studied by SEM shows the blisters formation of various sizes on electron irradiation and finally gradual degradation in the network structure was also observed. DSC studies showed a decrease in glass transition temperature and heat of fusion after irradiation due to chain-scission which indicates polymer was moving towards more disordered state. Thermal decomposition temperature of Lexan polycarbonate increases upon irradiation

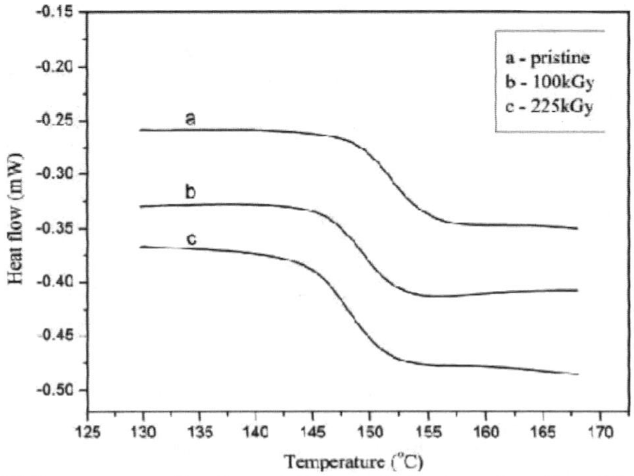

Fig. 9. DSC thermogram for the pristine and electron-irradiated Lexan films

The glass transition temperature (*Tg)* for the pristine sample was around150°C. At the dose of 100 kGy, the DSC curve showed a shift of *T*gto 146°C, and for 225 kGy, *T*gwas observed at 143°C. After irradiation, the shift of *T*gto lower temperature reveals that electronirradiation leads to chain scission and subsequently to reduction in molecular weight. As a result,the polymeric system was changing towards a more disordered state [12]

The a.c conductivity and dielectric constant of the Lexan film were found to increase with increase in electron dose due to the breakage of chemical bonds which increases the number of free radicals. The formation of defects and chromophores, and the presence of impurities, additives and unsaturation have been studied using photoluminescence technique

Several other collaborative research activities in the area of materials science, fisheries, radiobiology, radio-protectiveness of various plant extracts, radiation processing of ayurvedic medicines for shelf life increasal and other related aspects are also being carried out.

3. Program for users

Workshops, Seminars and Users Meeting on different areas of research are being organized to encourage both in building up the laboratory facilities and to use the facility for carrying out front-line research in the field.

4. Future plans

Up-gradation of the existing accelerator facility with higher beam power to cater to the needs of the Microtron Users is planned with the support of BRNS. Detailed neutron based studies on various aspects would be taken up and ongoing activities would be strengthened. It is also planned to extend the Microtron facility to postgraduate students for their laboratory experiments / project works.

Acknowledgement

The author is thankful to the research team at the Microtron Centre, concerned scientific personnel at RRCAT, Indore and BRNS for their valuable help and support. The author gratefully acknowledges the contributions of concerned Microtron users to this paper.

References

[1] Siddappa K, Ganesh S, Ramamurthy S S, Soni H C, Srivastava P, Sheth Y and Hemnani R 1998, Journal of Rad.Phys and Chem., Vol.51, No.4-6, p.441
[2] Rajaprakash H G, Ganesh Sanjeev, Vijay Kumar K B, Siddappa K, Nayak B K, Saxena A, 2011, Radiation Measurements 46, 413-417
[3] Praveen Joseph , Santhosh Acharya, Ganesh Sanjeev , Bhat N N, Narayana Y, 2011, J Radioanal Nucl Chem., 290:209–214
[4] Sheeja Krishnan, Ganesh Sanjeev, Manjunatha Pattabi, Xavier Mathew, 2009, Solar Energy Materials & Solar Cells,93, 2-5.
[5] Haobo Bao, Yanjun Gong, Zhen Li, and Mingyuan Gao , 2004, Chem. Mater., 16 3853-3859.
[6] Manjunatha Pattabi, Rani M Pattabi, Ganesh Sanjeev, 2009, J Mater. Sci:Mater Electron, 20:1233-1238
[7] Manjunatha Pattabi, Gurumurthy S C, Ganesh Sanjeev, Gaikwad A B, 2011,J Mater Sci: Mater Electron , 22:1095–1100
[8] Manjunatha Pattabi, Gurumurthy S C, Ganesh Sanjeev, Gaikwad A B, 2011, Nuclear Instruments and Methods in Physics Research B 269 (2011) 1534–1539
[9] Hassan C. M., Ward T. H., Peppas N.A. 2000.,*Polymer*, 41, 6729-6739.
[10] Chithra Lekha P, Balaji M, Subramanian S, Subramanian E, Ganesh Sanjeev, Pathinettam Padiyan D, 2011, Proc. of NAARI conference, Mumbai
[11] Ismayil V. Ravindrachary, Bhajantri R. F., Harisha A and Praveena S D, 2010, solid state physics, proceedings of the 55th dae solid state physics symposium, AIP Conf. Proc. 1349, pp. 196-197
[12] Hareesh K, Ganesh Sanjeev 2011, Materials Sciences and Applications, 2, 1682-1687

New Approach to meet vacuum requirements in UHV/XHV systems by Non Evaporable Getter Technology

E Maccallini,[1] P Manini, A Conte, F Siviero and A Bonucci
SAES®[2] Getters S.p.A., viale Italia, 77, 20020, Lainate (Mi), Italy

E-mail: enrico_maccallini@saes-group.com

Abstract. UHV/XHV users are increasingly demanding vacuum pumps with high performances with smaller package and weight. This requires a more efficient combination/integration of the currently available pumping technologies and, possibly, the development of new approaches in vacuum pumping systems. Non Evaporable Getter (NEG) pumps represent one of the most appealing option to UHV/XHV pumping, thanks to his large pumping speed and sorption capacity for several active gases (H_2, H_2O, CO, O_2, N_2, ...). NEG pumps can also reduce the pump down and bake-out time and can keep a stable pressure level in UHV conditions even if the other pumping systems are switched off. However, NEG pumps are unable to sorb, in UHV/XHV systems, the small amount of noble gases and methane which, on the other hand, can be removed by sputter ion pump (SIP). For this reason, extensive studies have been carried out to combine NEG and SIP technologies. Some results of the vacuum characterization of such combination are discussed in this work and compared with traditional pumping approaches based on large SIP. A further step forward is the NEXTorr®[3] pump which is an integration of a NEG pump (pumping speed ranging between 100 to 500 l/s for hydrogen) with a SIP (pumping speed of 6 and 10 l/s for Ar and CH_4 respectively) into a single small package unit. Examples of applications of this new approach in vacuum technology will be given to demonstrate the simplification of the design and operation of UHV/XHV systems.

1. Introduction

The design and operation of UHV/XHV systems requires the development of vacuum groups able to deliver higher pumping performances with a smaller package. In this framework, Non Evaporable Getter (NEG) pumps can satisfy these requirements due to the ability to chemically and permanently remove active gases in vacuum system [1]. However NEG pumps are unable to adsorb methane and noble gases and for this reason they are used together with sputter ion pump (SIP) [2]. The classical approach is to combine a large SIP with a compact NEG pump of comparable pumping speed. However, the analysis of the residual gases in UHV systems shows that hydrogen is the main gas, while methane and other noble gases are several orders of magnitude smaller. Therefore the most efficient approach to achieve UHV condition may be based on the use of high efficiency pumps to remove hydrogen backed by smaller pumping systems to remove the small amount of methane or noble gases. The use of performing pumping system to remove hydrogen could bring additional

[1] To whom any correspondence should be addressed.
[2] SAES® is an International Trademark registered by the "Madrid System" property of SAES Getters S.p.A.
[3] NEXTorr® is an International Trademark registered by the "Madrid System" property of SAES Getters S.p.A.

benefits into the UHV system. In fact if NEG pumps are activated and heated at moderated temperature during the bake out process, a reduction of either the baking time or the baking temperature can be achieved, obtaining even better pressure level at the end of the process in a shorter time [3].
In this framework, recently a new combination pump has been developed: the NEXTorr® pump.
The NEXTorr® is the combination of a NEG pump (100÷500) l/s for hydrogen with a 5 l/s (N2) SIP pumping speed.
A critical overview of the pumping performances obtained by using several types of SIP and NEG pumps is presented. In addition, a characterization of the pump down properties of NEXTorr® will be given and compared with SIP.

2. Experimental details

Tests 1 and 2: the tests were carried out on an 8.8 l volume AISI 304 steel chamber having a total inner surface area of 3000 cm^2. The chamber was fitted with an extractor gauge (detection limit is 10^{-12} mbar) and was leak checked down to 10^{-10} mbar l / s using a helium leak detector. The extractor gauge and the RGA were calibrated using a spinning rotor gauge. Primary vacuum was achieved by means of a 60 l/s (N2) turbomolecular pump (TMP) backed by a 1.8 l/s scroll pump. In test 1, a 60 l/s (N_2) SIP was connected to the bottom flange of the chamber and the NEG pump (Capacitorr®[4] D 400-2) was integrated into the SIP. In test 2, the 60 l/s SIP was replaced with 10 l/s SIP fitted directly to the test chamber. The NEG pump was also connected to the chamber directly. In both tests, pressure profiles were achieved first with the SIP alone (baseline) and then adding the NEG pump. During the tests, the vacuum chamber was first pumped with the TMP to less than 10^{-6} mbar. The SIPs were then switched on and the bake-out started. The bake-out was carried out at 150 °C for 18 h, with 1 h to ramp up and 2 h to cool down. At the beginning of the bake-out, the NEG pump was activated and then kept at 200 °C. At the end of the baking, when the chamber was still moderately warm (60–80 °C), the NEG pump was re-activated.

Test 3 and 4. The two tests were carried out mounting two NEXTorr models (NEXTorr D100-5 and NEXTorr D200-5 respectively), in an AISI 304 test chamber equipped with a 210 l/s (N2) TMP and a $3m^3/h$ rotary pump. The chamber was fitted with an Extractor gauge (detection limit is 10^{-12} mbar) and a quadrupole mass spectrometer (0-100 a.m.u.), which was calibrated against a certified Bayard Alpert gauge for H_2, CO, N_2, CH_4, O_2, Ar, CO_2 and He. In the NEXTorr design (see figure 1), NEG cartridges are integrated with a small ion pump (red box in figure 1) into one, very compact, functional unit. The getter cartridge provides large pumping speed and capacity, acting as the main pump for the removal of active gases. The ion pump has the task of removing ungetterable gases like argon and methane. Tests were carried out as follows. The test chamber was pumped with the mechanical pumps for 30 minutes down to 10^{-6} mbar and helium leak checked. Baking of the chamber, including the NEXTorr pump was started. In the first test the baking was carried out at 170°C for 18 hours, while in the second test the baking conditions were 190°C for 18 hours. In both series, after one hour from the start of the bake, the NEG elements of the NEXTorr pumps were activated at 500°C x 1h. Afterwards the NEG element was kept hot at 300°C. Two hours after the end of the baking, while the system was cooling down, the ion element of the NEXTorr was degassed, the NEG pump re-activated, to recover its full pumping speed, and the ion element of the NEXTorr switched on. In both tests, pressure profiles were achieved first with the SIP alone, featuring 120 l/s (N_2) and 200 l/s (N_2) of pumping speed respectively (baseline).

[4] CapaciTorr® is an International Trademark registered by the "Madrid System" property of SAES Getters S.p.A.

Figure 1. Picture of NEXTorr® D100-5 (2011-2013 © SAES Getters S.p.A.)

3. Experimental results

The pump-down results in test 1 and 2 are showed in figures 2a and 2b. In both experiments, the achieved pressure values are lower than $3x10^{-11}$ mbar. Compared to the experiments performed with only the SIP (baseline), an improvement of the pressure by factors of 6 and 30, respectively, was obtained.

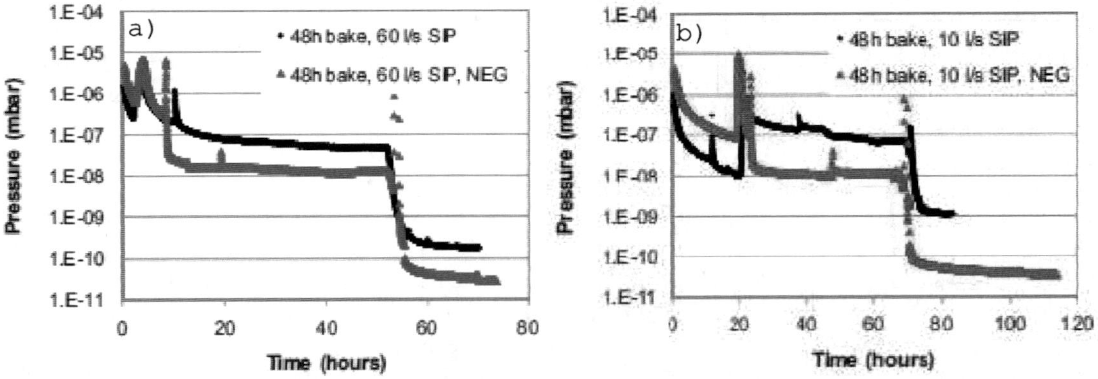

Figure 2. Pump down curves with (red curve) a without (black curve) the NEG pump in the chamber baked at 150°C for 48 h with SIP of a) 60 l/s and b) 10 l/s.

The improvement of the pressure in both experiments can be ascribed to 1) the activation of the NEG pump at the very beginning of the bake out process and 2) keeping the NEG at ≈300°C during the baking (conditioning mode). The conditioning temperature is sufficient to promote surface to bulk gas diffusion keeping a fresh getter surface which adsorbs all active gases degassed during the baking [4]. In addition, in conditioning mode, the pumping speed of the NEG pump will be higher, reducing in this way the achievable ultimate pressure. It must be noted that under the study conditions the used getter capacity during the bake-out process is a negligible fraction of the pump getter capacities (see table 1).

Table 1. Gas capacities of the NEG pump used into the test 1 and 2. The capacity of H2O, CO and N2 is relative at single run of adsorption. The sorbed quantity of each gas has been calculated considering the either the H2 or the H2O as the main gas species.

Gas	Capacity (mbar·l)	Sorbed capacity (mbar·l)
H_2	900	0.86
H_2O	80	0.57

This approach may be particularly advantageous in synchrotrons and accelerators, which can enhance the users operation. These results demonstrate that the getter can be used as the main UHV-XHV pump keeping the overall pump size sufficiently small. The weight vs pumping speed for NEG and SIP pumps is provided for reference in figure 3.

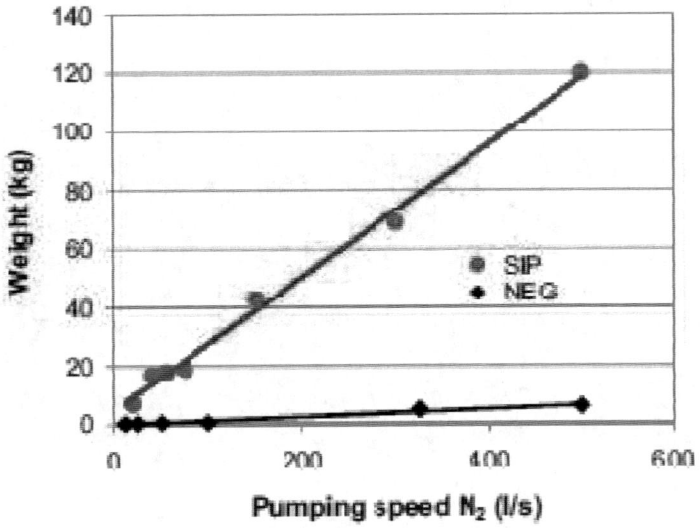

Figure 3. Weight comparison between NEG and SIP pumps as a function of their pumping speed for N_2.

This result is not totally new. In particular, Benvenuti et al. achieved [5, 6] XHV conditions ($<10^{-13}\div10^{-14}$ mbar) by using the NEG strip as the main system while the SIP is the secondary pumping system in order to remove the tiny amount of the other active gases that NEG pump are unable to adsorb.

The concept of combining large NEG with smaller SIP fully applies in the NEXTorr pump family. In particular in the present study, the NEXTorrD100-5 and NEXTorr D200-5 have been tested. The two pumps are a combination of a SIP of 5 l/s pumping speed and NEG pumps of 100 l/s and 200 l/s of pumping speed for H_2 respectively. They have been compared with 65 and 200 l/s(N2) SIP respectively (tests 3 and 4). The setups of the two experiments are shown in figures 4 and 5 respectively.

Figure 4. Experimental setup of tests 3 by using NEXTorr D100 (left) and SIP featuring 120 l/s of pumping speed (right).

Figure 5. Experimental setup of tests 4 by using NEXTorr D200 (left) and SIP featuring 200 l/s of pumping speed (right).

The base pressure achieved by NEXTorr D100-5 in the pump down experiment (test 3) was a factor 2-3 lower than that obtained by the ion pump (see figure 6). This is an important result considering the much smaller size and weight of the pump. Similar results have been achieved using the NEXTorr D200-5 (see figure 7). Interestingly, if the bake out time is reduced from 18 h to 7 h, the vacuum level obtained by using the NEXTorr D200-5 does not change Therefore even in the case of the NEXTorr pump, the use of the NEG element during the bake out process can allow to reduce either the ultimate pressure value and/or the baking time.

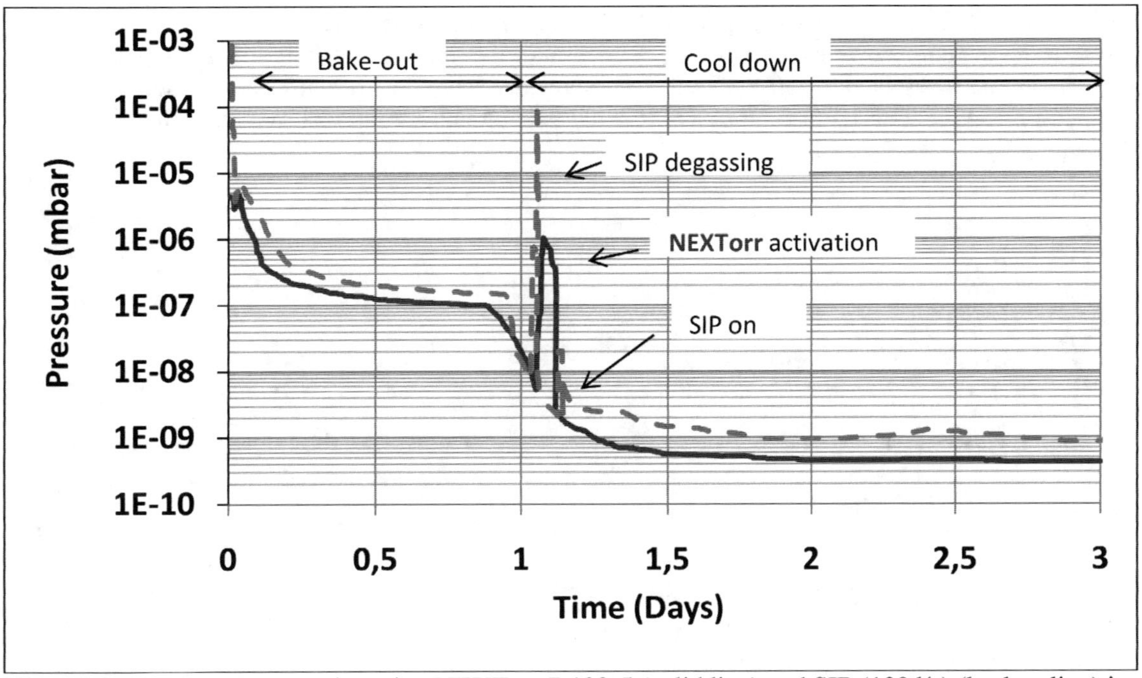

Figure 6. Pumpdown curves by using NEXTorr D100-5 (solid line) and SIP (120 l/s) (broken line) in the chamber baked at 170°C for 18 h.

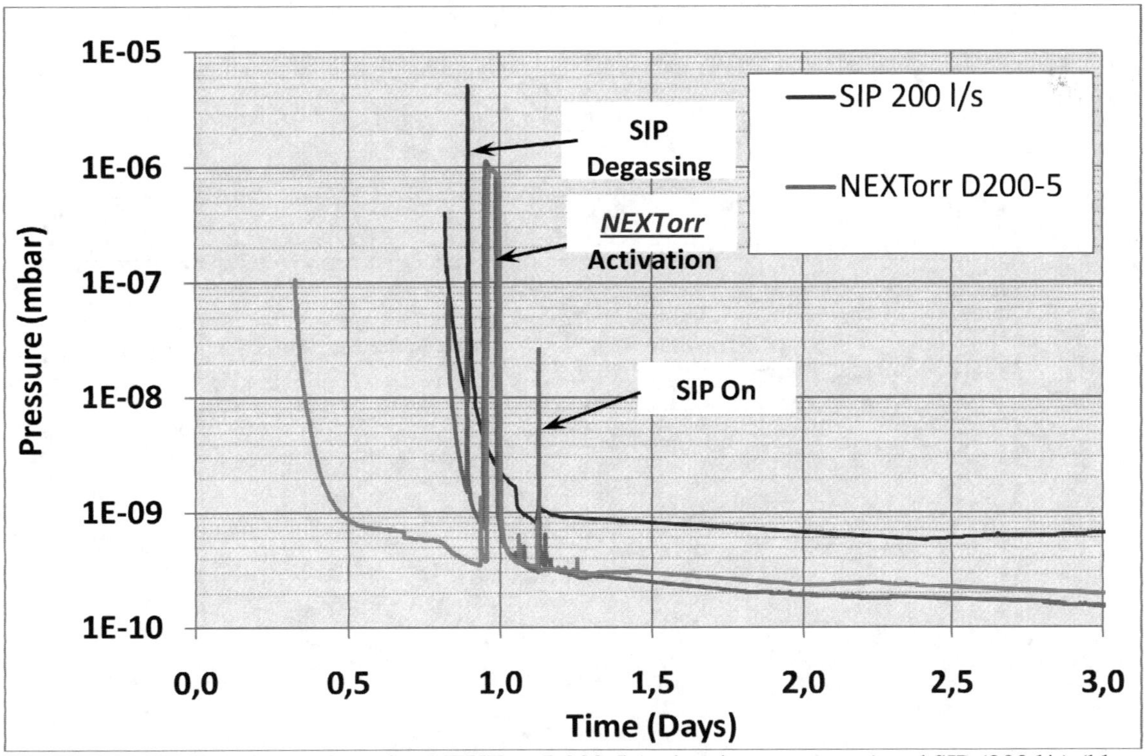

Figure 7. Pumpdown curves by using NEXTorr D200-5 (red and green curves) and SIP (200 l/s) (blue curve)

Journal of Physics: Conference Series **390** (2012) 012006 doi:10.1088/1742-6596/390/1/012006

The reasons of such results can be explained by considering the partial pressure profile of all active gases during the pumpdown experiments as a function of the time. In this experiment the bake out temperature has been fixed at 250 °C. Partial pressure curves of all active gases are smaller in the case of the test using the NEXTorr (see figure 8), mainly due to the large pumping speed of the NEG element.

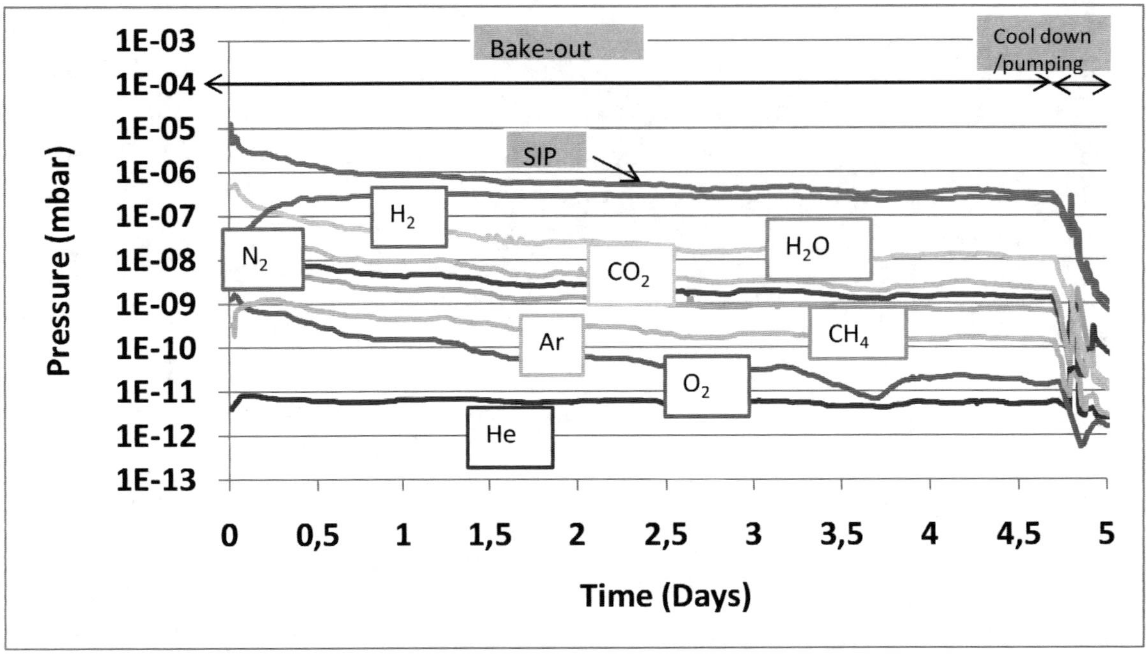

Figure 8. Partial pressure profile of all actrive gases by using SIP 120 l/s (N_2).

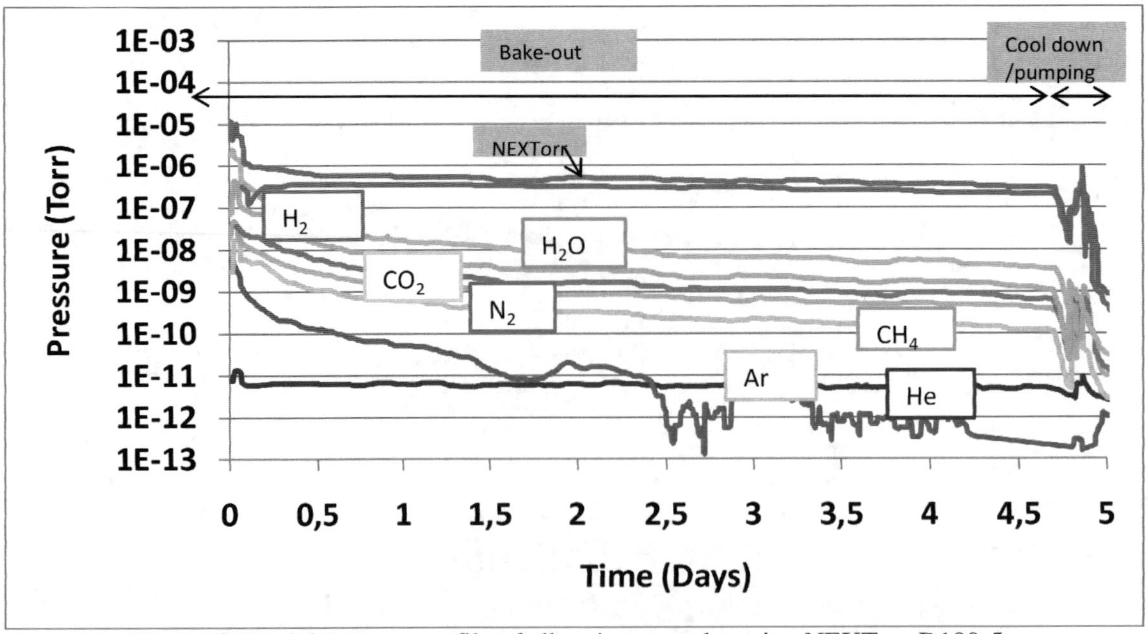

Figure 9. Partial pressure profile of all active gases by using NEXTorr D100-5.

4. Conclusions and perspectives

UHV-XHV systems can be effectively pumped using NEG and SIP pumps combined together. Being methane and argon a small fraction of the residual gas spectrum small ion pumps can be used keeping the NEG pump as the main pump to achieve the ultimate pressure. Such an approach is also used in the case of the NEXTorr a new combination pump integrating a NEG and SIP element in one small package pumping unit Results of pump down tests under different baking conditions shows that the NEXTorr compares favorably with larger and heavier Sputter ion pumps. Additional advantages, particularly appealing for application in large vacuum systems, are the ability of using the getter element of the pump to improve bake-out efficiency. This approach radically removes the main limitations of NEG pumps, providing in a very compact design a superior product in term of pumping speed, capacity, power consumption, reliability. A variety of UHV-XHV systems like surface science and MBE equipment, electron/ion spectrometers as well as high energy physics machines like synchrotrons and colliders can also benefit from this pump design. In fact these systems are generally very crowded with instrumentation, flanges and service ports. The NEXTorr can easily fit in a small space, freeing space for additional instrumentation and experiments.

References

[1] Manini P, Marino M, Belloni F and Porro M 1993 Proceeding IEEE, 3839.
[2] Benvenuti C and Chiggiato P 1996 J.Vac. Sci. Technol. A 14(6), 3278.
[3] Park CD, Chung SM and Manini P 2011 J.Vac. Sci. Technol A29(1), 011012-1.
[4] Kovac J, Sakho O, Manini P and Sancrotti M 2000 J. Vac. Sci. Technol. A 18, 2950.
[5] Benvenuti C and Chiggiato P 1996 J. Vac. Sci. Technol. A 14, 3278.
[6] Benvenuti C and Chiggiato P 1993 Vacuum 44, 511.

Journal of Physics: Conference Series **390** (2012) 012007 doi:10.1088/1742-6596/390/1/012007

PECVD/ECR/HWCVD Multichamber System with Robotic Substrate Handling System for deposition of Thin Film Electronic Devices

P. Rava

Elettrorava s.p.a. Via Don Sapino 176, 10078 Venaria, Torino Italy
tel +39-011-4240237, fax +39-011-4240364.

E-mail: paolo.rava@elettrorava.com

Abstract. The present work reports on progress in the design of modular UHV cluster tool multichamber systems. A wide range of processes has been implemented in the Deposition Process Chambers (DPC's), including rf-PECVD, vhf-PECVD, ECR-PECVD and HWCVD. A wide range of intrinsic and doped amorphous and microcrystalline silicon and silicon alloy materials have been produced and have been used in the fabrication of several types of electronic devices such as solar cells, Light Emitting Devices (LED's), Thin Film Transistors (TFT's), etc. using multichamber systems at several laboratories worldwide.

1. Introduction

Thin film electronic devices are formed by several layers of materials with different characteristics. The main layers in several types of devices are based on hydrogenated amorphous or microcrystalline silicon and its alloys such as silicon carbide and silicon nitride: these materials form the active layers in photovoltaic solar cells, sensors, detectors, LED's, thin film transistors, etc. These devices, in addition to p-doped, intrinsic and n-doped semiconductor layers, typically consist of several other layers, such as insulator layer, Transparent Conductive Oxyde (TCO) layer, and metal layer.

The technique of radio frequency plasma enhanced chemical vapour deposition (rf-PECVD) is commonly used for the deposition of hydrogenated amorphous or microcrystalline silicon and its alloys such as silicon carbide and silicon nitride; more recently other techniques such as vhf-PECVD, ECR-PECVD and HWCVD are being increasingly used for the deposition of these materials. The extensive work carried out both in research laboratories and in industry has resulted in the development of a wide range of deposition systems with widely varying features and many different reactor structures have been proposed to improve the quality of plasma deposited thin film materials and devices.

The main part of a deposition system consists in a vacuum chamber evacuated by a pumping unit to some appropriate vacuum level. After achieving the desired ultimate vacuum, the process gas mixture is introduced via Mass Flow Controllers and the pressure is kept constant by a variable conductance valve. Plasma is then ignited by an RF generator (or other source capable of generating high energy electrons); process gas is decomposed in ions and radical which are then deposited as thin films. The important figure of merit for any vacuum system, and in particular for a deposition

Journal of Physics: Conference Series **390** (2012) 012007 doi:10.1088/1742-6596/390/1/012007

chamber, is the total leak and degassing rate when the reactor is at deposition temperature: the ratio of this parameter to the process gas flow rate will largely determine the achievable purity of films deposited in that reactor.

In order to optimize the properties of each layer and therefore the characteristics of the electronic devices, it is desirable to deposit each layer, and in particular the active semiconductor layer, in high purity conditions; it is therefore necessary to minimize all external sources of contamination as well as cross contaminations among the different layers of the device.

Use of a UHV system which can achieve a good ultimate vacuum and therefore a low outgassing rate minimizes external sources of contamination, and therefore the incorporation in the deposited films of residual impurities such as oxygen, nitrogen and carbon, which can have a detrimental effect on the properties of silicon thin films. Deposition Process Chambers (DPC's) are constructed according to ultra high vacuum (UHV) standards (stainless steel construction and all metal seals) which allow ultimate pressures in the 10^{-9} mbar range at room temperature and in the 10^{-7} mbar range at process temperature to be readily reached with turbomolecular pumping .

For a UHV system to maintain a good base vacuum with as little contaminants as possible it is desirable to keep the process chamber always under ultra high vacuum (UHV) conditions, using a Load Lock chamber (LLC) isolated from the process chamber by a gate valve: only the Load Lock chamber is vented to atmosphere for substrate loading, and the substrate is introduced into the process chamber by means of a substrate handling system under vacuum.

Cross contamination is a very important factor for the deposition of devices, since these consist of several layers deposited from different gas mixtures: its effects may be detrimental especially in the case of boron and phosphorous doping, since a concentration of these dopants at the ppm level seriously degrades the optoelectronic properties of the intrinsic semiconductor layers. Cross contamination is a severe problem in the case of a system consisting of a single chamber in which several layers are deposited from different gas mixtures, due to outgassing from previous deposits on fixturing and chamber walls.

Cross contamination among different layers is avoided using a system configuration with multiple separate deposition process chambers (DPC's) for the deposition of each layer with a specific gas mixture. Several configurations of multichamber systems have been used. The in-line system consists of multiple reaction chambers configured in line, including entrance and exit load lock chambers. The cluster multichamber configuration consists of multiple reaction chambers configured in a circular geometry; the contamination at the interface between layers is avoided using a central vacuum chamber connected to all process chambers to transfer the substrate from one process chamber to another without venting it, named transfer chamber (TC). As a result of the avoidance of cross contamination, the reproducibility of each layer can be better guaranteed and therefore the device properties can be better controlled even if a variety of devices are deposited using a different sequence of layers.

Multichamber cluster systems with a vacuum transfer chamber are commonly used for research and production of thin film devices. The vacuum transfer chamber is kept under vacuum during normal operation and a Load Lock chamber is used for substrate loading. The handling system to transfer substrates under vacuum to the process chambers is the key technological component of multichamber systems and different technical solutions have been used in the multichamber systems installed (Fig.1)

The deposition of good quality materials requires also a good control of the plasma parameters during the deposition process in each DPC. The independently variable parameters of the plasma deposition process include the flow rate of the gases, the total pressure of the gas, the interelectrode distance, the substrate temperature and the electrical excitation parameters such as power and frequency. A combination of appropriate values for these parameters has to be chosen in such a way as to obtain the desired film properties, such as film thickness uniformity, high deposition rate, good adhesion, good electrical and mechanical properties. Reactors capable of depositing good quality materials should feature a good control by means of instruments with feedback loops of all the deposition parameters described above; automatic control and supervision of the whole system by means of a personal computer allows an improved control and repeatability of all deposition parameters, particularly in the case of devices with a large number of layers.

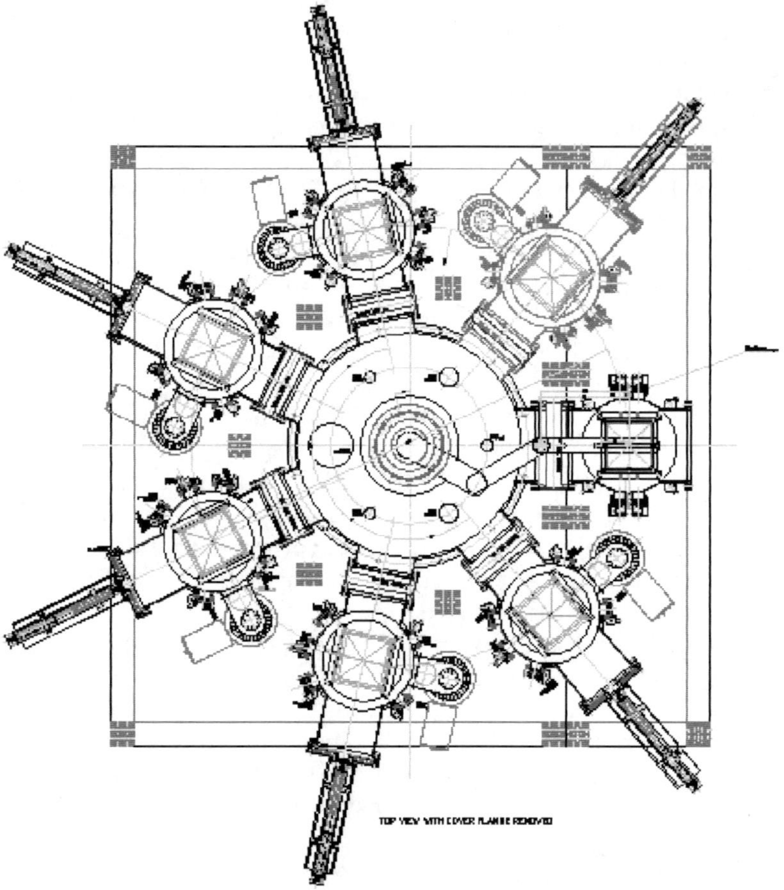

Fig.1. Multichamber system schematics

2. Deposition of the cluster tool multichamber systems

Several UHV cluster multichamber systems in different configurations have been installed at several locations worldwide. These include the IMM-CNR laboratory in Bologna and Politecnico di Torino in Italy, the University of Utrecht, the Technical University of Delft and Energy Centre Nederland (ECN) in The Netherlands, University of Barcelona and ITMA in Spain, Philips Research Laboratories in the UK and LNT at Vietnam National University. The deposition process technologies used in the DPC's of these systems are rf PECVD, vhf PECVD, HWCVD, ECR PECVD.

A five chamber cluster system for 156mm x 156mm substrates using rf PECVD in four process chambers (DPC1 for intrinsic silicon, DPC2 for p-doped silicon, DPC3 for n-doped silicon, DPC4 for silicon alloys), and plasma etching in DPC5 has been manufactured and installed at the ECN laboratories in The Netherlands (Fig.2). The system features complete control by PC with capability of simultaneous independent operation of each DPC and will be used for R&D on various types of thin film electronic devices.

Fig.2. Multichamber system installed at ECN

The main features of the UHV cluster multichamber system for deposition of high quality thin semiconducting films designed by Elettrorava are described in the following.

The system consists of five Deposition Process Chambers (DPC's) positioned around a central Transfer Chamber (TC) which contains the robotic transport arm.

The substrate handler used in the transfer chamber in the multichamber system is based on a reliable and field proven compact robotic manipulator which transfers under vacuum the necessary movements of the substrate carrier. The following movements are performed by the arm during the substrate transfer operations: rotary to allow positioning of the arm in front of each DPC; linear horizontal using a highly stable and reliable two or four sector arm to allow extension into the DPC's; linear vertical to allow engage and release of the substrate carrier by means of a pan shaped end effector.

An additional Load Lock Chamber (LLC) with easy access is connected to the front of the TC for front loading of substrates. Several substrates (156mm x 156mm) are first placed in the LLC for introduction into the TC; the substrates are then introduced into one of the DPC's to perform the relative process; after the process the substrate is extracted from the DPC and can be introduced into any other DPC for another process or taken into the LLC for unloading. The DPC's and the LLC are separated from the TC by means of gate valves which are open only to allow substrate insertion and extraction from the DPC's and LLC. This configuration allows the deposition of layers in any sequence to produce any type of multilayer device, avoiding cross contamination between different layers.

The DPC's are cylindrical chambers constructed according to ultra high vacuum (UHV) standards (stainless steel construction and all metal seals) which allow ultimate pressures in the 10^{-9} mbar range at room temperature and in the 10^{-7} mbar range at process temperature to be readily reached; dedicated turbomolecular pumping for each chamber is used both during pump down and during process, in order to avoid all possible contaminations from the pumping system; all pumps are purged by nitrogen to protect some of their most sensitive parts and to avoid possible accumulation of toxic or explosive components when reactive gases are used. The substrate is positioned by the robotic arm in each DPC on a holder facing downwards to avoid accumulation of dust particles on its surface; a heater placed outside vacuum (to avoid contamination by this source) heats the substrate by irradiation up to a maximum temperature of about 500°C; each DPC has a dedicated gas manifold for delivery of process gases; high vacuum is measured prior to process by an ionization gauge and process pressure is controlled by a throttle valve connected to a capacitance manometer; a dedicated rf generator with relative automatic matching network is connected to the electrode in each DPC for plasma excitation; the distance between the substrate and the electrode can be adjusted externally from 1 cm to 5 cm; the flange where the electrode is installed is removable and can be readily replaced with components for different deposition technologies, such as ECR and HWCVD. Optionally high temperature capability (1000°C) is available.

The multichamber system is controlled entirely by PC, which allows stable and repeatable control of all deposition parameters and make the system particularly suitable for the deposition of multilayer devices.

3. Experimental results

State of the art amorphous and microcrystalline semiconducting materials have been deposited using the multichamber systems installed by Elettrorava, research programs have been carried out on several types of devices.

The extensive solar cell research program carried out at Utrecht University and Technical University Delft has allowed to deposit devices with efficiencies in excess of 10% (1). More recently the addition of HWCVD chambers on the multichamber system has allowed to optimize deposition of amorphous and polycrystalline silicon thin films by HWCVD and to produce highly stable pin devices.

Extensive research programs have been carried out on the system installed at the IMM-CNR laboratory in Bologna and have led to scientific results reported in many papers. The main topics investigated and the main results obtained are summarized here.

Deposition of silicon carbide alloys using rf PECVD : the optoelectrical properties of amorphous silicon carbide have been optimized for films deposited in gas mixtures of silane and methane/acetylene both undiluted and with the addition of hydrogen (hydrogen diluted gas mixtures). It has been shown that the energy gap can be varied in a wide range and that hydrogen dilution improves the optoelectrical properties for films with high values of the energy gap (2, 3, 4, 5, 6, 7, 8, 9, 10, 11, 12, 14, 17, 19, 20, 21, 22, 25, 26, 27, 29, 33).

Deposition of silicon nitride alloys using rf PECVD : the optoelectrical properties of amorphous silicon nitride have been optimized for films deposited in gas mixtures of silane and ammonia both undiluted and with the addition of hydrogen (hydrogen diluted gas mixtures). It has been shown that the energy gap can be varied in a very wide range and that hydrogen dilution improves the optoelectrical properties for films with high values of the energy gap (15, 16, 24, 28, 30, 34).

Deposition of silicon and its alloys using vhf PECVD: amorphous silicon, silicon carbide and silicon nitride alloy films have been deposited at 100 MHz. These films are deposited at a high deposition rate and at the same time show good optoelectrical properties.

Deposition of multilayer structures: multilayer structures consisting of alternate high energy gap and low energy gap layers have been deposited; these structures show good photoluminescence properties (30, 32).

Deposition of heterojunction solar cells: intrinsic and doped amorphous silicon layers have been deposited on crystalline silicon wafers in order to produce photoelectrically active pin devices with good optoelectrical quality (13, 23, 31).

Deposition of LED's: LED's using as active layer amorphous silicon carbide and silicon nitride layers as well as multilayers have been deposited; these devices show good emission in the visible spectrum (18,32).

Extensive research programs are being carried out on the PECVD multichamber system installed at ECN: optimization of deposition processes for amorphous and microcrystalline silicon, silicon oxide and silicon carbide is in progress

4. Conclusions

Cluster tool multichamber systems have been used for the deposition of a wide range of semiconducting materials and devices with state of the art properties. The design of the systems has been improved with a highly reliable substrate transport arm and with complete computer control of the whole system.

References

[1] A.Madan, P.Rava, R.E.I.Schropp, B.Von Roedern. "A new modular multichamber Plasma Enhanced Chemical Vapor Deposition system". Applied Surface Science 70/71, 716-721, North-Holland (1993).

[2] J. Daey Owens, R.E.I. Schropp, C.H.M. van der Werf, M.B. von der Linden, C.H.M. Maree, W.F. van der Weg, P. Rava, F. Demichelis, C.F. Pirri, E. Tresso. "Effects of electrode spacing and hydrogen dilution on a-SiC:H and a-Si:H layers". Amorphous Silicon Technology-1993; E.A. Schiff, M.J. Thompson, A. Madan, K. Tanaka, P.G. LeComber Editors, Materials Research Society Symposium Proceedings, vol.297, p.61, Pittsburgh (PA) (1993)

[3] T. Pisarkiewicz, T. Stapinski, F. Demichelis, C.F. Pirri, E. Tresso, P.Rava. "Undoped and phosphorus doped μ-SiC:H films: investigation of electrical properties and Hall effect". Amorphous Silicon Technology-1993; E.A. Schiff, M.J. Thompson, A. Madan, K. Tanaka, P.G. LeComber Editors, Materials Research Society Symposium Proceedings , vol.297, p.431, Pittsburgh (PA) (1993)

[4] F. Demichelis, G. Crovini, C.F. Pirri, E. Tresso, L. Battezzati, P. Rava. "Structure and morphology of μ-SiC:H films produced by PECVD". Microcrystalline Semiconductors: Materials Science and Devices; F.M. Fauchet, C.C. Tsai, L.T. Canham, I. Shimizu and Y. Aoyagi Editors, Materials Research Society Symposium Proceedings, vol.283, p.543, Pittsburgh (PA) (1993)

[5] F. Demichelis, G. Crovini, C.F. Pirri, E. Tresso, R. Galloni, R. Rizzoli, C. Summonte, F. Zignani, G. Amato, P. Rava, A. Madan. "Optimization of optoelectronic properties of a-SiC:H films". Amorphous Silicon Technology-1993; E.A. Schiff, M.J. Thompson, A. Madan, K. Tanaka, P.G. LeComber Editors, Materials Research Society Symposium Proceedings, vol.297, p.681, Pittsburgh (PA) (1993)

[6] F. Demichelis, G. Crovini, F. Giorgis, C.F. Pirri, E. Tresso, G. Amato, H. Herremans, W. Grevendonk, P. Rava. "Electronic density of states in a-SiC:H films" Journal of Non-Crystalline Solids 164/166, 1015, (1993).

[7] F. Demichelis, G. Crovini, C.F. Pirri, E. Tresso, G. Amato, U. Coscia, G.Ambrosone, P. Rava. "Optimization of a-$Si_{1-x}C_x$:H films prepared by ultra high vacuum plasma enhanced chemical vapour deposition for electroluminescent devices ". Thin Solid Films 241, 274, Elsevier Sequoia (1994).

[8] F. Demichelis, G. Crovini, C.F. Pirri, E. Tresso, R. Galloni, R. Rizzoli, C. Summonte, F. Zignani, P. Rava, A. Madan. "The influence of hydrogen dilution on the optoelectronic and structural properties of hydrogenated amorphous silicon carbide films" (Invited paper) Philosophical Magazine B69, 377, (1994).

[9] R. Galloni, R. Rizzoli, C. Summonte, F. Demichelis, F. Giorgis, C.F. Pirri, E. Tresso, G. Ambrosone, C. Catalanotti, U. Coscia, P.Rava, G. Della Mea, V. Rigato, A. Madan. "Defect distribution and bonding structure in high band gap a-$Si_{1-x}C_x$:H films deposited in H_2 dilution". Amorphous Silicon Technology-1994 ; E.A. Schiff, M. Hack, A. Madan, M. Powell, A. Matsuda Editors, Materials Research Society Symposium Proceedings, vol 336, p.517, Pittsburgh (PA), (1994)

[10] R. Rizzoli, R. Galloni, C. Summonte, F. Demichelis, C.F. Pirri, E. Tresso, G. Crovini, P. Rava, F. Zignani. "Boron and phosphorous ion implantation in a-Si_xC_{1-x}:H thin films". Amorphous Silicon Technology-1994; E.A. Schiff, M. Hack, A. Madan, M. Powell, A. Matsuda Editors, Materials Research Society Symposium Proceedings, vol 336, Pittsburgh (PA) (1994)

[11] R. Rizzoli, R. Galloni, C. Summonte, F. Demichelis, C.F. Pirri, E. Tresso, G. Crovini, P. Rava, A. Madan "Recent improvements on a-SiC:H films deposited by UHV-PECVD". Proceedings of the 12th E.C. Photovoltaic Solar Energy Conference, Amsterdam, April 1994, p.366; H.S. Stephens and Associates, Bedford, UK, (1994).

[12] F. Demichelis, G. Crovini, C.F. Pirri, E. Tresso, R. Galloni, R. Rizzoli, C. Summonte, P. Rava. "Optimization of relevant deposition parameters for high quality a-SiC:H films". Solar Energy Materials and Solar Cells 37, 315, Elsevier Science (1995)

[13] F. Zignani, R. Galloni, R. Rizzoli, C. Summonte, A. Parisini, A. Armigliato, P. Rava. "Studio della eterogiunzione silicio amorfo\silicio cristallino e sua applicazione alla realizzazzione di celle solari" Atti II Convegno Scientifico Consorzio Interuniversitario Nazionale per la Chimica dei Materiali, Firenze, Febbraio 1995

[14] F. Demichelis, G. Crovini, F. Giorgis, C.F. Pirri, E. Tresso, R. Galloni, R. Rizzoli, C. Summonte, F. Zignani, P. Rava. "High quality a-SiC:H films and their application in p-i-n structures for optoelectronic devices". Vuoto Scienza e Tecnologia XXIV, 61, Pàtron (Bologna), (1995).

[15] C. Summonte, R. Rizzoli, R. Galloni, R. Pinghini, E. Centurioni, C.F. Pirri, E. Tresso, F. Demichelis, G. Crovini, F. Giorgis, P. Rava. "Study of amorphous hydrogenated silicon nitrogen alloys for photovoltaic applications". Proceedings of the 13th E.C. Photovoltaic Solar Energy Conference, pg. 199, Nice, October 1995, H.S. Stephens and Associates, Bedford, UK, (1995).

[16] F. Giorgis, P. Rava, R. Galloni, R. Rizzoli, C. Summonte, G. Crovini, F.Demichelis, C.F. Pirri, E. Tresso, V. Rigato. " Compositional, optoelectronic and structural properties of amorphous silicon nitrogen alloys deposited by plasma enhanced chemical vapour deposition". Journal of Non-Crystalline Solids 198-200, 596-600, Elsevier (1996)

[17] M. Fathallah, R. Gharbi, G. Crovini, F. Demichelis, F. Giorgis, C.F. Pirri, E. Tresso, P. Rava. "Light soaking in a-SiC:H films grown by PECVD in undiluted and hydrogen diluted SiH_4+CH_4 gas mixtures". Journal of Non-Crystalline Solids 198-200, 490-494, Elsevier (1996)

[18] R. Rizzoli, C. Summonte, R. Galloni, M. Ruth, A. Desalvo, F. Zignani, P.Rava, F. Demichelis, C.F. Pirri, E. Tresso, G. Crovini, F.Giorgis, A. Madan. "Brightness degradation controlled by current induced metastable defect creation in a-SiC:H based light emitting diodes" Amorphous Silicon Technology-1995, M. Hack, E.A. Schiff, A. Madan, M. Powell, A. Matsuda Editors, Materials Research Society Symposium Proceedings, vol 377, p.809, Pittsburgh (PA) (1995)

[19] P. Rava, G. Crovini, F. Demichelis, F. Giorgis, R. Galloni, R. Rizzoli, C.Summonte. "Power dissipation in PECVD for SiH_4-CH_4-H_2 gas mixtures". Journal de Physique IV, Colloque C5, suppl. Journal de Physique II, edt. by G.A.Battiston, R. Gerbasi, M. Porchia, vol.5, p. C5-1125, Les Editions de Physique, Les Ulis Cedex (1995)

[20] F. Demichelis, G. Crovini, F. Giorgis, C.F. Pirri, E. Tresso, V. Rigato, U. Coscia, G. Ambrosone, S. Catalanotti, P. Rava. "Effects of power density and molecule dwell time on compositional and optoelectronic properties of a-SiC:H alloys". Solid State Communications, vol.98, no.7, p.617, Pergamon Press, Elsevier Science (1996).

[21] P. Rava, G. Crovini, F. Demichelis, F. Giorgis, C.F. Pirri. "Characterization of the effect of growth conditions on a-SiC:H films". Journal of Applied Physics 80, no.7, 4116, (October 1996)

[22] A. Desalvo, R.Galloni, F. Giorgis, C.F. Pirri, P. Rava, R. Rizzoli, C. Summonte, E.Tresso. "Optoelectronic properties, defect structure and composition of a-SiC:H films grown in undiluted and H_2 diluted silane-methane plasma". Journal of Applied Physics (1996).

[23] F. Zignani, R. Galloni, R. Rizzoli, M. Ruth, C. Summonte, R. Pinghini, Q.Zini, P. Rava, A. Madan, Y.S. Tsuo. "Study of a-Si:H/c-Si Heterojunctions for PV applications". Amorphous Silicon Technology-1996, M. Hack, E.A. Schiff, S. Wagner, A. Matsuda, R. Schropp Editors, Materials Research Society Symposium Proceedings, vol. 420, pg. 45, Pittsburgh (PA) (1996)

[24] F. Giorgis, C.F. Pirri, E. Tresso, V. Rigato, S. Zandolin, P. Rava "Wide band gap amorphous silicon based alloys" Physica B, 229, 233-239, (1996)

[25] U. Coscia, C Catalanotti, G. Ambrosone, F. Demichelis, F. Giorgis, C.F. Pirri, E. Tresso, P. Rava. "Study of gap states and photoelectrical properties of a-SiC:H films deposited by PECVD in SiH_4+CH_4 gas mixtures". Vuoto Scienza e Tecnologia XXV, 38, Pàtron (Bologna), (1996).

[26] F. Demichelis, G. Crovini, F. De Zan, F. Giorgis, C.F. Pirri, E. Tresso, G. Della Mea, V. Rigato, S. Zandolin, P. Rava. "Optimization of a-SiC:H films for optoelectronic applications". Vuoto Scienza e Tecnologia XXV, 44, Pàtron (Bologna), (1996).

[27] F. Giorgis, C.F. Pirri, E. Tresso. P. Rava "a-SiC:H films deposited by PECVD from silane + acetylene and silane + acetylene + hydrogen gas mixtures". Diamond and Related Materials

[28] F. Giorgis, F. Giuliani, C.F. Pirri, E. Tresso, C. Summonte, R. Rizzoli, R. Galloni, A. DeSalvo, P. Rava "Optical, structural and electrical properties of device quality hydrogenated amorphous silicon-nitrogen films deposited by plasma-enhanced chemical vapour deposition". Phil. Mag. B, 77, no.4, 925-944, (1998)

[29] F. Giorgis, C.F. Pirri, P. Rava, E. Tresso "Correlation between optoelectronic properties and structure of a-SiC:H films grown from C_2H_2 source" Phil. Mag. B, 75, no.4, 471-483, (July 1997)

[30] F.Giorgis, F. Giuliani, C.F. Pirri, E. Tresso, R. Galloni, R. Rizzoli, C. Summonte, A. Desalvo, F. Zignani, P. Rava, F. Caccavale. "Photoluminescence and optical characterization of a-Si_xN_{1-x}:H based multilayers grown by PECVD". Amorphous Silicon Technology-1997, S. Wagner, M. Hack, E.A. Schiff, , R. Schropp, I. Shimizu Editors, Materials Research Society Symposium Proceedings, vol.467, Pittsburgh (PA) (1997)

[31] R. Rizzoli, R. Galloni, C. Summonte, R. Pinghini, E. Centurioni, F. Zignani, A. Desalvo, P. Rava, A. Madan "Study of a-Si:H/c-Si Heterojunctions for PV applications". Amorphous Silicon Technology-1997, S. Wagner, M. Hack, E.A. Schiff, , R. Schropp, I. Shimizu Editors, Materials Research Society Symposium Proceedings, vol.467, p.807, Pittsburgh (PA) (1997)

[32] C. Summonte, R. Rizzoli, R. Galloni, F. Giorgis, F. Giuliani, C.F. Pirri, E. Tresso, A. Desalvo, F. Zignani, P. Rava. "Photoluminescence and electroluminescence properties of a-Si_xN_{1-x}:H based superlattice structures". Journal of Non-Crystalline Solids, 227-230, 1127-1131, Elsevier (1998)

[33] F. Giorgis, F. Giuliani, C.F. Pirri, P. Mandracci, P. Rava, R. Reitano, L. Calcagno, P. Musumeci. " Carbon rich a-$Si_{1-x}C_x$:H films: an investigation on radiative recombination properties" Amorphous Silicon Technology-1998, S. Wagner, M. Hack, E.A. Schiff, , R. Schropp, I. Shimizu Editors, Materials Research Society Symposium Proceedings, Pittsburgh (PA) (1998)

[34] F. Giorgis, G. D'Agnelli, F. DeZan, F. Impavido, C.F. Pirri, P. Rava, E. Tresso. "Structural and optoelectronic properties of a-SiC:H and a-SiN:H films for applications as wide band gap semiconductors". Vuoto Scienza e Tecnologia, Pàtron (Bologna) (1997)

[35] P. Rava, F. Giuliani, F. Giorgis, C.F. Pirri, E. Tresso, P. Mandracci, C. Summonte, R. Rizzoli, A. Desalvo. "Amorphous silicon nitrogen alloys deposited by PECVD under hydrogen dilution conditions". Physics of Semiconductor Devices, V. Kumar and S.K. Agarwal (Eds.), Narosa Publishing House, New Delhi, India (1998)

Open loop, auto reversing liquid nitrogen circulation thermal system for thermo vacuum chamber

M.C.A.Naidu, Dinesh Nolakha, B S Saharkar, KM Kavani, D R Patel

Space Applications Centre, ISRO, Ahmedabad-380 015

E-mail: mcanaidu@sac.isro.gov.in

Abstract: In a thermo vacuum chamber, attaining and controlling low and high temperatures (-100 Deg. C to +120 Deg. C) is a very important task. This paper describes the development of "Open loop, auto reversing liquid nitrogen based thermal system". System specifications, features, open loop auto reversing system, liquid nitrogen flow paths etc. are discussed in this paper. This thermal system consists of solenoid operated cryogenic valves, double embossed thermal plate (shroud), heating elements, temperature sensors and PLC. Bulky items like blowers, heating chambers, liquid nitrogen injection chambers, huge pipe lines and valves were not used. This entire thermal system is very simple to operate and PLC based, fully auto system with auto tuned to given set temperatures. This system requires a very nominal amount of liquid nitrogen (approx. 80 liters / hour) while conducting thermo vacuum tests. This system was integrated to 1.2m dia thermo vacuum chamber, as a part of its augmentation, to conduct extreme temperature cycling tests on passive antenna reflectors of satellites.

1. Introduction

As a part of augmentation of existing 1.2m dia thermo vacuum chamber, which is about 35 years old, to conduct extreme temperature cycling tests on passive antenna reflectors up to 0.9m dia, this thermal system was developed, integrated and tested successfully. Also, this system is used for active sub-system (with power dissipation) testing. Existing twin lobe compressor based gaseous nitrogen type thermal system was replaced by the proposed system. Liquid nitrogen was used for cooling and IR lamps were used for heating the panel.

Even though the entire vacuum, thermal and control systems were augmented, this paper discusses only thermal system part applicable to shroud. Auto reversing system was implemented as the required temperature gradient (±7.5 K) on shroud was not obtained in one-way open loop system (±16 K). Some advantages of this system are: (i) It occupies less space, as there are only two pipe lines one, vacuum jacketed (25NB inner pipe) line as inlet line and another one, armaflex insulated (25 NB pipe) line as outlet, on the ground (ii) Operation of the system is silent, as there are no noise making blowers (iii) Low power consumption (4.5 KW) (iv) Nominal liquid nitrogen consumption (approx. 80 liters/hour) etc.

2. System details

Schematic of open loop auto reversing thermal system with liquid nitrogen flow path is shown in fig. 1. Thermal panel means double embossed panel (shroud). Size of the developed view of the panel is 2420 mm X 1080 mm. The panel is in "D" shape. There are four solenoid operated cryogenic valves (CV). Nine numbers of IR lamps were distributed on the shroud. IR lamps are given power through solid state relays. Cryogenic valves are controlled by set point of temperature sensors (PT100) located at inlets of the shroud. IR lamps are controlled by set points of respective temperature sensors. Entire operation and logic is fully auto mode with PID loops and PLC based, with an option to switch on to manual mode. Shroud has two inlets and one outlet. Inlets are at the bottom and outlet at the top. In auto reversing mode inlets become outlets and outlet become inlet. Entire outside surface of shroud was covered with two layers of radiation shields, except at inlet / outlet areas.

2.1 The sequence of operation is as follows:

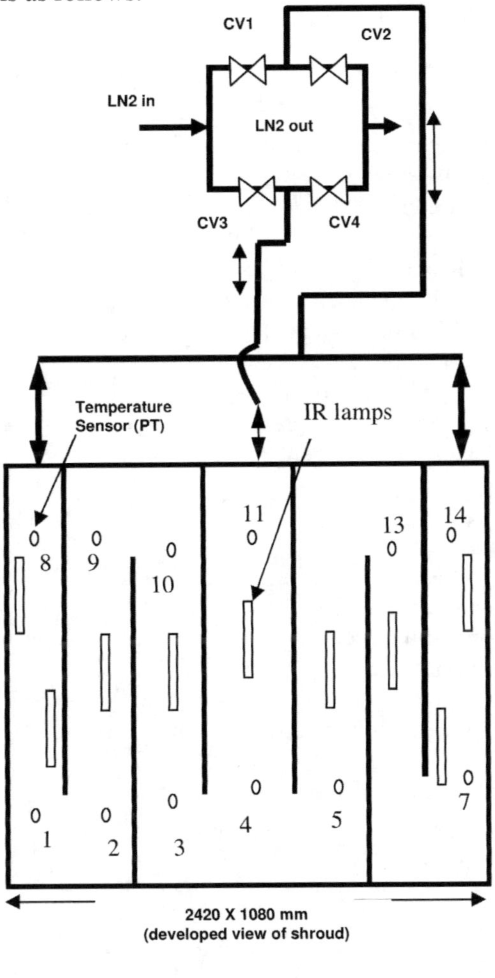

Fig.1. Schematic of open loop auto reversing thermal system

Ref. Fig.1

2.1.1. When CV1, CV4 are open, CV2, CV3 are closed.
2.1.2. In reverse flow, when CV2, CV3 are open, CV1, CV4 are closed.
2.1.3. The set point temperature of temperature sensor at respective inlets (PT8, 11, 14) in fig.1 will control the ON/OFF operation of CV1, CV4 or CV2, CV3.
2.1.4. During the initial cooling cycle, liquid nitrogen will flow from bottom inlets (Near PT8, 14).
2.1.5. The temperature difference between inlet and outlet is set as a differential set point.
2.1.6. Liquid nitrogen will flow from the bottom inlets, till a differential set point is reached.
2.1.7. Then the liquid nitrogen will flow from top inlet (Near PT11), as a reverse flow.
2.1.8. This way the operation of valves takes place till the set temperature is reached.
2.1.9. Once the set point temperature is reached, the valves remain closed for few minutes.
2.1.10. To maintain the set temperature, frequent on / off of valves will take place.
2.1.11. During off condition of valves, the liquid nitrogen in the pipe lines gets warmed up and become gas.
2.1.12. At this time, if any valve opens, gaseous nitrogen will flow inside the shroud and disturb the temperature.
2.1.13. To avoid this, auto flushing system is incorporated. The gas will be flushed out automatically for 10 seconds and then start the operation of inlet valve. This is electronically incorporated.
2.1.14. To maintain the temperature gradient (+/-7.5 Deg. C as specification) on shroud, IR lamps will be occasionally "ON" with variable power input.
2.1.15. It was observed that the IR lamps were not operated during -100 deg. C steady state.
2.1.16. First, to go to auto mode, auto mode is to be activated by pressing a dedicated switch.
2.1.17. In auto mode operation, PID loops will take care of opening and closing liquid nitrogen valves and IR lamps according to their given set points.
2.1.18. In case, manual mode to be activated the dedicated switches for liquid nitrogen valves for top entry, bottom entry and IR lamps, to be pressed.
2.1.19. In manual mode, all IR lamps can be controlled by manually entering the percentage value. Refer fig. 2 & 3.

3. Instrumentation and control

Instrumentation and control system consists of a dedicated PLC, PID controller, Temperature sensors, solenoid operated liquid nitrogen valves and IR lamps. In PLC program, proportional derivative integral control logic is developed. Temperature sensors reading is fed as input to PID, PID output is sent to solenoid operated liquid nitrogen valves through contactors.

IR lamps are controlled with individual temperature sensors through SSRs..

Fig.2 shows, the "ON /OFF" status of liquid nitrogen valves and IR lamps along with % power of IR lamps.

Fig.3 shows, parameters for manual mode of operation.

PLC (Micrologix 1200 base module with 14 DI / 10 DO) with: a) 12 channel RTD input module, b) 12 channel analogue output module, c) 6 channel digital output modules was added to existing PLC (Allen Bradley) which is used for vacuum system operation and temperature data monitoring of device under test.

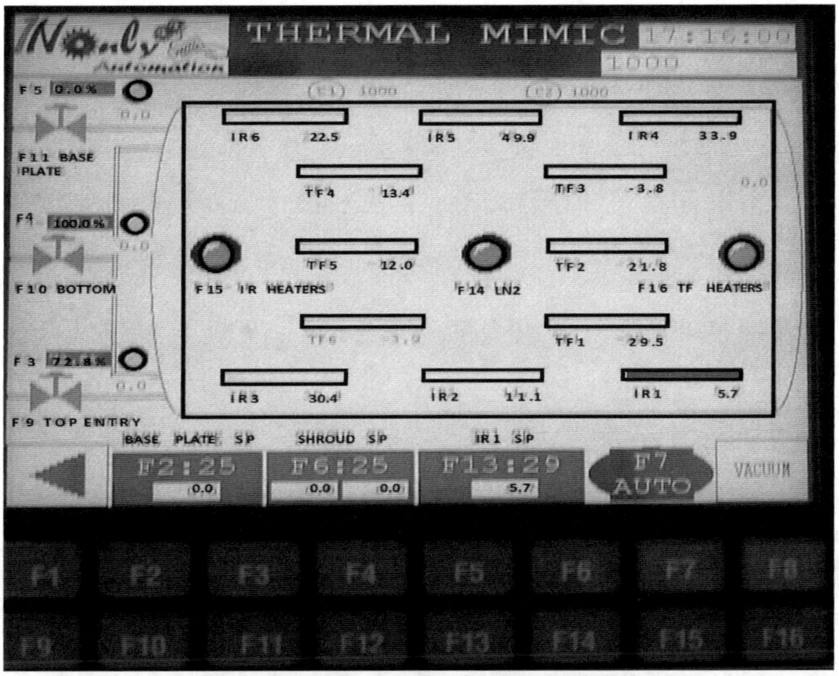

Fig.2. Operation status of valves and IR lamps

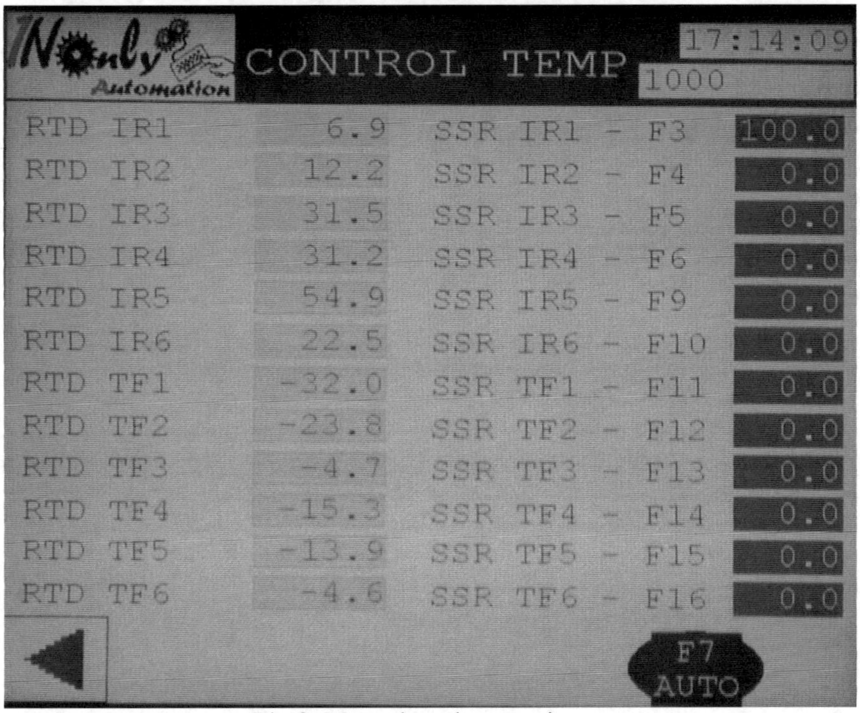

Fig.3. Manual mode operation screen

Journal of Physics: Conference Series **390** (2012) 012008 doi:10.1088/1742-6596/390/1/012008

4. Tests and results

Acceptance test document was prepared. Prior to acceptance test, experiments were carried out to know:

4.1 The variation of temperature along the length of bottom inlets by placing 5 temperature sensors in the same line,

4.2 Identified the conductive sensitive locations on the shroud and obtained information on the fluid flow pattern. Ref. Fig.1.

4.3 Mapping of temperature profile of bottom inlet and top inlet was carried out and fixed the locations of sensors.

Temperature sensors location is shown in Fig.1. Temperature distribution on shroud is shown in Fig.4

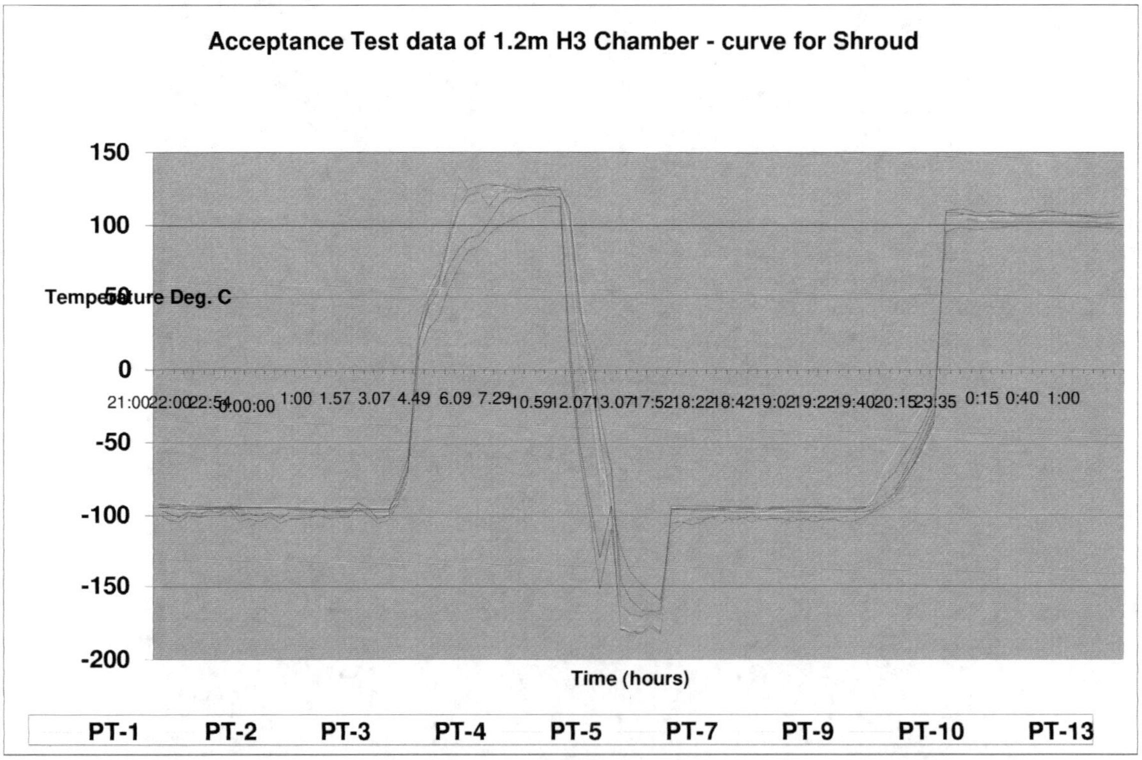

Fig.4. Temperature distribution on shroud

Maximum gradient of +/-6.5 deg.C was achieved @ -100 deg.C, with a control accuracy of +/- 1.5 deg. C. Maximum gradient of +/- 7.5 deg.C was achieved @ +120 deg.C, with a control accuracy of +/- 1 deg. C. The test was extended to see the lowest attained temperature on shroud which was seen as -186 deg.C Subsequently, 0.8m antenna reflector made of CFRP material was tested in the same chamber. Test setup is shown in Fig. 5 and Temperature Vs Time plot is shown in Fig. 6.

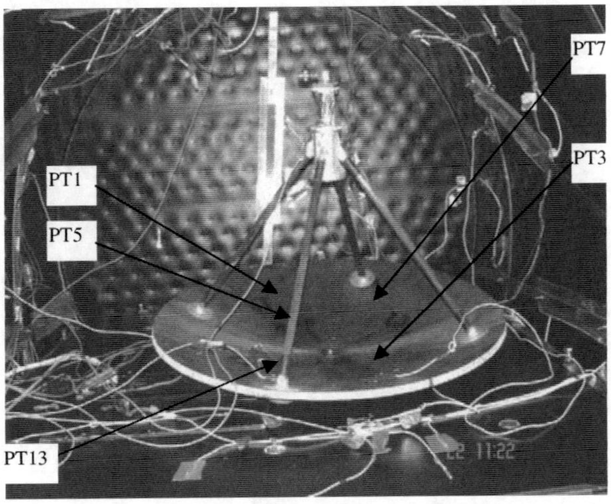

Fig. 5. 0.8 m Antenna reflector under thermo vacuum test.

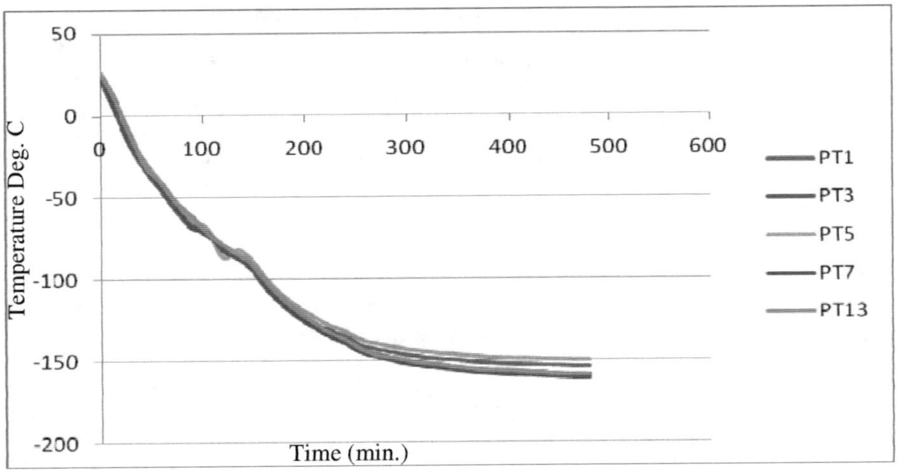

Fig.6. Temperatute (^{0}C) vs Time (min.) plot for antenna reflector

Acknowledgement

The authors thank Shri R K Arora, Deputy Director, ESSA for his continuous encouragement to bring out this technical paper. The authors are also thankful to Dr. R.R Navalgund, Director, SAC for permitting to present the paper. The authors acknowledge M/S Shree Raghavendra Technical Services Pvt Ltd, Ahmedabad for realization of this system successfully.

Journal of Physics: Conference Series **390** (2012) 012009 doi:10.1088/1742-6596/390/1/012009

Deuterium Gas Analysis by Residual Gas Analyzer

B K Das[*,1], R Shukla[1], R Das[1], A Shyam[1] and A D P Rao[2]

[1]Energetics & Electromagnetics Division, Bhabha Atomic Research Centre, Autonagar, Visakhapatnam, India

[2]Nuclear Physics Deaprtment, Andhra University, Visakhapatnam, India

Corresponding author's e-mail: dasbabu31@gmail.com

Abstract. Hydrogen gas is generated by electrolysis method in a compact hydrogen generator. A simple procedure reduces handling and storage of hydrogen cylinders for laboratory applications. In such a system, we are producing deuterium gas from heavy water by electrolysis method. After production of the deuterium gas, we have checked the purity level of the outgoing deuterium from the electrolyser. The test was carried out in a high vacuum system in which one residual gas analyser (RGA) was mounted. The deuterium gas was inserted by one manual gas leak valve in to the vacuum system. In this study, the effect of the emission current of the RGA on the detection of the deuterium was performed. In this paper, we will discuss the detail analysis of the deuterium gas and the effect of the emission current on the partial pressure measurement.

1. Introduction

Use of modern available technology for the basic requirements in research and development laboratories becomes important due to various safety reasons. One of the most important safety issues is handling of pressurized gas cylinders specifically for the flammable gases. Handling and use of hydrogen cylinders in laboratories requires lot of precautions. To avoid such safety issues related to handling, storage and use of hydrogen cylinders, we are working on one electrolysis system [1] for generation of hydrogen isotopes. Deuterium gas is generated by this electrolysis system from heavy water. The purity of the gases produced by this system is important for our application. The purity of the deuterium gas produced by an electrolysis system in our laboratory was checked in a high vacuum system by residual gas analyzer. In the next sections we will discuss about the detail experimental set up and the results.

2. Experimental set up

We are using one compact electrolysis system purchased from M/s. PCI Analytical Pvt. Ltd [2]. The photograph of the system is shown in figure 1. The electrolysis flow diagram is shown in figure 2. Potassium Deuteroxide (KOD) is used as electrolyte and heavy water. The usage of voltage across the electrolyte results in hydrolysis, breaking down the heavy water molecule into deuterium and oxygen gas, which are separated by the gas permeable membrane. To achieve the desired level of purity, the deuterium gas goes through series of purification and moisture removal systems while the oxygen gas is being discharged into the atmosphere. The electrolytic membrane technology has its own benefits over alternative hydrogen gas generator techniques as it is clean and requires minimum maintenance.

The deuterium gas thus produced by the electrolysis system is tested by one residual gas analyzer (RGA), Hiden Analytical Ltd., UK [3] make, mounted on a high vacuum system pumped by turbo molecular pump as per the figure 3. A vacuum in the order of 3.5E-7 Torr was achieved without baking. The deuterium gas was inserted into the chamber by one manual gas leak valve. The basic principle of the RGA is based on ionize and measurement [4]. RGA ionizes the neutral gases and then measure the ion current. The ion current directly corresponds to the number of neutral molecules of that particular ion. Neutral molecules present inside the vacuum chamber are ionized by one electron impact ion source present inside the RGA assembly. Once the gas is ionized they are focused to one filter system made of four electrodes know as quadruple. The quadruple filters the ions on the basis of their mass to charge ratio. The quadruple is powered by RF and DC voltages. The RF power makes the ions to oscillate inside the quadruple where as the DC voltage absorbs the unwanted ions when they strike any of the electrode surfaces. The ions those are passed through the quadruple are detected at the end either by one faraday cup or by channel electron multiplier. The RGA was operated in profile mode and MID mode. In profile mode, RGA does series of scans for each mass number. The scans were performed for different emission current of the filament. The emission current decides the ionization probability of the neutral gas molecules in the vicinity of the ion source.

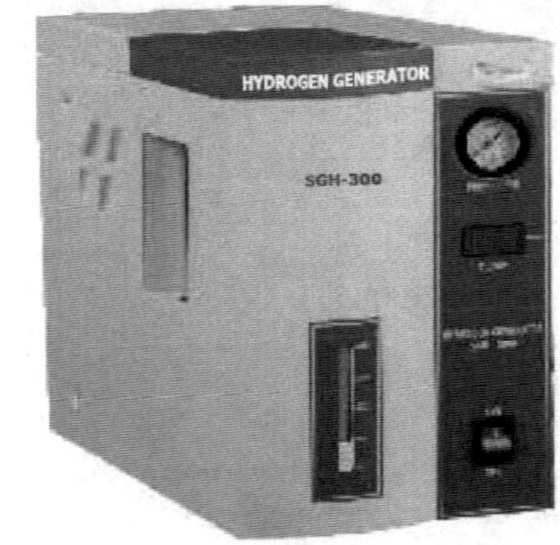

Fig. 1. Photograph of Hydrogen Generator

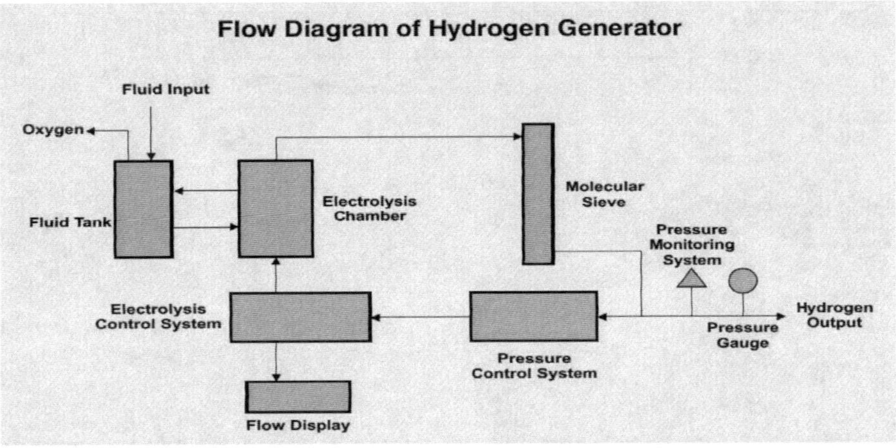

Fig. 2. Flow Diagram of Hydrogen Generator

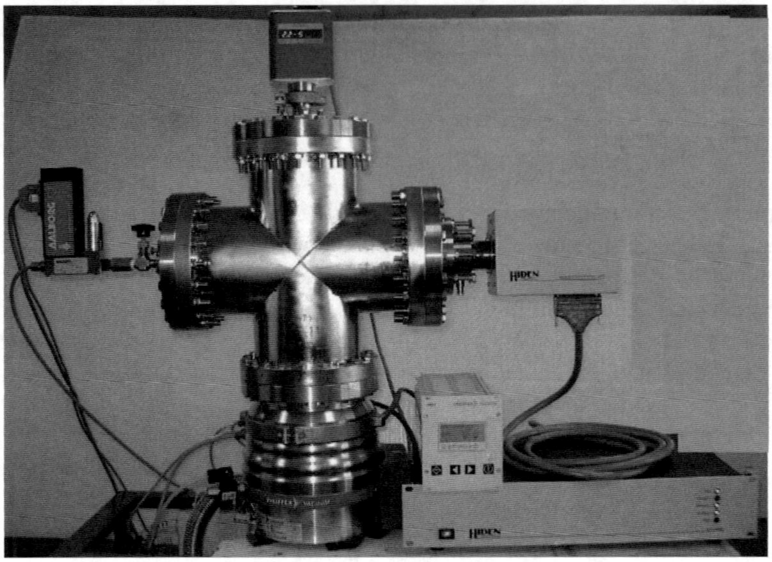

Fig. 3. Photograph of Experimental Set Up

3. Results

With the above experimental set up different spectrum were obtained from RGA for deuterium concentration, it's stability and the effect of emission current on the deuterium measurement. The spectrum obtained with the background i. e. without the filling of deuterium is shown in figure 4. From this figure, it could be observed that the vacuum system is free from leaks. The Oxygen signal at mass number 32 is absent, the nitrogen signal at mass number 28 is very less and there is no peak at mass number 14. The peak at mass number 28 is a contribution from carbon monoxide, carbon dioxide and nitrogen. As the vacuum system is not exposed to any heat, presence of water vapor in the system is unavoidable. The peak at mass number 18, 17, 2 and 1 are due to the presence of water vapor in the system.

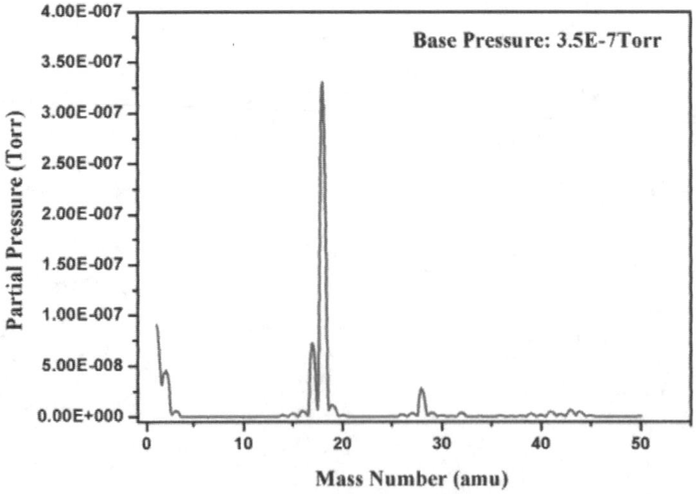

Fig. 4. Spectrum of profile scan of RGA for background

The spectrum shown in figure 5 is obtained with the deuterium filled pressure at 2.5E-5 Torr. The peak at mass number 4 is due to deuterium. In the spectrum there is no other significant peak is shown. There is a very little peak at mass number 3, which is due to formation of DH (Deuterium Hydride). This may be due to the presence of residual hydrogen left behind while the system was tested for hydrogen production. The formation of DH molecule may disappear after long use of the system for the production of deuterium. Another very little peak at mass number 18 is due to the presence of water vapor in the vacuum chamber even before filling of the deuterium gas. The spectrum shows free from any shorts of impurities.

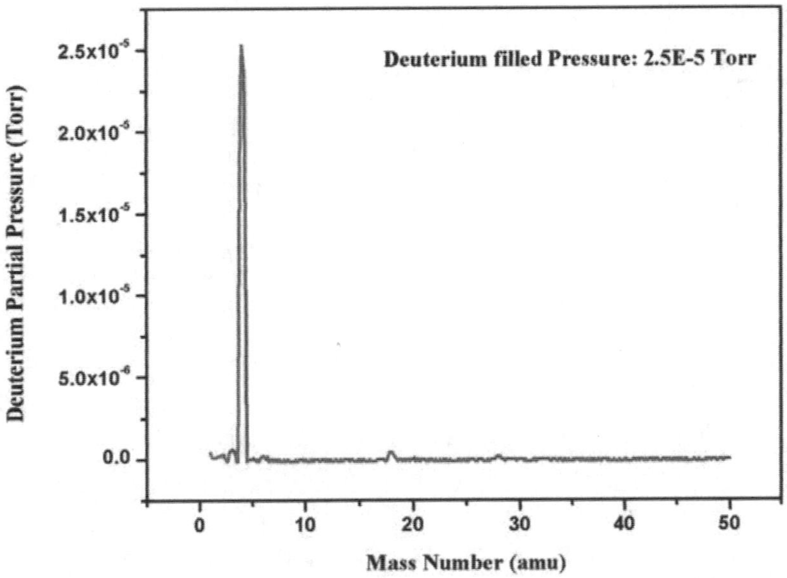

Fig. 5. Spectrum of profile scan of RGA with deuterium filled pressure

The electrolysis system continuously generates deuterium. To check the stability of the production of the deuterium, one MID scan was performed with the RGA. In MID scan, RGA continuously scans the particular mass number for given period of time. Such a scan is performed simultaneously for deuterium, oxygen, nitrogen and water vapor. Deuterium is for it's stability check and other molecules for their appearance during deuterium production either from electrolysis or from any other sources. Such a spectrum is shown in figure 6. The red line corresponds to deuterium where as the other colored lines for rest of the molecules. The red line once it is stable after production is steady throughout the measurement period for duration of ten minutes.

In an effort to check the effect of electron emission current on the partial pressure measurement, one measurement was done for different emission current. The electron emission current helps for the efficient ionization of the neutral gases in RGA. The spectrum shown in figure 7 represents the effect of emission current on the partial pressure measurement of deuterium. Though here, deuterium gas was taken as the test gas, the result may be applicable to any other gases. There is an exponential growth of the partial pressure with the emission current. The measurement was done at filled pressure of 4.5E-5 Torr. It is observed that the more emission current, i. e. higher side of the instrument specification is giving correct measurement, where as the lower side of the emission current is giving more incorrect value. The idea of using the instrument with lower emission current may be helpful for suppression of background.

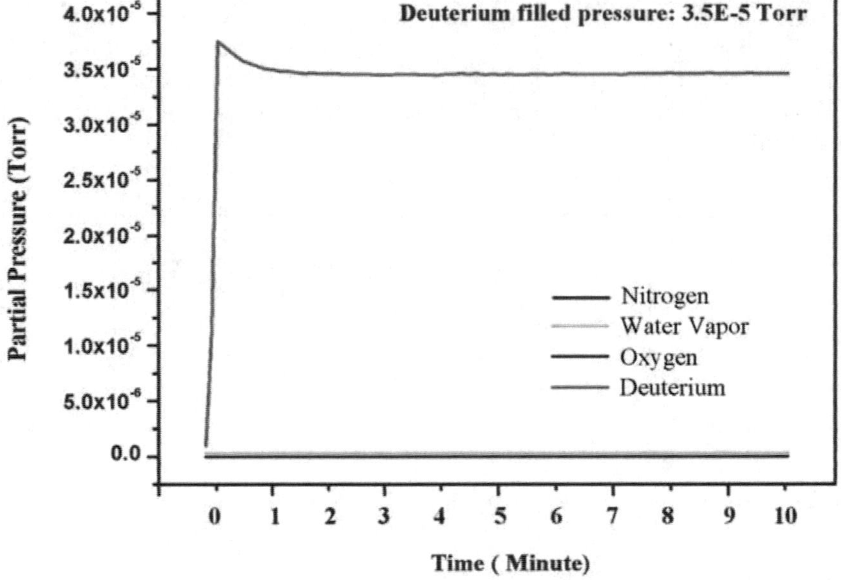

Fig. 6. MID scan of RGA for stability test of the deuterium production

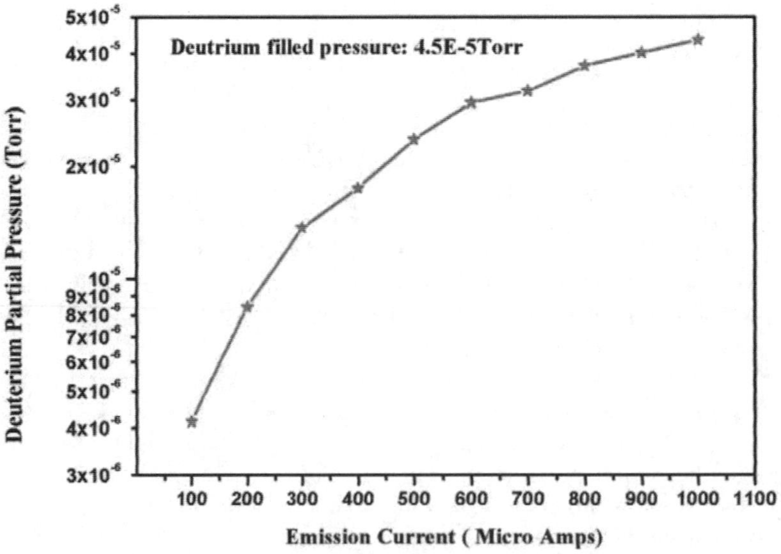

Fig. 7. Deuterium partial pressure Vs electron emission current

Conclusion

In this work, the production of deuterium gas by compact electrolysis system was carried out. The purity of deuterium is checked by residual gas analyzer. The output deuterium from the electrolysis system was found to be ultra pure and stable. Residual impurities are absent. This type of system for deuterium/hydrogen production in laboratories is useful from safety and logistic point of view. The emission current of RGA should be at the specified value according to the supplier's specification. Any lower value may give incorrect measurement.

References

[1] http://en.wikipadia.org
[2] http://www.pcianalytics.in
[3] http://www. hidenanalytical.com
[4] P. A. Redhead et al, The physical basis of ultrahigh vacuum,1993, NewYork

Design, fabrication, room temperature RF test of 1050 MHz, β = 0.49 single cell large grain niobium cavity

J. Mondal[1], T.K.Saha[2],S.Sarkar[1], S.B.Jawale[3],R. S. Vohra[3], A.V.Bapat[2], K.C.Mittal[1]

[1]Accelerator & Pulse Power Division, Bhabha Atomic Research Centre, Trombay, Mumbai – 400 085, India
[2]PBEDS, Bhabha Atomic Research Centre, Trombay, Mumbai – 400 085, India
[3]Centre for Design and Manufacture, Bhabha Atomic Research Centre, Trombay, Mumbai – 400 085, India

email: jmondal@barc.gov.in

Abstract. BARC is developing a technology for the accelerator driven subcritical system (ADSS) that will be mainly utilized for the transmutation of nuclear waste and enrichment of U233. Design and prototyping of a superconducting medium velocity cavity has been taken up as a part of the ADSS project. The cavity design for β =0.49, f = 1050 MHz has been optimized to minimize the peak electric and magnetic fields, with a goal of 5 MV/m of accelerating gradient at a Q > $5x10^9$ at 2 K. After the design optimization one single cell cavity was fabricated from large grain (RRR>99) Niobium material. This report presents the design, fabrication, electron beam welding and RF measurements at room temperature before and after electron beam welding.

1. Introduction

In the high energy section (> 100 MeV) of an ADSS accelerator, superconducting elliptical cavities will be used to accelerate protons up to 1 GeV [1]. After the successful completion of the Spallation Neutron Source (SNS) in the USA, the superiority of superconducting (SC) linacs over normal conducting ones is rather well accepted in the accelerator community to efficiently accelerate proton beams with velocity relative to speed of light, β, of about 0.5 [2]. Short, independently phased SC resonators with large velocity acceptance are also ideal for accelerating efficiently beams with different mass-to-charge ratio. SC cavities are ideal for low current beams, where efficiency is dominated by rf losses. In the high beam loading limit (~100mA), NC and SC linacs have comparable efficiency. Even for low- β proton beams SC linacs can be cost effective. In the studies for the EURISOL, 5 mA cw proton driver [3], a comparison between NC and SC option between 5 and 85 MeV was done, showing a similar construction cost.

For this purpose we have studied the RF properties of a β =0.49, 1050 MHz single cell superconducting elliptical cavity.

Section 2 of this report describes the design of β =0.49, f=1050 MHz single cell elliptic cavity. Section 3 describes the cavity fabrication. Section 4 presents the RF test results before EB welding.

Published under licence by IOP Publishing Ltd

2. Cavity design

RF parameters of 1050 MHz, $\beta = 0.49$ single cell Elliptical cavity has been optimized by means of 2D cavity tuning code SUPERFISH and 3D High Frequency Simulation code CST Microwave Studio for possible use in a High Current Proton Accelerator [4]. The low energy section of the ADS accelerator is being designed for 350 MHz and hence it is decided to build the superconducting accelerator section with the third harmonic of 350 MHz i.e. 1050 MHz.

It was found that Epk/E0 has a minimum at a particular value of the iris-ellipse aspect ratio a / b and around the same point Bpk/Epk has a maximum. A drawing of the single-cell cavity of 1050 MHz, $\beta = 0.49$ is shown in Figure 1, along with the electric field and magnetic field distribution calculated by SUPERFISH and CST-MWS. Our design is compared with the INFN-TRASCO and RIA design [5]. The INFN-TRASCO design adopts an elliptical iris and an elliptical equator shape where as in our case equator shape is circular and the iris shape is elliptical similar to RIA deign.

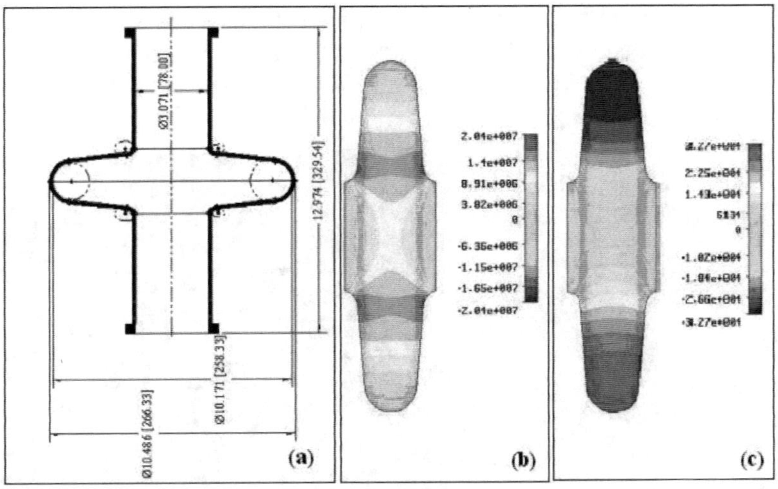

Fig. 1. (a)Drawing of the 1050 MHz, β=0.49 cavity, (b) Electrical field distribution in TM_{010} mode, (c) Magnetic field distribution in TM_{010} mode.

The main cavity parameters of our design are summarized in Table-1 and compared with the INFN-TRASCO and RIA design.

Table 1: 1050 MHz, β = 0.49 Single Cell Cavity Parameters

Parameters	Present Design	TRASCO Design	RIA Design
Frequency (MHz)		1050	
Geometrical β		0.49	
Iris Radius (cm)		3.9	
Cavity Diameter (cm)		25.833	
Dome Aspect ratio	1	1.6	1.0
Iris Aspect Ratio	1.43	1.3	1.45
Wall Angle (Deg.)	6.5	5.5	6.5
Wall Distance [mm]	7.37	7.0	8.5
r/Q (Ω)	9.19	9.46	9.04
Geometry Factor G (Ω)	141.91	147.387	135.675
E_{peak}/E_{acc}	4.26	4.44	3.98
B_{peak}/B_{acc} (mT/(MV/m)	8.02	7.78	8.37

The above table shows the design comparison with RIA and TRASCO. Actual RIA design is with frequency 805 MHz and geometrical beta 0.47 and TRASCO is with frequency 704.4 MHz and geometrical beta 0.47. In the above table both the RIA and TRASCO design were scaled for frequency 1050 MHz and geometrical beta 0.49, keeping the iris radius same as our present design i.e. 3.9 cm. The present design which is very close to the design of those other two cavities whose prototypes reached their performance specs and the ratios of the peak fields to E*acc* in our design is in between the values of the TRASCO cavity, designed to minimize B*p*/E*acc*, and the RIA cavity, designed to minimize E*p*/E*acc*

3. Fabrication

One prototype single cell Niobium cavity has been fabricated with the designed parameters mentioned in Table 1. The cavity is made from the large grain niobium of RRR ≥ 117 received from CBMM, Brazil [6]. The disks for the large grain cavity are sliced directly from the ingot by wire electro-discharge machining. A suitable set of dies has been made out of 6061 aluminium to form the cavity half cells from sheet niobium of thickness 4 mm. The schematic and the parts of the different components of the dies are shown in Figure 2.

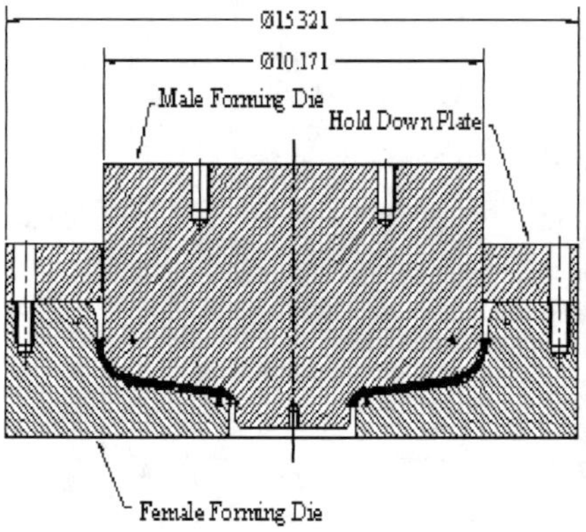

(a)

(b)

Fig. 2. (a) Schematic of the forming die for the 1050 MHz, 4mm thick cavity, (b) Different parts of the forming die.

The half cells are deep drawn and then the required machining at the equator and the iris region are carried out. All the defects, machine marks are removed by mechanical buffing followed by ultrasonic degreasing and buffered chemical polishing (BCP) ($HF:HNO_3:H_3PO_4$ = 1:1:1 (by volume)) to remove about 30μm from the inner surface of the cavity half cells. The beam tubes for both cavities are fabricated from a fine grain Niobium sheet of low RRR by rolling and electron beam welding (EBW) along the length of the tube. Then the half cells are EB welded with the respective beam tubes. No difficulties observed in welding of fine grain beam tubes with the large grain half cells. Finally equator welding is carried out to form the single cell. All the welded joints are leak tested by MSLD. In all the EB welded joints have shown a leak rate better than $1X10^{-9}$ std. cc/sec. The EB welded niobium large grain cavity is shown in Figure 3.

Fig. 3. f = 1050 MHz, β = 0.49 EB welded Niobium Large Grain cavity.

360^0 view of the EBW from inside of the Nb large grain cavity at the equator region is shown in Figure 4. Each picture shows about 40^0 sections in linear scale. The picture shows full penetration weld around the equator line except some regions. There is a clear evidence of grain growth in the welding heat affected zone.

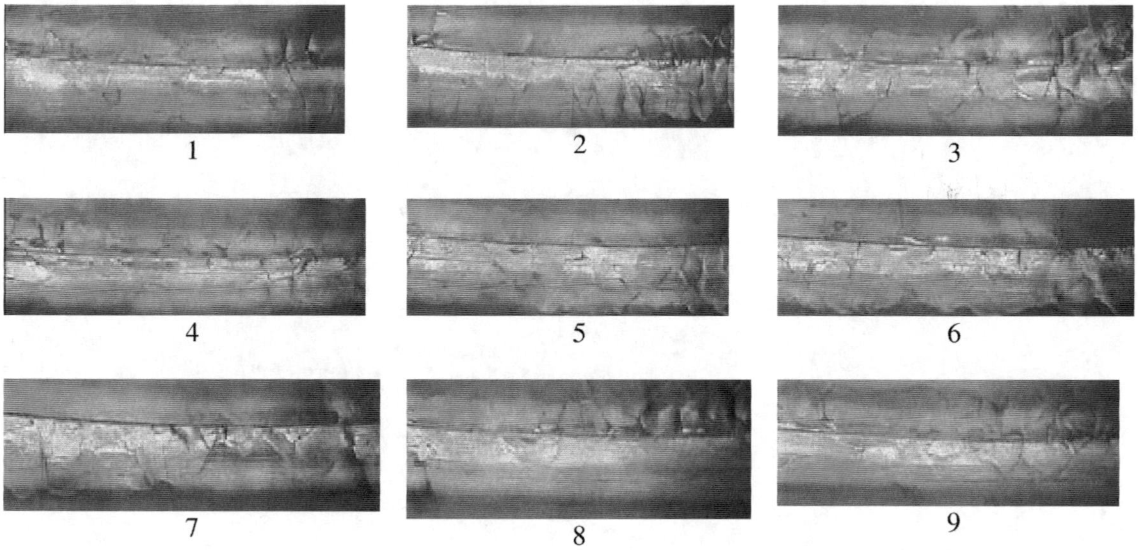

Fig. 4. Inner view of the equator weld joint

4. Results and discussions

4.1. Measurement of resonant frequency and Qo

To measure the resonant frequency and unloaded Q, Q0 of the cavity, the coupling coefficient β due to the input probe is taken into account where as the coupling coefficient of the transmitted probe is negligible. The coupling coefficient of the in put probe was 0.22. The measured frequency before EBW was 1026 MHz. The frequency bandwidth at 3-dB points was 2 MHz. Thus loaded Q, Q_L was 376. Therefore, $Q0 = Q_L(1+\beta) = 459$. After initial test it is found that the contact between the half cells are not proper and then the equator region is pressed with indium wire to have better RF contact. With the indium contact the measured frequency was 1048 MHz, Q_0 was 6832.

The measured f_0 and Q_0 after EBW were 1036.507 and 8076. Table 2 shows the details of the measurement results.

4.2. Measurement of geometrical shunt impedance, R/Q

Any object placed inside the cavity volume will disturb the field inside the cavity that will change the resonant frequency of the cavity. Experimentally R/Q can be evaluated by measuring the frequency shift due to perturbing object. The final relationship of R/Q with the frequency shift is given by,

$$R/Q = \left(ds^2 \sum_{i=1}^{n} \Delta f_i \right) \Big/ \pi \varepsilon k \Delta v f_0^2$$

Where, ds — bead displacement, Δv — volume of bead, $k = 1$ (assumed as bead is much smaller than the cavity dimensions), df —frequency shift, f_0 —resonant frequency, $\varepsilon = \varepsilon_0 \left(\varepsilon_r - 1/\varepsilon_r + 1 \right)$, ε_0 — permittivity of free space, ε_r — dielectric constant (= 2:1 for teflon and tends to infinity for metallic bead).A teflon bead of radius 5mm and length 5mm is made to pass through the central axis of the cavity, a change in resonant frequency is noted at the various positions of the bead with a step (=ds) of 10 mm. Figure 5 and 6 shows the measured axial electric field with theoretical on axis field before and after EBW. Table 2 shows the measured and theoretical values of shunt impedance and R/Q values.

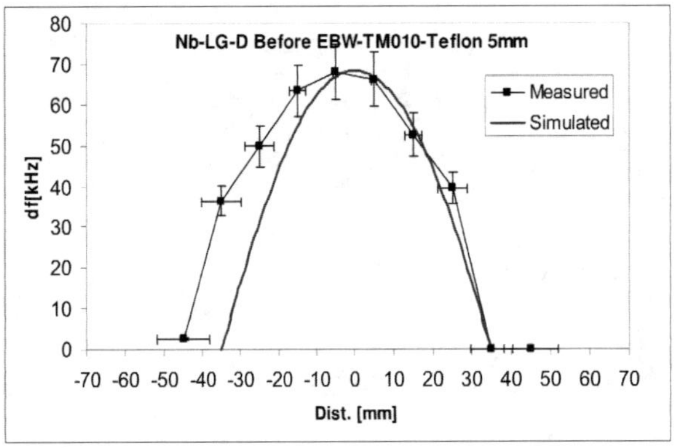

Fig. 5. Axial electric field distribution of TM_{010} mode.

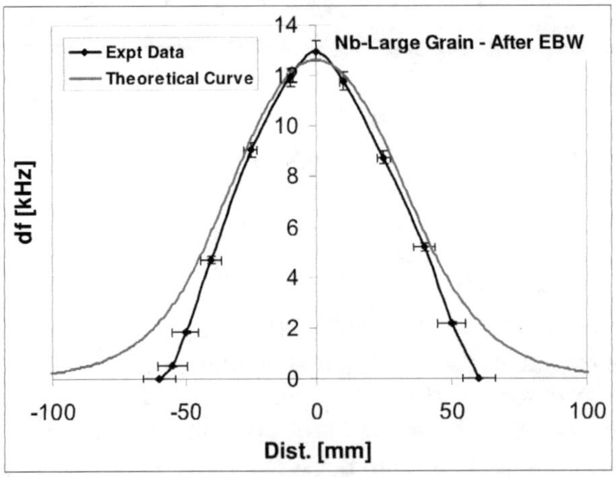

Fig. 5. Axial electric field distribution of TM_{010} mode.

Table 2 RF measurements of the cavity before EB welding and after EB welding with Vector network analyzer

Parameters	Simulated value without Beam tube	Measured value before EBW without beam tube	Simulated value with Beam tube	Measured value after EBW with Beam tube
Frequency [MHz]	1050	1048	1036.856	1036.507
Q Value	17000	459	17032	8076 (not cleaned after EBW)
R/Q [Ω]	9.225	8.59	9.675	9.0

5. Conclusion

A single cell large grain niobium cavity indigenously developed and tested at room temperature. The initial results are encouraging and matches with the simulated value within 5%. Soon the cavity will be tested at 2K and subsequent results will be reported.

References

[1] P. Singh et. al., Accelerator development in India for ADS programme, Pramana **68** (2007) 331.
[2] R. L. Kustom. "An Overview of the Spallation Neutron Source Project", Linac 2000, August 2000.
[3] http://www.ganil.fr/eurisol/Final_Report.
[4] "RF Properties of 1050 MHz, β = 0.49 Elliptical cavity for High Current Proton Acceleration" Amitava Roy , J. Mondal, K. C. Mittal, Journal of Instrumentation, 3 P04002, 2008 (IOP Publication).
[5] "Design Criteria for Elliptical Cavities" C. Pagani, D. Barni, A. Bosotti, P. Pierini, G. Ciovati, Proc. of the 10th SRF Workshop,Tsukuba, Japan , (2001) 115.
[6] J.Mondal, G.Ciovati, P.kneisel, K.C.Mittal, G.R.Myneni, Proc. of the 14th SRF Workshop,dBB Forum, Belin, Germany, September 2009, Paper THOAAU01.

Multipurpose Vacuum Induction Processing System

M Govindaraju[1], Deepak Kulkarni, K Balasubramanian,

Non Ferrous material Technology Development Centre (NFTDC), PO Kanchanbagh, Hyderabad- 500058, India, Ph: +919440727051, Fax: +91-40-24342567.

Email: govindaraju.rajusmg@gmail.com

Abstract. Multipurpose vacuum processing systems are cost effective; occupy less space, multiple functional under one roof and user friendly. A multipurpose vacuum induction system was designed, fabricated and installed in a record time of 6 months time at NFTDC Hyderabad. It was designed to function as a) vacuum induction melting/refining of oxygen free electronic copper/pure metals, b) vacuum induction melting furnace for ferrous materials c) vacuum induction melting for non ferrous materials d) large vacuum heat treatment chamber by resistance heating (by detachable coil and hot zone) e) bottom discharge vacuum induction melting system for non ferrous materials f) Induction heat treatment system and g) directional solidification /investment casting. It contains provision for future capacity addition. The attachments require to manufacture multiple shaped castings and continuous rod casting can be added whenever need arises. Present capacity is decided on the requirement for 10years of development path; presently it has 1.2 ton liquid copper handling capacity. It is equipped with provision for capacity addition up to 2 ton liquid copper handling capacity in future. Provision is made to carry out the capacity addition in easy steps quickly. For easy operational maintenance and troubleshooting, design was made in easily detachable sections. High vacuum system is also is detachable, independent and easily movable which is first of its kind in the country. Detailed design parameters, advantages and development history are presented in this paper.

1. Introduction Vacuum metallurgy:

Vacuum is misnomer relative term which is used to represent the meaning low pressure. Processing of selective materials (metals, alloys, intermetallic and refractory materials) like titanium in gaseous environment cause metallurgical problems or unwanted reactions. To avoid the reactions vacuum metallurgy was developed a century back. In addition, composition and partial pressures of the various components are carefully controlled in vacuum metallurgy. The pressure range in vacuum metallurgy is from sub-atmospheric to ultrahigh vacuum (less than 760 torr to 10^{-12} torr). The processes in vacuum metallurgy involve liquid/solid, vapor/solid, and vapor/liquid/solid transitions [1, 2].

The basic reasons for vacuum processing are elimination of contamination from the processing environment, reduction of the level of impurities in the product, and deposition with a minimum of impurities. In the vacuum process, impurities, particularly oxygen, nitrogen, hydrogen, and carbon, are released from the molten metal and pumped out during melting. There are numerous and varied application areas for vacuum metallurgy including special areas of extractive metallurgy, melting processes (Vacuum induction melting), casting of shaped products, degassing of molten steel, heat treatment, surface treatment, vapor deposition, space processing, joining(vacuum brazing, diffusion boning etc) processes, and testing of metals and alloys in controlled environments [2, 3].

Vacuum Induction Melting (VIM) is one of the important requirements for scientific research particularity in nonferrous metals and alloys arena. Oxygen Free Electronic (OFE) copper, copper-chromium alloys, silver containing copper, titanium alloys and super alloys are main materials among them. For International Thermonuclear Experimental Reactors (ITER), copper and copper-chromium alloys play important role. All these alloys are made by VIM and further processing [4, 5].

Several suppliers around the world offer processing equipments for vacuum processing like VIM. In many occasions, special purpose machines are built to satisfy the needs. But all these equipments are for particular applications or two similar applications. For example, systems built for vacuum induction melting cannot be used for surface treatment/heat treatment; Systems built for joining (vacuum Brazing) cannot be used for melting. Single operation oriented techniques (systems) like this are well suited for higher production. Butt research and development, pilot production, urgent requirement of some processing etc cannot afford the dedicated equipments for few experiments. So, a common platform is needed for the same which can fulfil various the needs with minimum effort and cost. This is similar to the assembly line of car factory. In this context, a multipurpose vacuum system was planned and built with operating vacuum ranging from 10^{-6} torr to 10^{-2} torr.

2.Design and Fabrication:

2.1. Main sections of the designed system are as below

Main sections of the vacuum induction system are listed below and shown in Figure. 1 A and B.

1. Vacuum systems
    - i. Pumps
        - a. Mechanical
        - b. Booster
    - ii. Pipe lines (detachable, provision for capacity addition)
2. Control pack (easy and user friendly, semi automatic with manual override)
3. Power pack
4. Induction Coil
5. Materials handling systems
    - i. Hot charge (for melting operations)
    - ii. Mould/crucible load/unload (material handling mechanism)
    - iii. Others
6. Chamber
7. Futuristic Ports for expansion/modifications
8. Subsystems which can be built if necessary
    - i. Resistance heating hot zone
    - ii. Mould chamber various outputs of induction melting

Possible operations, applications and methods:

2.2. Vacuum Induction Melting mode and refining pure metals and directional solidification of materials:

This process is suitable for the production of high-purity metals within an oxygen-free atmosphere. Refining of high purity metal and few alloys (eutectic alloys), electrode re-melting, investment casting, casting of aircraft engine components are few processes which can be carried out in this mode. For example, OFE Copper melting is done by using graphite crucible in VIM mode. It is melting, refining (keeping the liquid copper in vacuum for few hours to eliminate impurities and volatile matters), moving it slowly away from hot zone in such a way that the equilibrium solidifications starts at one end & ends at other end thus pushing all impurities to that end).

Journal of Physics: Conference Series **390** (2012) 012011 doi:10.1088/1742-6596/390/1/012011

2.3. Vacuum Induction Melting ferrous and non ferrous alloys

For many specialized materials, vacuum induction melting is indispensable, which must be melted under vacuum or in an inert gas atmosphere because of their reactivity with atmospheric oxygen. This limits the formation of non-metallic oxide inclusions. This process contains melting, holding, and casting by tilting the coil + assembly and pouring the melted liquid in moulds. Thus, it differs from the option 2.2; pouring in this option is done in a separate mould. When non ferrous materials are melted, a conducting crucible like graphite is used.

2.4. Vacuum heat treatment chamber for resistance heating:

Induction coil and its accessories (coil assembly) are designed in such a way that it can be disassembled and assembled in 2-3 hours easily. In addition the assembly is completely detachable package so the vacuum chamber is free from any fitments which occupy space. After removing the coil assembly, and installing resistance heat shield + hot zone set up, it can be used as resistance heating high vacuum furnace. This chamber and hot zone can handle huge size and weight (up to 4000 kg job size in one batch). Ports are designed to accommodate induction and resistance feed through. Making a separate furnace for such large job is uneconomical as the number of jobs in such system are very few and occasional.

2.5. Induction heating system

The system can be used as induction heating system of large components by providing a graphite suceptor (easily detachable) inside induction coil, which makes the hot zone for billet hating for forming (rolling, forging), homogenization annealing and induction hardening processes etc. Crucible of the induction coil is removed which makes way for susceptor fixing. Susceptor is conduction source for heating non ferrous materials in induction heating system that acts as uniform heat distribution source manly for non ferrous materials [6].

2.6. Continuous production/operation for melting copper and its alloys:

Continuous casting is possible by adding bifurcation valves and separating mould chamber and melt chamber with separate vacuum lines. The mould chamber can be opened to atmosphere by keeping melt chamber in vacuum, which enables to remove ingots in continuous production.

2.7. Provisions for future expansion:

Future expansions as following are possible with minimum effort and cost.

a. Power pack capacity and coil (increasing the capacity to 2 tons liquid copper from 1 ton) is possible with existing chamber, vacuum system. Feed through, coil support assembly can accommodate the increase.
b. Increasing the vacuum pumping system by adding additional diffusion pump/vapour booster pump + rotary pump. Vacuum ports are provided for the same. This can be useful for expanding the volume of chamber in horizontal direction, for which provision is available.
c. Extension of the vacuum chamber (design is in such a way that one side is fully detachable port)
d. Continuous casting arrangements can be fitted for horizontal continuous casting.

3. Conclusions:

1. A multipurpose vacuum processing system was designed and commissioned successfully.
2. It contains provision for vacuum induction melting, refining, investment casting, directional solidification and induction heat treatments.
3. It has provision for future expansion by easy additions.
4. It is possible for changing the system to resistance heating system by detaching induction coils and assembles, installing resistance heating hot zone.
5. Pumps, vacuum pipe line and chamber are easily detachable. This helps for easy servicing.

Journal of Physics: Conference Series **390** (2012) 012011 doi:10.1088/1742-6596/390/1/012011

It has provision for continuous casting of copper and its alloys by adding simple set up.

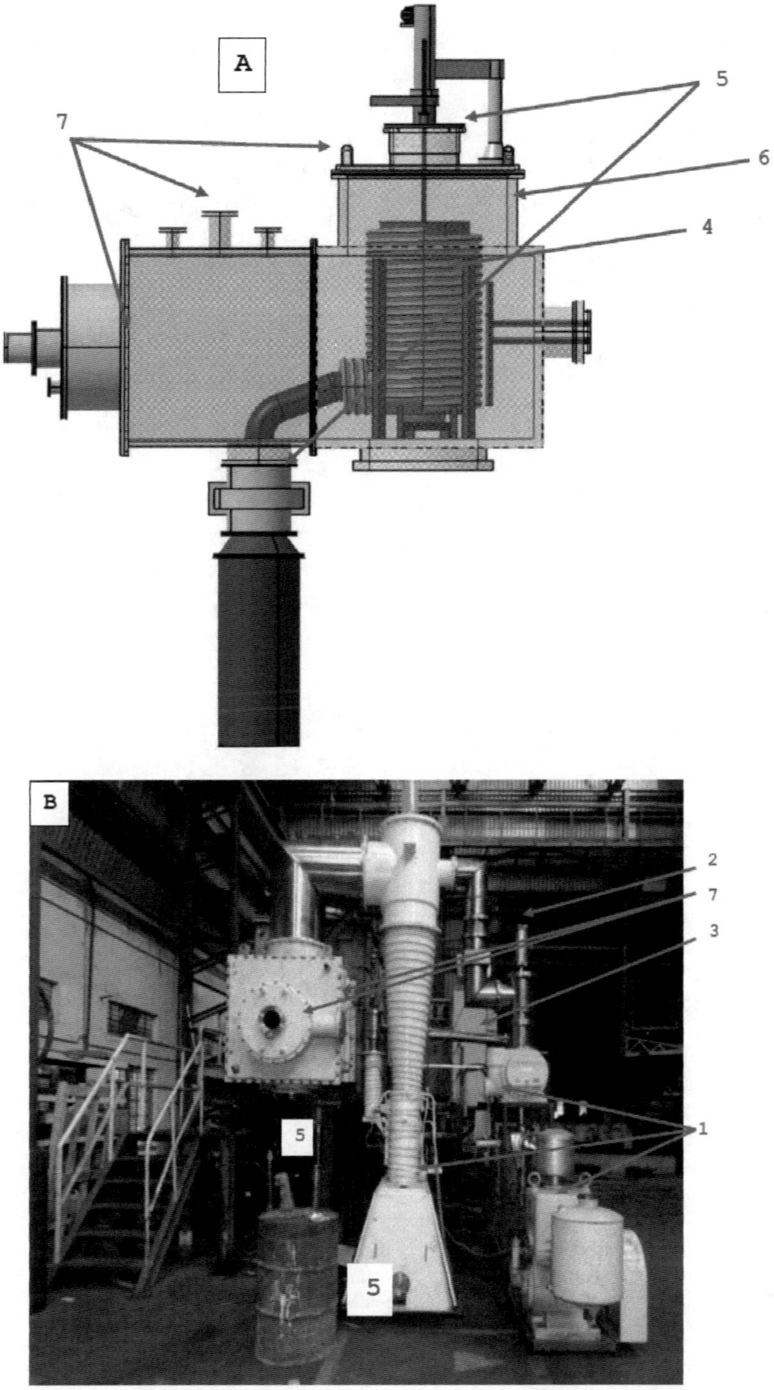

Fig. 1. Schematic view (A) and photograph (B) of multipurpose vacuum system

References:

[1] Chambers, Austin (2004). *Modern Vacuum Physics*. Boca Raton: CRC Press.
[2] Bunshah RF, 2008, *J. Vac. Sci. Technol.*, A12(4) 936–945.
[3] Choudhury A, 1990, *Vacuum metallurgy*, ASM International Materials Park, OH 88-91.
[4] Kenji Abiko, 1996 *Vacuum* 47 837-843. (From "Proceedings of the 13th International Vacuum Congress and the 9th International Conference on Solid Surfaces)
[5] Srinath Viswanathan et. all, *Casting*, 2008, ASM Handbook Volume 15, ASM International 409-410.
[6] J S Park, S Taniguchi and Y J Park, 2009 *J. Phys. D: Appl. Phys. 42*

Programmable Pulse Generator for Aditya Gas Puffing System

Narendra Patel, Chhaya Chavda, S B Bhatt, Prabal Chattopadhyay, Y C Saxena

Institute for Plasma Research, Nr. Indira Bridge, Bhat, Gandhinagar-382 428, Gujarat, India.

E-mail: ncpatel@ipr.res.in

Abstract. In the Aditya Tokamak, one of primary requirement for plasma generation is to feed the required quantity of the fuel gas prior to plasma shot. Gas feed system mainly consists of piezoelectric gas leak valve and gas reservoir. The Hydrogen gas is prior to 300ms loop voltage for the duration of 4 msec to 7 msec. Gas is puffed during the shot for required plasma parameters and to increase plasma density using the same system. The valve is controlled by either continuous voltage or pulses of different width, amplitude and delay with respect to loop voltage. These voltage pulses are normally applied through standard pulse generator. The standard pulse generator is replaced by micro controller based in housed developed programmable pulse generator system consists of in built power supply, BNC input for external trigger, BNC output and serial interface. This programmable pulse generator is successfully tested and is in operation for gas puffing during ADITYA Tokamak experiments. The paper discusses the design and development aspect of the system.

1. Introduction

Hydrogen gas is fed to vacuum vessel for generation of plasma. It is also puffed during the plasma discharge to reduce impurities hard X-ray etc. and to maintain the density. The gas feeding system consists of piezoelectric gas leak valve and gas reservoir. Gas is puffed during plasma shot using same gas feeding system to reduce hard X-ray and increase plasma density. Additional Gas puffing helps in sustaining the plasma.

For puffing gas at various instances during plasma discharge, either continuous voltage or voltage pulses of different amplitude and widths are applied to piezoelectric valve (model MV112) to get desired Plasma parameter. A piezoelectric gas feed valve is routinely used for feeding the fuel gas in Tokamak. The valve is operated linearly with applied voltage. The duration and the amplitude of applied pulse decide the opening of the valve i.e. valve will open more if the amplitude is more and will remain open for longer time for longer pulse width. Opening is varying depending upon voltage and the pulse width of the applied pulse. The valve opened fully in 2 msec.

The piezoelectric valve is controlled by the controls available either from CAMAC pulse generator or from standard function generator. An in-housed microcontroller based programmable pulse generator (PPG) is designed and developed to generate these pulses to control the valve. The output of the programmable pulse generator is amplified with the gain of 20 and applied to the piezoelectric valve to match the required potential to operate the valve.

Basic idea for the in-housed development is to make the control operation independent of the data acquisition and control system and work as stand-alone user-friendly system. User can set the parameters like number of pulses, amplitude, width and delay using PC serial interface.

2. Theoretical Basis

Microcontroller has two timer registers. One of the timers is used to generate pulse of fixed pulse width for "fixed pulse width (100μsec) High level, TH loop" while second timer is used to generate pulse of fixed pulse width for "fixed pulse width (1msec) Low level, TL loop".
TH = Number x TH loop
TL = Number x TL loop, Number is 8bit data fed by user.

On receiving the external trigger pulse, the program generates the desired delayed pulse. Data for pulse width, delay and Amplitude can be loaded using serial connectivity.

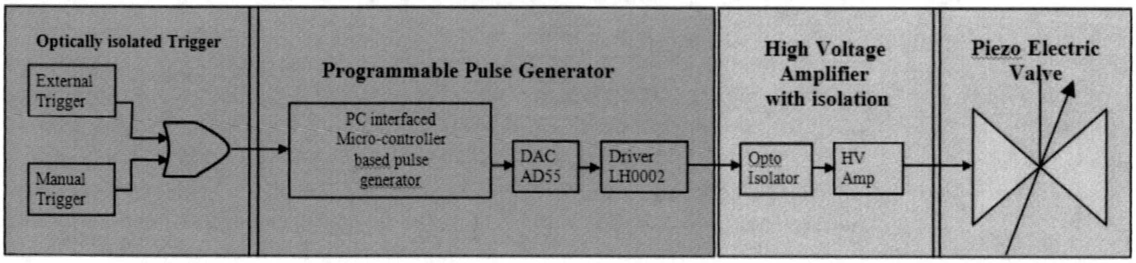

Fig. 1. Functional Block Diagram of PPG

As shown in the figure1 the system is developed in two halves. "Optically isolated external trigger and Programmable Pulse Generator" is one part which is installed in the Control room while second part "High Voltage Amplifier with isolation" is installed near piezoelectric Valve with Vacuum Control system.

Parameters are loaded to controller memory using serial text file transmitted using LabVIEW.

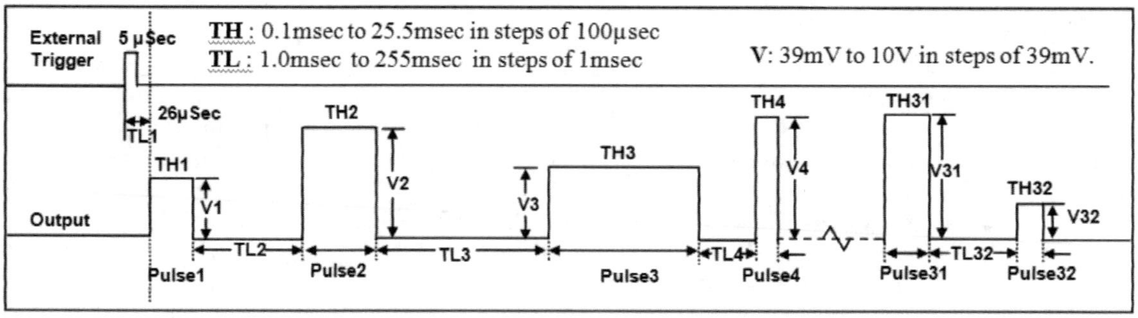

Fig. 2. Operational timing diagram of the system

To generate desired pulse following four main parameters are to be set: Total number of pulses (N), High time (TH) and Low time (TL) of each pulse and Amplitude (V) of each pulse.

As shown in the figure 2, the valve will be open for TH time and will close for TL time. Amplitude V will define opening or how much it will open and N will define for how many times the valve will be opened (gas will be puffed), Delay (TL1) will decide delay between Trigger and opening of Gate Valve.

3. Experimental set-up

In House developed programmable pulse generator is triggered by optically isolated signal conditioned loop voltage pulse. It generates pulses of different width (100μs to 25.5 ms in steps of 100 μs), amplitude (0-10V in steps of 39 mV) and delay (between external trigger to first output and between different output from 26 μsec to 255 ms in steps of 1 ms) using Digital to Analog converter. The pulse then applied to a high voltage amplifier with a gain of 20 to generate required potential to operate the valve. The amplified output is applied to the piezoelectric leak valve, to control the opening of the valve. Minimum delay from the external trigger pulse to the output is 26 μsec.

Fig. 3. Front Panel

As shown in the figure 3, the front panel has 'Ready" LED for indication of readiness of the system for data transmission or waiting for external trigger and a push button for Manual trigger. It has one SPST key for switching On/Off power supply. Programmable Pulse Generator output and optically isolated external trigger input are available on BNC connector while manual trigger is given by push button switch.

It has in-built regulated D C power supply which can be controlled by power ON switch on front panel-(figure 4). Programmable Pulse Generator is designed using AT89C52 micro controller, AD558 digital to analog converter, LH0002H output driver, IC 4N35 opto-coupler and MAX 232 serial converter. The parameter data text file is loaded to the controller memory through 9 pin D type serial connector and PC.

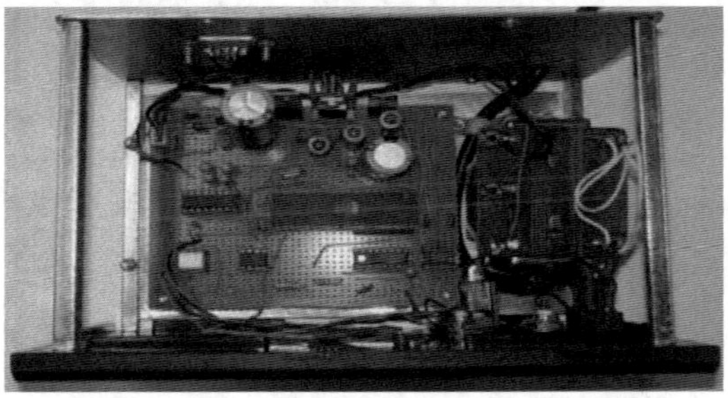

Fig. 4. System Circuit Assembly

4. Experimental results and Discussions

The experiment needs two stages

1) Sequence for setting parameters and
2) Receiving External trigger

4.1 Sequence for setting parameters

Generate an Excel file as shown in figure5. Column A (Type of parameter) describes name of parameter. Column B (Input data) is desired parameter. Column C is Converted data which will go to microcontroller memory.

For Offtime (TL), conversion is (data x 1).

For Ontime (TH), conversion is (data x 10).

For Amplitude (V), conversion is ((data x 1000) / 39) considering the DAC 558 parameter.

	A	B	C		A	B	C		A	B	C
1	**DATA ID**	#	#	35	TH - 1 ms	0.1	1	67	Volt Pulse - 1	5	128
2	Pulse no.	32	32	36	TH - 2 ms	0.1	1	68	Volt Pulse - 2	3	77
3	TL - 1 ms	1	1	37	TH - 3 ms	0.1	1	69	Volt Pulse - 3	8	205
4	TL - 2 ms	1	1	38	TH - 4 ms	0.1	1	70	Volt Pulse - 4	4	103
5	TL - 3 ms	1	1	39	TH - 5 ms	0.1	1	71	Volt Pulse - 5	2	51
6	TL - 4 ms	1	1	40	TH - 6 ms	0.1	1	72	Volt Pulse - 6	9	231
7	TL - 5 ms	1	1	41	TH - 7 ms	0.1	1	73	Volt Pulse - 7	4.5	115
8	TL - 6 ms	1	1	42	TH - 8 ms	0.1	1	74	Volt Pulse - 8	3.5	90
9	TL - 7 ms	1	1	43	TH - 9 ms	0.1	1	75	Volt Pulse - 9	1.5	38
10	TL - 8 ms	1	1	44	TH - 10 ms	0.1	1	76	Volt Pulse - 10	1	26
11	TL - 9 ms	1	1	45	TH - 11 ms	0.1	1	77	Volt Pulse - 11	9.5	244
12	TL - 10 ms	1	1	46	TH - 12 ms	0.1	1	78	Volt Pulse - 12	1	26
13	TL - 11 ms	1	1	47	TH - 13 ms	0.1	1	79	Volt Pulse - 13	2	51
14	TL - 12 ms	1	1	48	TH - 14 ms	0.1	1	80	Volt Pulse - 14	3	77
15	TL - 13 ms	1	1	49	TH - 15 ms	0.1	1	81	Volt Pulse - 15	4	103
16	TL - 14 ms	1	1	50	TH - 16 ms	0.1	1	82	Volt Pulse - 16	5	128
17	TL - 15 ms	1	1	51	TH - 17 ms	0.1	1	83	Volt Pulse - 17	6	154
18	TL - 16 ms	1	1	52	TH - 18 ms	0.1	1	84	Volt Pulse - 18	7	179
19	TL - 17 ms	1	1	53	TH - 19 ms	0.1	1	85	Volt Pulse - 19	8	205
20	TL - 18 ms	1	1	54	TH - 20 ms	0.1	1	86	Volt Pulse - 20	9	231
21	TL - 19 ms	1	1	55	TH - 21 ms	0.1	1	87	Volt Pulse - 21	9.5	244
22	TL - 20 ms	1	1	56	TH - 22 ms	0.1	1	88	Volt Pulse - 22	9	231
23	TL - 21 ms	1	1	57	TH - 23 ms	0.1	1	89	Volt Pulse - 23	8	205
24	TL - 22 ms	1	1	58	TH - 24 ms	0.1	1	90	Volt Pulse - 24	7	179
25	TL - 23 ms	1	1	59	TH - 25 ms	0.1	1	91	Volt Pulse - 25	6	154
26	TL - 24 ms	1	1	60	TH - 26 ms	0.1	1	92	Volt Pulse - 26	5	128
27	TL - 25 ms	1	1	61	TH - 27 ms	0.1	1	93	Volt Pulse - 27	4	103
28	TL - 26 ms	1	1	62	TH - 28 ms	0.1	1	94	Volt Pulse - 28	3	77
29	TL - 27 ms	1	1	63	TH - 29 ms	0.1	1	95	Volt Pulse - 29	2	51
30	TL - 28 ms	1	1	64	TH - 30 ms	0.1	1	96	Volt Pulse - 30	1	26
31	TL - 29 ms	1	1	65	TH - 31 ms	0.1	1	97	Volt Pulse - 31	3.5	90
32	TL - 30 ms	1	1	66	TH - 32 ms	0.1	1	98	Volt Pulse - 32	5.5	141
33	TL - 31 ms	1	1								
34	TL - 32 ms	1	1								

Fig. 5. Excel Parameter file

4.1.1 Load parameter as per following sequence

1) Load '#' as Data ID for valid data confirmation while serial transmission (row1).
2) Load 'Pulse No' as total no of pulses to be generated (row2).
3) Load "Offtime" or TL data first. "Off time data" or "TL "is the delay between pulses. TL1 is the delay between external trigger and first pulse, TL2 is delay between first and second pulse and so on (row 3 to row 34).
4) After loading delay information for all 32 (TL1-TL32) pulses, start loading data for pulse width "On time data", TH. So the 33rd data (TH1) is the pulse width for the first pulse, 34th data (TH2) is the pulse width for the second pulse and so on till data for 32nd pulse width (TH32) (row 35 to row 66).
5) After loading pulse width for all the channels, start loading data for "Amplitude"(V). Thus 65th data is the amplitude information (V1), for the first pulse. 66th data is amplitude information (V2) for 2nd pulse and so on till data is loaded for all 32 pulses (V1-V32) (row67 to row 98).
6) Save Excel file as text file. Run Gas Puff_32.vi. It will ask for the file to transfer.
7) Load the text file, confirm the data and transfer via serial port as shown in the figure 6.
8) This data will be stored in controller RAM and on board serial EEPROM so even if supply is switched off, last loaded will be available. So once the data is loaded till next modification, data will remain stored in the on board EEPROM. The system, then wait for Trigger pulse. On receiving the trigger pulse, output is available on Output BNC as per the loaded data.

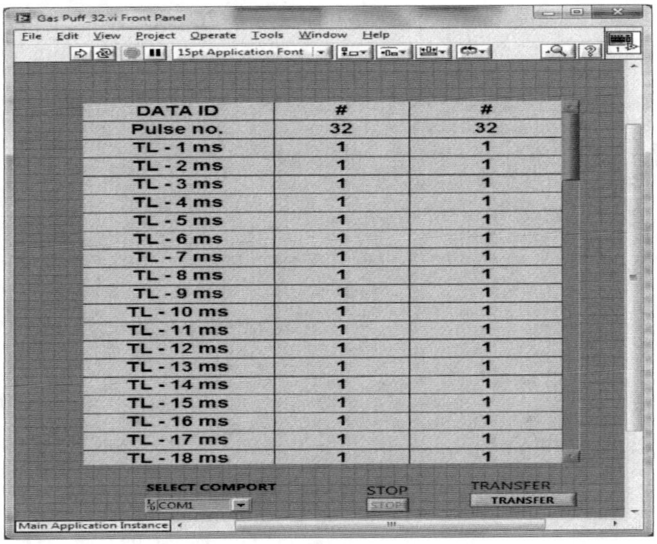

Fig. 6. Gas Puff 32.vi

Front Panel LED will glow when system power is switch 'On'. While transferring the parameter via serial port the front panel LED will be 'Off', indicating the transfer is going on. After transmission over, the LED will be 'On' again, indicating the system readiness for trigger.
AD558 is 8-bit, 0-10V digital to analog converter. The resolution is 39 mV. Actual input is converted using conversion factor (data x 1000) / 39 and data is rounded off for one decimal point.
System works in two different modes: Manual trigger or External Trigger mode. The trigger input (TTL) is isolated from the enclosure and is applied using BNC connector. The manual trigger is using Push button switch.

4.2 Receiving the external trigger
To generate delay or low time, "TL loop" will be executed, while to generate width or high time "TH loop" will be executed. On receiving trigger either manually or from external trigger, TL 1 loop will be executed. Then V1 data will be loaded to DAC and TH1 loop will start executing. After TH 1 loop will be over, DAC will be reset. So output will be low. Output will remain low till TL 2 loop will finish execution. Again V2 data will be loaded to DAC and TH 2 loop will start execution and so on until all pulses will be generated according to loaded "pulse number".

5. Test Results

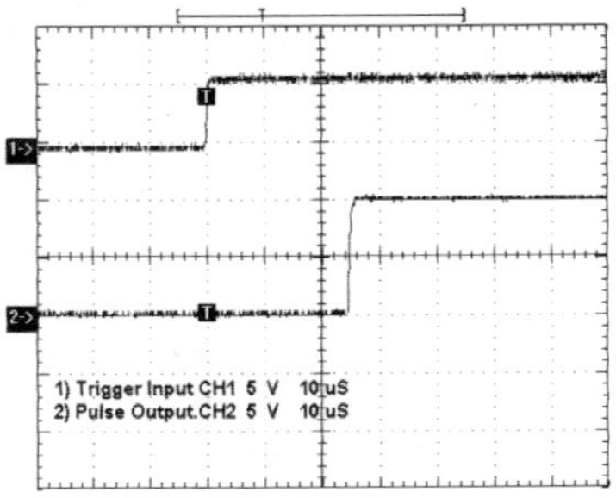

Fig. 7. Delay between Trigger & output 1st Pulse

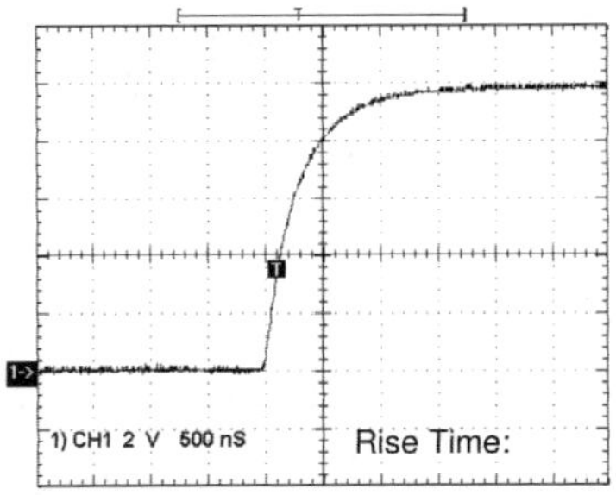

Fig. 8. Rise time of 900nsec

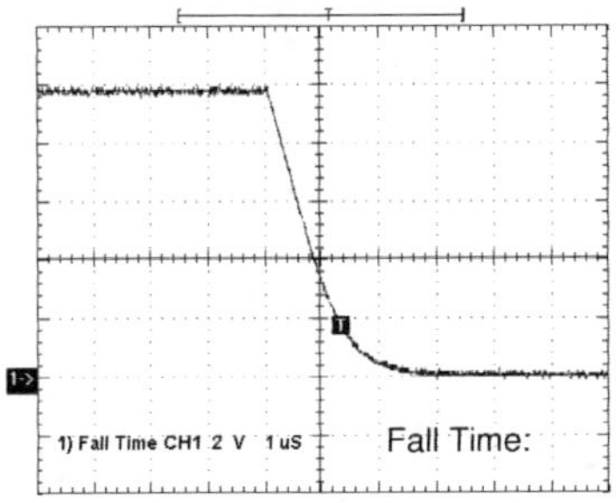

Fig. 9. Fall time of 800nsec

When TL 1 = 0, delay between trigger and output 1st Pulse is 26 μsec see the figure 7. The rise and fall time at any instance are 900nsec and 800nsec respectively, (figure 8 and figure 9).

6. Actual results

A typical recorded data of timing sequences during Aditya operation for 32 different pulse and for 10 different pulse output are shown figure 10 and 11.

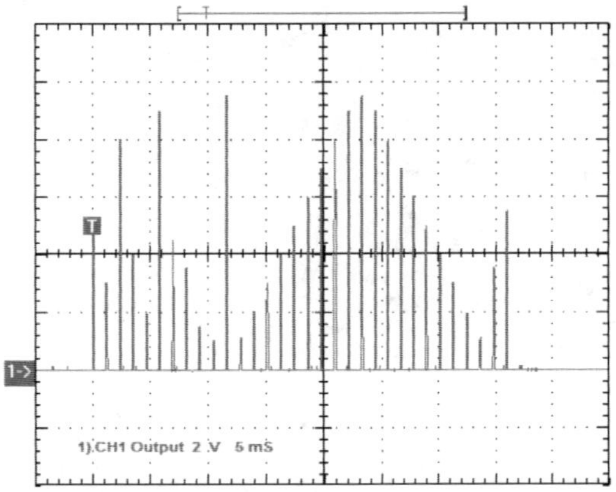

Fig. 10. Actual data for 32 pulses

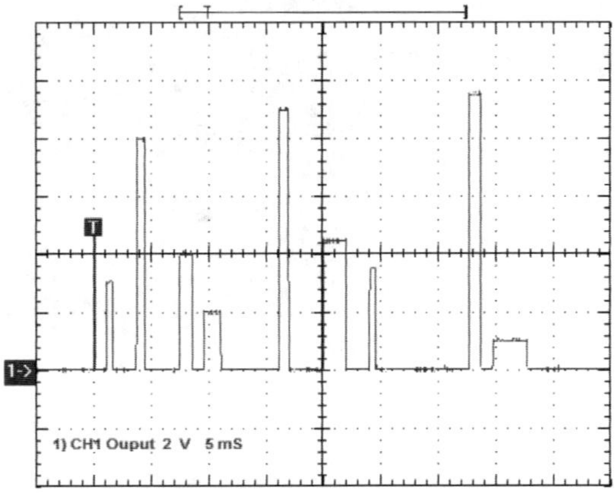

Fig. 11. Actual data for 10 pulses

The detailed parameter for figure10 is as shown in Excel file shown in the figure 5.

7. Conclusion

In house developed microcontroller based Programmable Pulse Generator is successfully working in gas puffing system in Aditya Tokamak. The CAMAC based controller system is replaced by developed stand alone system.

Acknowledgement

We would like to thank the members of Aditya Vacuum Group and Aditya Team for their support during the testing and implementation of the in house developed Programmable Pulse Generator.

Journal of Physics: Conference Series **390** (2012) 012013 doi:10.1088/1742-6596/390/1/012013

Establishment of a force balanced piston gauge for very low gauge and absolute pressure measurements at NPL, India

D Arun Vijayakumar, Om Prakash and R K Sharma

Vacuum and Pressure Standards, National Physical Laboratory, Dr K S Krishnan Marg, New Delhi 110 060, India

E-mail: arun@nplindia.org

Abstract: National Physical Laboratory, the National Metrology Institute (NMI) of India is maintaining Primary standards of pressure that cover several decades of pressure, starting from 3.0E-06 Pa to 1.0 GPa. Among which a recent addition is a Force Balanced Piston Gauge, the non-rotating piston type, having better resolution and zero stability compared to any other primary pressure standards commercially available in the range 1.0 Pa to 15.0 kPa (abs and gauge). The characterization of this FPG is done against Ultrasonic Interferometer Manometer (UIM), the National Primary pressure standard, working in the range 1.0 Pa to 130.0 kPa (abs and diff) and Air Piston Gauge (APG), a Transfer Pressure Standard, working in the range 6.5 kPa to 360 kPa (abs and gauge), in their overlapping pressure regions covering both absolute and gauge pressures. As NPL being one of the signatories to the CIPM MRA, the Calibration and Measurement Capabilities (CMC) of both the reference standards (UIM & APG), are Peer reviewed and notified in the Key Comparison Data Base (KCDB) of BIPM. The estimated mean effective area of the Piston Cylinder assembly of this FPG against UIM (980.457 mm^2) and APG (980.463 mm^2) are well within 4 ppm and 10 ppm agreement respectively, with the manufacturer's reported value (980.453 mm^2). The expanded uncertainty of this FPG, Q(0.012 Pa, 0.0025% of reading), evaluated against UIM as reference standard, is well within the reported value of the manufacturer, Q(0.008 Pa, 0.003% of reading) at k=2. The results of the characterization along with experimental setup & measurement conditions (for gauge and absolute pressure measurements), uncertainty budget preparation and evaluation of measurement uncertainty are discussed in detail in this paper.

1. Introduction

The force-balanced piston gauge is a primary vacuum standard developed by Ooiwa A [1] during 1990s. This new non-rotating force-balanced piston gauge is based upon a mass comparator to determine the force applied to a nominal effective area of 980 mm^2. The major difference from traditional rotating piston gauges is that the FPG measures the force generated from a given gas pressure against a force balanced load cell to which the piston is attached. Here the attachment has two chambers, one for high pressure (p_h) and the other for low/reference pressure (p_r), separated at the middle of the piston and the "lubricating gas flow" enters the chambers at this point. The non-rotating piston is connected to an electronic dynamometer and centered by means of transient gas flow in the tapered gap between piston and cylinder. The load cell is zeroed with high and low pressure chambers connected. In absolute mode, a precision capacitance diaphragm gauge measures the reference pressure at the low pressure chamber of the FPG. Pressure of gas in the upper chamber is adjusted by an automated very low pressure controller (VLPC) that is equipped with two parallel mass flow controllers for coarse and fine adjustment and a PC, which calculates and keeps generated pressure on a chosen value.

2. Experimental set up

The setup shown in figure 1 is used for our pressure measurements, except for the dashed line used only for gauge mode measurements. The valves V_9 to V_{12} that are associated with the FPG are pneumatically operated for which a continuous supply of 7.0 bar dry air is required; so a separate pneumatic supply is arranged for the same. Since the UIM at NPL, India is a mercury Manometer, a 1.0 torr differential CDG is used as an isolator cum null indicator (CDG#01) to avoid mercury contamination at FPG Piston Chambers as discussed by Hendricks et al [2]. Other added advantages of this isolation CDG is that it keeps away the necessity of thermal transpiration correction arising due to difference in temperatures of UIM and FPG if directly connected and prevented the FPG's VLPC from setting large pressure changes in large volumes of UIM manifold.

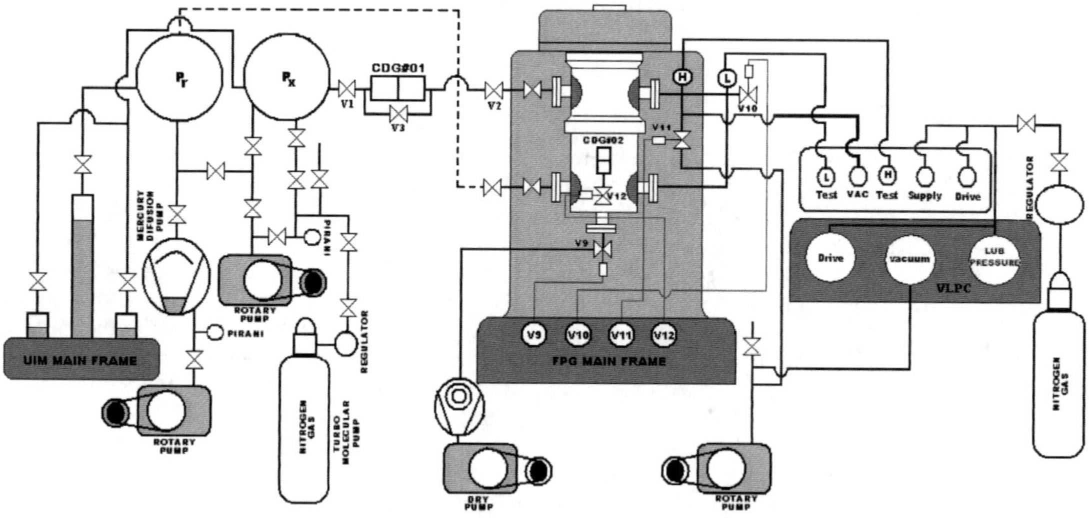

Fig. 1. Measurement Set Up (Dashed line is used only for gauge mode measurements)

All connections are made and secured as shown in figure 1. The UIM and FPG system are initialized for the required mode of measurement. After the whole system attained stabilization, FPG is recalibrated and zeroed. Subsequently the CDG#01 is also zeroed with V_2, V_3 open and V_1 closed.

With V_3 closed and V_1, V_2 open condition, both UIM and FPG (VLPC enabled) are pressurized to a selected nominal pressure. After establishing the pressure in the UIM and FPG through continuously monitoring and adjusting the CDG#01 reading close to null, the measurement system was allowed to stabilize. After stabilization, a set of 8 to 10 pressure readings were recorded for FPG, UIM, CDG#01 and drift in FPG zero pressure. Then FPG along with CDG#01 are isolated from UIM (V_1 closed) and zero-pressure readings of both were re-set (V_2, V_3 open) before proceeding to the next pressure point.

3. Results and discussion

The pressure exerted on the piston of the FPG from the force generated by the mass comparator is given by

$$p_{FPG} = \frac{K_{cal}\left(N+\delta N_1+\delta N_2+\delta N_3\right)}{A_{0_{20^oC}}\left\{1+\left(\alpha_P+\alpha_C\right)\left(t-20\right)\right\}}+p_r+p_{head} \tag{1}$$

Where δN_1, δN_2 and δN_3 are the force correction terms in counts, p_r is the reference pressure at the lower chamber (equal to zero when used in gauge mode), p_{head} is the pressure head, K_{Cal} is the calibration coefficient and N is the load cell reading in counts. The papers by P Delajoud and M Girard [3, 4] may be referred for detailed discussion.

From the data collected, as discussed by Jay Hendricks et al [2] the difference in corrected FPG and UIM reading for absolute and gauge mode are given by equation 2 and 3 respectively as

$$\langle p_{FPG_c} - p_{UIM_c} \rangle = (p_{FPG} - p_{ZC}) - (p_{UIM} + p_{vp_{Hg}} - p_{CDG\#01}) \quad (2)$$

$$\langle p_{FPG_c} - p_{UIM_c} \rangle = (p_{FPG} - p_{ZC}) - (p_{UIM} - p_{CDG\#01}) \quad (3)$$

The plot between measured pressure vs calibration factor (p_{FPG_c} / p_{UIM_c}) of all data collected in both absolute and gauge mode is depicted in figure 2.

3.1. Estimation of effective area of Piston cylinder assembly of FPG

The equation used for effective area estimation in gauge mode is given by

$$A_{0(20^o C)} = \left(\frac{m_{cal} g}{N_{cal}}\right)\left(1 - \frac{\rho_{lcal}}{\rho_{mcal}}\right)\left[\frac{(N + \delta N_1 + \delta N_2 + \delta N_3)}{(p_{Std} - p_{head})\{1 + (\alpha_P + \alpha_C)(t - 20)\}}\right] \quad (4)$$

Where p_{Std} in equation 4 may be UIM measured pressure at 20°C or Piston Gauge measured pressure at 20°C, as in our study both UIM and Air Piston Gauge were used as reference standards. The effective area data thus estimated in the overlapping pressure region (gauge mode) against UIM and Piston Gauge are depicted in figure 3.

The estimated mean effective area of the Piston Cylinder assembly of this FPG against UIM (980.456 mm^2) and APG (980.463 mm^2) are well within 4 ppm and 10 ppm agreement respectively, with the manufacturer's reported value (980.453 mm^2).

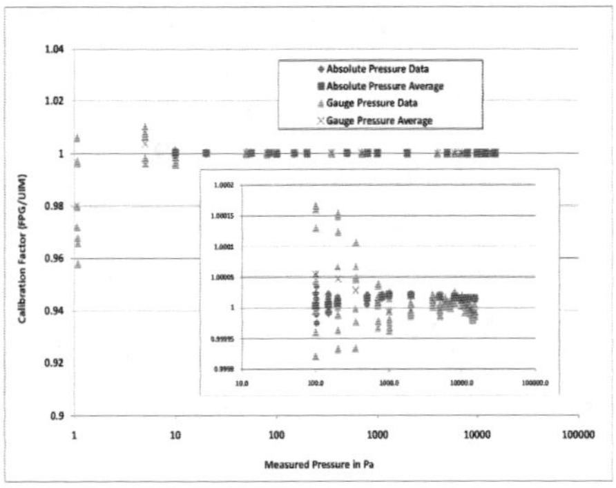

Fig. 2. Data collected in both absolute and gauge mode.

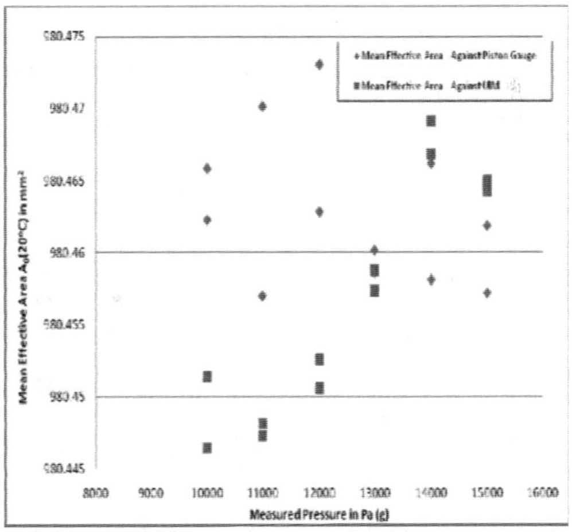

Fig. 3. Effective area of FPG estimated against UIM and APG.

3.1.1. Estimation of expanded uncertainty in Effective Area

Equation 4 is the model equation for the budget preparation. According to P Delajoud and M Girard [3, 4], the expanded uncertainty of effective area was worked out as ±26.0 ppm at k=2, where the uncertainty components due to force correction terms are ignored based on insignificant contribution. As the budget proposed by them is for the cross float method, the major uncertainty component is that of the reference standard, namely the Air Piston Gauge. But in our case, the reference standard is UIM and its measurement uncertainty is notified as Q(0.0092 Pa, 7.2 ppm) at k=2 in Appendix – C of KCDB of BIPM [5]. According to our budget prepared including the UIM uncertainty and repeatability, the expanded uncertainty of effective area is worked out to be ±19.98ppm at k=2.025 (effective degrees of freedom=74).

Quantity (X_i)	Estimate (x_i)	Uncertainty u_{xi}	Probability distribution	Type (A or B)	Degrees of Freedom	Uncertainty Contribution $u_i^2(y)$
Calibration Mass - m_{cal} (kg)	7.79E-01	3.90E-06	Normal	B	∞	2.40E-17
Accln. Due to Gravity g (m/s^2)	9.79E+00	1.96E-05	Normal	B	∞	3.85E-18
Lub air Density ρ_{lair} (kg/m^3)	1.61E+00	2.80E-03	Normal	B	∞	1.21E-19
Cal Mass Density ρ_{mcal} (kg/m^3)	7.90E+03	9.12E+01	Normal	B	∞	5.31E-18
Pressure Medium Density ρ_{mair} (kg/m^3)	4.83E-01	1.06E-07	Normal	B	∞	4.65E-20
Thermal coeff of Piston - α_P ($^oC^{-1}$)	4.50E-06	2.59E-07	Normal	B	∞	5.79E-19
Thermal coeff of Cylinder - α_C ($^oC^{-1}$)	4.50E-06	2.59E-07	Normal	B	∞	5.79E-19
Temp.diff from the ref. temp. $(t-t_{ref})$ (°C)	3.00E+00	8.10E-05	Normal	B	∞	5.11E-25
Measured Pressure- p (Pa)	1.50E+04	5.42E-02	Normal	B	∞	1.25E-17
Fluid Head Correction (Pa)	4.73E+00	1.65E-06	Normal	B	∞	1.17E-26
Verticality (Pa)		1.20E-03	Normal	B	∞	6.15E-21
System Stability (Pa)	0.15	4.33E-02	Rectangular	B	∞	8.01E-18
Load Cell Precision (Pa)		1.52E-02	Normal	B	∞	9.88E-19
Resolution (Pa)	1.00E-03	2.89E-04	Rectangular	B	∞	3.56E-22
Std Dev of Mean Eff Area (m^2)	7.80E-09	2.25E-09	Normal	A	11	5.07E-18
Repeatability (m^2)	1.80E-08	5.69E-09	Normal	A	9	3.24E-17
					Total Variance	9.35E-17
			Overall standard uncertainty (m^2)			9.67E-09
			Effective Degrees of Freedom			74
		Relative expanded uncertainty (ppm) at k=2.025				19.98

3.2. Estimation of expanded uncertainty of FPG
From the raw data collected, the repeatability for each nominal pressure point is worked out through normalization process and the uncertainty component of repeatability arrived at for absolute and gauge mode measurements are depicted in figures 4. The expanded uncertainty of FPG estimated including the repeatability component for absolute and gauge mode measurements are depicted in figure 5.

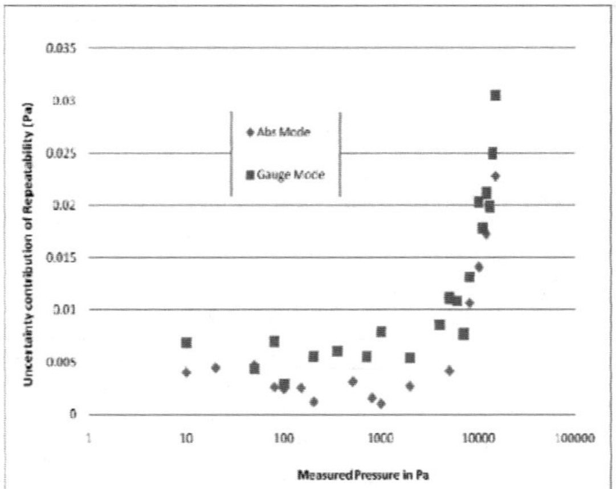

Fig. 4. Uncertainty contribution of repeatability - absolute and gauge mode

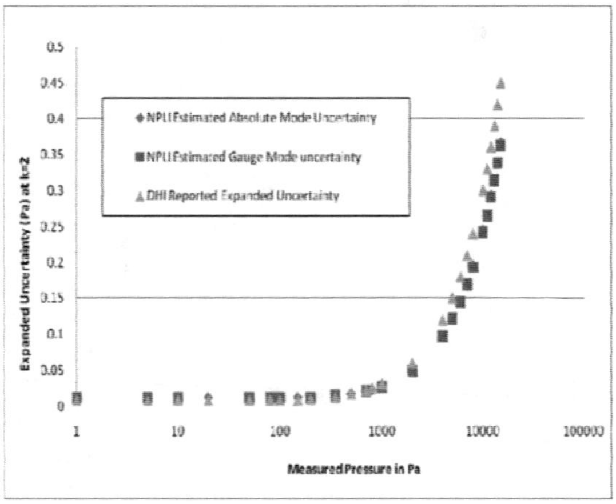

Fig. 5. Expanded uncertainty of FPG - NPLI estimated and DHI reported

Uncertainty budgets were prepared for two different methods namely, theoretical method and the comparison method proposed by D Arun Vijayakumar [6]. The budget prepared using theoretical and comparison methods for absolute and gauge mode are reproduced in tables 2 and 3 respectively.

Table 2: Theoretical method - expanded Uncertainty Budget of p_{FPG} at 15.0 kPa

Quantity (X$_i$)	Estimate (x$_i$)	Uncertainty u_{xi}	Type (A or B)	Degrees of Freedom	Uncertainty Contribution $u_i^2(y)$ Independent	Dependent
Calibration Mass - m_{cal} (kg)	7.79E-01	3.90E-06	B	∞		5.63E-03
Accln. Due to Gravity g (m/s^2)	9.79E+00	1.96E-05	B	∞		9.00E-04
Lub air Density ρ_{lair} (kg/m^3)	1.61E+00	2.80E-03	B	∞		2.83E-05
Cal Mass Density ρ_{mcal} (kg/m^3)	7.90E+03	9.12E+01	B	∞		1.24E-03
Pressure Medium Density ρ_{mair} (kg/m^3)	4.83E-01	1.06E-07	B	∞		1.09E-05
Thermal coeff. of Piston - α_P (^{o}C^{-1})	4.50E-06	2.59E-07	B	∞		1.36E-04
Thermal coeff. of Cylinder - α_C (^{o}C^{-1})	4.50E-06	2.59E-07	B	∞		1.36E-04
Temp.diff from the ref. temp. (t-t$_{ref}$) (^{o}C)	3.00E+00	1.35E-06	B	∞		3.32E-14
Effective Area- A_0 (m^2)	9.80E-04	9.80E-09	B	∞		2.25E-02
Fluid Head Correction (Pa)	4.73E+00	1.23E-06	B	∞		1.51E-12
Ref Pressure (Pa)		**3.35E-03** **0**	**B**	∞	**1.12E-05 (a)** **0 (g)**	
Verticality (Pa)		1.20E-03	B	∞		1.44E-06
System Stability (Pa)	0.15	4.33E-02	B	∞		1.88E-03
Load Cell Precision (Pa)		2.50E-03	B	∞	6.25E-06	1.00E-12
Resolution (Pa)	1.00E-03	2.89E-04	B	∞	8.33E-08	8.33E-08
Repeatability (Pa)	**7.21E-02** **9.64E-02**	**2.28E-02** **3.05E-02**	**A**	**9**		**5.20E-04 (a)** **9.30E-04 (g)**
				Total Variance	1.76E-05	3.30E-02
				Overall standard uncertainty (Pa)	4.19E-03	1.82E-01
				Effective Degrees of Freedom	**3.62E+04 (a)** **1.16E+04 (g)**	
					Absolute Value (Pa)	Relative Value (ppm)
				Expanded uncertainty at k=2	**8.38E-03** **7.64E-03**	**24.2 (a)** **24.4 (g)**

Table 3: Comparison Method – Uncertainty budget of p_{FPG_c} at 15.0 kPa

Quantity (X$_i$)	Estimate (x$_i$)	Uncertainty u_{xi}	Type (A or B)	Degrees of Freedom	$u_i(y)^2$ Independent	Dependent
Repeatability (Pa)	0.0721 0.0964	2.28E-02 3.05E-02	A	9		5.20E-04 (a) 9.29E-04 (g)
Resolution (Pa)	0.009	2.50E-03	B	∞	6.25E-06	
Mean Calibration Factor	1.000015 1.000009	1.09E-05 1.13E-05	A	14 19		2.66E-02 (a) 2.88E-02 (g)
Normalized Ref. Std. Reading (Pa)	14999.775 14999.865	5.42E-02	B	∞	2.12E-05	2.92E-03
				Total Variance	2.74E-05	3.00E-02
					$u_i(y)$	
					5.24E-03	1.73E-01

Absolute Mode Standard Measurement Uncertainty = Q(0.0052 Pa , 0.00116 % of reading)
Gauge Mode Standard Measurement Uncertainty = Q(0.0061 Pa , 0.0012 % of reading)

Estimated Effective Degrees of Freedom	18 24
Estimated Coverage Factor (k) at Degrees of Freedom =18 and 95.45 % Confidence level (a)	2.15
Estimated Coverage Factor (k) at Degrees of Freedom =24 and 95.45 % Confidence level (g)	2.11

$U_i(y)$	
1.13E-02	3.73E-01
1.29E-02	3.81E-01

Expanded Measurement Uncertainty = Q(0.0113 Pa , 0.00248 % of Reading) for k=2.15 (a)
Expanded Measurement Uncertainty = Q(0.013 Pa , 0.00254 % of Reading) for k=2.11 (g)

The mean differences of measured absolute pressure between FPG and UIM with error bars especially in the low pressure regions are plotted in between the expanded uncertainty bands of NPL estimated and manufacturer reported expanded uncertainty and the same are depicted in figure 6 and 7 for absolute and gauge mode measurements respectively.

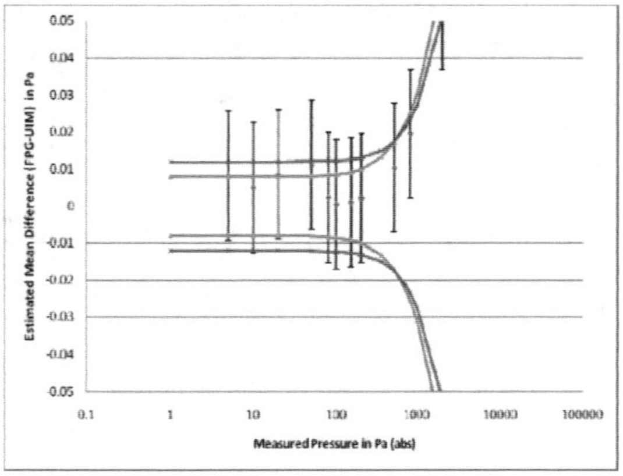

Fig. 6. Estimated mean difference of FPG and UIM with error bars – Absolute Mode

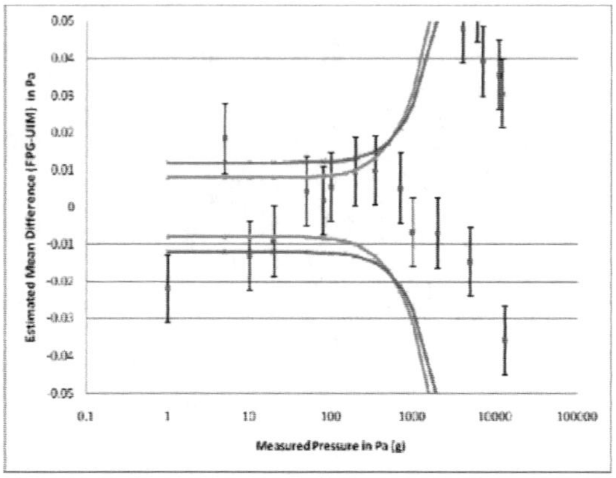

Fig. 7. Estimated mean difference of FPG and UIM with error bars – Gauge Mode

4. Conclusion

The expanded uncertainty of piston cylinder assembly of FPG is evaluated as ±19.98 ppm at k=2.025 (calculated effective degrees of freedom is 74), which is less compared to the manufacturer's reported value of ±26.0 ppm at k=2. This may be attributed to the choice of reference standard, namely the UIM, instead of a simple air piston gauge.

The expanded uncertainty of FPG is evaluated using theoretical method is Q(0.0084 Pa, 24.2 ppm of reading) at k=2 for absolute mode and Q(0.0076 Pa, 24.4 ppm of reading) at k=2 for gauge mode measurements.

The expanded uncertainty of FPG is evaluated using comparison method is Q(0.0113 Pa, 24.8 ppm of reading) at k=2.15 (calculated effective degrees of freedom is 18) for absolute mode and Q(0.013 Pa, 25.4 ppm of reading) at k=2.11 (calculated effective degrees of freedom is 24) for gauge mode measurements.

The expanded uncertainty evaluated through the budgets prepared using theoretical and comparison methods are found to be very much comparable for both absolute and gauge mode measurements, even when there is some difference observed in the estimated coverage factors. This proves the validity of the comparison method proposed [6].

Acknowledgement

The authors are highly thankful to Dr D. R. Sharma, Scientist – G (Retired) and Prof. R. C. Budhani, Director, National Physical Laboratory, New Delhi for their constant encouragement and permission to publish this work.

References

[1] Ooiwa A 1994 *Metrologia* **30** 607-610.
[2] Jay H Hendricks and Douglas A Olson 2009 *Measurement* **43** 664-674.
[3] Delajoud P, Girard M 2002 NCSLI Conference, San Diego.
[4] Delajoud P and Girard M 2003 *Vakuum in Forschung und Praxis* **15** 24–29.
[5] http://kcdb.bipm.org/appendixC, Oct 2009.
[6] Arun Vijayakumar D 2006 *MAPAN* **21(1)** 23-36.

Design, fabrication, and performance testing of a vacuum chamber for pulse compressor of a 150 TW Ti:sapphire laser

P.K.Tripathi[1,*], Rajvir Singh[2], V.K.Bhatnagar[2], S.D.Sharma[2], Sanjay Sharma[2], B. Sisodia[2], K.Yedle[2], R.P.Kushwaha[1], S.Sebastin[1], and G.Mundra[2]

[1]Laser Plasma Division, Raja Ramanna Centre for Advanced Technology, Indore - 452013, INDIA
[2]Accelerator Components Design and Fabrication Section, Raja Ramanna Centre for Advanced Technology, Indore - 452013, INDIA

E-mail: pkt@rrcat.gov.in

Abstract. A vacuum chamber, to house the optical pulse compressor of a 150 TW Ti:sapphire laser system, has been designed, fabricated, and tested. As the intensity of the laser pulse becomes very high after pulse compression, there is phase distortion of the laser beam in air. Hence, the beam (after pulse compression) has to be transported in vacuum to avoid this distortion, which affects the laser beam focusability. A breadboard with optical gratings and reflective optics for compression of the optical pulse has to be kept inside the chamber. The chamber is made of SS 316L material in cuboidal shape with inside dimensions 1370x1030x650 mm^3, with rectangular and circular demountable ports for entry and exit of the laser beam, evacuation, system cables, and ports to access optics mounted inside the chamber. The front and back sides of the chamber are kept demountable in order to insert the breadboard with optical components mounted on it. Leak tightness of $9x10^{-9}$ mbar-lit/sec in all the joints and ultimate vacuum of $6.5x10^{-6}$ mbar was achieved in the chamber using a turbo molecular pumping system. The paper describe details of the design/ features of the chamber, important procedure involved in machining, fabrication, processing and final testing.

1. Introduction

In this paper, we report the design, fabrication and performance testing of a vacuum chamber, which is to be used to house optical pulse compressor of 150 TW Ti:sapphire laser system. The pulse compressor of this laser consists of two optical gratings and reflective optics. The chirp pulse from the system compresses the optical pulse in the time to get back the femto second laser pulse. After the pulse compression, the intensity of the laser pulse becomes very high and this results phase distortion in the laser beam due to air. Therefore, the beam has to be in vacuum to avoid this. The optical gratings and relevant reflective optics are to be mounted on a bread board, which is to be housed in bottom of the chamber. The optical gratings and relevant reflective optics are to be positioned with the help of vacuum compatible positioners to be operated from outside the chamber through feed-throughs. The chamber is meant for obtaining dry vacuum of the order of 10^{-6} mbar.

2. Features

The chamber has a cuboidal geometry having rectangular and circular demountable ports with an overall volume of approximate 930 litres. The chamber has seven demountable rectangular ports of size 250 mm x 500 mm with aluminum covers. The overall size of the chamber is 1420 mm (L) x 1200 mm (W) x 820 mm (H) which is made from SS 316L material. The available space inside the chamber is 1370 mm (L) x 1030 mm (W) x 650 mm (H). Front and back sides of the chamber can be opened (opening:1030 mm x 650 mm) in order to insert the breadboard of size 1250 mm x 950 mm duly optical gratings and other associated optical components mounted laterally from any side of the chamber. These optical gratings and other optics need to be aligned during operation. Both the sides are having demountable covers of aluminum in order to make them light weight, as these are to be handled manually. One side cover plate have entry beam port and three ports with DN 63 ISO-K flanges, which are to be used as feed-through ports for system control cables as well as to operate the positioners from out side the chamber. The other side plate have two measurement ports of 160 mm inside diameter ports and a small port for vacuum measurement. The exit beam port of 160 mm inside diameter has been provided at desired place at right angle to the entry of laser beam. The output laser beam is to be transported from this port to the Laser Plasma Interaction Chamber through 160 mm ID beam line. The top of the chamber has two demountable rectangular ports for the purpose of access to optics mounted on breadboard from top and one demountable evacuation port with DN 200 ISO-F flange to couple the TMP via high vacuum gate valve. The front and back sides are also having three and two demountable rectangular ports respectively of same size for the purpose of accessibility to optics etc. All the demountable joints are made with the help of Viton 'O' ring sealing of suitable cross section.

3. Design

The design of the chamber is based on theory of plates. The chamber is designed rigid enough to withstand the external pressure equal to atmospheric pressure as well as keeping deflection with in limit for sealing the demountable joints with the help of 'O' rings reliably. In this, we have designed separate plates using theory of plates with end conditions as fixed ends in order to keep conservative approach. The ASME code Section VIII, Div 1 has also been referred for design of ports / flanges as well as to cross check the design from stability point of view. Stiffeners are attached with walls of the chamber in order to reduce the wall thicknesses as well to keep deflections within desired limit. All the groves are designed very carefully with proper groove dimensions to fulfill the sealing requirements i.e, appropriate surface contact and surface finish of the order of 1.6 micron. All permanent joints are designed in such a way that no crevices are left, thus avoiding trapped volumes.

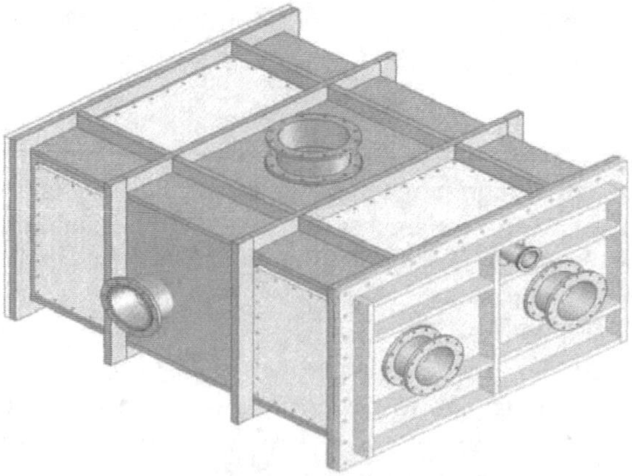

Fig. 1. 3-D view of the chamber

The vacuum pumping system had been selected to obtain vacuum better than 1 x 10^{-5} mbar within half an hour time after starting the same. The chamber was designed for a leak tightness better than 1 x 10^{-8} mbar lit./sec (He). The vacuum system was selected based on gas load in the chamber and ultimate vacuum requirement in stipulated period. A gas load of the order of 6 x 10^{-3} mbar lit./sec is calculated based on surface area exposed to vacuum and outgassing rate of construction materials. An additional gas load of 1 x 10^{-3} mbar lit./sec is expected due to out gassing of breadboard optical arrangement. A pumping system of 1250 lit /s speed was found suitable to evacuate the chamber into desirable vacuum within stipulated time considering conductance of connecting port and gate valve etc.

4. Fabrication

Manual GTAW welding process was used for welding using AWS 5.9 ER 316L filler wire and ultra pure argon gas under controlled RH (< 40%) in clean room. As higher coefficient of thermal expansion of austenitic stainless steel leads to unacceptable distortion during welding, the proper fixturing was used to avoid the same. The typical welding parameters used are as follows:

Welding current : 120-150 Amps,
Filler wire : ER 316L (dia. 2.5 mm and 3.15 mm.)
Shielding gas : 99.999% pure Argon
Shielding gas flow rate: 8 Ltrs./min.
Relative humidity : 35-40% @ 22° C

Fig. 2. Photograph during welding

The chamber was machined on CNC Horizontal boring machine for its desired tolerances. Surface finish of the order of 1.6 micron was achieved on sealing surfaces of the components using pocket milling cycle on CNC machine. Surface finish of all the 'O' ring grooves is better than 1.6 micron with no scratches and dents. All crevices in trapped volume had been removed to facilitate evacuation of the chamber. The stainless steel central chamber and other stainless steel parts are chemically cleaned and electro-polished in order to reduce the outgassing from them, while all aluminum covers plates are cleaned thoroughly to do the same.

5. Pumping system

The pumping system for the above Pulse Compressor Vacuum Chamber consists of combination of a turbo molecular pump with pumping speed of 1250 lit/sec and a dry backing pump with pumping speed of 50 M^3/Hr. The TMP is mounted in the top of the chamber in the vertical orientation via pneumatically operated gate valve. The chamber was rigidly mounted on a Mild Steel stand with levelling provision.

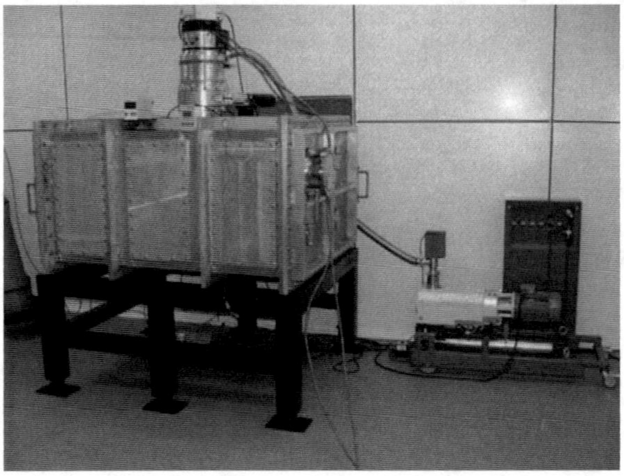

Fig. 3. Photograph of the chamber along with vacuum pumping system

6. Testing

Helium MSLD was used for carrying out leak testing. All permanent and demountable joints of the chamber have been tested and a leak tightness better than 9 x 10^{-9} mbar lit. per sec is found for all the joints. A vacuum of 6.5 x 10^{-6} mbar was achieved in less than two hours pumping time using a TMP of 1250 lit/sec pumping speed.

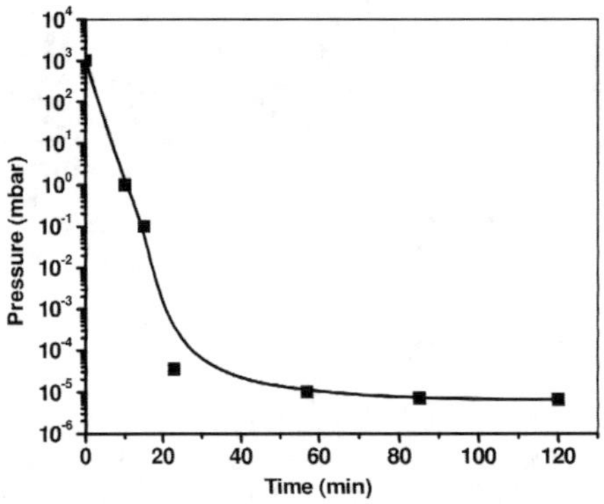

Fig. 4. Pump down curve

Acknowledgements

Authors sincerely acknowledge support and encouragement from Dr. P.A.Naik, Head, Laser Plasma Division. Thanks are due to Shri L.V.Thorat and Shri Sobhan Choudhary, ACDF Section, Shri P.Ram Sankar, CTF and Shri B.K.Sindal, UHV Division for drafting, fabrication, chemical treatment and leak testing work respectively. The authors are also thankful to other team members of ACDF Section for technical discussions and involvement in fabrication.

References

[1] Roarks formula for stress and strains by : Warren C Young.
[2] ASME boiler & pressure vessel code (Section VIII, Div. 1&2)
[3] Vacuum Engineering calculations by : Armand Berman.

Journal of Physics: Conference Series **390** (2012) 012015 doi:10.1088/1742-6596/390/1/012015

Fabrication of niobium superconducting accelerator cavity by electron beam welded joints

T.K.Saha[1], J. Mondal[2], K.C.Mittal[2], K.G.Bhushan[3] and A.V.Bapat[1]

[1] Power Beam Equipment Design Section
[2] Accelerator and Pulsed Power Division
[3] Technical Physics Division
Bhabha Atomic Research Centre, *Mumbai-400085, India*

Email. tanmay@barc.gov.in

Abstract. Fabrication of superconducting cavities has been taken up as a part of the development of accelerator driven sub critical system (ADSS) by Bhabha Atomic Research Centre. Large grain (RRR>99) pure niobium was chosen as the material for the cavity. Niobium,for its application as superconductor requires extremely high quality joints, feasible only by electron beam welding at high vacuum environment. An indigenously developed 100kV, 4kW high vacuum electron beam welding machine has been utilized to carry out the welding operations. Planning of the weld sequences was chalked out. Holding fixtures for the cavity, consists of seven numbers of joints have been fabricated beforehand. A few coupons were welded for optimization of the weld parameters and for inspection of the weld purity by indigenously developed secondary ion mass spectroscopy. The report describes the welding equipment and the stage wise joining operations of the cavity in details and also discusses the qualification testing of the welded cavity.

1. Introduction

High purity large grain bulk niobium is the most suitable material used in fabrication of SRF cavities for its excellent superconducting properties [1]. Although niobium is a refractory material, it exhibits strong affinity towards gases and react with oxygen, hydrogen, nitrogen at elevated temperature. Inclusion of these gases exceeding few tens to few hundred parts per million in the cavity material is detrimental to its quality [2]. Electron beam welding has been chosen for welding of cavity as the process is carried out under high vacuum, which is being considered to be the best inert environment for welding [3].

2.1 Description of the Equipment:

The welding has been carried out by a 100kV 4kW electron beam welding machine developed in house in early eighties [4]. After extensively being utilized for Nuclear and Defence applications for more than two and half decades in BARC, the equipment has been relocated in Nashik under MOU between BARC and M/s Precise Vacuum System Pvt Ltd to explore its industrial applications.

General specifications of the equipment are:

Accelerating voltage:	0-100 kV
Beam current:	0-40mA

Work Chamber size:	1500mm × 1000mm × 1000mm.
Work table size:	800mm × 500mm.
Table displacement:	700mm at X and 400mm at Y direction
Table speed:	20mm/min to 2000mm/min, in the X and Y directions and 0.5-20 rpm for the rotary table. All motions are programmable.
Gun vacuum:	2×10^{-6} mbar
Chamber vacuum:	5×10^{-5}mbar

2.2 Description of the Cavity

A single cell prototype elliptical RF cavity has been designed at f_0=1050 MHz, β = 0.49. Our design is based on the circular equator and elliptic iris for ease of mechanical fabrication. The schematic of the cavity dimensions is shown in Figure 1. The basic geometrical and RF parameters of the cavity are shown in Table1. Design and mechanical fabrication of the cavity in details has been reported elsewhere [5].

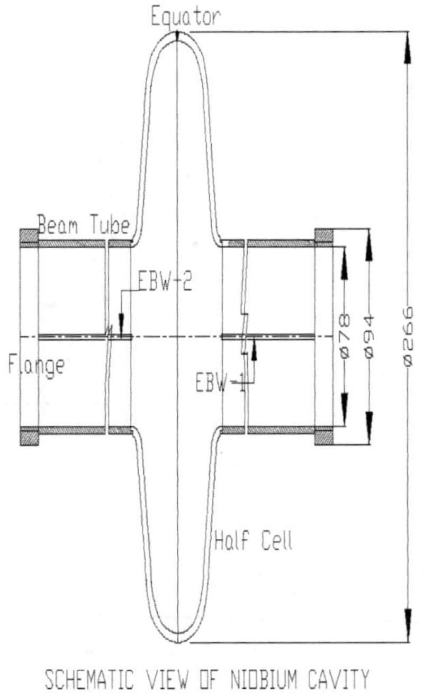

Fig. 1. Schematic drawing of designed Cavity

Table 1: 1050 MHz, β = 0.49 Single Cell Cavity Parameters

Parameters	Present Design
Frequency (MHz)	1050
Geometrical β	0.49
Iris Radius (cm)	3.9
Cavity Diameter (cm)	25.833
Dome Aspect ratio	1
Iris Aspect Ratio	1.43
Wall Angle (Deg.)	6.5
Wall Distance [mm]	7.37
r/Q (Ω)	9.19
Geometry Factor G (Ω)	141.91
E_{peak}/E_{acc}	4.26
B_{peak}/B_{acc} (mT/(MV/m)	8.02

3. Electron Beam Welding:

Niobium single cell cavity consists of 7 nos. weld joints in total. Fixtures for electron beam welding of various parts of the cavity have been designed and fabricated in advance. Sequence for welding is shown in Fig.2. All the weld parameters were optimised by welding trial on dummy coupons having similar contour and thicknesses as actual joints. The actual weld parameters used for joining of different components of cavity are tabulated in **Table-II.**

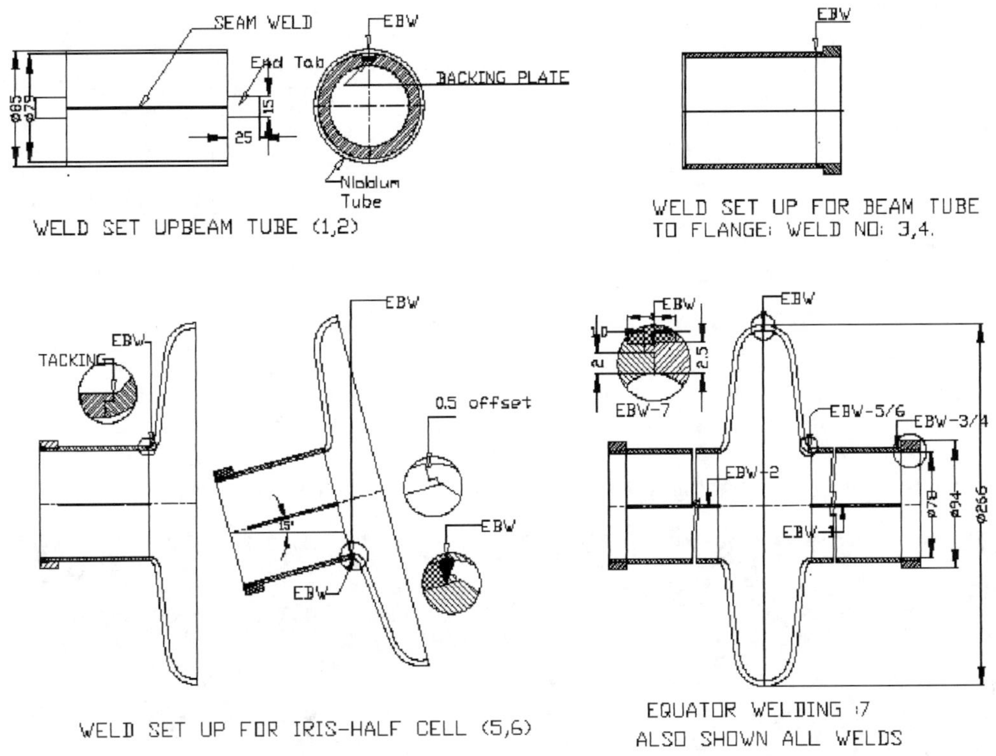

Fig.2. Sequence of electron beam welding of the components of the Cavity

Table-II: Weld Parameters for the Components

Name of the components And joint type	Accelerating Energy (kV)	Beam Current (mA)	Linear weld speed (mm/min)	Chamber vacuum (mbar)	Observations
Beam tube: Seam joint (right):	100	22	600	$5x10^{-5}$	Full penetration
Beam tube: Seam joint (left):	100	22	600	$5x10^{-5}$	Full penetration
Beam tube to Flange Step joint (right)	100	18	600	$5x10^{-5}$	Smooth joint Uniform U/B
Beam tube to Flange Step joint (right)	100	18	600	$5x10^{-5}$	Smooth joint Uniform U/B
Half cell to Iris (right) Tacking vertically	80	10	_	$8x10^{-5}$	Tacked at 12 places
Half cell to Iris (right) Inside weld at an angle	100	20	600	$5x10^{-5}$	Smooth bead
Name of the components And joint type	Accelerating Energy (kV)	Beam Current (mA)	Linear weld speed (mm/min)	Chamber vacuum (mbar)	Observations

Half Cell- Iris (right) Outside cosmetic pass	100	10	600	$5x10^{-5}$	Smooth bead
Half cell to Iris (left) Tacking vertically	80	10	–	$8x10^{-5}$	Tacked at 12 places
Half cell to Iris (left) Inside weld at an angle	100	20	600	$5x10^{-5}$	Smooth bead
Cosmetic pass (outside)	100	10	600	$5x10^{-5}$	Smooth bead
Equator welding : Tacking	100	10		$5x10^{-5}$	Tacked at 16 places
Equator weld	100	32	600	$5x10^{-5}$	Shrinkage of the lip took place on some portion.
Equator weld cosmetic pass at defocused beam	100	22	600	$5x10^{-5}$	Good upper bead.

3.1 Beam tubes (weld-1 and 2):

Beam tubes have been formed by rolling of 3.2mm niobium plates. Butt welding along seam of the tubes was carried out. Fig.3 shows photograph of one of the two welded beam tubes.

3.2 Beam tube to flange (Iris): Photograph of welded Iris (beam tube to flange) is shown in Fig.4.

Fig 3. Photograph of Seam welded Beam Tube

Fig.4. Photograph of welded Iris

Fig.5.Half cell to beam tube assembly

Fig. 6. EB welded Cavity (after final welding)

3.3 Half Cell to Iris

Welding has been carried out from inside at an angle of 15° with the axis of rotation and was followed after weld tacks. Fig.5 shows photograph of welded half cell to Irish in which very smooth weld is noticeable.

3.4: Equator Joint

Joint contour was step butt with 0.5mm thick upper and 2.5mm thick lower step. Actual joint line was along the lower step which was 1mm offset from upper joint line [fig.2]. A notch was provided on top to locate the lower step. 16 nos. weld tacks were provided on top joint to resist distortion of the joint due to asymmetric thermal expansion. It has been observed that shrinkage of the lip was taken place during welding at two places (25mm long each) along the weld periphery. Although full penetration was felt with finger, a cosmetic pass (table-II) was provided after machining of the deformed upper step.

4. Results and Discussions:

4.1. Leak Test:

All welded components of the cavity were leak tested by in MSLD process and found acceptable. Results of MSLD are shown in Table-III.

Mass Spectrometer Leak Detection results (TABLE-III)

Sl no	Name of the components	He leak rate (mbar-litre/sec)
1	Beam tube after seam welding (left)	$< 1x10^{-9}$
2	Beam tube after seam welding (right)	$< 1x10^{-9}$
3	Beam tube to flange welded (left)	$< 1x10^{-9}$
4	Welded beam tube to flange (right)	$< 1x10^{-9}$
5	Beam tube assembly to half cell welded (left)	$< 1x10^{-9}$
6	Beam tube assembly to half cell welded (right)	$< 1x10^{-9}$
7	Equator assembly after welding	$< 1x10^{-9}$

4.2. Micro Hardness Measurement

Micro hardness was measured across the weld pool of electron beam welded niobium sample. The measurement was taken at three stages (1,2,3) along the depth started on parent metal advanced to weld pool and stopped on parent metal (Fig.7). 25 gm load has been used with 5 sec dwell time. The hardness profile near top and middle of the weld profile showed homogenous mixture and more uniform values across the weld zone. Profile near the root of the weld shows a fluctuation of the values.

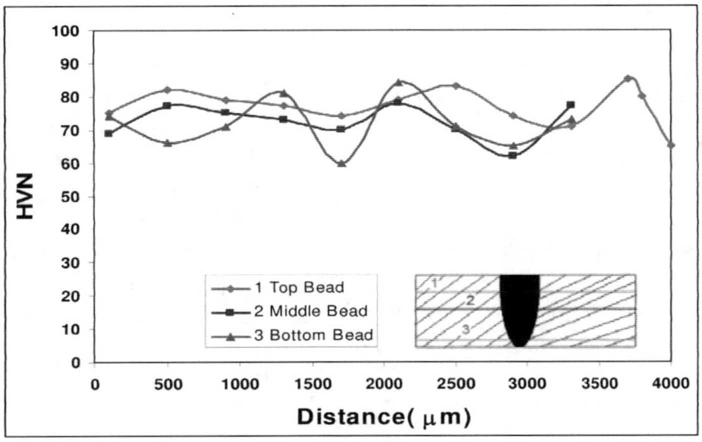

Fig.7. Micro-hardness profile of EB welded niobium sample

4.3. RF measurements and Electric field distribution

RF measurement of the cavity before EB welding and after EB welding with vector network analyzer is shown in **Table-IV.** The values measured after welding has been found at per with the values before welding. Electric field distribution before and after welding has shown in fig.8 & fig.9.

Table-IV: RF Values measured before and after EB weld

Parameters	Simulated value without Beam tube	Measured value before EBW without beam tube	Simulated value with Beam tube	Measured value after EBW with Beam tube
Frequency [MHz]	1050	1048	1036.856	1036.507
Q Value	17000	459	17032	8076 (not cleaned after EBW)
Effective Shunt Impedance [M□/m]	4.469	0.175	4.711	3.61 (not cleaned)
R/Q [□]	9.225	8.59	9.675	9

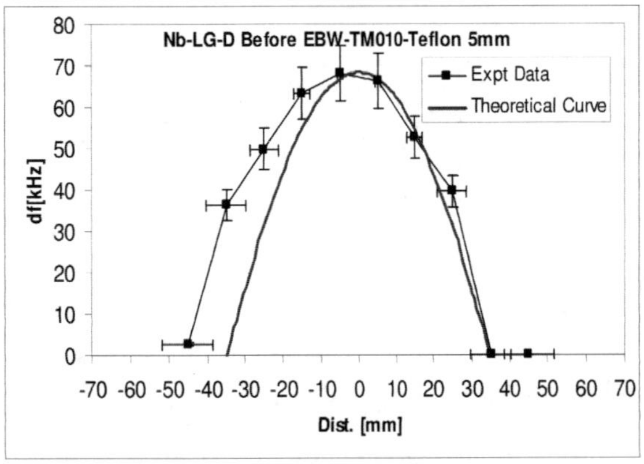

Fig.8. Axial electric field distribution before welding

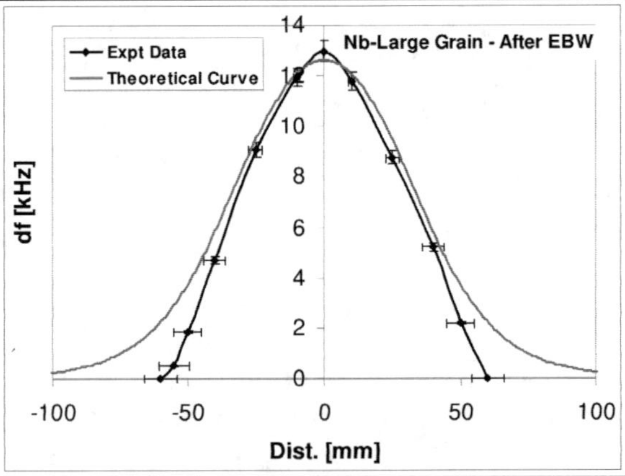

Fig. 9. Axial electric field distribution after welding

4.4: SIMS analysis of the weld profile

EB welded coupon was analyzed with the time of flight secondary ion mass spectrometry equipment (TOF-SIMS) developed in Technical Physics Division, BARC. A primary beam of mono-isotopic Gallium ions at an energy of 25 KeV is allowed to bombard the sample surface. Secondary ions emitted from the surface are mass analyzed with the help of reflectron TOFMS

The mass spectra were collected across the weld profile of the niobium starting from parent metal to weld profile to parent metal of the coupon [same as fig.7]. The relative concentration of NbH, NbO and NbOH levels were mapped and are shown in fig-10. Similarly evolved oxygen and hydrogen profiles were mapped and are shown in fig-11. It can be seen from the figs-10 and 11 that the oxygen and hydrogen profile in the weld zone has increased over the back ground i.e. parent metal.

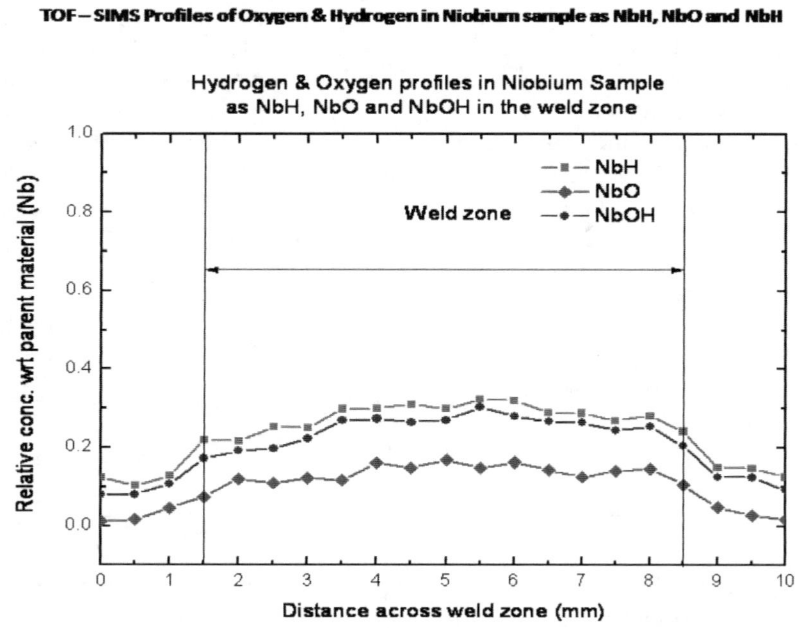

Fig.10. Time of flight secondary ion mass spectroscopic profiles on EB welded Nb specimen

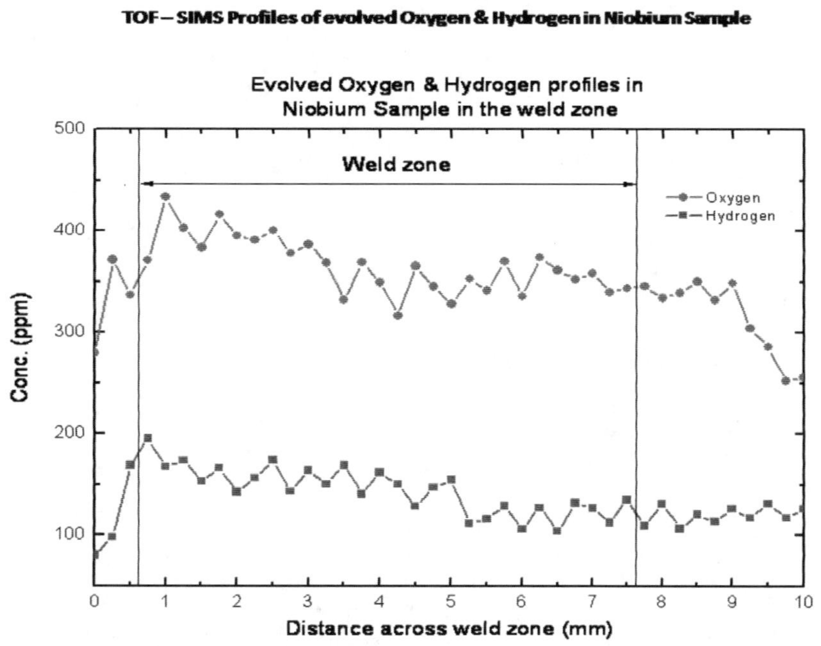

Fig.11.Time of flight secondary ion mass spectroscopic profiles on EB welded Nb specimen

5. Conclusion:

Welding of a single cell niobium superconducting cavity has been carried out by one of our in house developed electron beam welding machines. Tests like MSLD, electric field distribution and RF measurements at room temperature carried out on the welded component shown encouraging result. Analysis of the welded specimen showed an alarming level of oxygen and hydrogen pick up in weld pool. Chemical cleaning and vacuum heat treatment of the cavity is expected to reduce these levels to a significant extent. Results of the evaluation of the cavity at 2K are awaited. Depending on the above results desired modifications in welding environment or technique will be taken place for future cavities.

Acknowledgement:

Authors are thankful to Shri H.Badkas, Shri Shyam Amode and Shri Bhavsare of M/s Precise Vacuum Systems Pvt Ltd for the full fledging co-operation for fabrication of this job. Thanks are due to Dr.L.M. Gantayet, Director, Beam Technology Development Group, BARC for his constant encouragement and valuable advices for this development of activity.

References:

[1] "Ingot niobium RF cavity design and development at BARC": Mittal KC et al: AIP Conf proceedings1352.100-115 (2011)
[2] "RF properties 1050 MHz, β=0.49 elliptical cavity for high current proton accelerator" Roy Amitava et al: Journal of Instrumentation, P04002, 2008 (IOP Publication)
[3] "Vacuum-The ideal environment for welding of Reactive and Refractory materials" by Saha TK et al: Journal of Physics: 114(2008)02047, (IOP Publication)
[4] "Industrial perspectives of electron beam welding" by Saha TK et al proceedings of SEBTA-2005 p-283-292 held during Sept 28-30, 2005.
[5] "Design ,fabrication, room temperature RF test of 1050 MHz β=0.49 single cell large grain niobium cavity" by Mondal J: will be reported in same conference IVS-2012.

Comparative study of evaporation using DC and AC filament electron guns

Sutanwi Lahiri[1], G K Sahu, S Baruah, B Jana, A R Dixit, R L Bhardwaj, R C Das, R Kalra, V Kaushik, A Majumder, S Mohapatra, B Dikshit, K K Mishra, M S Bhatia, A V Bapat, V K Mago, K B Thakur, A K Das and L M Gantayet*

Laser & Plasma Technology Division,

*Beam Technology Development Group, Bhabha Atomic Research Centre, Mumbai-400085, India

E-mail: sutanwi@barc.gov.in

Abstract. Electron beam assisted physical vapour deposition (EB-PVD) and purification of metal by repeated melting using electron guns is a well-established technique in industrial metallurgy. Strip electron gun is considered a cost effective alternative to multiple pencil guns for handling of large size substrates. In the electron guns, the thermionic emission of the electrons from a filament is achieved by using AC or DC filament heating. A study of their relative merits and demerits was conducted for the both types of electron guns. Due to finite length of the filament, the magnetic field generated around the filament by heating current drops down towards ends. The DC filament heating results in electron beam with a comet shape having high power density hot spot at one end with low power density tails. With AC filament heating, electron beam oscillates with the frequency as that of heating current. The study of vapour flux distribution using DC gun revealed that highly directional vapour evolution takes place from a smaller hot spot whereas with AC gun vapour evolution occurs from an oscillatory 2D-evaporating source. The vapour deposit on substrate indicated that evaporation using DC gun caused splashing and granular deposit due to volumetric melting and evaporation from the ingot. This is contrary to the AC filament heating wherein quiet evaporation was observed due to surface melting and evaporation. The experimental results are critically reviewed to decide the configuration of electron guns for large-scale evaporation.

1. Introduction

Electron beam physical vapour deposition (EB-PVD) is a well-established industrial scale process for material processing [1,2]. In order to handle large surfaces, multiple pencil electron guns and oscillatory electron beams were used. However these techniques have their own limitations. Uniform power density can be achieved only when beam diameter is close to oscillation amplitude [1]. To overcome this disadvantage, the oscillating strip electron gun is considered the best alternative [3]. Filaments of refractory materials such as tantalum and tungsten are used in strip electron guns, based on their low work function, ease of shaping and high operating temperature. The electrons obtained from the thermionic emission caused due to resistive heating of the filament current are accelerated by applying a negative voltage of the order of 50 to 60 kV to the filament and grid cup assembly with respect to the water cooled anode. The sheet of electron beam is focused using anode-cathode configuration to form a parallel beam focused at 270^0 using magnetic field in field free region with an electromagnet in Helmholtz's configuration.

For generation of thermionic electrons, filaments can be heated using DC or AC. Both these have their merits and demerits. Due to high current passing through the conductor, the magnetic field generated in the vicinity of the filament has a deciding role on various parameters such as residence time, power density distribution and dimension of the beam, etc. [4].

In case of DC filament heated gun, the electron trajectories are permanently shifted at one end and rarefied at the other end depending on the polarity of the DC. This results in highly concentrated hot spot and rarefied tails, similar to that in a comet. Hence, these electron guns provide electron beam with non-uniform power density resulting in trench formation and smaller molten pool on the substrate surface.

In case of AC heated electron gun, the polarity of magnetic field due to current carrying conductor changes with time. This results in oscillation of strip electron beam on the target surface. Hence, the non-uniformity in current/power density gets smeared out to give uniform temperature, larger molten pool but less residence time. Due to the finite length of the filament, the magnetic field tapers off on either side. This end-effect can be overcome by using the dummy filament concept. The dummy filaments are thick filaments of the same material that carry same amount of current but do not contribute to the thermionic emission due to lower temperature. The dummy filament shifts the tapering of the magnetic field so that bunching and rarefaction of electron trajectories are avoided [4].

In this article we present comparative studies of copper evaporation using DC and AC heated electron guns.

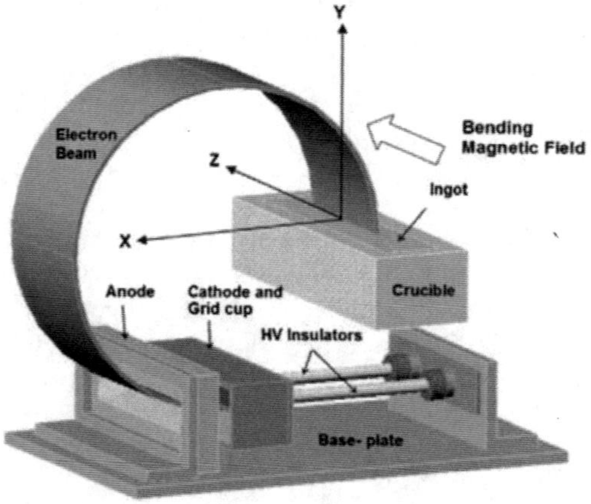

Fig. 1. Schematic diagram of the experimental setup.

2. Experimental Setup

The experimental studies with the DC and the AC heated filament electron guns were carried out in a water-cooled double-walled vacuum chamber operating at 5 x 10^{-6} mbar. The details of the 100 kW evaporator, along with its accessories are reported elsewhere [3]. The AC/DC heated strip electron guns are identical in design except the filament heating current. The schematic of the strip electron gun is given in Fig 1. It consists of segmented tantalum filament mounted on the filament holder assembly encapsulated inside the grid cup insulated to – 60 kV potential with respect to ground using alumina insulators. The sheet of electrons emerging out of the anode is focused on the copper ingot by deflecting the electron beam through 270^0 (radius of deflection R = 260 mm) in the presence of a magnetic field using electromagnetic coils arranged in Helmholtz configuration. The bending of the beam is necessary to prevent metal vapour from entering into anode–cathode space. In case of DC filament heated electron gun, segmented tantalum had length of 110 mm whereas in case of AC filament heated gun, the filament had a length of 130 mm and the dummy filaments on each side were 30 mm long.

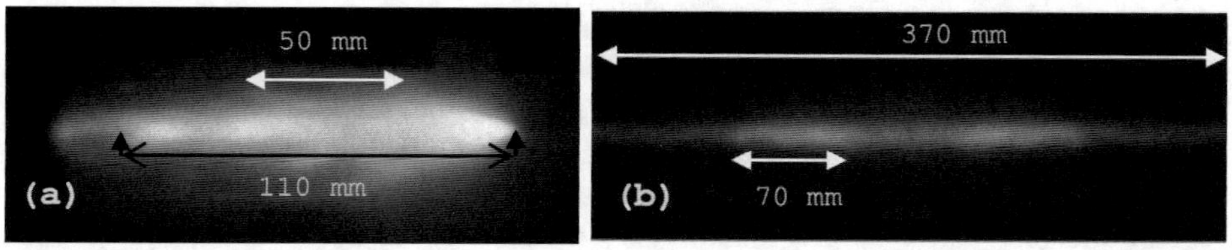

Fig. 2. Incident beam on the copper ingot in case of: (a) DC filament heated electron gun (b) AC filament heated electron gun

The performance of the AC/DC filament strip electron guns was tested using copper as the target material. The photograph of the evaporating source with DC and AC filament heated electron guns are shown in figure 2. The electron guns were operated with radius of deflection 260 mm. The DC evaporating source was observed to have an average dimension of 110 mm x 10mm with high intensity zone on one end and low intensity comet-like tails on other end. In case of AC filament heating evaporating source the average dimensions were 130 mm x 5 mm with oscillating amplitude of 370 mm.

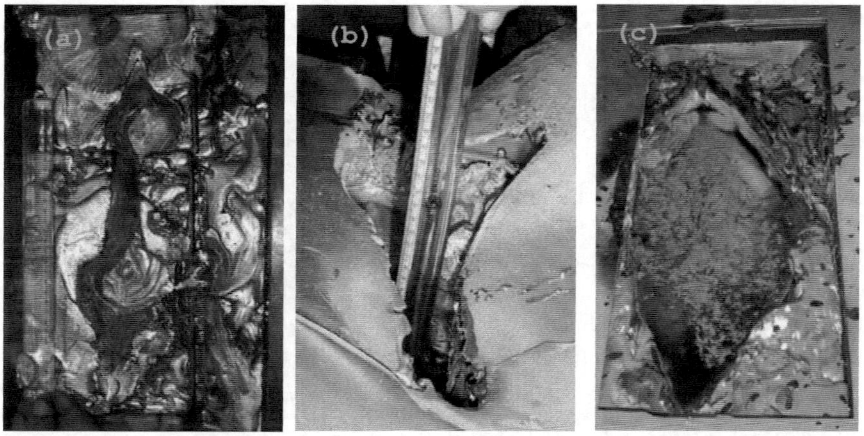

Fig. 3. Images of frozen molten pool in case of: (a) DC filament heated electron guns at 45 kW each (b) DC filament heated electron guns at 70 kW each and (c) AC filament heated electron guns at 85 kW each.

The molten pool produced with two DC guns at 45 and 70 kW each and two AC guns at 85 kW each are presented in figure 3. It can be seen that the molten pool during DC filament heating was much smaller compared to that formed with AC filament heating. Also there is a trench formation at the ends due to high intensity spots.

Fig. 4. Deposit collected in case of: (a) DC filament heated electron gun (b) AC filament heated electron gun.

In order to study the evaporation from a source generated using AC/DC filament heated electron gun, the vapour is allowed to deposit on collectors. Cumulative deposits were analyzed. The evaporation using a DC electron gun resulted in the formation of granules on the deposit formed on the SS target, placed at a height of 550 mm above the source, as is evident in figure 4(a). The deposit thickness was monitored at regular intervals along the length of the deposit.

In case of AC heated gun generated evaporating source, evaporating source was maintained at 1980 K with Knudsen number 0.06 and evaporation rate 580 g/h. The self-sustaining copper layer was taken for deposit analysis. The eighteen tokens (6.5 cm^2 each) were cut with known coordinates with respect to evaporating source at a height of 680 mm from the source. The uniform coating on the collector hardly showed any trace of granules, clearly indicating atomic layer deposition (figure 4(b)).

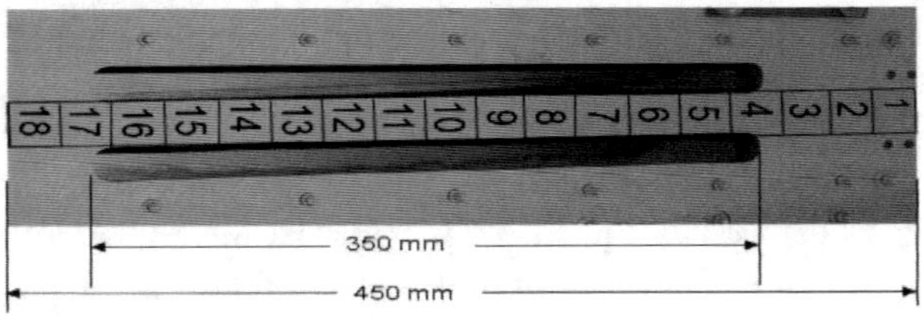

Fig. 5. Tokens cut from the collector for analysis of the deposit.

3. Results and Discussion

3.1 Shape of the incident beam

During filament heating for thermoionic emission, a magnetic field is generated around the filament. Due to finite size of the filament, magnetic field tapers off on either side. As a result of this, electron trajectories not only shifts permanently away from filament but also gets concentrates on one end and rarefy on other side causing non-uniform current density distribution within the beam [4].

In case of electron guns using DC heated filaments, the direction of the magnetic field is decided by the polarity of the DC. Thus electron trajectories are shifted in one direction with non-uniform current density distribution. As a result, evaporating source generated has non-uniform power density on the target surface with high power density on one side as can be seen in figure 2(a).

On the other hand, AC changes its polarity in each half cycle. The frequency of oscillation is same as that of the supply frequency, i.e., 50 Hz. As a result, the electron beam oscillates on the target surface with amplitude proportional to the magnitude of filament heating current and radius of deflection. Power deposition depends on the residence time of the electron beam at different location within the oscillation amplitude and non-uniformity in the electron beam gets smeared out as can be seen in figure 2(b).

Non-uniform current density distribution in electron beam due to magnetic field end effect caused by the AC/DC filament heating current can be overcome using a dummy filament [4]. As a result of incorporation of dummy filament, the non-uniformity in the magnetic field is done away with and bunching and rarefaction of electron trajectory is minimized. The AC filament electron gun results presented here are with incorporation of dummy filament.

3.2 Molten pool

During evaporation with DC heated gun, electron beam is stationary on the surface. Two DC filament heated guns placed in line and operated at 45 kW each are used to study the molten pool as can be seen in figure 3(a). The frozen profile of the target shows non-uniform power density distribution leading to trench formation, volumetric melting with smaller molten pool size. At higher power (70 kW each), the two trenches merge together to form a molten pool with greater depth but similar width as shown in figure 3(b).

In case of AC filament heated gun, the electron beam covers larger target surface area during its oscillation and temperature on the target depends on residence time of the electron beam. With present 50 Hz oscillations and 370 mm oscillation amplitude, the residence time is such that it does not allow solidification at any location within the oscillation amplitude maintaining molten pool size of a length more than the oscillation amplitude. We have used two AC filament heated guns placed in line and operated at 85 kW each to study the molten surface as can be seen in Fig 3(c). The frozen profile of the target shows larger molten pool with uniform depth.

3.2 Vapour flux distribution

In order to study vapour flux distribution, two DC filament heated guns were operated at 70 kW each. The copper vapour generated was allowed to deposit on a SS collector placed at 550 mm above the evaporating source. The vapour deposited was fitted using the generalized vapour flux distribution from a 2D evaporating source defined as

$$\left((n_1 + n_2 + 2)/4\pi\right)\cos^{n_1}\theta\cos^{n_2}\phi \tag{1}$$

where, 'θ' is the angle between normal plane passing through strip source and plane containing the collector and 'ϕ' is the angle between the line joining an elemental source and a collector area with respect to the normal to the elemental source area in any given 'θ' plane. The co-ordinate system used for analysis is reported earlier [5] and shown in figure 6. The exponent n_1 is the beaming coefficient along the width direction of the evaporating source corresponding to 'θ' and the exponent n_2, is the beaming coefficient along the length direction of the evaporating source corresponding to'ϕ',

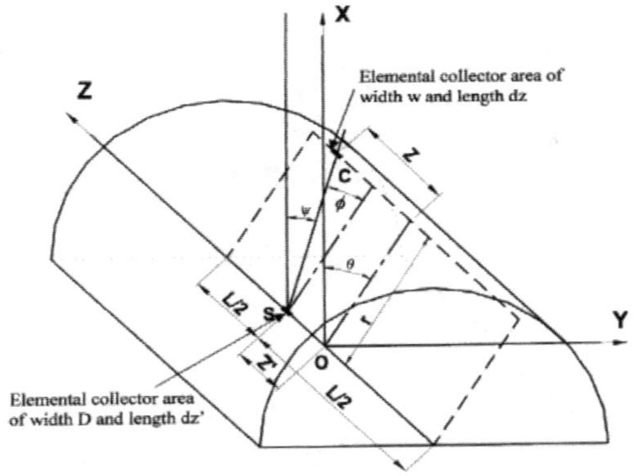

Fig. 6. Coordinate system along with the elemental source area (D x dz') at a distance z' and the elemental collector area (W x dz) at a distance z from central plane passing through source (D x L)

respectively.

With this general vapour flux distribution function [5], the vapour flux (atoms/sec) on a horizontal collector of area 'a_c' in *SC* direction (having length *SC*) from an elemental source area, 'a_S' is given by

$$F = \frac{Q}{DL} a_s \frac{(n_1 + n_2 + 2)}{4\pi} \cos^{n_1}\theta \cos^{n_2}\phi \frac{a_c \cos\theta\cos\phi}{SC^2} \tag{2}$$

where Q is the total evaporation rate (atoms/sec) from the source of dimension ($D \times L$).

The deposit obtained on the SS collector due to evaporation using the DC gun was analysed by measuring the thickness of deposit along the edge of the shutter, which is parallel to the length of the linear beam. During this fit, the vapour emitting source was assumed to have a high temperature central region of 1 cm x 2.4 cm embedded on a low temperature evaporating source of 1 cm x 30 cm at 1600 K. From earlier measured value of power vs. temperature [6], the central hot spot was assigned a temperature of 2000K. The beaming coefficient, length of the hot spot and the temperature difference between high & low evaporating area were kept as variable parameters while fitting the experimental data. The measured thickness could be fitted with error <14% with beaming coefficient as high as 15 with high evaporating source of 1cm x 2.4 cm with temperature 2000 K embedded on 1 cm x 30 cm evaporating source of 1600 K.

The vapour flux distribution using AC filament heated strip electron gun was carried out using single gun operated at 116 kW with copper as a target. The evaporating source was maintained at 1980 K (Knudsen number=0.06). The collection run was conducted by opening the shutter for 5160 sec. The vapour deposited on SS substrate placed at 680 mm above the evaporating source was analyzed by weight difference technique. The data was fitted using above mentioned generalized vapour flux distribution function for 2D source using Eqn. 2. The beaming coefficients were found to be: $n_1=2$, $n_2=3.5$. The error of fitting was found to be 13%.

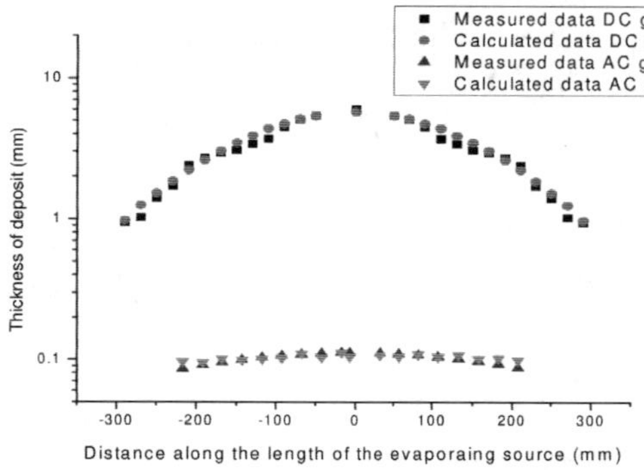

Fig. 7. Variation of the measured thickness of the deposit along the centreline of the collector during evaporation using DC/AC heated electron guns.

The deposition using DC/AC filament heated strip electron gun is presented in Fig 7. The difference in the deposition thickness is due to difference in collection time. The drastic fall of thickness of deposit along the length and the high beaming coefficient during DC filament heated strip electron gun evaporation clearly indicate highly collimated atomic beam from a hotspot embedded in a low temperature evaporating source generated from a comet-shaped electron beam having high power density spot with low intensity tail. Thus evaporation from a DC strip electron gun can be considered as volumetric with splashing. The uniform deposits along the length with moderate beaming coefficient during evaporation with AC heated electron gun clearly indicate quiet and uniform evaporation from the surface of the molten pool of the target.

4. Conclusion

It can be concluded that the high evaporation rates can be achieved using DC filament heated strip electron gun from a smaller molten pool. One has to be careful about power density distribution in the electron beam, as it is likely to result in hotspot formation, splashing and volumetric evaporation. It is possible to achieve uniform deposition over large substrate length from quiet surface evaporation using AC filament heated gun.

Acknowledgement

Authors would like to acknowledge technical discussion and suggestion by D Das. We are grateful to M Sunkanna, R Satpute, Lijeesh K, H Singh and R P Singh for their support in conducting the experiments.

References

[1] Schiller S, Heisig U, and Panzer S, *Electron Beam Technology* (Wiley New York, 1982).

[2] Rykalin N, Uglov A, Zuev I and Kokara A Laser and Electron Beam Material Processing *Handbook* (MIR Moscow, 1988)

[3] Thakur K B, Sahu G K, Tamhankar R V and Patel K 2001 *Rev. Sci. Instrum.* **72** 207–15

[4] Sahu G K, Baruah S, Thakur K B 2010 *Vacuum* **85** 236-244

[5] Thakur K B, Sahu G K 2004 *Vacuum* **75** 283–292

[6] Thakur K B and Sahu G K 2003 *Proc. 7th Int. conf. Electron Beam Technologies* (Varna, Bulgaria) p 56

Journal of Physics: Conference Series **390** (2012) 012017 doi:10.1088/1742-6596/390/1/012017

Design of large vacuum chamber for VEC superconducting cyclotron beam line switching magnet

Sumantra Bhattacharya, Chinmoy Nandi, Subhasis Gayen, Suvadeep Roy, Santosh Kumar Mishra, Sanjay Ramrao Bajirao, Gautam Pal and C Mallik

Variable Energy Cyclotron Centre, 1/AF Bidhan Nagar, Kolkata 700 064, India

Abstract. VEC K500 superconducting cyclotron will be used to accelerate heavy ion. The accelerated beam will be transported to different beam halls by using large switching magnets. The vacuum chamber for the switching magnet is around 1000 mm long. It has a height of 85 mm and width varying from 100 mm to 360 mm. The material for the chamber has been chosen as SS304.The material for the vacuum chamber for the switching magnet has been chosen as SS304. Design of the vessel was done as per ASME Boiler and Pressure Vessel Code, Section VIII, Division 1. It was observed that primary stress values exceed the allowable limit. Since, the magnet was already designed with a fixed pole gap; increase of the vacuum chamber plate thickness restricts the space for beam transport. Design was optimized using stress analysis software ANSYS. Analysis was started using plate thickness of 4 mm. The stress was found higher than the allowable level. The analysis was repeated by increasing plate thickness to 6 mm, resulting in the reduction of stress level below the allowable level. In order to reduce the stress concentration due to sharp bend, chamfering was done at the corner, where the stress level was higher. The thickness of the plate at the corner was increased from 6 mm to 10 mm. These measures resulted in reduction of localized stress.

1. Introduction

VEC K500 superconducting cyclotron will be used to accelerate heavy ion beams of energy up to 80 MeV/A for light heavy ions and about 10 MeV/A for medium mass heavy ions. The accelerated beam will be transported to different beam halls by using large switching magnets. The vacuum chamber for the switching magnet is around 1000 mm long. It has a height of 85 mm and width varying from 100 mm to 360 mm. The material for the chamber has been chosen as SS304.

2. Selection of Material

The vacuum chamber is to be operated at a pressure of $1X10^{-7}$ mbar inside and atmospheric pressure outside. Hence, all the joints must have helium leak tightness @ $1X10^{-9}$ Std. cc/sec. Under this circumstance, metal gasket only to be used to connect the chamber in the beam line. The use of metal gaskets using the conflate flanges rules out the possibility of the using aluminum as material for the vacuum chamber. The material should be weldable to facilitate the fabrication. So, SS 304 is chosen as it has both the qualities, non-magnetic as well as weldable.

Journal of Physics: Conference Series **390** (2012) 012017 doi:10.1088/1742-6596/390/1/012017

3. Design procedure

Design of the vacuum chamber was done as per ASME Boiler and Pressure Vessel Code, Section VIII, Division 1. A uniform cross section equal to the largest cross section was considered for the design and the rules of Appendix 13 were applied. The material property values from ASME Boiler and Pressure Vessel Code, Section II, part D. The allowable stress limit for SS 304 was found to be 126 MPa. It was observed that primary stress values varied from 370 MPa to 166 MPa for plate thickness variation from 4 mm to 6 mm. as the primary stress values exceeds the allowable limit, the thickness of the plates needed to be increased. Since, the magnet was already designed with a pole gap of 90 mm; increase of the vacuum chamber plate thickness restricts the space for beam transport. So, the plate thickness more than 6mm was not practically feasible. Hence, design needed to be optimized.

4. Optimization of design

Design of the vacuum vessel was optimized using stress analysis software ANSYS. Modeling of the vacuum vessel was done with CATIA and the model was used in ANSYS for analysis. Solid element SOLID95, a 20 node structural element was chosen. This element has plasticity, creep, swelling, stress stiffening, large deflection, and large strain capabilities. It can tolerate irregular shapes without as much loss of accuracy. SOLID95 elements have compatible displacement shapes and are well suited to model curved boundaries. Smart meshing has been chosen. Smart element sizing is a meshing feature that creates initial element sizes for free meshing operations. The Smart Sizing algorithm first computes estimated element edge lengths for all lines in the areas or volumes being meshed. The edge lengths on these lines are then refined for curvature and proximity of features in the geometry. Since all lines and areas are sized before meshing begins, the quality of the generated mesh is not dependent on the order in which the areas or volumes are meshed. This feature, provides a range of settings (from coarse to fine mesh) for meshing the models.

Analysis was started using plate thickness of 4 mm for all the sides. The degrees of freedom for all the directions for the end plates have been put zero, as the chamber will placed in the beam line, and all the other areas are put under atmospheric pressure. The result of the analysis is shown in figure 1.

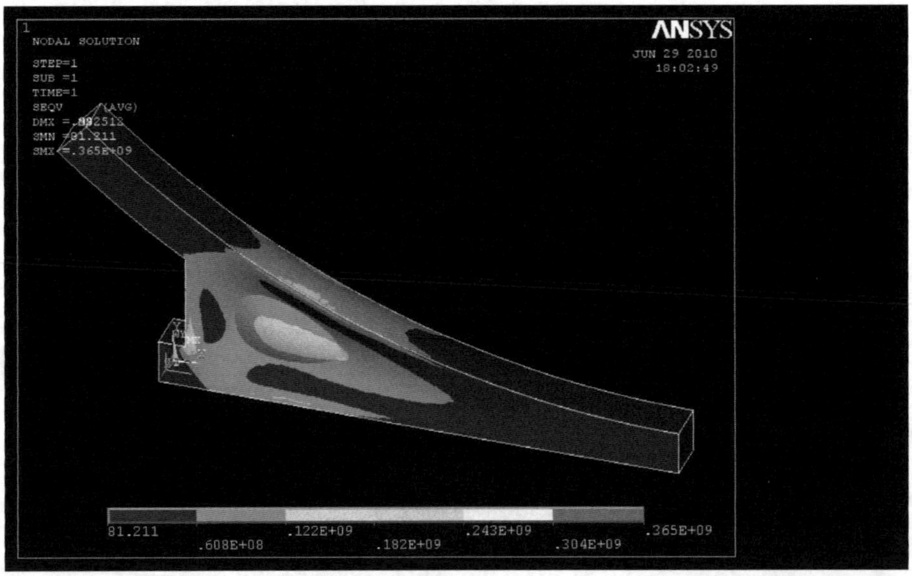

Fig.: 1. Stress Analysis of the Vacuum Chamber with 4 mm wall thickness.

After the analysis the stress level was found higher than the allowable level of 126 MPa. The localized stress at a corner was 365 MPa and the stress at the middle of the plate is also high between

Journal of Physics: Conference Series **390** (2012) 012017 doi:10.1088/1742-6596/390/1/012017

122 MPa to 182 MPa which is also higher than the allowable stress limits. Hence the thickness of the plates is needed to be increased.

The analysis was repeated by increasing plate thickness to 6 mm. The result is depicted in the figure 2.

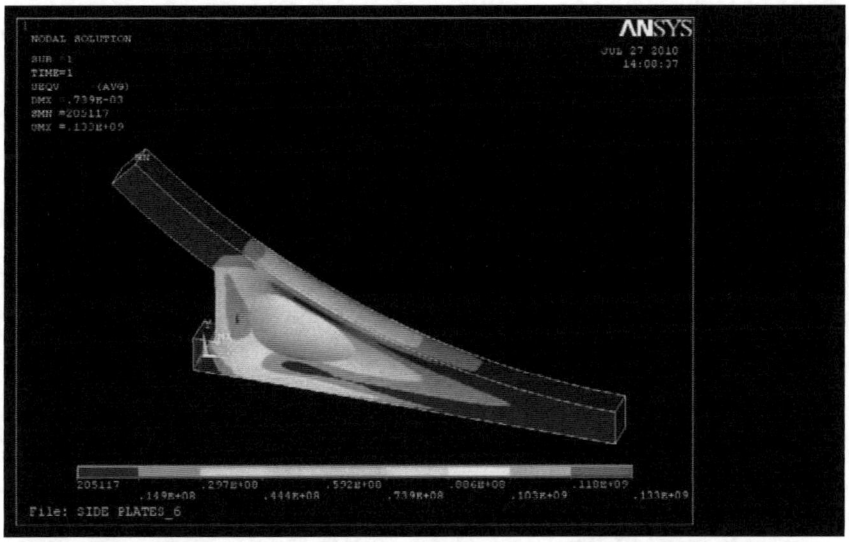

Fig. 2. Stress Analysis of the Vacuum Chamber with 6 mm wall thickness.

Figure 2 shows, the increase of thickness of the plates from 4 mm to 6 mm, resulting in the reduction of overall stress level below the allowable level. But the localized stress at a corner is still higher than the permissible limit.

In order to reduce the stress concentration at the corner due to sharp bend, chamfering was done at the corner, where the stress level was higher. The thickness of the plate at the corner was increased from 6 mm to 10 mm. The same analysis was repeated with this configuration. These measures resulted in reduction of localized stress. It was observed that the stress was 122 MPa, which is within acceptable limits of 126 MPa. The final shape of the vacuum chamber is shown in figure 3.

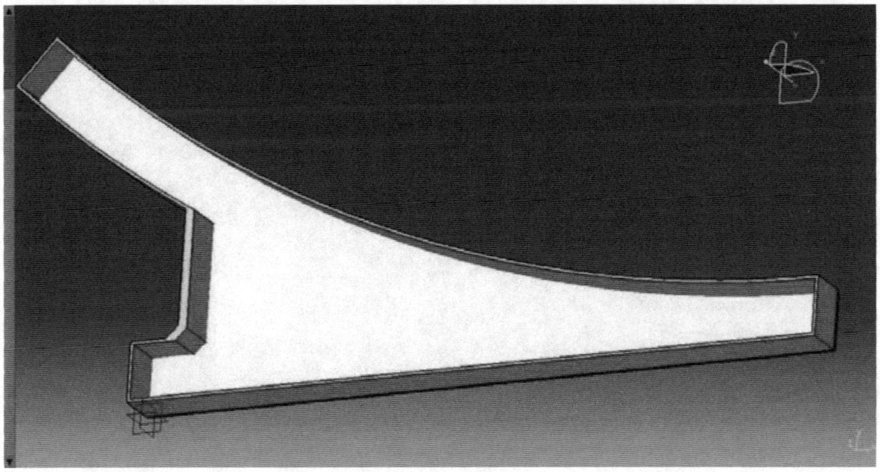

Fig. 3. The vacuum chamber for the switching magnet

Reference

[1] Mallik C, Bhandari RK 2010 first beam acceleration in Kolkata superconducting cyclotron and its present status *19th International Conference on Cyclotrons and their Applications*, CYCLOTRONS'10, Lanzhou, China, September 6-10, 2010.
[2] ASME Boiler and Pressure Vessel Code, Section VIII, Division 1
[3] ASME Boiler and Pressure Vessel Code, Section II, Part D
[4] Elements of Strength of Materials by S. P. Timoshenko, D. H. Young
[5] Strength of Materials, Part -1, by S. P. Timoshenko
[6] Strength of Materials, Part -2, by S. P. Timoshenko
[7] Theory and Design of Pressure Vessels by John F. Harvey
[8] The Finite Element Method and Applications In Engineering Using ANSYS by Erdogan Madenci, Ibrahim Guvan
[9] CATIA Software user manual

Journal of Physics: Conference Series **390** (2012) 012018 doi:10.1088/1742-6596/390/1/012018

Broad Band Antireflection Coating on Zinc Sulphide Window for Shortwave infrared cum Night Vision System

A S Upadhyaya and P K Bandyopadhyay

Instruments Research & Development Establishment, Raipur Road DehraDun 248008.

E-mail: asu@irde.drdo.in

Abstract. In state of art technology, integrated devices are widely used or their potential advantages. Common system reduces weight as well as total space covered by its various parts. In the state of art surveillance system integrated SWIR and night vision system used for more accurate identification of object. In this system a common optical window is used, which passes the radiation of both the regions, further both the spectral regions are separated in two channels. ZnS is a good choice for a common window, as it transmit both the region of interest, night vision (650 – 850 nm) as well as SWIR (0.9 – 1.7 µm). In this work a broad band anti reflection coating is developed on ZnS window to enhance the transmission. This seven layer coating is designed using flip flop design method. After getting the final design, some minor refinement is done, using simplex method. SiO_2 and TiO_2 coating material combination is used for this work. The coating is fabricated by physical vapour deposition process and the materials were evaporated by electron beam gun. Average transmission of both side coated substrate from 660 to 1700 nm is 95%. This coating also acts as contrast enhancement filter for night vision devices, as it reflect the region of 590 – 660 nm. Several trials have been conducted to check the coating repeatability, and it is observed that transmission variation in different trials is not very much and it is under the tolerance limit. The coating also passes environmental test for stability.

1.Introduction

Vision in dark is not only important but it is necessary toll for modern civil and allied operations. Night vision devices based on image enhancement and thermal imagers are already available for this purpose. Shortwave infrared (SWIR) imaging is quietly earning a growing place in night vision system, although thermal imaging has been well established for night vision. It is poised to make night vision capabilities even better than they are today. Even in moon less night, the short wave region of EM spectrum contains five to seven times more energy than any other spectral wavelength from star light and glow of the night sky [1]. SWIR camera is smaller and lighter then all thermal cameras, and cost far less than many of them. SWIR detectors can see through glass and also can see things intrinsically which thermal detectors cannot. Thermal imagers are good at spotting temperature differentials; but they have drawbacks that it does not work well at dawn or dusk, which are known as the crossover points. But SWIR do not have this problem which is a complement to this technology. As per surveillance experts SWIR works well round the clock and especially well at those crossover points. SWIR also produces images similar to a black and white visible camera, requiring almost no training to interpret. With SWIR, the pedestrian or middle size vehicle can get information of a distant vehicle, even on a moonless night where as interpretation of thermal images, by contrast, requires intensive training and, even then, does not permit definitive identifications.

To detect SWIR wave, different types of detector generally used depending on the different sub-band between 0.9µm to 3 µm range. They are NVG Gen III, Silicon CCD, and different InGaAs based detector. InGaAs based detectors have different types, depending on higher wavelength cut. InGaAs (1.7µm) detector based SWIR cameras operate at 27^{0} C, mechanical shutter or cooling is not required, it never need field Non Uniformity Corrections (NUCs) and also have high quantum efficiency [2]. As a result, InGaAs (1.7µm) SWIR cameras are as compact, versatile and simple to use as a commercial digital video camcorder. They provide analog video to standard TV or any commercial frame grabber card.

For SWIR camera with InGaAs (1.7µm) detector we require highest light gather efficiency of the glass based optics in 900nm to 1700nm band. Zinc sulphide's transmission spectrum from 0.4 lm to 11 lm is free from major absorption, with the material easily available in large quantity and high purity to suit for most electro optical applications. Investigation on general optical properties of zinc sulphide has been extensively carried out over the years by different authors [3,4]. Very good surface hardness, robustness and ease of fabrication makes zinc sulphide a strong candidate for optical elements in SWIR (0.9–2.5 µm). Refractive index of zinc sulphide is 2.2, so it transmits only 73% incident radiation in uncoated condition. So antireflection coating is a critical requirement for zinc sulphide substrate, to make it useful for any optical system.

For military applications, it has been seen that the reflectivity of green paint is more than that of green vegetation from 550 to 660 nm region. After 660 nm to the remaining portion of spectrum, green paint reflectivity becomes less than that of green vegetation. So a night vision device with a contrast reversal at 660 nm make easy to detect the green military target. So a state of art system for night surveillance can integrate night vision system as well as SWIR imaging system. The present paper describes design and development of antireflection coating on zinc sulphide substrates, which is effective from 650 nm to 1700 nm. This coating also works as contrast reversal filter for night vision system.

2. Coating Design

Design of the coating is done using flip flop method [4, 6]. In this method the overall thickness is to be chosen. We have fixed the overall thickness at 1.5 λ. It has been shown that the overall thickness between λ and 2 λ give the good result [7]. We have selected λ 1300 nm. Then low and high index materials are selected. We have selected SiO_2 and TiO_2 as two materials [8]. The total thickness is divided in very thin layers and fix the starting index of each layer either low or high index. Each thin layer index is interchanged and the better design is kept. This all is done with the help of the Macleod thin film design software. The final design is refined using simplex method [9]. We get the design with 7 layers, and the top SiO_2 layer is replaced by MgF_2. Index profile of the design is shown in figure 1.

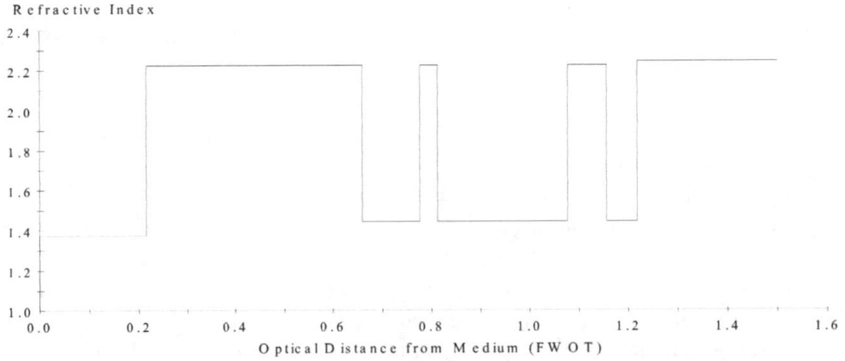

Fig. 1. Index profile of the coating design

3. Experimental

This was fabricated using electron beam gun evaporation technique on Hind High Vacuum coating unit BC 600 on Zincsulphide Substrate. Before deposition, substrate was cleaned ultrasonically in three stages: i) soap solution for ten minutes, ii) de ionized water for ten minutes, iii) isopropyl alcohol for 15 minutes. The substrates were then subjected to vapor degreasing in isopropyl alcohol for half an hour before mounting in vacuum chamber. Substrate was finally cleaned by high tension glow discharge for about fifteen minutes just before coating. Initially chamber was evacuated upto a pressure of $2x10^{-6}$ mbar then TiO_2 and SiO_2 were evaporated in oxygen atmosphere at a pressure of $6x10^{-4}$ mbar and last layer MgF_2 at $2x10^{-6}$ mbar with out oxygen percale pressure. The oxygen atmosphere requires for TiO_2 and SiO_2 for mentioning low absorption in the film. The substrate was heated to 300^0 C (with tolerance $\pm 5^0$C) by radiant heater before coating and the temperature was maintained during deposition. Substrates were rotated on flat plate holders, to ensure layer thickness uniformity.

All the deposition parameters i.e. glow discharge time, heating powers for coating materials, rate of deposition, thickness of various layers, layer sequence etc were controlled by microprocessor interfaced with the coating unit. Thickness and rate of deposition were monitored with quartz crystal monitor. Rate of deposition of TiO_2 was kept at 0.2nm/s for SiO_2 0.3nm/s and rate of deposition of MgF_2 was kept at 1.5 nm/s. The transmission of the coated substrate was measured with the help of Perkin Elmer UV – Vis – NIR double beam spectrophotometer model 'Lambda 950'. The transmission curve is shown in figure 2.

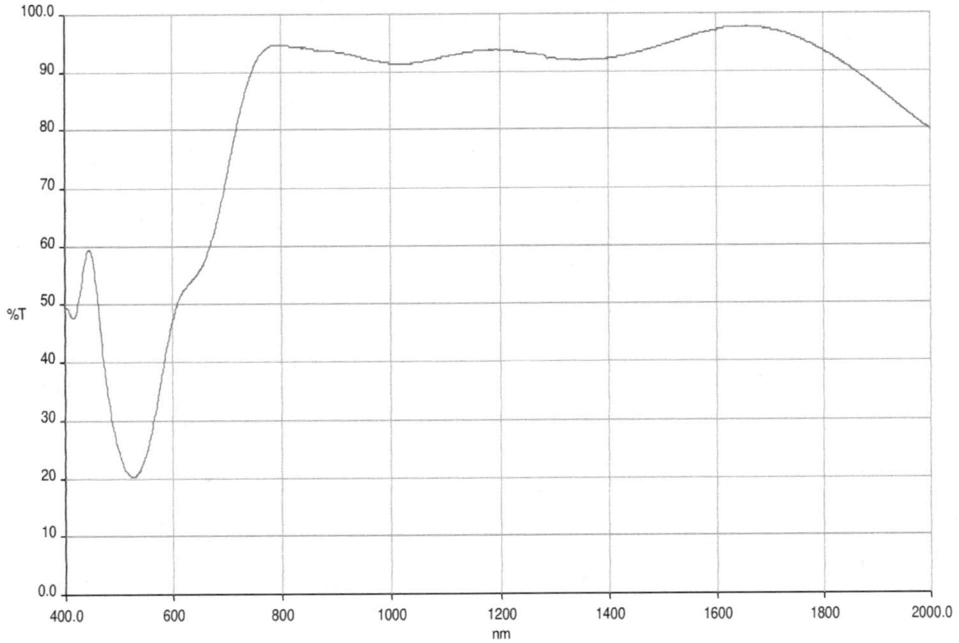

Fig. 2. Transmission variation of both side coated sample

4. Result

The practical transmission variation with wavelength is shown in figure2. The experimental transmission curve is measured in Perkin-Elmer lambda 950 spectrophotometer. Practically is 95% average transmission achieved for both side coated sample (sample thickness is around 3 mm). The average reflection from 590 to 660 nm is 50 %. This coating shows excellent stability and passes the entire MIL standard test as per MIL-F-48497.

Acknowledgment
The authors are thankful to Mr S S Sundaram, Director IRDE for his encouragement in carrying out the above work. The authors are also grateful to Mr Ikbal Singh, Associate Director IRDE for his guidance and valuable suggestions in carrying out the above work.

References
[1] L.V.Mirshri, "Atmospheric Optical Environment", R&D Technical Report ECOM-7023, (1972).
[2] Photonic Solution for the design engineer-"PTB Photonics Tech Brief" (2004).
[3] G.T. Petrovsky, Proc. SPIE 1540 (1991) 401.
[4] E.M. Gavrishckuk, E.V. Yashina, J. Opt. Technol. 71 (2004) 822.
[5] W H Southwell, App Opt. 24, (1985) 457.
[6] J A Dobrowolski Appl Opt 25, (1986) 1966.
[7] A Tikhonravov et al Proc SPIE 5250, (2004) 312.
[8] R R Willey 47 SVC conference, (2004) 322.
[9] J A Dobrowolski et al Appl Opt 29, (1990) 2876.

International Symposium on Vacuum Science & Technology and its Application for Accelerators IOP Publishing
Journal of Physics: Conference Series **390** (2012) 012019 doi:10.1088/1742-6596/390/1/012019

A study on brazing of Glidcop® to OFE Cu for application in Photon Absorbers of Indus-2

D P Yadav[1], R Kaul, P Ram Sankar, A Kak, P Ganesh, R Shiroman, R Singh, A P Singh, P Tiwari, L Abhinandan, L M Kukreja and S K Shukla

Raja Ramanna Centre for Advanced Technology, Indore - 452 013 INDIA

Email: dpyadav@rrcat.gov.in

Abstract: The paper describes an experimental study aimed at standardizing brazing procedure for joining Glidcop to OFE Cu for its application in upgraded photon absorbers of 2.5 GeV synchrotron radiation source, Indus-2. Two different brazing routes, involving brazing with silver base (BVAg-8) and gold base (50Au/50Cu) alloys, were studied to join Glidcop to OFE Cu. Brazing with both alloys yielded helium leak tight and bakeable joints with acceptable shear strengths.

1. Introduction

Indus-2, a 2.5 GeV synchrotron radiation source emits 187 kW of SR power from bending magnet at 300 mA [1]. Only 15% is channelled in beam lines while the rest is absorbed by 64 water-cooled photon absorbers [2-4]. In the upgraded design of photon absorber, the bottom part facing SR is made of low oxygen (LOX) grade Glidcop AL-15 while the intermediate part is made of OFE Cu which is to be joined to the bottom part. The dissimilar joint between Glidcop and OFE Cu is required to be leak tight (helium leak rate < 2 x 10^{-10} mbar.lit/s), bakeable upto 150°C and reasonably strong to withstand high pressure cooling water. The present study was undertaken to standardize brazing process to obtain helium leak tight, bakeable and strong brazed joint between Glidcop® to OFE Cu.

Glidcop represents an oxide dispersion strengthened (ODS) copper with uniform dispersion of sub-micron size Al_2O_3 particles in a fine-grained high purity copper matrix. Dispersion of hard and thermally stable Al_2O_3 particles, with low solubility in copper, provides effective barrier to the motion of dislocations and grain boundaries, thus preventing re-crystallization of the matrix [5-8]. The material offers a unique combination of high strength, high electrical and thermal conductivities, coupled with an ability to resist softening and retain a large portion of these properties even after exposure to elevated temperatures [9].

Fine grained of Glidcop is a disadvantage in the brazing operation as large number of grain boundaries

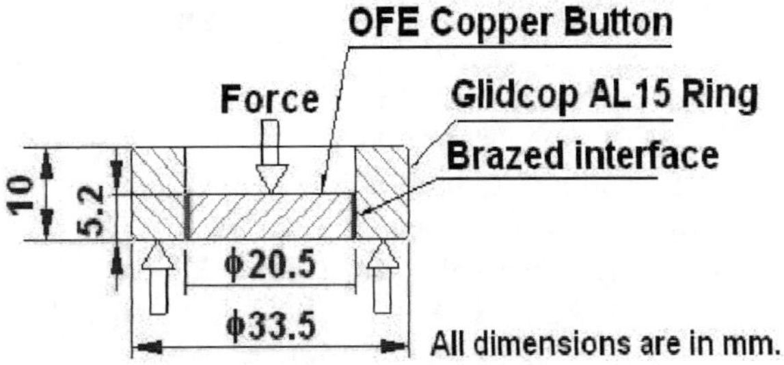

Figure 1. Details of specimen used for brazing experiments. The figure also provides schematic illustration of shear testing of the brazed joint.

permit rapid diffusion of constituents of the braze alloy, particularly Ag, along the grain boundaries [9-11]. The effect would leave the joint with less than adequate amount of braze alloy. The problem is addressed by electroplating Glidcop® part with Ni or Cu (particularly for Ag-bearing brazing alloy) prior to brazing. The plated layer acts as a barrier to the diffusion of Ag into Glidcop [10].

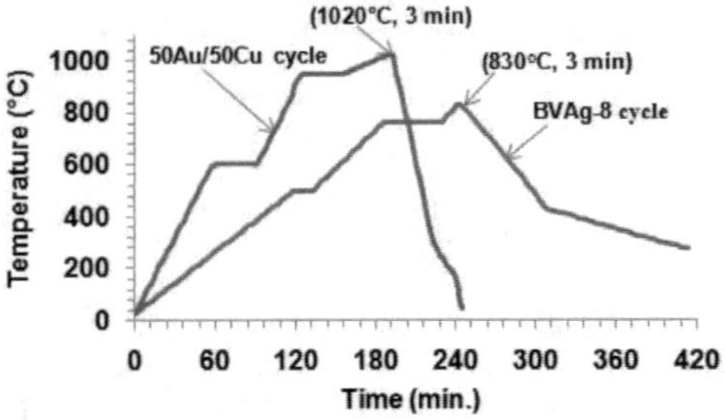

Figure 2. Thermal cycles used for brazing with BVAg-8 and 50Au/50Cu alloys.

2. Experimental Procedure

The study involved brazing of Glidcop ring to OFE Cu button, as shown in figure 1. The geometry of specimen facilitated leak and shear testing on the same specimen. The study was carried out in two sets viz. (i) Set 1: vacuum brazing with BVAg-8 alloy and (ii) Set2: brazing with 50 Au/50 Cu alloy in hydrogen furnace. Brazing with BVAg-8 alloy was performed in an in-house developed vacuum furnace [12]. For brazing with BVAg-8 alloy, Glidcop ring of the assembly (figure 1) was electroplated with three different types of coatings viz. Ni, Cu and Ni + Cu before loading the assembly in the furnace. Different electroplating routes were evaluated before brazing with BVAg-8 alloy. The methodology adopted for evaluating electroplating process involved metallographic examination of electroplated Glidcop specimens (10 mm x 10 mm x 3 mm) in as-plated and as-fired (at brazing temperature of 830°C for 3 mins.) conditions. The procedure adopted for electroplating Glidcop ring involved (i) preliminary cleaning, (ii) cleaning, (iii) activation and (iv) plating in the electrolytic bath. Four different cleaning treatments (step ii) viz. "A", "B", "C" and "D" were evaluated for Glidcop rings. On the other hand, different routes studied for electroplating Glidcop part were: (i) Ni electroplating in Watts' bath, (ii) Cu strike in cyanide bath followed by thick Ni plating in Watts' bath, (iii) Cu plating in cyanide/ acid baths, (iv) Ni strike in Watts' bath followed by Cu plating in acid/cyanide baths. Table 1 presents experimental electroplating parameters.

The electro-plated Glidcop ring was subjected to vacuum firing at brazing temperature before its brazing with mating OFE Cu part with BVAg-8 alloy in vacuum furnace. On the other hand, no coating was provided on Glidcop part while brazing with 50Au/50Cu alloy in hydrogen furnace. Figure 2 presents thermal cycles used for brazing Glidcop/OFE Cu assemblies with BVAg-8 and 50Au/50Cu alloys. Brazed specimens were subsequently characterized by helium leak test (HLT), thermal cycling test to evaluate its bakeability, shear test and metallographic examination under optical and scanning electron microscope with energy dispersive spectroscopy (EDS). Thermal cycling test involved subjecting the brazed specimen to multiple thermal cycles. Each cycle involved heating to 150 °C, soaking for 8 hours, followed by cooling to room temperature. After each thermal cycle, the brazed specimen was subjected to HLT to determine hermeticity of the brazed joint.

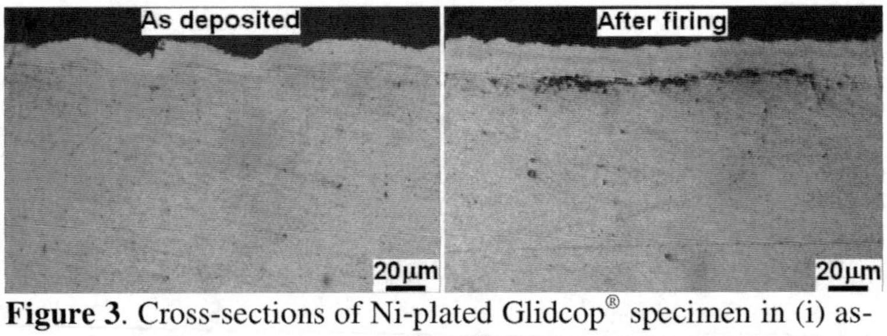

Figure 3. Cross-sections of Ni-plated Glidcop® specimen in (i) as-plated and (ii) as fired (at 830°C) conditions. Note blister formation at Ni/Glidcop interface after firing.

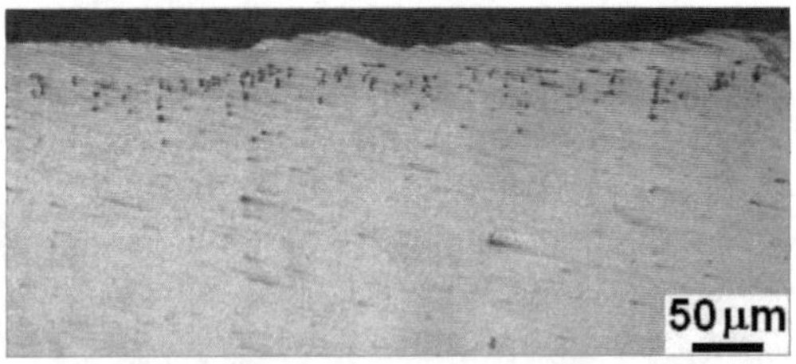

Figure 4. A row of gas porosities on the cross-section of Cu strike + Ni-plated Glidcop specimen after firing at 830 °C.

Table 1. Details of different steps involved in electroplating of Glidcop specimens

Step 1 - Preliminary cleaning:
(i) Ultrasonic cleaning in trichloroethylene for 15 mins.
(ii) Immersion in 50% v/v HCl (37%) for 2 mins. + water rinsing
(iii) Immersion in a soln. of chromic acid (50 gm/l) & H_2SO_4 (5 ml/l) for 2 mins. + water rinsing
(iv) Immersion in 50% v/v HCl for 2 mins. + water rinsing
Step 2 – Cleaning:
A: 1. Hot immersion cleaning in 10% w/v NaOH for 10 mins.
2. Cleaning in conc. HCl (37% w/w) for 40 sec.
B: 1. A1; **2.** A2; **3.** Bright dip in a soln. of conc. HNO_3 (1 part) & conc. H_2SO_4 (2 parts) for 1 min.
C: 1. A1; **2.** B3
D: 1. A1;
2. Dip in a soln. of 75% HNO_3 (69 % w/w) & 25% ortho-H_3PO_4 (85 % w/w) by volume for 20 sec at room temperature
3. Bright dip in a soln. of 55% ortho-H_3PO_4 (85% w/w), 20% HNO_3 (69% w/w) & 25% CH_3COOH by volume for 2 mins.
Step 3 - Activation: Immersion in 10 % v/v H_2SO_4 for about 30 sec.
Step 4 – Electroplating:
1. Nickel plating: Watts' bath - $NiSO_4.7H_2O$ (250 g/l) + $NiCl_2.6H_2O$ (65 g/l) + Boric acid (45 g/l)
pH: 3.8 - 4.2; Temperature: 55 -65 °C; Current density: 3 A/dm^2
2. Copper plating:
2.1 Cyanide Copper bath - NaCN (48 g/l) + CuCN (40 g/l) + Na_2CO_3 (30 g/l) + Rochelle salt (60 g/l)
pH: 10 - 10.5; Temperature: 55 - 55 °C; Current density: 2.4 A/dm^2 (2 mins.) + 1.2 A/dm^2 (10 mins.)
2.2 Acid Copper bath - $CuSO_4.5H_2O$ (200 g/l) + H_2SO_4 (38 ml/l) + chloride (30 - 40 mg/l)
Temperature: room temperature; Current density: 2 A/dm^2
Plating was performed with periodic reversal - forward cycle: 20 sec.; reversed cycle: 4 sec.

3. Results and Discussion

3.1 Set 1: Vacuum brazing with BVAg-8 brazing alloy

3.1.1 Evaluation of electro-plated coatings through metallographic examination

Metallographic examination of Ni and Cu plated Glidcop specimens, prepared under different conditions, exhibited dense coatings with sound bonding with underlying substrate. All four cleaning procedures were found to be effective for successful electroplating. However, it was noticed that

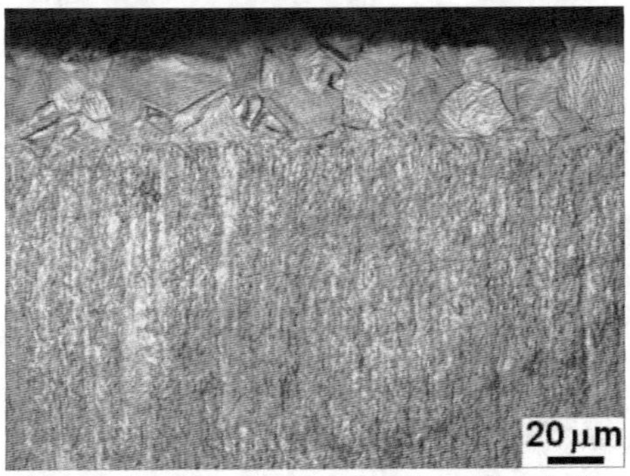

Figure 5. Cross-section of as-fired Cu-plated (cyanide bath) Glidcop specimen.

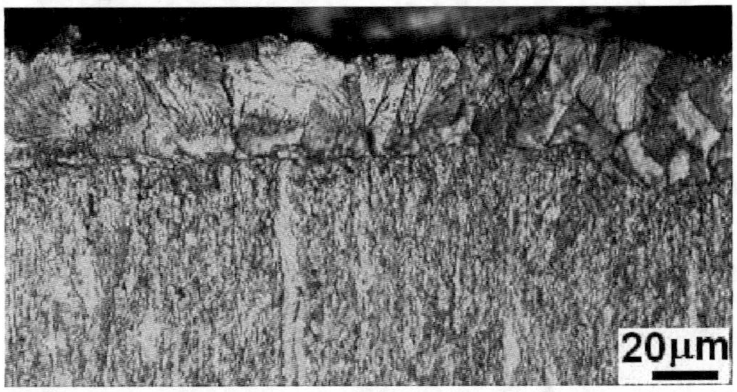

Figure 6. Cross-section of Ni-strike + Cu plated (acid bath) Glidcop specimen after firing at 830°C.

nickel plated specimens, with or without Cu-strike, when fired at 830°C, developed blisters/gas porosities at plating/substrate interface, as shown in figures 3 and 4. It is believed that these blisters are formed due to hydrogen charging during electroplating [13-15]. During firing at brazing temperature, atomic hydrogen dissolved in the electroplated deposits, diffuse to substrate/plating interface to form molecular hydrogen and resultant gas pressure led to the formation of blisters [15]. On the other hand, in Cu-plated Glidcop specimens, involving (i) plating in cyanide bath and (ii) Ni strike followed by Cu plating in acid bath, no porosities/blisters were seen in as fired condition (figures 5 and 6).

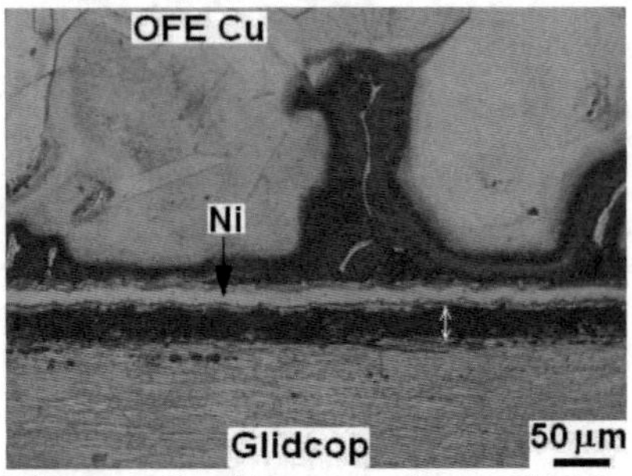

Figure 7. Cross-section of Ni-plated Glidcop/OFE Cu brazed specimen. White arrow marks decohesion at Ni/Glidcop interface.

3.1.2 Evaluation of brazed specimens

The specimens were prepared by vacuum brazing OFE Cu button with (i) Ni-plated and (ii) Cu-plated (Ni-strike + Cu-plating in acid /cyanide bath) Glidcop® rings, (refer figure 1). Vacuum brazed Ni-plated Glidcop /OFE Cu specimens exhibited helium leak rate of 10^{-4} - 10^{-5} mbar.l/sec. Metallographic examination of one of these specimens showed regions of decohesion at Ni plating/Glidcop interface, as shown in figure 7. The specimen exhibited intergranular penetration of brazing alloy into OFE Cu (refer figure 8) as well as into Glidcop. About 200 µm thick region of inter - granular penetration of brazing alloy into Glidcop is identified as "X" in figure 8. The figure also presents EDS concentration profiles of Ag, Cu and Ni across the brazed joint. Frequent peaks in Ag concentration profile represent Ag-rich grain boundary regions (in region "X") in Glidcop. Shear testing of two brazed specimens exhibited failure strength of 135-151 MPa. Cross-sectional metallographic examination of the shear tested specimens confirmed that the site of shear fracture was Glidcop/Ni plating interface, as shown in figure 9. On the other hand, vacuum brazed Cu-plated Glidcop (Ni strike + Cu plated)/OFE Cu specimens exhibited leak tightness better than 2 x 10^{-10} mbar.lit/s. The brazements remained leak tight even after undergoing 6 thermal cycles and thus established their bakeability for UHV application. Vacuum brazed specimen exhibited a sound brazed joint with about 30 µm thick layer of electro-plated Cu on Glidcop part (figure 10). The brazed specimens displayed shear strength of 136-144 MPa. Joint shear strength remained unaltered even after thermal cycling at 150°C. Cross-sectional metallography of the shear tested brazed specimen exhibited fracture at Cu-plating/Glidcop interface, as shown in figure 11.

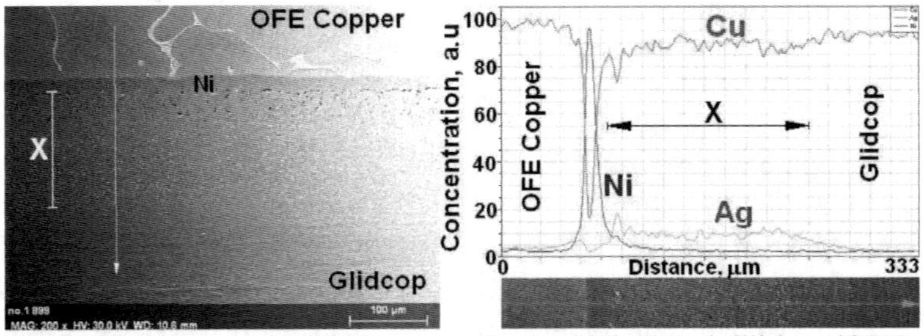

Figure 8. Cross-section of vacuum brazed Ni-plated Glidcop/OFE Cu specimen and EDS concentration profiles of Cu, Ag & Ni across the brazed joint. X – Ag-penetrated region in Glidcop.

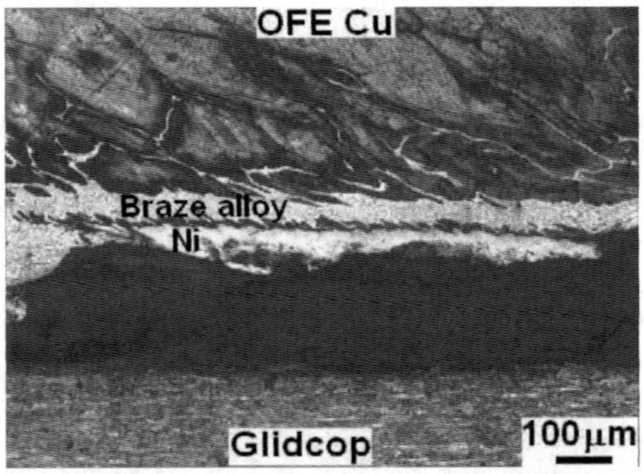

Figure 9. Cross-section of shear tested Ni-plated Glidcop/OFE Cu brazed specimen. Note fracture at Ni/Glidcop interface.

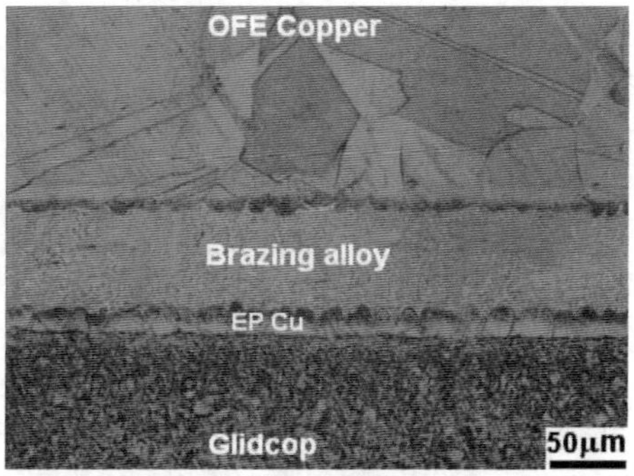

Figure 10. Cross-section of brazed Cu-plated Glidcop/ OFE Cu specimen. EP Cu: electroplated copper

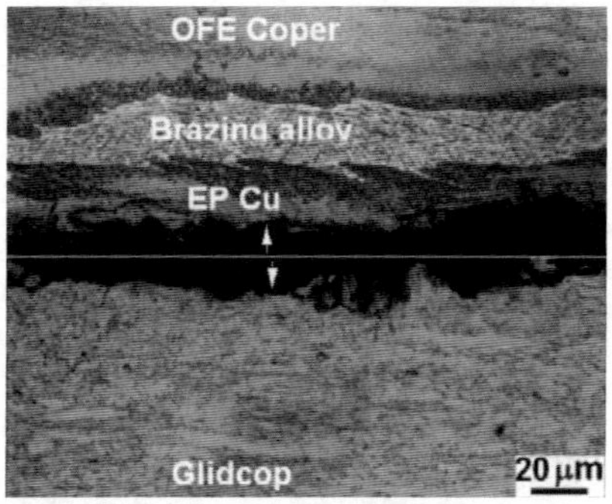

Figure 11. Cross-section of shear tested Cu-plated Glidcop/OFE Cu brazed specimen.

3.2 Set2: Brazing with 50 Au/50 Cu alloy in hydrogen furnace

Visual examination of the brazed specimen exhibited brazing alloy flowing down to the bottom of the specimen, indicating full penetration of the brazing alloy across the length of the joint. Helium leak testing of the brazed specimens exhibited a leak rate < 2 x 10^{-10} mbar.l/s. Leak tightness of the brazed joint remained intact after undergoing 6 numbers of thermal cycles at 150°C. This ensured bakeability of the brazed joint for UHV application. Metallographic examination of the cross-section of brazed specimen exhibited full penetration of the brazing alloy, as shown in figure 12. Due to brazing at a temperature close to melting temperature of copper, OFE Cu part of the brazed assembly exhibited significant grain coarsening while no such effect was noticed in Glidcop ring. The brazed joint, although associated with isolated pinhole porosities, exhibited good bonding with both Glidcop and OFE Cu parts, as shown in figure 13. No inter-granular penetration of braze alloy was observed in the two substrates. Final composition (in wt%) of brazing alloy (as determined by EDS) - Cu: 79.3/84.4; Au:20.8/15.3, is indicative of extensive dilution of brazing alloy by the two substrates. Figure 14 presents concentration profiles of Cu and Au across the brazed joint. Shear strength of the brazed joint was found to be 163 MPa and fracture of the specimen took place in the brazing alloy (figure 15),

Figure 12. Cross-section of specimen brazed with 50 Au/50 Cu alloy. Arrow marks full penetration of the brazing alloy.

Figure 13 Magnified view of the brazed joint made with 50 Au/50 Cu alloy.

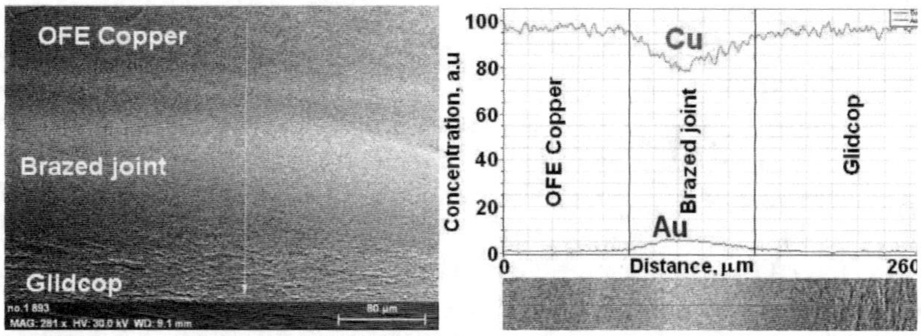

Figure 14. EDS concentration profiles across brazed joint made with 50 Au/50 Cu alloy.

indicating that the two associated interfaces (Glidcop/brazing alloy & OFE Cu/brazing alloy) were stronger than the brazing alloy.

4. Conclusions

In the light of the results of this investigation, two brazing routes were standardized with BVAg-8 and 50Au/50Cu alloys for obtaining leak tight, bakeable and strong brazed joint between OFE Cu and Glidcop. Due to blister formation in Ni-plated Glidcop part during its exposure to brazing temperature, Cu plating of Glidcop was preferred for brazing with BVAg-8 alloy. Standardized brazing procedure for brazing with BVAg-8 alloy involved (i) electroplating Glidcop part - Ni strike in Watts' bath followed by Cu plating in acid bath, (ii) pre-brazing firing of electroplated Glidcop part in vacuum furnace at 830 °C and (iii) vacuum brazing of Cu-plated Glidcop part with OFE part. In contrast, single step brazing with 50Au/50Cu alloy in hydrogen furnace produced satisfactory quality joint between OFE Cu and Glidcop.

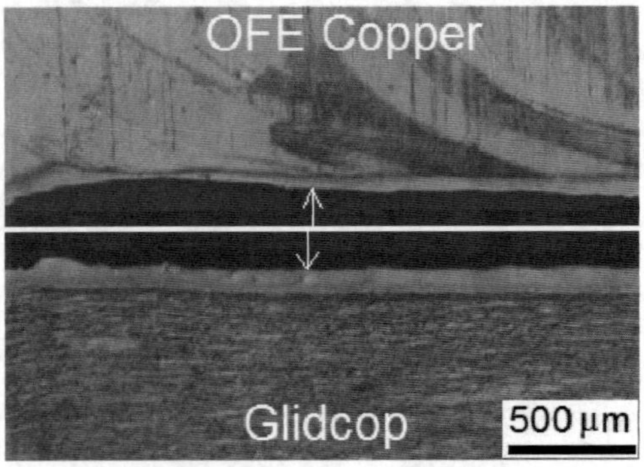

Figure 15. Cross-section of shear tested/ Glidcop specimen brazed with 50 Au/50 Cu alloy. Fracture edge is marked with arrows.

References

[1] Shukla S K, Sridhar R, Yadav D P, Patel R J, Rajarao A S 1998 *Proc. Asian Particle Accelerator* Yadav S K, Shiroman R, Shukla S K, Rajarao A S 1997 *Proc. Int. Conf. IVSNS* (Mumbai, India) *Conf.* (KEK, Tsukuba, Japan).

[2] Yadav S K, Shiroman R, Shukla S K, Rajarao A S 1997 *Proc. IVSNS-97* (Mumbai, India), 02, C-234.

[3] Yadav D P, Shiroman R, Shukla S K 2003 *Proc.Indian Particle Accelerator Conf.* (Mumbai, India), 44.

[4] Yadav D P, Kumar K V A N P S, Sindal B K, Awasthi S K, Shukla S K, Rajarao A S 2005 *Proc. Indian Particle Accelerator Conf.* (Kolkata, India).

[5] Nadkarni A V, Troxell J D, Verniers F 1989 *Glidcop: dispersion strengthened copper: an advanced copper alloy system for automotive and aerospace applications* (Cleveland, USA: SCM Metal Products Inc.).

[6] Nadkarni A V, Dispersion strengthened copper: properties and applications, Final Technical Report, SCM Metal Products Inc., Cleveland, OH, USA, 1985.

[7] Davies J R et al, eds. 1984 *Metals Handbook* Vol. 7 (OH, USA: American Society of Metals) p 711.

[8] Samal P K and Nadkarni A V 1984 *Modern Developments in Powder Metallurgy* **16** 341.

[9] Samal P K 1992 *The Metal Science of Joining* ed M J Cieslak et al (Minerals, Metals & Materials Society).

[10] Technical data sheet 703A 2005 Brazing of Glidcop North American Höganäs High Alloys LLC.

[11] Ryding D G, Allen D and Lee R 1996 *Proc. Int. Conf. on High Heat Flux Engineering (Denver, USA, Aug. 1996)*, SPIE 2855, **14**.

[12] Singh R, Yedle K, Jain A K 2005 *Proc. Int. Conf. IVSNS* (Gandhinagar, India).

[13] Mirkova L, Maurin G and Monev M 2003 *Journal of Applied* Electrochemistry **33** 93-100.

[14] Hydrogen embrittlement, www.tau.ac.il/~chemlaba/Files/Electrodeposition/13208_02.pdf,visited on 11.12.2011.

[15] Durney L J, ed 1996 *Electroplating Engineering Handbook Part I* (London, UK: Chapman and Hall) p 361.

Electron Beam Transport Analysis for Compact, High Power Sheet Beam Microwave Tube

P C Panda[1, 2], V Srivastava[1], A K Nehra[1], A Vohra[2]

[1]Microwave Tubes Division, CSIR-Central Electronics Engineering Research Institute, Pilani, Rajasthan, India
[2]Department of Electronic Science, Kurukshetra University, Kurukshetra, Haryana, India

E-mail: pc_panda.physics@yahoo.com, vs@ceeri.ernet.in, vohra64@gmail.com

Abstract. Sheet beam transport through closed periodic cusped magnet has been analysed. On the basis of analysis, closed PCM is designed to transport elliptical sheet beam of size 30 mm (width)×1.2 mm (thickness), beam voltage 120 kV, current density 600 A/cm^2. This beam is selected for 30 GHz high power sheet beam klystron which has potential application for high gradient LINAC. Selection of aspect ratio of magnetic tunnel and magnetic period is the critical aspect. When aspect ratio of magnetic period is 4:1 and magnetic period is 10 mm, horizontal magnetic force profile matches nearly with horizontal space charge field profile. However, to get desired peak value of magnetic field is very critical. When magnetic period is 10 mm, cross-sectional size of magnetic tunnel is 42.4 mm (width) ×10.6 mm (height), axial thickness of each magnets is 4.2 mm, transverse thickness is 20 mm, the simulated value of magnetic field is 720 G. Above sheet beam with current density 100 A/cm^2 transports up to 120 mm distance through this magnetic tunnel without any instability.

1. Introduction

Sheet electron beam microwave tubes have been emerging for high power microwave applications including directed energy weapon, power beaming, linear accelerators and muon accelerators [1-3]. Large cross section area of sheet beam solves the high current carrying capacity and high voltage break down problems faced by conventional round beam microwave tubes. However, sheet beam focusing is the major issue. Sheet beam transport under uniform magnetic field is unstable due to dicotron instability and edge effect which arise due to $\boldsymbol{E}\times\boldsymbol{B}/B^2$ velocity drift, where E and B are space charge field and applied axial magnetic field, respectively [4-6]. Due to these instabilities, kinks and vortices are formed on the beam and the beam power is dissipated as heat. Although sheet beam transport under periodic magnetic field is stable against diocotron mode, edge focusing is the major issue. This paper deals with suitable periodic cusped magnet design for stable transport of elliptical sheet beam of size 30 mm× 1.2 mm, beam voltage 120 kV, current density 600 A/cm2. These beam parameters are selected for 30GHz for very high power sheet beam klystron which has potential application for high gradient LINAC.

2. Analysis of space charge field

Space charge field is the self electric field created by electrons present in the beam due to which beam diverges. In open boundary, space charge field due to elliptical beam of width *w* and thickness *t* taken respectively along x and y- axes can be expressed as [4]:

$$E_x = \frac{\rho}{\epsilon_0}\frac{t}{w+t}x \quad (1.a)$$

$$E_y = \frac{\rho}{\epsilon_0}\frac{w}{w+t}y \quad (1.b)$$

Where, ρ is volume charge density, E_x and E_y are x and y components of space charge field, respectively.

Expressions (1.a) and (1.b) indicate that both *x* and *y*-components of space charge field vary linearly with *x* and *y*, respectively. In presence of drift tube, nature of the space charge field will different. Figure 1 shows the schematic view elliptical beam inside rectangular drift tube tunnel. It is very difficult evaluate space charge field analytically for this configuration. Hence, space charge field is numerically analysed using OPERA 3D software tool [7]. Specified current density and kinetic energy emitter is used. The parameters considered for analysis are shown in table 1 which can be suitable for high power microwave at 30 GHz frequency. Emitter surface coincides with z=0 plane. Space charge field is estimated at z=1 mm. Figure 2 shows the variation of x-component of space charge field with *x* and indicates that unlike theoretical value in open boundary, space charge field varies non linearly with *x* and increases sharply near beam edge in presence of drift tube tunnel. Figure 3 shows the variation of *x*-component of space charge field with *y*. Figure 4 shows the variation of *y*-component of space charge field with *y*. *Y*-component space charge field varies linearly with *y* like the theoretical results.

Table 1. Parameters considered for analysis

Parameters	Value
Beam width	30 mm
Beam thickness	1.2 mm
Beam kinetic energy	120 keV
Current density	600 A/cm^2
Drift tube tunnel width	40 mm
Drift tube tunnel height	2 mm

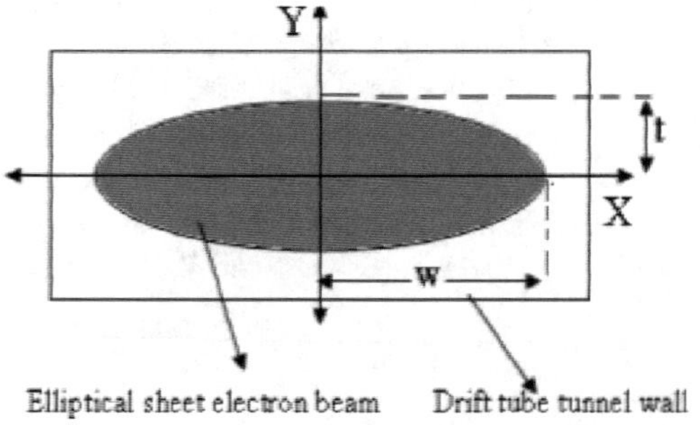

Fig. 1. Cross-sectional view of elliptical sheet beam through elliptical tunnel

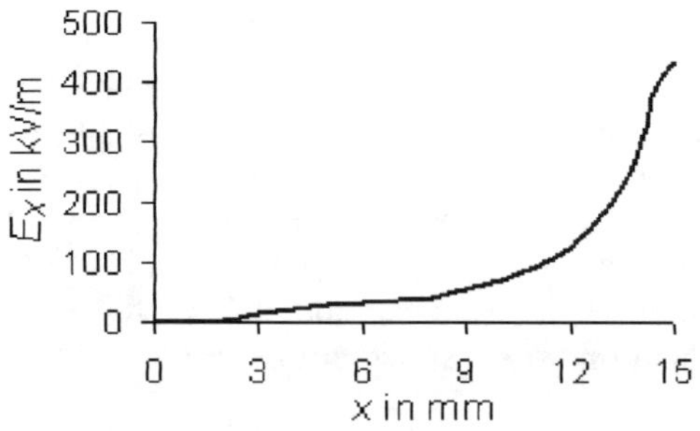

Fig. 2. Variation of *x*-component of space charge field with *x* at y=0

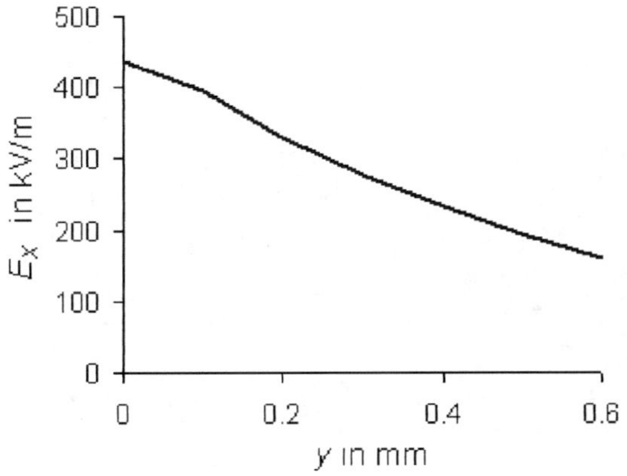

Fig. 3. Variation of *x*-component of space charge field at beam edge with y

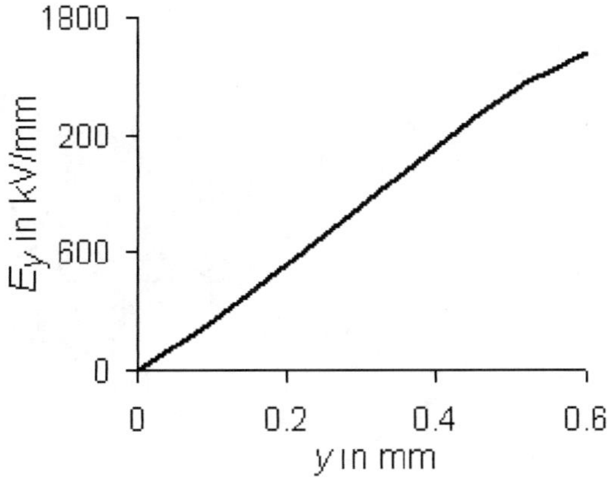

Fig. 4. Variation of *y*-component of space charge field with *y*

3. Closed PCM Design for Sheet Beam Transport

Schematic view of closed periodic cusped magnet (PCM) is shown in Figure 5. Magnetic field at any point inside closed PCM can be expressed as [4, 8]:

$$\boldsymbol{B_x} = -\left(\frac{k_x}{k_m}\right)\boldsymbol{B_0}\sinh(\boldsymbol{k_x x})\cosh(\boldsymbol{k_y y})\cos(\boldsymbol{k_m z}) \tag{2.a}$$

$$\boldsymbol{B_y} = -\left(\frac{\boldsymbol{k_y}}{\boldsymbol{k_m}}\right)\boldsymbol{B_0}\cosh(\boldsymbol{k_x x})\sinh(\boldsymbol{k_y y})\cos(\boldsymbol{k_m z}) \tag{2.b}$$

$$\boldsymbol{B_z} = \boldsymbol{B_0}\cosh(\boldsymbol{k_x x})\cosh(\boldsymbol{k_y y})\cos(\boldsymbol{k_m z}) \tag{2.c}$$

Where, B_0 is the peak value of magnetic field; $k_m=2\pi/l_m$; l_m is magnetic period; $k_x=k_m[1+(a_m/b_m)^2]^{-1/2}$; $k_y= k_m[1+(b_m/a_m)^2]^{-1/2}$; a_m and b_m are width and height of magnetic tunnel, respectively.

Horizontal and vertical component of magnetic force experienced by the sheet beam transporting through closed PCM can be expressed as [8]:

$$\boldsymbol{F_{xs}} = \left(\frac{e^2 B_0^2}{4m\gamma^2 k_m}\right)\left[\mathbf{1} + \left(\frac{a_m}{b_m}\right)^2\right]^{-3/2}\sinh(\boldsymbol{2k_x x_s}) \tag{3.a}$$

$$\boldsymbol{F_{ys}} = \left(\frac{e^2 B_0^2}{2m\gamma^2}\right)\left[\mathbf{1} + \left(\frac{b_m}{a_m}\right)^2\right]^{-2}\cosh^2(\boldsymbol{k_x x_s})\,y_s \tag{3.b}$$

Where, suffix *s* stands for slow time scale solution; F_{xs} and F_{ys} represent respectively horizontal and vertical component of average magnetic force exerted by beam.

Figure 6 shows the variation of normalized magnetic force and space charge force with *x* for different aspect ratio of magnetic tunnel and magnetic period and indicates that proper horizontal focusing is possible when magnetic period and aspect ratio of magnetic tunnel are 10 mm and 4:1, respectively. However getting required peak value of magnetic field is the critical issue. However, to get desired peak value of magnetic field is very critical. When magnetic period is 10 mm, cross-sectional size of magnetic tunnel is 42.4 mm×10.6 mm, axial thickness of each magnets is 4.2 mm, transverse thickness is 20 mm, the simulated value of magnetic field is 720 G. Beam transport analysis has been performed for variable current density with other parameters same as shown in table. CST Particle studio software tool [9] is used for simulation. Sheet beam with current density 100 A/cm^2 and other parameters transport up to 120 mm distance through the above magnetic tunnel without any instability which is shown in Figure 7.

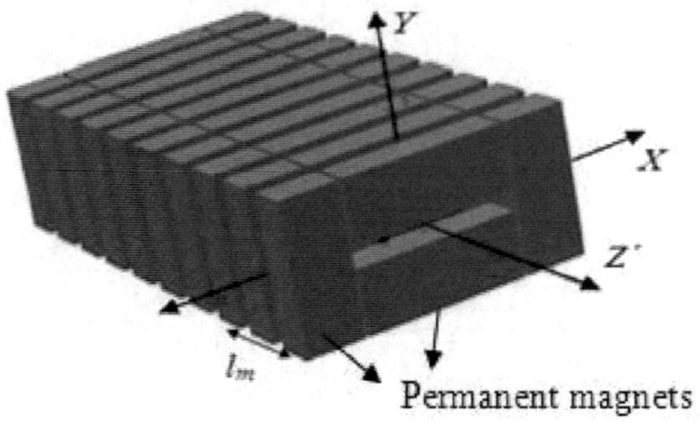

Fig. 5. Schematic View of closed PCM designed by CST particle studio software tool on the basis of reference [8]

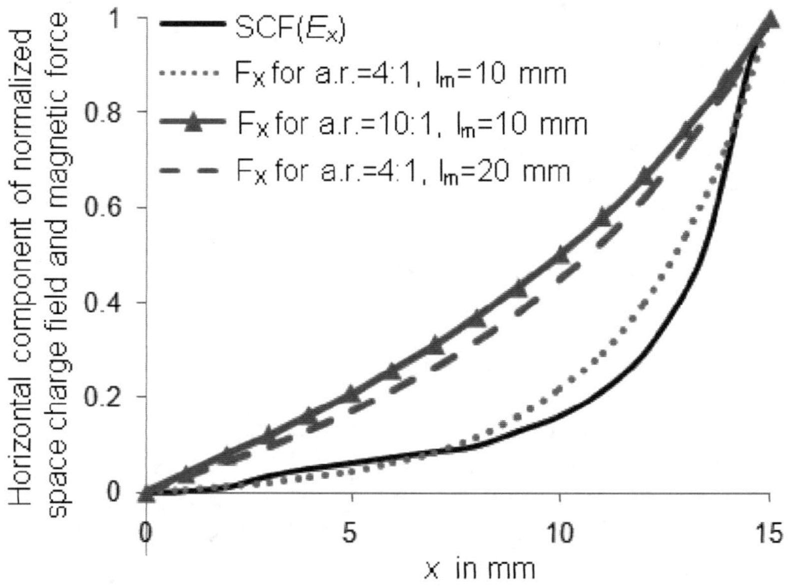

Fig. 6. Variation of normalized x-component space charge field and magnetic force for different magnetic period(l_m) and aspect ratio of magnetic tunnel(a.r.)(Both space charge field and magnetic force are normalized to their respective values at beam edge.

Fig. 7. 3D free view of elliptical beam transporting up to 120 mm distance under closed PCM of magnetic period 10 mm, peak value of field 720 G and aspect ratio of magnetic tunnel 4:1

4. Conclusion

Analysis of transport of sheet beam under closed PCM has been performed by selecting beam parameters suitable 30GHz high power sheet beam klystron. Closed PCM with magnetic period 10 mm, aspect ratio of magnetic tunnel 4:1 is suitable to provide proper horizontal focusing force. However, getting high peak value of magnetic field for these magnetic parameters is very difficult. Simulated peak value of magnetic field is 720 G when axial and transverse thickness of magnets is 20 mm and 4.2 mm respectively. Elliptical sheet beam of size 30 mm×1.2 mm, beam voltage 120 kV, current density 100 A/cm^2 transport up to 120 mm distance without any instability. To transport beam of current density higher than 100 A/cm^2 pole pieces should be introduced in order to increase peak value which is the future work of authors.

Acknowledgement

Funding for this research is provided by HRDG of Council of Scientific and Industrial Research (CSIR) through NET-SRF Scheme. Authors are thankful to Director, CEERI for providing required facility to do this work and allowing publishing this work.

References:

[1] Benford J 2008 IEEE Trans. Plasma Sci. **36(3)** 569
[2] Barker R J and Schamiloglu E 2001 *High Power Microwave Sources and Technologies* (III series, Inc., New York: IEEE press)
[3] Booske J H 2008 *Phys. Plasmas* **15** 055502-1
[4] Booske J H, McVey B D and Antonsen T M, Jr. 1993 *J. Appl. Phys.* **73(9)** 4140
[5] Qui J X, Levush B, Pasour J, *et al* 2009 *IEEE Microwave Magazine* **10(7)** 38
[6] Panda P C, Srivastava V and Vohra A 2011 *Indian J. Pure Appl. Phys.* **49(6)** 410
[7] OPERA 3D Modeller 13.0, Combham Technical Services, Vector Field Software, England
[8] Zhao D 2009 *Phys. Plasmas* **16** 113102-1
[9] CST Particle Studio, CST Studio Suite^{TM} 2010, Darmstad, Germany

Journal of Physics: Conference Series **390** (2012) 012021 doi:10.1088/1742-6596/390/1/012021

Design, Installation and Commissioning of new Vacuum chamber for Analysing Magnet of K-130 Cyclotron

Bidhan Chandra Mandal[1], S Saha, S C Sarkar, D Adak, T Viswanathan, B Hemram, P S Chakraborty, R C Yadav, C Mallik and R K Bhandari

Variable Energy Cyclotron Centre, 1/AF, Bidhannagar, Kolkata-700064.

E-mail: bidhan@vecc.gov.in

Abstract. In view of up-gradation of K-130 Cyclotron at VECC, Kolkata, we have designed a new Vacuum chamber to modify the existing vacuum chamber system. This new chamber is meant for C-shaped 1T dipole type 159.5^0 Analysing Magnet of 4710 OD x 2750 ID x 1075 mm tall in the RIB feeder beam-line. The welded type vacuum chamber is made of SS-304. The chamber with trapezoidal cross-section is of 4447 OD x 4057 ID x 61.5 mm average height. Pumping ports and modules are selected accordingly to ensure the required high vacuum for beam transport. The chamber improves the base vacuum and reduces the complicated O-ring replacement mandatory for existing chamber made of aluminium alloy. The new chamber is installed at site along with all the pumping module and beam line components. This paper presents the detailed design, installation and commissioning results.

1. Introduction

Analyzing magnet in the beam transport line (BTS) of K-130 cyclotron is used to bend the extracted beam selectively. It is a C-shaped, dipole magnet made of low carbon steel, with normal entry and exit and can produce 9 KG field maximum in an average pole gap of 63.5 mm. The beam path is confined in the vacuum chamber kept at a vacuum of the order of 2×10^{-6} mbar. At present, the vacuum chamber of analyzing magnet is nothing but the beam space enclosed by two vertical walls made of aluminium alloy against the upper and lower pole pieces. Both the vertical walls' ends are vacuum tight against pole cap by a large "O" ring. It is a very critical job to achieve required vacuum tightness of this chamber because of sealing groove configuration and limited accessibility. To assemble or disassemble, a series of laborious, exhaustive steps like removal and re-assembly of RIB laboratory, ECR laboratory, chilling plant as well as nos. of shielding plank weighing ~27 ton each and various power supply terminals etc. are mandatory. As a modification to the present vacuum chamber, a welded type stainless steel vacuum chamber has been designed, fabricated and installed to cut short these huge jobs as well as to reduce the down time coming out of frequent vacuum leaks. This chamber is consists of two vertical and two horizontal plates welded together that forms the 159.5^0 arc of the magnet. The trapezoidal chamber volume is of ~19 ltr. and weight is about 150 kg.

2. Design aspects and theory

The semi-circular vacuum chamber is trapezoidal shape (see Fig-1.) in cross section due to tapered pole pieces. Thereby, the 200 mm wide section is 59.94 mm and 63.06 mm in height at the ends respectively thus an average 51.5 mm beam space is to be ensured. Being the non-circular cross-section with an arc shape complex vacuum chamber i.e. vessel under external pressure, structural design is limited to carry out by the standard formulas or ASME –Section-VIII, Div-1. Thereby, a Finite element tool like ANSYS Multi-physics is used. Considering the circular length to a equivalent straight length of non circular cross section vacuum chamber , calculation through MathCAD as per ASME code seeks for higher thickness of plate for shorter side to 6.5mm as against 5mm as per FEM calculation.

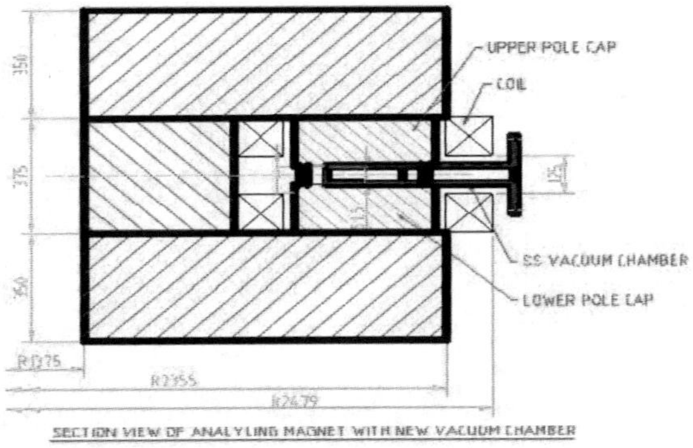

Fig. 1. Sectional view of Analysing Magnet

Equation or formulas used for calculation:-

- $$S_{mA} = \frac{P_e.h.H}{2.(t_1.h + t_2.H)} \qquad S_{mB} = \frac{P_e.h}{2.(t_1)}$$
- $$S'_{crA} = \frac{\pi^2.E_2.\left(\frac{t_1}{H}\right)^2}{12.(1-\upsilon^2)}.K_A \qquad S'_{crB} = \frac{\pi^2.E_2.\left(\frac{t_1}{L_v}\right)^2}{12.(1-\upsilon^2)}.K_B$$
- $$S''_{crA} = S_Y - \frac{S_Y^{\ 2}}{4.S'_{crA}} \qquad S''_{crB} = S_Y - \frac{S_Y^{\ 2}}{4.S'_{crB}}$$
- $$S_{crA} = S'_{crA} \quad if\ S'_{crA} \le \frac{S_Y}{2} \qquad S_{crB} = S'_{crB} \quad if\ S'_{crB} \le \frac{S_Y}{2}$$
- $$S_{crA} = S''_{crA} \quad otherwise \qquad S_{crB} = S''_{crB} \quad otherwise$$

- $2.\frac{S_{mA}}{S_{crA}} + 2.\frac{S_{mB}}{S_{crA}} \le 1.0$

For column stability

- $S_{ma} = \frac{P_e.h_o.H_o}{2.(t_1.H_o + t_2.h_o)}$ $\qquad M = P_e.h_o.H_o.y$
- $C_c = \sqrt{\frac{2.\pi^2.E_2}{Sy}}$ $\qquad S_b = \frac{M.c}{I_e}$
- $\frac{S_a}{F_a} + \frac{S_b}{\left(1 - \frac{S_a}{F'e}\right).S} \le 1.0$ where,
- $F_a = \frac{1 - \frac{\left(2.\frac{L_v}{R1}\right)^2}{2.C_c^{\,2}}}{\frac{5}{3} + \frac{3\left(2.\frac{L_v}{R1}\right)}{8.C_c} - \frac{\left(2.\frac{L_v}{R1}\right)^3}{8.(C_c)^3}}$ if $2.\frac{L_v}{R1} \le C_c$ other wise F'_e
- $F'_e = \frac{12.\pi^2.E_2}{23\left(2.\frac{L_v}{R1}\right)^2}$

Where, Pe-External pressure, t_1,t_2 - plate thickness for long and short side, K_A,K_B - plate buckling factor; h, H - heights respectively, i,e subscript - inner and outer side. S -allowable stress, S_{mA}, S_{mB} - compression stress at the short and long edge, S_{crA}, S_{crB} - plate buckling stress, I_e -moment of Inertia, R_l - least radius of gyration, υ - poison ratio, L - equivalent length, c - neutral axis to extreme fibre distance.

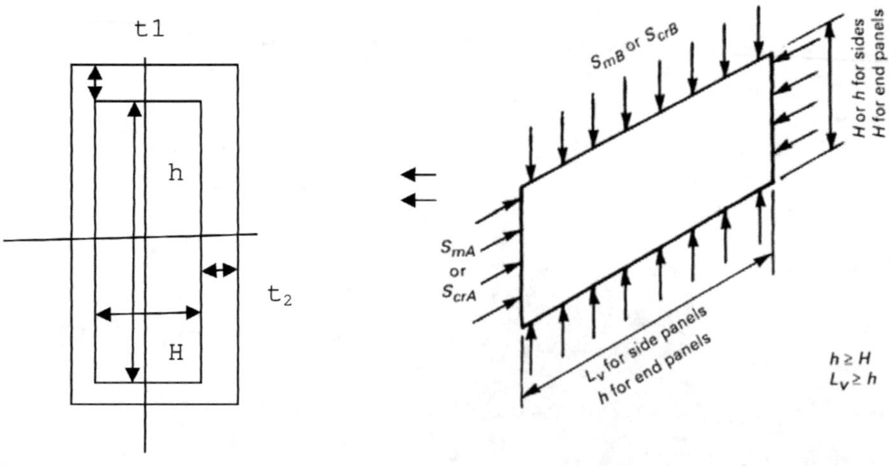

Fig.-2: Cross Section of non circular Vacuum chamber

3. **Materials and fabrication aspects**

The chamber material is required to be paramagnetic since they are placed inside the central magnetic field. Moreover, the material should have low out gassing rate and high modulus of elasticity. SS-304 grade stainless steel is used and the quality of the material used for fabrication is ensured through sample testing by induced field measurement method. The flatness and the surface finish, complex geometry as well He leak tightness (better than 1 x 10^{-9} mbar ltr/s) are enhanced strictly. The most difficulty in fabricating this chamber of this size was to maintain the required flatness after performing the circumferential TIG welding in all the four sides.

Table 1. *"SS-304"* Constant Properties

Name	Value
Density	7.75×10^{-6} kg/mm³
Poisson's Ratio	0.31
Tensile Yield Strength	172.37 Mpa
Tensile Ultimate Strength	482.63 Mpa
Young's Modulus	193,053.0 Mpa
Compressive Yield Strength	172.37 Mpa

Allowable stress for SS-304, S = min{Syt/1.5, Sut/4} = 150 MPa

4. **Analysis report**

The vacuum chamber may be placed between the spacers and pole tips. In this case, inside width of the cross section of the chamber made of SS-304 will be 210 mm. The thickness of the chamber wall should be as minimum as possible to provide the sufficient space for beam path. Initially, calculation was carried out based on ASME to find out the minimum wall thickness which can withstand the vacuum load for this chamber. It is assumed that the vessel of noncircular cross section is straight (equivalent length) instead of circular nature of the chamber. Wall thickness is required as per ASME / Section-VIII / Appendix-13 for Vessels of Non-circular cross subject to external pressure, to be found as 6.5 mm.

Table-2: Structural calculation results of vacuum chamber for various thicknesses

Sl. No.	Thickness (mm)	Maximum Deformation (mm)	Maximum Stress (Mpa)	F.O.S
a)	4	0.73	181.5	0.95
b)	5	0.38	64.43	2.67
c)	6	0.21	53.73	3.20
d)	7	0.13	52.01	3.31

To increase the beam path space, 3-D model analysis for external pressure is carried out on ANSYS Multi-physics platform. The bending stress distribution, deformation and buckling stability were checked for various wall thicknesses. The result of the analysis for 4 mm to 6 mm is given in the Table-2. The FEM result shows that the wall thickness 5 mm is safe for this annular vessel as the curvature of the inside and outside walls are imparting more strength than the straight one. Finally, the chamber has been fabricated as per the ANSYS Multi-physics calculation which conforms to structural as well as beam dynamics requirements.

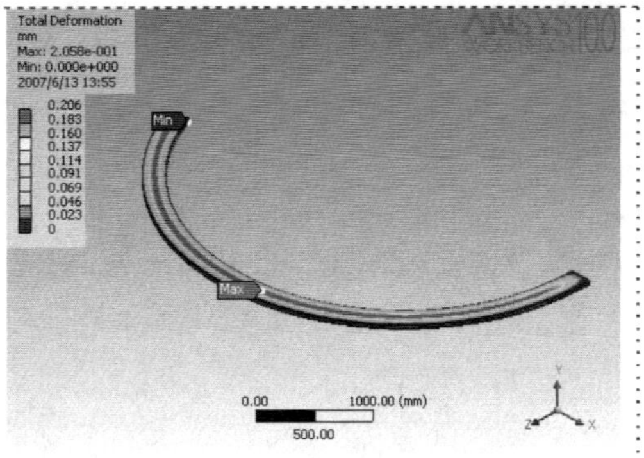

Fig. 3. Total deformation

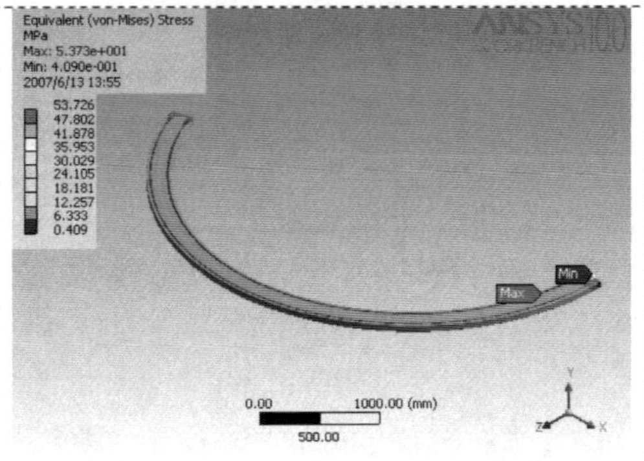

Fig. 4. Equivalent.(Von-Mises) Stress

5. Installation and commissioning report

The said chamber is installed at site on conforming all quality control as well assembly requirements.

Figure 5. Evacuation test of chamber prior to upper pole cap placement

Figure 6. Re installed AM Magnet with SS vacuum chamber

6. Final result:

Evacuation was carried out with two set of Rotary pump (14 m^3/hr) backed Diffusion pump (Diffstak DN 160 flange, 700 l/s N_2) and the system vacuum was achieved to 3.0 x 10^{-5} mbar after 24 hrs and 1.0 x10^{-6} mbar after 60 hrs smoothly with leak rate of < 2.0 x 10^{-8} mbar l/s.

Beam transport line with this Analysing Magnet (achromatic mode) in the channel-IV i.e. RIB feeder line was commissioned with 40 MeV alpha beam. Presently, the beam line is in use for various experiments.

Acknowledgement

The author acknowledges the help of Sri K P Basu in drafting.

References

[1] Roth A , Vacuum Technology
[2] R F Barron, Cryogenics System (Monograph on Cryogenics)
[3] Hahn U et al, 2000, *Nucl. Inst. and Methods in Phy. Research*, A 445(2000) 442-447, Design and performance of the vacuum chambers for the undulator of the VUVFEL at the TESLA test facility at DESY.
[4] ASME Code 2000, Design of Pressure Vessel
[5] Chavet I, 1969, *Vacuum, vol 19*, issue 12, pg 547-552
[6] ANSYS Multi-physics manual.

Development and UHV testing of LN_2 cooled Titanium Sublimation Pump

B K Sindal[1], K V A N P S Kumar, Tripti Bansod and S K Shukla

Ultra High Vacuum Technology Division, RRCAT, Indore - 452 013

E-mail: bksindal@rrcat.gov.in

Abstract. Titanium sublimation pumps (TSP) provide inexpensive large pumping speeds for active gases especially in UHV environment where residual gases are mainly H_2 and CO etc. The pumping speed of TSP is independent of pressure in this range. Sublimation of Ti on liquid nitrogen (LN_2) cooled surfaces is expected to provide large porosity in deposited film and this may also result in further enhancement of its pumping speed. Accordingly a liquid nitrogen cooled TSP was developed to achieve clean vacuum in 10^{-11} mbar range in combination of sputter ion pump which is mainly used to take care of inert gas load. The LN_2 shroud was designed to have 1.9 litre capacity and offered approximately 2000 cm^2 cooled surface area with approximately 250cm^2 area coated with titanium film. This paper describes the performance of the TSP at room temperature as well as after cooling with LN_2. Quadrupole mass analyzer was utilized to analyze residual gases. Simultaneously for pumping speed measurements for different gasses, a setup with known conductance was also appended.

1. Introduction

Titanium Sublimation Pumps (TSP) are mostly used to supplement the pumping of Sputter Ion Pumps. They have low cost, simple operation and provide very high pumping speed for active gases in UHV region. We have used commercially available 'U' shaped filaments made of Titanium-Molybdenum alloy (85%+15%) of Ø2.1 mm and having 35 mΩ resistance. Three filaments mounted on simple four pin low electrical resistance feed-through with the help of SS connectors (Figure 2). A high current passed through the filaments by a power supply causing its temperature to rise sufficiently and sublimate titanium from filament. This sublimated titanium then gets coat the on wall of the vacuum chamber, the film reacts with gas molecules within the system to form low vapour pressure compounds. The saturation time of film depends upon pressure and types of gasses pumped.

The pumping efficiency of TSP increased many times by cooling of coated wall to liquid nitrogen temperature, due to increase in sticking coefficient for gases on cooled surfaces especially for H_2 which shows highest partial pressure in UHV region. This paper describes the development works and pumping speed measurement for different gases (N_2, H_2 & CO) at various pressures and at 300K as well as 77K.

2. Development of UHV chambers

2.1. Design Titanium sublimation Pump body

Design of the vacuum chamber is governed by features like Geometry optimization, material selection, design for mechanical strength against various loadings like atmospheric pressure, gasket sealing load and dead weight, bakeability, minimum gas load and maximum effective pumping speed available at pumping port. TSP chamber is a triple layer, vertically mounted cylindrical chamber with flat top having a port of size DN 40 Con-flat flange for mounting of TSP filament cartridge and two tubes of Ø12mm for liquid nitrogen filling, The torispherical bottom having port of DN 160 Con-flat flange was used for connection with vacuum system. Overall dimensions of the vacuum chamber are Ø203mm OD X 525 mm height. Schematic orthographic front X-sectional view is shown in figure 3.

Fig. 1. Photograph of TSP.

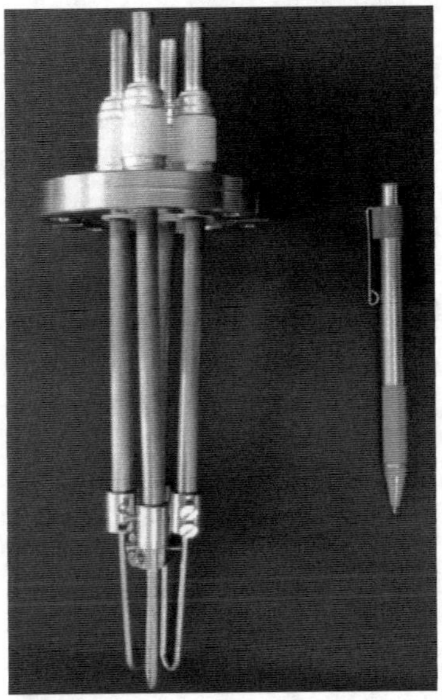

Fig. 2. Filament cartridge of TSP.

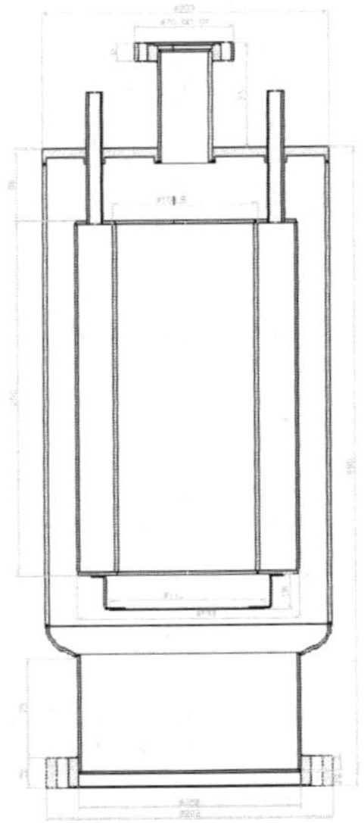

Fig. 3. Schematic front X- sectional view

The cylindrical shape for main shell was chosen to reduce wall thickness. Similarly bottom was having torispherical head because of higher strength-to-weight ratio facilitating light weight compared to flat plate construction. The design of main shell and torispherical head was based on the guidelines of ASME section-VIII, Div-I code. Optimization of their thickness was based on stability against elastic buckling due to external atmospheric pressure of 15 PSI. Iterative process floowed to arrive at minimum required thickness for this loading, using rules under subsection UG-28(C) and UHA-28 part-D for SS 304L. 10% torispherical head was designed for ease in fabrication. Nominal thickness of 2 mm was selected, for fabrication of outer pump body and 1.6 mm for liquid nitrogen shroud, material of construction was SS 3034L. Standard CF flange made by SS 316-L were used.

2.2. Design of Titanium Sublimation Pump

TSP with Liquid Nitrogen (LN_2) shroud was developed which provides approx. 2000 cm^2 cooled surface area with approx. 250 cm^2 Titanium film quoted area for pumping at 300K as well as 77K.

2.3. Fabrication

The main shell was made from 2mm thick seam welded pipe of Ø203 OD, and 10% torispherical dished end both were readily available which was made as per the guidelines of ASME Section IX. Standard available pipes were used to make liquid nitrogen shroud. All the pipes were faced with high precision lathe and ports opening were also machined in dished ends, standard CF flanges were machined on high precision lathe and boring operation was performed accordingly to accommodate pipe size. Prior to welding all the joints were degreased and cleaned properly. In order to give oxide free, smooth and regular surface all the welds were performed from inside. DC TIG welding was done autogenously without using any filler material.

2.4. Chemical treatment, Vacuum Degassing & Leak Detection

After fabrication entire chamber was subjected to various cleaning stages in order to have very clean and smooth surface with minmum concentration of out-gassing species. Following cleaning procedure was adopted; Ultrasonic cleaning and vapour degreasing in Tri-Chloro Ethylene, tap water and demineralised (DM) water rinsing, electro-polishing, neutralization dip, washing with tap water and DM water and drying with air. During electro-polishing the entire knife edges and sealing surfaces were masked by lacquering. After cleaning, chamber was packed nicely to avoid any contamination. Vacuum degassing was performed to remove the H_2 gas from the bulk material.

Leak detection with HMSLD was carried out after fabrication and cleaning, all the welding joints and leak tightness < $1x10^{-10}$ mbar-l/s, was ensured.

3. Assembly and UHV Testing

After chemical cleaning the chamber was assembled on a standard pumping speed measurement setup, consists of two test chambers, upper and lower with number of ports and a known conductance plate was assembled in between. In upper chamber TSP was assembled on its top, BA Gauge and Residual Gas Analyser (RGA) assembled on radial ports. Lower chamber also accommodate BA gauge, a right angle valve for roughing and pumping and a leak valve for gas purging. Sputter Ion Pump 140 l/s capacity assembled in lower chamber mainly pump inert gasses and CH_4. TSP filament cartridge was assembled in TSP port on top and TMP station was also connected using metal hose to system for roughing and pumping during baking. All CF flanges were tightened using standard torque values. The schematic of whole set-up is shown in figure 7.

Leak detection with HMSLD was carried out after assembly and all the joints were found leak tight within $5x10^{-11}$ mbar-l/s.

Initial prolonged pumping was carried out with TMP station for approximately 16 hours and pumping was shifted to SIP and $2x10^{-8}$ mbar vacuum was recorded after 12 hours and improved to $4.6x10^{-9}$ after 3 days pumping of SIP in unbaked condition. Liquid Nitrogen was poured in shroud to find out the integrity at low temperature before bake-out of the system. RGA plot recorded in unbaked condition during LN_2 filling, shown as figure 4

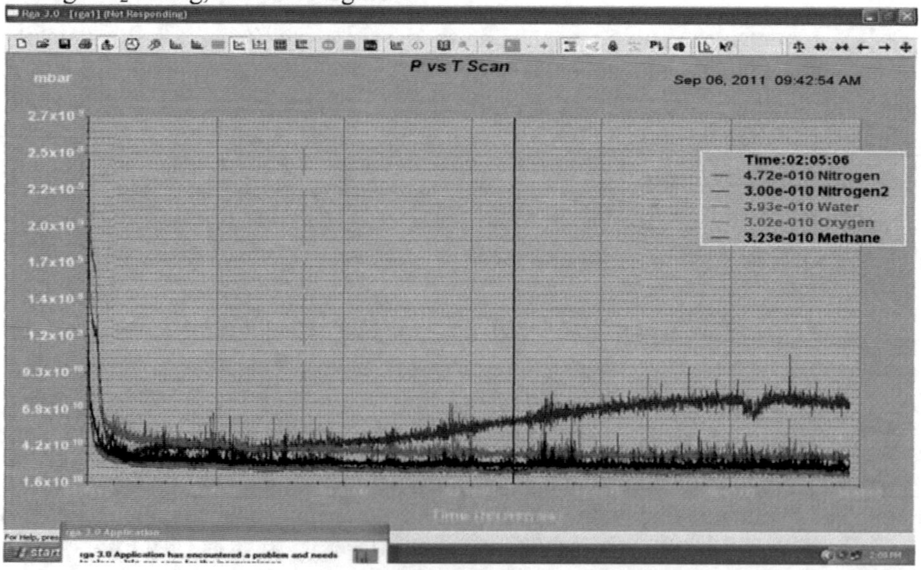

Fig. 4. Residual Gas Spectrum in unbaked condition at 77K

From the figure 4, it was evident that H_2O was dominating gas followed by N_2/CO and O_2 in unbaked condition, and then LN_2 was filled to see the performance of the shroud. A sudden fall in H_2O, N_2/CO and O_2 was observed as the shroud temperature reached 77K.

Entire set-up was baked at 250 °C for 48 hours and BA gauges, RGA and TSP filaments were degassed properly at the end of baking cycle. Ultimate vacuum was recorded after 24 hours pumping with only SIP. One TSP filament was fired with 42 amps current with the help of power supply for two pulses of 1 min each with interval, and 12 hrs delay time was set for next firing, the ultimate vacuum was noted after 24 hrs pumping of TSP at 300K. Simultaneously LN_2 was poured into shroud and TSP was fired, again ultimate vacuum noted after 24 hrs pumping. Ultimate vacuum recorded with different pumping conditions are shown in table 1.

Table-1

Pressure (mbar)	Time duration	Pump used	Temp.
$1.5x10^{-10}$	24 hrs.	SIP only	300 K
$7.3x10^{-11}$	24 hrs.	SIP + TSP	300 K
$2.6x10^{-11}$	24 hrs.	SIP + TSP	77 K

The pump-down details from atmospheric condition to ultimate vacuum achievement using different pump combinations are shown in figure 5.

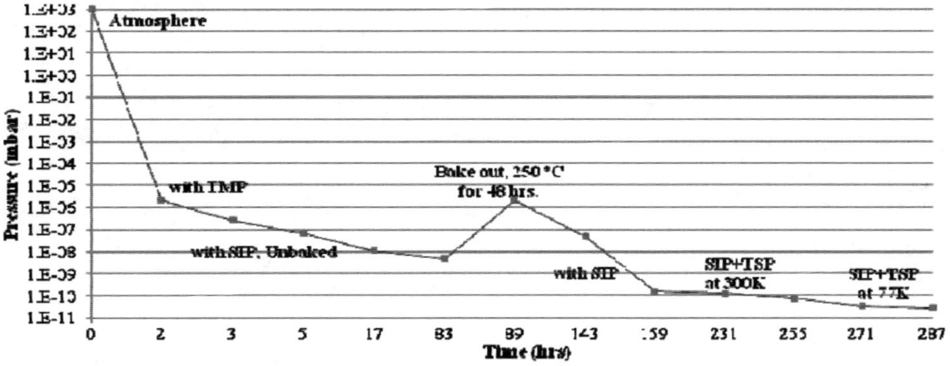

Figure 5. Pump-down curve of the TSP test set-up

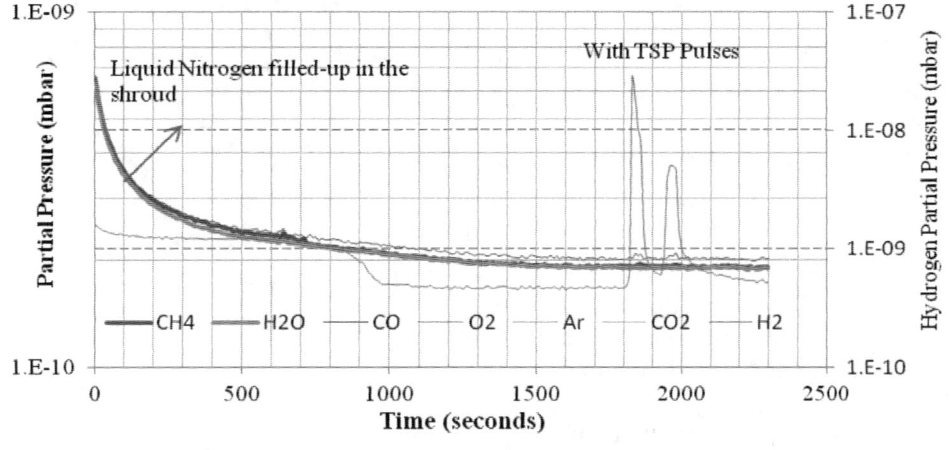

Fig. 6. Residual gas behaviour at 300K, 77K and TSP firing

At 77 K temperature, it was observed that partial pressure of H_2O, CO, CO_2 CH_4 and O_2 fall immediately and there was delay observed in H_2 gas on cool surface. The TSP was fired once to see the release of gasses from the filament, except H_2 no other gases were released. The ultimate vacuum achieved was 7x10^{-11} mbar with-out RGA and the ultimate vacuum was limited to 5x10^{-10} mbar with RGA filament on. RGA plot was shown in figure 6.

4. Pumping speed measurements

The pumping speed measurement of TSP was carried out using known conductance method as per AVS 4.1 standard for pumping speed measurement of vacuum pumps. Known conductance of 9.66 l/s (for N_2 at 20 °C) was introduced between two test chambers. Standard pumping speed measurement set-up for DN 160 size was utilised for pumping speed measurement. For pumping speed measurement gas was introduced in the lower chamber through leak valve and it was measured as P_1 with the help of BA gauge mounted in chamber, known quantity of gas flowing to the upper chamber through the conductance 'C' where TSP pump was connected and BA gauge measured pressure as P_2. The gas flow was controlled by a leak valve for the desired pressure reading P_2 in upper chamber.

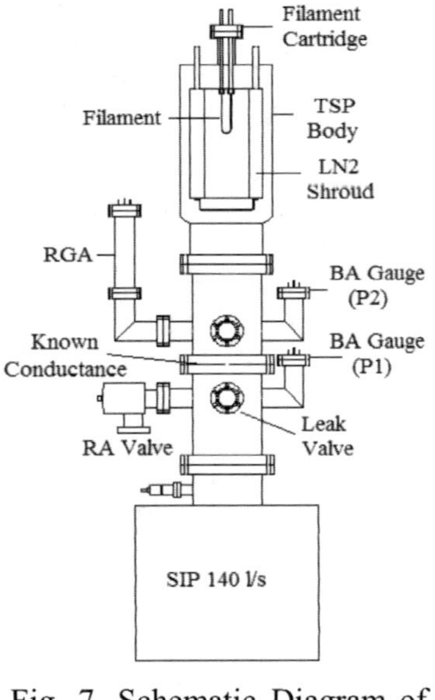

Fig. 7. Schematic Diagram of Test Set-up

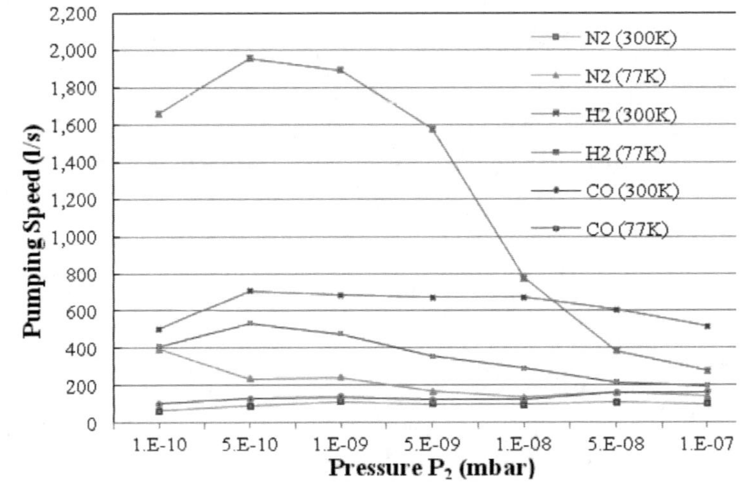

Fig. 8. Graph of Pumping Speed V/s Pressure

Following relation was utilised for calculation of pumping speed:

$$\mathbf{S = C \times [(P_1/P_2)-1]}$$

Where,

S - Pumping speed of TSP at P_2 pressure in l/sec
C - Conductance of orifice in l/sec
P_1 – Pressure at gas purging side in mbar
P_2 - Pressure at TSP Pump side in mbar

The pumping speed was measured with N_2, H_2 and CO at room temperature 300K and after liquid Nitrogen pouring 77K. For pumping speed measurements gas was purged through leak valve, pure Nitrogen was obtained by boiling LN_2 and H_2 and CO (ultra high pure) was purged from gas cylinders.

The graph of pumping speed for different gasses to pressure at different temperature is also shown in figure 8. The pumping speed of different gasses at 300 K and 77 K are shown in table 2.

Table-2

S. No.	Pressure P_2 (mbar)	Pumping Speed for N_2 (l/s)		Pumping Speed for H_2 (l/s)		Pumping Speed for CO (l/s)	
		300 K	77 K	300 K	77 K	300 K	77 K
1.	$1x10^{-10}$	64	396	506	1662	103	410
2.	$5x10^{-10}$	90	237	708	1956	130	533
3.	$1x10^{-9}$	113	243	686	1889	139	477
4.	$5x10^{-9}$	100	170	673	1577	123	357
5.	$1x10^{-8}$	98	135	673	776	127	291
6.	$5x10^{-8}$	109	161	609	385	161	216
7.	$1x10^{-7}$	100	142	519	278	166	197

5. Conclusion

A LN_2 cooled Titanium Sublimation Pump was successfully developed, specified design parameters were obtained and an ultimate vacuum of $2.6x10^{-11}$ mbar was achieved. The TSP has been characterised for its pumping speed with N_2, H_2 and CO at 300 K as well as at 77K successfully.

Acknowledgement

Authors would like to acknowledge technical help from members of UHVT Division specially Shri T.R. Meena, Shri Nitin More and Shri Vijay Singh Bais, help from Workshop-A for fabrication and vacuum degassing of the components and Chemical Laboratory for chemical cleaning work, is also acknowleged.

Reference

[1] B.K. Sindal, D.P. Yadav and S.K. Shukla, proceedings of the Indian vacuum society National Symphosium-2005

Evaluation of Ti-Zr-V (NEG) Thin Films for their pumping speed and pumping Capacity

Tripti Bansod*, BK Sindal, KVANPS Kumar and SK Shukla

UHV Technology Division, RRCAT- INDORE.

*E-mail: tripti@rrcat.gov.in

Abstract. Deposition of NEG thin films onto the interior walls of the vacuum chambers is an advanced technique to convert a vacuum chamber from a gas source to an effective pump. These films offer considerably large pumping speed for reactive gases like CO, H_2 etc. A UHV compatible pumping speed measurement system was developed in-house to measure the pumping speed of NEG coated chambers. To inject the fixed quantity of CO and H_2 gas in pumping speed measurement set-up a calibrated leak was also developed. Stainless steel chambers were sputter coated with thin film of Ti-Zr-V getter material using varied parameters for different compositions and thickness. Pumping capacity which is a function of sorbed gas quantities was also studied at various activation temperatures. In order to optimize the activation temperature for maximum pumping speed for CO and H_2, pumping speeds were measured at room temperature after activation at different temperatures. The experimental system detail, pumping performance of the NEG film at various activation temperatures and RGA analysis are presented.

1. Introduction

Non evaporable getters (NEG) are widely used to effectively capture molecules impinging on their active surface. Due to small vacuum chamber conductance and the particle induced desorption of gas molecules from chamber surfaces in the particle accelerators or storage rings, the required vacuum levels are difficult to maintain. An effective solution to overcome these problems is to coat the whole inner wall of the vacuum chambers with NEG film. On exposure to air, the NEG surface is covered by gases and become inactive. However, the surface may be activated by heating it. Thus the walls of the vacuum systems may be converted to a pumping surface and after insitu activation by baking these films provide a large pumping speed. The Ti-Zr-V coating has the attractive feature of a low activation temperature [1]. The NEG thin film was deposited using Ti-Zr-V getter material in a DC magnetron sputtering system developed in house[2]. From vacuum system designer's point of views, the pumping speed and pumping capacity are very important vacuum parameters and these studies were conducted for H_2 as well as CO. Two SS chambers, coated with different coating parameters, were studied.

2. Description of the pumping speed measurement system

The experimental set-up developed and used for pumping speed and pumping capacity measurement of NEG coated chambers is shown in figure 1. The NEG coated SS chamber was installed on the speed measurement setup. One end of the coated chamber was connected to the gas inlet chamber and other end was connected to another SS chamber and a sputter ion pump (SIP) of 70 l/s pumping speed, through the orifice. The orifice conductance was 3.27 l/s for CO gas and 12.2 l/s for H_2 gas. On both side of the orifice, the chambers had Bayard Alpert Gauges and residual gas analysers (RGA) for total and partial pressure measurements. To check the pumping performance, CO and H_2 gases were injected through a calibrated leak valve. In order to reduce any outgassing from the other components in the testing apparatus, pumping was provided by the turbo molecular pump (TMP) during baking. After baking, system was pumped by the SIP during activation and to achieve ultimate vacuum.

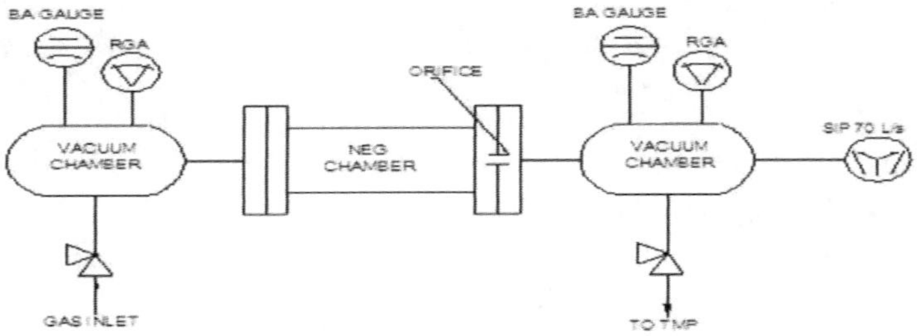

Fig. 1. Pumping speed measurement setup for NEG coated chamber.

The pumping performance of the Ti-Zr-V coated stainless steel cylindrical chamber was evaluated by pumping speed and pumping capacity [3]. The pumping speed measurement of the NEG coated chamber was studied for CO and H_2 gas. These gases were individually introduced from the calibrated leak valve. The pressure in the gas reservoir was kept constant at almost 200 mbar. The pressure of the NEG chamber P_1 and that on other side of orifice P_2 were recorded during injection of the gas. The pumping speed of the NEG coated chamber was calculated from the measured P_1 and P_2:

$$S_{NEG} = \{Q_L - C\,(P_1 - P_2)\} / A.\,P_1 \quad (\text{l/s-cm}^2)$$

Where,

P_1, P_2 are pressure on both side of orifice in mbar
C = Conductance of orifice
A = Coated surface area of the SS chamber (cm^2)
Q_L is the calibrated leak of the test gas which was kept at $2\text{x}10^{-6}$ mbar.l/s for CO and $2\text{x}10^{-5}$ mbar.l/s for H_2.

The total quantity of gas pumped Q_{sorbed} was calculated from the equation

$$Q_{sorbed} = \int (S_{NEG.}\, P_1)\, dt$$

Pumping capacity was evaluated by finding out the quantity of gas which was adsorbed at particular value of pumping speed.

3. Calibrated leak and gas inlet setup

The pumping speed measurement was done by injecting the gas through a calibrated leak valve at known leak rate which was developed for the test gases CO and H_2. The gas inlet system with calibrated leak is shown in figure 2. There is a gas reservoir container of nearly 2 litres volume, it had four ports to install Bourdon Gauge, leak valve, and two high vacuum right angle valves to connect TMP and gas cylinder. This bakable gas inlet setup was used to flush the system with test gas and also to inject the known quantity of gas into the pumping speed measurement system.

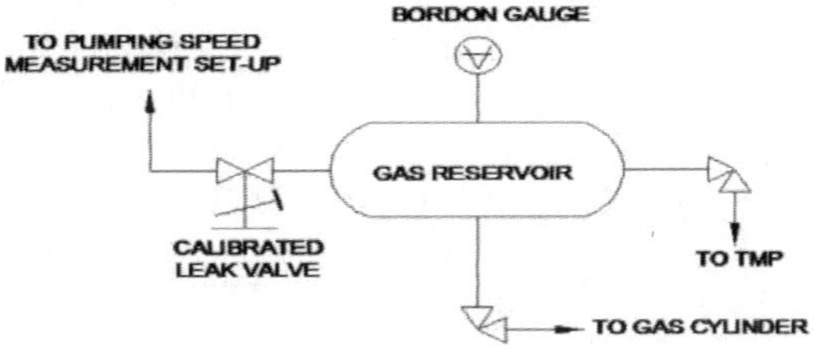

Fig. 2. Calibrated leak valve and gas inlet setup

Leak valve calibration - The test gas was filled in a reservoir at 200 mbar pressure which was measured by a Burdon Gauge. At a particular opening of the leak valve, gas was leaked from the reservoir to speed measurement setup which had an orifice of conductance C. Pressure p_1, p_2 on both side of the orifice were noted down at the time of injection of gas at particular opening of leak valve. During this measurement pumping was provided by the TMP. For a particular opening of the leak valve, specific leak Q_L was obtained which is given by $Q_L = C\ (p_1\text{-}p_2)$. For different opening of leak valve different specific leaks were measured as given in table 1. UHV leak valve was calibrated for CO and H_2 gas by filling the reservoir with CO and H_2 respectively. This experiment was also performed using the helium (He) gas. The reservoir was filled with the He gas at 200 mbar and He gas was injected through leak valve into a chamber whose other port was connected to leak detector calibrated with standard leak for He. The He leak rate measured by LD was used to match leak rate for CO with correction factor.

Table 1. Calibrated leak for CO and H_2

Leak valve opening	Specific leak (mbar-l/s)	
	CO	H_2
Mark-I	$4x10^{-7}$	$6x10^{-6}$
Mark-II	$2x10^{-6}$	$2x10^{-5}$
Mark-III	$7x10^{-6}$	$8x10^{-5}$

4. Experimental procedure for pumping speed measurement

The two SS chambers of 100 mm diameter and 400 mm length were coated in the DC magnetron coating system [4]. These were chemically cleaned, eletropolished and vacuum degassed. The chamber #D (CH #D) was coated with target made of Ti-Zr-V wires in ratio of 1:2:2 and the coating parameters were -600V discharge voltage, 200 gauss solenoid field, $2x10^{-2}$ mbar argon gas pressure, 150mA discharge current and 150°C substrate temperature. For CH #D, the film deposition rate was 0.2 μm /hr, film thickness formed was nearly 1.65 μm and the composition of the NEG film in atomic percentage was Ti-25%, Zr-27%, and V-48%. The chamber # E (CH #E) was coated with target made of Ti-Zr-V wires in ratio of 1:1:1 and the coating parameters were -600V discharge voltage, 200 gauss solenoid field, $1.2x10^{-2}$ mbar argon gas pressure, 100mA discharge current and 150°C substrate temperature. For CH #E, the film deposition rate was 0.15 μm /hr, film thickness formed was nearly 1.62 μm and the atomic composition was Ti- 35%, Zr-22%, V- 43%. The effective NEG coated area of both SS chambers was almost $1000cm^2$.

The NEG coated chamber was installed on the pumping speed measurement setup and leak tested. Initially the system was pumped by the TMP for 12 hrs and then whole system except the NEG coated chamber was baked at 250°C for 24 hrs. The temperature of coated chamber was maintained at 100°C to avoid NEG activation. This also avoided saturation of NEG from the gases released during bakeout. At the end of baking, as the temperature of system was reduced to 100°C the temperature of NEG coated chamber was increased to 150°C for 24 hrs while the temperature of the rest system was gradually decreased. Figure 3 shows the temperature profile. Gauges and RGAs were degassed at the end of the baking process, after baking system was pumped by the SIP. After cooling down to room temperature, the test gas was injected into the system for pumping speed measurement.

In a standard measurement cycle, the coated chamber was heated for 24 hrs at every activation temperature and the pumping speed for individual gases were measured after cooling down of system to room temperature. The pumping speed was measured at different activation temperatures in steps of 50°C from 150°C to 350°C for 24 hrs without air venting in between. During speed measurement, system pumping was provided by the SIP. After each activation when the coated chamber cooled down to ambient temperature, the test gas was introduced into the system and the pressures P_1, P_2 were recorded until the saturation of NEG film. To avoid accumulation of CH_4 and rare gases, the system was continuously pumped with SIP.

5. Results

5.1. Ultimate Vacuum & RGA of NEG chambers

During the CO pumping speed measurement of NEG coated chamber, the coated chamber was heated at different activation temperatures. After each activation and before injection of CO, the NEG coated chamber was tested for ultimate vacuum. As the activation temperature was increased to 200°C, there was a sharp improvement in the ultimate vacuum in the NEG coated chamber. The ultimate vacuum was achieved for CH #D and CH #E at different activation temperature is shown in figure 4. In the ultimate vacuum testing setup [5], the ultimate vacuum $7x10^{-12}$ mbar was achieved in CH #D, after activation at 300°C for 2hrs.

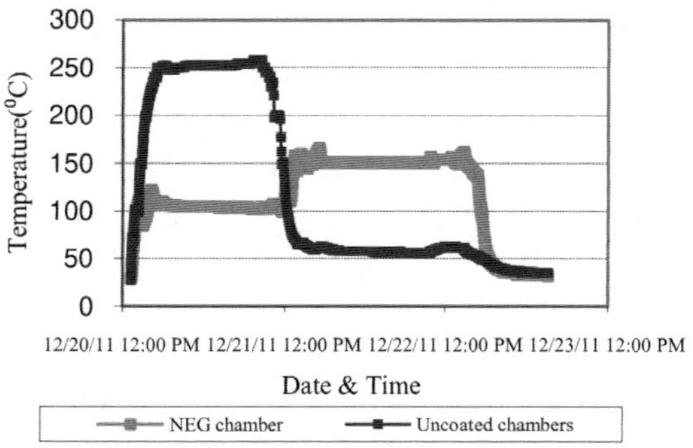

Fig. 3. Temperature profile during baking and activation.

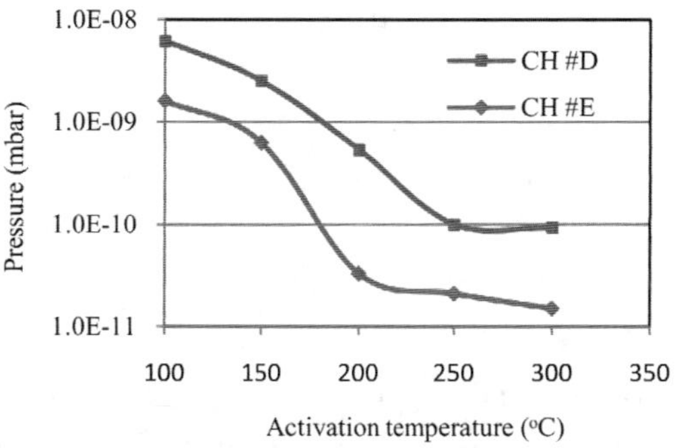

Fig. 4. Ultimate vacuum vs activation temperature.

After activation at 150°C and 250°C temperature, NEG chamber was cooled down to room temperature and then the residual gas analysis was also performed. The residual gases present in the NEG coated CH #D and for CH #E are shown in figure 5 and figure 6 respectively. After 150°C activation the CO partial pressure (PP) was large compared to methane peaks, which was nearly the same as that for uncoated well baked vacuum chamber. This showed that the NEG surface was not activated at 150°C. As the coated chamber was activated at 250°C and cool down to room temperature, the partial pressure of CO was reduced compared to PP of CH_4 and its value reduced to $3x10^{-12}$ mbar for CH #D and $1.2x10^{-12}$ mbar for CH #E. This shows the activation of the NEG chamber for CO and it also indicated that the NEG film has negligible pumping speed for CH_4. Except H_2 all peaks were nearly to $1x10^{-12}$ mbar for CH #E, in RGA spectra.

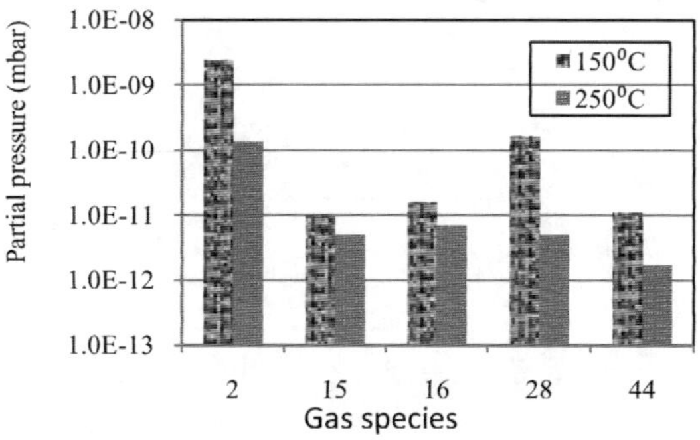

Fig. 5. RGA of NEG coated CH #D after activation at 150°C and 250°C.

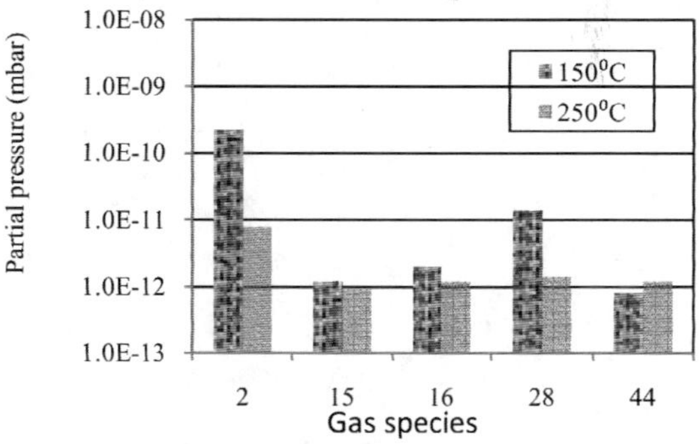

Fig. 6. RGA of NEG coated CH #E after activation at 150°C and 250°C.

5.2. Gas release during activation process

After CO pumping speed at particular activation, the system was pumped by the SIP for almost 24 hrs and then coated chamber was again heated to activate at next higher activation temperature. During the activation, the changes in the partial pressure of CO and H_2 are shown in figure 7. The initial PP of H_2 was $1.5x10^{-9}$ mbar and for CO it was $2x10^{-11}$ mbar. As the temperature was raised H_2 partial pressure continuously increased. But there was a small increase of CO partial pressure and as the temperature reached to 200°C the surface started pumping and PP of CO reduced further. At 250°C PP of CO increased slightly in 10^{-11} range, then as the chamber cooled down to room temperature it reduced to 10^{-12} mbar range. However during activation mainly H_2 gas was desorbed.

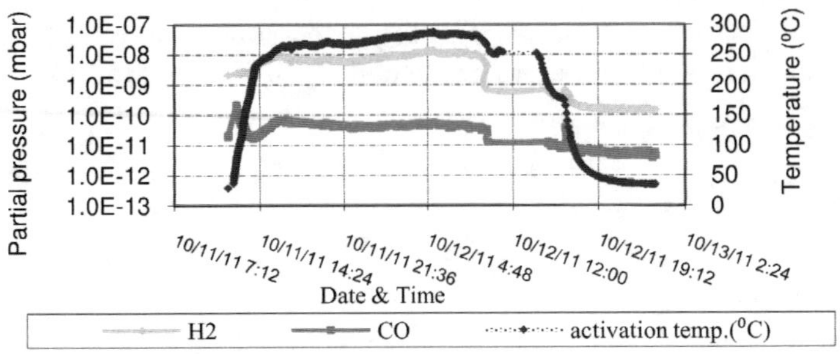

Fig. 7. Partial pressure of CO and H_2 during activation of NEG CH #D.

5.3 Pumping performance of CH #D.

At first the CO pumping speed of NEG chamber was measured at various activation temperatures and it was plotted as a function of the adsorbed gas load Q_{sorbed} is shown in figure 8. After activation at 250°C, a pumping speed of nearly 0.7 l/s/cm^2 for CO was achieved; this was comparatively higher than that at 200°C activation. It was also observed that as the activation temperature was increased the pumping capacity also increased as shown by the Q_{sorbed} quantity in figure 8. At 350°C pumping capacity slightly reduced because large quantity of CO gas was already pumped by the NEG film. The pumping speed remained nearly constant upto nearly $1x10^{-5}$ mbar.l/cm^2 of sorbed gas quantity. After that it reduced very fast.

After completing the pumping speed measurement for CO, the whole setup was baked for 48 hrs and then same procedure followed for pumping speed measurement using H_2 gas. Figure 9 shows the H_2 pumping speed after activation at different temperatures. Compared to CO, H_2 pumping speed was less. But the pumping capacity for H_2 was high; H_2 pumping speed was nearly constant upto nearly $5x10^{-3}$ mbarl/cm^2 of sorbed gas quantity after activation at 300°C. As shown in figure 8 and figure 9, pumping capacity for H_2 was larger than that for CO, due to different pumping mechanism for CO and H_2. CO molecules were only pumped by chemisorptions process onto the NEG surface and it was irreversible process. But the H_2 molecules were pumped by the diffusion process into the NEG film.

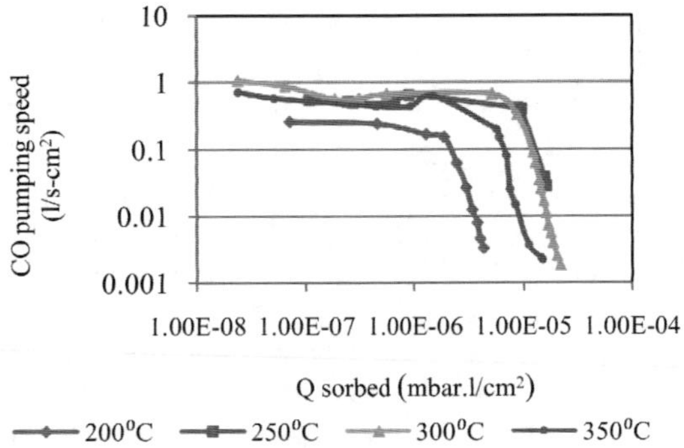

Fig. 8. CO pumping speed and capacity at various activation temperature Vs Q_{sorbed} for CH #D.

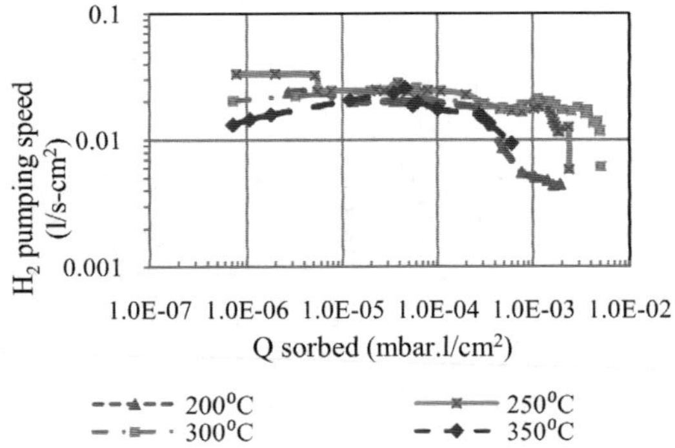

Fig. 9. H_2 pumping speed and capacity at various activation temperature Vs Q_{sorbed} for CH #D.

5.4 Pumping performance for CH #E.

For CH #E, the CO pumping speed of NEG chamber was measured at various activation temperatures. The CO pumping speed was plotted as a function of the Q_{sorbed} as shown in figure 10. After 300°C activation, CO pumping speed of 1.85 l/s-cm^2 was achieved. There was not much difference in pumping speed at 250°C and 350°C activation temperature. It indicated that the NEG surface was fully activated at 250°C. At the same time as the activation temperature was increased the pumping capacity also increased as in figure 10. The pumping speed remained nearly constant upto $2.3x10^{-5}$ mbarl/cm^2. After that it reduced very fast.

After CO pumping Speed measurement for CH #E, H_2 pumping speed was measured at different activation temperatures is shown in figure 11. The H_2 pumping speed was 0.22 l/s-cm^2 after 250°C activation and after 400°C activation it increased to 0.65 l/s-cm^2.

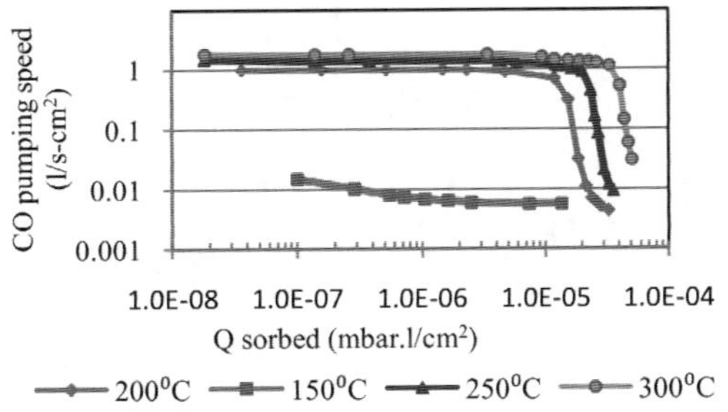

Fig. 10. CO pumping speed Vs Q_{sorbed} for NEG coated CH #E at different activation temperatures.

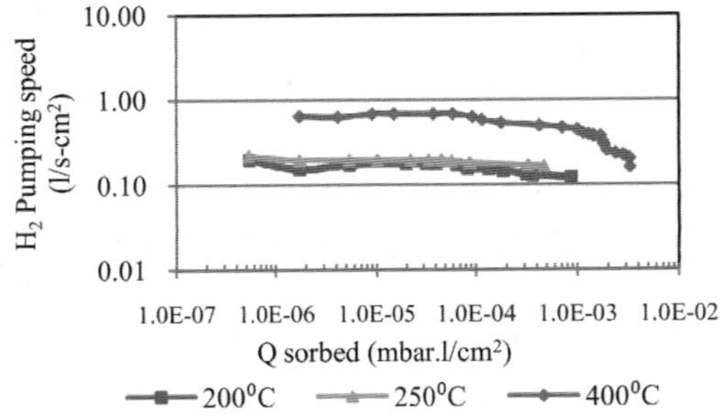

Fig. 11. H_2 pumping speed Vs Q_{sorbed} for NEG coated CH #E at different activation temperatures.

The comparison of initial CO and H_2 pumping speed for CH #D and CH #E at various activation temperatures is shown in figure 12. The CO pumping speed increased as the activation temperature was increased. Initially, the H_2 pumping speed increased marginally but after activation at 400°C there was large increase in H_2 pumping speed. NEG coated chambers have low pumping speed for H_2 compared to CO. The CO pumping speed and capacity for CH #E were large compared to those for CH #D at same activation temperatures.

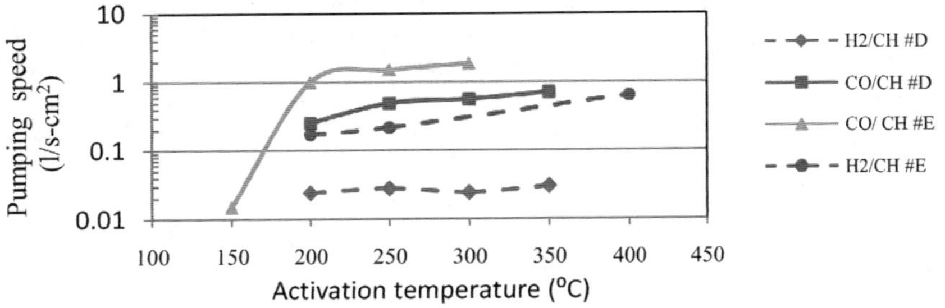

Fig. 12. CO and H_2 pumping speed for CH #D and for CH #E at different activation temperatures.

6. Conclusion

A UHV compatible speed measurement set-up was developed in house for studying important vacuum parameters like pumping speed and capacity for the NEG coated films. These measurements were successfully conducted for chambers coated with different coating parameters. The NEG coated chamber #E provided large pumping speed and capacity compared to CH #D. The observed activation temperature for the NEG coated chamber was 200°C to 250°C. It was found that the gases present in NEG pumping volume were mainly H_2 and large reduction of CO was observed. PP of CH_4 was high comparable to CO, as NEG had low pumping for it.

Acknowledgement

Authors would like to sincerely acknowledge the appreciation and encouragement from Director RRCAT, Indore. The authors would also like to sincerely acknowledge the contribution of all members of UHVTD.

Reference

[1] C. Benvenuti, *J.Vac.Sci.Techy.A* 19(6), Nov/Dec 2001.
[2] C. Benvenuti, *Vacuum* 53 (1999) 219-225.
[3] Yulin Li, *J.Vac.Sci.Techy.A* 16(3), May/Jun 1998
[4] SK Shukla, *presented at IVS – TFST, Mumbai,* 9-12 Nov 2011.
[5] Tripti Bansod, *presented at IVS – TFST, Mumbai,* 9-12 Nov 2011.

Journal of Physics: Conference Series **390** (2012) 012024 doi:10.1088/1742-6596/390/1/012024

Study of Residual Gas Analyser (RGA) Response towards Known Leaks

Firozkhan S Pathan, Ziauddin Khan, Pratibha Semwal, Siju George, Dilip C Raval, Prashant L Thankey, Himabindu Manthena, Paravastu Yuvakiran and Kalpesh R Dhanani

Institute for Plasma Research, Near Indira Bridge, Bhat, Gandhinagar – 382428, Gujarat, India

E-mail: firose@ipr.res.in

Abstract. Helium leak testing is the most versatile form of weld qualification test for any vacuum application. Almost every ultra-high vacuum (UHV) system utilizes this technique for insuring leak tightness for the weld joints as well as demountable joints. During UHV system under operational condition with many other integrated components, in-situ developed leaks identification becomes one of the prime aspect for maintaining the health of such system and for continuing the experiments onwards. Since online utilization of leak detector (LD) has many practical limitations, residual gas analyser (RGA) can be used as a potential instrument for online leak detection. For this purpose, a co-relation for a given leak rate between Leak Detector and RGA is experimentally established. This paper describes the experimental aspect and the relationship between leak detector and RGA.

1. Introduction

A leak means an unintended crack, hole or porosity in an enveloping wall or joint. The basic function of the leak detection is to localize and to quantify the leak size in a sealed enclosure or system. In spite of modern technologies, it is practically impossible to seal any enclosure or system that can guarantee a complete leak proof system. What maximum can be practically feasible is to minimize it to the best possible extents. For such situations, the helium leak detection plays a major role.

Mass spectrometer based leak detectors (MSLD) are most commonly used detectors and are very sensitive instruments for identifying very small leaks presence in vacuum systems. Mainly there are two broad categories of MSLD based helium leak testing – (1) Spray method in which the object under scanning is evacuated and connected to the MSLD through the roughing line of the pumping system. Helium gas is spread throughout the system and if there is a leak the Helium gas will penetrate into the system due to pressure difference and will be detected by the MSLD. (2) Sniffer method in which the object under testing is pressurized with helium gas to a desired value and a helium-sensing probe connected to the MSLD is moved as close as possible to the weld joints, de-mountable joints and the surface of the object. Both the techniques are briefly represented in the figure 1.

SST-1 Tokamak [1, 2] is a steady state super-conducting tokamak for plasma discharge of 1000 sec duration. Mainly it comprises of two vacuum chambers. One is Vacuum vessel, which is ultra high vacuum chamber and other is Cryostat, which is a high vacuum chamber. Vacuum vessel will be

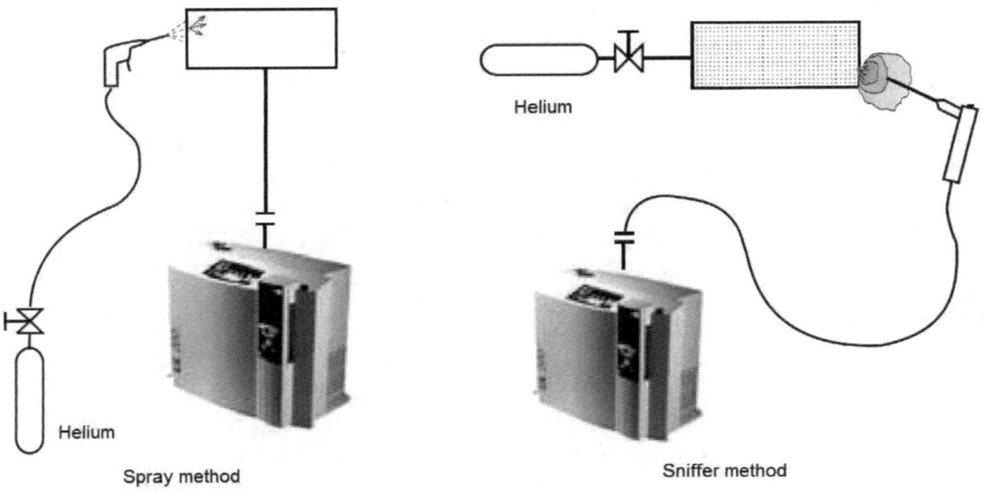

Figure 1. Different helium leak testing methods using MSLD.

pumped to < 1.0×10^{-8} mbar and Cryostat will be pump to < 1.0×10^{-5} mbar vacuum. Both this chambers are having pressurized subsystems inside, which if leaks will deteriorate the system vacuum. Figure 2 shows the schematic of SST-1 cryostat circuits.

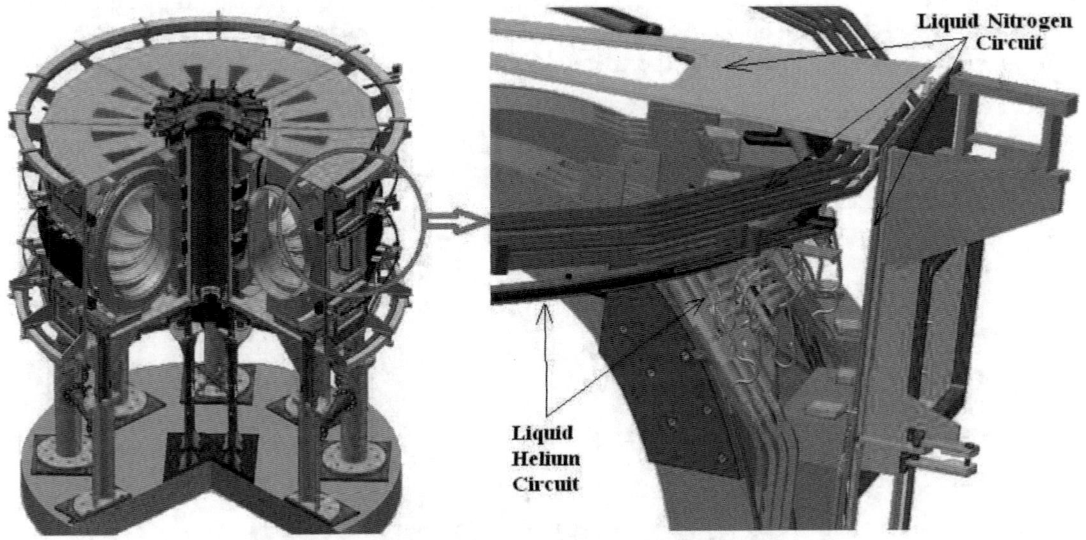

Figure 2. High pressure circuits of SST-1 Cryostat

For baking of vacuum vessel at 150 °C during bake out, U-shaped SS 304L channels of 2 mm thickness have been welded on the inner surfaces of the vessel wall. Hot Nitrogen gas will be passed through this baking channels at pressure of 4.5 bar (g). Any leak on this baking channel will affect the vacuum inside vessel. Vacuum vessel is also equipped with first wall components which are having copper tubing for cooling and heating purposes. Any leakage in these Cu tubing will also disturb the vessel vacuum.

Cryostat is having 80 K shields, TF & PF coils and their supply and return headers. 80 K shields are cooled with liquid nitrogen at 6.0 bar (g) whereas TF & PF coils are cooled with supercritical liquid helium at 4.0 bar (a) 4.5 K. Considering such different pressurized circuits with different gases, the smooth operation of vacuum system requires a kind of online tool which keep watch on leak integrity of all these circuits and can give indication of a leak for various types of gases without disturbing the pumping process. Leak detector can not be utilized for this purpose, because (i) it cannot detect multiple gases as per system requirement, (ii) it cannot be kept online during operation and (iii) it will disturb the pumping process. These critical requirements foresee to find best alternative solution which could be online residual gas analyzer (RGA). Hence, efforts were made for characterization of the response of RGA towards known leaks.

Residual gas analyzer (RGA) is used to measure the partial pressures of gases present inside a vacuum system. There are various types of RGAs based on different working principles. MKS make RGA used in this experiment, is capable of scanning masses up to 200 amu with a minimum detectable pressure of 2.0×10^{-11} torr [3]. If required, a particular mass can be selected to measure its partial pressure trend with time. Since a single mass is being scanned, the scanning rate and hence the response will be faster in this mode. This feature of RGA can be used for online leak detection in a large vacuum chamber like SST-1 Tokamak. With systematically pre-cleaned vacuum system, it is possible to detect the leaks smaller than 1.0×10^{-12} torr l/s with a RGA [4].

2. Experimental set-up and procedure

The experiment is carried out in a stainless steel ultra-high vacuum (UHV) chamber. Chamber is pumped to an ultimate pressure < 1.0×10^{-7} mbar using a 520 l/s (Nitrogen) turbo-molecular pump (TMP). Leak integrity of the entire system is ensured at a background of 1.0×10^{-9} mbar l/s. Vacuum measurement is carried out using B. A. gauge while the partial pressure is measured using MKS make RGA. Provisions are made to connect Standard leaks of different ranges to the system through a 25 KF right angle valves. Figure 3 shows the experimental set up used for the test.

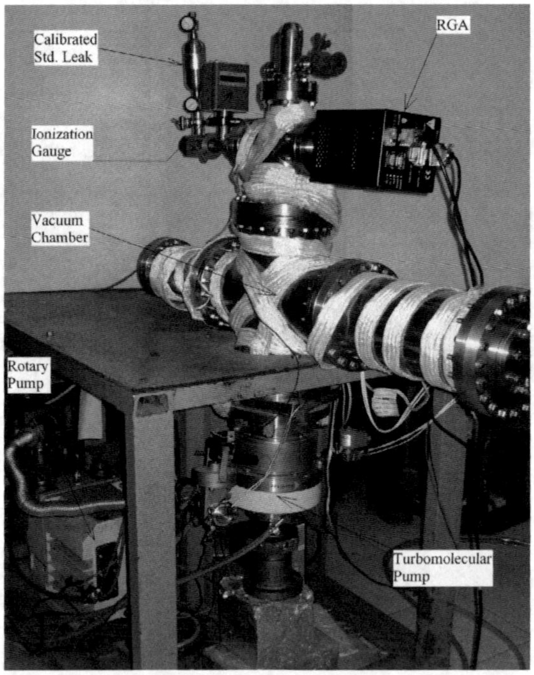

Figure 3. Experimental set-up for RGA characterization.

Different standard leaks are used and also their locations are varied to see the response of RGA. Initially it is installed in such a way that almost all the helium gas coming out of standard calibrated leak gets fully exposed to RGA. Initially the standard leak is connected in such a way that RGA will be directly exposed to the gases coming from standard leak and then observations are noted. In the similar fashion, the standard leak is kept nearly 3.0 meters away using 25 KF bellow and again the observations are noted. Finally the standard leak is shifted in such a way that it is directly exposed to TMP and RGA is exposed partially. In this condition the observations are also noted. Finally observation for air leak is carried out. All these observations are only taken when the system is in very stable condition. Also the entire experiment is carried out for three (03) times to ensure the repeatability.

3. Results

The leak rates are calculated using RGA partial pressures with effective pumping speed of 275 l/s (Helium) which is as per system configuration. The observed partial pressures of the helium gas, base partial pressure of the system, the standard leak values and the calculated leak values are tabulated in the table 1.

Table 1. Comparison of RGA based leak values with Standard leak.

Location of Standard leak	Partial pressure of He (mbar)	Base Partial pressure of He (mbar)	ΔP (He) (mbar)	RGA based leak value (mbar l/s)	Standard leak value (mbar l/s)
Directly exposed to RGA and very close to it	1.27×10^{-09}	3.00×10^{-11}	1.24×10^{-09}	2.45×10^{-06}	3.70×10^{-06}
	1.83×10^{-09}	3.00×10^{-11}	1.80×10^{-09}	3.58×10^{-06}	6.70×10^{-06}
	3.09×10^{-09}	3.00×10^{-11}	3.06×10^{-09}	6.15×10^{-06}	1.04×10^{-05}
Directly exposed to RGA and 3.0 m away from it	1.23×10^{-09}	3.00×10^{-11}	1.20×10^{-09}	2.41×10^{-06}	3.70×10^{-06}
	1.75×10^{-09}	3.00×10^{-11}	1.72×10^{-09}	3.46×10^{-06}	6.70×10^{-06}
	2.80×10^{-09}	3.00×10^{-11}	2.77×10^{-09}	5.57×10^{-06}	1.04×10^{-05}
Partial exposed to RGA	4.40×10^{-09}	3.00×10^{-11}	4.10×10^{-10}	8.24×10^{-07}	3.70×10^{-06}
	6.30×10^{-09}	3.00×10^{-11}	6.00×10^{-10}	1.21×10^{-06}	6.70×10^{-06}
	1.00×10^{-09}	3.00×10^{-11}	9.70×10^{-10}	1.95×10^{-06}	1.04×10^{-06}

While varying the standard leak values at given location, RGA snap shorts are taken in order to see the effects in the partial pressure variation which are shown in the figure 4 and figure 5.

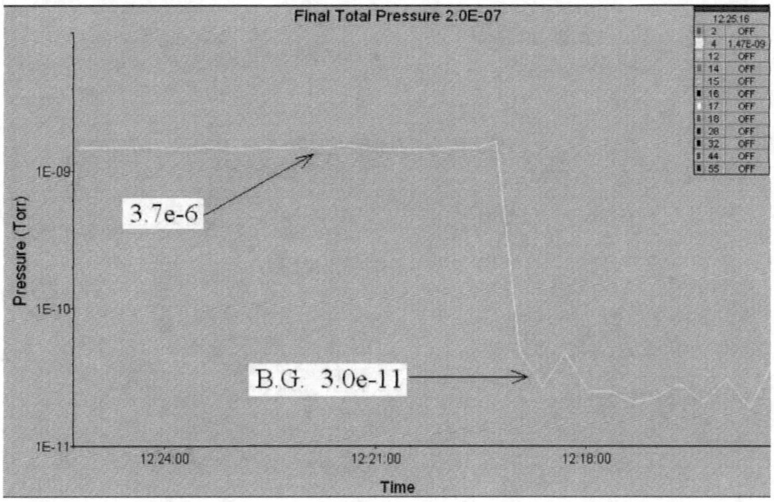

Figure 4. Screenshot of RGA scan with a single standard leak.

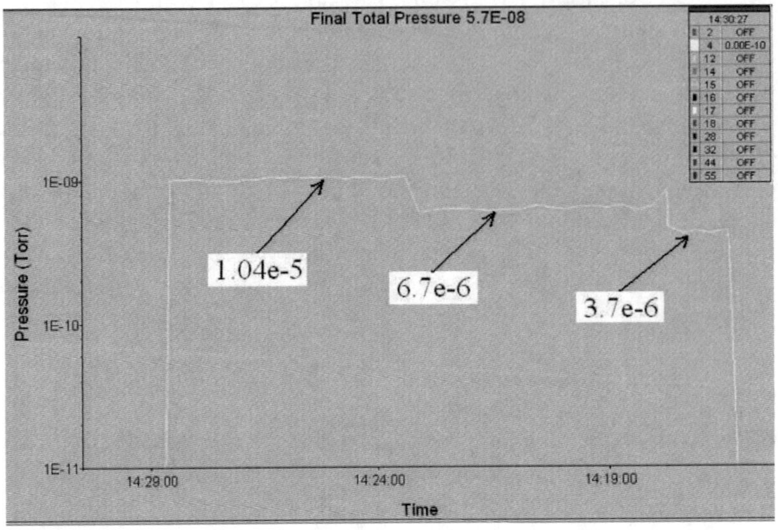

Figure 5. Screenshot of RGA scan with different standard leaks.

Table 1 shows that in directly exposed conditions, response of RGA is in accordance to standard leak value at lower leak rate and it is within 10 % variation. But as the leak value increases, difference between RGA and standard leak value differs more i.e. RGA shows less response. Even after varying standard leak distance, the same results are repeated. For partially exposed condition, the response of RGA is even more deteriorated.

Finally air leak is introduced and RGA scan is taken as shown in the figure 6. Data for air leak is tabulated in the table 2.

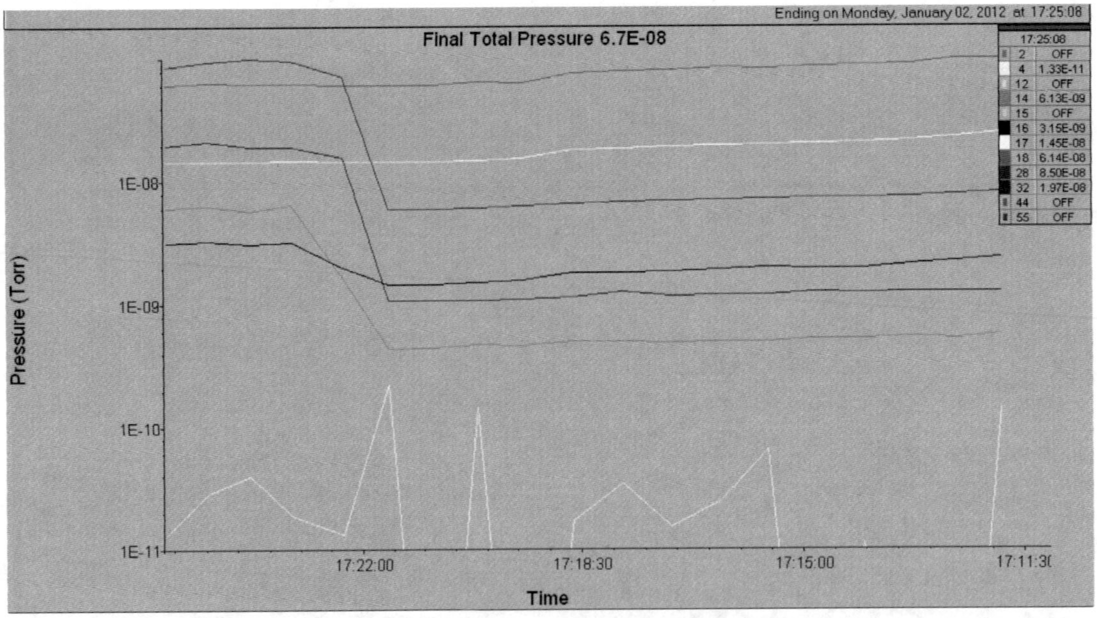

Figure 6. Screenshot of RGA scan with air leak.

Table 2. Experimental observation for air leak.

Partial pressure of mass 14 (mbar)	Partial pressure of mass 16 (mbar)	Partial pressure of mass 28 (mbar)	Partial pressure of mass 32 (mbar)	Ratio (14+28)/(16+32)
6.13×10^{-09}	3.15×10^{-09}	8.50×10^{-08}	1.97×10^{-08}	3.99

Table 2 shows a ratio of 3.99:1 for the partial pressure of Nitrogen to that of Oxygen which is same to the percentage composition of Nitrogen and Oxygen in air (4:1). Hence the air leak can be confirmed through RGA scan if this ratio appears.

4. Conclusion

RGA can be used as an online leak detector in a system like SST-1 tokamak with so many different high pressure components with different gases species. This will give an instant sense for a leak in a particular system.

References

[1] Despande S P and SST-1 Team 1997 *Proceedings of the 17th IEEE/NPSS Symposium on Fusion Engineering* (San Diego, CA) vol 1 p 227
[2] Pradhan S & SST-1 Mission Team 2010 *J. Plasma Fusion Res. SERIES* vol 9 p 650
[3] MKS Instruments Inc 2009 *Opertaing manual for RGA* (Andover, USA)
[4] Philip A and Ganapati M 2007 *J. Phys.Conference Series* 114 (2008) 012011

Vacuum Brazing of Accelerator Components

Rajvir Singh[1], KK Pant[2], Shankar Lal[2], DP Yadav[3], SR Garg[4], VK Raghuvanshi[4] and G Mundra[1]

[1]Accelerator Components Design and Fabrication Section,
[2]Free Electron Laser Lab,
[3]Ultra High Vacuum Technology Division,
[4]Indus Synchrotrons Utilization Division, Raja Ramanna Centre for Advanced Technology, Indore - 452013, INDIA

E-mail: rvsingh@rrcat.gov.in

Abstract. Commonly used materials for accelerator components are those which are vacuum compatible and thermally conductive. Stainless steel, aluminum and copper are common among them. Stainless steel is a poor heat conductor and not very common in use where good thermal conductivity is required. Aluminum and copper and their alloys meet the above requirements and are frequently used for the above purpose. The accelerator components made of aluminum and its alloys using welding process have become a common practice now a days. It is mandatory to use copper and its other grades in RF devices required for accelerators. Beam line and Front End components of the accelerators are fabricated from stainless steel and OFHC copper. Fabrication of components made of copper using welding process is very difficult and in most of the cases it is impossible. Fabrication and joining in such cases is possible using brazing process especially under vacuum and inert gas atmosphere. Several accelerator components have been vacuum brazed for Indus projects at Raja Ramanna Centre for Advanced Technology (RRCAT), Indore using vacuum brazing facility available at RRCAT, Indore. This paper presents details regarding development of the above mentioned high value and strategic components/assemblies. It will include basics required for vacuum brazing , details of vacuum brazing facility, joint design, fixturing of the jobs, selection of filler alloys, optimization of brazing parameters so as to obtain high quality brazed joints, brief description of vacuum brazed accelerator components etc.

1. Introduction

Indus-1 and Indus-2 are two storage rings presently operational at RRCAT. In these machines, electron beam from an injector microtron is accelerated to rated energies (450 MeV for filling Indus-1 and 600 MeV for filling Indus-2) in a booster synchrotron before being used to fill up Indus-1 and 2 synchrotrons to rated currents. In Indus-2, beam energy is further ramped up after filling to ultimately achieve the design energy of 2.5 GeV. Synchrotron Radiation (SR) from Indus-1 and Indus-2 is extracted through the different beamlines on these machines and used for a wide range of experiments in different areas of science and technology. The components and assemblies used between points of beam extraction and experimental area are called Front-End components. These include collimator, water cooled shutter, fixed mask, water cooled safety shutter etc made of stainless steel and OFHC copper. Main requirements of the above components are either removal of the heat or to check/contain radiation produced by travelling beams. These components have SS-to-copper and copper-to-copper joints. In most of the cases vacuum brazing of the joints of these components is carried out using Cusil BVAg-8 brazing alloy in the vacuum atmosphere of $5x10^{-5}$ mbar at temperatures ranging from 830°C to 860°C.

A Plane Wave Transformer (PWT) linac structure is another accelerating structure that has been brazed in the Vacuum Brazing Furnace at RRCAT. It involves brazing of copper to copper and copper to SS joints with minimum distortion and good electrical conductivity. Eight and twelve cell PWT linac structures have been successfully developed indigenously using Vacuum Brazing.

OFHC Copper is main engineering material used in the accelerators. It has relatively less strength which results into selection of its higher thicknesses. Glidcop is a new material to be used in lieu of OFHC/OFE Grade copper. It is similar to OFHC copper in all respects but with higher strength as compared to OFHC copper. Brazing process has been developed for the above material using Cusil BVAg-8 brazing alloy vacuum atmosphere. Brazed joints are developed using test samples.

2. Working Principle of Brazing Process

Joining of two pieces of base metal is performed when melted filler alloy flows across a very thin gap or clearance between them and on cooling it forms a solid bond. Brazing creates an extremely strong joint, usually stronger than the base metal pieces themselves, without melting or deforming the components. The principle by which the filler alloy is drawn through the joint to create this bond is capillary action. In a brazing operation, the heat is applied to the entire base material. The filler alloy automatically comes in contact with the heated parts. The filler alloy melts instantly by the heat available in the base metals and on melting it is drawn by capillary action completely through the joint. Thus, a brazed joint is made. In Vacuum brazing process all gases are essentially removed from the area surrounding the job. For this reason it is an expensive process. No reaction or oxidation takes place of the components being brazed as gases present removed vacuum pumping. Flow, wetting and spreading of liquid filler is affected by the presence of oxides and impurities present at the surfaces to be brazed. For removing the oxides proper cleaning methods are used. Most of the impurities and the remains are generally of volatile nature detrimental to joining and are cleaned by physical and/or chemical methods. Their presence in traces is easily removed by evacuating the surroundings of the job. Thus the surfaces are maintained clean. In some cases oxide present on the surfaces of the components being brazed dissociate themselves during heating cycles under vacuum environment. This results into self- cleaning of the components. Gases are desorbed from the components during heating cycles and these are also removed completely. In this way in vacuum brazing an environment is created for carrying out joining process in ideal conditions. The components are uniformly heated. All of the above factors in vacuum brazing result into several advantages which are not possible to get in other methods of joining and brazing.

3. Vacuum Brazing Furnace

The above facility consists of a double walled water cooled vacuum chamber having demountable type door ends with provision of vertical bottom loading. Job loading is motorized. The above furnace can accommodate jobs of 700mm diameter x 2000mm long sizes with job weight of 500kgs. The furnace was designed for a maximum temperature of 1250°C at a vacuum of $5x10^{-5}$ mbar. At present the furnace is in regular use for temperatures upto 1000 ^{O}C with vacuum of order of $2x10^{-5}$ mbar. Oil diffusion pumping system with a combination of rotary and mechanical booster pump have been employed for obtaining vacuum. Molybdenum heating elements, radiation shield of molybdenum and Stainless Steel Grade 304 have been used. The above furnace is computer controlled with manual over-ride facility. PLC and Pentium PC are integrated together to maneuver steps of operation and safety interlocks of the system. Closed loop water supply provides cooling to the system, **Figure 1** [4].

4. Main Steps in Vacuum Brazing

Following are the main steps to be followed in vacuum brazing –

4.1 Clean and Compatible Vacuum Furnace

The vacuum furnace in which vacuum brazing of components is scheduled should be absolutely clean. If it is a new furnace the hotbox assembly is required to be carried out in a dust free clean room with control on humidity. High humidity may spoil molybdenum components of hotbox assembly. In case the furnace is old and has been in use for a long time it requires cleaning of the hotbox and vacuum chamber surfaces exposed to vacuum. Vacuum degassed materials and oxides formed deposit on hotbox components. They are required to be removed chemically or by using mechanical methods. Thick black tar layer of contaminants deposited on the hotbox also requires its removal. Hydrogen purging at elevated temperature is also employed for removal of metal oxides from surfaces of the hotbox components. To know and ascertain the health of the vacuum furnace from cleanliness view point it will be ideal to identify the level of various contaminants by using Residual Gas Analyzer (RGA). The above practice has been followed in present facility.

4.2 Cleaning of the components/Assemblies

Proper cleaning of the components is equally important. Capillary action works properly if the surfaces are clean. The surfaces contaminated with oil, grease, rust, scale or dirt will form a barrier between base metal surfaces and the molten filler alloys. An oily base metal will result into poor wetting and spreading of the filler. The molten filler that oxidizes due to heat results into voids and in imperfect joint. Oils and grease carbonize the parts when heated and form a film over which the filler metal will not flow. Similarly brazing filler cannot bond rusty surfaces. Hence cleaning is essential part of vacuum brazing. Cleaning of base metals and non metals is a complex process. In general, most of the impurities are cleaned and washed with water and detergent soaps. Further cleaning is done with organic solvents. Rough surfaces are smooth buffed and acid pickling is done to remove buffing compounds from the surfaces. Good quality joints have been obtained following the above practices.

4.3 Joint Design and Types of Joints

Many types of joints are used in brazing similar to other brazing methods. However, two basic types of joints are commonly used. These are butt and lap types. Butt joints do not provide as much strength as lap joints. This is because more area is available for brazing in lap type joints. Main consideration in designing a joint is given to the strength of the joint. If high strength is required lap or scarf type joint is preferred rather than a butt. Strength depends on area of brazing of faying surfaces of the components. The joint design can further be improved by adopting a joint which can re-distribute stresses. This will reduce stress concentrations in the areas nearer to the joints, *Figure 6*. It also depends on gap or clearance between the components[4]. The recommended clearance for typical furnace brazing is given the Table 1 for particular type of brazing filler metals. Necessary care has been taken in designing of the components which had been vacuum brazed.

4.4 Good Fit and Proper Clearance

For getting a quality joint it is necessary to provide and maintain proper clearance between components of the base metal to allow capillary action to work most effectively. A close clearance is maintained during the brazing operation as the tensile strength of the brazed joint varies with the amount of clearance between the parts being joined. Joint strength is reduced if the clearance is either reduced or increased than a value which is specific to a particular base metal in given set of conditions. Figure 9. Typically these clearances range from 10 microns to 100 0f microns [4]. In joining dissimilar metals thermal coefficient of expansions should also be considered once deciding for the clearances. For this reason manufacturing tolerances and allowances in the design of the individual components and an assembly as a whole are required to be specified so as to meet the brazing process requirements, Figure 2.

4.5 Assembly and Loading of the Job

On accomplishment of cleaning and oxide removal the components are assembled in a clean room. Assembly of components requires holding in position. This is achieved by using fixtures. Filler metals/materials are cut or formed to the required sizes and are placed/sand witched between the components. It shall be ensured that the assembly remains in required position till it is taken out of the chamber on completion of the heating and cooling cycle. It should further be ensured that the assembly remains in correct and aligned position throughout the cycle. The simplest way of doing this is to hold the components together by gravity. For easy and successful brazing the joints surfaces are kept in horizontal plane for effective working of capillary action. Joints inclined to the horizontal plane are also brazed by holding fillers in position using special holding fixtures. For brazing several joints and/or subassemblies multi-stage brazing is carried out. The components received after cleaning should immediately be assembled and loaded in the vacuum furnace with minimum time gap so that the surfaces to be brazed are joined in clean and oxide free condition.

Fig. 1 Vacuum Brazing Furnace

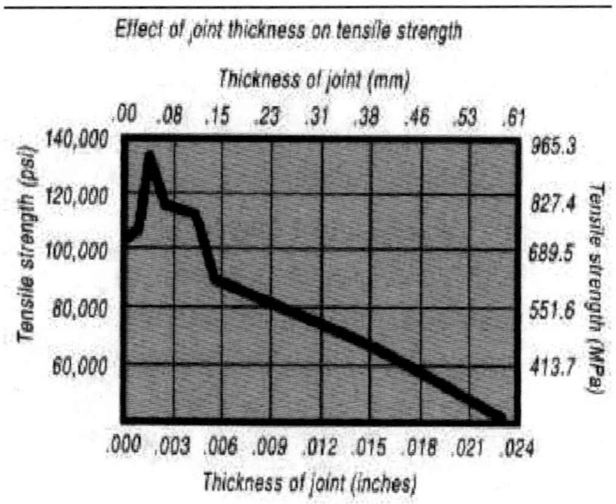

Fig. 2 Variation of Strength with thickness of joints

4.6 Heating Cycle of Furnace/Job

Once the job is loaded into the furnace, it is ready for starting the heating cycle. Cooling water, compressed air and power supply are checked before starting the furnace. Vacuum pumping system and all other sub-systems are also checked for their proper working. Initially the heating rate High purity alumina ceramics are kept from 3 deg. C to 5 deg. C per minute. For achieving temperature uniformity in the hot zone temperature soaking of adequate durations are given. For achieving a quality brazed joint heating is carried out to a temperature of 50 deg. C above the melting point of the filler alloy with duration of few minutes. Very low voltages Thyristor/SCR controlled power supply are used in the furnace. Operation & control of the furnace is computer controlled with manual over ride facility. PLC and SCADA are also used in the vacuum brazing furnaces to make them operator and user friendly. Operation and safety interlocks are provided for safe operation of the furnaces. Closed loop clean and cooled water supplied is used. Periodic maintenance and timely up-keeping of the sub-systems reduced down time to the minimum.

5. Brazing of Accelerator Components

Various components have been vacuum brazed successfully at RRCAT, Indore. These components as mentioned below are explored as better and suitable alternatives to the available components. Some of them are new developments. Brief description of brazed components is given below -

5.1 Brazing of a PWT linac structure

In order to support high electric field gradients $\geq$ 20 MV/m, linac structures are operated under high vacuum $\leq$ 1x10-7 mbar. Further, all joints should have very good electrical/rf conductivity to minimize losses. A 12-cell PWT linac structure that is capable of accelerating electron beam to > 10 MeV energy consuming ~ 8 MW of RF power has been successfully brazed and leak-tested to a leak rate $\leq 5 \times 10^{-10}$

mbar l/s. The brazed disk-array of the linac has 101 joints that are brazed in a single cycle. The disk array of the linac, at the axis of which the accelerating electric field is generated, has twelve disks machined from Class I Oxygen-free Electronic (OFE) Grade Copper, which are supported on four tubes of AISI Stainless Steel (SS) 316L plated on the outside surface with copper. These tubes serve the dual purpose of support and cooling since they carry water from a reservoir at one end of the array to all the disks. The total number of joints to be brazed in a single cycle is 101. The outer vacuum envelope of the linac (tank), into which the disk array is inserted, is made of AISI SS 316L and it has the necessary ports for vacuum pumping and for feeding RF. The RF port made of Class I OFE copper is brazed to the outer surface of the tank. In all these brazing cycles, the filler alloy used is eutectic Copper-Silver (72-28) and all brazed SS surfaces are electroplated with copper before the brazing cycle. Appropriate fixtures are employed during the brazing cycles to ensure alignment and precise spacing between the disks of the array. Figure 3 shows the brazed disk array and figure 4 shows the tank envelope of the 12-cell PWT linac structure.

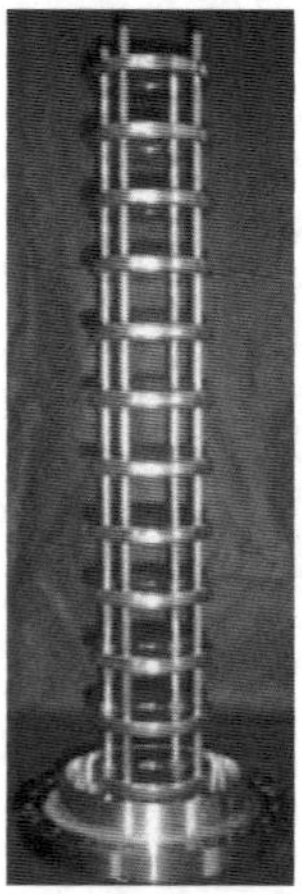

Fig.3: Disk-array of 12-cell PWT linac

Fig.4: Tank and Reservoir-flange assembly of 12-cell PWT linac

Fig.5: S-band pre-buncher structure

5.2 Brazing of an S-band pre-buncher structure:

An S-band pre-buncher structure has been designed to bunch an electron beam from a sub-harmonic pre-buncher before injection into the PWT linac structure for acceleration to rated energy. This pre-buncher is a single-cell RF structure with a geometry identical to the full cell of the standard BNL/SLAC/USLA type of laser photocathode RF guns. The half cell part of a photocathode RF gun has been modified to accommodate a beam port. The structure has two beam ports, two tuner ports, one vacuum pumping port and one RF port with a rectangular wave-guide section. Excepting the RF port, all ports are made of AISI SS 316L and electroplated with copper before brazing of the assembly using eutectic copper-silver/Cusil (72-28) filler/brazing alloy. The complete assembly has been vacuum brazed in two cycles with appropriate fixtures to maintain precise alignment of the ports. Figure 5 shows the S-band pre-buncher after the final brazing cycle. The above has been vacuum and leak tested to a leak tightness of $2x10^{-10}$ mbar-lit/sec with excellent quality brazed joints.

5.3 Front End Components

5.3.1 Collimator (CM)

Collimator is a beam-defining aperture made of OFE Copper solid block with rectangular tapered hole along its central axis. Collimator receive Synchrotron Radiation (SR) Beam of around 24 mrad horizontal divergence out of which only 7 mrad SR beam will pass through it resulting in heating the collimator. The water cooled copper block is vacuum brazed with Cu tube and two S.S. conflat flanges such that the water cooling is done inside UHV beam line chamber. Vacuum brazing is done by using BVAg-8 silver copper eutectic filler metal in foil form as per SFA-5.8 of ASME boiler and pressure vessel code Section II part C at the above mentioned vacuum brazing facility.

5.3.2 Water Cooled Shutter

Water-cooled shutter is used to completely intercept the SR beam and absorb the full thermal power of the beam to isolate downstream components from the thermal load. The shutter is made of a water cooled Cu plate which can be moved up & down pneumatically inside a vacuum chamber. The Cu tube is vacuum brazed with Cu plate and the conflat flange such that there is no direct vacuum to water joint. Vacuum brazing is done by using BVAg-8 silver copper eutectic filler metal in foil form as per SFA-5.8 of ASME boiler and pressure vessel code Section II part C.

Fig.6 Collimators

Fig. 7 Water Cooled Safety Shutters

5.3.3 Fixed Mask (FM)

Fixed mask is a final beam-defining aperture made of OFE Copper solid block with rectangular tapered hole along its central axis. Fixed mask receive S.R. beam of 7 mrad horizontal divergence. The purpose of FM is to provide the required beam size for a particular beam line. The water cooled copper block is vacuum brazed with Cu tube and two S.S. conflate flanges such that there is no direct vacuum to water joint. Vacuum brazing is done by using BVAg-8 silver copper eutectic filler metal in foil form as per SFA-5.8 of ASME boiler and pressure vessel code Section II part C at the above mentioned facility.

5.3.4 Safety Shutter

The purpose of safety shutter is to absorb Bremsstrahlung gamma radiation generated from electron beam scattering, absorb the SR beam and absorb the thermal load. The safety shutter is consist of a radiation absorber head made of densimet, a tungsten alloy, and a water-cooled copper plate which can be moved up & down inside a vacuum chamber by pneumatic cylinder. The Cu tube is vacuum brazed with Cu plate and the conflat flange such that there is no direct vacuum to water joint. Vacuum brazing is done by using BVAg-8 silver copper eutectic filler metal in foil form as per SFA-5.8 of ASME boiler and pressure vessel code Section II part C.

6. Water-cooled Four Blade slit system of XRF Microprobe Beamline

Water cooled four blade slit system will be used to precisely define the white beam size. Slit system is 4 blade type which is housed in UHV chamber. Each blade is independently actuated by linear motion system. The cooling of slit blade is done through a water-cooled Cu plate. The above copper plate is vacuum brazed to copper tube carrying cooling water. The Cu tube is vacuum furnace brazed with Cu plate and the conflat flange using BVAg-8 silver copper eutectic filler metal in foil form as per SFA-5.8 of ASME boiler and pressure vessel code Section II part C.

7. Vacuum Brazing of of Glidcop® to OFE Cu for application in Photon Absorbers of Indus-2

Indus-2, a 2.5 GeV synchrotron radiation (SR) source emits 187 kW of SR power from bending magnet source at the design current of 300 mA. Only 15% of this power is channeled in beam lines while the rest is absorbed by 64 water-cooled photon absorbers installed on 16 numbers of bending magnet chambers. The normal incident SR power density on the exposed surface ranges from 10 to 12 kW/cm2. In the upgraded design of photon absorber, the bottom part facing SR is made of low oxygen (LOX) grade Glidcop AL-15 (UNS C15715) while the intermediate part is made of oxygen-free electronic copper (OFE Cu - UNS C10100) which is to be joined to the bottom part. The dissimilar joint between Glidcop and OFE Cu is required to be hermetically sealed (helium leak rate < 2 x 10^{-10} mbar.lit/s), bakeable upto 150°C and reasonably strong to withstand high pressure of cooling water. The experimental study was done to standardize brazing process to obtain helium leak tight, bakeable and strong brazed joint between Glidcop® to OFE Cu. Vacuum brazing joint thus produced yielded helium leak tight and bakeable joints with acceptable shear strengths and the joints were comparable to the joints made using gold-50%-copper-50% alloy using hydrogen atmosphere in some other furnace[6].

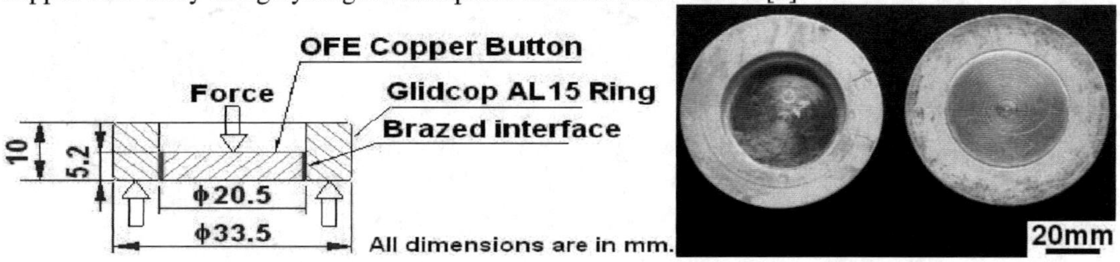

Fig. 8. Details of specimen used for brazing experiments.

8. Conclusion

Vacuum brazing of hundreds of components and assemblies have been successfully done using the vacuum brazing facility available at RRCAT, Indore. The vacuum compatible brazing assemblies are leak tested by using Helium leak detector. All the brazing joints are found leak tight with leak rate better than 10^{-10} mbar liter/sec. The above process of joining will be highly useful for developing UHV compatible accelerator components and systems in the future.

References

[1] Giles Humpston and David M. Jacobson, 1993, ASM International.
[2] D. Apelian, Introduction to Furnace Brazing, 2001, APD.
[3] Handy & Harman Brazing Book, 2001.
[4] Singh R, Development of Vacuum Brazing Techniques, Invited Talk 2005 *Proc. Int. Conf. IVSNS* (Gandhinagar, India).
[5] Singh R, Yedle K, Jain A K, Development of Vacuum Brazing Furnace, 2005 *Proc. Int. Conf. IVSNS* (Gandhinagar, India).
[6] Yadav DP, Vacuum Brazing of of Glidcop® to OFE Copper, unpublished.

Baking of SST-1 vacuum vessel modules and sectors

Firozkhan S Pathan, Ziauddin Khan, Paravastu Yuvakiran, Siju George, Gattu Ramesh, Himabindu Manthena, Virendrakumar Shah, Dilip C Raval, Prashant L Thankey, Kalpesh R Dhanani and Subrata Pradhan

Institute for Plasma Research, Near Indira Bridge, Bhat, Gandhinagar – 382428, Gujarat, India

E-mail: firose@ipr.res.in

Abstract. SST-1 Tokamak is a steady state super-conducting tokamak for plasma discharge of 1000 sec duration. The plasma discharge of such long time duration can be obtained by reducing the impurities level, which will be possible only when SST-1 vacuum chamber is pumped to ultra high vacuum. In order to achieve UHV inside the chamber, the baking of complete vacuum chamber has to be carried out during pumping. For this purpose the C-channels are welded inside the vacuum vessel. During baking of vacuum vessel, these welded channels should be helium leak tight. Further, these U-channels will be in accessible under operational condition of SST-1. So, it will not possible to repair if any leak is developed during experiment. To avoid such circumstances, a dedicated high vacuum chamber is used for baking of the individual vacuum modules and sectors before assembly so that any fault during welding of the channels will be obtained and repaired. This paper represents the baking of vacuum vessel modules and sectors and their temperature distribution along the entire surface before assembly.

1. Introduction

SST-1 vacuum vessel [1, 3] is fabricated as a continuous torus structure using non-magnetic SS 304L material. For easy fabrication and assembly point of view, SST-1 is divided into sixteen (16) parts out of which eight (08) of them are vessel sectors (VS) while other eight are vessel modules (VM). Vessel Sector (VS) is comprised of one number of radial port (RP), two numbers of vertical ports (top and bottom each) and one number of vessel sector ring whereas vessel module (VM) is made up of one number of VS and two numbers of inter connecting rings (ICR) on both sides. The thickness of vacuum vessel is 10 mm while that of the ports is 6 mm. Figure 1 shows the schematic of VS and VM.

Since an ultra-high vacuum (UHV) is the basic requirement for plasma discharge inside any tokamak, the baking facility to bake the main vacuum vessel has to be developed. Baking effectively reduces the net gas load due to out-gassing of the materials. Also it reduces the surface impurities like water vapor, heavy hydrocarbons and other gases get absorbed / adsorbed into the bulk of parent materials during fabrication and commissioning. These high-Z impurities, if present, will lead to plasma contamination during the plasma discharge. Further during the cool down of cryostat components like toroidal filed (TF) coils and 80 K thermal shields, there is a probability of the vacuum vessel attending low temperature due to conduction and radiation leading to water condensation. In order to avoid this condition, the vacuum vessel has to be kept at elevated temperature during cryostat components cool down.

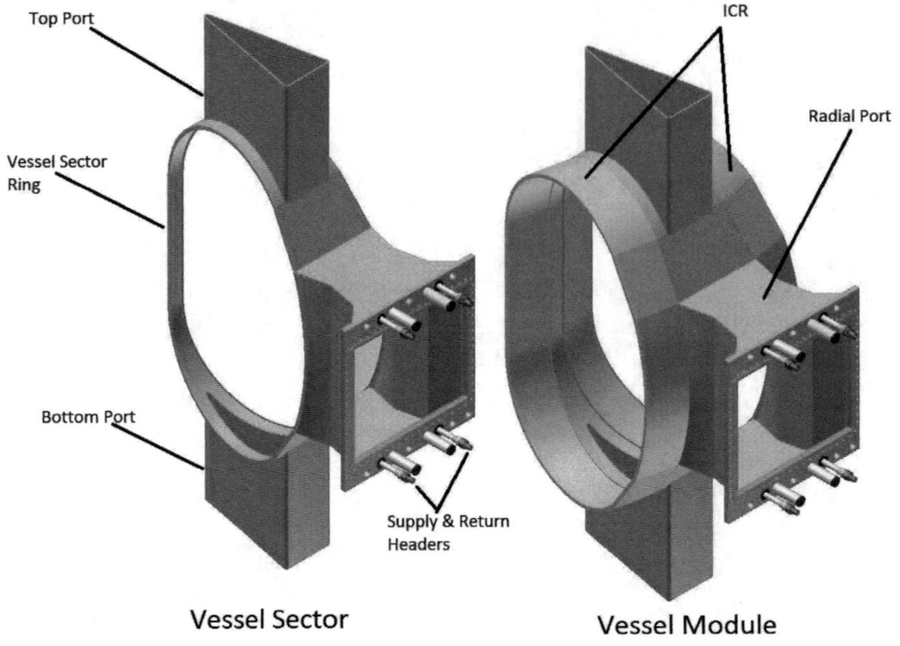

Figure 1. Schematics of SST-1 vessel sector (VS) and vessel module (VM).

2. SST-1 VS and VM baking

For steady state baking of vacuum vessel at 150 °C during bake out and to maintain the vacuum vessel at 50 °C during cool-down phase, U-shaped SS 304L channels of 2 mm thickness have been welded on the inner surfaces of the vessel wall. Figure 2 shows the baking paths of each vacuum vessel indicating the different numbers of bends, the details of which is given in the table 1. The internal baking layout as shown in the figure 2 is preferred due to constrains like space availability and the mounting of thermal shields over the outer surface of the vacuum vessel to protect superconducting TF coils from radiation. For such baking option, the gases like Oxygen, Nitrogen, Air or water can be used as a heating medium. Nitrogen gas [4] is preferred due to the following advantages over other gases like

- Non-corrosive and non-toxic nature.
- Non-combustible and non-supporting of combustion.
- Does not combine with other gases under normal conditions.

Table 1. Different bends of each vessel sector baking layout.

Paths	No of bends			
	Meter	90°	Interconnecting	Closed loop
Radial port	8	12	0	0
Top vertical port	19	7	0	0
Bottom port	19	8	0	0
ICR	2	0	2	2

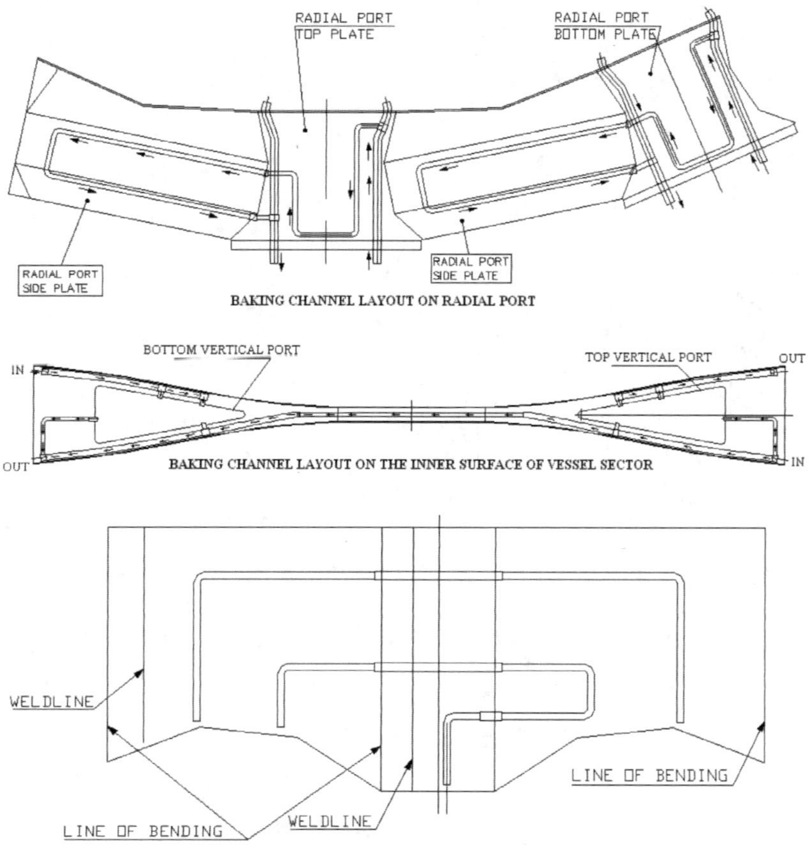

Figure 2. Baking lay-out of SST-1 vacuum vessel.

For the baking purposes, all sixteen-vessel sectors will be connected in parallel with the main header of hot nitrogen gas heating and supply system. Supply and return main headers are of four inch size while sub-headers are one inch size. All these supply and return lines are thermally insulated using perlite material to reduce the radiation heat loss. Each vessel sector / module has two numbers of diagonally opposite supply and return lines provided on radial port as shown in figure 1. Metal braided flexible stainless steel hoses along with indigenously developed demountable type bull nose coupler are used to connect this header to SST-1 tokamak.

This hot nitrogen gas heating and supply system comprises of series connected positive displacement blowers to boost the inlet gas pressure to required value. It has 150 kW capacity shell and tube type heater to heat this pressurized nitrogen gas. Schematic layout of the same is shown in figure 3. Nitrogen cylinders are connected in parallel to the cylinder manifold which contains pressure regulator, control valve and non-return valves. Dry nitrogen gas are fed to the suction chamber of higher capacity up to a pressure of 3.5 bar (a) through this manifold. During operation if the suction chamber pressure reduces below a set point, a low-pressure switch mounted on suction chamber will open the manifold inlet valve to compensate the pressure drop. A safety valve is also provided with the suction chamber to avoid any pressure rise above the set point. All the above functions are controlled via the control panel and PLC based system. Control panel consists of volt meter, ammeter, flow indicator and controller, temperature indicator and controller, pressure indicator, AC drive and ON / TRIP indicator lamps. All the necessary safeties for temperature, pressure and high current etc are provided in the system to ensure the safe functioning. These safety arrangements will take care of high and low pressures, high current and over loading, proper water cooling and flow rate etc.

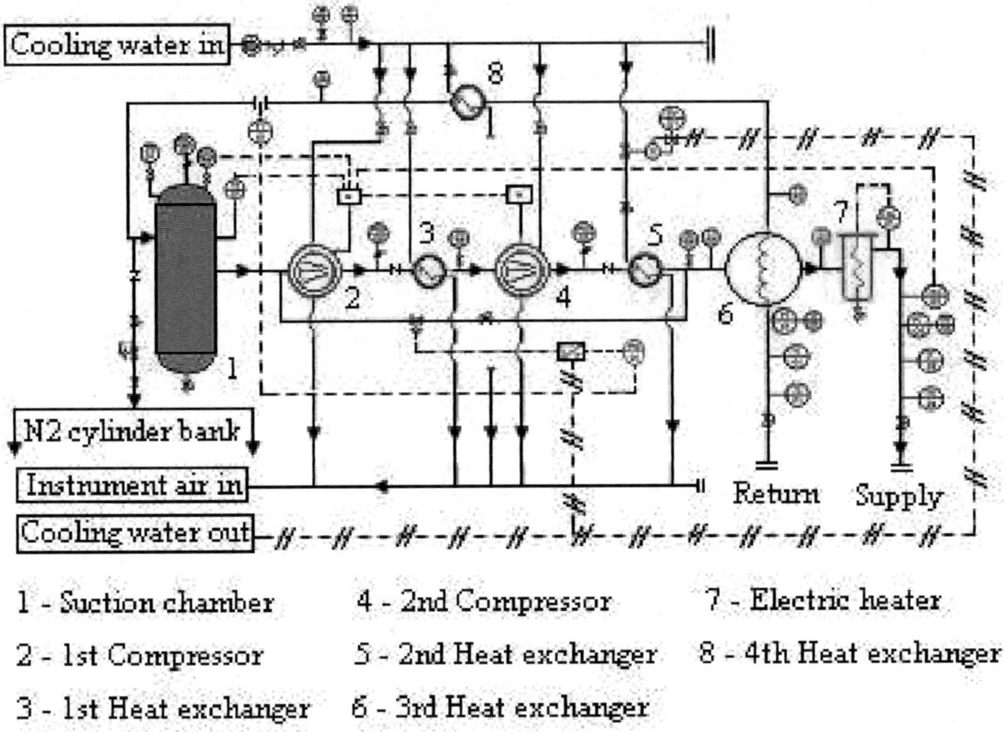

Figure 3. Schematic layout of Nitrogen Gas Heating and Supply System.

3. Experimental set-up and Results

A dedicated high-vacuum (HV) chamber of 2.4 meter diameter and 3.5 meter height is fabricated for baking these sixteen numbers of VS and VM. Since this chamber has very larger dimension, double O-ring configuration was utilized along with inter-space pumping in order to ensure the better leak tightness. Two numbers of 500 ISO-F flanges are provided on this chamber, out of which one is used for pumping purpose while other is used for human entry. Four numbers of 63 CF flanges are provided for connecting supply and return lines of VS and VM with hot nitrogen gas supply and return lines as shown in the figure 4.

Figure 4. Experimental set-up for VS and VM baking.

System was run via PLC program in auto mode with predefined set parameters. Various interlocks are incorporated into the PLC program to ensure maximum safety to plant operation. Before starting the system it checks for all interlock conditions like water temperature, water flow, reservoir tank pressure, etc. It starts the system cycle after pre-filling system reservoir tank with nitrogen gas up to 1.0 bar (g) and steady state flow condition was established. After confirming the normal system operation by running system for few minutes, the heater was switched ON in Warm UP mode. In this mode system passes 10-15 % (adjustable) of full current into the heater for 15-20 (adjustable) minutes. After completion of Warm Up mode, heater enters run mode. In this mode, current supply of heater is increased gradually depending on ramp rate. Normally ramp rate of 50 °C /Hr is used. Heater current is controlled based on feedback from supply gas temperature. K-type Thermocouples are installed at different locations in the system. PLC gets supply gas temperature from thermocouple and pass on it to PID controller which in turn controls 4-20mA output for AC drive. AC drive passes current based on 4-20 mA current input from PLC. Pressure transmitter is installed on reservoir tank to give pressure information to PLC. Based on this feedback, system controls the switching of solenoid valve to maintain the pressure within specified range. Orifice based mass flow meter is installed in return line for measuring the mass flow rate.

K-type Kepton insulated thermocouples are mounted at the different locations of VS / VM using high temperature Ceramabond glue from M/s. Aremco. Both these thermocouples and glue can withstand the temperature up to 500 °C. Also these thermocouples are calibrated using high temperature hot bath at 100 °C, 150 °C and 250 °C within the accuracy of ± 1 °C. Temperature signals from vacuum chamber are acquired through de-mountable type 36-pins feed-through using MASIBUS make DATA logger. Temperature data from this DATA logger is stored in PC using CITECT SCADA based software via RS-485 transfer protocol. All sixteen VM and VS are baked up to 150 ^{0}C for a flattop of eight hours. Cool down of VS / VM is completed by flowing room temperature nitrogen gas through them. Figure 5 shows the temperature profile at different locations of one of the vessel module.

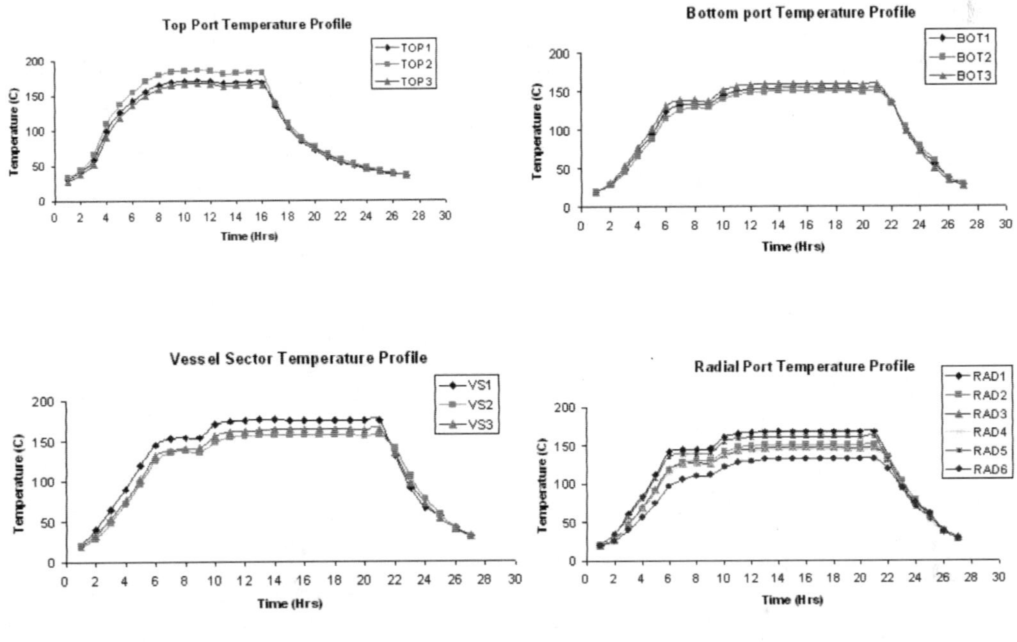

Figure 5. Temperature profile at different locations of one of the vessel module.

4. Conclusion

Nitrogen gas heating and supply system for SST-1 Tokamak is demonstrated successfully in its present status for baking of vessel modules and sectors. We are further planning to upgrade the system equipped with more reliable blower system so that desired mass flow rate of 1.1 kg/s at 4.5 bag (a) can be achieved in order to baking the entire SST-1 vacuum vessel. Further the baking channels layout inside the vacuum chamber is demonstrated for their performance as per design requirement.

References

[1] Despande S P and SST-1 Team 1997 *Proceedings of the 17th IEEE/NPSS Symposium on Fusion Engineering* (San Diego, CA) vol 1 p 227
[2] Saxena Y C and SST-1 Team 2000 *Nuclear Fusion* vol 40 issue 6 p 1069
[3] S. Pradhan & SST-1 Mission Team 2010 *J. Plasma Fusion Res. SERIES* vol 9 p 650
[4] Greenwood Norman N and Earnshaw Alan 1984 *Chemistry of the Elements* (Oxford: Pergamon Press. ISBN 0080220576)

Spinning rotor gauge based vacuum gauge calibration system at the Institute for Plasma Research (IPR)

Pratibha Semwal, Ziauddin Khan, Kalpesh R Dhanani, Firozkhan S Pathan, Siju George, Dilip C Raval, Prashant L Thankey, Yuvakiran Paravastu and Himabindu M

Institute for Plasma Research, Near Indira Bridge, Bhat, Gandhinagar – 382428, Gujarat, India

E-mail: pratibha.semwal@gmail.com

Abstract. The Steady-state Superconducting Tokamak (SST-1) is an indigenously built medium sized fusion device at IPR designed for plasma duration of 1000 seconds. It consists of two large vacuum chambers – Vacuum Vessel (16 m^3) and Cryostat (39 m^3) which will be pumped to UHV and HV pressures respectively using a set of turbo molecular pumps, Cryo-pumps and Roots pumps. The total as well as the partial pressure measurement in these chambers will be carried out using a set of Pirani gauges, Bayard Alpert type gauges, Capacitance manometers and Residual Gas Analyzers (RGA). A reliable and accurate pressure measurement is essential for successful operation of SST-1 machine. For this purpose a gauge calibration system is set up in SST-1 Vacuum laboratory based on Spinning Rotor Gauge which can measure absolute pressure in the range 1.0 mbar to 1.0×10^{-7} mbar. This system is designed to calibrate up to five gauges simultaneously for different gases in different operating pressure ranges of the gauges. This paper discusses the experimental set-up and the procedure adopted for the calibration of such vacuum gauges.

1. Introduction

Vacuum System is one of the main subsystems of SST-1tokamak [1, 2] and it consists of (i) Vacuum vessel, (ii) Cryostat (iii) Pumping system and (iv) Gas feed system. Vacuum vessel is the main chamber which will be pumped down to $< 1.0 \times 10^{-8}$ mbar while the cryostat is a high vacuum chamber with the base pressure $< 1.0 \times 10^{-5}$ mbar. Plasma discharge will be carried out in vacuum vessel while the cryostat encloses magnet system and radiation shields. The pumping system comprising of Roots pumps, Turbo-molecular pumps and Cryo-pumps will be used for evacuating these two main vacuum chambers. Total pressure measurement will be done using a set of Convectron gauges, Capacitance manometers (CM) and B. A. gauges while the partial pressure will be measured using Residual Gas Mass Analysers (RGA).Gas feed system will be used during plasma operation for the pre-filling of vessel at required gas pressure, density build up phase and during wall conditioning.

Vacuum gauges mounted on SST-1will play a passive role because the entire pumping system and vacuum DATA acquisition will be remotely operated using PXI based system which has both the software and hardware interlocks for any un-eventual failures. All the feedback control like pumps operation, gate vales interlock and will be based on the gauges reading. Also the gas feed and density control completely depends on the gauges reading. In either case, the accuracy in the pressure measurement is very important to achieve the desired results in SST-1 experiments. Hence all the pressure gauges must be calibrated in their respective measuring ranges before mounting into SST-1. Also the ageing effect and contamination of the gauges lead to a change in the gauge sensitivity, therefore periodic calibration of the gauges must be carried out with a reference gauge. For this purpose a Spinning rotor gauge (SRG) calibration system has been set-up.

2. Calibration Methods

Pressure measuring instruments are classified as the primary and the secondary gauges. Primary gauges measure pressure simply by the direct application of the definition of pressure as a force acting on a surface. On the other hand, the secondary gauges measure pressure indirectly by exploiting a gas property that depends on pressure. The secondary gauges therefore require to be calibrated against a primary standard. Additionally the sensitivity of the secondary gauge depends upon the type of gas and therefore they must be calibrated for a specific gas. Due to advancement in technologies, nowadays secondary gauges are available in a wider range from atmospheric pressure (1013 mbar) to less than 10^{-10} mbar. Still not a single gauge can measure the entire ranges with considerable accuracy and one has to use a combination of secondary gauges to cover the complete pressure range.

The calibration process must be carried out in the entire measurement range of the vacuum gauges since many times the gauges exhibit non-linearity in their response with pressure. Thus, ideally a primary gauge is required in the entire measurement range of the gauges for the calibration purpose. In case where primary gauge is not available, an appropriate vacuum must be created in the calibration chamber in a primary way using basic gas laws. A brief overview of the calibration processes in different pressure ranges is given below.

2.1.Direct comparison with a primary gauge

It is the comparison of a gauge reading with a primary gauge where both the gauges are mounted to a calibration chamber and are exposed to the gas at identical gas densities or pressures. The primary gauges available for this type of calibration are mainly the liquid column manometers. The pressure in a liquid column is given by the well-known formula

$$P = \rho\, g_1 \Delta h + p_r \tag{1}$$

where ρ is the density of the liquid used (kg m^{-3}),
g_l is the local acceleration due to gravity (m s^{-2}),
Δh is the level difference in the two columns of the U tube and
p_r is the residual pressure in the reference side.

McLeod gauge belongs to this category. Despite its wide use, McLeod gauge is now considered to be obsolete because of its complicated operation and the numerous uncertainty sources involved which cannot always be analysed exhaustively or reduced. The major limitations of the McLeod gauges and liquid manometers are the accurate measurement of the liquid level.

2.2. Static/dynamic expansion systems (below 10^{-3} mbar)

For pressures lower than 10^{-3} mbar, the primary gauges are not available. In such case, the gauge can be mounted on a special calibration chamber in which an accurately known pressure has been established by applying the gas laws. Such calibration systems are essentially of two types, i.e. (i) Static expansion systems and (ii) Dynamic expansion systems.

In case of static expansion systems, gas at known pressure high enough to be measured by a primary instrument is expanded from a volume into a larger one, both volume values being determined with primary methods. If the temperature and number density are constant, the pressure value is obtained by applying Boyle's law (PV = constant).

In the dynamic expansion method, the pure gas whose flow rate Q is to measure is caused to flow through a known conductance C separating the measurement and the pumping chambers. The pressure for such case is represented as

$$P = \frac{Q}{\left[C\left(1-\frac{1}{r}\right)\right]} \tag{2}$$

where r is the ratio between the pressure values in the calibration and pumping volumes.

This formula is applicable only if the temperature of the entire system is constant. The gas flow rate can be determined in different ways but the most widely used method is the constant pressure method [3-7].

2.3. High Accuracy Gauge

Another popular method of vacuum gauge calibration is to use a high accuracy gauge such as Spinning Rotor Gauge (SRG) which offers < 1% calibration accuracy over an extended range of 1.0 mbar to 1.0 × 10^{-7} mbar. The gauge calibration system described in this paper is based on this method.

3. The Spinning Rotor Gauge

The spinning rotor gauge is based on the principle of momentum transfer. It can measure the pressures in the range from of 1.0 mbar to 1.0 × 10^{-7} mbar. A small steel sphere (ϕ = 4.5 mm) is axially suspended in vacuum inside a metal tube by means of a permanent magnet as shown in the figure 1. The magnetic forces balance the weight of the sphere. A series of electromagnets stabilize the position of unstable equilibrium. The sphere is accelerated to a working frequency of 340 Hz by two pairs of guiding coils. The guiding field is then stopped and the rotational frequency of the sphere will vary depending upon the number of collisions of the residual gas molecules against the sphere surface.

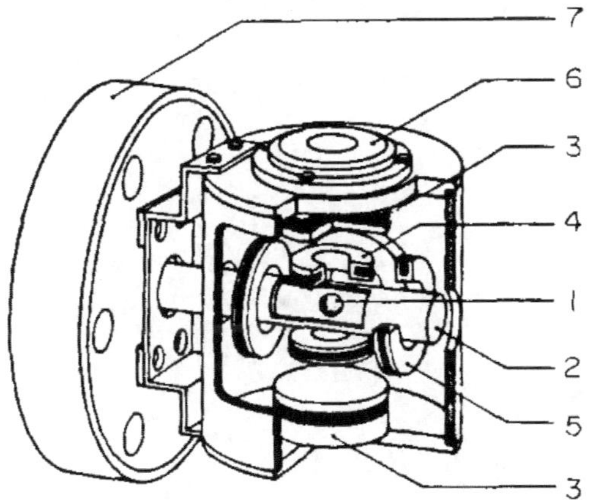

1. Steel sphere,
2. Evacuated tube,
3. Permanent magnets,
4. Vertical stabilization coils,
5. Driving coils,
6. Water level,
7. Flange

Figure 1. Schematic of spinning-rotor gauge.

The rotational frequency is measured using inductive coils and the rate of its decrement is then converted to pressure using the formula given as

$$P = \frac{\pi v^{*} a \rho \varpi}{10 w} = \left(\frac{\pi v^{*} a \rho}{10 \sigma}\right) \times \left(\frac{t_{n} - t_{n-1}}{t_{n} \times t_{n-1}}\right) \tag{3}$$

where $v^{*} = \sqrt{\frac{8RT}{\pi M}}$ is the mean molecular velocity,

M is the average molar mass of the gas,
a is the sphere diameter,
ρ is the density of the gas and
σ is the viscosity coefficient or the momentum transfer coefficient.

Thus the pressure is given by

$$P = \left(\frac{K}{\sigma}\right) \times \left(\frac{t_n - t_{n-1}}{t_n \times t_{n-1}}\right) - ofs \qquad (4)$$

with $K = \pi\rho a\sqrt{\frac{8RT}{\pi M}}$ which must be calculated for each type of gas and sphere at the operating temperature. The factor σ can only be measured which varies between 0 (for an ideal smooth surface) to 1.27 (for a very rough surface), depending on the mass of the gas used. The term 'ofs' refers to the residual magnetization [3]. Offset value of an SRG must be determined every time the gauge is switched on.

4. Experimental Set-up

A schematic of the calibration set-up is shown in the figure 2. The calibration chamber is fabricated using SS 304L material in a spherical form of diameter ϕ = 30 cm. Six (06) numbers of DN 63 CF ports are provided symmetrically around the circumference of this chamber with a dimensional accuracy of 0.01 mm. An MKS make SRG is connected to one of these ports while five different gauges under calibration can be mounted on the remaining ports. The pumping port is provided at the bottom of the system and is connected to a Varian make 500 l/s Sputter ion pump. The initial roughing of the system to a base pressure < 5.0×10^{-5} mbar is achieved using a 500 l/s Pfeiffer make portable turbo-molecular pump. Further pumping down to a pressure < 1.0×10^{-7} mbar is achieved using the ion pump with 5 hours of baking at 150 °C and long duration of pumping of about 8 hours. A Varian make sapphire gas feed valve is provided on the top of the chamber to maintain the desired pressure in the chamber with a specified gas with respect to which the secondary gauge has to be calibrated.

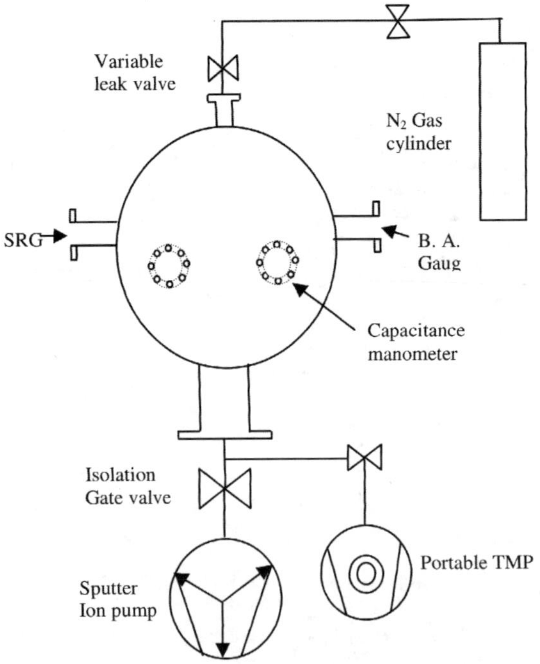

Figure 2. Schematic diagram of the calibration set-up.

5. Calibration Procedure and Results

The entire system is leak tested in vacuum mode using a Mass Spectrometer Leak Detector (MSLD) and is found to be leak tight below 1.0×10^{-9} mbar l/s. With net pumping speed of 100 l/s inside the chamber and considering the out-gassing rate of the electro-polished stainless steel material to be 1.0×10^{-10} mbar l/s/cm^2 [8], the system pressure is excepted below 1.0×10^{-7} mbar. After achieving the system pressure $< 1.0 \times 10^{-7}$ mbar, SRG offset value is measured and stored in the control unit. Nitrogen gas of 99.95 % purity is fed into the chamber using the leak valve and a desired pressure is maintained in the system.

RGA scans of the system for its background and with nitrogen gas feed are shown in the figure 3 and figure 4 respectively. It can be seen that the partial pressure of nitrogen (base peak at 28 amu, secondary peak at 14 amu) is at least one decade more than the remaining gas components. Once the required pressure is stabilised, the readings of the gauge and the SRG are noted and the calibration factor is determined. This process is carried out in the entire pressure range of the gauge to be calibrated and the calibration curve is plotted. The calibration factor is determined for different pressure ranges and the correction factor is incorporated directly into the Data acquisition program.

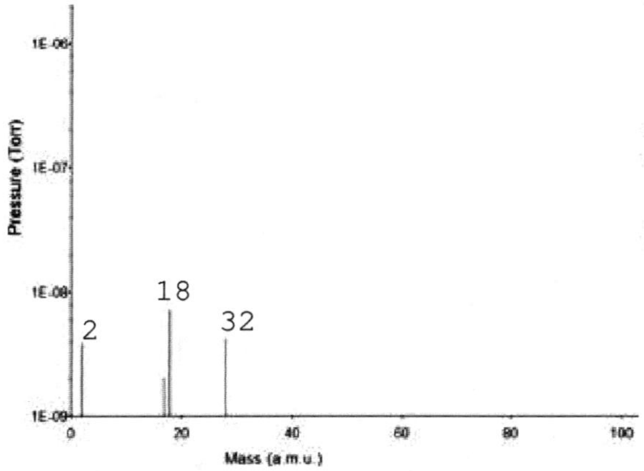

Figure 3. RGA scan after baking and 24 hours pumping.

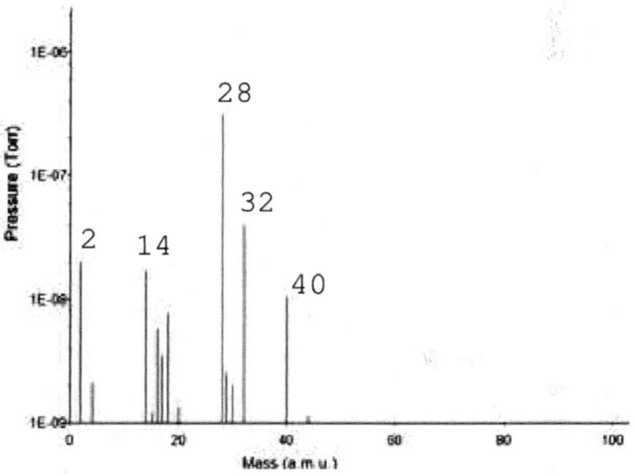

Figure 4. RGA scan with Nitrogen gas feed.

The calibration of a B.A. gauge with measuring range from 1.0×10^{-10} mbar to 1000 mbar and a Capacitance manometer with measuring range from 1.0×10^{-5} mbar to 0.1 mbar is carried out simultaneously. The calibration plot for these two gauges is shown in the figure 5. The capacitance manometer readings show a good match with the SRG reading in the range of 0.1 mbar to 1.0×10^{-4} mbar with percentage error less than 5 % whereas in the range from 1.0×10^{-4} mbar to 1.0×10^{-5} mbar this error increases up to 25 % . This could be due to the offset correction in the two gauges which becomes significant in this range as it is of the same order as the actual pressure value in this range.

In the case of B. A. gauge the error is within ± 10 % in the measurement range from 1.0×10^{-7} mbar to 1.0×10^{-3} mbar whereas in the range of 1.0×10^{-3} mbar to 0.1 mbar, this error increases up to 30 %. The reason for this could be due to the measurement principle of thermal conductivity of the gas where inaccuracy increases due to the various types of thermal losses such as the lead loss at the filament ends. The percentage error of both of these gauges at different pressure ranges is tabulated in the table 1.

Table 1. Percentage error of B. A. Gauge and Capacitance manometer (N_2 gas)

SRG Reading (mbar)	B.A. Gauge (mbar)	Capacitance Manometer (mbar)	Percentage error (B.A.Gauge) %	Percentage error (CM) %
1.3×10^{-6}	1.2×10^{-6}	-	7.7	-
2.2×10^{-5}	2.4×10^{-5}	1.7×10^{-5}	9.0	22.7
1.1×10^{-4}	1.0×10^{-4}	8.7×10^{-5}	9.0	20.9
1.0×10^{-3}	9.2×10^{-4}	1.0×10^{-3}	8.0	0
8.1×10^{-3}	7.8×10^{-3}	5.6×10^{-3}	30.9	3.7
7.9×10^{-2}	7.4×10^{-2}	5.3×10^{-2}	32.9	6.3

Again the calibration process is repeated for Hydrogen gas of 99.9995% purity. The calibration curve B. A. Gauge is shown in the figure 6. In normal condition the sensitivity of the B. A. Gauge is set for nitrogen gas. However, since the ionisation cross-section for Hydrogen gas is different from Nitrogen gas, an appropriate correction factor is applied to obtain the Hydrogen gas equivalent pressure.

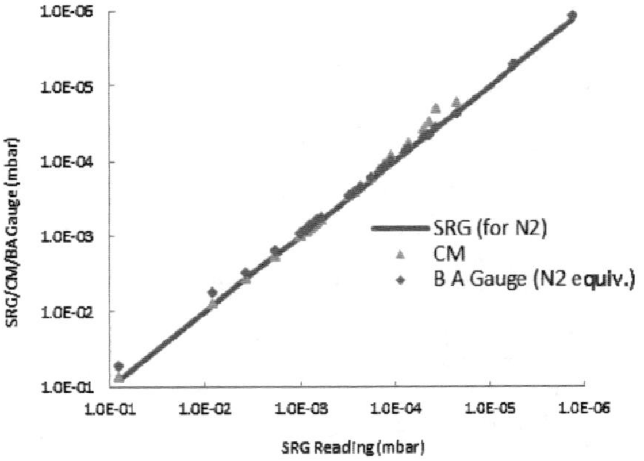

Figure 5. Calibration curve for CM and B.A. Gauge (for Nitrogen).

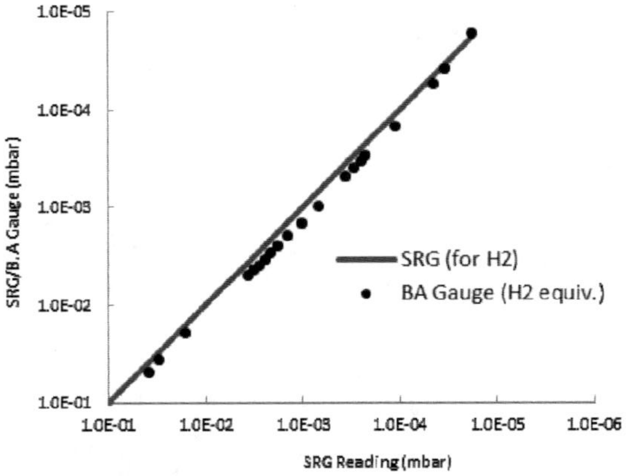

Figure 6. Calibration curve for CM and B A Gauge (for Hydrogen).

6. Conclusion

An SRG based vacuum gauge calibration system has been commissioned and is in the operation for the calibration of five numbers of different gauges at a time in the range of 1.0 mbar to 1.0×10^{-7} mbar for different gases.

References

[1] Saxena Y C and SST-1 Team 2000 *Nuclear Fusion* vol 40 issue 6 p 1069
[2] Pradhan S & SST-1 Mission Team 2010 *J. Plasma Fusion Res. SERIES* vol 9 p 650
[3] Calcatelli Anita 1989 *Vacuum* 39 p 799
[4] Steckelmacher W 1987 *Vacuum* 37 p 657
[5] Fitch R K 1987 *Vacuum* 37 p 637
[6] Mohan P 1998 *Vacuum* 51 p 69
[7] H F Dylla1982 *J. Vac. Sci. Technol.* 20
[8] Strausser, Yale, Review of out-gassing results, Varian Associates, Palo CA.

Study of Hydrogen Pumping through Condensed Argon in Cryogenic pump

K A Jadeja and S B Bhatt

Institute for Plasma Research, Nr. Indira Bridge, Bhat, Gandhinagar-382 428, Gujarat, India.

E-mail: kumarpal@ipr.res.in

Abstract. In ultra high vacuum (UHV) range, hydrogen is a dominant residual gas in vacuum chamber. Hydrogen, being light gas, pumping of hydrogen in this vacuum range is limited with widely used UHV pumps, viz. turbo molecular pump and cryogenic pump. Pre condensed argon layers in cryogenic pump create porous structure on the surface of the pump, which traps hydrogen gas at a temperature less than 20° K. Additional argon gas injection in the cryogenic pump, at lowest temperature, generates multiple layers of condensed argon as a porous frost with 10 to 100 A° diameters pores, which increase the pumping capacity of hydrogen gas. This pumping mechanism of hydrogen is more effective, to pump more hydrogen gas in UHV range applicable in accelerator, space simulation etc. and where hydrogen is used as fuel gas like tokamak. For this experiment, the cryogenic pump with a closed loop refrigerator using helium gas is used to produce the minimum cryogenic temperature as ~ 14° K. In this paper, effect of cryosorption of hydrogen is presented with different levels of argon gas and hydrogen gas in cryogenic pump chamber.

1. Introduction

In cryogenic technology, pumping through cryosorption is widely used, to achieve high pumping speed and capacity of non condensable gases (H_2, D_2, He, Ne). Cryosorption also provides the highly cleaned vacuum condition to achieve extreme vacuum. The adsorbents like active carbon (charcoal), zeolite molecular sieves are mostly used for cryosorption pumping in cryogenic pumps, i.e. Refrigerator cooled cryopump, Bath cryopump, Supercritical helium cooled cryopump. These adsorbent materials have large surface area to capture non condensable gases below 20°K temperature. But all cryogenic pumps are designed with limited adsorbent material for particular application. Thus these adsorbents have limited capacity to pump non condensable gases at saturation level. To improve this pumping limitation, A Cryosorption by condensed gases like Ar, CO_2, SF_6 below 20°K temperature also used as creative adsorbents to increase pumping speed and capacity of non condensable gases. Argon gas is more preferable for condensed gas adsorbent than other gases, CO_2, SF_6. As an adsorbent, Argon has effective properties for cryosorption, like high thermal conductivity; non reactive, inert gas etc. There are two types of gas filling of adsorbent gases in cryopump, as pre condensed (frost) and continuous condensed. In pre condensed [1], adsorbent gas is filled sufficiently in cryopump to create porous structure before actual pumping of main vacuum chamber. It is, useful to achieve extreme vacuum condition and option of saturated static adsorbents (activated charcoal, molecular sieves) to increase pumping capacity and speed. Gradually, the pumping efficiency is reduced with time in pre condensed technique due to filling of pores with fresh gas load. In continuous condensed [2], adsorbed gas is filled continuously in cryopump with actual pumping of the vacuum chamber to get effective pumping speed of non condensable gases. Continuous condensed of adsorbent gas is used to determine cryosorption pumping effect on non condensable gases and also provides continuous pumping of these gases with fresh frost.

ADITYA [3] is ohmically heated limiter based, moderate size Tokamak. The ADITYA vacuum vessel is pumped through three turbo molecular pumps and a refrigerator cooled cryopump. Hydrogen is the fuel gas to produce various discharges like high temperature plasma discharge, glow discharge cleaning (GDC) [4] etc. For wall conditioning, Glow discharge cleaning is operated with high pressure of hydrogen and for long duration for many hours. The resultant of GDC, hydrogen pressure is increased significantly in vacuum vessel, which perturbs the high temperature plasma discharge as recycling of hydrogen in hot plasma. Thus, the experimental study of cryosorption using argon adsorbent to increase pumping speed and capacity of hydrogen has been carried out to improve pumping efficiency of ADITYA vacuum vessel.

In this experiment, the refrigerated cooled cryopump has been used at cryogenic temperature 14°K of 2nd stage as a manufactured parameter.

2. Experimental set-up

The schematic view of this experimental setup is described in Figure 1. The refrigerated cooled cryopump of 250 torr.liter crossover evacuates the SS304 vacuum chamber to achieve ultra high vacuum. Before cryopump operation, rotary pump is used to evacuate the cryopump and vacuum chamber simultaneously to get crossover pressure. For total and partial pressure measurement, Ionization gauge and residual gas analyzer are used. Two gas leak valves are used to fill gases as argon and hydrogen.

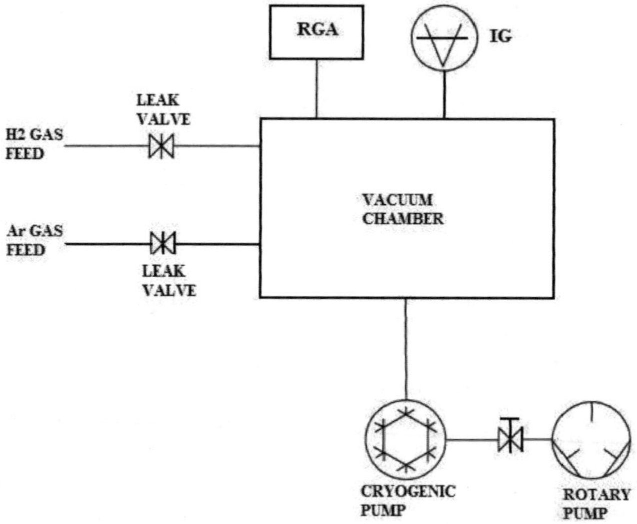

Fig. 1. Schematic view of the experiment setup to study of the hydrogen pumping through condensed argon in cryogenic pump.

Other important parameters of this cryopump are: (1) pumping speed (liters/sec): 9500 for water vapor, 4000 for nitrogen, 3500 for argon, 6500 for hydrogen (2) crossover: 250 Torr liters for N_2 (3) capacity: condensable gases greater than 2200 standard liters at 5 x 10^{-6} torr and hydrogen 30 standard liters at 5 x 10^{-6} torr (4) maximum argon throughput at 20°K: 11 torr.liter/sec (5) minimum temperature on 2nd stage: 14°K

3. Theoretical Basis

In cryogenic pumping technology, Effectiveness of cryo pumping is significantly dependent on sticking co-efficient (α), which is the ratio of number of particles stick on cryo surface to number of particles impinge on cryo surface. This ratio stays between zero (no pumping) to one (100 % pumping of impinging particles) for cryo surface. Sticking coefficient (α) derives from equation (1) [5] for commercially available refrigerator cooled cryopump with two components (baffle and cryo panel) connected in parallel.

$$1/\alpha = 1/c - 1/w + 1 \tag{1}$$

Here, *c* is capture coefficient, which characterize the cryo pump's pumping speed. Capture coefficient (*c*) is the ratio of actual pumping speed (S) to theoretical black hole pumping speed (S_{id}) and derives from equation (2) [6].

$$S = c.\ S_{id} = c.\ A\ \ [R_0T/2\pi M]^{1/2} \tag{2}$$

Thus, the pumping speed of cryo pump depends on geometry as inlet area A and types of gas being pumped as mass M. Here R_o is gas constant and T is temperature as 273.15ºK [6]. The pumping speed (S) is also measured using general formula of throughput in vacuum system.

$$S_{measured} = Q/P \tag{3}$$

Throughput or gas influx (Q) is adjusted using gas flow controller and measured pressure of particular gas or total pressure for temperature 293ºK. In equation (1), *w* is transmission probability, which reduces the ideal pumping speed of cryopump, due to *w* depends on baffle designed to restrict gas particles movement to cryo panel. In this experiment, baffle design data of cryopump is not available. Thus, transmission probability is taken by Ref [6] as 0.25 for hydrogen pumping of standard refrigerator cryopump. The hydrogen sticking coefficient (α_{H2}) on condensed argon can be determined for 14ºK temperature of refrigerator cooled cryopump using above equations and the measured pumping speed of hydrogen with different gas load of argon.

4. Experimental results and Discussions

The cryosorption effect of continuous condensed argon on hydrogen at temperature 14°K is shown in figure 2. Hydrogen is pre filled (before argon filling) in cryopump chamber with high quantity as more than 10 liters to achieve saturated condition of cryosorption of activated charcoal on 2nd stage of the cryopump. This affects hydrogen background level stable on 5 x 10^{-7} torr and this background pressure is enough to realize the pumping effect of continuous condensed argon. During continuously increasing gas feed of argon, when argon pressure is reached at 1 x 10^{-5} torr (point A in figure 2), the hydrogen pumping is started on enough argon frost structure. Gradually, the hydrogen pumping is increased with argon pressure increasing. When argon gas feed is stopped at point E, the hydrogen pressure is increased suddenly and stable on 1 x 10^{-7} torr, which is reduced in factor of 5 from the initial saturated pressure.

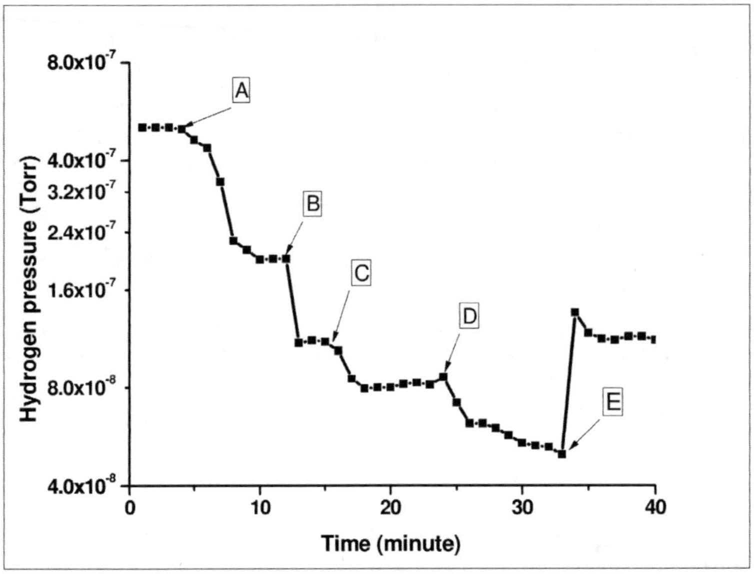

Fig. 2. Cryosorption effect of continuous condensed argon on saturated level of hydrogen, at 14°K temperature in cryopump. Various pressure levels of argon are shown as points A to E, where argon pressure at (A) 1 x 10^{-5} torr (B) 5 x 10^{-5} torr (C) 6 x 10^{-5} torr (D) 8 x 10^{-5} torr (E) Argon gas feed off.

The cryosorption effect of condensed argon on hydrogen during simultaneously filled argon and hydrogen in cryo chamber at temperature 14°K is shown in figure 3. In this experiment, continuous filled hydrogen throughput is steadied on pressure 1.3 x 10^{-7} torr. Gradually increasing of argon pressure reaches at 1.3 x 10^{-5} torr, steady filled hydrogen pressure starts to reduce.

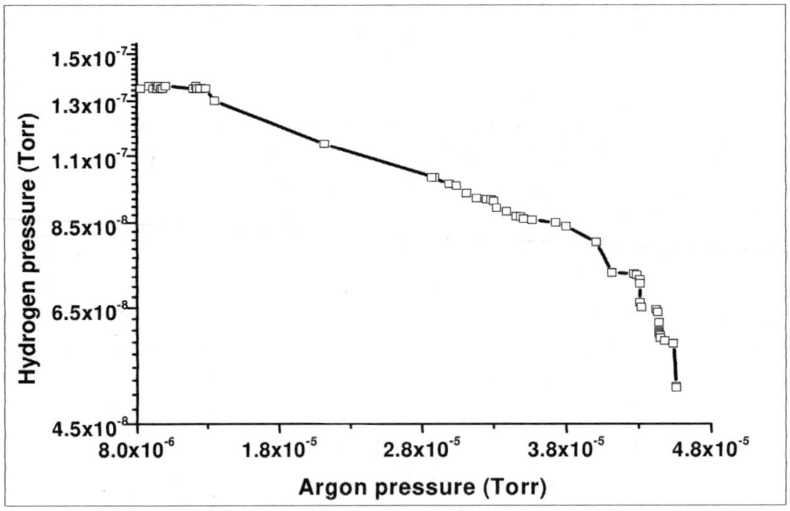

Fig. 3. Cryosorption effect of argon adsorbent, the continuous increasing of argon, which effect on continuous filled hydrogen at 1.3 x 10^{-7} torr with 14°K temperature

This effect indicates the cryosoption due to sufficiently filled argon creates condensed porous structure to trap hydrogen molecules. With argon pressure increasing in 10^{-5} torr range, hydrogen is reduced

Journal of Physics: Conference Series **390** (2012) 012028 doi:10.1088/1742-6596/390/1/012028

linearly up to pressure 5 x 10^{-8} torr. The argon filling is limited due to operation limit in 10^{-4} torr range of residual gas analyzer (RGA). Thus, we cannot distinguish limitation of hydrogen pumping with continuous increasing of argon at temperature 14°K.

The sticking coefficient is a universal parameter to measure cryo-pumping effect. Using experimental data of figure 3 and theoretical basis, the sticking coefficients of hydrogen (α_{H2}) are calculated for different levels of argon gas load on cryopump at temperature 14°K. In figure 4, the sticking coefficients of hydrogen (α_{H2}) are plotted with various argon gas loads in torr.liter/cm^2. The hydrogen sticking coefficient (α_{H2}) is calculated maximum as 0.34 on argon gas load 3.2 x 10^{-6} torr.liter/cm^2 and minimum as 0.14 on argon gas load 1.4 x 10^{-6} torr.liter/cm^2. The maximum value 0.34 of hydrogen sticking coefficient (α_{H2}) can be increased with argon gas load increasing. It is limited due to measurement limit of RGA in this experiment.

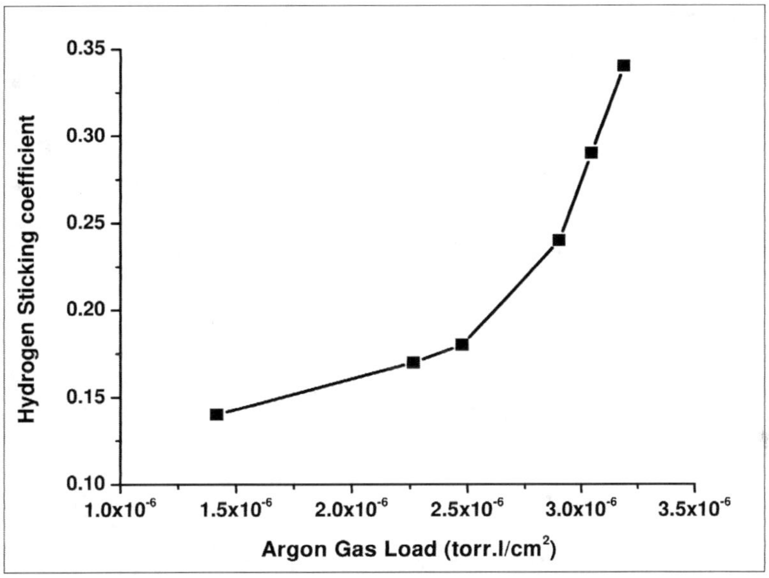

Fig. 4. The sticking coefficients of hydrogen (α_{H2}) with various argon gas loads in torr.liter/cm^2 at temperature 14°K through steady gas feed of hydrogen as 1.3 x 10^{-7} torr.

4. Conclusion

The initial result of continuous condensed argon adsorbent effect on hydrogen at temperature 14°K of refrigerator cooled cryopump has been concluded. The cryosorption effect of argon frost on hydrogen is started on argon gas feed pressure near 1 x 10^{-5} torr in this cryopump system. The sticking coefficient of hydrogen (α_{H2}) at temperature 14°K is determined to be 0.34 for continuous condensed argon on argon gas load 3.2 x 10^{-6} torr. liter/cm^2 and steady gas feed of hydrogen at pressure 1.3 x 10^{-7} torr. The final value of sticking coefficient of hydrogen (α_{H2}) at temperature 14°K in this cryopump can be determined with changing in experimental setup of measuring instruments. It is noticed that the second stage temperature of this cryopump is decreased up to 12°K from its standard manufactured value as 14°K after sufficiently filled argon (pre condensed effect). This effect is useful to increase pumping capacity and speed of non-condensable gases. These experiments more depend on cryo surface temperature, so experiments should be carried out at lower temperature of cryo pump. This initial study is motivated us to further experimental studies of pre condensed argon effect, increasing hydrogen pumping in actual vacuum vessel of ADITYA Tokamak.

Acknowledgement

The authors wish to thank, Mr. P.M. Chavda and Mr. T.P. Purabia for their help in many stages during this work.

References

[1] J. Kim, J. Vac. Sci. Technol. A 8 (1990) 3084
[2] J. Hangevoss, Trans. 3rd AVS vac. Symp., (Pergamon press, Newyork, 1966), Vol.I,p.51.
[3] S.B. Bhatt et al, *Ind. j. of Pr. and appl. Phys*. Vol. 27, Sept.-Oct. Pg. 710-742 (1989).
[4] H.A Pathak et al, *J. Nuclear Materials,* 220 -222 (1995) 708- 711
[5] Haefer RA. Cryopumping theory and practice. Oxford: Clarendon Press, 1989.
[6] C. Day, "Basics and applications of cryopumps" CERN Accelerator School Proceedings on Vacuum in Accelerators' 2006

High-vacuum compatibility tests of SST-1 superconducting magnets

Prashant L Thankey, Ziauddin Khan, Siju George, Firozkhan Pathan, Kalpesh R Dhanani, Yuvakiran Paravastu, Himabindu Manthena, Dilip C Raval and Subrata Pradhan

Institute for Plasma Research, Near Indira Bridge, Bhat, Gandhinagar – 382428, Gujarat, India

E-mail: plt@ipr.res.in

Abstract. SST-1 Tokamak is under commissioning at Institute for Plasma Research in mission mode. It comprises of a toroidal doughnut shaped plasma chamber, surrounded by liquid helium cooled superconducting magnets and LN_2 thermal shields, housed inside the cryostat chamber. The superconducting magnet system of SST-1 consists of toroidal field (TF) magnets and poloidal field (PF) magnets and will be operated at internal supercritical helium pressure of 4.5 bar (a) under very low temperature of 4.5 K and carrying a DC current of 10 kA. High-vacuum compatibility up to low-pressure $\leq 1 \times 10^{-5}$ mbar is one of the most essential features of these superconducting magnets in order to avoid the heat losses due to conduction and convection. This paper describes the extensive tests carried out under representative conditions to ensure the high-vacuum compatibility of the SST-1 magnets before assembly to the main system.

1. Introduction

Steady state Superconducting (SST-1) Tokamak [1, 2] is in refurbishment state at Institute for Plasma Research (IPR) for Double Null configuration plasma discharge. This double null configuration is provided by the magnet system. Magnet system is one of the main subsystems of SST-1 and it consists of sixteen (16) superconducting D-shaped Toroidal Field (TF) coils and nine (09) superconducting Poloidal Field (PF) coils [3] together with a pair (02) of resistive PF coils. The TF magnets are designed to generate 3.0 T field at the major radius of 1.1 m while the PF magnets are meant for the plasma equilibria with wide range of elongation and triangularity. The base conductor for these magnets is an NbTi based cable-in-conduit conductor (CICC) [4, 5]. The CICC consists of 135 NbTi/Cu strands of 0.86 mm diameter each, with a high copper to superconductor ratio of ≥ 4.9: 1. These strands are twisted in four stages before being jacketed inside a conduit made of stainless steel having a cross-section of 14.8 mm × 14.8 mm and a void fraction of 40 % inside the cable space for liquid helium to flow. The entire magnet systems are cooled with supercritical helium at 4 bar (a) and 4.5 K, which is fed at the high field region in the middle of each of the double pancake over a hydraulic path length of 47 m.

Each TF coil is made up of six (06) double pancakes, each pancake having nine turns. The complete double pancakes winding pack is shrunk fitted into a stainless steel (SS 316L) casing, which supports most of the electromagnetic loads. For cool down of casing, each TF coil is also fitted with 5 K bubble type thermal shield. Figure 1 shows a 3D model of a pair of TF coils along with some associated components and hydraulics. Due to such pancake configuration, six numbers of low resistance lap inter-pancake joints within and inter-coil joints between the coils are made for current transportation. These joints are fabricated by inserting unconduited cable of each double pancake into a hollow copper block and by filling silver solder (95Sn-05Ag) in the voids between inserted cable and copper termination block. Two adjacent such prepared cable terminations are joined by 0.1 mm thick soft solder (60Sn-40Pb) layer to form a joint. This coil joint is jacketed using especially joint boxes made of SS 304 L. The schematic of these joint boxes along with weld joint locations is shown in the figure 2.

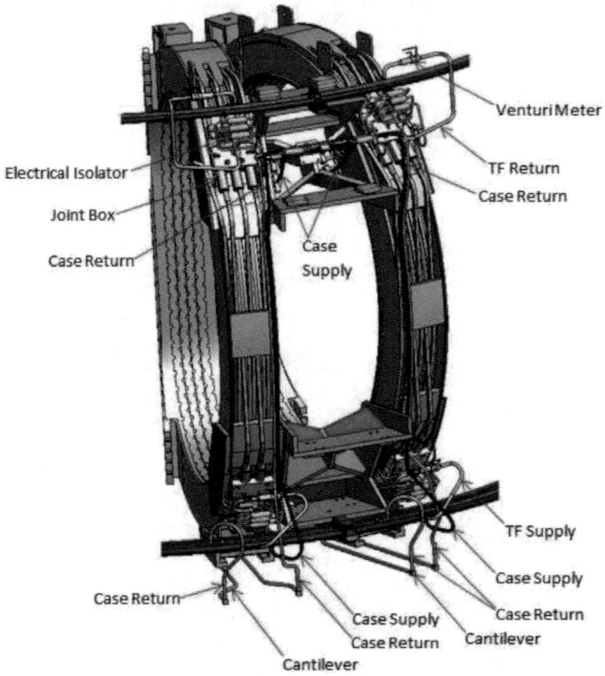

Figure 1. 3D model of a pair of TF coils along with some associated components and hydraulics

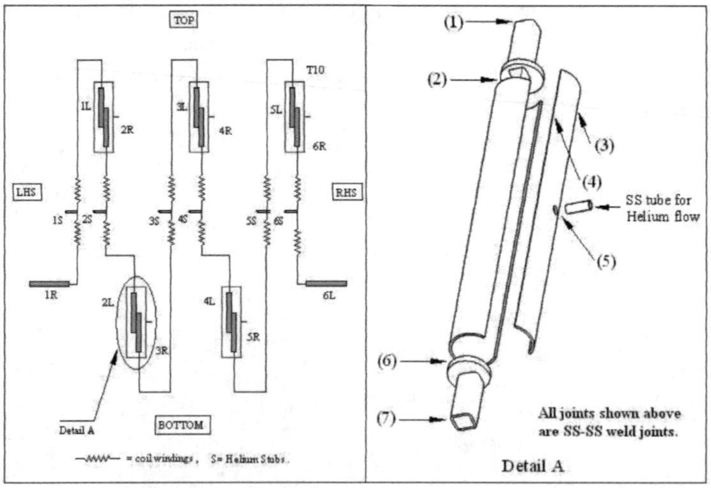

Figure 2. Schematic of the CICC joint boxes along with weld joint locations

The entire SST-1 TF coils system has such 96 numbers of such joint boxes enclosing the super-conducting transition joints. These transition joints with such a large numbers of SS to SS welding are forced flow cooled and are exposed to thermal stresses during cool down to 4.5 K and electromagnetic stresses during 10 kA current charging. Hence these weld joints have to be rigorously tested and qualified for the leak tightness both at RT and cold conditions before taken for assembly into the tokamak as there is almost zero accessibility after assembly.

Since these TF coils are connected electrically in series and each of them are fed separately for cool down, their inlets and outlets are electrically isolated from liquid helium supply and return headers using isolator manifolds. Each TF coil has 02 numbers of isolator manifolds containing 06 numbers of isolators each as shown figure 3. There are 32 pairs of isolator manifolds. Isolators used here are fabricated using specially grade G-10 materials, SS 304 L tube and epoxy. During cool down of TF coils and operation at 10 kA, each manifold assembly, figure 3 and joint boxes will be subjected to very high thermal as well as electrical stresses. Due to dissimilar materials presence in isolators, these are more prone towards the leak development. Here also these manifolds joints along with isolators have to be rigorously tested and qualified for the leak tightness both at RT and cold conditions before taken for assembly into TF coils.

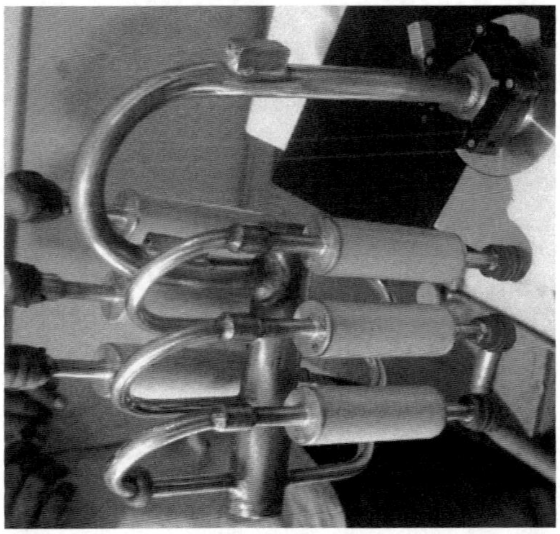

Figure 3. Isolator manifold consisting of 6 numbers of isolators

Since the vacuum < 1.0×10^{-5} mbar inside the cryostat is basic criterion in order to prevent any shot of discharge occurring during charging of TF coils at 10 kA apart from other concerns, so each and every weld joints has to be qualified for the leak tightness individually as well as at their operational conditions below certain defined values.

2. Experimental set-up and procedure

A dedicated ultra-high-vacuum (HV) chamber of 2.4 meter diameter and 3.5 meter height is fabricated from SS 304 L material for testing all 16 numbers of TF coils. Due to very larger dimension, double O-ring configuration is utilized along with inter-space pumping in order to ensure the better leak tightness. Since these TF coils are cool down to 4.5 K, copper thermal shields are used in between TF coil and test chamber inner surface. Also a proper fixture arrangement is done in order to sustain the forces due to charging of TF coils at 10 KA current.

For roughing, two numbers of rotary pumps are connected, each having a pumping speed of 120 m^3/hr. For high vacuum, one turbo-molecular pump of 5000 l/s pumping speed (N_2 equivalent) is connected to the chamber with effective pumping speed of 2300 l/s. Due to very high magnetic field produced during the charging, the pumping system is mounted at far location where the magnetic field value was < 5 mT as per pump specification in order to avoid the pump failure. To measure the vacuum inside the chamber, two numbers of BA gauges were mounted. One is mounted at the bottom port of the test chamber while other is mounted nearer to pump so that during charging of TF coil this gauge could be used for vacuum measurement. Also one MKS RGA is mounted at the bottom port of the chamber. The entire pumping system and vacuum DATA acquisition is remotely operated using PXI based system which has both the software and hardware interlocks for any un-eventual failures. The schematic of the test facility is shown in the figure 4.

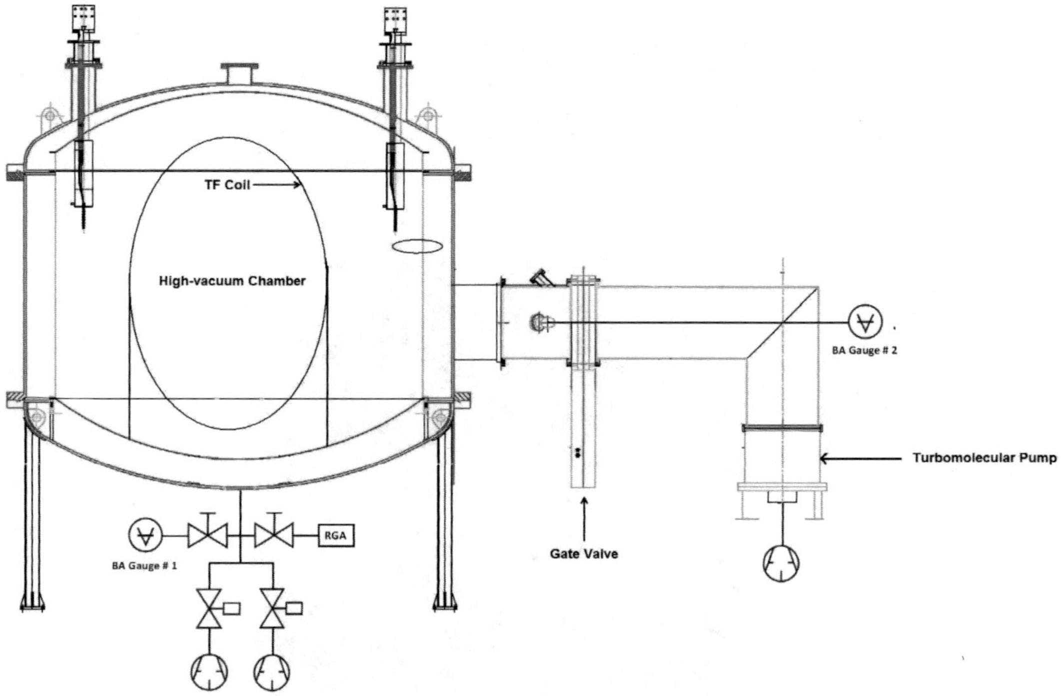

Figure 4. Schematic of the TF coil test facility

At the component level, 5 K panels, individual isolators and isolator manifolds are leak tested before and after thermal shock at the background < 5.0×10^{-9} mbar l/s for their initial qualifications. Then these components are mounted to the TF coils followed with other diagnostics. TF coil as an assembly then leak tested in the vacuum mode at the background < 5.0×10^{-9} mbar l/s and then leak tested in sniffer mode at the background < 1.0×10^{-6} mbar l/s with internal helium pressure of 12.5 bar (g). Then TF coil assembly was inserted into the test chamber as shown in the figure 5. Pumping of the test chamber was carried out and after achieving the vacuum < 1.0×10^{-4} mbar, cool down of TF coil assembly was started at the rate of 5-6 K/hr. The vacuum inside chamber during the entire cool down and during the charging of one of the TC coil is shown in the figure 6. During cool down, it is observed that as soon TF coil reaches below 273.15 K, water vapor gets condensed and at ≤ 63 K, a sudden vacuum improvement is observed indicating the cryo-condensation of nitrogen gas. Also the test chamber vacuum along with TF coil temperature status during the magnet charging is shown in the figure 7 where the magnetic field effects on the vacuum gauge is clearly seen.

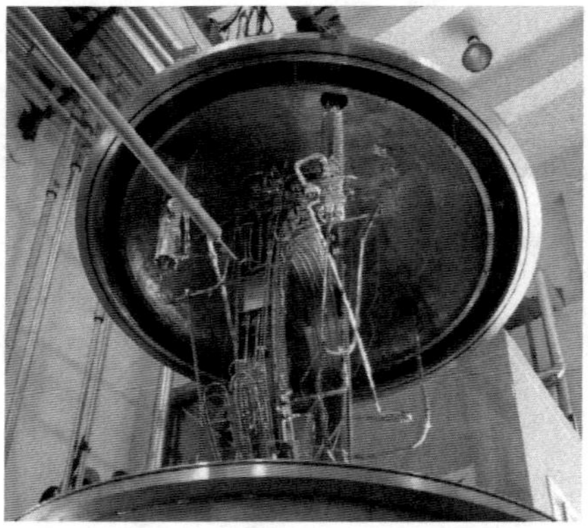

Figure 5. TF coil insertion into the experimental setup

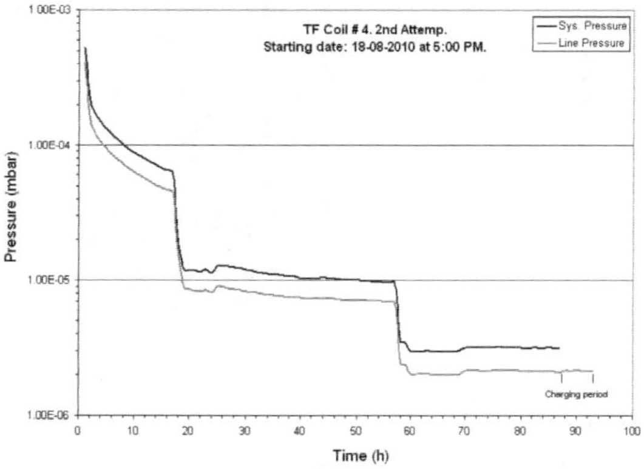

Figure 6. Pressure verses time curve during cool down and charging of TF coil

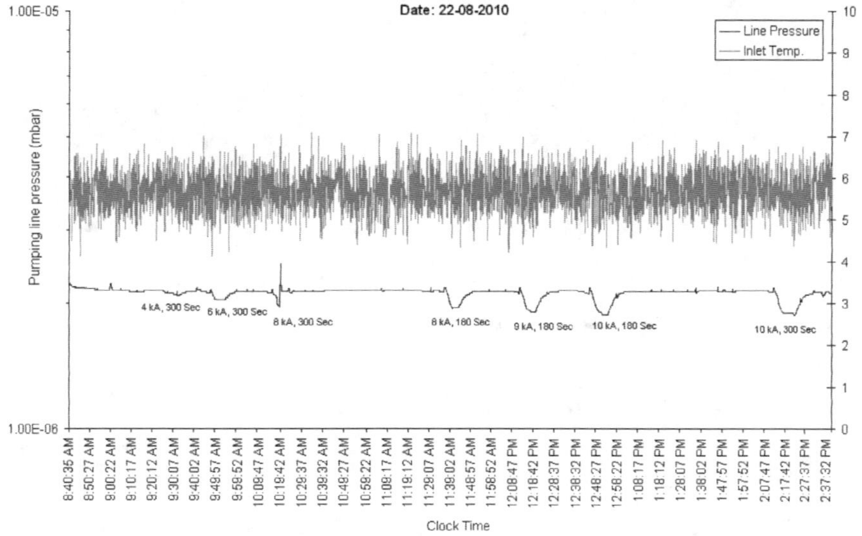

Figure 7. Test chamber vacuum during the charging of TF coil

3. Results

After achieving a pressure < 1.0×10^{-4} mbar in the chamber, RGA scan is taken to see the status of leak tightness of the TF coil by pressurizing it with helium at an internal pressure of 12.5 bar (g) at room temperature. Then, after the leak tightness is ensured, the cool down of the TF coil is started by passing liquid helium. RGA scans are taken time to time throughout the cool down up to 4.5 K, to ensure the leak tightness of the TF coil. The scans are also taken after the charging of the TF coil and during warm up of the TF coil. After warm up, the TF coil is tested for its leak integrity.

Figure 8 shows the RGA scans before and after the passing of current in a leak-tight TF coil, both showing negligible (~ 1.3×10^{-10} mbar) partial pressure of helium (yellow graph).

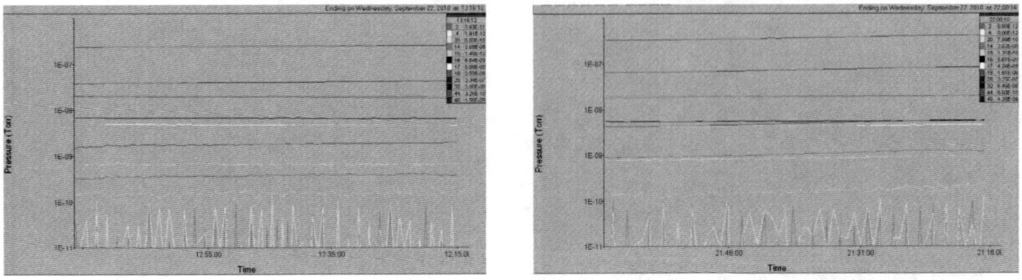

Figure 8. RGA scans before and after passing of current in a leak-tight TF Coil

Figure 9 shows the RGA scans before and after the passing of current in a TF coil in which a leak developed in the helium circuit during the passing of current through it. The scans show a rise in the partial pressure of helium (yellow graph) from ~ 1.0×10^{-11} mbar to up to ~ 5.0×10^{-9} mbar.

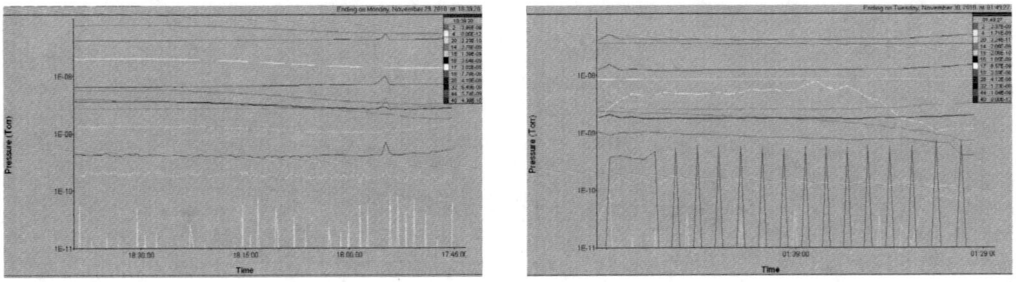

Figure 9. RGA scans before and after passing of current in a TF Coil in which a leak developed during passing of current

In this particular coil, the helium leak testing by sniffer method at room temperature and 12.5 bar (g) internal helium pressure revealed that a leak of 1.0×10^{-4} mbar l / sec had developed in the edge of the CICC of a winding pack.

4. Conclusion

The probability of leak development in the weld joints of TF coils and 5 K thermal shields are eliminated ensuring leak tightness of the system at RT and at their operating conditions. Also the helium isolators which are the most venerable to leaks in cold are ensured for leak tightness.

References

[1] Saxena Y C and SST-1 Team 2000 *Nuclear. Fus.* vol 40 1069
[2] Pradhan S. & SST-1 Mission Team 2010 *J. Plasma Fusion Res. SERIES* vol 9 p 650
[3] Pradhan S et al 1997 *Proc 17th IEEE/NPSS Symposium on Fusion Engg* (San Diego, CA) vol 1 p 665
[4] Pradhan S et al 1999 2nd *IAEA Technical Committee Meeting on Steady State Operation of Magnetic Fusion Devices* (Fukuoka, Japan) vol 2 p 482
[5] Anaskin O P et al 2002 *IEEE Trans Appl Superconductivity* vol 12 p 567

PXI Based Vacuum Control for Testing Various Components of SST-1

Kalpesh R Dhanani, Ziauddin Khan, Dilip C Raval, Prashant L Thankey, Firozkhan S Pathan, Siju George, Yuvakiran Paravastu, Pratibha Semwal and Himabindu Manthena

Institute for Plasma Research, Near Indira Bridge, Bhat, Gandhinagar – 382428, Gujarat, India

E-mail: kdhanani@ipr.res.in

Abstract. Vacuum system of steady-state Superconducting Tokamak (SST-1) has a very essential role during SST-1 plasma operation. For this purpose, its data acquisition and control system should be reliable and accurate. The PXI based faster real time data acquisition and control system were used for performing various operations like online data measurements, control, display, status indication in form of graphical visualization and storing the data for future analysis. We developed such PXI based vacuum control and implemented to our ongoing experimental set-up. This paper will describe the detailed information and guidance on PXI based platform for data acquisition and control used during the campaign.

1. Introduction

Vacuum chamber of SST-1 Tokamak [1, 2] comprises of two vessels. One is vacuum vessel and another one is cryostat. The vacuum vessel will be used for the plasma confinement while the cryostat will be used to provide the operational environment for superconducting coils and LN_2 thermal shields. The superconducting coils are cooled down to 4.5 K and are charged at 10 kA current for long time. These coils have to be protected from the heat radiation coming from baked vacuum vessel and room temperature cryostat surfaces by using LN_2 radiation shields all around them. Apart from this, the cryostat is maintained at a high vacuum $< 1.0 \times 10^{-5}$ mbar in order to prevent the plasma discharge inside the cryostat during operation. Also vacuum vessel is maintained at ultra-high vacuum $< 1.0 \times 10^{-8}$ mbar. Ultra-high vacuum as well as High vacuum can be maintained by applying the reliable pump operation, data acquisition and control system which is possible with proper functioning of vacuum pumps, gate valves and other related vacuum equipments. Hence the faster real time based data acquisition and control system is used for performing various operations such as online data measurements, controlling the various parameters, display, status indication in form of graphical visualization and data stores for future analysis. Such PXI based vacuum system is designed and implemented during the testing of various SST-1 components like TF coils, PF coils and 5 K thermal shields.

2. PXI based Data Acquisition and Control System

Data acquisition systems, as the name implies, are the products and/or processes used to collect information to analyze some phenomenon. The purpose of data acquisition and control is to measure an electrical or physical phenomenon such as voltage, current, temperature, pressure or vibration. PC-based data acquisition uses a combination of modular hardware, application software and a computer interface to take measurements. PXI stands for PCI extensions for instrumentation, which is a rugged PC-based platform that offers a high-performance, complete system solution for measurement and automation systems. It consists of a central host, remote units and a communications network followed with in-house developed software. It collects data from various sensors located at different places and then stored this data in the local server. At the same time, this data is also shared with main central control, which manages and distributes the data with other sub-systems. The schematic diagram of this PXI Based vacuum control with signal description is shown in the figure 1.

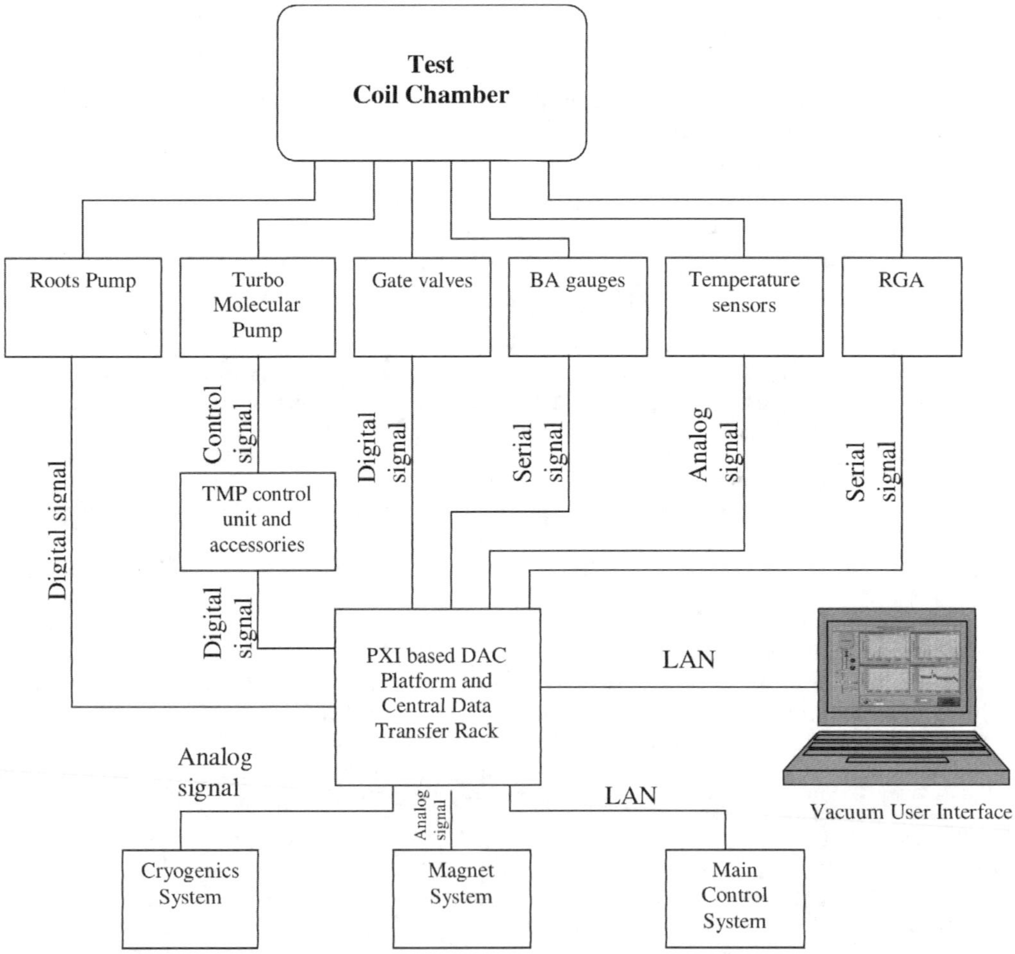

Figure 1. The Schematic diagram of PXI Based vacuum control signal description.

The schematic diagram representing the major components of the vacuum data acquisition and control system is shown in the figure 2.

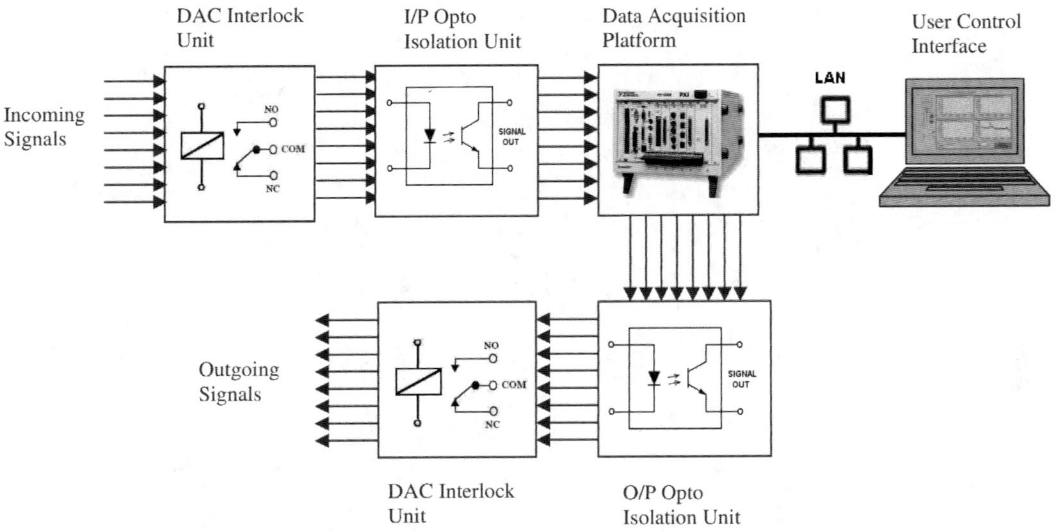

Figure 2. Major Components of vacuum data acquisition and control system.

3. System Description

The test facility is developed for testing the various SST-1 components as described earlier. This facility is a large ultra-high vacuum (UHV) chamber having a turbo-molecular pump (TMP), two numbers of Roots pumps, three numbers of electro-pneumatic gate valves, temperature sensors, B-A gauges and residual a gas analyzer (RGA). Using PC based user interface, the complete pumping system is controlled. Also a provision is adapted for the switch over to manual mode in case emergency. TMP has its own hardwired safety interlocks like speed, vibration and temperature parameters to ensure safe operation. Further TMP can also be manually switched ON/OFF from its respective control unit whenever it is required.

The test facility is pumped initially with two numbers of Root pumps. After achieving the pressure inside the chamber $< 1.0 \times 10^{-3}$ mbar, Root pumps are isolated by closing their respective gate valves. Once the isolation of Roots pumps is ensured, TMP is used for further pumping by opening its gate valve. Vacuum inside the chamber is monitored continuously and in the event of increase in pressure above the critical limit, the gate valve of TMP will be closed and Root pumps will be switched ON for the rough pumping by opening their respective gate valves to maintain the chamber vacuum. Graphics User Interface of the application software developed in LabVIEW is shown in the figure 3. It shows the status of TMP ON / OFF condition, TMP speed, Roots pumps ON / OFF condition and their respective gate valves OPEN / CLOSE position etc in graphical mode. The coil inlet temperature and LN_2 temperature can be seen here in the graphical mode. Also the vacuum status of the chamber and the pumping line can be seen.

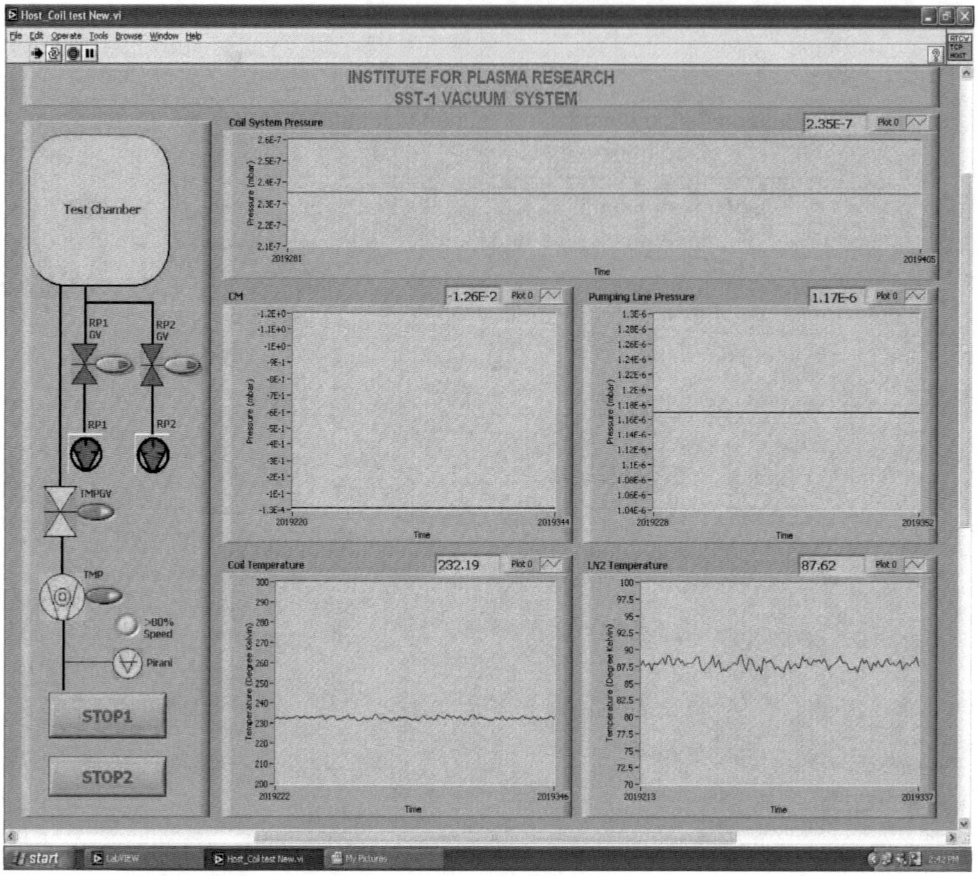

Figure 3. The schematic of Graphical user interface and vacuum status of the test facility.

Electro-pneumatic gate valve with its hardwired safety interlocks to ensure minimum required conditions for opening/closing is mounted over TMP for its safe operation. For pressure measurement, monitoring and operation, the number of UHV gauges are mounted into the vacuum chamber and pumping line. Status of UHV gate valves (open/close) is transmitted to vacuum measurement system. A large number of data is generated from different pressure measuring equipments, controller of the pump, valves, temperature sensor connected to the TF coil and RGA during the operation. Data varies in terms of its nature and is used to monitor the functionality of the different equipments. Also it is used to diagnose the reason for malfunctioning of any unit. All the parameters used for the operation and monitoring will be stored in the hard disk for post analysis purpose. Figure 4 shows the vacuum status and control parameters stored in Excel format. Excel format is chosen because of its wide spread availability and user friendly operating environment for analysis and plotting. Also one can easily access the data for a give time and date whenever required. Data acquisition of the vacuum system should be capable to carry out all these functions which can be possible by using PXI bus based Hardware. In front end electronics isolation cards are used to isolate the PXI system from external disturbances like high voltage spikes.

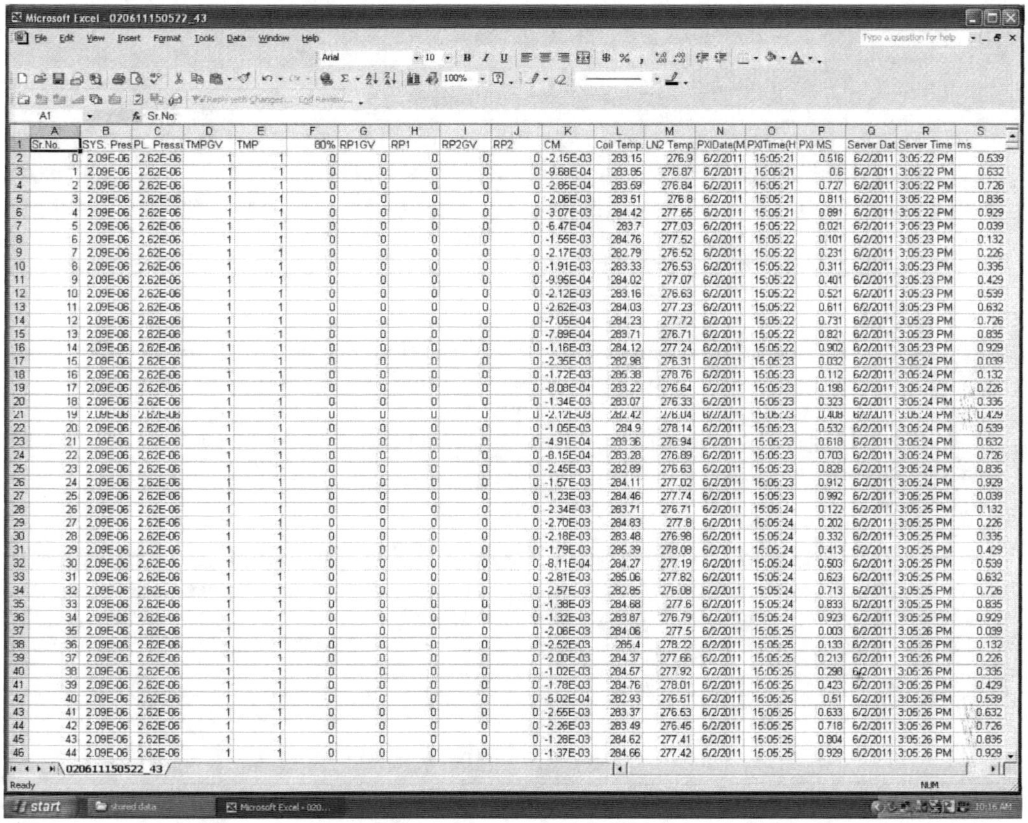

	A	B	C	D	E	F	G	H	I	J	K	L	M	N	O	P	Q	R	S
1	Sr.No.	SYS. Pres	PL. Pressu	TMPGV	TMP	80%	RP1GV	RP1	RP2GV	RP2	CM	Coil Temp.	LN2 Temp.	PXIDate(M	PXITime(H	PXI MS	Server Dat	Server Time	ms
2	0	2.09E-06	2.62E-06	1	1	0	0	0	0	0	-2.15E-03	283.15	276.9	6/2/2011	15:05:21	0.516	6/2/2011	3:05:22 PM	0.639
3	1	2.09E-06	2.62E-06	1	1	0	0	0	0	0	-9.68E-04	283.85	276.87	6/2/2011	15:05:21	0.6	6/2/2011	3:05:22 PM	0.632
4	2	2.09E-06	2.62E-06	1	1	0	0	0	0	0	-2.85E-04	283.69	276.84	6/2/2011	15:05:21	0.727	6/2/2011	3:05:22 PM	0.726
5	3	2.09E-06	2.62E-06	1	1	0	0	0	0	0	-2.06E-03	283.51	276.8	6/2/2011	15:05:21	0.811	6/2/2011	3:05:22 PM	0.835
6	4	2.09E-06	2.62E-06	1	1	0	0	0	0	0	-3.07E-03	284.42	277.65	6/2/2011	15:05:21	0.891	6/2/2011	3:05:22 PM	0.929
7	5	2.09E-06	2.62E-06	1	1	0	0	0	0	0	-6.47E-04	283.7	277.03	6/2/2011	15:05:22	0.021	6/2/2011	3:05:23 PM	0.039
8	6	2.09E-06	2.62E-06	1	1	0	0	0	0	0	-1.55E-03	284.76	277.52	6/2/2011	15:05:22	0.101	6/2/2011	3:05:23 PM	0.132
9	7	2.09E-06	2.62E-06	1	1	0	0	0	0	0	-2.17E-03	282.79	276.52	6/2/2011	15:05:22	0.231	6/2/2011	3:05:23 PM	0.226
10	8	2.09E-06	2.62E-06	1	1	0	0	0	0	0	-1.91E-03	283.33	276.53	6/2/2011	15:05:22	0.311	6/2/2011	3:05:23 PM	0.335
11	9	2.09E-06	2.62E-06	1	1	0	0	0	0	0	-9.95E-04	284.02	277.07	6/2/2011	15:05:22	0.401	6/2/2011	3:05:23 PM	0.429
12	10	2.09E-06	2.62E-06	1	1	0	0	0	0	0	-2.12E-03	283.16	276.63	6/2/2011	15:05:22	0.521	6/2/2011	3:05:23 PM	0.539
13	11	2.09E-06	2.62E-06	1	1	0	0	0	0	0	-2.62E-03	284.03	277.23	6/2/2011	15:05:22	0.611	6/2/2011	3:05:23 PM	0.632
14	12	2.09E-06	2.62E-06	1	1	0	0	0	0	0	-7.05E-04	284.23	277.72	6/2/2011	15:05:22	0.731	6/2/2011	3:05:23 PM	0.726
15	13	2.09E-06	2.62E-06	1	1	0	0	0	0	0	-7.89E-04	283.71	276.71	6/2/2011	15:05:22	0.821	6/2/2011	3:05:23 PM	0.835
16	14	2.09E-06	2.62E-06	1	1	0	0	0	0	0	-1.16E-03	284.12	277.24	6/2/2011	15:05:22	0.902	6/2/2011	3:05:23 PM	0.929
17	15	2.09E-06	2.62E-06	1	1	0	0	0	0	0	-2.35E-03	282.98	276.31	6/2/2011	15:05:23	0.032	6/2/2011	3:05:24 PM	0.039
18	16	2.09E-06	2.62E-06	1	1	0	0	0	0	0	-1.72E-03	285.38	278.76	6/2/2011	15:05:23	0.112	6/2/2011	3:05:24 PM	0.132
19	17	2.09E-06	2.62E-06	1	1	0	0	0	0	0	-8.08E-04	283.22	276.64	6/2/2011	15:05:23	0.198	6/2/2011	3:05:24 PM	0.226
20	18	2.09E-06	2.62E-06	1	1	0	0	0	0	0	-1.34E-03	283.07	276.33	6/2/2011	15:05:23	0.323	6/2/2011	3:05:24 PM	0.336
21	19	2.09E-06	2.62E-06	1	1	0	0	0	0	0	-2.12E-03	282.42	276.04	6/2/2011	15:05:23	0.408	6/2/2011	3:05:24 PM	0.429
22	20	2.09E-06	2.62E-06	1	1	0	0	0	0	0	-1.05E-03	284.9	278.14	6/2/2011	15:05:23	0.532	6/2/2011	3:05:24 PM	0.539
23	21	2.09E-06	2.62E-06	1	1	0	0	0	0	0	-4.91E-04	283.36	276.94	6/2/2011	15:05:23	0.618	6/2/2011	3:05:24 PM	0.632
24	22	2.09E-06	2.62E-06	1	1	0	0	0	0	0	-8.15E-04	283.28	276.89	6/2/2011	15:05:23	0.703	6/2/2011	3:05:24 PM	0.726
25	23	2.09E-06	2.62E-06	1	1	0	0	0	0	0	-2.45E-03	282.89	276.63	6/2/2011	15:05:23	0.828	6/2/2011	3:05:24 PM	0.835
26	24	2.09E-06	2.62E-06	1	1	0	0	0	0	0	-1.57E-03	284.11	277.02	6/2/2011	15:05:23	0.912	6/2/2011	3:05:24 PM	0.929
27	25	2.09E-06	2.62E-06	1	1	0	0	0	0	0	-1.23E-03	284.46	277.74	6/2/2011	15:05:23	0.992	6/2/2011	3:05:25 PM	0.039
28	26	2.09E-06	2.62E-06	1	1	0	0	0	0	0	-2.34E-03	283.71	276.71	6/2/2011	15:05:24	0.122	6/2/2011	3:05:25 PM	0.132
29	27	2.09E-06	2.62E-06	1	1	0	0	0	0	0	-2.70E-03	284.83	277.8	6/2/2011	15:05:24	0.202	6/2/2011	3:05:25 PM	0.226
30	28	2.09E-06	2.62E-06	1	1	0	0	0	0	0	-2.18E-03	283.48	276.98	6/2/2011	15:05:24	0.332	6/2/2011	3:05:25 PM	0.335
31	29	2.09E-06	2.62E-06	1	1	0	0	0	0	0	-1.79E-03	285.39	278.08	6/2/2011	15:05:24	0.413	6/2/2011	3:05:25 PM	0.429
32	30	2.09E-06	2.62E-06	1	1	0	0	0	0	0	-8.11E-04	284.27	277.19	6/2/2011	15:05:24	0.503	6/2/2011	3:05:25 PM	0.539
33	31	2.09E-06	2.62E-06	1	1	0	0	0	0	0	-2.81E-03	285.06	277.82	6/2/2011	15:05:24	0.623	6/2/2011	3:05:25 PM	0.632
34	32	2.09E-06	2.62E-06	1	1	0	0	0	0	0	-2.57E-03	282.85	276.08	6/2/2011	15:05:24	0.713	6/2/2011	3:05:25 PM	0.726
35	33	2.09E-06	2.62E-06	1	1	0	0	0	0	0	-1.38E-03	284.68	277.6	6/2/2011	15:05:24	0.833	6/2/2011	3:05:25 PM	0.835
36	34	2.09E-06	2.62E-06	1	1	0	0	0	0	0	-1.32E-03	283.87	276.79	6/2/2011	15:05:24	0.923	6/2/2011	3:05:25 PM	0.929
37	35	2.09E-06	2.62E-06	1	1	0	0	0	0	0	-2.06E-03	284.06	277.5	6/2/2011	15:05:25	0.003	6/2/2011	3:05:26 PM	0.039
38	36	2.09E-06	2.62E-06	1	1	0	0	0	0	0	-2.52E-03	285.4	278.22	6/2/2011	15:05:25	0.133	6/2/2011	3:05:26 PM	0.132
39	37	2.09E-06	2.62E-06	1	1	0	0	0	0	0	-2.00E-03	284.37	277.66	6/2/2011	15:05:25	0.213	6/2/2011	3:05:26 PM	0.226
40	38	2.09E-06	2.62E-06	1	1	0	0	0	0	0	-1.02E-03	284.57	277.92	6/2/2011	15:05:25	0.298	6/2/2011	3:05:26 PM	0.335
41	39	2.09E-06	2.62E-06	1	1	0	0	0	0	0	-1.78E-03	284.76	278.01	6/2/2011	15:05:25	0.423	6/2/2011	3:05:26 PM	0.429
42	40	2.09E-06	2.62E-06	1	1	0	0	0	0	0	-5.02E-04	282.93	276.51	6/2/2011	15:05:25	0.51	6/2/2011	3:05:26 PM	0.539
43	41	2.09E-06	2.62E-06	1	1	0	0	0	0	0	-2.65E-03	283.37	276.53	6/2/2011	15:05:25	0.633	6/2/2011	3:05:26 PM	0.632
44	42	2.09E-06	2.62E-06	1	1	0	0	0	0	0	-2.26E-03	283.49	276.45	6/2/2011	15:05:25	0.718	6/2/2011	3:05:26 PM	0.726
45	43	2.09E-06	2.62E-06	1	1	0	0	0	0	0	-1.28E-03	284.62	277.41	6/2/2011	15:05:25	0.804	6/2/2011	3:05:26 PM	0.835
46	44	2.09E-06	2.62E-06	1	1	0	0	0	0	0	-1.37E-03	284.66	277.42	6/2/2011	15:05:25	0.929	6/2/2011	3:05:26 PM	0.929

Figure 4. Vacuum status and control parameters stored in Excel format.

To ensure uninterrupted operation and measurement, the system is provided with emergency power. Also a battery back up is provided for smooth transition from main power to emergency power incase of interruption. The PXI system [3] is useful to control different components of vacuum system by software programming using interlock logic. Apart from software interlocks, hardware interlocks are also implemented in vacuum system for the safety of crucial vacuum system components like vacuum pump, gauges etc. The PXI system is useful to control different components of vacuum system by software programming using interlock logic. Apart from software interlocks, hardware interlocks are also implemented in vacuum system for the safety of crucial vacuum system components like vacuum pump, gauges etc.

For implementation of hardware interlocks Relay box has been developed. Relay box detects conditions like vacuum pump ON/OFF, speed of pump and accordingly operates the Gate valves. The digital output lines from PXI system are connected to the relay box so that gate valves can be operated remotely by using software. PXI System can be used in both Window and Remote RT modes. In Window mode, PXI controller is loaded with Windows-2000 operating system while in Remote RT mode, PXI system runs on Real time operating system LabVIEW RT. National Instruments LabVIEW with the LabVIEW Real-Time Module and RT Series hardware delivers deterministic, real-time performance for data acquisition and control systems. Window mode can provide a general platform for developing and running any non-critical measurements and control applications. It is not an ideal platform for running the applications that require precise timing or extended up time. It never follows the program priorities strictly. Also it takes a variable amount of time to respond to a given interruption. While RT mode can be used to run the applications with very precise timing and high degree of reliabilities. The priority of the tasks can be defined using this RT mode. RT mode places the high priority tasks in time-critical codes while the low priority tasks such as storage, network communication etc are placed to a less important codes. Also real-time operating systems guarantee that all interrupts will be performed within a given time frame. Using LabVIEW graphical programming, LabVIEW Real-Time embedded control application is developed and executed the program to run on an independent hardware target. LabVIEW Real-Time targets include an embedded processor running on RTOS, onboard memory and local storage. Currently we have tested all the functions from PXI System in RT mode.

4. Conclusion

PXI based vacuum control for Remote RT modes for remote control of a single TMP, Roots pumps and their respective gate valves for opening and closing is successfully developed and demonstrated during SST-1 components testing. This work will be very useful for SST-1 pumping system automation. The above-developed system is scalable and it can be expanded and integrated for future needs for automation.

5. Future works

In SST-1 Tokamak, twenty one numbers of TMPs, three numbers of Roots pumps, two numbers of Cryo-pumps will be used for pumping purpose. Also a large number of Piezo-electric valves will be used for gas feed. For such purposes, more numbers of channels for these large numbers of TMP ON/OFF, Root pump ON/OFF, respective gate valves OPEN / CLOSE etc will be incorporated for smooth and remote operation of SST-1 vacuum system.

References

[1] Saxena Y C and SST-1 Team 2000 *Nuclear Fusion* vol 40 issue 6 p 1069
[2] Pradhan S & SST-1 Mission Team 2010 *J. Plasma Fusion Res. SERIES* vol 9 p 650
[3] National Instruments Corporation 2003 *Development Guidelines* (LabVIEW7 Express, USA)

Journal of Physics: Conference Series **390** (2012) 012031 doi:10.1088/1742-6596/390/1/012031

Preparation of magnetron sputtered ZrO_2 films on Si for gate dielectric application

P. Kondaiah[1] *, G. Mohan Rao[2] and S. Uthanna[1]

[1]Department of Physics, Sri Venkateswara University, Tirupati – 517 502, India

[2]Department of Instrumentation and Applied Physics, Indian Institute of Science, Bangalore – 560 012, India

* E-mail: paruchurikondaiah@gmail.com

Abstract. Zirconium oxide (ZrO_2) thin films were deposited on to p - Si and quartz substrates by sputtering of zirconium target at an oxygen partial pressure of $4x10^{-2}$ Pa and sputter pressure of 0.4 Pa by using DC reactive magnetron sputtering technique. The effect of annealing temperature on structural, optical, electrical and dielectric properties of the ZrO_2 films was systematically studied. The as-deposited films were mixed phases of monoclinic and orthorhombic ZrO_2. As the annealing temperature increased to 1073 K, the films were transformed in to single phase orthorhombic ZrO_2. Fourier transform infrared studies conform the presence of interfacial layer between Si and ZrO_2. The optical band gap and refractive index of the as-deposited films were 5.82 eV and 1.81. As the annealing temperature increased to 1073 K the optical band gap and refractive index increased to 5.92 eV and 2.10 respectively. The structural changes were influenced the capacitance-voltage and current-voltage characteristics of $Al/ZrO_2/p\text{-}Si$ capacitors. The dielectric constant was increased from 11.6 to 24.5 and the leakage current was decreased from $1.65x10^{-7}$ to $3.30x10^{-9}$ A/ cm^2 for the as-deposited and annealed at 1073 K respectively.

1. Introduction

Zirconium oxide (ZrO_2) thin films find potential applications in various fields like optical coatings due to transparency over a wide spectral range from ultraviolet to mid-infrared regions, and protective coatings on devices working at high temperature, corrosive and other critical environments [1,2]. Because of its large band gap, good thermal stability and reasonable band gap offset with silicon it is used as high dielectric constant material for future complementary metal oxide semiconductor devices [3,4]. ZrO_2 thin films can be deposited by different methods such as electron beam evaporation, sol-gel process, pulsed laser deposition and sputtering [5-8]. Among these methods, sputtering has advantage in deposition of uniform films on large area substrates. Post deposition annealing plays an important role to modify the structure hence control over the interfacial layer between the Si substrate and the ZrO_2 film. Hence in this investigation, ZrO_2 thin films were deposited on unheated quartz and silicon substrates by using DC reactive magnetron sputtering method. The as-deposited films were annealed in air at different temperatures and studied their structural, optical and dielectric properties.

2. Experimental Details:

Zirconium oxide thin films were deposited using DC reactive magnetron sputtering onto un-heated p-(100) silicon and quartz substrates. The 100 mm diameter zirconium target was used as sputtering. Before deposition of ZrO_2 films, the vacuum chamber was evacuated to a base pressure of $4x10^{-4}$ Pa using combination of rotary and diffusion pumps. The sputter zirconium target (99.99 %) was cleaned by pre-sputtering for 5 min at sputter power of 100 W and argon pressure of 0.4 Pa. The substrates were shielded during pre–sputtering of the target. The ZrO_2 films were deposited at fixed oxygen partial pressure of $4x10^{-2}$ Pa. The as-deposited films were annealed in air at temperatures (T_a)= 873 K and 1073 K for one hour. The deposition parameters maintained for the growth of ZrO_2 films are given in table 1.

Table.1. Deposition parameters for the preparation of ZrO_2 films.

Deposition method	: DC reactive magnetron sputtering method
Sputter target	: Zirconium (100 mm dia, 3 mm thick)
Base pressure	: $4x10^{-4}$ Pa
Oxygen partial pressure	: $4x10^{-2}$ Pa
Sputtering pressure	: 0.4 Pa
Sputter power	: 140 W
Substrate temperature	: 303 K
Deposition time	: 30 min

The thickness of the as-deposited films was determined by using Dektak depth profilometer. The crystallographic structure of the films was investigated by X-ray diffractometer. The chemical bonding configuration of the films was determined by using Fourier transform infrared spectroscopy (FTIR). The optical transmittance of the films formed on quartz substrates was recorded in the wavelength range 185 – 800 nm using UV–Vis–NIR double beam spectrophotometer. The electrical properties such as capacitance - voltage (C-V) and current - voltage (I-V) of the fabricated Al/ZrO_2/p-Si structure were measured by using MIOKI (model 3532-50) LCR meter and Hewlett Packard (model hp 4140B) pA meter respectively. The top Al electrodes on ZrO_2 films were deposited by thermal evaporation method.

3. Results and discussion:

The thickness of the as-deposited ZrO_2 films determined with Dektek depth profilometer was 75 nm.

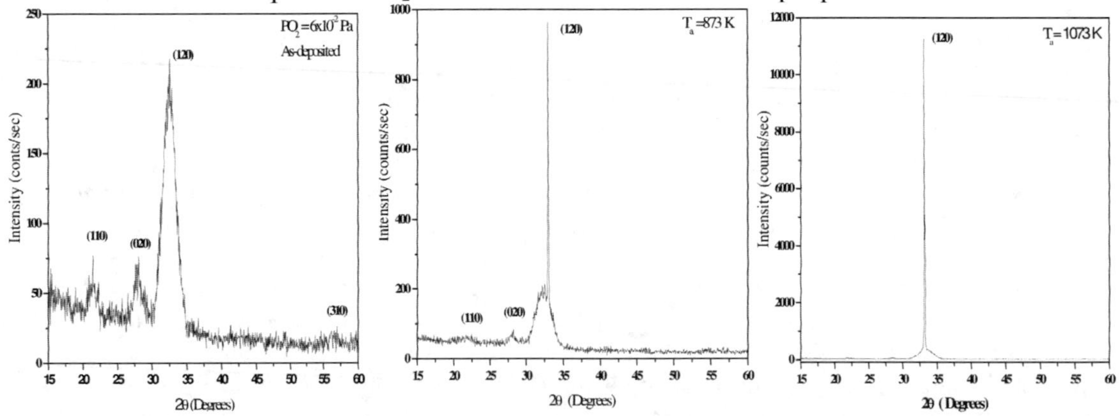

(a) (b) (c)

Fig.1. XRD patterns of ZrO_2 films: (a) as-deposited,(b) annealed at 873 K and (c) annealed at 1073K.

Figure 1 (a) to (c) shows the XRD patterns of the as-deposited ZrO_2 films and the films annealed at 873 and 1073 K. For the as-deposited films (Fig.1.a) the observed reflections at $2\theta = 21.46^0$, 28.06^0 and 32.58^0 were related to the (110) (020) and (120) planes of orthorhombic phase of ZrO_2 [JCPDS- 83- 0810]. Another reflection seen at 56.98^0 connected to the (310) plane of monoclinic phase of ZrO_2 [JCPDS- 88- 2390]. It indicates that the as-deposited films were mixed phase of orthorhombic and monoclinic ZrO_2. It can be seen that the intensity of the (120) reflection was increased at annealing temperature of 873 K and the shift in position of reflection is due to strain developed in the films (Fig. 1.b). At annealing temperature of 1073 K the weak (310) reflection disappear and the intensity of (120) reflection becomes stronger (Fig. 1.c). It is interesting to note that the phase transformation from mixed phase of monoclinic and orthorhombic to single phase of orthorhombic takes place at annealing temperature of 1073 K. Zhou et. al. [9] also observed the transformation of monoclinic ZrO_2 to single phase of ZrO_2 with increase of substrate temperature. The full width at half maximum of the (120) reflection was found to be decreased with increase of annealing temperature. The crystallite size of the films was calculated from Debye scherrer's relation. The crystallite size of the as-deposited films was 15 nm. As the annealing temperature of films increased to 1073 K the crystallite size was increased to 55 nm due to improvement of the crystallinity of the films.

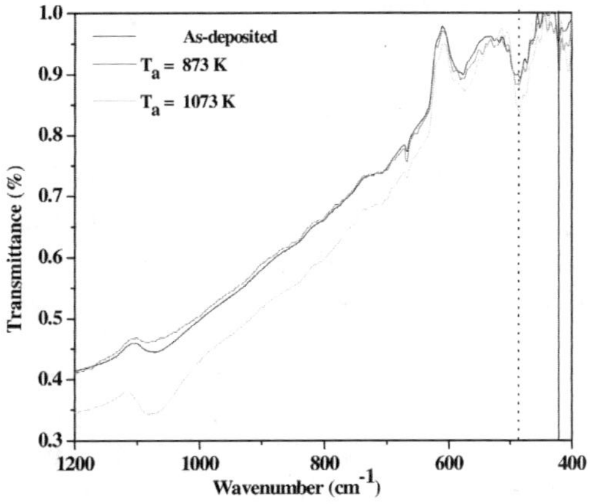

Fig. 2. FTIR spectra of the as-deposited and the ZrO_2 films annealed at 873 and 1073 K.

The Fourier transform infrared transmittance spectra of the ZrO_2 films measured in the wave numbers in the range 1200 - 400 cm^{-1} were shown in Figure 2. It can be seen from the Fig 2 that all the films exhibited strong absorption peaks located near 486 and 420 cm^{-1} are correspond to Zr–O vibrational modes as noticed by Li et al. [10]. As the annealing temperature increased to 1073 K, the appearance of broadness of the band 1073 cm^{-1} was suggests a wide dispersal of vibrational states which are characteristic of SiO_2 in amorphous matrix. At higher annealing temperatures, the oxygen diffused to the surface of the silicon substrate and results the formation of SiO_x layer at the interface between ZrO_2 film and the Si substrate. Busch et. al. [11] observed the substantial exchange of oxygen diffusion along the grain boundaries of ZrO_2 films under various annealing conditions.

The optical transmittance of the zirconium oxide films formed on quartz substrates were recorded as a function of photon energy in the wavelength range 185 – 500 nm. Figure 3 shows the wavelength dependence of optical transmittance spectra of as-deposited ZrO_2 films and the films annealed at temperatures of 873 and 1073 K. The average optical transmittance of the films was about 85 % above the wavelength of 250 nm. Sharp absorption edge was observed around the wavelength of 230 nm. The optical absorption edge of the films shifted towards lower wavelengths with the increase of annealing temperature. The optical absorption coefficient (α) of the films was calculated from the optical transmittance (T) data where the reflection losses were taken into consideration. The optical band gap (E_g) of the ZrO_2 films was determined by extrapolating the linear portion of the curves $(\alpha h\nu)^2$

versus photon energy (h□) to (□h□)^2 = 0. The optical band gap of the as-deposited films was 5.82 eV. After annealing of the films to 1073 K there was a blue shift in the band gap to5.93 eV which may be due to recovery of defects in zirconium oxide films caused by more incorporation of oxygen at higher annealing temperature [12]. According to Pauling's theory, the increased E_g value is due to increased stoichiometric in the ZrO_{2-x} film because of the oxidation of suboxidized zirconium oxide at higher annealing temperature [13].

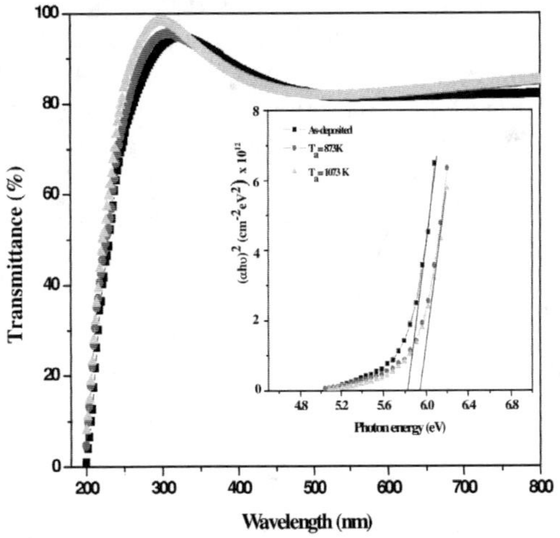

Fig. 3. Optical transmittance spectra of as- deposited ZrO_2 films and the films annealed at 873 and 1073 K and optical band gap (inset).

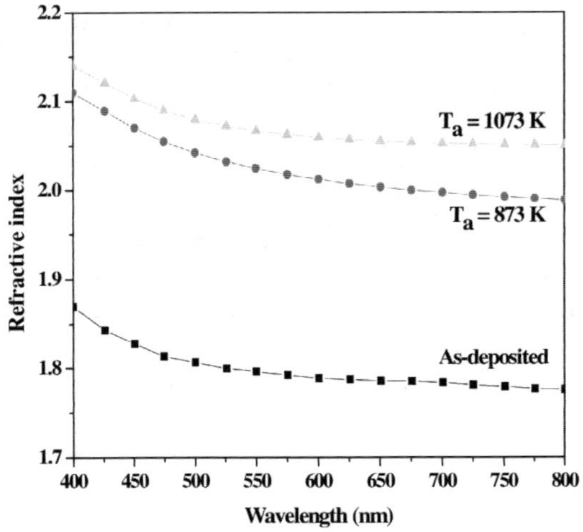

Fig. 4. wavelength dependence of refractive index of ZrO_2 films.

Figure 4 shows the wavelength dependence refractive index of as-deposited and air annealed films. The refractive index of the films decreased with increase of wavelength up to 550 nm thereafter it remains almost constant. At a fixed wavelength of 500 nm the refractive index of the as-deposited films was 1.80 and it increased to 2.07 with increase of annealing temperature to 1073 K [12]. The increase of refractive index with increase of annealing temperature was due to the increase of packing density of the films. As the annealing temperature increases the oxygen vacancies decreased there by increase in the packing density of the films. Figure 5 shows the capacitance - voltage curves of as-deposited and air annealed ZrO_2 films. The gate voltage was applied from accumulation to inversion

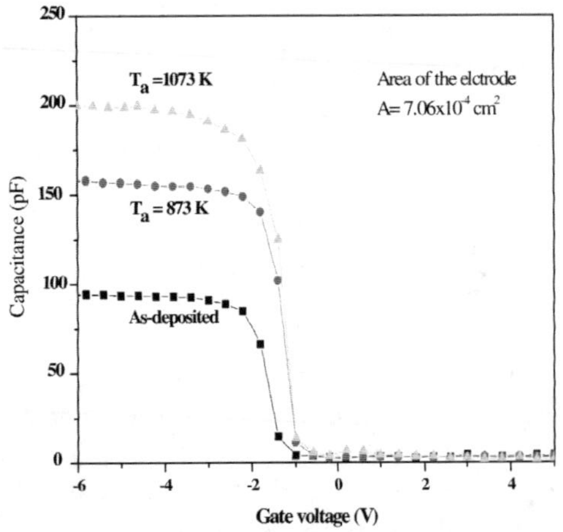

Fig. 5. Capacitance – voltage curves of $Al/ZrO_2/$ Si capacitors.

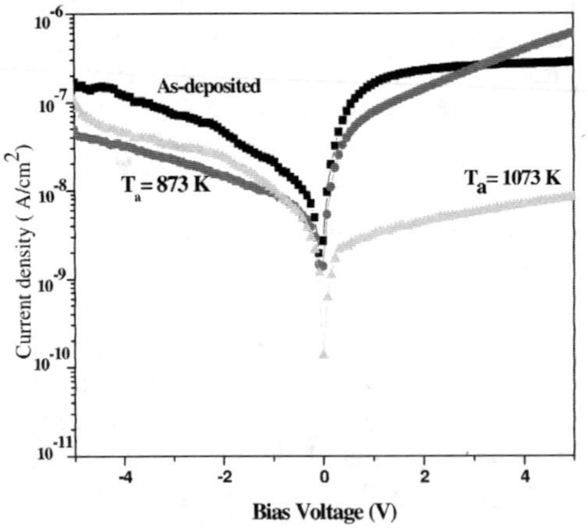

Fig. 6. Current- voltage curves of $Al/ZrO_2/Si$ capacitors.

with a sweep rate of 0.1 V/s at fixed frequency of 1 MHz. The accumulation capacitance of as-deposited films was 95 pF and it increased to 200 pF for the films annealed to 1073 K. The slope of the C-V curves at depletion region increases with increase of annealing temperature, which may be due to decrease of defects and interfacial trap density at higher annealing temperatures [14]. Moreover, the dielectric constant of the as-deposited films was 11.6 and it increased to 24.5 in the films annealed at 1073 K. The increase in the dielectric constant with annealing temperature could be due to decrease in structural defects and increase of the crystallite size. Figure 6 shows the leakage current density versus electric field (*J–E*) characteristics of the Al/ZrO_2/p-Si structures formed by as-deposited and the films annealed at different temperatures. The current density decreased with the annealing temperature at an applied electric field -1 V. The leakage current density was 1.65×10^{-7} A/cm^2 for as-deposited films and it decreased to 3.30×10^{-9} A/cm^2 after annealing at 1073 K. This is because of an interfacial layer of SiO_2 was formed between the interface of Si and ZrO_2 films.

4. Conclusion

Zirconium oxide films (75 nm thick) were deposited on p - silicon and quartz substrates held at room temperature by DC reactive magnetron sputtering of zirconium target at a fixed oxygen partial pressure of $6\text{x}10^{-2}$ Pa and sputtering pressure of 0.4 Pa. The as-deposited films were annealed in air at temperatures of 873 and 1073 K for one hour and studied the influence of annealing temperature on the structural, dielectric and optical properties. The capacitance - voltage and current - voltage of the Al/ZrO_2/p-Si fabricated capacitors were systematically studied. The X-ray diffraction results revealed that the as-deposited films were mixed phase of orthorhombic and monoclinic ZrO_2. At annealing temperature of 1073 K only orthorhombic ZrO_2 was observed with enhancement in the crystallanity of the films. The Fourier transform infrared absorption spectra of the as-deposited films showed absorption bands at 420 and 489 cm^{-1} were related to Zr-O vibrations, while those annealed at 1073 K showed an additional band at 1073 cm^{-1} indicated the growth of SiO_2 layer between the Si substrate and ZrO_2 oxide film. The leakage current density of the films was decreased from $1.65\text{x}10^{-7}$ to $3.30\text{x}10^{-9}$ A/cm^2 (at 1 V gate bias voltage). The dielectric constant was increased from 11.6 to 24.5 for the as-deposited and the films annealed at 1073 K. The average optical transmittance of the films was 85 % in the wavelength > 200 nm. The optical band gap of the as-deposited films was found to be 5.82 eV and it increased to 5.93 eV at the annealing temperature of 1073 K. The refractive index enhanced from 1.80 to 2.07 for the as-deposited and the ZrO_2 films annealed at 1073 K due to improvement in the packing density of the films.

Acknowledgement

One of the authors P. Kondaiah, thankful to UGC for the awardee of UGC-RFSMS fellowship to carry out the present work.

References

[1] Giuri C, Perrone M R and Piccinno V 1997 *Appl.Opt.* **36** 1143
[2] Al-Kuhaili M F 2004 *Opt. Mater.* **27** 383
[3] Wilk G D, Wallace R M and Anthony J M, 2001 *J. Appl. Phys.* **89** 5243
[4] Gusev E P, Cartier E, Buchanan D A, Gribelyuk M, Copel M, Okorn-Schmidt H and D'Amico C, 2001 *Microelectron. Eng.* **59** 341
[5] Jeong S H, Bae I S, Shin Y S, Lee S B, Kwak H T and Boo J H 2005 *Thin Solid Films* **475** 354
[6] Yanming S, Shuying S, Hua Y, Zhengxiu F, Hongbo H and Vianda S 2007 *Appl. Surf. Sci.* **254** 552
[7] Hwang S M, Lee S M, Park K, Lee M S, Joo J, Lim J H, Kim H, Yoon J J and Kim Y D 2011 *Appl. Phys., Lett.* **98** 022903
[8] Tang W T, Ying Z F, Hu Z G, Li W W, Sun J, Xu N and Wu J D 2010 *Thin Solid Films* **518** 5442
[9] Zhou Y, Kozima N and Sasaki K 2008 *J. Phys. D: Appl. Phys.* **41** 175414

[10] Li W, Liu X, Huang A and Chu P K 2007 *J. Appl. Phys.* **40** 2293
[11] Busch B W, Schulte W H, Garfunkel E and Gustafsson T 2000 *Phys. Rev. B* **62** 3290
[12] Zhu L Q, Fang Q, He G, Liu M and Zhang L D 2006 *J. Phys. D: Appl. Phys.***39** 5285
[13] Pauling L 1960 The Nature of the Chemical Bonds (Ithaca, New York: Cornell University press)
[14] Cho B, Wang J, Sha L and Chang J P 2002 *Appl. Phys. Lett.* **80** 1052

Effect of Aluminum concentration on structural and optical properties of DC reactive magnetron sputtered Zinc Aluminum Oxide thin films for transparent electrode applications

B.Rajesh Kumar and T.Subba Rao

Materials Research Lab, Department of Physics, Sri Krishnadevaraya University, Anantapur-515055, A.P, India

Email: rajphyind@gmail.com

Abstract. Zinc Aluminum Oxide(ZAO) thin films were deposited on glass substrates by DC reactive magnetron sputtering in an Ar+O_2 gas mixture using commercial available Zn metal (99.99% purity) and Al (99.99% purity) targets of 2 inch diameter and 4 mm thickness. The films were characterized and the effect of aluminum (Al) concentration (2 at %-6 at %) on the structural and optical properties was studied. The average crystallite size obtained from Scherer formula is in the range of 32-44nm. Microstructural analysis using Scanning Electron Microscope (SEM) supplemented with EDS is carried out to find the grain size as well as to find the composition elemental data of prepared thin films. Optical study is performed to calculate the extinction coefficient (k), absorption coefficient (□), optical band gap (E_g) using transmission spectra obtained using UV-VIS-NIR spectrophotometer. There was widening of optical band gap with increasing aluminum concentration. ZAO film with low resistivity 3.2 X 10^{-4} □ cm and high transmittance of 80% is obtained for 3at% doped Al which is crucial for optoelectronic applications.

1. Introduction

Doped Zinc Oxide thin films such as Zinc aluminum oxide (ZAO) are being widely used as transparent conductive oxide films in many applications due to their numerous advantages over indium tin oxide (ITO) films [1-4]. ZAO films have many industrial applications, such as in the fabrication of thin films solar cells, smart windows, silicon, hetro-junction, high transparency and wide band-gap. Various deposition techniques have been utilized for ZAO films such as direct current(DC) and radio frequency(RF) magnetron sputtering[5], pulsed laser deposition[6], chemical vapor deposition[7], spray hydrolysis[8], and sol-gel[9]. In order to satisfy the demands of good film quality, high–rate and large-area deposition, and low cost of equipment for commercialization of all transparent conductive oxide films, DC magnetron sputtering has been regarded as one of the most attractive and effective fabrication techniques in the mass production of ZAO films. The main purpose of this work is to demonstrate the feasibility of using these ZAO films in photovoltaic devices as front electrode.

Journal of Physics: Conference Series **390** (2012) 012032 doi:10.1088/1742-6596/390/1/012032

2. Experimental details

Zinc Aluminum Oxide (ZAO) thin films were prepared by DC reactive magnetron sputtering. High purity of Zinc (99.999%) and Alumina (99.99%) targets with 2 inch diameter and 4mm thickness are used for deposition on glass substrates. The base pressure in chamber was $3X10^{-4}$Pa and the distance between target and substrate was set at 60mm. The glass substrates were ultrasonically cleaned in acetone and ethanol, rinsed in an ultrasonic bath in deionized water for 15 min, with subsequent drying in an oven before deposition. High purity (99.99%) Ar and O_2 gas was introduced into the chamber and was metered by mass flow controllers for a total flow rate fixed at 25sccm. Deposition was carried out at a working pressure of 1Pa after pre-sputtering with argon for 10min. The depositions were carried out at different doping concentrations of aluminum (Al). Film thickness was measured by Talysurf thickness profilometer. The resulting thicknesses of the films were found to be ~300nm respectively. XRD (Philips: PW1830) was used to analyze the crystalline orientation and structure of ZAO thin films. The sheet resistance (R_s) of the film was measured with a four-point probe method. The resistivity of the film (ρ) was calculated using the simple relation $\rho = R_s.d$, where d is the film thickness. Surface morphology of the samples has been studied using HITACHI S-3400 Field Emission Scanning Electron Microscope (FESEM) with Energy Dispersive Spectrum (EDS). EDS is carried out for the elemental analysis of prepared thin films. Optical transmittance measurements were carried out using Cary 5E UV-VIS-NIR spectrophotometer.

3. Results and discussion

XRD measurements were performed to investigate the structural properties of the deposited films. Figure 1 shows the XRD pattern of the thin films deposited at 2, 3, 4 and 6 at % of Al doped ZnO (ZAO) films. As it can be observed, the patterns were clearly dominated by a peak around 34.4°, corresponding at the (0 0 2) reflection. The position of the peak was very close to the preferred orientation of the standard d powder ZnO crystal (2θ=34.45°). This fact confirms the preferential c-axis orientation as reported by other authors [10]. It is observed that all XRD peaks were influenced by doping process where the peaks intensity reduced with doping concentrations. The peaks at doping concentration of 4 and 5 at% were more reduced to indicate higher doping concentration heavily affects ZAO thin film structural properties. The result suggested that the crystallinity of ZAO thin films decreased with increase of dopant concentrations. The aggravation of ZAO crystallinity with doping concentration might be from the formation of stress by the difference in ion size between Zn and Al. At higher doping concentration, due to nuclear charge of Al^{3+} is larger than Zn^{2+}, extrinsic Al^{3+} will captured more oxygen in completion of Zn^{2+} which decrease crystallinity and hexagonal structure of the thin films. The average crystallite size obtained from Scherer formula [11] is in the range of 32-44nm.

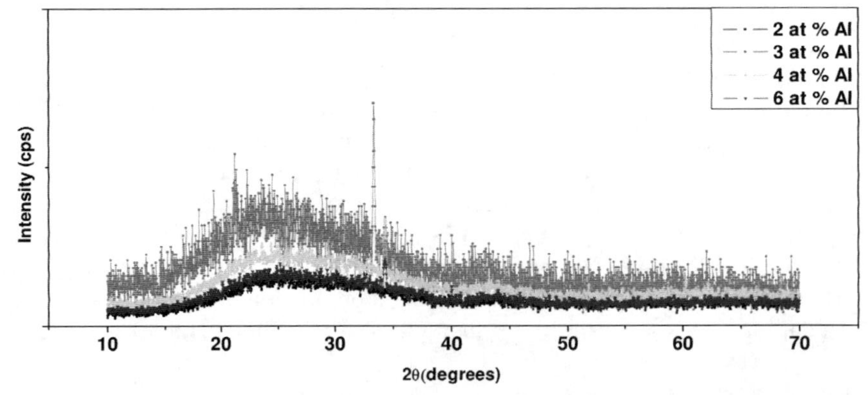

Fig. 1. XRD patterns of ZAO thin films at different Al doping concentrations.

Figure 2(a)-2(d) shows Scanning electron microscopy (SEM) images of ZAO thin films deposited on glass substrate doped with 2, 3, 4 and 6 at % of Al. The SEM images reveal that the films are uniformly deposited with nanoscale particles. It could be observed that grain size tends to reduce with increment of Al doping. The average grain size from SEM images have 53nm, 42nm, 34nm and 28nm for 2 at %, 3 at%, 4 at % and 6 at % Al doped ZnO. Energy Dispersive Spectrum (EDS) of ZAO thin films with its compositional elemental data are shown in figure 3(a)-3(d).

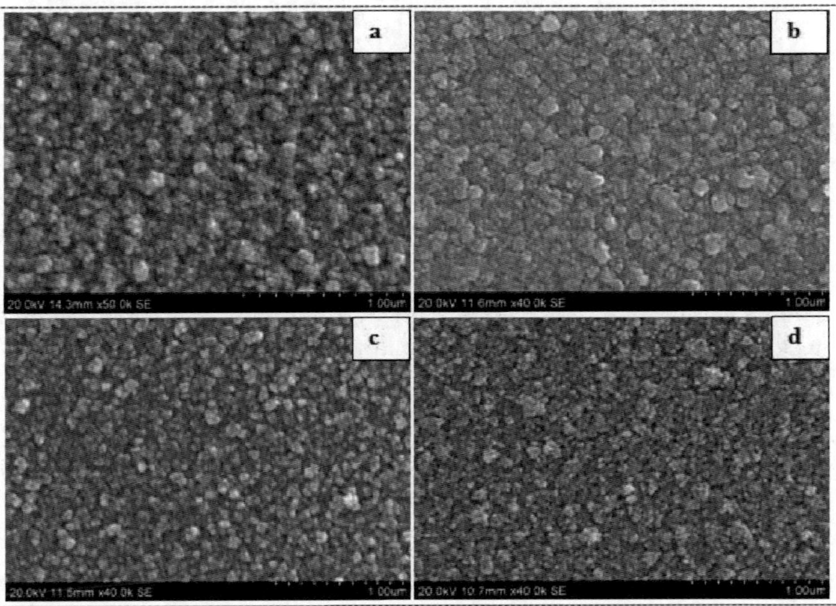

Fig. 2. Surface morphology of ZAO thin films doped with (a) 2 at %, (b) 3 at % (c) 4 at % and (d) 6 at % of Al.

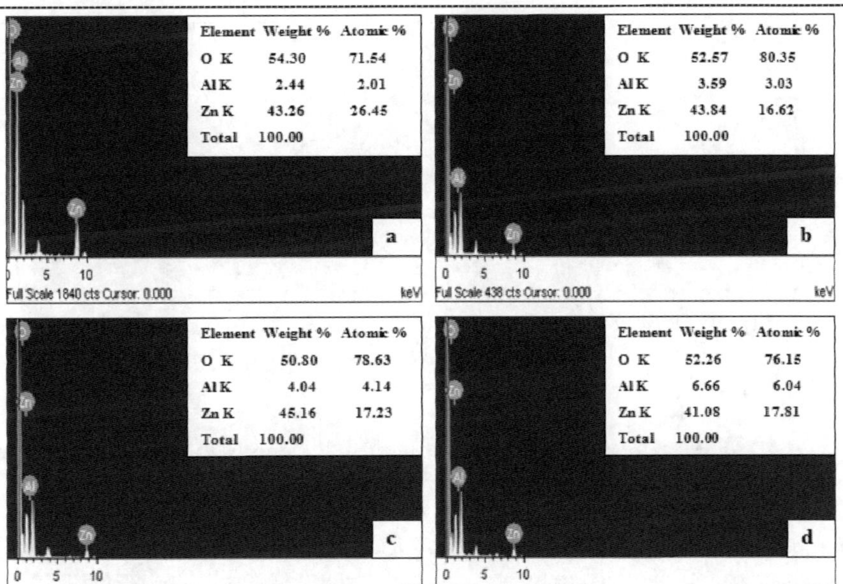

Fig. 3. EDS of ZAO thin films doped with (a) 2 at %, (b) 3 at % (c) 4 at % and (d) 6 at % of Al.

Journal of Physics: Conference Series **390** (2012) 012032 doi:10.1088/1742-6596/390/1/012032

The electrical resistivity of the ZAO films was investigated by four-point probe method at room temperature. The electrical resistivity is in the range of $3.48X10^{-4}$ Ω cm – $5.2X10^{-4}$ Ω cm. The lowest resistivity of $3.2X\ 10^{-4}$ Ω cm is obtained for 3 at % doped Al. The electrical properties of ZAO films were analyzed because of the strong influence of Al dopant, as they exert on the final performance of the optoelectronic devices such as solar cells.

The transmittance spectra of ZAO thin films are shown in figure 4. The film deposited with 3 at% doped Al exhibits maximum transmittance of 80%. Optical parameters such as absorption coefficient (□), extinction coefficient (k) and optical band gap (E_g) were determined from the transmission data. The extinction coefficient (k) can be calculated by using the relation [12].

$$k = \frac{\alpha\lambda}{4\pi} \qquad (1)$$

where □ is the absorption coefficient and is calculated from the relation

$$\alpha = \frac{1}{d}\ln\left(\frac{1}{T}\right) \qquad (2)$$

where d is the thickness of the film and T is the transmittance [13].

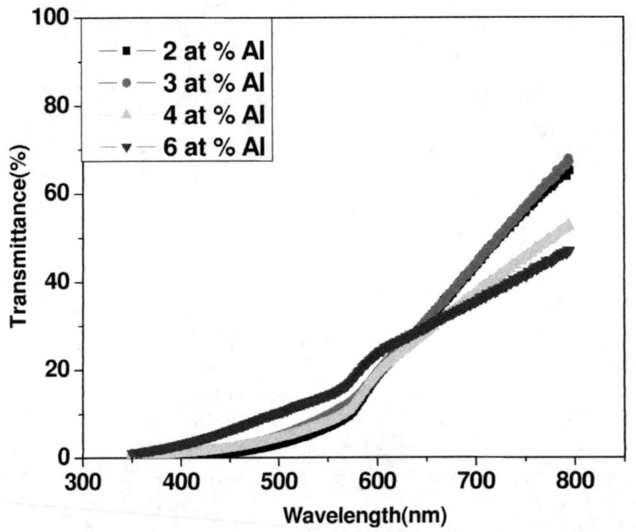

Fig. 4. Optical transmittance spectra of ZAO films deposited at different atomic weight % of Al.

Absorption coefficient (□) decreases with increase of wavelength (□). The absorption coefficient (□) is found to be in order of 10^6 cm^{-1} at shorter wavelength regions. For higher values of photon energy, the energy dependence of absorption coefficient (□ > 10^4 cm^{-1}) suggests the occurrence of direct electron transitions. The extinction coefficient increases with the increase in energy of the incident beam. The values of extinction coefficient (k) with wavelength (□) are found to be in the range of 0.02-0.58.

The optical band gap, E_g is determined from the dependence of absorption coefficient values (□) on the photon energy, using Tauc's relation [14]

$$(\alpha h\nu) = B(h\nu - E_g)^n \tag{3}$$

where B is a parameter that depends on the transition probability, E_g is the optical band gap energy of the material, hυ is the photon energy and n is an index that characterizes the optical absorption process and is theoretically equal to 2 and ½ for indirect and direct allowed transitions respectively. The optical band gap of the films was evaluated from the Tauc's plot of $(\alpha h\upsilon)^2$ versus hυ shown in figure 5. The optical band gap is in the range of 3.2-3.46eV, the contribution from Al^{3+} ions on substitution sites of Zn^{2+} ions and Al interstitial atoms determines the widening of the band gap caused by increase in carrier concentration , which is believed to be Burstein-Moss effect because of doping by Al[15].

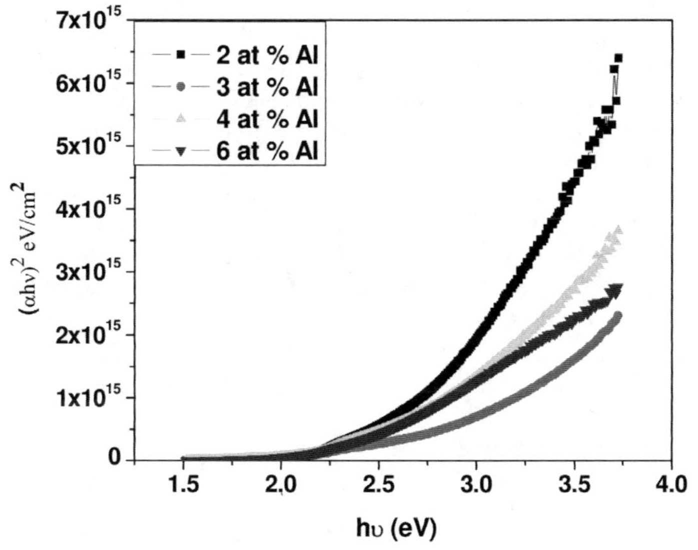

Fig. 5. Tauc's plot of ZAO films deposited at different atomic weight % of Al.

The figure of merit F is introduced in order to quantify the optoelectronic properties of transparent oxides. It is defined as [16]

$$F = \frac{-1}{[R_s \ln(T_{av})]} \tag{4}$$

where R_s is sheet resistance and T_{av} is average visible transmittance of the films.

The best combination of low resistivity and high transmission, ie., the highest figure of merit F = 4.87 X 10^{-1} Ω^{-1} was obtained for the ZAO thin film doped with 3 at % of Al with $R_s = 21.7\ \Omega/\square$.

4. Conclusions

ZAO thin films have been grown by DC reactive magnetron sputtering technique on glass substrates at different doping concentrations of Al. The deposited ZAO films have a preferred crystalline orientation of (0 0 1) direction. The conduction characteristics of ZAO films are dominated by O^{2-} vacancies, Zn interstitial atoms, Al ions on substitutional site of Zn ions and Al interstitial atoms. ZAO films with low resistivity 3.2 X 10^{-2} □ cm and high transmittance over 80% is obtained for Al concentration of 3 at %. This indicates that these films satisfy basic requirement as transparent electrodes for display and solar cell applications.

Acknowledgements

The authors are thankful to UGC, New Delhi, India for financial support under the major research project (F.NO.37-346/2009, SR).

References

[1] Lin Y C, Li H Y,Yen W T 2008 *Appl.Surf.Sci.* **254** 3262
[2] Lin Y C, Jian Y C, Jiang J H 2008 *App. Surf. Sci.* **254** 2671
[3] Rim Y S, Kim S M, Choi H W, Park S J, Kim K H 2008 *Colloids Surf. A.* **313** 461
[4] Moon C S, Chung Y M, Jung W S, Chan J G 2007 *Surf.Coat.Technol.* **210** 5035
[5] Agashe C, Kluth O, Schope G, Siekmann H, Hupkes J, Rech B 2003 *Thin Solid Films* **334** 263
[6] Agura H, Suzuki A, Matsushitam T, Aoki T, Okuda M 2003 *Thin Solid Films* **445** 263
[7] Sato H, Minami T, Takata S, Miyata T, Ishii M *Thin Solid Films* **236** 14
[8] Nunes P, Fortunato E, Martins R 2001 *Thin Solid Films* **383** 277
[9] Jimenez-Gonzalez A.E, Urureta J.A.S, Suarez-Parra R 1998 *J.Crsyt. Growth* **192** 430
[10] Fujikura N, Nisihhara T, Goto S, Xu J,Ito T 1993 *J.Crst. Growth* **130** 269
[11] Cullity B D 1959 *Elements of X-ray diffraction* (Addition-Wesley London) p 99
[12] Marquez E, Ramirez J, Villares P, Jimenez R, Ewen P.J.S, Owen A.E. 1992 *J.Phys.D: Appl.Phys.* **25** 535
[13] Priyamvada Bharadwaj, Shishodia P K, Mehra R M 2001 *J.Optoelectron.Adv.Mater.* **3** 319
[14] Tauc J, Grigorovici R, Vancu A 1966 *Phys. Status Solidi* **15** 627
[15] Burstein E 1982 *Phys. Rev.* **25** 7826
[16] Halina Czternastek 2008 Vacuum **82** 994

Microwave emission from an AXIAL-Virtual Cathode Oscillator driven by a compact pulsed power source

R.Shukla. S.K.Sharma, P.Banerjee, P.Deb, T. Prabaharan, R. Das, B.K.das, B. Adhikary, R.Verma, A.Shyam

Energetics & Electromagnetics Division, Bhabha Atomic Research Centre, Visakhapatanam, Andhra Pradesh, India, 530012

rohit.ipr@gmail.com

Abstract. For the generation of microwaves, Electron beam devices operating in vacuum are most widely used. For pulsed and high power microwave generation, Virtual cathode oscillators (VIRCATORs) are said to be simple in operation and construction. They are generally driven by a pulsed power source which gives high input powers to the Vircator connected as load. Vircator, depending upon its efficiency, converts the electrical input power to the microwave power. We are presenting the results of an axial Vircator operating in $2X10^{-4}$ mbar vacuum and is driven by a compact pulsed power source. The energy source and pulse compression is realized in very user friendly approach to run the system. The radiating system presently runs at relatively low powers but has the scope of reaching to high power by a logical improvement. A study of effect of collapsing diode impedance, of the vacuum field emission diode of the Vircator, on the microwave emission is presented in the paper. We are also presenting the microwave emission measurement conducted in the given system. Effect of vacuum is also studied to the extent of present experimental limits.

1. Introduction

Axial Virtual cathode oscillators are the devices which are recently being studied by various groups-worldwide [1-2] as a source of high power microwave radiations operating with relatively much ease as compared with other conventional and well established high-power-microwave generating devices. This device requires huge amount of electrical power, in the form of power of generated Electron beam inside it, which is converted to microwave power in the geometry of Vircator. To the best knowledge of the authors it is first system of its kind using coaxial cable based system driven by the compact pulsed power source to drive the complete assembly. Because of its compact nature, the system has potential of operating at repetitive mode also. The results of a bulkier pulsed power system driving a solid dielectric PFL and Vircator connected to it have been reported previously [3].

The principle of operation of Vircator is as given under. A high voltage source is connected with the Electron beam diode which is part of Vircator geometry inherently. In the case of axial Vircator, which is in fact the case of present study, the electron beam diode is formed of cylindrical cathode and mesh/thin foil anode. The high power Electron beams are emitted from the surface of cathode mainly by two emission mechanisms viz. Field Emission and Explosive Electron Emission depending upon the material and surface properties of the cathode. Once high power electron beam is generated it is accelerated towards anode by the same electric field which is responsible for its generation. As the electrons of the emitted beam reach the anode, some of them pass through anode depending upon the transparency for the anode. These passed electrons are forced to propagate inside a drift tube of Vircator which interacts electromagnetically with the beam passing through it. The space charge effects causes the bending of cylindrical beam towards axis of drift tube and this results in the formation of electron cloud of certain charge density inside the drift tube. This charge cloud is formed only when the electron beam current is higher than space charge limiting current of this drift tube.

Figure 1 : Axial Virtual Cathode Oscillator System

This charge cloud is also called virtual cathode. After formation of virtual cathode some part of the successive electrons emitted from cathode are reflected back by the virtual cathode. Because of this they get trapped between cathode and virtual cathode and keep on oscillating in the space between cathode and virtual cathode till they are lost either by diffusion or by hitting the foil/mesh. These oscillating electrons radiate the radiations depending upon their oscillating frequency which again depends upon the A-K gap and electron velocity considerably.

Figure 2. Stainless Steel Cathode

Figure 3. Punctured RG218 Cable

Other than the radiations caused by oscillating electrons between cathode and virtual cathode, there is another mechanism which contributes to the total microwave radiations from the virtual cathode oscillator. The virtual cathode which is an electron cloud also keeps on oscillating in space and in time.

The oscillation of the virtual cathode oscillator in space and time also causes the emission of electromagnetic radiation to take place. The frequency of the radiation from virtual cathode oscillations is dependent on the electron plasma frequency of the charge cloud. Moreover there is always a loss of the electrons from the virtual cathode and electron beam in the form of drift and diffusion loss of electrons. This loss of electrons from the virtual cathode is compensated by the incoming electron beam which is emitted from the cathode.

There is formation of plasma on the cathode surface and also on the anode surface caused by the impact of energetic electrons of the beam on it. This plasma also expands in all free directions and occupies the space in between the region of anode cathode. Once the plasma is filled in this region the electric field between anode cathode gap regions diminishes to a value where no further field emission of electrons takes place. The event is called diode closure. This requirement makes the pulse compression and pulse shaping necessary before applying it to the Vircator.

As discussed above the Vircator requires high electrical powers for its operation and only for very short durations. This is done by pulse-compression and pulse-shaping. An attempt has been made to do pulse-compression and pulse-shaping by compact pulse compression system mentioned in the details given under. The experimental results are also presented after the system details.

2. Experimental Setup

The Vircator system is shown in figure 1. The Virtual cathode oscillator is a vacuum chamber with a measured vacuum inside the chamber of $2X10^{-4}$ mbar. The vacuum is created with a diffusion pump of 500liter/ second pumping speed backed by a rotary pump. The Vircator chamber is shown in figure 1 and is made by stainless steel. The cathode is also made of S.S. electrode of 60mm diameter and is shown in figure 2. The anode is made of S.S. mesh of 80% optical transparency. In the present study at different PFL voltages, the anode cathode spacing is fixed at 4mm. The drift tube diameter is 140mm and estimated space charge limited current in this geometry is less than 1kA.

The Vircator is operated using a solid dielectric pulsed forming line (PFL) made of four pieces of 50-ohm, RG-218 coaxial cables connected in parallel to make a net estimated impedance of 12.5 ohms. The length of each cable is 12 meters and hence the estimated pulsed-width which is twice of its transit-time, is expected to be nearly 140nS.

The Pulsed forming line is being charged by the compact pulsed transformer [4] which has its primary capacitor bank made in the shape of its primary. The capacitors used in the bank have shown very high power delivery driving a plasma focus device [5]. The connection between pulse transformer and pulse forming line is done using specialized connectors at the ground part in order to reduce the electric filed strength caused by sharp braid wires to such levels where the breakdown at feed point (as is shown in figure 3) does not take place which, otherwise, happens only at 20kV otherwise. The improved connections between the transformer and Pulsed forming line result in charging of PFL to still higher voltages reaching few hundreds of kV. The compact transformer with the combination of Pulsed forming line is having 1:20 Voltage gain when compared with primary charging voltage and PFL peak voltage at its second peak. The capacitor bank –cum- transformer, when first charged to 10kV and then discharged, delivers 200kV peak voltage in the second peak in 2.0 microseconds at the Pulsed forming line which is connected to the Virtual cathode oscillator system using a self triggered spark gap switch. A conical horn antenna is also designed for the system with input waveguide matching with the dimensions of drift tube.

As far as diagnostics are concerned, the standard diagnostics are used for the measurement of the PFL discharge current, PFL charging voltage, and microwave (power envelop and frequency). For the measurement of Pulse forming line charging voltage we have used 10000X resistive voltage divider. For the measurement of currents we have used a 17nS rise time current transformer with standard sensitivity of 50V/kA. The current transformer output is connected to the oscilloscope using standard 20dB attenuator in order to reduce the voltage to the oscilloscope input requirements for the given current pulse. The current transformer is placed inside the spark gap chamber. The microwave signal is recorded on a 10GHz, 40GSa/sec oscilloscope using a double-ridged horn antenna having bandwidth of 800MHz to 18GHz. The receiving antenna is kept at 1.2 meters from cathode location facing the drift tube which has no radiating antenna. The receiving antenna is connected to the oscilloscope using high Bandwidth cable LMR-400 having a cut-off at 16GHz. This signal coming directly from the receiving antenna to the oscilloscope is analyzed for its FFT to give frequency spectrum of radiated microwave signal. To see the power envelop of radiated microwave, the output of antenna is bifurcated and passed through microwave detector diode Agilent-423B and its output is smoothened to average out the acquisitions over 1ns (40samples) comparable to the rise time of diode.

3. Experimental Results

Initially when the spark gap connecting the PFL with Vircator was filled in air, the peak voltages possible were not exceeding 160kV on PFL and consistent breakdowns took place at the output side. The PFL charging voltages under these circumstances are shown in figure 4 collectively. Irrespective of the primary charging voltages which were varied from 8kV to 12kV in steps of 2kV the output voltage could not exceed 160kV.

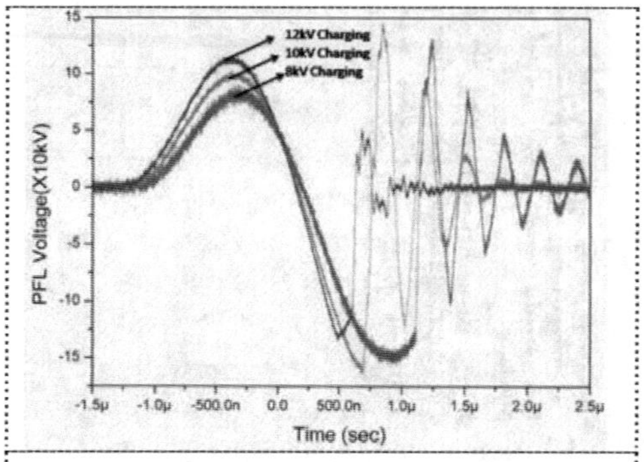

Figure 4 : Voltage time history **of** PFL (when spark gap is in air)

To enhance the Voltages on PFL, the spark gap was filled with transformer oil and the gap was reduced to hold a maximum of 200kV which is also shown in figure 5

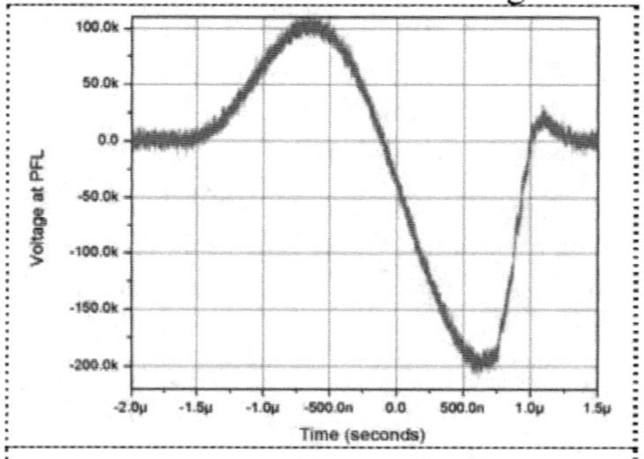

Figure 5. Voltage time history of PFL when spark gap is in oil

Once a peak voltage of 200kV was achieved on the pulsed forming line the Vircator which was connected to its output was operated in vacuum. It was noted that when the Vacuum was not there inside the Vircator, the high frequency components (of higher that 2.5 GHz) in the FFT of antenna signal were not present.

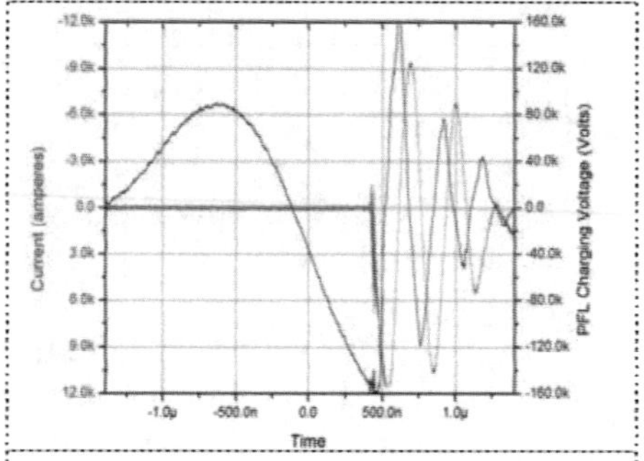

Figure 6. : PFL Charging Voltage and PFL Discharge Current

The microwave experiments with Vircator were conducted at two different PFL-charging-voltages of 160kV and 180kV decided by the spark gap settings. Figure 6 shows the voltage measured at PFL charging end and discharge current of PFL measured using Current Transformer (CT). The peak discharge current reaching in the system is 12kA. The peak PFL charging voltage is 160kV and the small Anode cathode gap leads to fast reduction of diode impedance. The net impedance of the complete discharge-circuit comes out to be nearly 13ohms.

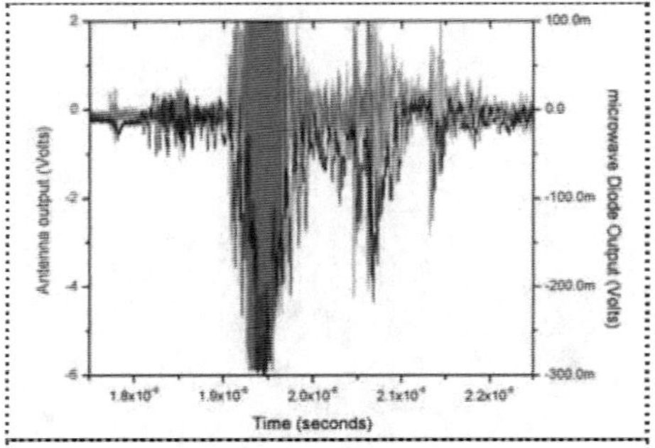

Figure 7. : Receiving Antenna (kept at 1.2Meters from Cathode) signal and envelop seen by the microwave detector diode (Vircator operating with PFL at 160kV)

Figure 7 represents the microwave signals received from the antenna and fed to the oscilloscope at 50ohm input impedance. The microwave pulse record of not more than 50nanosecond also suggest that fast closure of diode is taking place because the applied high voltage pulse is nearly 140ns in duration. The reflections of voltage and current also suggest the same sequence of events to be happening in the Vircator.

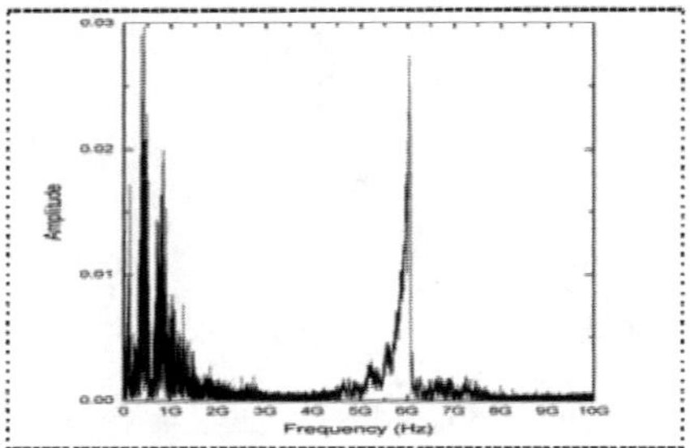

Figure 8. FFT of receiving antenna signal (Vircator operating at 160kV PFL Voltage)

Figure 8 is more informative in terms of the frequency content of radiated signal from the Vircator. It is clearly visible from the spectrum that high frequency of higher than 5GHz is present in the signal. The Vircator has successfully demonstrated the high frequency radiations from the compact pulsed power system.

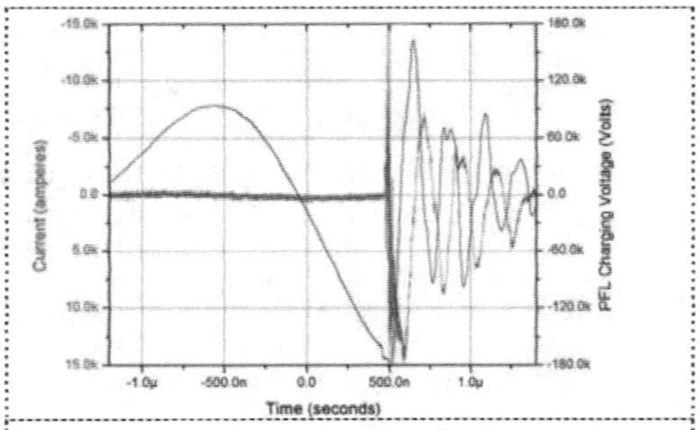

Figure 9. : PFL Charging Voltage and PFL Discharge Current

The other set of results achieved at 180kV of Vircator operation are also highly interesting. Figure 9 is the record of PFL charging voltage and discharge current reaching 180kV and 14kA peak current again demonstrating the same 13 ohm impedance of the discharge circuit. The microwave signal is shown in figure 10.

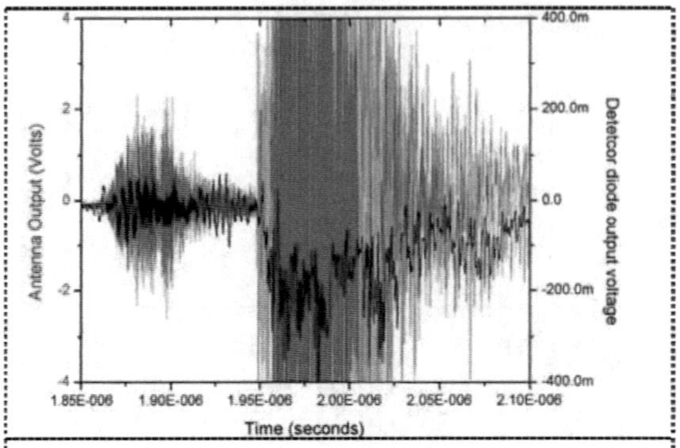

Figure 10. : Receiving Antenna (kept at 1.2Meters from Cathode) signal and envelop seen by the microwave detector diode (Vircator operating at 180kV PFL Voltage)

The microwave envelop is higher as compared with the signal at 160kV indicating a higher emisision of microwave at higher voltage.

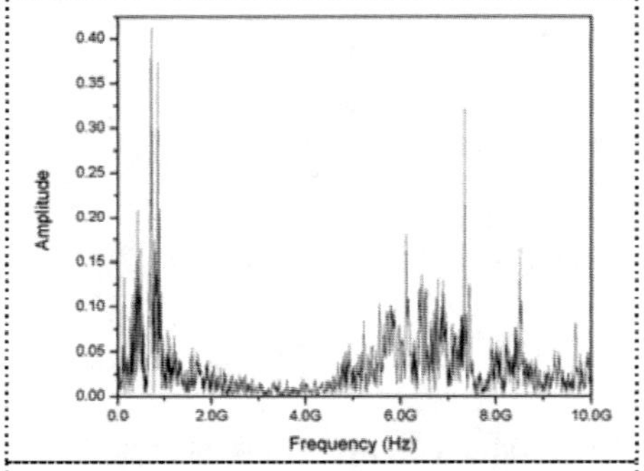

Figure 11. : FFT of Receiving antenna signal (Vircator operating at 180kV PFL Voltage)

Figure 11 represents the FFT of radiated fields indicating that a frequency spectrum of greater than 5GHz is present as a result of Vircator operation.

4. Conclusion

A successful operation of Vircator is demonstrated in the article using a compact pulse power source. The energy involve in each experiment was 125Joules making the system efficient in terms of power compression and shaping and also for a futuristic view of achieving repetitive operation.

Acknowledgement

Authors wish to thank Director, E&I Group for his consistent encouragement for this work.

References

[1] A Roy et al 2010 *IEEE Trans. Plasma Sci.* **7** p 1538-1545
[2] Queller T. Et al 2010 *J.Appl.Phy.* **10** 103302
[3] R. Shukla et al April 2010 *J. of Nepal Phys. Soc.* **26**, No. 1, p 29-34
[4] R Shukla et al 2011 *Rev. Sci. Instr.* **10** 106103
[5] R Shukla et al 2010 *Rev. Sci. Instr.* **8** 83501

Development of High vacuum facility for baking and cool down experiments for SST-1 Tokamak components

Ziauddin Khan, Firozkhan S Pathan, Paravastu Yuvakiran, Siju George, Himabindu Manthena, Dilip C Raval, Prashant L Thankey, Kalpesh R Dhanani, Manoj Kumar Gupta and Subrata Pradhan

Institute for Plasma Research, Near Indira Bridge, Bhat, Gandhinagar – 382428, Gujarat, India

E-mail: ziauddin@ipr.res.in

Abstract. SST-1 Tokamak, a steady state super-conducting device, is under refurbishment to demonstrate the plasma discharge for the duration of 1000 second. The major fabricated components of SST-1 like vacuum vessel, thermal shields, superconducting magnets etc have to be tested for their functional parameters. During machine operation, vacuum vessel will be baked at 150 °C, thermal shields will be operated at 85 K and magnet system will be operated at 4.5 K. All these components must have helium leak tightness under these conditions so far as the machine operation is concerned. In order to validate the helium leak tightness of these components, in-house high vacuum chamber is fabricated. This paper describes the analysis, design and fabrication of high vacuum chamber to demonstrate these functionalities. Also some results will be presented.

1. Introduction

SST-1 Tokamak is designed for 1000 sec long pulse hydrogen plasma discharge with 220 kA current [1, 2]. The main subsystems of SST-1 tokamak consist of vacuum chamber, magnet systems and LN_2 radiation shields as shown in the figure 1. Vacuum chamber of SST-1 has been divided into two parts such as vacuum vessel and cryostat. The vacuum vessel will be used for plasma confinement while the cryostat will be used to provide the operational environment for super-conducting coils. These super-conducting coils have to be cooled up to 4.5 K using super-critical helium gas and they will be charged at 10 kA current for longer time period. For this purpose, they have to be protected from heat radiation coming from room temperature cryostat surfaces and vacuum vessel surfaces maintained at 50 °C during entire campaign by using LN_2 radiation shields around them. The cryostat will be maintained at high vacuum $\leq 1.0 \times 10^{-5}$ mbar.

Nearly 162 numbers of 10 verities of different shape and size of bubble type of LN_2 thermal shields are used. These panels are cooled using Liquid Nitrogen (LN_2) maintained at 80 K uniform temperature at 6.0 bar (g) internal pressure. Since these panels will be operated for much longer time period at higher internal pressure, it is very essential to qualify all the weld joints of these panels for their leak tightness at room temperature (RT) and at their operational conditions. For plasma confinement, the main vacuum vessel will be maintained at ultra-high vacuum $\leq 1.0\times 10^{-9}$ mbar.

The complete SST-1 vacuum vessel structure consists of eight numbers of vessel modules (VM) and eight numbers of vessel sectors (VS). In order to achieve such ultra-high inside the chamber, it is very much essential to bake the vacuum vessel at 150 °C for longer duration to remove water vapor and other gases which are absorbed and adsorbed. Further during cool down phase and plasma operations, the vacuum chamber will be maintained at 50 °C under evacuated condition to avoid the cool down of vessel towards low temperature due to radiation loss. For baking purpose, rectangular channels (16 mm × 8 mm) of 2 mm thickness are welded on the inner surface of the vacuum vessel as per finite element analysis [3].The high pressure dry nitrogen gas at 4.5 bar (a) heated up to 250 °C is passed through these rectangular channels of the vacuum vessel for baking. Since a large amount of welding are exposed directly to the UHV portion and will be under high pressure at high temperature, the qualification of these welds in their operation conditions has to be established.

Before assembling these VM & VS into the system, the individual VM and VS are baked up to steady state temperature of 150 °C for eight (08) hours in a dedicated high vacuum (HV) baking facility developed in-house. After baking, the leak testing of VM and VS baking channels are carried out and qualified at the leak rate < 1.0×10^{-8} mbar l/s in vacuum mode and at < 1.0×10^{-6} mbar l/s in sniffer mode at 6.0 bar (g) helium gas pressure. Similarly LN_2 radiation shields are cool down to 85 K in the same chamber and integrated leak testing is carried out in sniffer mode using helium gas at a pressure of 7.5 bar (g) using RGA.

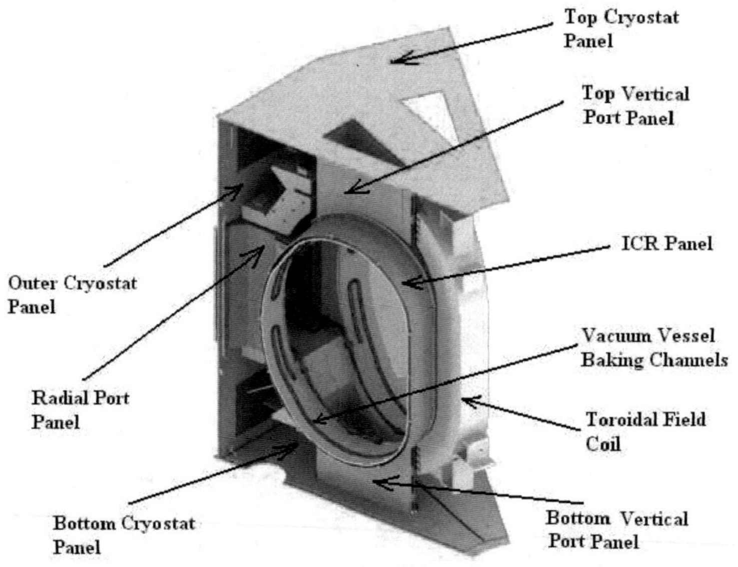

Figure 1. Schematics of SST-1 Tokamak with main sub-systems.

2. Design of high vacuum chamber

In order to accommodate VM and cryostat LN_2 shields (Dimensions: h = 2.5 m, w = 1.5 m) for their qualification along with additional structures necessary for their support, a dedicated chamber with a diameter of ϕ = 2.4 m, height of h = 3.45 m and wall thickness of t = 10.0 mm is designed and fabricated. The chamber is fabricated using SA 516 Gr-70 material, the out-gassing rate of which is 2.0×10^{-9} mbar l/s.cm^2 [4]. It is electro-plated with chrome to reduce the out-gassing rate to 1.0×10^{-9} mbar l/s.cm^2. The chamber is fabricated in two halves based on finite element analysis (FEA) using ANSYS software [5]. Model of this high vacuum chamber has been generated using CATIA software. The model is transferred to ANSYS through CATIA interface for the analysis considering the parameters as shown in the Table 1.

Journal of Physics: Conference Series **390** (2012) 012034 doi:10.1088/1742-6596/390/1/012034

Table 1. Parameters considered for FEA analysis.

Analysis type	Structural
Element used	Solid 186
Material properties	
Density	7850 Kg/m^3
Young's modulus	190 GPa
Possion's ratio	0.27
Boundary conditions	
Inside surface pressure	1.0×10^{-5} Pa
Outside surface pressure	1.013×10^5 Pa
Displacement of lower elliptical end	x, y, z - displacements = 0

The element Solid 186 is a higher order 3-D solid element that exhibits quadratic displacement behavior. The element is defined by 20 nodes having three degrees of freedom per node. The element supports plasticity, hyper-elasticity, creep, stress stiffening, large deflection, and large strain capabilities. Considering the above parameters of table 1, the deformation and von-mises stress results obtained after ANSYS analysis are shown in the figure 2. Analysis results show that maximum deformation of 0.2 mm and maximum stress of 21.9 MPa will be developed due to pressure difference which is less than the allowable tensile strength of 205MPa.

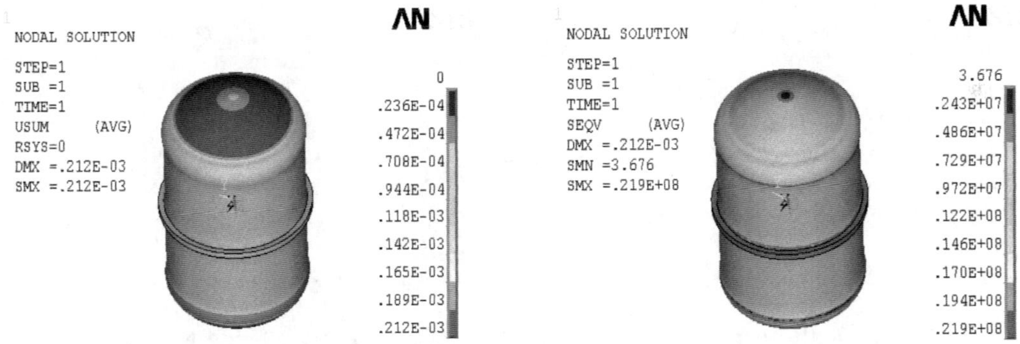

Deformation result Stress result

Figure 2. ANSYS results for deformation and stress values.

3. Experimental set-up and procedure

Due to larger dimension of the vacuum chamber, double O-ring configuration was utilized along with inter-space pumping in order to ensure the better leak tightness. Two numbers of 500 ISO-F flanges are provided on this chamber, out of which one is used for pumping purpose while other is used for human entry. Four numbers of 63 CF flanges are provided for connecting supply and return lines of VS and VM with hot nitrogen gas supply and return lines. Also two numbers of 300 class flanges are provided for connecting LN_2 supply and return lines for cool down of thermal shields as shown in the figure 3.

For rough pumping of the vacuum chamber, two numbers of rotary pumps are connected, each having a pumping speed of 35 m^3/hr. For high vacuum, one turbo-molecular pump of 5000 l/s pumping speed (N_2 equivalent) is connected to the chamber with effective pumping speed of 3550 l/s. Combined ionization gauge is used for the measurement of chamber vacuum. MKS make RGA is used to measure the partial pressure of different gases inside vacuum chamber.

Figure 3. High vacuum chamber with vacuum accessories.

Before insertion of VM and VS into the chamber, the complete baking channel is leak tested for the acceptable leak rate. Twenty eight (28) numbers of K-type Kapton insulated thermocouples calibrated within the accuracy of ± 1 °C are mounted at the different locations of VM / VS using high temperature Ceramabond glue from M/s. Aremco which can withstand the temperature up to 500 °C. Temperature data are acquired through de-mountable type 36-pins feed-through using MASIBUS make DATA logger and stored using CITECT SCADA based software via RS-485 transfer protocol. After insertion of VM / VS, the chamber is closed and pump down to the pressure < 1.0×10^{-5} mbar and then baked for steady temperature of 150 °C for eight (08) hours by passing hot nitrogen gas at 250 °C with the ramp rate of 50 °C/hr. Then VM / VS is cooled down naturally and baking channels are leak tested.

Similarly, the thermal shields are leak tested in a component labels for the acceptable leak rate. Then these thermal shields are fixed with VM / VS and are kept inside this chamber with their common supply and return headers. A provision for inter-face between the vacuum side headers and the atmospheric side is provided with special type of connector. Also the provision for pressurizing the panels at room temperature and cold condition is provided using valve arrangement at supply end. Eleven numbers of Pt-102 temperature sensors are mounted to these panels at midpoint of each bobbles. Once again the entire configuration is leak tested in vacuum mode at the background < 1.0×10^{-8} mbar l/s and then tested in sniffer mode at the background < 1.0×10^{-6} mbar l/s with internal helium pressure of 8.0 bar (g). After qualification test, the vacuum chamber is pumped down to the vacuum < 1.0×10^{-5} mbar. QMA was switched ON and the panels are pressurized to 8.0 bar (g) with helium gas. Panels were kept under pressurized condition for few minutes and a QMA scan was taken. Leak tightness of the panel was confirmed by observing the Helium partial pressure in the QMA scan. Then, the panels were de-pressurized to atmosphere and Liquid nitrogen was passed through the panels to cool them up to 80 K. The same procedure was repeated after achieving 80 K temperature on panels to see the leak tightness at cold condition.

4. Experimental results

The test chamber vacuum during baking along with temperature profiles at different locations of one of the VM is shown in the figure 4. Also the thermocouples mounting locations at the radial port and vertical ports with saturated temperature values are shown in the figure 5.

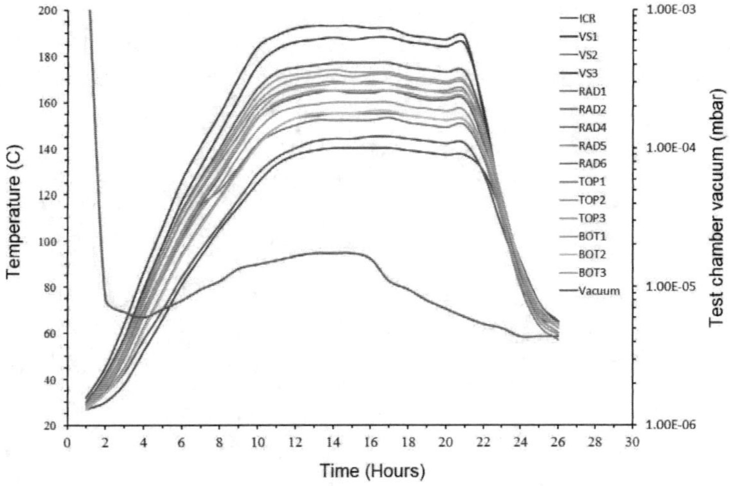

Figure 4. Test chamber pressure and temperature profiles of one of the VM.

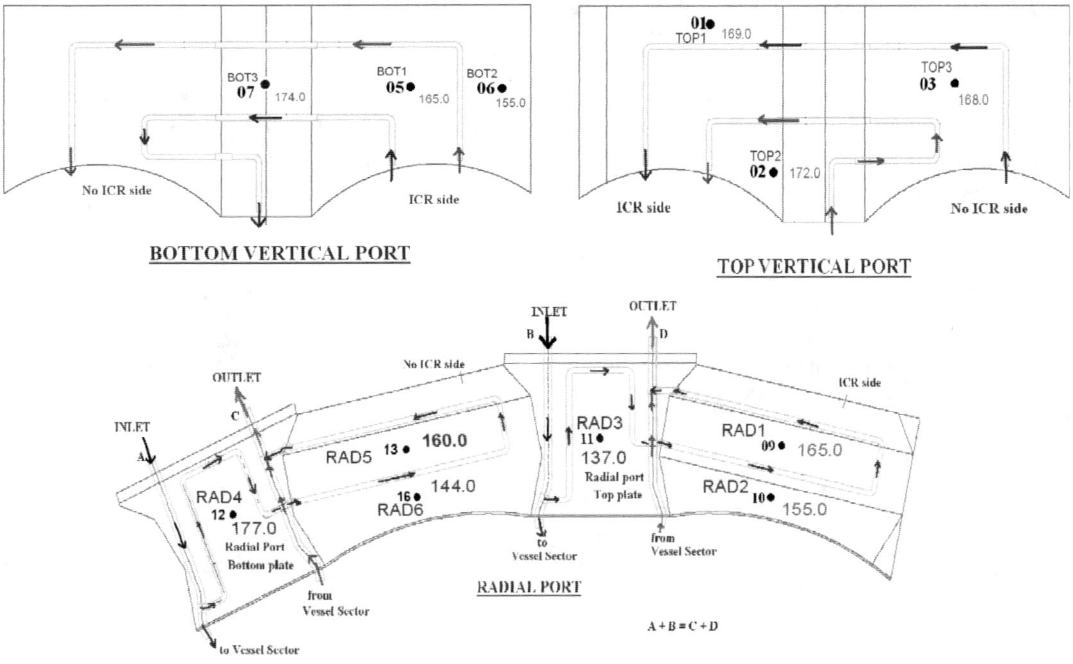

Figure 5. Temperature sensor locations of vertical and radial ports of one of the VM.

After baking and cool down, the leak testing of VM / VS baking channels is carried out. In few VM / VS, some leaks are detected ranging from 10^{-7} mbar l/s to 10^{-3} mbar l/s. These leaks are repaired and further leak tested and baked. This process is repeated till, the leaks are completely removed.

The cool down of thermal shields of two numbers of VS in series with time is shown in the figure 6. After achieving 80 K, QMA scan was taken as shown in the figure 7. This scan indicate no traces of helium gases confirming that there was no leak developed due to cool down. After integrated leak test, these panels are warmed up and again leak tested in integrated form. No leak found the background < 1.0×10^{-8} mbar l/s in vacuum mode and in the background < 1.0×10^{-6} mbar l/s with internal helium pressure of 8.0 bar (g) in sniffer mode.

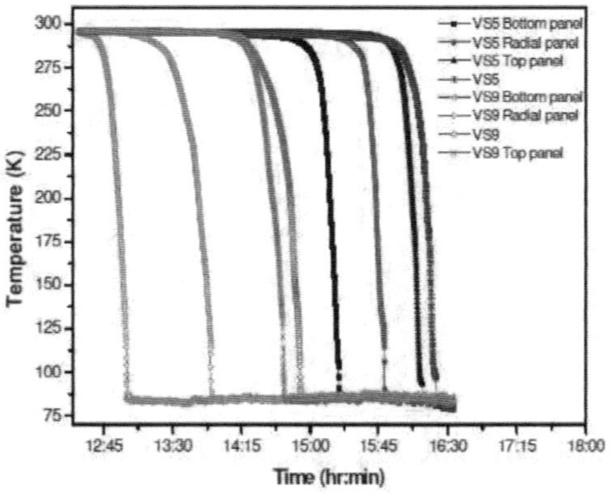

Figure 6. Vessel panels temperature with time.

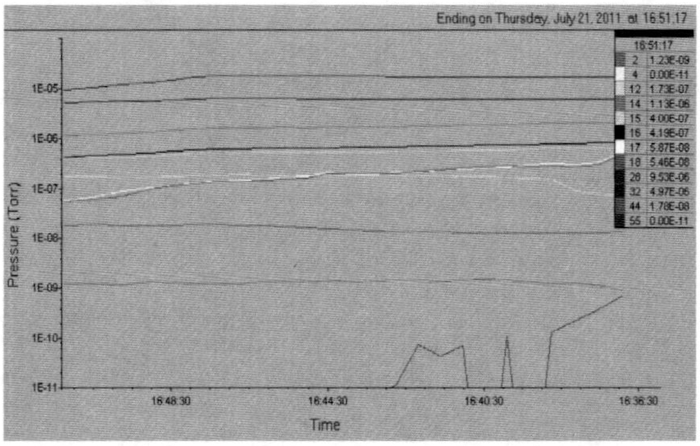

Figure 7. RGA Scan after cool down of shields.

5. Conclusion

The probability of leak development in the weld joints of VM /VS baking channels and in the thermal shields is significantly eliminated ensuring leak tightness of the system. Also their performances at their operating conditions are well established assuring their integration with SST-1 Tokamak.

References

[1] The SST-1 Team 1995 *16th IEEE/NPSS symposium on fusion Engineering* University of Illinois Urbana-Champaign **1** 481

[2] S. Pradhan & SST-1 Mission Team 2010 *J. Plasma Fusion Res. SERIES* vol 9 pp 650

[3] Rajendra Kumar E, Nagabhushana S, Pathak H A, Panigrahi S, Ranga Nath T, Santhana babu A V, Gangradey R, Patel R J and Saxena Y C 1998 *Fusion Technology* vol 2 p 1737

[4] Rao V V, Ghosh T B and Chopra K L 1998 *Vacuum science and Technology* (Allied Publisher limited)

[5] ANSYS Inc 2009 *ANSYS finite element analysis code Release 12.0.1* (Canonsburg, Pennsylvania, USA)

Compact pulsed electron beam system for microwave generation

S K Sharma[1], P Deb[1], R Shukla[1], P Banerjee[1], T Prabaharan[1], B Adhikary[1], R Verma[1], A Sharma[2] and A Shyam[1]

[1]Energetics & Electromagnetics Division, BARC, Visakhapatnam, A.P 530012

[2]Accelerator & Pulsed Power Division, BARC, Mumbai 400085

Email: surender80@gmail.com

Abstract. A compact 180 kV electron beam system is designed for high power microwave generation. The electron beam system is consists of a secondary energy storage device, which can deliver energy to the load at faster rate than usual primary energy storage system such as tesla transformers or marx generator. The short duration, high voltage pulse with fast rise time and good flattop is applied to vacuum diode for high power microwave generation. The compact electron beam system is made up of single turn primary tesla transformer which charges a helical pulse forming line and transfers its energy to vacuum diode through a high voltage pressurized spark gap switch. We have used helical pulse forming line which has higher inductance as compared to coaxial pulse forming line, which in turns increases, the pulse width and reduce the length of the pulse forming line. Water dielectric medium is used because of its high dielectric constant, high dielectric strength and efficient energy storage capability. The time dependent breakdown property and high relative permittivity of water makes it an ideal choice for this system. The high voltage flat-top pulse of 90 kV, 260 ns is measured across the matched load. In this article we have reported the design details, simulation and initial experimental results of 180 kV pulsed electron beam system for high power microwave generation.

1. Introduction

The pulsed power field is driven by weight, size and volume constraints nowadays. In both the military and commercial areas it is needed to provide more and more capability in smaller and lighter systems. The need for higher energy density, power density, reliability, and efficiency is driving progress in the field. Some applications also require technology that can be deployed in smaller spaces under stressful environments. The pulsed electron beam has utilization in various applications such as flash X-ray generation [1], Compact laser drivers[2], high power microwave generation [3] etc. The electron beam generator system consists of primary energy storage system, high voltage generator, pulse compression system, high voltage switch and vacuum field emission diode. .

2. Compact Pulsed Power System

The compact pulsed power system is consists of primary energy storage capacitor, charged with high voltage dc power supply and discharges in to the primary of tesla transformer, which produces high voltage pulse with the rise time of microsecond duration and charges a helical pulse forming line. The block diagram of the system is shown in Figure.1. The inductance of helical pulse forming line is

Published under licence by IOP Publishing Ltd

higher than normal coaxial pulse forming line so it produces longer duration high voltage pulse. The helical pulse forming line transfers its energy through high voltage self breakdown

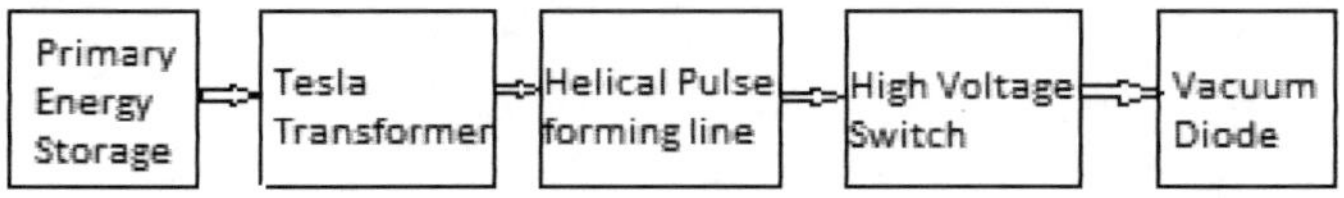

Figure 1. Block Diagram of electron beam system

pressurized spark-gap switch to the vacuum field emission diode.

2.1 Primary Energy Storage Capacitor

The capacitor is charged with 25 kV high voltage DC power supply. Electrical energy is electrostatically stored in the capacitor bank which discharges into the primary of the tesla transformer. The specifications of capacitor bank are
Energy stored – 400 J
Capacitance - 1.29μF
Maximum Charging Voltage – 25kV
Peak Current – 125kA
Inductance - <50nH
Four numbers of coaxial cable (RG 213) are used to keep the inductance low and connected to the primary of tesla transformer through a pressurized spark gap closing switch. The capacitor is charged with unregulated 25 kV DC power supply.

2.2 High Voltage Tesla Transformer

The high voltage tesla transformer is a primary and secondary circuit coupled with air core. The optimum design of tesla transformer depends upon the coefficient of coupling, turning ratio and dielectric medium [4,5]. The secondary capacitance of tesla transformer is determined by the capacitance of the helical pulse forming line. The primacy capacitance is the capacitance of the energy storage capacitor.
The specifications of tesla transformer are
Primary Inductance, L_p = 145 nH
Secondary Inductance, L_s = 150 uH
Coupling coefficient, k = 0.56

2.3 Helical Pulse Forming Line and Spark-gap Switch

The transmission line characteristics of helical line can be used for generating rectangular high voltage pulse across the vacuum diode. The helical pulse forming line produces longer pulse compared with conventional coaxial transmission line [6-8] as it has higher inductance compared to coaxial pulse forming line.
The specifications of helical pulse forming line are
Capacitance of helical pulse forming line, C_h = 5.8 nF
Impedance of helical PFL Z_0 = 21
Pulse width = 240 ns
It should be ensured that no breakdown of dielectric occurs between the conductors. The radial electric field E_r between the conductors is given by $E_r = V_0 / r \, . \ln (b/a)$. The electric field stress is maximum on the surface of the inner conductor and it should not be more than the breakdown strength of the medium. The electrical field stress calculated on inner conductor is 61kV/cm, which is less than the breakdown strength of the medium. The helical pulse forming line transfers peak power to the vacuum diode through a high voltage self breakdown pressurized spark-gap switch.

Journal of Physics: Conference Series **390** (2012) 012035 doi:10.1088/1742-6596/390/1/012035

2.4 Vacuum Field Emission Diode

The vacuum field emission diode is made of graphite cathode and SS mesh anode. The electron beam is generated by application negative high voltage across the cathode. Microscopic protrusions exist on all surfaces and when negative voltage of > 100 kV/cm is applied between the cathode and anode the electric field at the tip is enhanced many times and electrons are emitted from the surface of the cathode [9].

The current density for a planer diode is given by

$J_{CL} = (4\varepsilon_0 / 9).\ (2e / m)^{1/2}\ .\ V^{3/2} / d^2$

And impedance of system is given by

$Z= V/ I = (9 / 4\pi\varepsilon_0 r^2).\ (2e / m_0)^{-1/2}\ .\ V^{-1/2}\ d^2$

Helical Pulse Forming Line Voltage = 180 kV
Impedance of the vacuum diode = 20 ohm
Voltage across diode matched load, V = 90 kV
Electron beam current = 4.5 kA & Pulse width = 240 ns

3. PSPICE Modelling of Electron Beam Generator

The modelling of electron beam system is done with PSPICE software. The primary energy storage capacitor discharges into the primary of tesla transformer (Figure. 2). The tesla is modelled as an equivalent air core pulse transformer having primary and secondary inductances of 145 nH and 150 uH respectively with coupling coefficient of 0.56. Helical pulse forming line is modelled as ideal transmission line with characteristic impedance of 21 ohms and one-way transit time of 120 ns. It

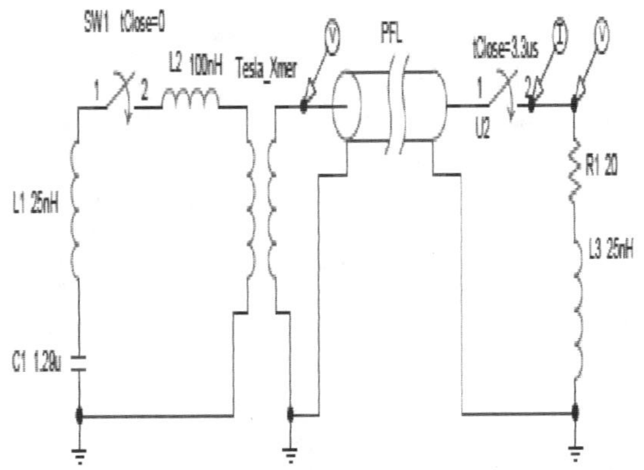

Figure. 2. PSPICE schematic circuit

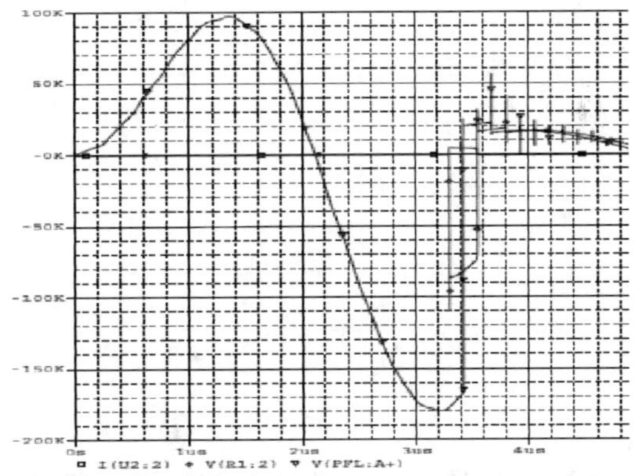

Figure. 3. PFL Voltage, Load Voltage & Load Current

discharges into a 20 ohm load. For simulating practical condition inductance of 25 nH on load side is included and capacitor inductance of 25 nH and connecting cable inductance of 100 nH is taken on primary side. The simulation result when the primary is charged with 18 kV is shown for Helical pulse forming voltage, voltage across diode & electron beam current (Figure. 3).

4. Experiments

The primary energy storage capacitor was charged to 18 kV and discharged into the primary of tesla transformer using pressurized spark-gap switch (Figure. 4). The secondary of the tesla charges the helical pulse forming line to 180 kV. To measure the pulse duration across the load the

Figure. 4. Compact electron beam system

PFL was discharged into matched load of 20 ohms through a pressurized self breakdown spark-gap switch and the voltage is measured across the PFL & the load (Figure. 5). The vacuum field emission diode was connected after the self breakdown pressurized spark-gap switch. The switch closes at 180 kV and the voltage across the PFL and electron beam current is measured (Figure. 6). The neon bulb

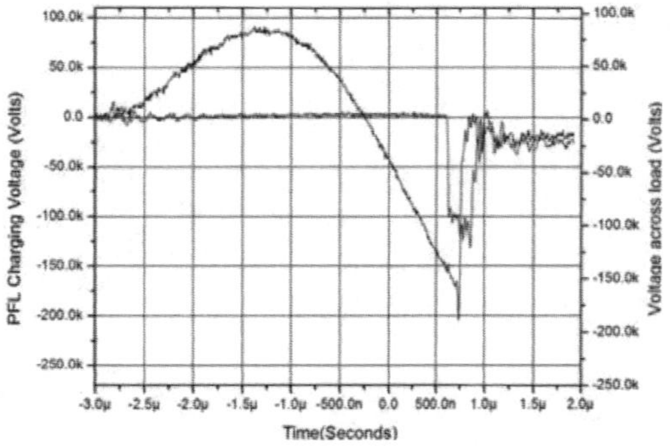

Figure. 5. PFL Voltage & Load Voltage

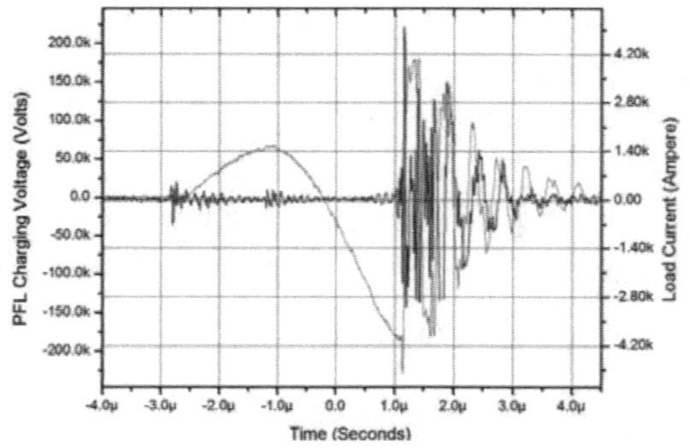

Figure.6. PFL Voltage & Electron beam current

was kept at a distance of 15 cm and the glow was observed (Figure. 7). The microwave signal is

Figure. 7. LED glow with microwave

measured using an antenna at a distance of 50 cm. The microwave detector diode output (Figure. 8)

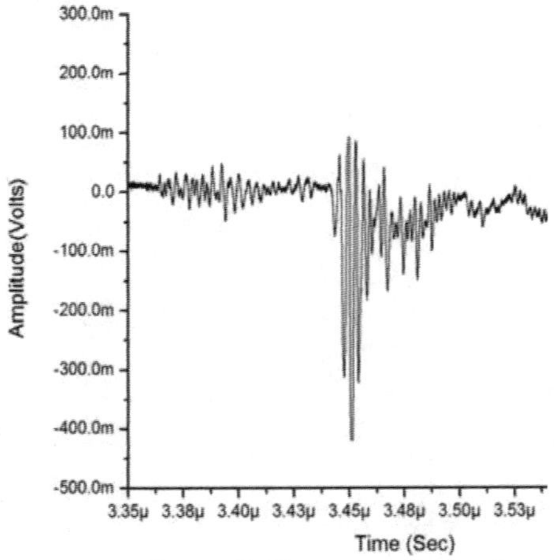

Figure.8. Detector diode signal

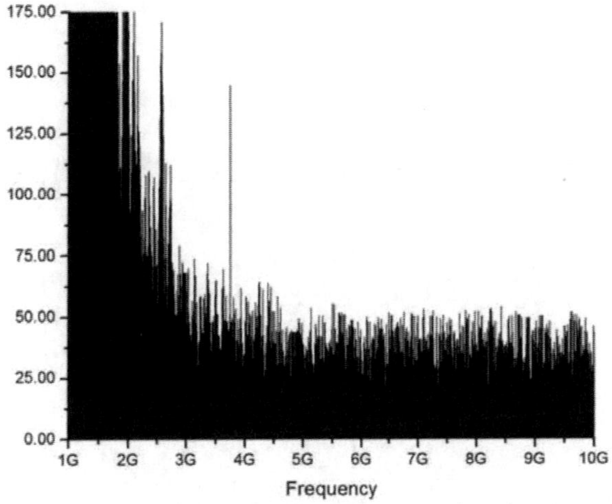

Figure.9. FFT of microwave signal

and the FFT of signal (Figure. 9) is done to determine the dominant frequency components.

5. Results & Conclusion

A compact electron beam is designed and fabricated. The pulse forming line was charged up to 180 kV and discharged into a matched load and voltage of 90 kV, 240ns has been measured. It is seen that helical pulse forming line produces longer duration high voltage pulse and makes the system compact as compared to coaxial pulse forming line, which requires longer system for producing same pulse width. The field emission diode was connected after the PFL and switch. The peak voltages of 90 kV have been applied across the diode & peak electron beam current of 4.5 kA has been measured. The microwave signal is measure at a distance of 50 mm and detector diode measure negative 425 mV signal and microwave pulse duration of 30 ns. The FFT of the microwave signal is done and the dominant frequency components of 2.5 GHz & 3.6 GHz are seen.

Acknowledgement

We would like to acknowledge Shri S Bhattacharya AD(T), E&I Group and Shri G P Srivastava Director, E&I Group for their continuous encouragement and their valuable suggestions.

References

[1] Yusuke Sakai, Shnsuke Takahashi,Takanonri Komatsu, Inho Song, Masato Watanable and Eiki Hotta, Review of Scientific Instruments 81, 013303 (2010)
[2] E. Spahn, G. Buderer and W. Wenning, IEEE Transaction on Magnetics, 35, 378 (1999)
[3] Y J Chan, A A Neuber, J Mankowski, J C Dickens, M Kristiansen and R Gale, Review of Scientific Instruments 76, 104703 (2005)
[4] C R J Hoffmann. Review of Scientific Instrument, Vol. 46 , pp 1 – 5 (1975).
[5] K K Jain, D Chennareddy, P I John and Y C Saxena (Feb 1986), Sadhana, Vol. 9, (1986)
[6] G S Kino, S F Paik, Journal of Applied Physics, 33, 3002 (1962)
[7] Liu Zhenxiang, Zhanng Jiande, Plasma Science & Technology, Vol.8, No.5. pg 596-599(2006)
[8] S F Paik, IEEE Transaction on Electron Devices, ED-16, 1010 (1969)
[9] R B Miller, Intense Charged Particle Beams, New York: Plenum (1982)

SST-1 Gas feed and Gas Exhaust system

Dilip C Raval, Ziauddin Khan, Prashant L Thankey, Kalpesh R Dhanani, Firozkhan S Pathan, Pratibha Semwal, Siju George, Paravastu Yuvakiran, Himabindu Manthena and Subrata Pradhan

Institute for Plasma Research, Near Indira Bridge, Bhat, Gandhinagar – 382428, Gujarat, India

E-mail: raval@ipr.res.in

Abstract. SST-1 tokamak is a long pulse tokamak designed for the plasma operation up to 1000 sec duration. Gas feed system and gas exhaust management will play a very crucial role during plasma discharge. During the different type of operations of tokamak like wall conditioning, diverter operation and neutral beam injection, a large amount of gas will be fed into the vacuum chamber at different locations. Also during plasma operations, the gas will be fed both in continues and pulse mode. Gas feed will be carried out mainly using piezo-electric valves controlled by PXI based data acquisition and control system. Such operations will lead to a huge amount gas exhaust by the main system which requires good exhaust facility to searches, great care should be taken in constructing both. Also initial pumping of cryostat and vacuum vessel of SST-1 will release a large amount of gas. Exhausted gases from SST –1 will be Hydrogen, Nitrogen, Mixture gases or some toxic gases. Dedicated exhaust system controlling the different gases are installed. Special treatment of hazardous/explosive gases is done before releasing to the atmosphere. This paper describes design and implementations of the complete gas feed and exhaust system of SST-1.

1. Introduction

Vacuum chamber of steady state superconducting (SST-1) tokamak comprises mainly of two vacuum chambers, (1) Vacuum Vessel in which plasma will be produced and confined and (2) Cryostat which will provide working environment for all superconducting magnetic coils [1, 3]. Vacuum vessel and cryostat are designed and fabricated for ultra-high vacuum and high vacuum applications respectively. Ultimate base pressure in vacuum vessel will be $\leq 1.0 \times 10^{-8}$ mbar while that in cryostat will be $\leq 1.0 \times 10^{-5}$ mbar. For tokamak experiments, the gas puffing is carried out to have an initial break down. After break down, the plasma fuelling plays an important role to maintain and control the plasma density. For the tokamak operating in steady state and in divertor mode, the steady state removal of heat and exhaust particles are much required to maintain the plasma density for longer periods. A number of techniques like external gas feed, neutral beam injection and pellet injection etc. are being implemented for fuelling the plasma. Out of these techniques, the gas feed system is the most basic and primary need for tokamak experiment during plasma shot.

Main functions of the gas feed system are (a) the pre-filling of the vessel to a required pressure, (b) feeding of required amount of gas during the plasma current and the density build up phase, (c) feeding of amount of gas to maintain the plasma density at required value, (d) feeding of gases into the vessel during wall conditioning and (e) feeding of required amount of gases in divertor region to remove the heat and exhaust particles for plasma diagnostic. For all these functions during plasma shot, a very efficient gas feed system is essential.

Further due to very large volumes of the vacuum vessel (16 m^3) and the cryostat (39 m^3), during initial pumping of both of them, a large amount of gases (mainly air) will be extracted and released through the pumps. Also during plasma discharge operation of tokamak, a large amount of hydrogen gas will be released through the pumps as the vacuum vessel's pressure will be maintained $\leq 1 \times 10^{-4}$ mbar during entire plasma discharge of 1000 sec. Special gases like helium, neon and other special gases will also be used for vessel wall conditioning, divertor operations, plasma diagnostics, Neutral Beam Injection etc. So an effective gas exhaust handling system for SST-1 vacuum system as well as other systems is required.

2. SST-1 gas feed system

For proper fulfilment of SST-1 tokamak requirements, the gas feed system is divided into two sub-systems. First sub-system will be used for plasma fuelling in which wall conditioning, pre-filling and the density control will be carried out. Second sub-system will be used for divertor fuelling. Total twenty-two numbers (22) of piezoelectric gas feed valves will be used in SST-1. Piezoelectric gas feed valve works on 100 VDC and starts functioning above 30 VDC. By varying applied voltage, the required gas throughput from this valve can be varied. Gas feed pulse shape can also be adjusted as per requirement using PXI based control system.

2.1. Plasma fuelling

Eight (08) numbers of piezoelectric valves will be mounted diagonally opposite to each other at four radial ports to maintain the overall balance of gas distribution inside the vessel over a given period of time. These valves will be used for wall conditioning, boronization, pre-filling and density control. During all above phases, hydrogen gas of very high purity of ~ 99.9999 % will be injected into the vessel through these valves. The hydrogen gas of such purity will be obtained by feeding the supplied gas of 99.95 % purity through a hydrogen purifier of the same order. This ultra-high pure gas cannot be fed into the vessel on line due to variation of gas requirements in different phases. Also the function of hydrogen purifier, incorporated with a palladium cell consisting of palladium alloy tubing which acts as the diffusion element, depends on the inlet and outlet pressures.

SST-1 tokamak is a long pulse tokamak with maximum plasma duration of 1000 seconds and the amount of high purity hydrogen gas required during single plasma shot will be approximately 80,000 torr litre. To fulfil these requirements, one buffer chamber is used for storage of ultra-high pure hydrogen gas. As the outlet pressure of the hydrogen purifier is 5 bar, the buffer chamber will be maintained at the same pressure. At room temperature and 5 bar, a buffer chamber of 30 litre will be sufficient for storage of gases required for at least 2 cycles. During different phases, the amount of gas feed into the vessel should be known so that the behaviour of the graphite surface absorption can be determined for next plasma discharge. For this purpose, a mass flow meter for hydrogen gas is being used for measuring the amount of gas puffed during operation. Since these values are operated using DC voltage, they should be isolated from both the high-pressure side and the low-pressure side of the vacuum vessel by using vacuum isolators to avoid the grounding, which may lead to change in the amplitude of the applied signal. Before the piezoelectric valve, a know conductance and a pressure gauge will be mounted to measure the quantity of gas being puffed to the vessel by each piezoelectric valve.

During wall conditioning, the mixture of hydrogen and helium gas will be puffed to the vessel. The mixture of required percentage of hydrogen and helium gases will be carried in the 10 litre chamber by puffing these gases into this chamber through mass flow meters for each of them. The entire installation of the gas feed line will be carried out by the polished SS 316 tubes and swagelock type quick connectors. Since the entire line will be bifurcated towards the different radial ports as per requirement, the ON/OFF valves will be provided at each line for functioning of the respective line whenever required. Before operation, the entire gas feed line will be pumped to a pressure below 10^{-3}

mbar using the dry rotary pumps, followed by the baking of line at 150° C for few hours. The schematic of plasma fuelling is shown in the figure 1.

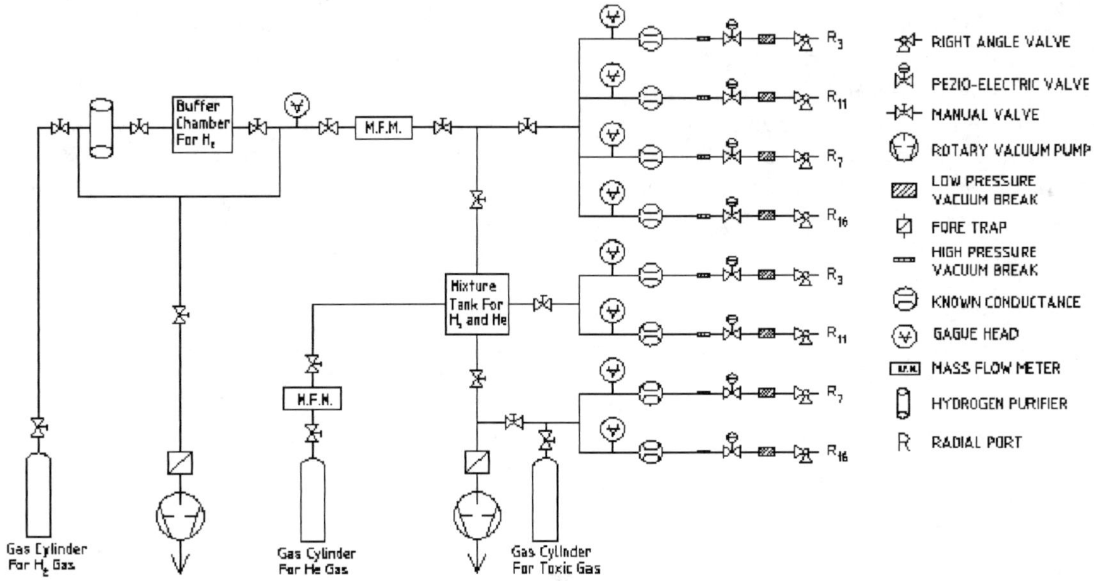

Figure 1. Schematic for SST-1 Gas-feed System

2.2 Divertor fuelling

Divertor fuelling schematic of SST-1 is shown in the figure 2. The integration of this supply line will be carried out in the same manner that to previous one. Here each of eight piezoelectric valves will be connected to the top and bottom vertical ports respectively. In both top and bottom regions, the feeding will be performed through eight outer ports and four inner ports. The amount of gas required will be supplied from another 10 litre chamber. If mixture of gases is desired for puffing in divertor region, then mixing of gases can be done in the same chamber using mass flow meter.

3. SST-1 Gas Exhaust System

Apart from SST-1 main vacuum system, a large number of other subsystems will also be connected to SST–1 vacuum vessel for plasma heating, plasma current drive, plasma diagnostics etc. Many of these subsystems will be using their own dedicated vacuum systems. So, an effective gas exhaust handling system for SST–1 vacuum system as well as other subsystems is required.

3.1 Design Calculation

One roots rotary pumping station of 1800 m^3/hr will be used for initial roughing of the vacuum vessel from atmosphere to 1.0×10^{-3} mbar. Two numbers of turbo-molecular pumps, each having 5000 l/s (N_2) pumping speed and two numbers of closed cycle cryo-pumps (each of 10,000 l/s pumping speed for water vapor) will be used to pump down vessel from $1 .0 \times 10^{-3}$ mbar to $\leq 1.0 \times 10^{-8}$ mbar. Sixteen (16) numbers of similar turbo-molecular pumps will be used to pump the divertor region during steady state (1000 sec) plasma discharge. Vacuum vessel pumping system will have to remove about 60 mbar l/s hydrogen gas during this phase of operation.

Two numbers of roots rotary pumping stations, each of 1800 m^3/hr will be used for initial rough pumping of cryostat from atmosphere to 1.0×10^{-3} mbar. Three numbers of turbo-molecular pumps each having 5000 l/s pumping speed will be used to pump the cryostat from 1.0×10^{-3} mbar to $< 1.0 \times 10^{-5}$ mbar. During normal operation of cryostat at pressure $< 1.0 \times 10^{-5}$ mbar, liquid nitrogen (LN_2)

panels will be cooled down to 80 °K and super-conducting coils will be cooled down to 4.5 K. These cold surfaces will provide large pumping speed for all gases except hydrogen and helium. So, the gas load will be very small during normal operation of cryostat. However, during rough pumping as well as thermal run away (cold surfaces getting warmed up accidentally), very large amount of gases will be exhausted from cryostat. Total gas load from all other systems connected with SST–1 (e.g. plasma diagnostic etc.) is assumed to be much less than the gas loads from vacuum vessel and cryostat.

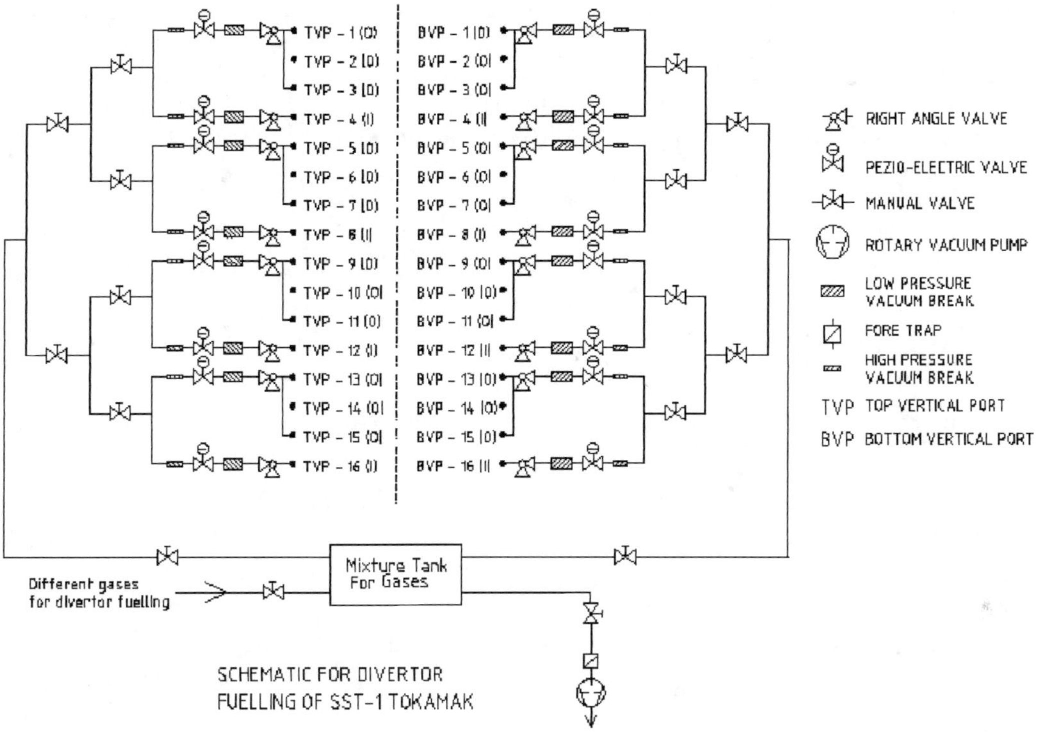

Figure 2. Schematic for SST-1 Divertor Fueling System

As a design criterion, it is important to ensure that gas exhaust system removes the gas from the exhaust port of the rotary pump with a speed such that pressure more than 1.2 bar (a) is not built up at the exhaust port of the pump. This is because feather valve of the rotary pump exhaust port is designed to operate efficiently at this pressure on its exhaust side. The volumetric flow rate (speed) for different inlet pressure is given in the table 1 for these root pumps.

Table 1. Volumetric flow rate of root pumps at different pressures

Pressure (mbar)	Volumetric flow rate (m^3/hr)	Total throughput Q (mbar l/s)
1.0×10^3	250	69.42×10^3
1.0×10^2	320	8.88×10^3
10	1.0×10^3	2.77×10^3
1	1.7×10^3	4.72×10^2
1.0×10^{-1}	1.8×10^3	50.00
1.0×10^{-2}	1.6×10^3	4.47
1.0×10^{-3}	1.1×10^3	0.305

The maximum throughput that will be exhausted by a single pump in SST–1 vacuum system is Q = 69.42 × 10^3 mbar l/s. Thus the net gas throughput that will be exhausted to the exhaust line by all three pumps is Q = 2.08 × 10^5 mbar l/s. For proper functioning of rotary pump, the pressure at its exhaust should not exceed 1.2 bar (a). This is possible by the use of pipeline of suitable conductance. The conductance needed to maintain the pressure difference of Δp = 0.2 bar for above throughput is C = Q / Δp = 1.041 × 10^3 l/s. This is the minimum required conductance that the line should have to maintain 0.2 bar pressure difference between inlet point of gas exhaust line (i.e. outlet of rotary pump) and outlet of gas exhaust line (i.e. 1 bar (a)).

The relation between the throughput (mbar l/s) and the diameter (cm) of the pipe for turbulent flow (Q > 196 d) [4] shows that for the tube up to 806 cm diameter, we will get turbulent flow. Again the conductance of a circular tube of diameter (d) cm and length (l) cm in viscous flow is given by

$$c = 182\left(\frac{d^4}{l}\right)\left(\frac{p_1 + p_2}{2}\right) \quad \text{l/s} \qquad (1)$$

where p_l = 1.2 bar is the pressure at entrance of the pipe,
p_2 = 1.0 bar is the pressure at exit of the pipe.

The velocity of exhaust gas for a given diameter d (m) is given by

$$v = \frac{4v'}{\pi d^2} \quad \text{m/hr} \qquad (2)$$

where v' is the volumetric flow rate in m^3/hr.

According to the possible lay out of the gas exhaust line in SST–1 hall, the effective length of the exhaust line (including bends etc) in the design is 80 meter. Thus for this length and different diameter of the pipe, the conductance of the pipe is as follows

(1) For d = 6.5 cm, C = 0.340 × 10^5 l/s
(2) For d = 10 cm, C = 0.830 × 10^5 l/s
(3) For d = 10 cm, C = 2.026 × 10^5 l/s
(4) For d = 12.5 cm, C = 4.947 × 10^5 l/s
(5) For d = 15.0 cm, C = 1.025 × 10^6 l/s

The above calculations and comparison between two conductance show that the pipe of 10.0 cm diameter is suitable for exhaust line.

3.2. Gas Exhaust Line Layout

SST–1 gas exhaust system is designed to remove (a) air during normal pumping of vacuum vessel and cryostat, (b) hydrogen during plasma operation, (c) some special gases for plasma diagnostics and wall conditioning. Considering this, three exhaust lines are designed for SST–1 tokamak. First line will be used only for main vessel and cryostat pumping system of SST–1 tokamak. Second one will be used by all other sub-systems connected to vacuum vessel. Third exhaust line will be used for special gases. These three lines are labeled as (1) VACUUM SYSTEM EXHAUST, (2) DIAGNOSTIC SYSTEMS EXHAUST and (3) SPECIAL GASES EXHAUST respectively. A standard rigid PVC pipes are selected for different exhaust lines. Vacuum System exhaust line has 104 mm internal diameter for vessel and cryostat pumping system due to very high gas loads. Exhaust line for diagnostics and other subsystems attached to SST–1 vacuum vessel has 84 mm internal diameter because it has relatively low gas load as compare to main pumping system. Special gases exhaust line has 57 mm internal diameter. Each line will have a number of ports to connect the exhaust line to the exhaust port of rotary pump. Each port will have cap to close the port when it is not being used to stop gas from re-entering in SST–1 hall. All three lines will be taken into exhaust gas treatment room and connected to the centrifugal blowers. Special gases exhaust line can be converted into toxic gas exhaust line, if required, by connecting toxic gas treatment chamber in its series. After commissioning, all three gas exhaust lines will be helium leak tested for leak rate $\leq 1.0 \times 10^{-5}$ mbar liter/sec at 1.2 bar (a) internal pressure with sniffer probe.

A number of sensors for hydrogen, helium and toxic gas (if required) will be deployed in SST–1 hall as well as exhaust gas treatment room to detect the leakage of these gases either from supply lines or from exhaust lines. Oxygen level sensors will also be deployed in SST–1 hall and exhaust gas treatment room to monitor safe working level of oxygen concentration.

4. Conclusion

Gas feed system for SST-1 is commissioned and tested for its functionality for 1st first plasma operation i.e. for circular plasma operation. This gas feed system will be extended further for divertor fuelling in future. Also exhaust system commission is over. The probability of back trust of exhaust gases from pumps is eliminated by use of higher capacity bowers operation.

References

[1] Bhise M K et al 1996 *Fusion Technology* vol 1 p 805
[2] Ranga Nath T et al 1997 *17th SOFE* vol 1
[3] Pradhan S & SST-1 Mission Team 2010 *J. Plasma Fusion Res. SERIES* vol 9 p 650
[4] Armand Berman 1992 *Vacuum Engineering calculations, Formulas, and Solved Exercises* (Academic Press Inc, New York) p 81

Design and Simulation of Electron Gun for a Multibeam Klystron

A K Nehra[1,2], R K Gupta[1], S M Sharma[1], P C Panda[1],Y Choyal[2] and R K Sharma[1]

[1]Microwave Tubes Division, CSIR-Central Electronics Engineering Research Institute, Pilani-333031, Rajasthan, India
[2]School of Physics, Devi Ahilya University, Indore-452001, M.P, India

E-mail: ashokkumarnehra@gmail.com

Abstract. This paper represents the design of multi-beam (sixty-beam) electron gun and focusing system for high power, compact klystron. The beam voltage is 4 kV with a total beam current of 513 (8.55 x 60) mA which is equally divided among sixty-beam. Each beam has a perveance of 0.033 μP making a total gun perveance of 2.02 μP corresponding to a total beam power of more than 2 kW. The cathode radius is7 mm and individual emitter radius is 0.2 mm having current density 6.7 A/cm2. The design has been accomplished using OPERA 3D code. All beamlets have individual anode as well as BFE and a common focusing system. Potential difference between cathode and anode is 4 kV. A magnetic field of 1200 Gauss is applied along the beam axis. A major challenge for the development of multi-beam klystron is design and technology for the focusing of off-axis beamlets because off-axis beams are at various azimuths.

1. Introduction

The basic principle of operation of any microwave tube is the effective interaction between the electron beam and the electromagnetic wave in an interaction region. Therefore electron gun plays a vital role in the design and development of any microwave tubes. For new generation microwave tubes like multi-beam klystron electron beam of desired shape, energy and current density and its focusing is really a challenging task. Since the drawbacks of the single beam klystron like high operating voltage, narrow bandwidth and more weight cannot be eliminated due to some physical limitations or technical restrictions. Therefore, concept of multi-beam electron gun has been introduced. The multi-beam klystrons (MBKs) are the only microwave amplifiers that fulfil the most of present demand of users, such as high power, high efficiency, large bandwidth, low noise and low-voltage operation, as compared to conventional single-beam klystrons. Due to these features MBK is used for ground-based, airborne and space-based radars, radio-communication and high energy particle accelerators [1]-[7].

In an MBK each beamlet propagates along its own individual path through the RF section consisting of two or more cavity resonators. The most important advantage of an MBK is that the current and perveance per beamlet are low while total current and perveance remain high in this device. For higher electronic efficiency and gain of device bunching should be more, which can be achieved by low perveance per beamlet. Further, the overall higher perveance allows a lower voltage operation yielding higher beam power and RF power as compared to conventional single beam klystron. In addition, the reduced operating voltage significantly decreases the cost of the power supply system and reduces the device length thereby making the device more compact. Moreover, the multiple-beam devices are designed with lower cavity shunt impedance as compared to single-beam devices resulting in a broader bandwidth [2], [8].

Figure 1 An example of 4 kV sixty-beam electron gun consists of sixty emitters (radius = 0.2 mm) on a common plate with sixty electrically isolated BFE (again a common plate) as well as sixty individual anodes constructed on a common plate has been taken for study and focused with a common solenoid.

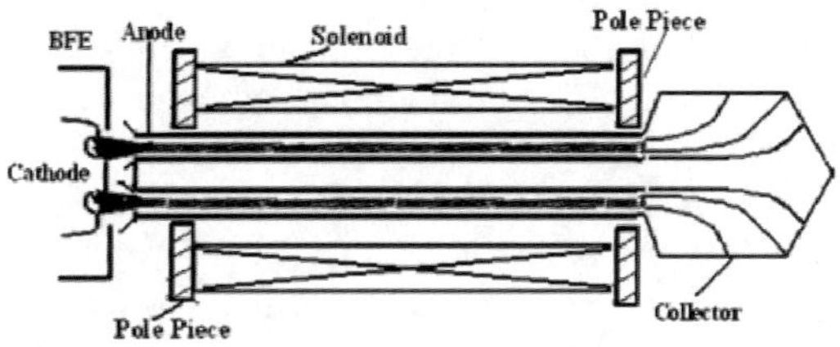

Figure 1. Schematic of sixty beam stick tube

Table 1. Gun design parameters

Parameter	Value
Cathode voltage	-4 kV
No. of beamlets	60
Total beam current	513 mA
Total beam perveance	2.02 μP
Individual beam perveance	0.033 μP
Cathode diameter	14.4 mm
Cathode loading	6.7 A/cm^2
Beam waist radius	0.15 mm
Beam throw	2.75 mm

2. Electron Gun Design

In this section we discuss how to format the title, authors and affiliations. Please follow these instructions as carefully as possible so all articles within a conference have the same style to the title page. This paragraph follows a section title so it should not be indented.

2.1 Objective of the sixty beam electron gun

The main objective of designing the multi beam electron gun is to replace a single beam high perveance (1.2 μP), high voltage (6 kV) electron gun with low perveance (0.033 μP) and low voltage (4 kV) compact multi beam electron gun. An effort has been made to design sixty beam low perveance electron gun using 3D code OPERA. Initially solenoid focusing has been planned to achieve the sixty-beam focusing with minimum beam scalloping and at later stage permanent magnet can be tried.

2.2 Sixty-beam gun topology

The gun design parameters of multi-beam electron gun are shown in Table 1. A sixty beam electron gun with planer cathode has been simulated using OPERA 3D code [9]. The electron gun employs a concentric packing topology with four PCDs 12.8 mm, 9.6 mm, 6.4 mm, and 3.2 mm having emitters 24, 18, 12, and 6 respectively as shown in figure 2(a) and is expected to produce 513 mA at its nominal operating voltage of 4 kV. The cathode radius is 14.4 mm and individual emitter radius is 0.2 mm having current density 6.7 A/cm2. All beamlets have individual anode as well as BFE and a common focusing system. Potential difference between cathode and anode is 4 kV. A magnetic field of 1200 Gauss is applied along the beam axis. Figure 2 (b) illustrates one quadrant of the basic topology of the optimized sixty beam electron gun design, which includes sixty planer emitters set within a cathode mask, an sixty-aperture focus electrode, and an sixty-aperture anode.

Figure 3(a) shows the beam trajectories in a full gun geometry and beam tunnels with a guiding magnetic field of 1200 Gauss. Total beam current is 513 mA (8.5 mA×60) at a voltage of 4 kV corresponding to a total beam power of more than 2 kW and magnified view of their corresponding graph trajectories on different PCDs are shown in figure 3(b). The computed beam-throw and beam-waist radius are 2.75 mm and 0.15 mm, respectively. Confined flow focusing system has been chosen to have better focusing, resulting in higher value of output RF power. The end-on views of electron trajectories corresponding to figure 3(a), is shown in figure 4(a). It has been observed that the beamlets are laminar and within the initial beam-to-tunnel fill factor at the beam waist. The achievement of this sixty beam electron gun is that the beam cut off in single beam takes place when BFE is biased with negative 300 V with respect to cathode while in case of multi-beam it takes place at negative 100 V as shown in figure 4(b) [10].

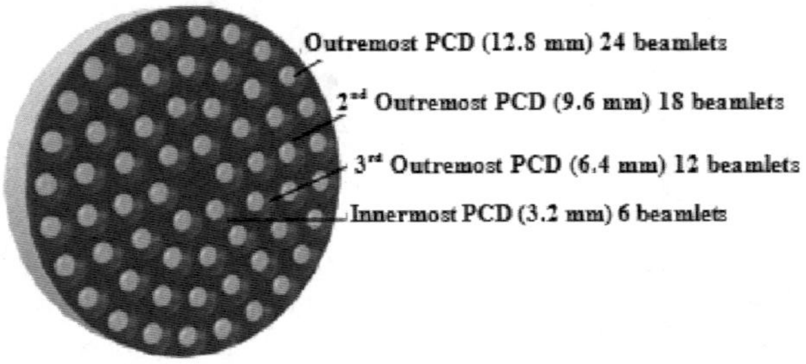

Figure 2(a). Sixty emitters and BFE on a common plate

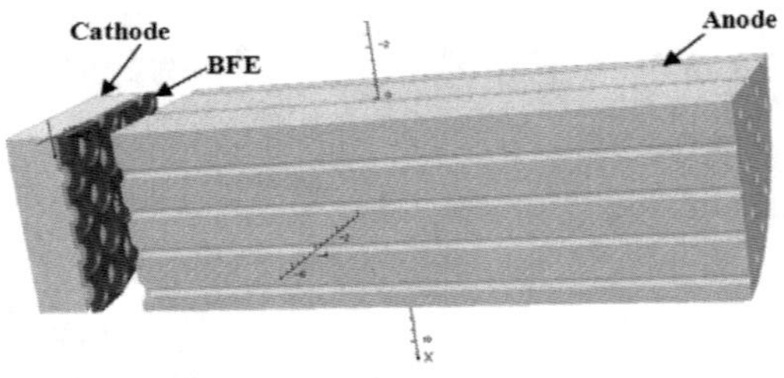

Figure 2(b). Quarter section of multi-beam electron gun

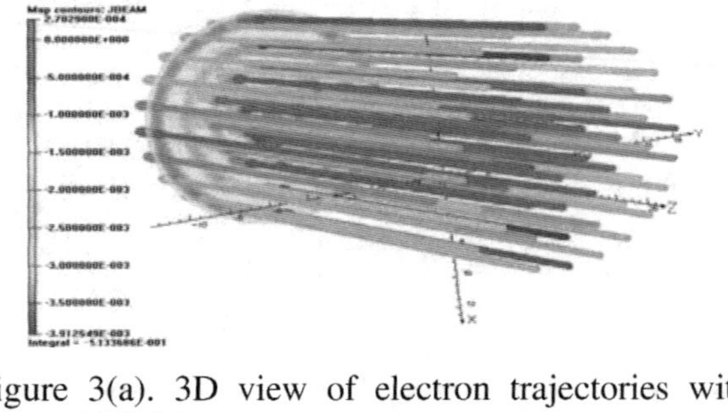

Figure 3(a). 3D view of electron trajectories with magnetic field

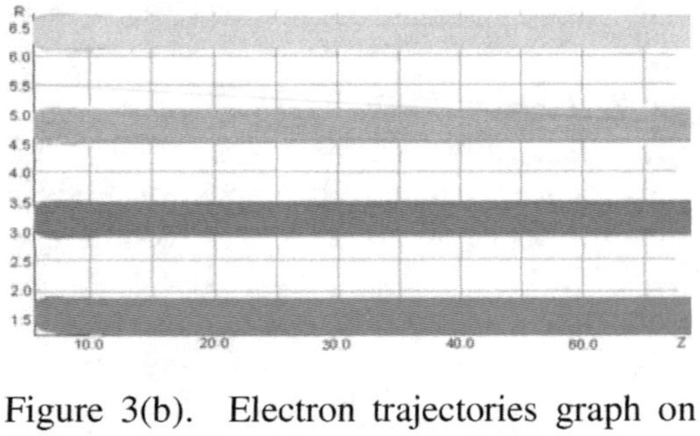

Figure 3(b). Electron trajectories graph on different PCDs with magnetic field

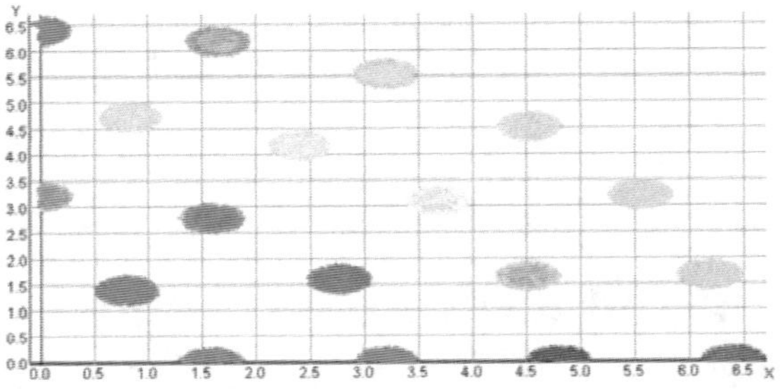

Figure 4(a). 3D view of electron trajectories with magnetic field

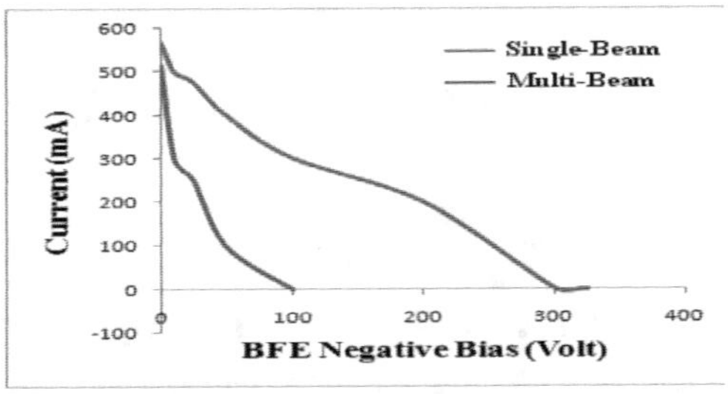

Figure 4(b). Electron trajectories graph on different PCDs with magnetic field

3.Conclusion

A sixty beam electron gun with individual beam perveance 0.033 □P and total beam perveance 2.02 □P has been designed to replace the single beam electron gun of 1.2 □P, operating at beam voltage 6 kV and 564 mA current. It has been observed that emission of electrons take place from the edge of emitter but it can be controlled by applying negative potential at beam focusing electrode and at a particular potential beam completely cut off or no emission of electrons takes place.

Acknowledgement

The authors are thankful to Dr Chandershaker, Director, CEERI, Pilani, for granting permission to publish this paper. They owe their gratitude to Dr SN Joshi, Dr V Shrivastava, and other team members for the constant encouragement and useful suggestions. Thanks are due to CSIR, New Delhi for funding the project.

References

[1] Korolyov A N, Gelvich E A, Zhary Y V, Zakurdayev A D, and Poognin V I, Jun. 2004, IEEE Trans. Plasma Sci., vol. 32, no. 3, pp. 1109-18.

[2] Nguyen K T, Abe D K, Pershing D E, Levush B, Wright E L, Bohlen H, Staprans A, Zitelli L, Smithe D, Pasour J A, Vlasov A N, Antonsen T M, Jr., Eppley K, and Petillo J J, Jun. 2004, IEEE Trans. Plasma Sci., vol. 32, no. 3, pp. 1119-35.

[3] Abe D K, Pershing D E, Nguyen K T, Myers R E, Wood F N, and Levush B, May 2007 *IEEE Trans. Electron Devices*, vol. 54, no. 5, pp. 1253–58.

[4] Gelvich E A, Borisov L M, Zhary Y V, Zakurdayev A D, Pobedonostsev A S, and Poognin V I, Jan. 1993, *IEEE Trans. Microwave Theory Tech.*, vol. 41, pp. 15–19.

[5] Korolyov A N, Gelvich E A, Zhary Y V, Zakurdayev A D, and Poognin V I., Dec. 2001, *IEEE Trans. Electron Dev.*, vol. 48, pp. 2929–37.

[6] Ding Y, Shen B, Shi S, Cao J, May 2005, IEEE Trans. Electron Dev., vol. 52, no. 5 pp. 889-94.

[7] Nguyen K T, Pershing D E, Abe D K, Levush B, Wood F N, Calame J P, Pasour J A, Petillo J J, Cusick M, Cattelino M J, and Wright E L, Jun. 2004, IEEE Trans. Plasma Sci., vol. 32, no. 3, pp. 1212-22.

[8] Gilmour A S, Jr, 2011 Boston, Artech House, Klystrons, Traveling Wave Tubes, Magnetrons, Crossed-Field Amplifiers, and Gyrotrons,

[9] Opera-3D, Cobham technical services, Vector fields software, 24 Bankside Kidlington Oxford OX5 1JE England.

[10] Sharma R K, Sharma S M, Sinha A K, Gupta R K, Kumar L, and Joshi S N, Dec. 1999, *IETE, Technical Review*, vol. 16, pp. 487-96.

General and crevice corrosion study of the in-wall shielding materials for ITER vacuum vessel

K S Joshi[1], H A Pathak[1], R K Dayal[2], V K Bafna[3], Ioki Kimihiro[4], V. Barabash[4]

[1] ITER-India, IPR, A-29, GIDC, Sector-25, Gandhinagar:382025, India
[2] IGCAR, Kalpakkam: 603102, India (retired),
[3] TCR Eng. Pvt. Ltd., New Mumbai: 400705, India
[4] ITER Organization, Route de Vinon sur Verdon, 13115, Saint Paul, Lez Durance France

E-mail: kaushal.joshi@iter-india.org

Abstract. Vacuum vessel In-Wall Shield (IWS) will be inserted between the inner and outer shells of the ITER vacuum vessel. The behaviour of IWS in the vacuum vessel especially concerning the susceptibility to crevice of shielding block assemblies could cause rapid and extensive corrosion attacks. Even galvanic corrosion may be due to different metals in same electrolyte. IWS blocks are not accessible until life of the machine after closing of vacuum vessel. Hence, it is necessary to study the susceptibility of IWS materials to general corrosion and crevice corrosion under operations of ITER vacuum vessel. Corrosion properties of IWS materials were studied by using (i) Immersion technique and (ii) Electro-chemical Polarization techniques. All the sample materials were subjected to a series of examinations before and after immersion test, like Loss/Gain weight measurement, SEM analysis, and Optical stereo microscopy, measurement of surface profile and hardness of materials. After immersion test, SS 304B4 and SS 304B7 showed slight weight gain which indicate oxide layer formation on the surface of coupons. The SS 430 material showed negligible weight loss which indicates mild general corrosion effect. On visual observation with SEM and Metallography, all material showed pitting corrosion attack. All sample materials were subjected to series of measurements like Open Circuit potential, Cyclic polarization, Pitting potential, protection potential, Critical anodic current and SEM examination. All materials show pitting loop in OC2 operating condition. However, its absence in OC1 operating condition clearly indicates the activity of chloride ion to penetrate oxide layer on the sample surface, at higher temperature. The critical pitting temperature of all samples remains between 100° and 200°C.

1. Introduction

One of the key issues of study the corrosion attack on IWS materials during operation phase of ITER vacuum vessel. As IWS block assemblies will be placed between shells of vessel and will be immersed in water which will be used for baking and cooling of vacuum vessel. The IWS block assemblies comprises of number of plates stacked together with spacers in between the plates. Thus there will be numbers of crevices will be form and it may be cause rapid and extensive crevice corrosion. Also there will be high possibility of galvanic corrosion due to various IWS materials in same electrolyte [1]. In addition local accumulation of the aggressive compounds like chloride and other anions might be higher for ITER condition due to the different operational modes [2].

The schematic diagram of IWS assemblies between inner shell and outer shell of vacuum vessel are shown in figure 1.

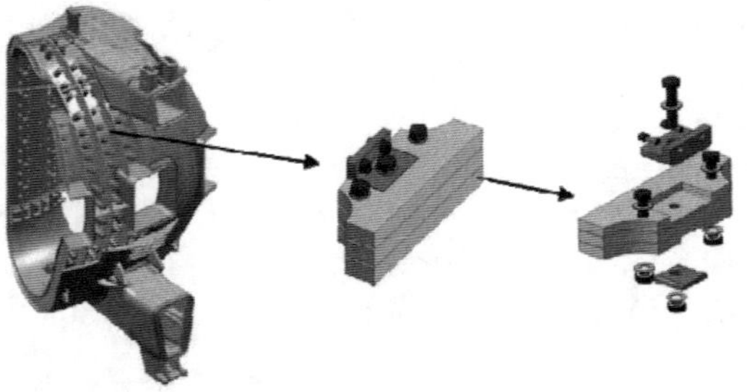

Figure 1 schematic of In-wall shielding block between double shell vacuum vessel

The materials of IWS that used for corrosion study:

(1) Austenitic steel 316L(N) – IG – ribs, brackets, washers, fixing plates;
(2) Borated steel 304B7 (~ 1.75-2.25 % wt. B) – plates for inboard region;
(3) Borated steel 304B4 (~ 1 – 1.24 % wt. B) – plates for outboard region;
(4) Ferritic steel 430 – plates for outboard region;
(5) Austenitic steel XM-19 (bolts designation B8R) – bolts

The corrosion properties of IWS materials were studied by (1) Immersion technique and (2) Electro-chemical Polarization techniques

2. Corrosion study

2.1. Corrosion study by Immersion method

Three custom designed Autoclaves from SS 316L material were used to conduct this study. The eight test coupons of each material were placed in autoclaves and heated to three different operating conditions for a specific time period as per table-1. The Operating cycle were repeated for 5 times in a sequential manner [4].

Table 1 Operating cycle for immersion method

1	OC1	Water at a temperature of 100 °C and the pressure of 1.1MPa	5 Days
2	OC2	Water at temperature of 200 °C and the pressure of 2.4MPa	5 Days
3	OC3	Draining out water from the autoclave and then Using hot Nitrogen gas at 150 to 200 °C	5 Days

- OC1 = Water temperature 100° C, pressure 1.1 MPa
- OC2 = Water temperature of 200° C, pressure 2.4 MPa.
- OC3 = Hot Nitrogen gas 150-200 degree C after draining out the solution from the autoclave

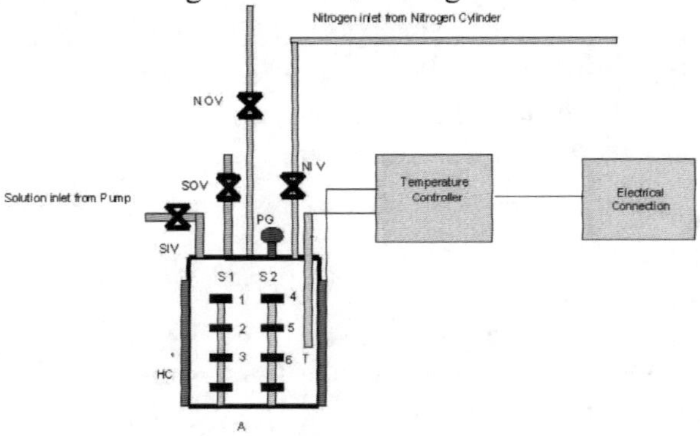

Figure 2 Schematic of test autoclave

2.1.1. Testing procedure for Immersion test in W1000 HP water chemistry

These tests were carried out with the Chloride concentration level of 1000ppb and Hydrogen Peroxide of 2ppm (referred as W1000HP water) in each of the autoclaves. The test coupons of IWS sample materials were subjected to a series of examinations before and after the Immersion Test as follows:

- Loss or gain of weight of sample
- Micro structural analysis using optical microscope and SEM
- Analysis of elements on area to be covered by crevice former before the test
- Analysis of elements over corrosion affected area.
- Measurement of surface profile and hole diameter of test coupons near the crevice location before and after the test
- Material hardness of each sample for (i) as received and (ii) after preparing and (iii) after completion of test
- Oxygen concentration in water was measured before and after each cycle of OC1 and OC2.

The test specimens clamped by XM-19 crevice former and M10 nuts at the torque of 80 N-m. All the crevice assembly was kept on their respective stand and placed the stand into their respective autoclaves. Water chemistry of W1000HP was checked before starting the experiment, prior to filling the water into the autoclaves. To minimize the oxygen concentration and chloride concentration in water, Argon gas was purged for 48 hours at 150cm3/min before start of testing (temperature rising of autoclave). The water chemistry achieved as described in table 2.

Table 2 W1000HP water chemistry

Required Operational Parameters	Test Result of initial water	Test results after Argon purging
Conductivity (at 20° C) μS/cm	At 27.8°C, 199.800 μS/cm	At 27.8°C , 199.800 μS/cm
Oxygen, μg/kg	6500 μg/kg	1080 μg/kg
Chloride μg/kg	5800 μg/kg	1000 μg/kg
Sulphate, μg/kg	200 μg/kg	200 μg/kg
Copper, μg/kg	Not Detectable	Not Detectable
Iron, μg/kg	15000 μg/kg	15000 μg/kg
Hardness (Ca, Mg etc.) μg/kg	20900 μg/kg	20900 μg/kg
Oil products, organic, μg/kg	Nil	Nil
pH at RT	6.6	6.6

All the autoclaves were then cycled through following Operating conditions described in table 3.

Table 3 Operation cycle

Autoclave	Immersion Test Sample	No. of Samples	Operating Conditions		
1	ITS 1: SS 304B4 + CF1 +CF1	8	OC 1	OC 2	OC 3
2	ITS 2: SS 304B7 + CF1 +CF1	8	OC 1	OC 2	OC 3
3	ITS 3: SS 430 + CF1 +CF1	8	OC 1	OC 2	OC 3

2.1.2. Observations

(a) Material hardness of test samples

Material hardness was measured at the surface of the specimen with 2.5 mm diameter tungsten ball at 187.5 kg load. Two samples were taken before the Immersion test to generate the base line data and compared with the hardness after test.

Table 4 Material hardness of samples

Material	Average Hardness Before Immersion Test (HB 2.5/187.5)	Average Hardness After Immersion Test (HB 2.5/187.5)
SS 304B4	176	190
SS 304B7	195.5	221
SS 430	145	167

(b) Weight loss/gain measurement

Each sample was weighed before and after the test to find out % weight loss due to corrosion. The marginal weight loss observed in SS 430 coupon material and hence the corrosion rate calculated by following formula.

Rate of corrosion (mm/year) = (87480 x W) / (A x t x D)

Here W = Weight loss, A = Area under crevice, t = time of exposure and D = Density

The corrosion rate calculation is shown in table 5.

Table 5 Weight measurement of samples

Specimen No.	Cross Sectional Area of entire Specimen (cm2)	Weight loss (gm)	Corrosion Rate in mm/Yr	Average Corrosion Rate in mm/Yr
0	37.441	0.0156	0.00261	0.00267
5	37.572	0.0155	0.00259	
6	37.609	0.0153	0.00255	
7	37.696	0.0153	0.00254	
8	37.572	0.0152	0.00254	
9	37.528	0.0175	0.00293	
10	37.572	0.0177	0.00296	
11	37.528	0.0159	0.00266	

(c) Optical Microscopy

The samples removed from the experiment were followed by diamond polishing and optical observation at both side. The crevice corrosion has found on both sides of samples. Following table 6 shows the images of crevice at 100X and 200X magnification showing presence of crevice corrosion.

Table 6 Microscopic images of samples

	SS 304 B4	SS 304 B7	SS 430
100x			
200x		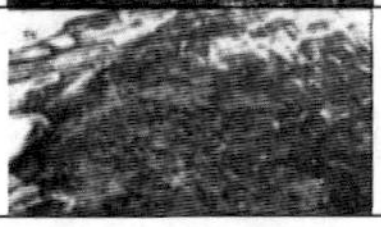	

(d) Scanning Electron Microscopy (SEM)

The SEM analysis was carried out on samples of each IWS materials before immersion which considered as base line data. After test, two samples of each material were polished by 600 grit polishing paper and followed by SEM analysis. Following table 7 shows images of SEM analysis.

Table 7 SEM images of samples

	SS 304 B4	**SS 304 B7**	**SS 430**
Before test	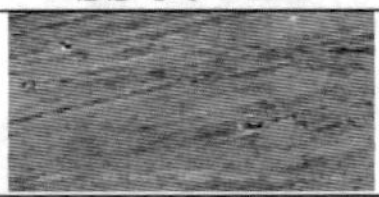		
After Test	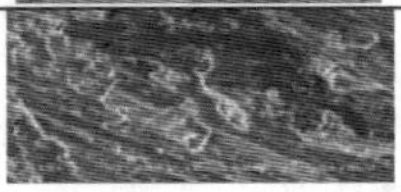	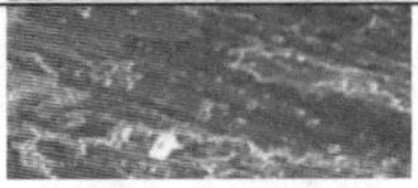	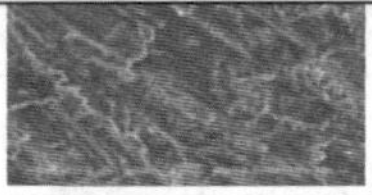

(e)Microstructure Examination

Microstructure examination was carried out at sample surface for all the materials to collect the evidences of the corrosion under optical metallography. The testing was carried out on two samples of each material to collect the baseline data of each material type. Two samples from each material were considered for microstructure after the test from top and bottom position in the autoclave as per the testing procedure.

Table 8 Microstructure of samples

	SS 304 B4	**SS 304 B7**	**SS 430**
Before test	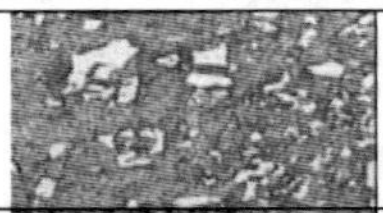	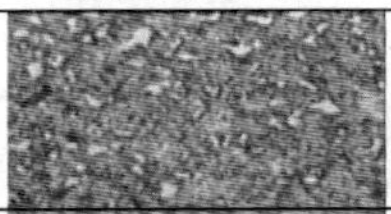	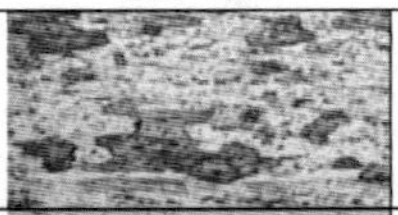
After Test	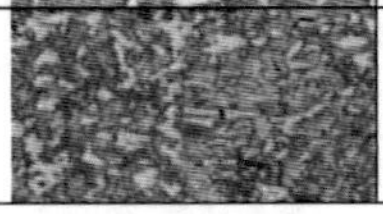	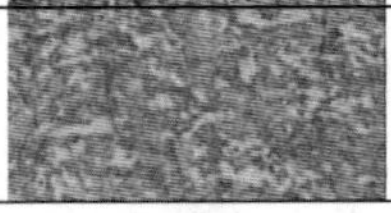	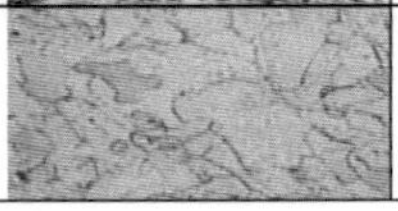

(f) Conclusion of Immersion testing:

- On visual observation, along with SEM and metallography of the crevice former area and nut tightening area of the specimen of all material type showed pitting corrosion attack for W1000HP water condition. Therefore, immersion test and observation was repeated for operating condition with W100HP water and found that all material specimens showed pitting attack.
- The materials SS304 B4 and SS 304 B7 showed marginal weight gain after crevice corrosion experiment. The marginal weight gain is due to a thin layer of oxidation on the surface after exposure.
- SS 430 material after crevice corrosion experiments showed weight loss of very negligible quantity indicating that the material in a given environment showed mild effect of general corrosion. The corrosion rate is calculated as 0.00267 mm/year.

2.2. Corrosion study by Electro-Chemical Polarization method

The susceptibility of various materials for galvanic/general corrosion was studied as part of the Electro-Chemical Polarization test method. Test specimen materials considered for the Electro-Chemical Polarization test study included SS 304 B4, SS 304 B7, Ferritic SS 430, XM-19, SS 316LN(IG) and three samples (SS 304 B4, SS304 B7 and SS 430), which were exposed for immersion testing. One custom designed autoclave from SS 316L material was used to conduct this study as shown in figure 3.

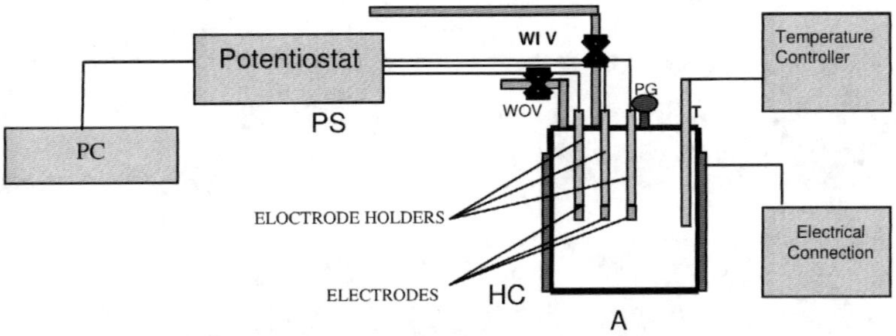

Figure 3 Schematic of test autoclave

These testing were carried out with the water having a chloride concentration level of level of 1000ppb and hydrogen peroxide of 2ppm (referred as W1000HP water) and chloride concentration level of level of 100ppb and hydrogen peroxide of 2ppm (referred as W100HP water) in the autoclave. Following table-9 shows the test matrix of test sample, water chemistry and operating condition.

Table 9 Test matrix of electro-chemical polarization test

Sr. No.	Test	Material	Water chemistry	Operating condition
1	ET3	SS 304 B4	W1000H	OC1
2	ET4	SS 304 B4	W1000H	OC2
3	ET7	SS 304 B7	W1000H	OC1
4	ET8	SS 304 B7	W1000H	OC2

The following observations were made for each test listed as per the above table:

- Cyclic Potential measurement before start of polarization.
- Polarization curve (Tafel slope and corrosion rate will be calculated).
- Reverse curve for Pitting potential and behavior.
- Difference between pitting potential and protection potential.
- Critical anodic current density for passivation and passivation potential.
- Corrosion potential.
- Micro structural analysis using optical microscope and SEM before and after the test.
- Oxygen concentration in water will be measured before and after each test.

2.2.1. *Test Procedure*

The entire specimen polished with 120 grit emery sheet and ensures that it does not contaminate the test solution. Specimens of all eight materials observed using Optical Microscope at the magnification of 100X, 800X and SEM. The sample were inserted from flange of the autoclave and fixed the test specimen and platinum electrodes in their respective position. Autoclave filled with the test solution W1000HP or W100HP and temperature increased to 100 ± 10 or 200 ± 10 degree C. Test was executed by the control and measured software in computer and measured Open Circuit Potential (OCP) till steady graph is observed.

By Filling-in experimental parameters in software, the temperature and pressure values were measured till the experiment is complete.

2.2.2. *Observations*

(a) Micro Structure examination of samples

Microstructure examination was carried out on all materials surface to collect base line information. It was carried out on one sample of each material for the baseline data. All samples were considered for microstructure at cross section after the electrochemical test as per the procedure.

Table 10 Microstructure images of samples

	SS 304 B4	SS 304 B7	SS 430	SS 316L(N) IG	XM-19
After OC1	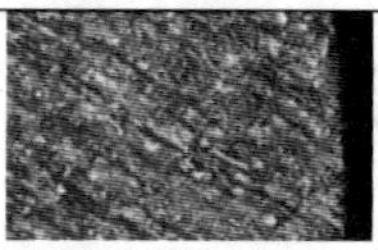	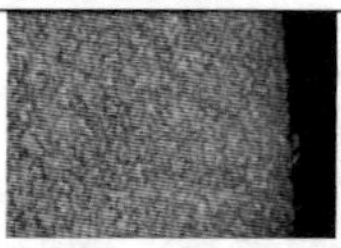	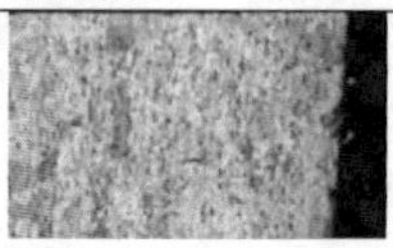	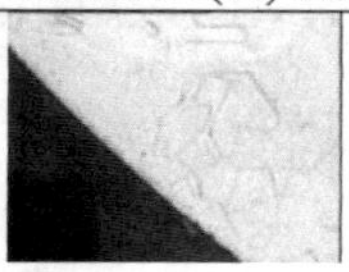	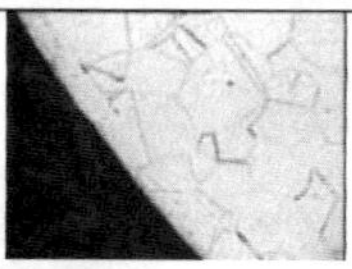
After OC2	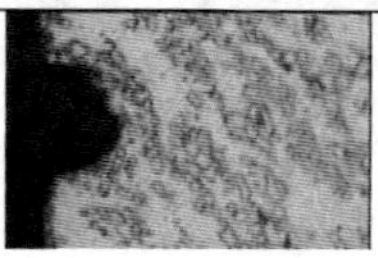	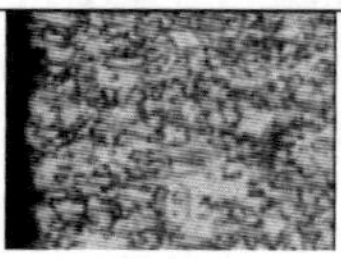	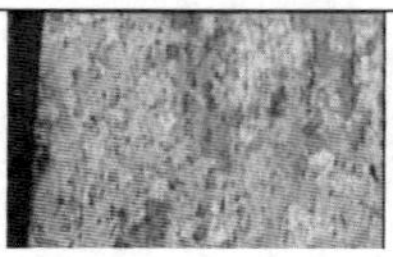		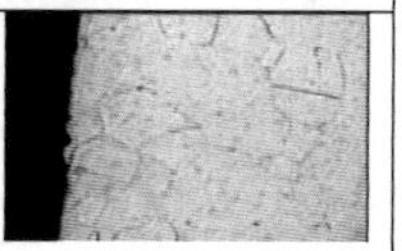

(b) Scanning electron Microscopy (SEM)

Scanning Electron Microscopy was carried out at sample surface for all the materials to collect the detailed evidences of the corrosion initiation sites and its nature. SEM was carried out on one sample of each material before test to generate the baseline data. Also, SEM analysis was carried out on samples which showed pitting behaviour in electrochemical tests. SEM examination before test was carried out on surface of the sample prepared with 600 grit polishing paper. Following are the images of Sample surface at 1000X, as there was no evidence of pitting in Electro-Chemical test in OC1 condition whereas OC2 Condition indicated pitting hence SEM was carried out after OC2 cycle and compared with baseline data.

Table 11 SEM images of samples

	SS 304 B4	SS 304 B7	SS 430	SS 316 L(N) IG	XM-19
Before Test	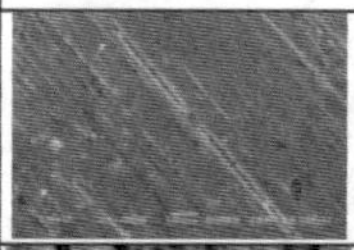	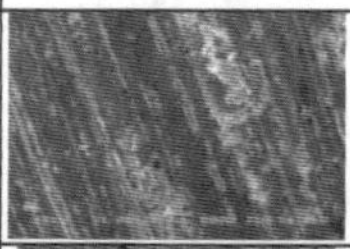	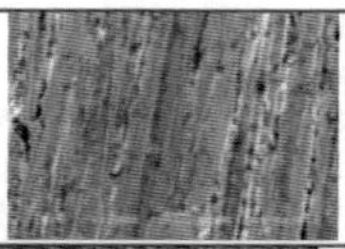	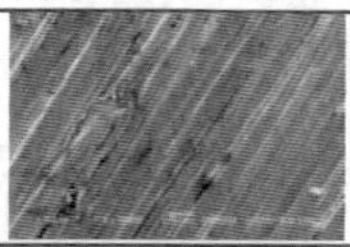	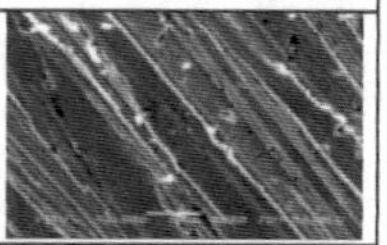
After OC2 Cycle	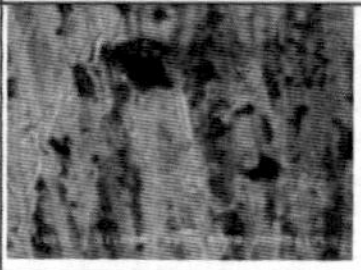	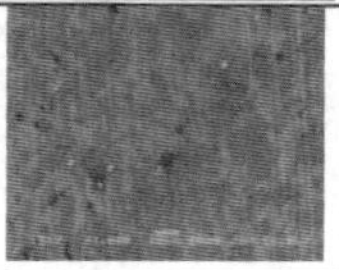	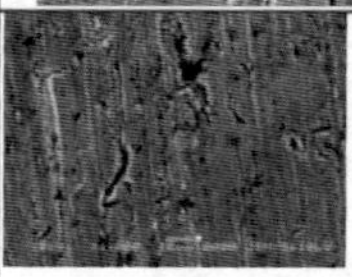		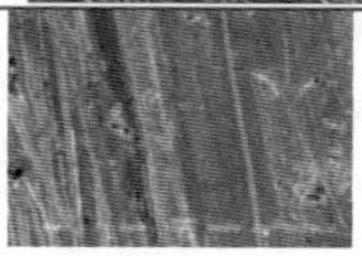

(c) Electro-Chemical test data

Electro-Chemical tests were conducted on all materials including coupons from immersion test for

OC1 and OC2 operation condition with W1000HP water chemistry. The following parameters were measured by software.

- Open Circuit Voltage: to study steady state polarization.
- Tafel Plot: To measure the general corrosion rate of the material in a particular test condition.
- Cyclic Polarization Curve: To characterize the susceptibility of material to pitting for a given test condition.

Table 12 Tafel plots of samples

	SS 304 B4	SS 304 B7	SS 430	SS 316L(N) IG	XM-19
OC1	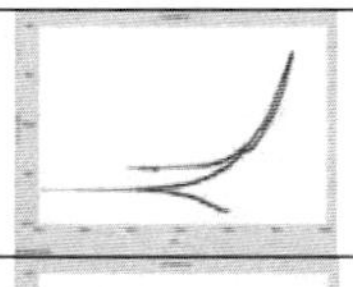	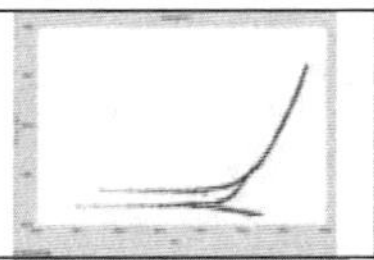	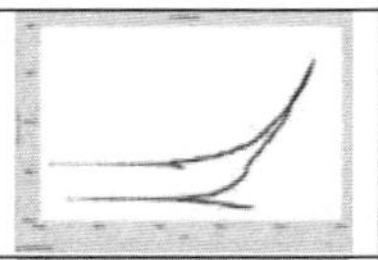	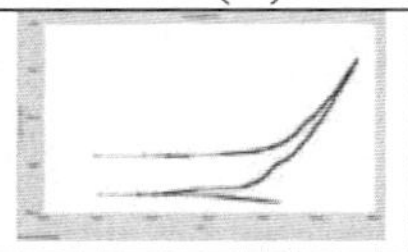	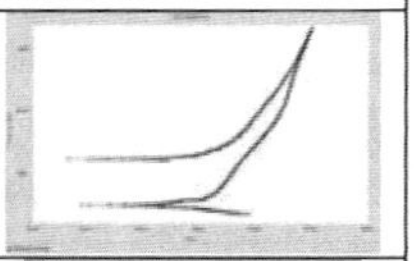
OC2			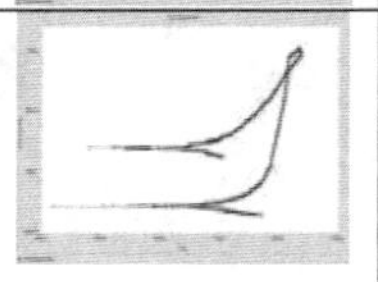	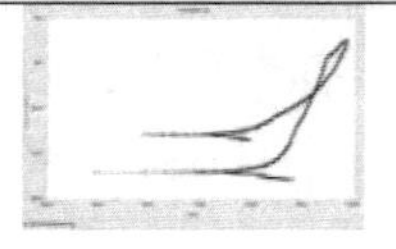	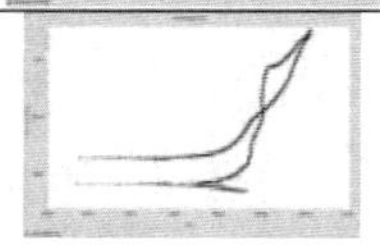

(d) Result summary

❖ OC1 Condition:

Pitting loop was not observed in OC1 operating condition at 100 C in all the samples. ICC value of XM-19 sample was lowest (1.85 E-03) and that of SS 430 sample was the highest (0.00149 E-03). Lower ICC value denotes easy passivation. Metallography and SEM did not show any corrosion damage on the surface when compared to before and after exposure to the OC1 condition.

❖ OC2 Condition:

Pitting loop was observed in all experiments as per OC2 operating conditions. In OC2 operating condition, the ICC value of SS 304 B7 was the highest (6.7038E-03). Under OC2 conditions, Eb – Ep value was the lowest in SS 430 sample. Where this value is high, pitting can take place more easily. SEM, EDS and metallography results showed pitting at all samples tested at OC2 condition.

(e) Conclusion of Electro-chemical potential test

❖ General corrosion Rate of the samples:

The General corrosion rate under OC1 conditions is lowest with SS 304 B7 and highest for SS430 material in as supplied conditions. Under OC2 conditions SS 304 B7 showed highest corrosion rate and XM19 showed lowest in as supplied conditions. Overall the materials evaluated in the OC2 conditions showed higher corrosion rates then the OC1 conditions. This can be explained by the fact that Increase in temperature would increase the corrosion rate.

❖ Pitting corrosion behaviour:

Absence of pitting loop in OC1 operating condition, and its presence in OC2 operating condition, clearly indicates the activity of chloride ion to penetrate oxide layer on the sample surface, at higher temperature. The results show that critical pitting temperature of all samples remains between 100° and 200° C. At higher temperature, reaction rate (corrosion rate) of alloy with water increases, and kinetic energy of aggressive ions like chloride increases. In the absence of dissolved oxygen, this severely affects the passive layer break-down frequency, and at OC2 operating conditions, it becomes evident that the rate of oxide film breakdown becomes more than the rate of film build-up.

3. References

[1] Ioki K "Detailed Design Document 1.5 Vacuum vessel", ITER IDM ref. IDMUID 22FPWQ

[2] Fontana Mars G, Greene Norbert D "Corrosion Engineering (Materials Science and Engineering)" (ISBN: 0071003606 / 0-07-100360-6)

[3] Kaushal Joshi, H.A.Pathak and V.K.Bafna "Final report of corrosion testing of ITER In wall Shielding material", ITER IDM Ref. II-2PPDOWL-v1_0.

[4] Kaushal Joshi, H.A.Pathak and V.K.Bafna, "Final completion report of crevice corrosion study-Immersion test of ITER In wall shield materials." ITER IDM Ref. II-CRY8X0L_v1.0.

Profiling of back-scattered electrons in opposed magnetic field of a Twin Electron Beam Gun

S. Sethi, Anchal Gupta, V. Dileep Kumar, Jaya Mukherjee and L.M. Gantayet

Laser &Plasma Technology Division, Beam Technology Development Group, Bhabha Atomic Research Centre, Mumbai 400 085

Email address: s_sethi@barc.gov.in

Abstract: Electron gun is extensively used in material processing, physical vapour deposition and atomic vapour based laser processes. In these processes where the electron beam is incident on the substrate, a significant fraction of electron beam gets back-scattered from the target surface. The trajectory of this back scattered electron beam depends on the magnetic field in the vicinity. The fraction of back-scattered depends on the atomic number of the target metal and can be as high as ~40% of the incident beam current. These back-scattered electrons can cause undesired hot spots and also affect the overall process. Hence, the study of the trajectory of these back-scattered electrons is important. This paper provides the details of experimentally mapped back-scattered electrons of a 2x20kW Twin Electron Beam Gun (TEBG) in opposed magnetic field i.e. with these guns placed at 180° to each other.

1. Introduction

Electron Beam (EB) evaporation is useful for atomic vapour based laser separation process, where a metallic charge contained in water cooled copper crucible, is evaporated in high vacuum environment [1, 2]. When an electron beam is incident on the charge (metal/alloy), temperature of the substrate under the beam rises causing a molten pool contained in solid skull of metal/alloy. The evaporation rate depends on the surface temperature which is governed by the power density of the electron beam, while the molten pool size depends on the total power of the electron beam and the effective heat transfer coefficient at the surface of water cooled crucible. Two electron beams focused on a single ingot are useful for independently controlling the evaporation rate and the molten pool size. One beam is focused to attain the required power density and the second defocused electron beam supplies the balance power for attaining fully molten condition of the charge. In EB evaporation a part of the incident beam gets back scattered and its trajectory is dictated by the surrounding magnetic field, while the fraction depends upon the atomic number of the charge. It can be as high as 40% for element with atomic number more than ninety [3]. The trajectories of the back-scattered electrons should be contained to ensure (i) no local heating of surfaces and (ii) these electrons do not land in the process zone.

The trajectory of backscattered electrons for 2x20KW twin electron beam gun (TEBG) system, in opposed magnetic field, i.e., with these guns placed at 180° to each other has been experimentally determined. A number of experiments have been conducted to identify the limits of operation of the TEBG, for minimizing its interference with the process. The experimental details and observations have been presented in this paper.

2. Experimental Set-up

TEBG has a water cooled cooper crucible with a cavity of 156cc for holding the charge to be evaporated. The two electron guns are configured opposite to each other, i.e., at 180° orientation as shown in the top view of crucible in Fig 1(a). The magnetic shunt pieces are meant for altering the magnetic field, to increase the extent of sweep of the electron beam in oscillating mode. Each of 270° bend electron gun has independent 12kV, 1.7A power supply for its use up to maximum of 20kW. The copper crucible and the assemblies of both the electron guns have separate water cooling circuit as depicted in the side view of crucible in Fig 1(b). A water cooled copper shroud around the crucible protects the crucible and the assemblies of electron gun from undesirable heating. The entire assembly is placed in a high vacuum water cooled jacketed chamber as shown in Fig 2. This jacketed vacuum chamber (W 550mm, L 550mm, H 750mm) is made of stainless steel. A vacuum of $3x10^{-6}$ mbar is maintained using 3000lps diffusion pump, backed by a $60m^3$/hr rotary pump. The Pirani and Penning gauges were used to measure the vacuum in the chamber.

(a) Top View of TEBG (b) Side View of TEBG

Fig 1: Top and side views of a 2 x 20kW Twin Electron Beam Gun (TEBG) with 156cc cavity

The orientation of the magnetic fields in the longitudinal direction i.e. along the path of primary electron beam and in direction normal to longitudinal i.e. transverse direction is shown in the Fig 3. The magnetic field of the permanent magnets has been mapped at the surface of crucible in both the directions. The profile of magnetic field for the cases of with shunt and without shunt is given in Fig 4.

Fig 2: TEBG assembled in a vacuum system

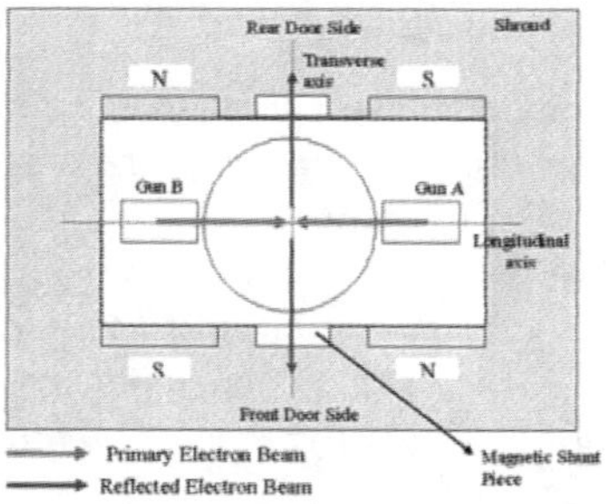

Fig 3: Configuration of magnetic field in TEBG and expected directions of primary and back-scattered electron beam

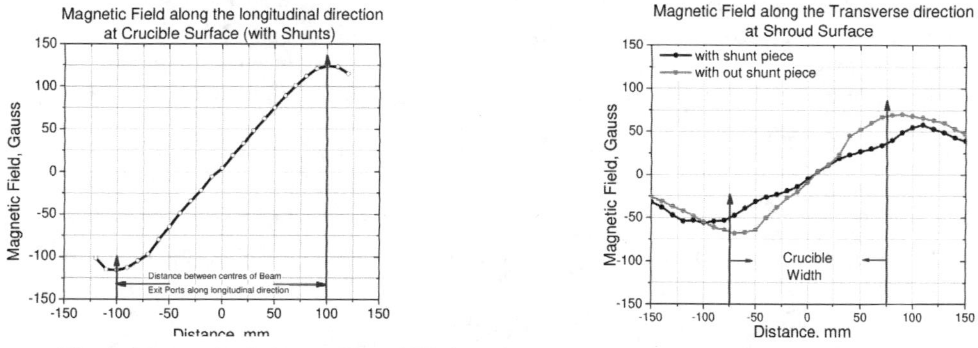

Fig 4: Magnetic field profile of TEBG along longitudinal and transverse directions

3. Experimental Procedure and Results

3.1. Identification of position of primary electron beam

The permanent magnetic field around the electron beam ensures that the primary electron beam always falls within surface of cavity of crucible. However, the magnetic field due to electromagnet is used for fine adjustment of the location of the incident electron beam. The correspondence between the magnetic field and the location of incident beam was initially obtained by the puncturing a metal sheet placed on the surface of ingot at low beam power. The calibration of actual location of beam spot corresponding to the reading on sweep control unit is given on Fig 5. This information is useful in correlating the profile of back-scattered electron with the position of beam.

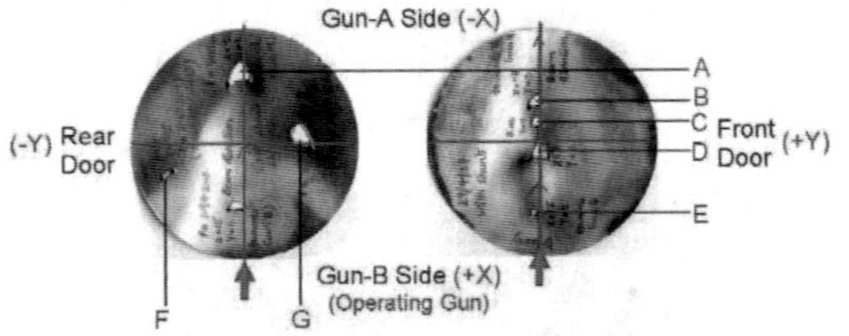

Position Number	As indicated on sweep control unit		Measured Coordinates of puncture (mm)
A	-15, 0	Longitudinal	-28, -1
B	-7, 0		-16, -1
C	0, 0		-8, 0
D	7, 0		3, 1
E	15, 0		27, -2
F	0, -15	Transverse	13, -21
G	0, 15		18, 21

(a) with magnetic shunts

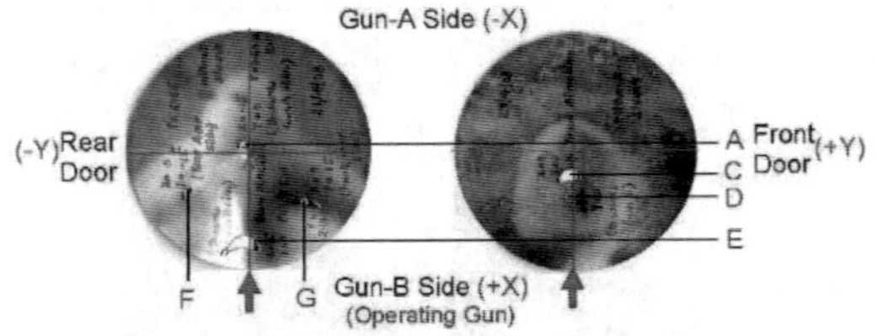

Position Number	As indicated on sweep control unit		Measured Coordinates of puncture (mm)
A	-15, 0	Longitudinal	-3, -1
B	-7, 0		----------
C	0, 0		8, -1
D	7, 0		16, 2
E	15, 0		32, -2
F	0, -15	Transverse	12, -29
G	0, 15		4, 21

(b) without magnetic shunts

Fig 5: Measured beam position of operating gun of TEBG on cavity surface vis-à-vis as indicated on sweep control unit for with and without magnetic shunts

3.2. Profiling of back-scattered electron

It can be seen in seen in Fig 4 that the magnetic field has a null point at the centre of surface of ingot. In case of profile along longitudinal direction the magnetic field peaks near the electron beam exit port in the crucible and in the transverse direction the magnetic field peaks at the edge of the copper crucible due to the pole pieces of the permanent magnet. It can be expected that the back-scattered electrons from the centre of the ingot may sense minimum or nil magnetic field for its bending and hence may go in vertically upward direction. Whereas, the back scattered beam away from the centre of ingot is expected to bend on both sides along the transverse direction due to significant magnetic field in this direction. Therefore, experiments were carried out in three stages to make an assessment of direction and profile of the back-scattered beam as a function of spot location of primary beam on the ingot. Firstly, the trajectory of back-scattered electron was determined just near the ingot over a cylindrical surface with radius of 75mm. Thereafter, these electrons have been profiled at various distances (0mm, 50mm and 100mm) from the side of the crucible i.e. starting from 75mm from the centre of ingot. And finally, the total current of the backscattered electrons is measured at various heights (150mm, 200mm and 250mm) in vertically upward direction from the surface of ingot. All these results are summarized as follows.

3.2.1. Profiling in cylindrical coordinates with fixed radius

The semi-cylindrical structure with 15 numbers of 12 x 270 mm^2 stainless steel strips as electron collector at a radius of 75mm is shown in Fig 6. Each of these collectors is electrically insulated from the frame and is shorted to ground through a resistance. The voltage drop across the resistance provides the information of current collected by respective collector.

Fig 6: Semi-cylindrical structure used for profiling back-scattered electron beam with 15nos. of strip shape electron collectors insulated from support structure

One of the electron guns was located with and without magnetic shunts and the current collected by each collector is measured. The electron beam power was 600W (12kV, 50mA). The Fig 7(a) and Fig 7(b) provide the plots of electron current of these strip electron collectors, for various locations of the primary incident beam along longitudinal direction, with and without the magnetic shunt pieces respectively. The Fig 7(c) is a plot of electron current collected by collectors for changing location of primary incident beam along transverse detection and without magnetic shunt pieces. These results clearly show that the electron current in vertical direction is in general more in case of use of magnetic shunt pieces in comparison to without use of magnetic shunts and this current drastically reduces if the position of electron beam is kept towards the operating gun side. The movement of primary beam along transverse direction only shifts the location of peak of back-scattered electrons but does not change its magnitude.

3.2.2. Profiling at various distance from the sides of crucible

The set-up for profiling the back-scattered electron at various distances on the side of crucible consist of a number of electron collectors with a diameter of 25mm and with a pitch of 33mm mounted with electrical insulation on a 300mm wide plate. Fig 8(a) shows this plate placed on the edge of the side of crucible. The back-scattered current on the disc collectors in each row and each column have been recorded for electron beam power of 1200W (12kV, 100mA) and for the distance of 0mm , 50mm and 100mm from the side of crucible.

Table-1 & 2 show the consolidated results of cumulative current collected by each row and each column of these disc collectors for all the cases of distances and for with and without magnetic shunt pieces. These results clearly indicate that the back-scattered electrons from the centre travel longer radius. A major fraction of these electrons is contained in bottom three rows, i.e., up to height of ~120mm and with in middle three columns i.e. 70 to 80mm measured in the transverse direction.

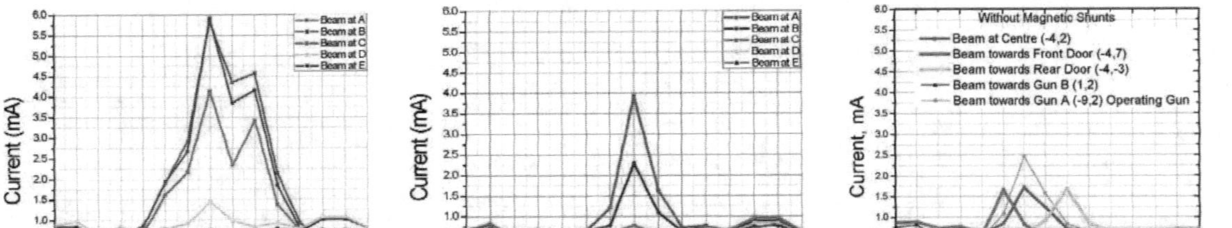

(a) for various positions of EB along longitudinal direction (with magnetic shunt pieces)
(b) for various positions of EB along longitudinal direction (without magnetic shunt pieces)
(c) for various positions of EB along transverse direction (without magnetic shunt pieces)

Fig 7: Profile of current collected on 15 numbers of 12 x 270mm^2 strip shape electron collectors on cylindrical surface with diameter of 75mm radius

(a) Typical orientation used at the side of crucible

(b) Typical orientation used at position vertically above the crucible

Fig 8: Experimental set-up with an array of electrically insulated disc for profiling back-scattered electron with a diameter of 25mm and pitch of 33mm

3.2.3. Profiling at various heights from the surface of crucible

The same structure with an array of φ25 disc collectors was used for measuring the electron current in vertically upward direction from the surface of the crucible. However, in this case only the total current of all of these collectors have been recorded for the location of primary beam near centre on the ingot. This structure, holding the disc collectors, is mounted on a linear motion feed through from the top of chamber as shown in Fig 8 (b). This mechanism helps in changing the height of structure and hence for recording the current of back-scattered electron at various heights. Fig 9 (a) shows the current of back scattered electrons collected for 50-600mA primary beam at 150, 200 & 250mm height from the crucible surface, when only one of the electron gun is operated. It clearly shows (i) there is a significant back-scattered current in the vertical upward direction above the crucible, (ii) current of back-scattered electron collected at various heights is, as expected, proportional to the primary beam current, and (iii) there is no change in the current collected by array of disc collectors on 170x300 mm^2 plate as the bending of back-scattered electron, due to magnetic field is only below this height. Fig 9 (b) shows the current of back scattered electrons collected for the cases when only one gun is operated and also when both the guns are operated with 50-600mA current of primary beam. The beam positions, as indicated on the control panel (-5, 0) for beam A and (0, 0) for beam B, are at nearly same location and is around the centre of the ingot. The results clearly show that behavior of back-scattered electron, for both the guns, is the same and linear with respect to the current of primary beam.

Table 1 : Back-scattered electron current collected by each row of discs at various side of crucible at electron gun power of 1200W (12kV, 100mA)

Row Number (with height of each row from crucible surface)	Distance of collector discs from the side of crucible					
	0 mm		50 mm		100 mm	
	With Shunt	Without Shunt	With Shunt	Without Shunt	With Shunt	Without Shunt
R-8 (Height 286)	0.13	0.12	0.15	0.12	0.17	0.13
R-7 (Height 253)	0.16	0.17	0.19	0.17	0.2	0.16
R-6 (Height 220)	0.25	0.25	0.29	0.23	0.28	0.22
R-5 (Height 187)	0.37	0.35	0.41	0.33	0.37	0.28
R-4 (Height 154)	0.51	0.49	0.53	0.43	0.46	0.34
R-3 (Height 121)	0.79	0.69	0.77	0.68	0.55	0.50
R-2 (Height 88)	1.20	1.13	1.09	0.94	0.69	0.54
R-1 (Height 55)	1.56	0.69	1.09	1.02	0.66	0.55
Total of all Discs	4.97	3.89	4.52	3.92	3.88	2.72

Table 2 : Back-scattered electron current collected by each column of discs at various side of crucible at electron gun power of 1200W (12kV, 100mA)

Column Number (with distance from centre of crucible)	Distance of collector discs from the side of crucible					
	0 mm		50 mm		100 mm	
	With Shunt	Without Shunt	With Shunt	Without Shunt	With Shunt	Without Shunt
C 1 (-135mm)	---	---	---	---	---	---
C 2 (-102mm)	0.06	0.05	0.06	0.05	0.06	0.05
C 3 (-69mm)	0.14	0.08	0.16	0.10	0.20	0.10
C 4 (-36mm)	1.15	1.00	1.08	0.53	0.69	0.53
C 5 (-3mm)	1.73	2.10	1.96	2.42	1.70	2.42
C 6 (30mm)	1.44	1.56	1.21	0.75	0.64	0.75
C 7 (63mm)	0.49	0.12	0.11	0.10	0.15	0.10
C 8 (96mm)	0.06	0.06	0.06	0.05	0.07	0.06
C 9 (129mm	---	---	---	---	---	---

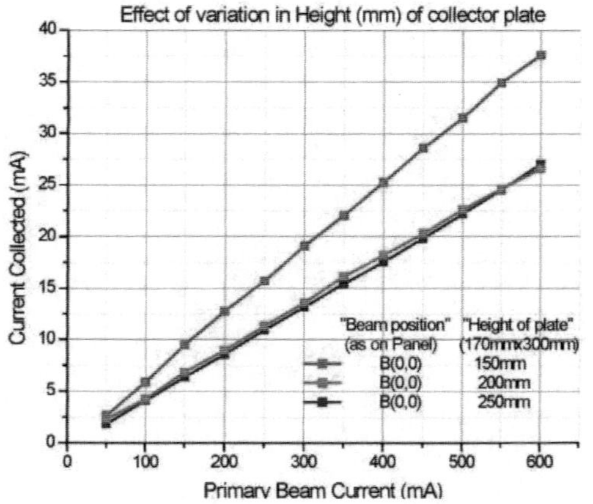

(a) current of back scattered electrons collected at various heights

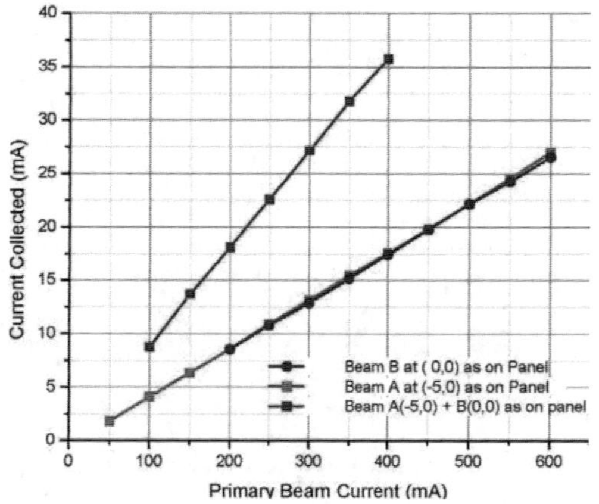

(b) current of back scattered electrons collected from single/both guns of TEBG at 250mm height

Fig 9: Total current of back-scattered electrons collected by 45 nos. of discs mounted on 170x300^2 mm plate for various primary beam currents and at various heights

4. Conclusions

The trajectories of TEBG suggest that the distribution of the backscattered electrons is concentrated at certain locations according to the resultant magnetic fields in the vicinity of electron beams. These electrons can also interfere with the process zone if the EB incidents on the ingot towards opposite side gun. To confine the back-scattered electrons, the primary electron beam should not cross the mid plane along longitudinal direction, as it leads to enhanced back-scattering in vertically upward direction. Moreover, magnetic shunt pieces are not useful in TEBG with 180° orientation as shunt pieces are meant for increasing the extents of oscillating beam in order to cover a larger area over the crucible surface.

Acknowledgement

Authors are grateful to Dr. A.K. Das, Head, Laser & Plasma Technology Division, BARC for his support and encouragement for carrying out this study. The help provided by Shri Rahul Mishra, L&PTD in the fabrication and assembly of experimental set-ups is also acknowledged.

References

[1] L. M. Gantayet and S. Sethi, "Beam material interactions - Role of melt pool", Proc. PBEAM-02.

[2] S. Sethi, Anupama P.; Manisha Gupta, Jaya Mukherjee; R.K. Singh and H.S. Kushwaha, "Modeling of Vapour Generator for Cleanup Separator Module", Proc. INSAC 2000

[3] Schiller, S.; Ullirich, H. and Siegfried, P., "Electron beam technology", John Wiley & Sons, NY., 1982.

Thermo-mechanical induced deformation simulation studies for metal gaskets for UHV application

B Ramesh Kumar [1], **S Purohit**

Institute for Plasma Research, Bhat, Gandhinagar-382428.

[1]E-mail: *buddu@ipr.res.in / brkumar75@gmail.com*

Abstract. In vacuum technology, metal gasket seals are extensively employed to achieve a UHV with reduced contamination considering the pressure and temperature variations as it performs a static seal between two stationary members of a mechanical assembly. The optimum sealing is attained over the balancing of the forces effective, which are function of temperature, governs the surface deformation for the metal gasket seal follows into degradation in the leak tightness at elevated temperatures. The prime component exerting the most deformation force over metal gasket seals, gasket seating force is a constant value generated by the bolting of the stationary members of a mechanical assembly. The paper address to metal gasket seals, copper and aluminum, behavior under thermo-mechanical load is analyzed (simulation), with ANSYS platform, workbench. The major concern is to investigate the typical deformation behavior as a function of thermal variation, baking/ cooling. For copper and Aluminum gasket seals, 16mm to 250mm internal diameter, exposed to pre-established gasket seating force under wide temperatures range. The deformation, average and the deformation range, observed to move in a very specific manner and runs to a wide range for a given material and size. The data reported here deserves to be substantial enough to establish the prediction of thermal behavior of metal gasket seals for standardization.

1. Introduction

Invented by Austrian engineer Richard Klinger about 100 years ago gaskets are employed to achieve static seal between two stationary members of a mechanical assembly, harboring surface imperfection induced probable leaks, under extremes of temperature and pressure. Metal seals are the crucial components in UHV systems for achieving the required leak tightness and to avoid any kind of catastrophe for optimum performance[1]. The mechanical behavior between the knife-edge-type flange and gasket material involves various types of complicated mechanisms such as elasto-plastic deformation, large deflection, thermal creep, and contact-friction. Metallic gaskets seals are generally used where applications in which joint might be exposed to UHV conditions with baking requirements and higher leak tightness over wide range of temperature operation. Also in addition, these metal seals exhibit low degassing, permeation and long term radioactive resistance. The metal seal capable of withstanding very high temperature is possible if the joint is designed to yield locally over a narrow location with application of bolt load. UHV systems in general use Aluminium and Copper as sealing materials over wide range of temperature due to their better mechanical properties and low outgassing rate. Metal gasket sealing, in particular the Copper and Aluminium materials are

[1] To whom any correspondence should be addressed.

Published under licence by IOP Publishing Ltd

the choice of experimentalists for attaining the suitable leak- tightness which are widely used in vacuum systems in various applications like laboratory plasma vacuum systems, and many other industrial vacuum processing production systems. Some typical applications like large vacuum systems like fusion devices, stellarators, accelerators etc will use custom designed seals based on required dimensions. On applying a load, bolting of the flanges containing the metal gasket holds the sealing to attain the required tightness. The bolting of flanges exerts mechanical force over the gasket, induces the local surface deformation, in case of knife edge flange system. An established experience reveals baking, as in integral part of number of experiments to withstand the degree of cleanliness, degassing for vacuum, adds to the gasket a nonzero deformation. The deformation information in terms of magnitude and dynamics is highly significant for gasket designing and selection for the specified experiment with a satisfactory performance. Attempts to understand the mechanical loading over gasket, mathematical modeling [2,3], based on theory of elasticity, have been made to understand the contact stress distribution for high pressure pipeline along with rotational flexibility [4], holding nonzero probability. H.Fend has given attempt to reduce the seating force on metal seals by modified mechanical design of seal and the same were validated with experiments for different kinds of UHV applications[5] and further this analysis revealed that the seating force applied has been reduced to 50% in comparison with regular metal standard CF design and less material usage.

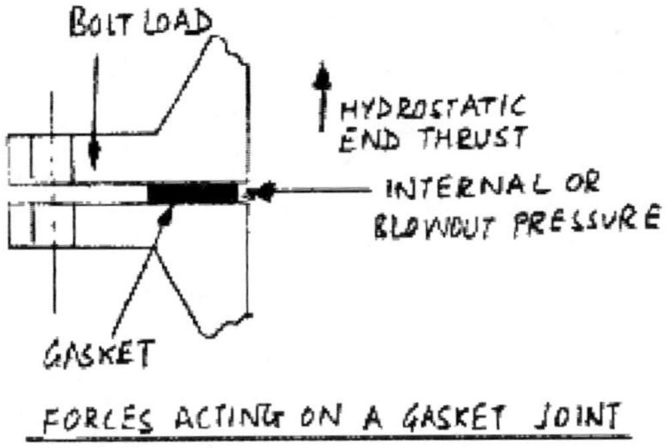

Fig. 1. Example of Load onto Seals

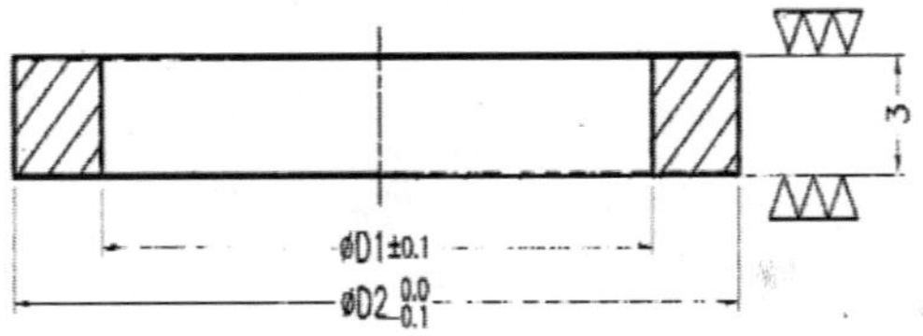

Fig. 2. Schematic of Seal dimensions used

The present study, simulation attempt has been carried out with standard mechanical metal gasket of Copper and Aluminium seal materials over the wide range of dimensions and thermal baking conditions. The thermo- mechanical load on metal seals which in similar to bolt tight gasket in a two side knife edge flange assembly deployment under variable backing conditions, for range gasket material and sizes are analyzed.

Table 1. ID/OD dimension for the gasket specification considered

Standard CF range	D1(mm)	D2(mm)
16	16.2	21.2
35	38.0	47.7
63	63.6	82.2
100	101.7	120.2
150	152.5	171.0
200	203.3	222.0
250	254.0	272.5

The gasket and flange system under operation are exposed to a complex force assembly having peculiar formation as shown in Figure 1. The acting forces are balanced as to restore the highest leak-tightness achievable. The total force compressing the gasket to create a seal, mechanical loading is referred as seating force. The mechanical characteristics exerting pressure against the gasket to unseal gasket joint and against the flanges holding the gasket in place, named hydrostatic end force. The mentioned forces are indicative of the mechanical loading, creates deformation as effective over in the sealing assembly. The temperature also adds to the deformation, creates thermo-mechanical effects induces the promotion of "creep relaxation", a permanent strain or relaxation quality of soft materials under stress .As a rule, the higher the temperature, the more critical becomes the selection of the proper gasket. The assurance of perfect sealing will be given by the nonzero residual force, difference of gasket seating force and hydrostatic end force.

2. Simulation and Analysis

The simulation for gasket deformation under thermo-mechanical load is performed on ANSYS Workbench software followed by IDL analysis of the data for the deformation studies. The applied load in the form of uniform distributed torque is assumed in all the analysis procedure. The details of the adopted procedures are given in the following.

2.1 Gasket design

Standard metal gasket design/dimension was incorporated in the simulation as per Table 1. Imagery dimensions were considered for both copper and aluminum. The details of the structural material Copper and Aluminum properties of which the gaskets are made up of are employed, as per the Table 2, were the standard ANSYS library which is found to be constant.

Table 2. Material properties considered for the analysis

Properties	Aluminum	Copper
Young's Modulus	7.1 e+10 Pa	1.1e+11 Pa
Poisson's Ration	0.33	0.34
Density	2770Kg/m^3	8300Kg/m^3
Thermal Expansion	2.3e-5 1/^{o}C.	1.8e-5 1/^{o}C.
Tensile Yield Strength	2.8e+8 Pa	2.8e+8 Pa
Compressive Yield Strength	2.8e+8 Pa	2.8e+8 Pa
Tensile Ultimate Strength	3.1e+8 Pa	4.3e+8 Pa

2.2 Solid and finite element (FE) model of Gasket

The thermo – mechanical structural analysis of metal gaskets was carried out using finite element package ANSYS work bench [6]. A gasket is deployed in a flange assembly (considered to be CF type) offers a two side knife edge contact. When bolted the gasket experiences bolting load via knife edges at a well defined confined surface. This enhances the deformation probability as the effective area decreases.

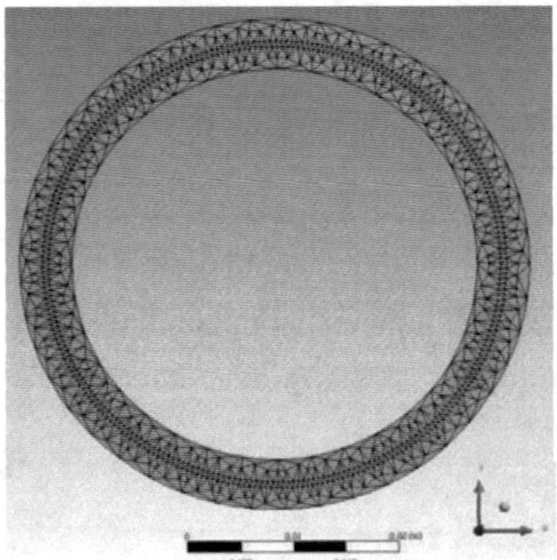

Fig. 3. The mesh model created in ANSYS

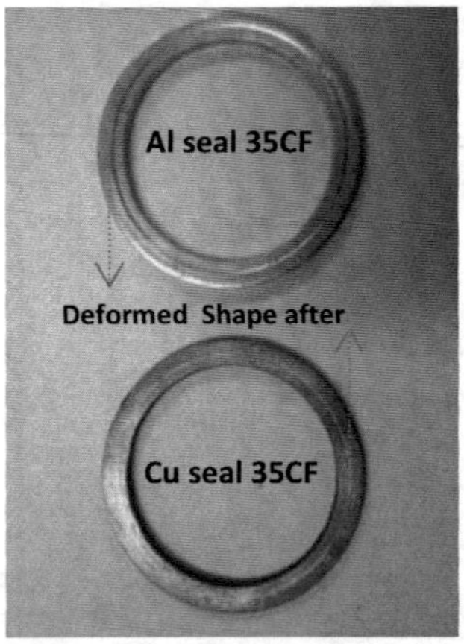

Fig. 4. Al and Cu seals deformation

The symmetric circular gasket with symmetric loadings (defined connection area), is modelled for coupled thermo mechanical structural analysis. The first demanded for simulation is for a mesh model, shown in Figur3**.** The differential meshing is considered for the gasket as the criticality of the physical phenomenon is unevenly distributed. The area of contact where the knife edge is effective will be expected to exhibit higher deformation probability. Some images of 35CF flange Al and Cu seals are shown in Figure 4 for deformed shape due to mechanical loading after removed from the UHV baked systems.

2.3 Boundary and loading conditions

The next step for the simulation is to set the environment parameters over the 3D geometry in to the when imported to the simulation section of ANSYS. The assumption was taken in order to check the standard practices followed in UHV laboratory systems such that the validation of the analysis can be done.

2.3.1Mechinical Loading.

The seating force which is applied on bolting is the primary binding of the assembly, induces substantial deformation. The binding force is calculated [2,7] for each and every dimension and material which then provided to the ANSYS simulation as the effective force. The force is given by the following equation.

$$\frac{\pi}{4}(OD^2 - ID^2)Y - \frac{\pi OD^2}{4}P \leq \frac{\pi}{4}(OD^2 - ID^2)Pm \quad \ldots\ldots\ldots\ldots(1)$$

(Gasket Seating force – Hydrostatic force = Residual gasket force)

As per equation 1, OD is outer diameter, ID inner Diameter, Y is minimum Seating stress, P maximum Internal pressure, m is Gasket factor which is ratio of contact pressure with contained pressure for the applicable seal materials. The gasket designer use the above formula for the seal design practices . A nonzero magnitude of residual gasket force ensures the leak tightness for the assembly. The calculation of seating force was carried out by a dedicated IDL codes so as to have the precise values of the force magnitude acting when flanges are bolted.

2.3.2 Thermal Loading.

The thermal loading is a potential contributor to significant deformation over the gasket. The simulation considers whole volume of the gasket on a constant and uniformly distributed temperature, pre decided input. As mentioned, for our studies the mechanical loading is constant all across the observation with a range of thermal loading, temperature range 10^0C to 250^0C. The temperature range serves different operating condition of the gasket, till backing or degassing, common practice for UHV applications.

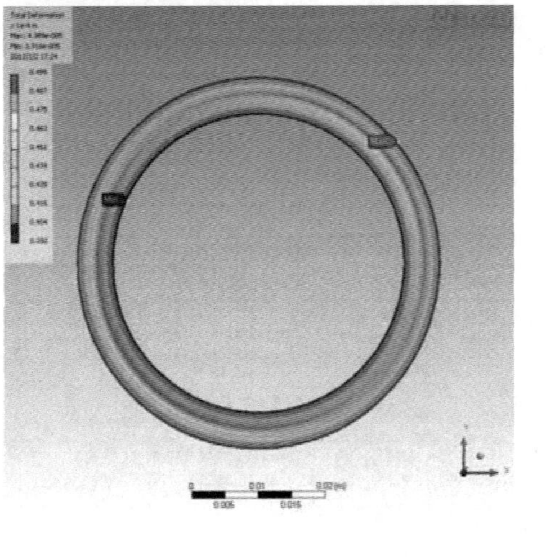

Fig. 5(a). Deformation with 35CF Al @100^0C seal

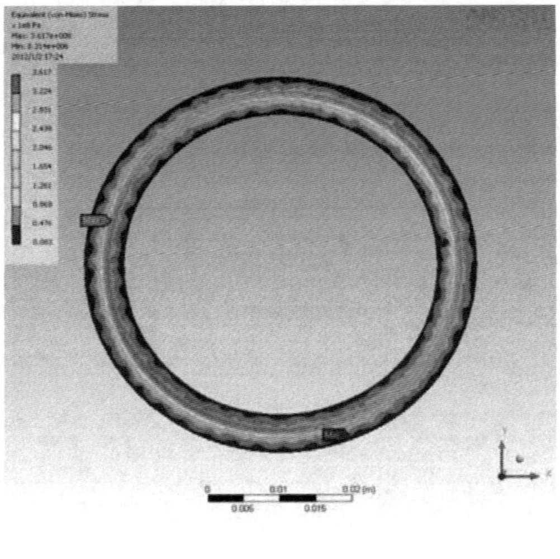

Fig. 5(b). Stresses for 35 CF Al seal @100^0C

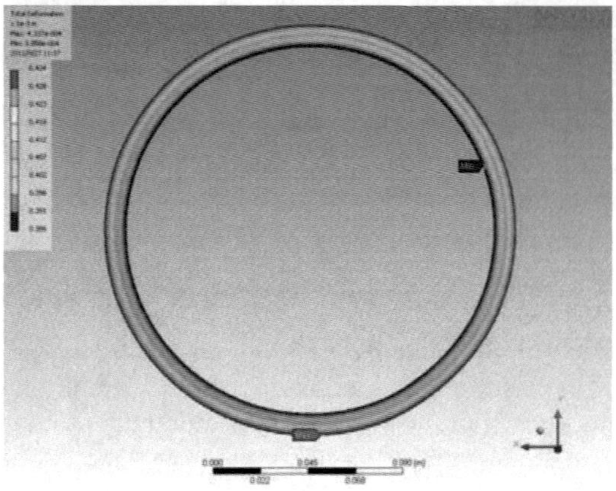

Fig. 6(a). Deformation with 150 CF Al seal @230^{0}C

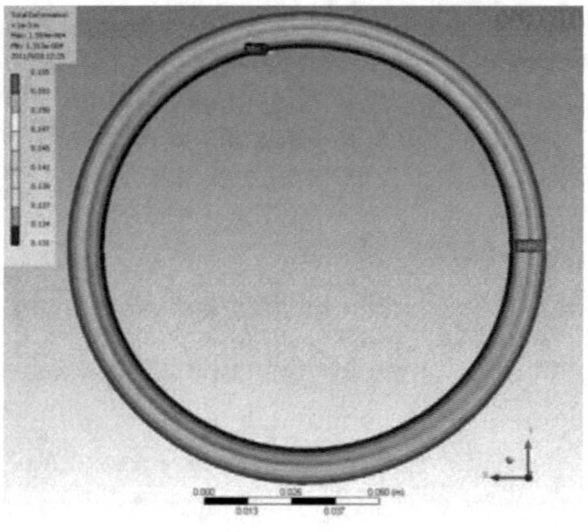

Fig. 6(b). Deformation with 100 CF Cu seal @150^{0}C

2.4 Simulation output

Simulation was conducted to know the deformation information for different gasket size as a function of time and material. Figure 5 and Figure 6 shows the typical output from the simulation made. Figure 5(a) describes the Al seal which has gone under stress with 35CF standard flange gasket load (seating stress of 35kN) conditions and with applied temperature of 100^{0}C. The deformation with knife edge dislocation mechanism are shown for reference. Figure 5(b) describes the stress levels on the deformed shape with Aluminium seal. In figure 6(a) deformation levels of 150CF Al seal with 230^{0}C was shown with applied seating force of 285 kN. Similarly in figure 6(b), deformation level was shown with 100CF Cu seal with constant seating force 288 kN and at 150^{0}C.

3. Results of the analysis

The Thermo-Mechanical load analysis cases for Al and Cu metal seals for range of inner diameter (ID) have been considered for the quantification of deformation. ANSYS simulation provides the maximum and minimum values for deformation. In order to compare different dimensions and material of gasket average deformation and the deformation range at a particular temperature are measured. This allows uniform and easy comparison between different cases. Figure 7 and Figure 8 shows the average deformation over the temperature scale for different ID for Al and Cu.

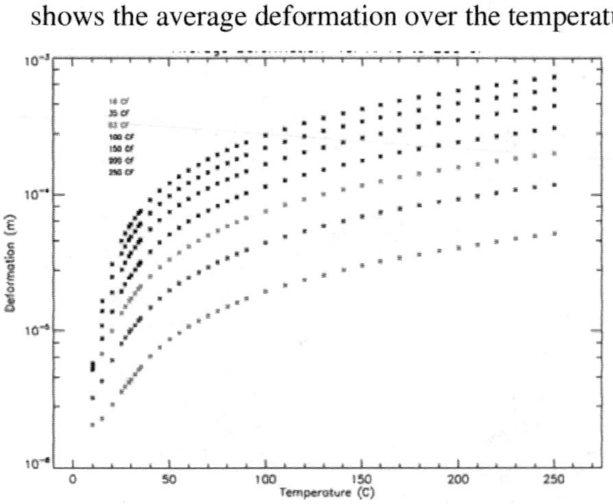

Fig. 7. Average Deformation Al seals Over wide temperature and different

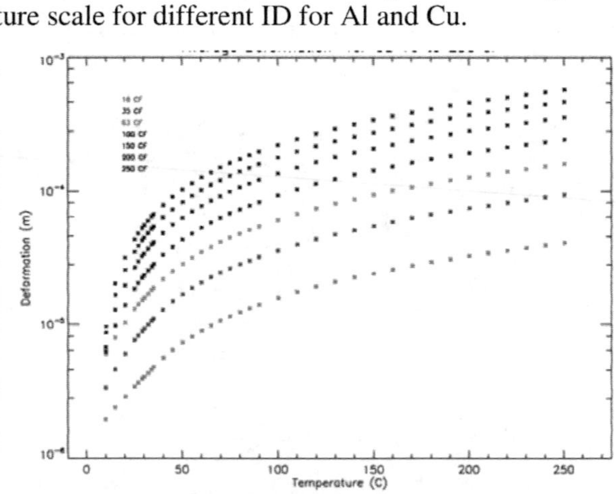

Fig. 8. Average Deformation Cu seals Over wide temperature and different seals

The average deformation is plotted on the y axis (log scale) as a function of temperature (^{0}C), exhibits that the change in the average deformation is higher at lower temperature with respect to the higher side of the temperature range. Figure 8 is shown with the average deformations analyzed with reference to the measured values in range of temperature from 10^{0}C to 250^{0}C temperature with Cu gasket seals which are regularly applied in UHV systems. Similarly for Al gasket seals deformations analysed with are shown in figure 8.

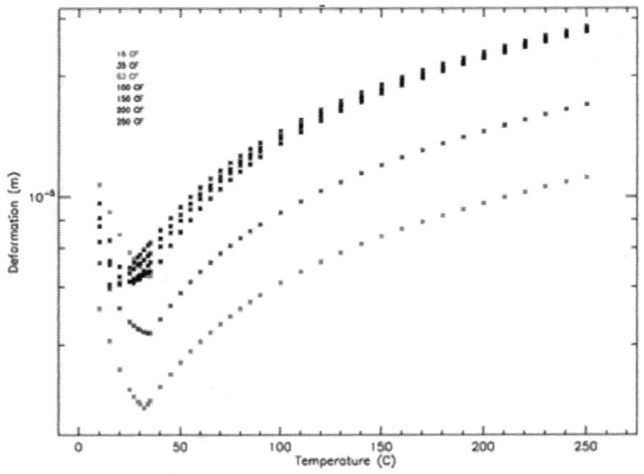

Fig. 9. Differential . deformation with Cu seals

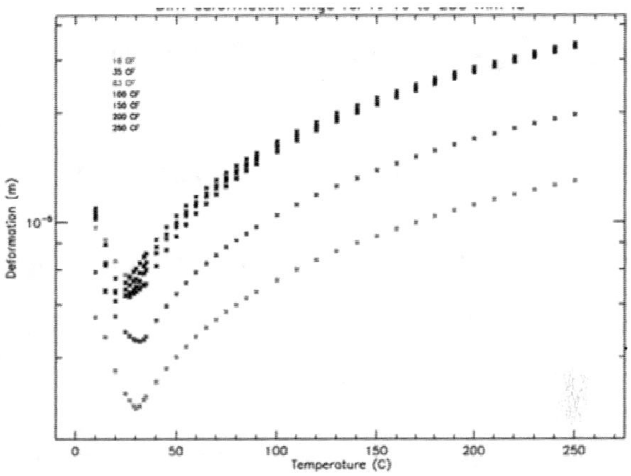

Fig. 10. Differential deformation with Al seals

4. Summary

Thermo-mechanical analysis of the metal vacuum seals is carried out with standard laboratory dimensions regularly used in UHV applications. The analysis for Cu and Al seal materials has been carried out with uniform seating load and wide temperature range from 10^{0}C to 250^{0}C with different flange seals like from 16 CF to 200 CF compatible flanges. The simulation analysis revealed that the deformations at the early temperatures are quiet significant when compared with the higher temperature range in both materials with standard applied seating load on to the seals. The analysis procedure will be helpful for the estimation of deformations in working conditions like mechanical and thermal loads.

Reference

[1] A.K. Roth, 1994 *Vacuum Sealing Techniques*,(AVS Classics in Vacuum Science and Technology)

[2] Lloyd E. Brownell, Edwin H. Young, 1959 *Process Equipment Design*,(John Wiley &Sons.)

[3] Antonio Carlos Guizzo, Determination of design gasket assembly stress with new constants, *6th Annual Fluid Sealing Association Technical Symposium, Houston, Texas,(1996).*

[4] W.Steckelmacher, Seals and Gaskets for Ultra High Vacuum Systems, *Vacuum*, **12**,(1962), 109- 113.

[5] H Fend, Progress in Metal Sealing techniques, *Vacuum*, **47**(1996) 527-529.

[6] ANSYS User's Manual, ANSYS Inc. PA, and USA

[7] Przemyslaw Lutkiewcz , Ph.D thesis, *The tightness conditions of UHV all metal seals subjected to the radiation and inelastic deformations.*

Thin films of Ti–Nb–Zr as non-evaporable getter films

R.K. Sharma, Jagannath, S. Bhattacharya, S.C. Gadkari, R. Mukund and S. K. Gupta

Technical Physics Division, Bhabha Atomic Research Centre, Mumbai 400085, India

Email: rajendra@barc.gov.in, Tel: +91 22 25592246

Abstract. Thin films of Non-evaporable getter (NEG) Ti–Nb–Zr, have been deposited using a DC magnetron sputtering technique. The NEG films were produced from a Ti–Nb–Zr target in the form twisted wires of pure individual metals combined to give the appropriate stoichiometry. The performance and morphology dependence on deposition pressure, sputtering conditions and substrate surface roughness are investigated. Energy dispersive X-ray and X-ray photoelectron spectroscopic techniques are utilized to characterize and study the activation behaviour of the NEG thin films. After characterizing with surface techniques the activation temperature of this alloy is found to be in the range 250 – 300°C for 2 h heating with well defined composition. It is seen that film topography and density depend largely on the substrate surface roughness, while the activation temperature depends almost exclusively on the final stoichiometry of the thin film.

1. Introduction

Production of extreme high vacuum (XHV) (pressure $\leq 10^{-10}$ Pa) is a technologically challenging task. Many of the modern particle accelerators as well as conventional analytical instruments have been pushing the limits of producing XHV that gives several advantages in terms of clean vacuum conditions[1-4]. Surfaces coated with Non evaporable getter(NEG) films have been known to selectively adsorb hydrogen and are being utilized either in the form of stand-alone pumps or more recently as thin film coated vacuum chambers that effectively reduce the partial pressure of hydrogen at very low pressures[5,6].

It is well known that the lifetime of a charged particle beam in storage rings depends mainly on the presence of residual gas molecules and beam instabilities. The conventional method of reducing residual gas molecules is through vacuum pumps located discretely at chosen positions in order to reduce interference with the charged particle beams. The disadvantage of this method is the lack of uniformity of vacuum conditions between two pumps especially in the case of long pipe sections. In order to overcome this disadvantage, it has been proposed to coat the inner walls of the vacuum chambers or pipes with thin films of suitable NEG materials. Active gases such as CO, CO_2, H_2O, N_2, and O_2 are chemisorbed on the NEG surface while H_2 diffuses into the getter bulk, thereby acting like a long continuous pump. Noble gases and CH_4 are not pumped by NEG materials. After in situ activation through the conventional baking procedure, these films provide large pumping speeds and also much lower dynamic and static degassing compared to traditional structural materials.

Activation of the NEG proceeds by heating to a temperature where the native oxide layer on the surface is dissolved into the bulk [7]. The suitability of the NEG for coating on the inner walls of vacuum chambers is given by its lower activation temperature and adhesion of the NEG films on to the substrate. Thus by effectively reducing surface outgassing and creating additional pumping, production of very low pressures is achieved[8]. Further, the desorption caused by the synchrotron radiation is very low in case of NEG coated surfaces compared to uncoated stainless steel surfaces. However, the sorption capacity for gases, except H_2, is limited to between 1 and 10 mL. This means that the materials suitable for use as an NEG coating should have large oxygen diffusivity, high oxygen solubility at bakeout temperature. The material should also have high sorption capacity and low dissociation probability for H_2. Alloys of Ti, Zr, and Nb fulfil all these requirements. Ti and Zr have an oxygen solubility of greater than 10%, but they have low oxygen diffusion coefficients. Niobium is the opposite case. Hence alloying Nb with any of other elements either in binary or ternary form is most efficient.

In this paper we present the characterization studies of a alloy getter material Ti-Nb-Zr deposited on stainless steel substrate to get a better understanding of the pumping properties of these coatings. The film stoichiometry, activation temperature of the NEG is also reported. It is seen that film morphology depends critically on deposition parameters such as pressure, sputtering conditions, and the substrate surface roughness.

2. Experimental details

We have used DC sputter deposition technique to deposit thin films of NEG material, since sputtering is simple and applicable to a wide range of the materials and alloys as well preserving the stoichiometry. Sputtering also allows uniform and distributed coating of long narrow vacuum chambers and also good adhesion. By employing composite cathodes it is possible to produce thin films of alloys/compounds of different materials. This feature is very attractive since elemental materials are easily available, and also coatings of pyrophoric materials could safely be produced from suitable targets. Once deposited as a thin film the danger of pyrophoricity is circumvented by the thermal capacity of the substrate, which limits the getter heating which results from exothermic gas absorption. Schematic of the DC magnetron sputtering deposition system has been shown in Fig.1

Here, we describe our indigenously fabricated cylindrical type DC magnetron sputtering system specially developed for depositing thin films of NEG materials on long thin SS sheets. Fig.1 shows a schematic of our experimental setup. In order to deposit NEG materials on long thin SS sheets, a substrate holder in a cylindrical form of 100mm diameter and length 300mm was utilized. A multi-turn, high current cooled solenoid was utilized to produce the magnetic field required for confining the plasma discharge. By varying the current through the solenoid, it was possible to produce uniform magnetic fields in the range of 100 – 300 gauss. The NEG target was prepared from a set of inter-twisted wires of diameter 1 mm from high purity (purity > 99.9%) materials of Ti, Zr and Nb in ternary form. Prior to initiating deposition, surface cleaning of the target and the substrate was performed by sonication with soap solution, followed by rinsing in clean distilled water.

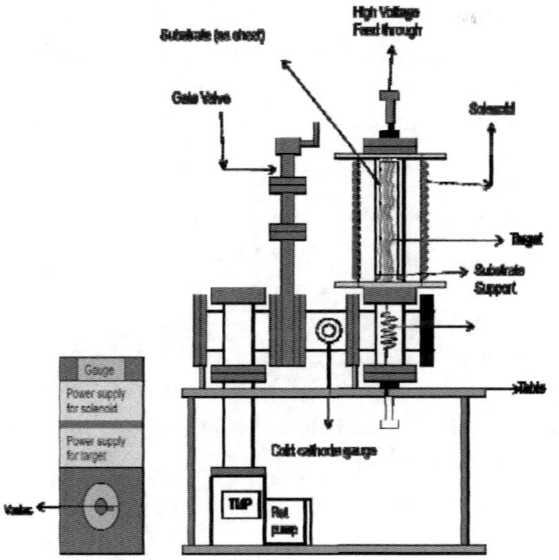

Fig. 1. Schematic of the DC magnetron sputter deposition system

The entire setup along with the target and the substrate was initially pumped down using a 500 l/s turbomolecular pump to a base pressure of 2×10^{-4} Pa. High purity, Lab grade argon was introduced through a leak valve to initiate the discharge. The pressure of sputtering gas Ar was about 1.1Pa. Discharge was initiated in the argon environment using a current stabilized high voltage power supply.

The sputtering parameters were optimized to 200V, 30mA and 0.02Tesla. The distance between the target and the substrate (sheet of 304L stainless steel of 100mm dia and length 300 mm) was 50 mm over the entire length of the target. In order to avoid the sagging of the target due to heating, the target was tensioned with the help of a spring on one side. Under these conditions, a coating rate of about 1 Å/sec was achieved. The film thickness as monitored by quartz crystal thickness monitor was 1 micron(approx.) for 3 hrs of deposition.

3. Characterization

An Oxford Instruments made Energy dispersive X-ray(EDX) instrument was utilized to identify the surface composition of the thin film is shown in table.1 indicating an elemental composition of (Ti 20Zr25Nb55 at.%.).

X-ray Photo electron spectroscopy (XPS) measurements were conducted for characterizing the activation properties of the NEG film. Characterization of Ti-Zr-Nb film was performed by (SPECS make XR 50 twin X-ray source). During NEG activation under ultra high vacuum conditions, the surface oxygen content is progressively reduced and surface starts pumping. Activation is completed when the oxygen surface content reaches a minimum and the pumping speed reaches its maximum value.

An Mg K_α X-ray source with 1253.6 eV was used with a detection resolution of concentration of 1 at.%. The detectable signal depth of the sample was better than 5 nm. The base pressure of the experimental chamber was 1.5×10^{-8} mbar during the measurement. XP spectra were obtained with a VG Scienta made spectrometer with single channeltron detector having a resolution of < 1 eV. A Thermo Vacuum Generator make sample manipulator (model TX 4570) with heating (upto 1000°C) and cooling down to liquid nitrogen temperature (77K) facilities was employed for controlled sequential heating of the samples. The samples could be rigidly clamped on to the sample holder for good thermal contact with the help of suitable clamps.

The temperature of the sample holder was measured by two K-type thermocouples. In order to study the activation temperature of the NEG films, controlled heating of the sample holder was done through suitable resistive heating element under ultrahigh vacuum conditions. The following procedure was adopted to exactly understand the activation mechanism of the NEG thin film.

4. Results

The high resolution photoelectron spectra allow the metal oxide reduction as a result of heating to be followed. Fig. 2 – 5 show the XP spectra for the Ti-Zr-Nb NEG recorded after heating the sample to different temperatures (25, 100, 120, 140, 180, 200 250 and 300 °C). The spectrum at 25°C is the 'as-deposited' condition.

Niobium is also the metal which is very easily reduced. The shift in Niobium peak starts at 140 deg C heating , indication the onset of oxide reduction already at this temp. Fig. 2 shows the decomposition of C(1s) to graphitic (atomic) carbon at 285 eV, CO bond at 287 eV, and carbide C-metal bond at 282 eV.

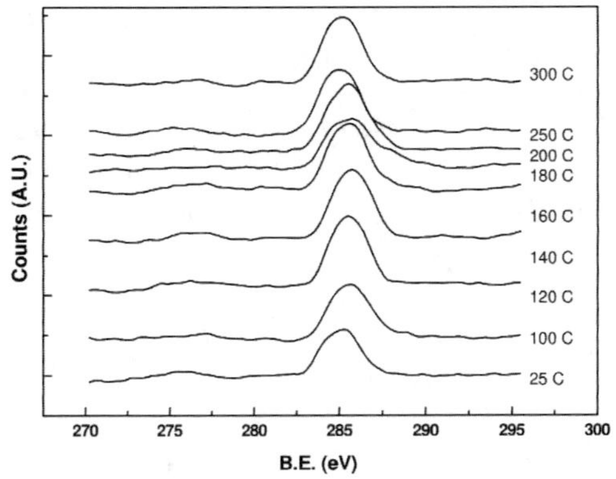

Fig. 2. Carbon 1s peak at different temperatures

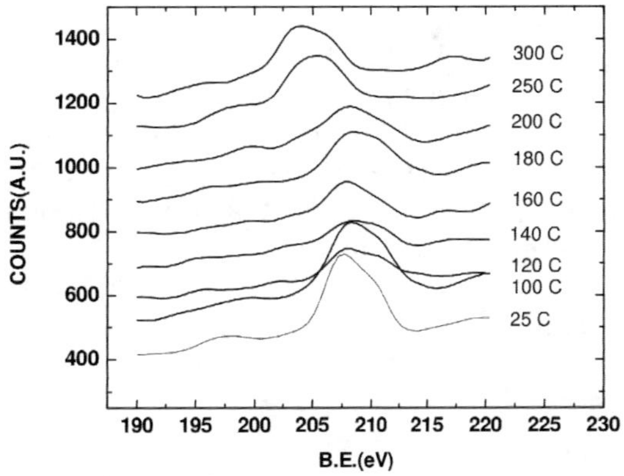

Fig. 3. Niobium $3d_{5/2}$ 1s peak at different temperatures

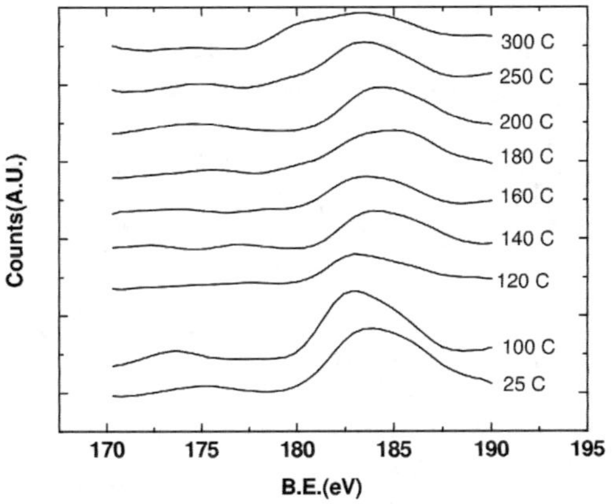

Fig. 4. Zirconium $3d_{5/2}$ 1s peak at different temperatures

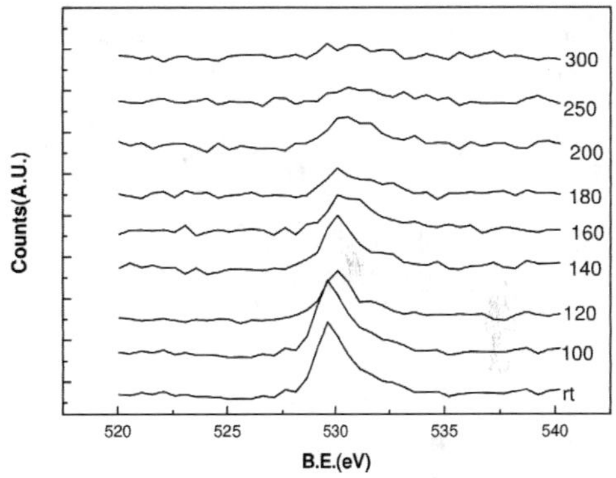

Fig. 5. Oxygen 1s peak at different temperatures

The change of the oxidation state of Ti-Zr-Nb can be seen in the spectra which indicates a transformation from a higher oxidation to lower oxidation state and then to the metallic state. This is clearly seen in Fig. 3 for Niobium. The $3d_{5/2}$ peak of Nb shows a clear shift of 2.5eV at a temperature of 250°C. The shift in the peak corresponds to lower oxidation state (NbO_2). This peak further shifts to a lower value at a temperature of 300°C.

Fig. 4 shows the evolution of $Zr(3d_{5/2})$ peak at different temperatures. There is significant shift in the oxidation state of the $Zr(3d_{5/2})$ after 2 h of heating at different temperatures indicating that

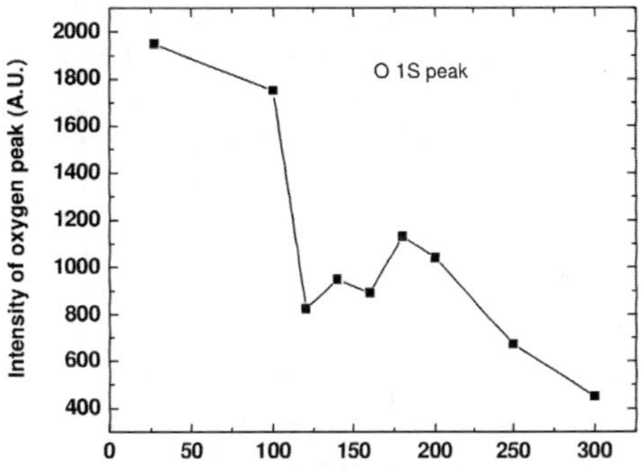

Fig. 6. Plot of Intensity of oxygen versus temperature

the peak corresponds to ZrO_2 in Ti-Zr alloy[6] n the as-deposited film which broadens with temperature. Similar behaviour was found for titanium indicating a Ti – Zr matrix.
The oxygen peak in Fig.5 is composed of two elemental intensities at 531 eV and 530 eV. These peaks correspond to different metallic oxides of Ti, Zr and Nb. In Fig.6, the plot between the intensity of the oxygen peak versus temperature using XPS technique shows the oxygen (O1s) evolution with temperature. It can be concluded that in activation of the film surface oxygen diffuses inside the bulk and leave the surface active.
From the above observations, it can be concluded that the activation temperature of the NEG coating(Ti-Nb-Zr) will be is in the range of 250 - 300°C.

5. Conclusion

Coatings of NEG materials in the form of ternary alloy of Ti-Zr-Nb were deposited on stainless steel substrate by using a DC magnetron sputtering technique. The XPS studies shows the lowest achievable activation temperature is in the range of 250°C to 300°C for 2 h of heating for Ti-Zr-Nb films. The Ti-Zr-Nb films can be used as NEG to achieve lower pressures in extreme high vacuum applications. Ternary elements Ti-Zr-Nb in the form of alloy was seen to proceed activation by the subsequent transition of oxides (TiO_2, ZrO_2 and Nb_2O_3) to residual zirconium and niobium sub-oxides mixed with metallic Ti, Zr and Nb phases even at 250°C. This oxide reduction leads to formation of active surface for H_2 and CO to form hydrides and results into the lowering of the pressure inside sealed vacuum chambers. The EDX spectrum of the film provides the surface stoichiometry of ($Ti_{20}Zr_{25}Nb_{55}$ at.%.).

All elementary XPS peaks are shifted by about 2 eV to lower BE for Zr($3d_{5/2}$), 2eV for Ti($2p_{3/2}$) after heating the sample to 250°C while niobium (Nb $3d_{5/2}$) B.E. peak is shifted by 3 eV after 250°C.

Acknowledgements
The authors are thankful BARC for giving the support.

Reference

[1] C Benvenuti, P Chiggiato, F. Cicoira, and V Ruzinov, Vacuum **50**, 57(1998).
[2] C Benvenuti and F Francia, J. Vac. Sci. Technol. A **6**, 2528 (1988).
[3] C Benvenuti and F Francia, J. Vac. Sci. Technol. A**8**, 3864 (1990).
[4] C Benvenuti, P Chiggiato, F Cicoira, Y Aminot, and V Ruzinov,Vacuum **73**, 139 (2004).
[5] M Sjöström, E Wallén, M Eriksson, and L-J Lindgren, Nucl. Instrum. Methods Phys. Res. A **601**, 229 (2009).
[6] J D Herbert, O B Malyshev, K J Middleman, and R J Reid, Vacuum **73**, 219 (2004)
[7] Z A Munir, Journal of material Science **14**(1979) 2733-2740
[8] C Benvenuti, P Chiggiato, A Mongelluzzo, A Prodromides, V Ruzinov, C Scheuerlein, M Taborelli, and F Levy, J. Vac. Sci. Technol. A **19**, 2925(2001).

Design and construction of a target chamber and associated equipments for the BARC Charged Particle Detector Array

Bency John[1], S. Sunil Kumar[2], Manish Kumar[2], R.V. Jangale[1], A.L.Inkar[1] and Lalit Kinage[1]

[1]Nuclear Physics Division, Bhabha Atomic Research Centre, Mumbai-400085, India
[2]Mechanical Design & Prototype Development Division, Bhabha Atomic Research Centre, Mumbai-400085, India

bjohn@barc.gov.in, bencyv.john@gmail.com

Abstract. A 60 cm diameter spherical high-vacuum target-chamber with side-opening hemispherical-lids, two ancillary-chambers, beam-line-tubes, tees and other high-vacuum components, and chamber-lid handling systems have been designed, constructed and installed for the Charged Particle Detector Array in BARC-TIFR Pelletron-LINAC Facility, Mumbai. This array of several tens of Si-CsI detector modules and other ancillary-detectors will be used for investigations in fusion-fission dynamics, nuclear structure at elevated temperatures and angular momenta, exotic nuclear clusters and related fields. This paper describes the unique features of the system that aid different coincidence experiments, the chamber fabrication experience and the pump-down characteristics with a turbo molecular pump. Unlike many other target chambers in use, this chamber allows multiple overall geometrical configurations to be set to reach experimental goals. For instance, by replacing a hemispherical-lid from one side with a flat-lid, the overall configuration becomes hemispherical. This way, high geometrical efficiency can be provided to an ancillary gamma detector array by allowing it to move close to target from the flat-lid side, although with some sacrifice of geometrical efficiency for charged particles. In experiments where a further improvement of geometrical efficiency for a gamma array is desired, a third compact-cylinder configuration can also be arrived at. Thinned portion of the lids of the chamber also allow neutron coincidence measurements with charged particles and gamma rays.

1. Introduction

A charged particle detector array (CPDA) which can provide Z and energy information of nuclear reaction products is being installed in the recently upgraded BARC-TIFR Pelletron-LINAC Heavy Ion Facility, Mumbai. This array will be used for investigations in fusion-fission dynamics, nuclear structure at elevated temperatures and angular momenta, exotic nuclear clusters, and related fields. These are some of the fields the nuclear physics community has taken special interest since they permit unraveling of complex nuclear structural and dynamic phenomena. This paper describes the unique features of the target chamber of the array that aid different coincidence experiments in the above fields, other facilities in the beam line, the chamber fabrication experience and the pump-down characteristics with a turbo system.

The standard investigative approach in heavy ion reaction physics is the measurement and analysis of the collision effects of a heavy ion beam over a target foil. Depending on the entrance channel parameters, the nuclear collisions produce a number of particles of different mass, charge, energy, and

Published under licence by IOP Publishing Ltd

angle of emission. Some of the studies require only simple measurements with a few detectors, while many other studies require more complete measurements with arrays of different types of detectors. First generation of detector arrays came in to existence around 1990 as array of Compton suppressed Ge detectors. With the availability of compact, convenient and reliable charged particle detectors and advances in data acquisition instrumentation, a second generation of such arrays came in to existence as in the following examples: INDRA-GANIL (1992), 8πLP-LNL (1997), NIMROD-TAMU (1999), CHIMERA-LNS (2000). With the ever increasing sophistication, reliability and convenience in detector and data acquisition instrumentation, third generation of arrays are also being built in advanced laboratories around the world.

For a comparison of charged particle detector array capabilities, the main characteristics of second generation of such arrays can be classified as: 1) Large 4π arrays for multi fragmentation studies that requiring complete exclusive reconstruction of emitting sources, e.g., INDRA, CHIMERA, NIMROD and FOPI, 2) 4π charged particle detector array as an ancillary detector for gamma spectroscopy, e.g., NORDBALL, GASP, GAMMASPHERE and EUROBALL, 3) Arrays for special purposes, e.g., MEGHA, AMPHORA and ISOLDE Silicon Ball.

The third category– where the present CPDA also belong– in most cases are upgrades of detector systems that cater to variety issues in nuclear dynamics and structure, and they are not necessarily (close to)4π systems as arrays in the first two categories. For example, the MEGHA array has 44 three-detector-telescopes (gas-silicon-CsI) mounted suitably for nuclear cluster-breakup studies. In the first two categories, the geometrical configuration of charged particle detectors is more or less singular (being 4π setups) where in the third, multiple configurations *can be* set to reach experimental goals, provided the scattering chamber is suitably designed to accommodate multiple configurations.

2. Conceptual design of the CPDA

2.1. Charged particle detector module

Having resolved the type of CPDA to be built, the next issue is selection of charged particle detectors. Ion implanted silicon detectors made by BARC at BEL-Bangalore facilities is a natural choice due to qualities silicon possess as a charged particle detection medium and other advantages since the detectors are indigenously produced[1]. Since BARC-BEL detectors have thickness about 300 μm, it will not stop all energetic charged particles of our interest, therefore, it is necessary to have a stop detector in a □E –E configuration and, therefore, commercially available CsI detectors were selected as stop detectors. In the mechanical design work for the array, one of the first tasks was to design a detector module mount which will hold the Si and CsI detectors in a transmission and stop detector configuration (see Fig.1(a-b)). The active area of the Si detector is a square of approximately 400 mm^2 and it is segmented in to four equal squares of 100 mm^2. The CsI detector is having an active area of approximately 500 mm^2 square and overall thickness 25 mm with crystal thickness equal to 10 mm. The Si detector is in a precision machined detachable insert which is a foldable structure with a rectangular opening. One side of the insert is made with stainless steel and the other with Derlin. The metal part gives strength and act as a heat sink and the Derlin side protects the conducting tracks in the ceramic substrate from shorting. The detachable design allows one to use CsI detector alone in a module when need arises. It was decided that the detector module should be movable to a desired lab angle but there will not be any movable arms inside the chamber in the first phase. Therefore, graduated rails are provided to fix the detectors at desired angles. A part of such rail is shown in Fig.1(a-b) with arrangement for fixing the module. Since we have option for a large number of detector modules to cover sufficient solid angle, the routine for chamber opening, fixing, closing and then evacuation can be minimized to one or two in an experiment to economize in the machine time. The angular opening of one segment in the detector module will depend on the distance from target at which the module is fixed. It was desired that this opening be around 3° in the array and accordingly the target to first detector distance was selected as 18.5 cm. It followed that the diameter of the spherical chamber required is around 60 cm.

2.2. Target chamber

The main task was to design a versatile high vacuum target chamber made of stainless steel(SS304) which is sufficiently flexible to allow different types of experiments[2] and there are no preceding chambers of this type in our laboratory. The variety of experiments we are interested in can be noted from our recently published works carried out at Pelletron Linac Facility using different simple cylindrical chambers and beam lines. A small cylindrical (20 cm dia., 4 cm height) scattering chamber and a gamma ray multiplicity setup consisting of 14 BGO detectors were used in one of the works. Coincidences between charged particles measured using silicon detectors placed inside the chamber, in particular alpha particles, and gamma rays detected in the BGO array placed outside the chamber were recorded in the experiment. An evaporation residue(ER) detector would have been very useful to improve accuracy of the measurements, however, due to lack of space inside the compact chamber and other reasons, this option was not available in the experiments. The total solid angle subtended by charged particle detectors was only 11 msr, which is rather a low value for this type of work. Another experiment is a first of its kind heavy cluster knockout reaction $^{16}O(^{12}C, 2^{12}C)^{4}He$ where two energetic ^{12}C ions where detected in coincidence using a pair of silicon telescopes of 4.5 msr solid angle at fixed angles of 41 and 45 degrees. For better event statistics, it would have been better to cover more □ angles at selected scattering angles(□) with more detector telescopes. A third kind of experiments is pre-scission charged particle emission where again a large number of charged particle detectors at selectable 'odd' angles are required. Details on the physics interests in the above experiments can be obtained from the published articles[3,4]. It is necessary to eliminate above mentioned drawbacks to the extent possible and facilitate several more types of experiments with the new target chamber.

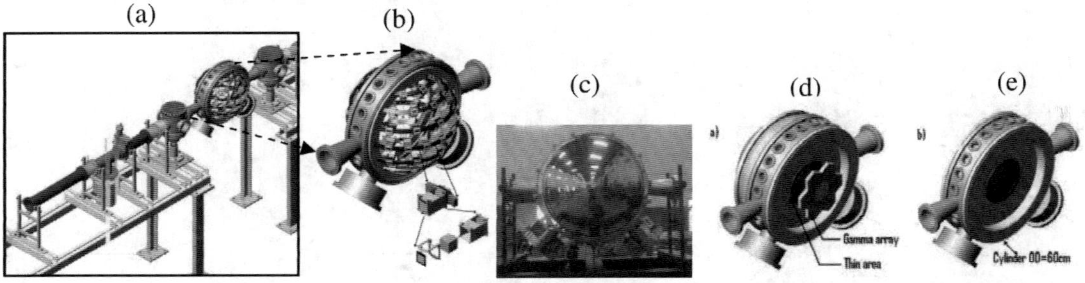

Fig.1: (a)CAD diagram of the CPDA beam-line part. (b)Expanded view of the target chamber. Full assembly of a single detector module is shown in the inset below. Components of the detector module such as four pad silicon detector, hinged holder for the silicon detector, CsI detector and its holder are shown in a exploded view(bottom). (c) Photograph of the target chamber after fabrication. (d) CAD diagram of hemispherical and (e) compact cylindrical configurations of the CPDA.

There are three distinct configurations possible for the target chamber. One of them, the spherical configuration of diameter 60 cm, is shown in Fig.1(a)-(c). Thin walled hemi-spherical lids and flange cover plates, that are not shown in Fig.1(a), will ensure vacuum. Expanded views of the spherical mount and the detector module are shown in Fig1(b) to illustrate the design concepts. Detector modules can be held on to circular rails by brackets and the modules can be fixed at any angle within the range provided by the rails. Detector modules can be held similarly inside the central cylindrical part(see Fig.2). As stated earlier, first element of the detector module will be at a distance of 18.5 cm from the center. The range of scattering angles that can be covered is from 10° to 170°. Maximum solid angle that can be covered in this configuration is 1520 msr (12% of 4π). If needed, the total solid angle coverage can be increased by bringing the rails closer by doing a rework on the rails.

By removing a part of the hardware a second configuration namely hemisphere can be arrived at as illustrated in Fig.1(d). In this figure, charged particle detector modules are covered by hemispherical and flat lids. The flat lid has a central thin convex circular area of 25 cm diameter made of 2mm stainless steel. This configuration will have a maximum solid angle coverage of only 930 msr for charged particles. Reconfiguring this way has a major advantage that a gamma detector array can be brought closer to target position while still maintaining a reasonable solid angle coverage for charged particle detectors. In experiments where a further improvement of geometrical efficiency for a gamma array is desired, a third configuration can be arrived at as shown in Fig.1(e). Here, charged particle detectors are retained only in the compact-cylindrical portion and the flat lids cover both sides of the chamber. Gamma detector arrays can be brought closer from either sides to double their total geometric efficiency.

3. Construction

Installation aspects of the fabricated compact cylindrical chamber are described first. This part of the chamber is vertically placed and it defines a vertical 'reaction' plane(Fig.2). It is supported from below at three places, at beam inlet and outlet by coupling to sturdy ancillary chambers, and in the middle by a single pillar fixed on the floor. Two stainless steel rings of width 24 mm and thickness 9 mm were welded inside the cylindrical surface so as to provide good reinforcement for the cylindrical structure and they also act as holders for the hemispherical rail structure as well as a place to mount detector modules. This way of chamber alignment can avoid broad stands which usually support horizontally-aligned chambers. Such broad stands may create problems for bringing close bulky gamma detector arrays from the sides.

(a)

Fig.2: (a)Photographs of the compact cylindrical chamber with one of the flat lids open. The opened lid is hanging in the foreground. Central pillar and the second lid-handling stand with crane are partially visible. (b) Same chamber with both lids closed.

Fine adjustments for alignment of cylindrical part with respect to beam line could be made by movements on vertical bolts and in horizontal cuts made in the base plates of ancillary chambers. Chamber lids are bought from two sides by two mini-cranes, one for each side, for closing. Detachable semi-circular hooks have been fabricated for lifting the lids by cranes. The motor driving the crane is fixed on top of table-cum crane stand on a linear motion platform. The linear motion provided by these platforms and the vertical motion by the cranes can align the lids to the cylindrical part of the chamber. Use of the guide pins provided at the chamber periphery facilitates the lid alignment to the chamber. Accompanying photograph(Fig.2(a)) shows the vertically aligned compact cylindrical chamber with one lid moved to the foreground for display purpose. Ancillary chambers and the central pillar can also be seen in their aligned positions. One of the lid-handling table-cum-stand with crane attached on top can be seen in the background. The target chamber with both lids closed and the table-cum-stand retracted can be seen in photograph Fig.2(b). The suitably designed table-cum-stands have storage mechanism for storing the hemispherical and flat lids when not in use and have table-tops for users to work on.

Some more details and dimensions of the chamber and associated equipments can be obtained from Fig.3. The multiple target holder contains 6 targets, any one of which may be used without breaking the vacuum. There are total 18 ports on the cylindrical part of which 2 are for 200CF flange fitted with ten numbers of 25-pin D-type electrical feed-throughs. Rest 16 are custom sized ports of OD 70 mm distributed at various angles. These ports are of multi-purpose, including those for insertion of the target ladder, view ports, and electrical feed-throughs. Expanded view of one of the 70 mm port, the one at the bottom, is given in the top right corner inset of Fig.3. A section of the center pillar can also be seen here since the hollow-center-pillar support the main cylinder by surrounding the bottom most port. A thick washer made of Teflon prevents electrical contact between the chamber and the pillar.

As per the design, particles emitted in ±6° most forward and most backward angles will enter in the ancillary chambers placed after and before the target chamber, respectively. These chambers can house detectors for measuring charged particles emitted at these angles, for example, evaporation residue detectors at most forward angles for measuring ERs. Ports are provided at the sides and on top (100 CF and 200 CF size, respectively) of the ancillary chambers for facilitating these works. Construction of these chambers are such that they and the target chamber are electrically insulated from ground. (Suitably designed nylon washers as shown in the left-top inset of Fig.3 help to achieve this insulation.) At the same time the ancillary chamber fixtures provide necessary mechanical support to the target chamber and avoid mechanical stress on beam line insulators. The turbo pumping station (TPU 1201 P with control unit TC 750 from Pfeiffer Vacuum) and high vac gauge(PKR 251) are located upstream and a beam-line insulator and an electrically actuated gate valve isolate the ancillary chamber and the pump station. In the down-stream line also there is a provision for a pump station after a gate-valve, however at present no second pump station is connected. Although the positions of pumps are not at good gas conductance, the same helped in achieving electrical isolation for the chambers. At the end of the beam-line, after a beam-line insulator and a drift tube of length 75 cm, a Faraday cup(FC) is connected for measuring beam current.

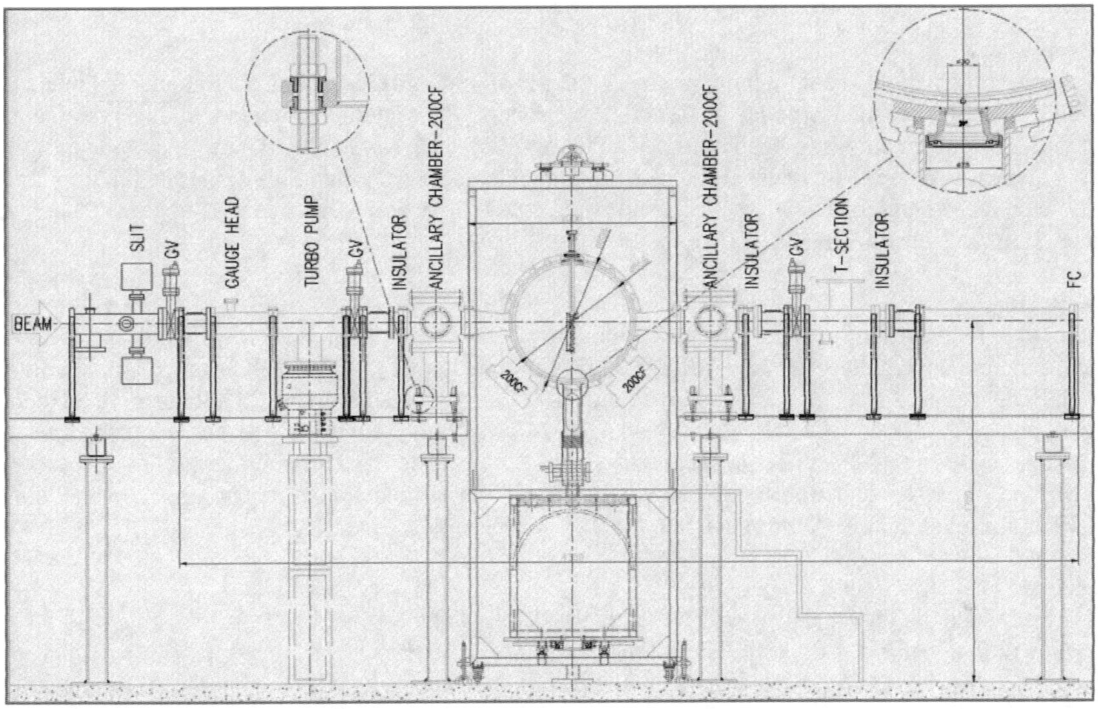

Fig.3: Schematic of the CPDA beam-line. Target chamber is in the middle.

The hemispherical mounts(Fig.4) and corresponding lids have also been installed and tested for vacuum and for operations with the lid-handling crane and storage table. The alignment of this part was separately tested using guide blocks and laser pointers. Overall alignment at the target position is better than ±2 mm. For beam collimation there are two provisions, the primary one is an electrically actuated x-y slit at the position indicated in Fig.3. In addition a collimator disc can be placed at the entrance of the cylindrical chamber on a removable retainer ring.

Pump-down characteristics of the compact cylindrical chamber and beam line is shown in Fig.5. Pump-down time from atmospheric pressure to working pressure $2x10^{-6}$ mbar is 2 hours. This was measured with relatively small gas load from detectors, cable etc. Detector response has also been measured[5] in an actual heavy ion reaction experiment for confirmation. The response was satisfactory and detailed analysis of the same will be communicated separately.

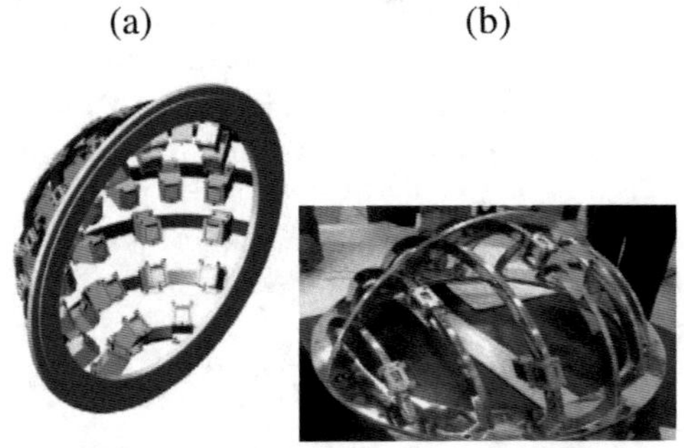

Fig.4: Hemispherical rail structure for holding the detector module. (a) CAD diagram (b) partially assembled after fabrication.

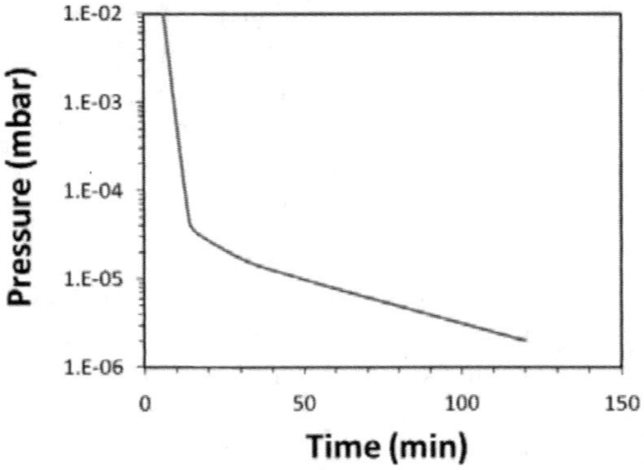

Fig.5: Pump-down characteristics.

4. Summary

Design and construction of a new high vacuum target chamber and associated equipments for nuclear reaction and structure studies at BARC-TIFR Pelletron-LINAC Facility is described. Experimental work with this beam line has been initiated. This facility will house the BARC charged particle detector array that comprising of several tens of Si-CsI detector modules and it got unique features that aid coincidence measurements between the array of charged particle detectors and arrays of neutral particle (gamma, neutron) detectors.

Acknowledgements

Authors wish to thank Dr.S.Kailas, Dr. R.K.Choudhury and Shri.A.W.Matkar for helpful discussions in finalizing the design. Several useful discussions with many colleagues of Pelletron LINAC Facility and MDPDD, most particularly Drs.D.C.Biswas, B.K.Nayak, Suresh Kumar, S.C.Sharma, A.Saxena, A.Agarwal, A.Chatterjee, P.K.Nema, V.Nanal and R.G.Pillay, have helped in the fabrication, installation, and tests. M/s Vacuum Technique Pvt. Ltd carried out the fabrication of the target chamber and M/s LNCNC Consultants carried out fabrication of mechanical parts of detector modules and table-cum-crane stands. Authors wish to thank all of them as well.

References

[1] Bency John *et al.*, Nucl. Inst. And Meth. Phys. Res. A **609** (2009) 24
[2] Bency John *et al.*, DAE Symp. Nucl. Phys. **54** (2009) p.698
[3] Y.K.Gupta *et al.*, Phys. Rev. C **78**, 054609(2008), *ibid* Phys. Rev. C **84**, 031603(2011).
[4] Bhushan Joshi *et al.*, Phys. Rev. Lett. **106**, 022501(2011)
[5] Bency John *et al.* (to be published).

Preliminary Design of the Vacuum System for FAIR Super FRS Quadrupole Magnet Cryostat

J. Akhter, G. Pal, A. Datta, P. R. Sarma, U. Bhunia, S. Roy, S. Bhattacharyya, C. Nandi, C. Mallik and R. K. Bhandari

Variable Energy Cyclotron Center, Kolkata, India

javed.akhter@vecc.gov.in

Abstract. The Super-Conducting Fragment Separator (Super FRS) of the Facility for Antiproton and Ion Research (FAIR) at GSI Darmstadt is a large-acceptance superonducting fragment separator. The separator consists of large dipole, quadrupole and hexapole superconducting magnets. The long quadrupole magnet cryostat houses the helium chamber, which has the magnet iron and NbTi superconducting coil. The magnet weighs about 30 tons. The helium chamber is enclosed in vacuum inside the magnet cryostat. Multilayer Insulation (MLI) will be wrapped around the thermal shield to reduce radiation loss. Polyster of MLI comprises the major component responsible for outgassing. In order to reduce outgassing, pumping at elevated temperatures has to be carried out. In view of the large size and weight of the magnet, a seal off approach might not be operationally feasible. Continuous pumping of the cryostat has also been examined. Pump has been kept at a distance from the magnet considering the effect of stray magnetic fields. Oil free turbo molecular pump and scroll pump combination will be used to pump down the cryostat. The ultimate heat load of the cryostat will be highly dependent on the pressure attained. Radiation and conduction plays an important role in the heat transfer at low temperatures. This paper presents the vacuum design of the long quadrupole magnet cryostat and estimates the heat load of the cryostat.

1. Introduction

The Superconducting Fragment Separator (Super FRS) of the Facility for Antiproton and Ion Research (FAIR) at GSI in Darmstadt is a large-acceptance superconducting fragment separator to efficiently separate rare isotopes. The seperator consists of large dipole, quadrupole and hexapole superconducting magnets.

The long quadruple magnet [1, 2] along with helium chamber is housed inside a vacuum chamber. The pump has been kept at a distance from the magnet considering the effect of stray magnetic fields. Oil free turbo molecular pump and scroll pump combination will be used to pump down the cryostat. Figure 1 shows the overall view of the quadrupole magnet cryostat.

Heat transfer through Multilayer Layer Insulation (MLI) in cryostats has been elaborately studied [3-5]. C. K. Krishnaprakas [3] has considered four models for the calculation of heat flux through MLI: (1) conductance model, (2) effective emittance model, (3) conduction radiation model, (4) Connington and Tien model [4]. These models use average properties to obtain the overall heat transfer. Cunnington has considered heat transfer through each layer of MLI.

Heat transfer between the vacuum vessel at 300K and the radiation shield at 80K through each layer of MLI and the requirement of the pumping speed has been evaluated in this paper. The cryostat vacuum system is designed to satisfy the vacuum requirement inside the cryostat and the overall pumping system is proposed.

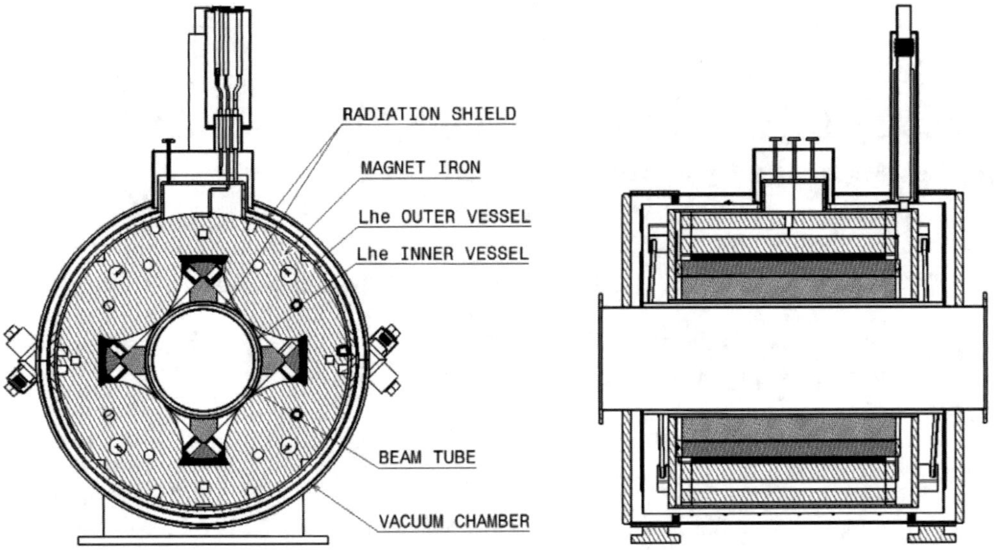

Fig. 1. Cross sectional view of the FAIR Super FRS quadrupole cryostat

2. Heat Transfer between cryostat and the radiation shield

In the quadrupole magnet, the coil will be at liquid helium temperature. A liquid nitrogen cooled radiation shield will be placed between magnet coil at 4.2 K and cryostat at 300 K. The radiation shield will be covered with MLI for reducing heat load on the liquid nitrogen system and the liquid helium system. Heat leak to the radiation shield is by radiation, gas conduction and conduction through the supports.

2.1. Numerical Model

The heat transfer due to gas conduction in molecular flow region and radiation is calculated using equations (1) [5, 6] and (2) [7]. A MATLAB code is written to solve equations (8) and (9). The code uses iterative method to calculate the resistance to heat transfer using assumed temperatures of the MLI layers. These calculated resistances were then used to find the total heat transfer rate and corrected temperatures. The calculation proceeds till there is no appreciable correction in the temperatures.

$$Q_{cond} = \alpha\left(\frac{\gamma+1}{\gamma-1}\right)p_{ins,i}\sqrt{\frac{R}{8\pi T_{ins}}}\left(T_{i+1}-T_i\right)A_{i+1} \tag{1}$$

$$Q_{rad} = \sigma A_i\left(T_{i+1}^4 - T_i^4\right)\Big/\left[\frac{1}{\varepsilon_i}+\frac{A_i}{A_{i+1}}\left(\frac{1}{\varepsilon_{i+1}}-1\right)\right] \tag{2}$$

$$\alpha = \frac{\alpha_i\alpha_{i+1}}{\alpha_{i+1}+\alpha_i\left(1-\alpha_{i+1}\right)A_i/A_{i+1}} \tag{3}$$

where α is the accommodation coefficient for copper thermal shield and cryostat wall, γ is c_p/c_v, the specific heat ratio of the gas, $p_{ins,i}$ is the interstitial pressure of insulation gas pressure, T_{ins} is the effective temperature of the non equilibrium gas [5]. The subscripts '*i*' ranges from 1 to *N*+2, where *N* is the number of layers of MLI. Subscript 1 will refer to radiation shield and subscript *N* refers to cryostat inner wall. $p_{ins,i}$ is obtained using equation (4), P_c and T_c being the chamber pressure and warm chamber temperature, respectively. This model takes into account the effect of thermal transpiration.

$$p_{ins,i} = p_c\left(\frac{T_{ins,i}}{T_c}\right)^{1/2} \tag{4}$$

where,

$$T_{ins,i} = \frac{T_i + T_{i+1}}{2}$$

The gas heat conduction in viscous and transition region is calculated using equation (5) [8].

$$Q_{cond,i} = \frac{\kappa(T_{i+1} - T_i)}{(D + 2L)} \tag{5}$$

where, κ is the heat conductivity of the gas, *D* is the typical distance between two surfaces and *L* is the characteristic distance relating the mean free path of the gas of the gas molecules. The total heat transfer through the layers is obtained by adding the heat transfer by gas conduction and radiation.

The following assumptions have been made:

- The emissivities of stainless steel cryostat wall, copper radiation shield and MLI are 0.1, 0.15 and 0.07 respectively.
- Accommodation coefficient varies linearly from 1.0 at 80K to 0.8 at 300K for air [7].
- Area of MLI layer closest to the radiation shield is equal to the area of the radiation shield itself and for the successive layers of the MLI the area increases marginally and is same as the area of the first layer of MLI.

Radiation shape factors and thermal conductance are used for evaluating radiation heat transfer and gas conduction heat transfer.

$$F_{i,i+1} = \frac{1}{\left[\frac{1}{\varepsilon_i} + \frac{A_i}{A_{i+1}}\left(\frac{1}{\varepsilon_{i+1}} - 1\right)\right]} \tag{6}$$

$$\Lambda = \left(\frac{\gamma+1}{\gamma-1}\right)\sqrt{\frac{R}{8\pi T_{ins}}} \tag{7}$$

The radiation heat transfer and gas conduction heat transfer equations can then be written as in equation (11) and (12). The total heat transfer through i^{th} layer will be sum of the two modes of heat transfer.

$$Q_{rad,i} = \sigma A_i F_{i,i+1}\left(T_{i+1}^4 - T_i^4\right) \tag{8}$$

$$Q_{cond}, i = \Lambda p_{ins,i}\left(T_{i+1} - T_i\right)A_{i+1} \tag{9}$$

Fig. 2 shows the calculated variation of total heat transfer with the chamber pressure for 0, 10, 20 and 30 layers of MLI considering the different heat transfer mechanisms considered in equations (5), (8) and (9). It is observed that by the use of MLI the total heat transfer is reduced by up to 27 times. Figure 3 shows the variation of rate of heat transfer with number of MLI layers at a pressure of 0.1 Pa for the different heat transfer mechanism: radiation, gas conduction and combined gas conduction and radiation. Figure 4 shows the temperatures of the thirty layers of MLI for the different modes of heat transfer. Temperatures of the upper curve are calculated by restricting the heat transfer to radiation, lower curve are calculated by restricting the heat transfer to gas conduction. The middle curve is obtained by considering both radiation and gas conduction. The temperature gradient is high near the cold boundary; consequently the gas conduction increases and the radiation heat transfer decreases when approaching the cold wall. The highest temperature difference and consequently the highest heat transfer rate is calculated between the radiation shield and the adjacent MLI layer.

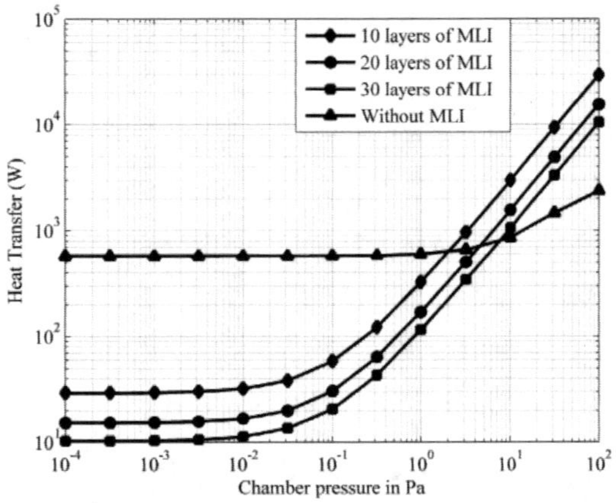

Fig. 2. Relationship between Heat Transfer rate and the chamber pressure for 0, 10, 20, 30 layers of MLI

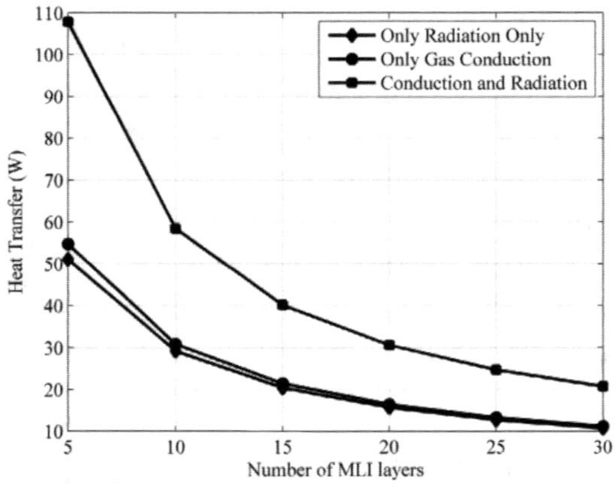

Fig. 3. Calculated Variation of heat leak with number MLI layers

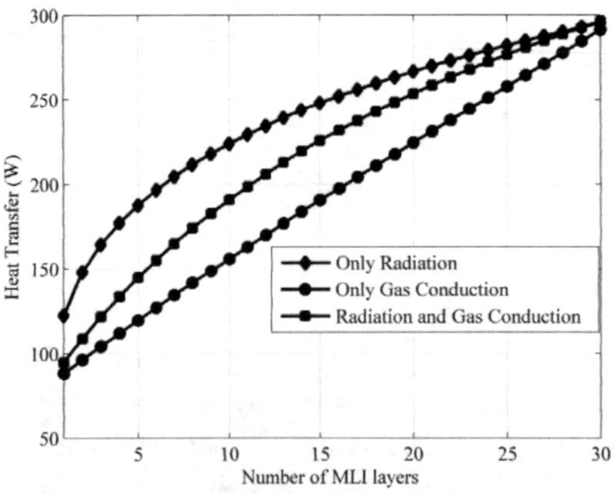

Fig. 4. Calculated temperatures of 30 layers of MLI considering different heat transfer mechanism

3. Vacuum in the Cryostat

From the results of heat transfer in the cryostat (Fig. 2) it is seen that the upper limit of the pressure in the cryostat can be set at 0.1 Pa as the total heat transfer rate with 30 layers of MLI is about 21 Watts. In view of the large size and weight of the magnet, a seal off approach might not be operationally feasible and the option of continuous pumping of the cryostat is considered. The internal volume of the cryostat is 1.84 m^3. Total internal surface area of the cryostat, the helium vessel, the beam tube, and other SS elements sums up to 39 m^2. The surface area of the thermal shield (both internal and external surfaces) is 37 m^2. The G10 epoxy material surface area is 1.9 m^2. Table 1 tabulates the area and the outgassing rates of the materials after 1 hr in vacuum. The outgassing rates are taken from the references [9, 10]. The total outgassing load was found to be 6.74 Pa l/sec.

Table 1: Outgassing rates of material inside the cryostat

	Material	Surface Area (m^2)	Outgassing rate (Pa.l/sec/cm^2)	Estimated gas load (Pas.l/sec)
Cryostat+Beam Tube+He Vessel	SS	39	1.20E-05	4.68
Radiaion Shields	Copper	37	5.30E-06	1.96
Support Links	G10	1.9	1.33E-06	2.49×10^{-02}
MLI	Mylar	888.68	8.50E-09	7.55×10^{-02}

An effective pumping speed of 90 liter/sec is expected to achieve a pressure better than the operating pressure. However, degassing/ vacuum baking of the system is required in order to achieve a better vacuum in short time. Taking the outgassing rates for the degassed materials from [9], a pressure of 1×10^{-3} Pa can be achieved.

4. Preliminary Design of the Vacuum System for Vacuum System for FAIR Super FRS Quadrupole Magnet Cryostat

Considering the effect of stray magnetic field the pumping system for the FAIR Super FRS Quadrupole Magnet system will be located at a distance of 1.5 meter from the central position and connected to the cryostat via a 100 mm diameter vacuum duct. The conductance of the vacuum duct will be 121 liter/sec. A dry mechanical pump of speed 210 liter/min will be used for the rough pumping of the cryostat from atmosphere. A turbomolecular pump of 350 liter/sec pumping speed will be used for attaining the base pressure. The expected pump down time from the atmosphere to 0.01 Pa by the pumping system is evaluated by solving mass balance equation (10). It was found that a pumping time of about 2 hr and 25 minutes will be required. Figure 5 shows the expected pump down curve for cryostat vacuum system.

$$\overline{Q} = SP + V\frac{dP}{dt} \tag{10}$$

Where $\overline{Q}$ is the gas load to the cryostat vacuum system obtained by adding the outgassing contributions from the different materials inside the cryostat, S is the effective pumping speed, V is the volume of the vacuum space and P is the pressure inside the cryostat. The outgassing from the surface varies depends of times after the start of pumping, expressed as in equation (11) [11].

$$\overline{Q} = K_1 t_h^{-\lambda}$$

Where K_1 is the outgassing rate at one hour.

Figure 5 shows the expected pump down curve for unbaked cryostat vacuum system. To reduce the pump down time rough pumping of the cryostat at elevated temperatures is required.

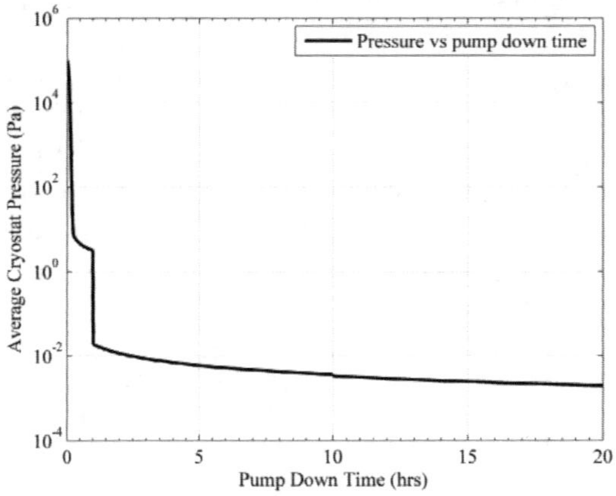

Fig 5. Expected Pump Down curve the Quadrupole Magnet Cryostat Vacuum system

5. Summary

The heat transfer in the FAIR Super FRS Quadrupole magnet cryostat was discussed and the required vacuum was evaluated. The vacuum system that will achieve the required vacuum and hence minimize the heat transfer has been discussed. Since the outgassing within the cryostat cannot be negligible, the pressure in the system may be determined by the gas load. Hence it is important to determine the outgassing rates of the materials present inside the cryostat at 300K. Further the need for degassing the cryostat is discussed for improving the pump down time of the cryostat.

References

[1] G. Pal, U. Bhunia, J. Akhter, C. Nandi, A. Datta, P.R. Sarma, S. Roy, S. Bajirao, S. Bhattacharyya, T.K. Bhattacharyya, M.K. Dey, C. Mallik and R.K. Bhandari. Superferric quadrupoles for FAIR Super FRS energy buncher, Cryogenics, ACASC 2011 Conference Proceedings in print.

[2] "Design of Outer Vacuum Chamber for Long Superconducting Quadrupoles for Fair Super FRS Energy Buncher", J. Akhter, G. Pal, C. Nandi, S. Roy, U. Bhunia, A. Datta and C. Mallik, presented at ACASC 2011, New Delhi. Indian Journal of Cryogenics in print.

[3] C. K. Krishnaprakas, K. Badari Narayan, Pradip Dutta. Heat Transfer Correlations for Multilayer Insulation system. Cryogenics Volume 40 (2000) 431-435.

[4] Cunnington GR, Tien, CL. A study of heat transfer processes in multilayer insulation. In: Bevans JT, editor. Thermophysics: application to thermal design of spcaecraft, AIAA progress in astronautics and aeronautics, vol. 23. New York: Academic Press; 1970. O. 111

[5] R. J. Corroccini. Gaseous Heat Condcution at low pressures and Temperatures, Vacuum, Vol. 7 and 8; 1957-1958.

[6] T. Nast. Radiation Heat Transfer in Handbook of Cryogenic Engineering: Taylor and Francis; p 186-202.

[7] R.B. Scott, Cryogenic Engineering, Met-Chem Research, 1963.

[8] A. Roth. Vacuum Technology. North-Holland Publishing Company. 1976

[9] R. J. Elsey. Outgassing of vacuum materials-II, Vacuum, Vol. 25, Number 8, 1975.

[10] A. P. M. Glassford and C. K. Liu. Outgassing rate of Multilayer Insulation, Lockheed Palo Alto Research Laboratory

[11] O'Hanlon, John F., A User Guide To Vacuum Technology, John Wiley and Sons, 1980

Influence of Wall Conditioning on ADITYA Plasma Discharges

R L Tanna, K A Jadeja, S B Bhatt, P S Bawankar, C N Gupta, Y S Joisa, P K Atrey, R Manchanda, Nilam Ramaiya, J Ghosh, D Raju, P K Chattopadhyay, R Jha and the ADITYA team

Institute for Plasma Research, Bhat, Gandhinagar-382 428, Gujarat, India.

Email: rakesh@ipr.res.in, rtan.ipr@gmail.com

Abstract. ADITYA (R0 = 75 cm, a = 25 cm), an ohmically heated circular limiter tokamak is regularly being operated to carry out several experiments related to controlled thermonuclear fusion research. In recent operational campaign, various experiments have been carried out to enhance the discharge performance as well as improve the plasma parameters. A comparative plasma discharges study with SiC and Graphite limiter was carried out to increase the plasma heating and reduce runaways. Excellent plasma heating has been observed in many discharges using Graphite limiter. Good repeatability of low hard X-rays, high temperature discharges was obtained. The control of plasma impurities and hydrogen recycling is very much essential for high performance discharges. The wall conditioning in ADITYA tokamak is carried out by hydrogen glow discharge cleaning (GDC), Pulse discharge cleaning and electron cyclotron resonance (ECR) discharge cleaning techniques with and without lithium wall coating. GDC assisted Lithiumization was found to be the most effective technique for substantial reduction in Hα and low Z (CIII & O-I) impurities. The partial pressure of mass number 18 (H2O) and 28 (N2/C2H4/CO) were regularly monitored before plasma discharge operation. Furthermore, experiment on optimization of pulse gas feed was helped in reducing wall loading and recycling. However, hard X-rays suppression with the application of multiple gas puff has been successfully achieved during negative converter operation, which led to the extension of plasma pulse length up to ~ 250 ms. All the supporting facts and operation aspects are reported.

1. Introduction

ADITYA is ohmically heated limiter based, moderate size tokamak and is regularly being operated with the transformer converter power supply [1]. During recent operation campaign, the toroidal field of (B_T = 0.75-0.84 T), the peak loop voltage of (V_L ~ 20 V), the plasma current of (Ip = 70- 90 kA) and duration of 100-270 ms have been achieved in ADITYA tokamak. The line average electron density (n_e) is in the range of 1.5–2.0 x 10^{13} cm^{-3} and the electron temperature (T_e) is in the range of 300-600 eV. The plasma is generated in a stainless steel vessel evacuated to a base pressure of 1.0 x 10^{-7} torr. The working plasma fueling gas is hydrogen filling at a pressure of 8 x 10^{-5} to 1 x 10^{-4} torr. The pre-ionization filament kept on with filament current 19 A and bias voltage 150 V during the plasma experiments.

Recent upgradation in ADITYA performance has been carried out with various experiments including recycling study with SiC and graphite limiter, low hard X-ray, high temperature plasma discharge operation, application of gas-puff on plasma discharges and long pulse plasma discharges with negative converter have been performed to enhance the plasma discharge performance. In past, the discharge failure has been restricted and the repeatability as well as reliability of successful plasma discharges has been obtained after optimizing the method of gas fuelling [2].

2. Experimental set-up

In ADITYA, all the magnetic coils are powered by computer-controlled thyristor based pulsed power supply and discharge data are stored in CAMAC based data acquisition system. The main set of diagnostics used in these experiments are namely, Rogowski coils for plasma current and VF current measurements, voltage loop for loop voltage measurements, Optical detectors for impurity radiation including H_α measurements, microwave interferometer for electron density, Scintillation detector with NaI (Tl) source for hard X-ray measurements, soft X-ray detectors for electron temperature measurement, magnetic probes garland for magnetic fluctuation measurement, impurity radiation power using bolometry and the position probes are used for plasma position measurements.

2.1 Gas fueling systems at ADITYA tokamak

The Piezo-electric valve is installed to control the fuel gas pressure. Two different methods have been used for gas fueling viz. static (continuous) gas feed and pulsed (Pre-fill) gas feed [3]. In continuous gas feed, the fuel gas is introduced in the vessel by applying DC voltage just few seconds before the loop voltage is established. Whereas, in the pulsed gas feed, a square pulse is applied through pulse generator with the time-delay of few ms before the loop voltage is established.

2.2 Wall conditioning techniques at ADITYA tokamak

Various wall-conditioning techniques are used all over the world in tokamak operation to reduce the impurity influx from the wall and enhance the plasma performance. To achieve better wall conditioning in ADITYA, DC glow discharge cleaning (GDC), pulse discharge cleaning (PDC), electron cyclotron resonance (ECR) discharge cleaning as well as superimposing a PDC on the ECR background techniques [4] with and without lithium coating are carried out with hydrogen plasma as per experimental requirements. The partial pressure of various Mass numbers like O_2, CO, N_2 CH_4, and H2O etc. were monitored with quadruple mass analyzer (QMA).

GDC system consists of two UHV bellow driven movable electrodes, placed diametrically opposite at the center of the vessel [5], which have a positive potential act as anodes and vacuum vessel being kept as negative potential act as a cathode. The discharge voltage is about 350 V to 400 V and discharge current is of the order of ~3.5 A. The GDC is carried out at higher hydrogen gas pressure of ~ 8×10^{-4} to 1×10^{-3} torr and automated for maximum 12 h in absence of toroidal magnetic field. The electron temperature during GDC is of the order of ~5 eV. During PDC, the ohmic coils are powered by a capacitor bank power supply to generate the loop voltage required for the gas breakdown. The toroidal field is of the order of 500 G during combined ECR + PDC, whereas it is 900 G for alone PDC. The hydrogen gas is fueling at the pressure of ~2 to 3×10^{-5} torr. The measured electron density is ~1×10^{11} cm^{-3}. The pulse repetition rate during PDC is ~ 900 pulses/h of 4 ms duration. Microwaves of 2.45 GHz frequency is produced by a Magnetron of 1 kW power level are launched towards the plasma chamber during ECR. The anode current is 200 mA. The ECR operation duty cycle is about 66%. The maximum electron density (measured using Reflectometry) during combined ECR+PDC is of the order of 6×10^{11} cm^{-3}. The effect of solid target Lithiumization has been performed in ADITYA with various methods. Initially, two lithium rods of diameter 12 mm, toroidally 180^0 apart, were inserted at 20 mm inside the plasma volume. The evaporation of lithium in vacuum vessel has been done with ECR, PDC and GDC as well as with actual tokamak discharges.

3. Experimental results and Discussions

3.1 Recycling study with SiC and Graphite limier

The optimal choice of limiter material for plasma-facing components have major role on plasma discharge behavior and it was experimentally studied in ADITYA tokamak. Similar studies have been reported in TCA, JET-2 [6] and ALCATOR C tokomaks [7]. Silicon Carbide (SiC) coated graphite limiters have been chosen in ADITYA being low Z materials, higher thermal conductivity and lower vapor pressure. This includes lower chemical and sputtering yield, better oxygen gettering and lower hydrogen recycling. Till Dec. 2009, ADITYA has been operated with poloidal graphite limiter with radius of ~ 25 cm. The main problem with this limiter was that discharges were not consistent at higher operating pressure (≥ 1 X 10^{-4} torr) and lower (E/P ~ 400 V $cm^{-}1$ $torr^{-1}$) because of more carbon impurity and its insufficient burn through in the start-up. Therefore, the replacement of graphite limiter to SiC was envisaged. During vessel opening in Dec. 2009, the SiC limiter was introduced. It consists of 20 nos. of SiC coated graphite tiles mounted on SS support structure in poloidal direction. The limiter radius of ~ 24 cm was kept to increase the gap between vessel wall and limiter. The thickness of SiC coating was about 100 μm [8].

During recent operation campaign (15th October to 3rd December 2010), discharges with SiC coated graphite limiter showed (1) Increase in H_α and C-III, O-I impurity line radiation restricts the electron temperature rise (2) Hard X-rays could not be controlled even with maximum gas pressure of 1 x 10^{-4} torr and lowest E/P up to 425 V/torr-cm. This was probably due to the lowering the radius of limiter. Furthermore in last week of Dec 2010, the original graphite limiter of radius ~ 25 cm was installed back to identify the reason. The time evolution of ADITYA discharges comparison between SiC coated graphite and pure graphite limiters for similar operating parameters are shown in 'figure 1'.

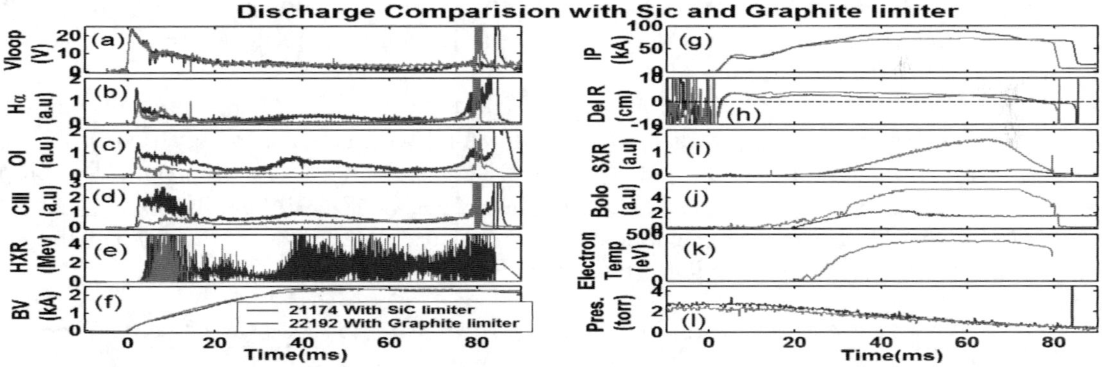

Fig. 1. Time series of *(a)* Loop voltage, *(b)* H_α, (c) O-I, *(d)* C-III, *(e)* Hard X-rays, *(f)* VF current, *(g)* plasma current, *(h)* Radial position (i) Soft X-rays, (j) bolometer (k) electron temperature and (l) Pressure (value x calibration factor 4) with SiC and graphite limiter.

The experimental results show significant reduction in H_α and C-III, O-I impurity line radiation signals and huge reduction in hard X-rays (except for first 10 ms) are observed in discharges with graphite limiter. This is also reflected in the plasma current driven by runaways. The excellent plasma heating is observed in many discharges with graphite limiter. The electron temperature goes up to 450 eV for shot 22192 in 'figure 1' measured with Soft X-ray diagnostic.

3.2 Low hard X-ray, high temperature plasma operation

In order to obtain low hard X-ray, high temperature discharges with better repeatability, the experiment was carried out with total 25 h of limiter baking at maximum 100° C temperature out of that 12 h limiter baking was carried out with GDC. The variation in plasma electron temperature and wall conditiong progress after installation of graphite limiter is tabulated in Table 1.

Table 1. The wall conditioning and plasma temperature progress after installation of graphite limiter

Shot series	Wall conditioning	Partial pres. (M # 18) (torr)	Partial pres. (M # 28) (torr)	Base Pres. (torr)	Electron temp. (eV)
22124 – 22148	12 h GDC	6.7×10^{-9}	5.6×10^{-8}	3.6×10^{-7}	300
22149 – 22172	12 h GDC	5.8×10^{-9}	2.5×10^{-8}	3.4×10^{-7}	400
22173 – 22196	12 h GDC	4.4×10^{-9}	2.4×10^{-8}	3.1×10^{-7}	450
22210 – 22237	12 h GDC	4.4×10^{-9}	2.1×10^{-8}	2.7×10^{-7}	500
22238 – 22263	12 h GDC	3.0×10^{-9}	2.1×10^{-8}	2.9×10^{-7}	600

The time evolution of typical ADITYA discharges for plasma electron temperature progress after installation of graphite limiter is shown in 'figure 2'.

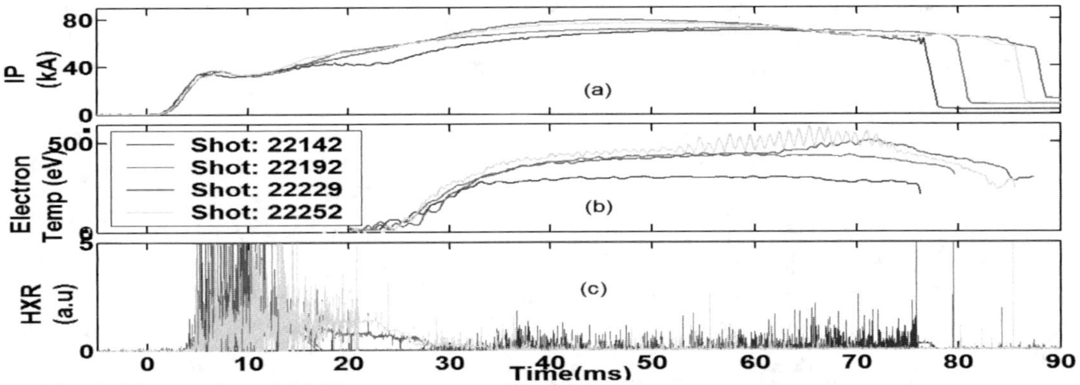

Fig. 2. Time series of (*a*) Plasma current, (*b*) electron temperature, (c) hard X-rays

In addition to that the ADITYA discharges repeatability was established with the application of fast feed back position control and excellent wall conditioning with graphite limiter backing. We have achieved 13 repeatable discharge of around 100 ms duration with good heating and stable electron temperatures of around 350 eV. The time evolution of typical ADITYA discharges repeatability is shown in 'figure 3'. The sufficient pumping time was provided between two shots and partial pressure of water vapour and N_2 / CO is monitored before each shot.

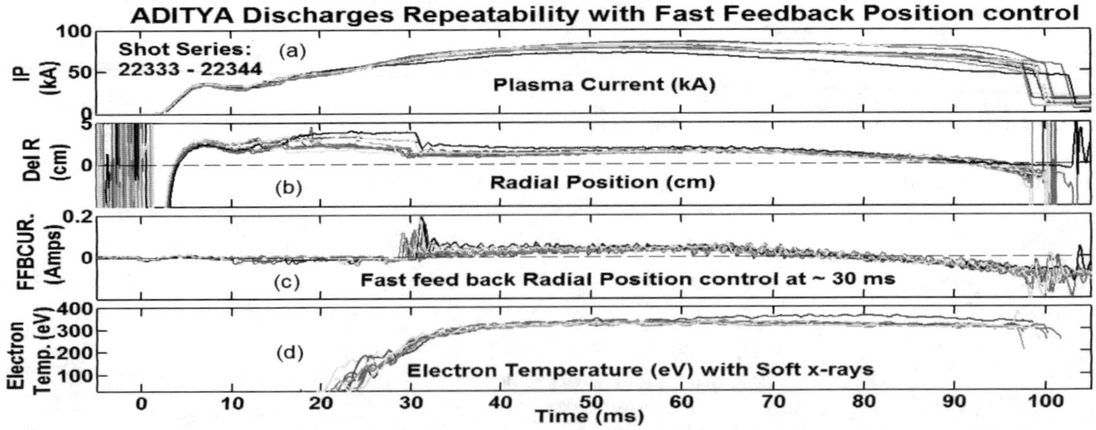

Fig. 3. Temporal evolution of (*a*) Plasma current, (*b*) radial position, (c) FFB current, (d) electron temperature describes discharges repeatability.

3.3 Application of Gas-puff on ADITYA plasma discharges

The behavior of density-limit disruptions caused by different gas fuelling methods like Molecular Beam Injection (MBI), gas-puff (GP) have been already studied in ADITYA tokamak discharges [9]. The edge fluctuation suppression with the application of gas-puff of working gas fuel has also been studied in ADITYA tokamak [10]. Apart from these studies, recently we have studied the runaway suppression by enhancing the line average electron density as well as MHD mode suppression with the application of multiple gas-puff of working gas fuel (hydrogen). The similar experiments on runaways suppression by strong helium gas-puff has been studied in TEXTOR, ASDEX and JET tokamak [11].

The multiple gas-puff is introduced in the vessel by using a Piezo-electric valve (500 SCCM at 100 V). The valve is mounted on the bottom port and located 10 cm radially (on the bottom port) and 90° toroidally from the limiter. The pulse widths timing and voltage level, time (T) for gas-puff to start, number of pulses and the time gap between the pulses were varied with pre-programmed gas-puff according to the discharge requirements. The time evolution of typical ADITYA discharge parameters comparison with and without multiple gas-puff is shown in 'figure 4'.

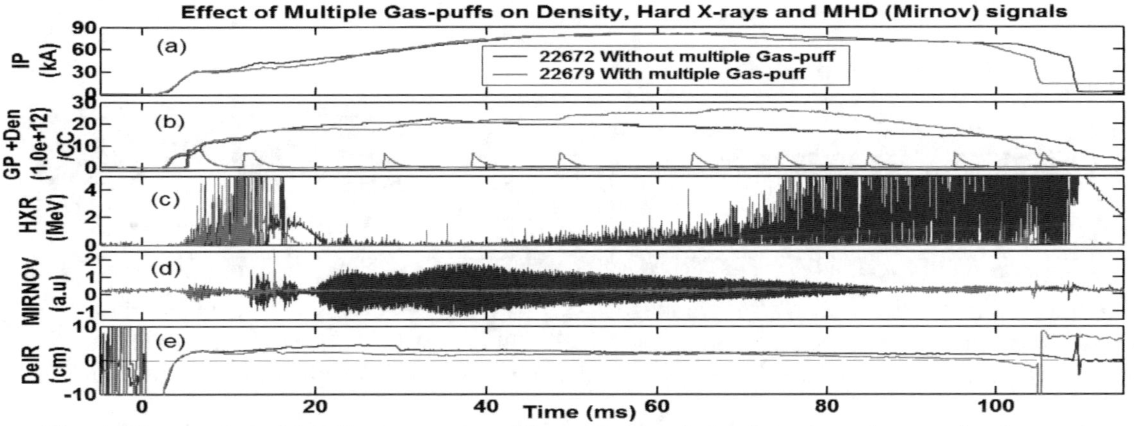

Fig. 4. Time series of (*a*) Plasma current, (*b*) gas-puff and chord average electron density, (c) Hard X-rays, (d) Mirnov signal and (e) Radial position with & without multiple gas-puff.

Figure 4 shows that in absence of multiple gas-puff (Shot: 22672), Plasma column shifted towards outer wall interacts with impurity that strikes the MHD activities. Mode analysis results describes in 'figure 5' shows first m=4 mode, later m=3 that converted into m=2 at the end. Chord average electron density falls, which brings lots of runaways.

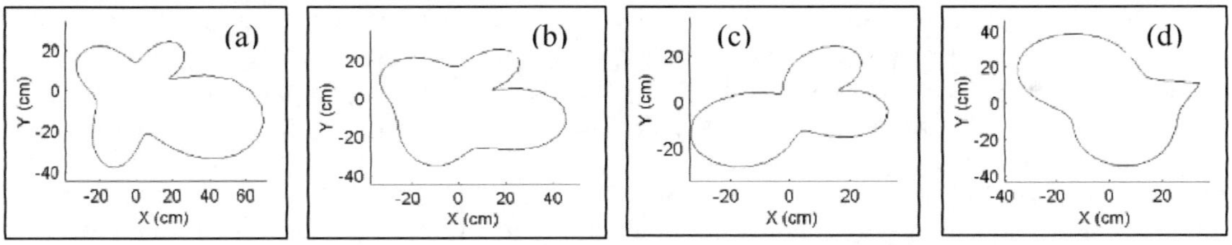

Fig. 5. MHD mode analysis with Mirnov signal for shot: 22672 (*a*) 32 –34 ms, m=4 mode, (*b*) 46 – 48 ms m=4 mode, (c) 72 –74 ms, m=3 mode and (d) 82 –84 ms m=2 mode.

Hydrogen gas pulse of 10 nos. with typically pulse width of ~ 0.2 ms and amplitude of ~ 100 V at different time interval were introduced in discharge (shot: 22679), which raised the electron density up to maximum of 2.8 x 10^{13}/cc and control MHD activities as well as hard X-rays. Stable plasma equilibrium towards the center was observed in discharge with multiple gas-puffs.

3.4 Pulse length enhancement of plasma current with negative converter operation

In order to increase the volt-sec, ohmic current swing was extended in the negative direction (+11 kA to –7.5 kA) to generate loop voltage up to ~ 275 ms. This will provide constant flattop loop voltage of ~ 2 V during negative converter phase. The maximum discharge duration of ~ 270 ms was obtained through the tuning of operating parameters. The time evolution of ADITYA discharge parameters with negative converter operation is shown in 'figure 6(a)'. The multiple gas-puff were introduced during the middle of the discharge that enhance the plasma density and helps in control the hard X-ray up to some plasma length is shown in 'figure 6(b)'. The maximum electron density of the order of 3 x 10^{13} /cc and electron temperature of the order of 500 eV was observed.

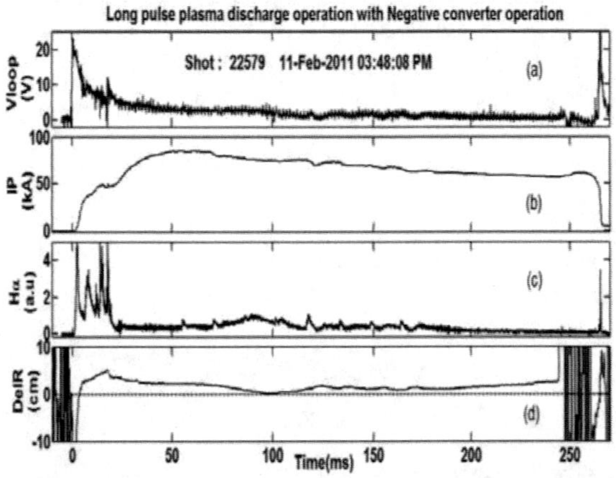

Fig. 6(a). Time evolution of (*a*) Loop voltage, (b) Plasma current, (*c*) H_α and (d) Radial position with negative converter.

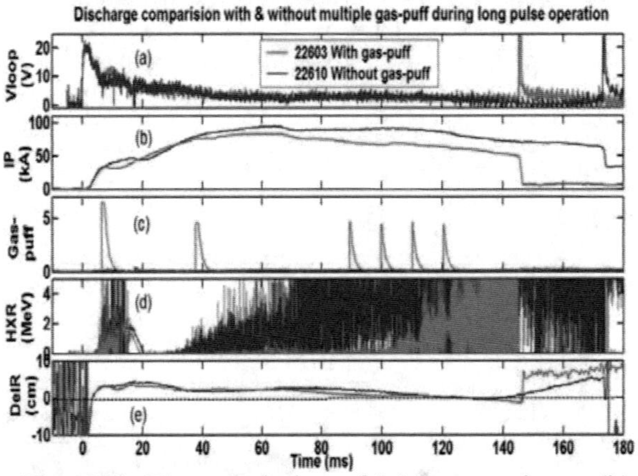

Fig. 6(b). Time evolution of (*a*) Loop voltage, (b) Plasma current, (*c*) Gas-puff and (d) Hard X-ray (e) Radial position with negative converter.

The plasma current starts falling during auxiliary phase (~ 100 ms) because of decreasing loop voltage during that phase and remains at lesser value during negative converter phase.

4. Conclusion

In recent operation campaign, ADITYA machine has been successfully operated for various performance improvement experiments. In this article, we have reported some of the relevant results of recent experiments. The experiment on recycling study with SiC coated graphite limiter and graphite limiter, revels significant reduction in H_α and CIII, O-I impurity line radiation in discharges with graphite limiter. The excellent plasma heating improves the electron temperature up to 500 to 600 eV. Optimal operation parameters tuning, better wall condition and Fast Feedback position control provides excellent repeatability of ADITYA discharges. The runaways present in the discharge are successfully suppressed using multiple gas-puff of hydrogen gas of optimum pulse width at different time interval by enhancing line average electron density. This could also suppress MHD activities and maintains plasma equilibrium towards the centre. The plasma pulse length is enhanced beyond 250 ms with the application of negative converter. The drops in plasma current after 100 ms because of slightly lowering of loop voltages in auxiliary converter phase will be resolved in near future.

Acknowledgement

The author would like to acknowledge Mr. V K Panchal of data acquisition group and Mr. Nirav Mecwan, Mr. Pintu Kumar, Mr. M B Kalal and Mr. D S Varia of ADITYA operation group for their help and support provided during experiments.

References

[1] S.B. Bhatt et al, *Ind. j. of Pr. and appl. Phys*. Vol. 27, Sept.-Oct. Pg. 710-742 (1989).
[2] R. L. Tanna, et al, *21st Nat. Symps. on Plasma Sc. & Tech.*, Dec. 19-22, 2006, Jaipur, India.
[3] R.J. Hawryluk et al, *J. Nuclear Fusion* Vol. 16 (1976), Page. 775-781
[4] Ram Prakash et al, *J. Appl. Phys*. 97, 043301 (2005)
[5] H.A Pathak et al, *J. Nuclear Materials,* 220 -222 (1995) 708- 711
[6] Yoshio Gomay, et al, *Japan. J. Appl. Phys., Vol. 18, No. 7, July 1979, pp. 1317 -1324.*
[7] E. Marmar et al., *J. Nuclear Materials* 121 (1984) 69 -74
[8] S.B. Bhatt et al, *12th Int. Workshop on PFMC* 11 – 14 May 2009, Julich, Germany
[9] S.B. Bhatt et al, *Journal of Fusion. Eng. & Design*, 75 – 79 (2005), 655 -661.
[10] R. Jha et al, *Plasma Phys. Control Fusn.* 51 (2009), 095010 (17pp)
[11] H.R.koslowski et al, *IPP/FZJ/Julich/Scientific report*-2001/Vol.5/page.1-5.

Removal of water from unbaked vacuum system

G Pal, RC Yadav, J Akhter, T Das, A Sarkar, C Mallik and RK Bhandari

Variable Energy Cyclotron Centre, I/AF Bidhannagar, Kolkata 700064, INDIA

Email: gautam.pal@vecc.gov.in

Abstract: The water vapour contained in the atmosphere contaminates a vacuum systems, when it is exposed to air. The degree of contamination will depend on the surface characteristics, treatment of the system walls, the relative humidity and the turbulence of air, the amount of time the system is opened to the atmosphere, etc. If the temperature of the exposed surface is low, possibility of water condensing on the surface exists. In order to study the effect and optimise procedure for achieving high vacuum in a vacuum chamber initially containing water, a vacuum chamber has been fabricated. Pump down time has been measured for different amount of water placed in the vacuum chamber with varying pumping speeds. The paper presents a review of similar activities and results of the experimental study.

1. Introduction:

The beam chamber of VEC superconducting cyclotron [1] has a number of penetrations in the median plane in between the upper and lower superconducting coils to accommodate extraction components and beam diagnostic components. Because of the compact structure, adequate space was not available at the penetrations for placing a thermal shield. The temperature at these penetrations is quite low. It is observed that moisture from air condenses there and considerable time is taken to achieve good vacuum, specially, in monsoon. In the case of water leakage in the system, considerable amount of water gets accumulated in the beam chamber. Because of the complex geometry of the cyclotron, it is quite difficult to remove water from the restricted areas. It is also not possible to bake the acceleration chamber to enhance the rate of moisture removal.

An experimental set-up has been fabricated to study the pump down characteristics in the presence of water in a vacuum enclosure. The paper presents a literature survey on work carried out on similar lines and results of experiments carried out.

2. Review of similar activities

Chan 2010 [2] examined the effect of environment humidity on the rate of thermal outgassing from an aluminium chamber. An aluminium glove box and an air shower, which provided dehumidified environments with water vapor concentrations of 0.1 ppm and 5 ppm respectively, were utilized to assess the effect of environmental humidity on the rate of thermal outgassing. Chan observed that only 4 h of evacuation was required to yield a small rate of the order of 1×10^{-13} mbar L s^{-1} cm^2 of thermal outgassing upon filling with super-dry nitrogen without baking. A smaller rate of thermal outgassing was obtained using a drier venting gas, or by exposure to a drier environment. A thermal outgassing rate of the order of 1×10^{-11} mbar L s^{-1} cm^2 was achieved by venting the glove box with super-dry nitrogen inside and controlling of environmental humidity at 0.1 ppm.

Yamazaki 2010 [3] carried out control of water vapor in a nitrogen gas purge line in addition to surface treatments of chambers using buff polishing and electrolytic polishing, followed by measurement of outgassing rate of the chambers. He observed that with proper surface treatment and introduction of well controlled nitrogen gas to the vacuum system which was not baked out reduces the pump down time considerably.

Liapis 2008 [4] constructed expressions for determining the exergy inputs and exergy losses due to heat and mass transfer in the primary and secondary drying stages as well as in the water vapor condenser and vacuum pump of the process involving freeze drying of pharmaceuticals in vials on trays. The exergy expressions can be used for operational control policies to minimize the irreversibilities occurring in the operations of a freeze drying system of a given design, thus enhancing the efficiency of energy utilization.

Chen 2007 [5] studied the morphological changes of water in a vacuum cooling system in the course of vacuum chamber pressure step down. The experiment result shows that steam adhered to vacuum chamber due to the atmospheric pressure evaporates as the chamber pressure decreases and produces a great amount of smoke after the vacuum pumping system is initiated for 20 s. After 115.2s, the liquid water in the vessel begins to produce tiny bubbles and then large bubbles at 121.2 s, and boils at 126.7 s. After the rapid boiling is over, the liquid water enters the freezing stage at 181.5 s. When the vacuum pumping system is stopped at 480 s, the liquid water inside the vessel is frozen into solid ice which is consisted of two layers, irregular porous layer on the top and dense layer on the bottom. The cooling rate of liquid water in this experiment is between 0.33–0.38 ^{0}C/s, and 0.04–0.07 ^{0}C/s during the freezing process.

Caleman 2006 [6] studied the evaporation of pure water clusters under vacuum. He simulated the evaporation of water molecules. The average number of evaporated water molecules per surface area is 0.9–1.0 molecule/nm^2. The evaporation becomes slow, once the temperature reaches 240 K. We find that as the evaporation ceases, the temperature falls to about 215 K. He observed that the cooling rates are in good agreement with experimental results, and evaporation rates agree well with a phenomenological expression based on experimental observations.

Dostal 2004 [7] has developed a simple mathematical model of the vacuum cooling process which enables to predict a temperature evolution regarding an equipment size, vacuum pump parameters and properties of the cooled liquid. Real thermophysical properties of the cooled liquid were considered in the model along with the assumption that the main resistance against mass transport is situated on the side of liquid phase. Parameter identification of mass and heat transfer coefficients were based on available experimental data.

Song 2002 [8] developed a numerical code to predict vacuum freeze drying processes in trays and vials using a finite volume method to discretize the governing partial differential equations. Along with the finite volume method, a moving grid system was adopted to handle irregular and continuously changing physical domains encountered during the primary drying stage. To show the validity of the present calculation scheme, freeze drying in a tray was simulated and the results were compared with available experimental data.

Akaishi 1997 [9] experimentally investigated the effect of helium glow discharge cleaning as a wall conditioning technique for the production of ultrahigh vacuum in an unbaked vacuum system. A test chamber made of 304 stainless steel was constructed. The chamber was initially exposure to air before pump-down. Discharge cleaning was done at the constant helium ion fluence of 0.5 Coulomb/cm^2. It was observed that the pumping time to attain the pressure of order of magnitude of 10^{-8} Torr is shortened after the discharge cleaning by one fourth compared with the pumping time without discharge cleaning. But the ultimate pressure at the pumping time of 72 h after the discharge cleaning is rather the same as that without discharge cleaning.

Porta 1996 [10] has recommended the use of getters for removal water in sealed devices for semiconductor processing, vacuum insulated vessels and other applications

Berman 1995 [11] examined the problem associated with water vapour in vacuum system and methods to remove it. In order to reduce the desorption rate and remove large amounts of water vapour Berman suggests the use of methods for conditioning the construction materials and transfer energy to the adsorbed gases to break their bonds. Berman observes that traditional vacuum bakeout cannot be used with some applications such as accelerators, or fusion devices. Other methods such as flushing with cold or hot neutral gas, glow discharge, or either electron, ion, or ultraviolet radiations have been successfully used.

Novikov 1968 [12] examined the mechanism of vacuum sublimation of ice with experimental data and theoretical results. He observed that in the vacuum chamber very favourable conditions are created for external mass transfer, when small fluctuations of pressure can cause mass transfer velocities of several meters per second. In the presence of an intense source of mass release (sublimation of ice in a vacuum) and in the absence of leakage the total pressure in the chamber may be assumed equal to the saturation pressure and the chamber may be considered to contain vapor only. In this case, the rate of mass transfer is determined by the vapor pressure drop.

3. Experimental set-up

A vacuum system (Fig. 1) has been set up to study removal of water from the enclosure and optimisation of the procedure for removal of water and subsequent achievement of high vacuum in it. The vacuum chamber is connected to a scroll pump with pumping speed 7 l/sec through a 25 mm port. A turbo molecular pump with pumping capacity 210 l/sec is also connected to the vacuum system using another 25 mm port. The effective pumping speed at the vacuum chamber for the scroll and turbo molecular pump was evaluated and found to be about 0.7 l/sec and 0.6 l/s respectively. Vacuum gauges G1 and G2 are placed close to the vacuum chamber and the turbo molecular pump respectively. An extension having a glass view port has been added to visualise removal of water from a glass beaker partially filled with water. The total volume of the chamber is about 7 litres. Pneumatic valves GV1 and GV2 and used to isolate scroll pump and turbo molecular pump respectively.

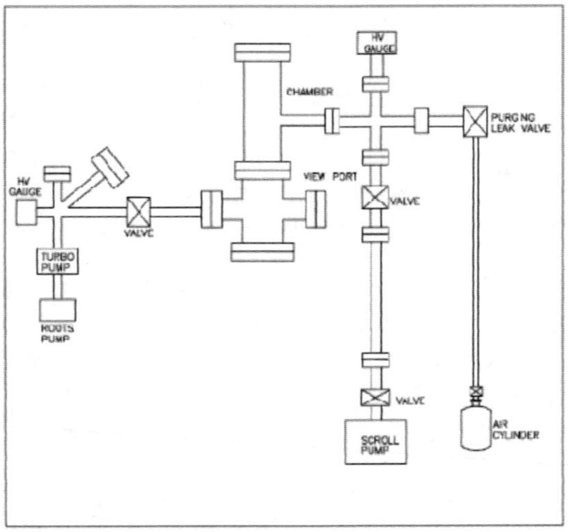

Fig. 1: Schematic diagram of vacuum system

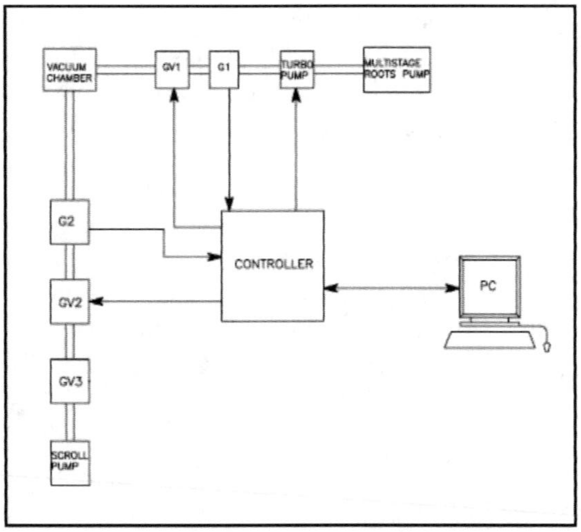

Fig. 2: Control diagram for vacuum system

A control circuit (Fig. 2) is used to sequentially operate the valves, backing and high vacuum pump. Two compact gauges monitor the vacuum in the system. The valves and vacuum pumps are operated using the relays available in the gauge controller. The scroll pump is initially operated till the vacuum in the vacuum reaches 2 x 10^{-2} mbar. The high vacuum pump is then connected to the vacuum system. The gauges are connected to a computer for data acquisition. Safety interlocks have also been incorporated to isolate the turbo molecular pump in case of power failure and accidental pressure rise in the vacuum chamber.

4. Results

The different stages of pump down was observed through the view glass. A webcam was placed in front of the view glass and the evacuation of the chamber containing a glass beaker of water was observed. It was observed that as pumping progresses vigorous boiling of water is observed. After some time, the boiling phenomena stops, the water solidifies and ice is formed. The amount of ice reduces slowly. Pressure data in the vacuum chamber is stored on-line in a computer. The quantity of water was varied from 2.5 gram to 20 gram. Fig. 3 shows the variation of pressure with time. In order to assess the effect of exposed surface, the experiment was carried out by varying the exposed area from 11 sq. cm to 50.9 sq. cm (Fig. 4). The effect of speed of pumping was evaluated by changing the conductance of the scroll pump connected to the vacuum system. The variation of pressure with time for orifice diameters 2.5mm to 9 mm is given in Fig. 5. The orifice diameter is chosen to reduce the conductance by 25%, 50% and 75% respectively. Fig. 6 shows the effect of pumping with the scroll pump.

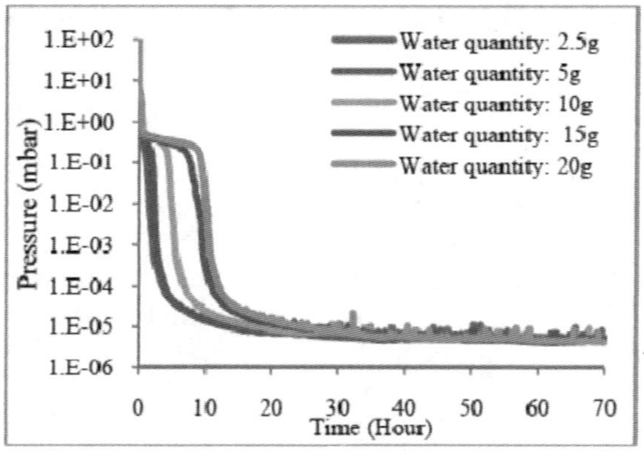

Fig. 3: Pump down-different mass

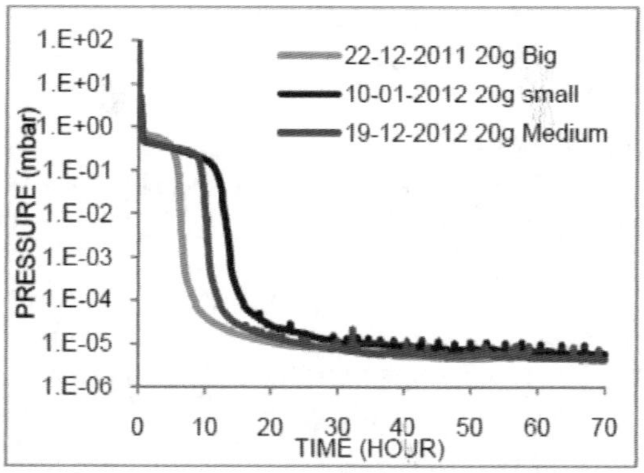

Fig. 4: Pump down-different area

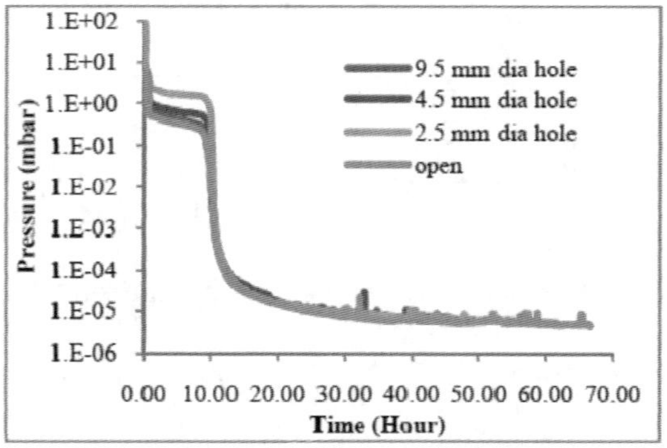

Fig. 5: Pump down-different conductance

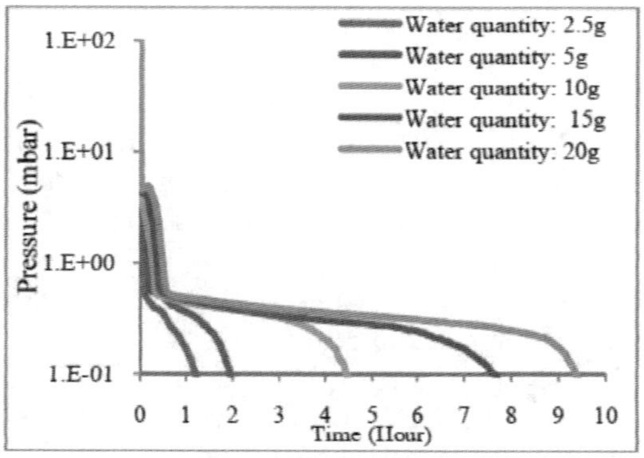

Fig. 6: Pump down – initial period

It was observed from the experiments (Fig. 3) that after the end of 70 hours of pumping, the pressure in the system was about 5.52 x 10^{-6} mbar, 4.47 x 10^{-6} mbar, 4.64 x 10^{-6} mbar, 5.1 x 10^{-6} mbar and 4.32 x 10^{-6} mbar for initial water of 2.5 g, 5 g, 10 g, 15 g and 20 g water respectively. The initial rate of water removal is highly dependent on the surface area, but the effect after about 70 hours is not so prominent (Fig. 4). The pressure in the system was about 7.3 x 10^{-6} mbar, 6.54 x 10^{-6} mbar, 5.96 x 10^{-6} mbar and 4.32 x 10^{-6} mbar respectively for initial water of 20 g water after the end of 70 hours of pumping, when the conductance was 25%, 50%, 75 % and 100%. The time taken to achieve a vacuum of 0.5 bar was about 8 min, 30 min, 30 min, 42 min and 50 min for initial water of 2.5 g, 5 g, 10 g, 15 g and 20 g water respectively (Fig. 6).

5. Discussion

Removal of water using vacuum has been studied for a long time. It has application in electronic industry [10], food processing [8,13,14], food preservation [5, 7], desalination [15], pharmaceutical industry [16], accelerators [17], tokomaks [18], etc. Different forms of heating, viz., microwave, thermal, dry gas purging, glow discharge, etc. have enhanced removal of water.

The literature survey and experiment have shown that during the evacuation process, initially free water evaporates at the surface of water. As the pressure reduces, the boiling point of water reduces and boiling of water is observed. Boiling continues and the liquid transforms to vapours. The corresponding amount of heat, the latent heat, is continuously supplied from the system. The value of latent heat is rather large (order of MJ/kg) and it play a substantial role in the energy balance of liquid phase. The balance includes the flow of water vapours which leave the liquid surface and which actually contain the energy of the latent heat. This energy is, therefore, taken off the liquid phase and in the absence of a heat input the liquid temperature decreases. The temperature of the vessel is high. So, different modes of boiling can be observed. The lower temperature would reduce the partial pressure of vapours below the total pressure and the boiling process would terminate. Once, water converts to ice, removal of water is by sublimation.

6. Conclusions

It was observed from the experiments that the long term performance of the vacuum is not highly dependent on the initial amount of water. The rate of removal of water is dependent on the surface area and conductance.

It is further studies have to be continued. It is planned to add a residual gas analyser to the vacuum system and assess the effect of residual amount of H_2O after about 70 hours for different amount of initial water. A vacuum gauge for accurately measuring the pressure from 1 bar to about 0.01 bar is also planned to minutely observe the initial pressure change.

Reference

[1] Mallik C, Bhandari R K 2010 First beam acceleration in kolkata superconducting cyclotron and its present status *Proceedings 19th International Conference on Cyclotrons and their Applications CYCLOTRONS'10* (Lanzhou China) p 292-4

[2] Yamazaki Ko, Shike Junichi, Yamagata Motoi, Kitano Masahiro, Nishiwaki Michiru, Kato Shigeki 2010 High-speed pumping to UHV *Vacuum* **84** 756-59

[3] Chan Che-Kai Chan, Cheng Yu-Tsun, Hsiung Gao-Yu, Chan Bei-Jing, Ou Yi-Ching, Yang Chia-Ying, Chen June-Rong 2010 Effect of environmental humidity on rate of thermal outgassing *Vacuum* **84** 747-50

[4] Liapis A I, Bruttini R 2008 Exergy analysis of freeze drying of pharmaceuticals in vials on trays *International Journal of Heat and Mass Transfer* **51** 3854–68

[5] Cheng H P, Lin C T 2007 The morphological visualization of the water in vacuum cooling and freezing process *Journal of Food Engineering* **78** 569-76

[6] Caleman Carl, Spoel David van der 2006 Temperature and structural changes of water clusters in vacuum due to evaporation *The Journal of Chemical Physics* **125** 154508

[7] Dostal M, Petera K 2004 Vacuum cooling of liquids: mathematical model *Journal of Food Engineering* **61** 533-9

[8] Song C.S., Nam J.H., Kim C.J., Ro S.T., 2002 A finite volume analysis of vacuum freeze drying processes of skim milk solution in trays and vials, *Drying Technology* **20(2)** 283–305

[9] Akaishi K, Ezaki K, Motojima O, Nakasuga M 1997 Production of ultrahigh vacuum by helium glow discharge cleaning in an unbaked vacuum chamber *Vacuum* **48(7-9)** 767-70

[10] Porta P della 1996 Gas problem and gettering in sealed-off vacuum devices *Vacuum* **47(6-8)** 771-7

[11] Berman A 1995 Water vapour in vacuum system *Vacuum* **47(4)** 327-32

[12] Novikov P A, Vagner N A 1968 Inzhenerno- Fizicheskii Zhurnal **15(5)** 788-93

[13] Sun Da-Wen, Wang Lijun 2006 Development of a mathematical model for vacuum cooling of cooked meats *Journal of Food Engineering* **77** 379–85
[14] Sun Da-Wen, Hu Zehua 2003 CFD simulation of coupled heat and mass transfer through porous foods during vacuum cooling process *International Journal of Refrigeration* **26** 19–27
[15] Gude Veera Gnaneswar, Nirmalakhandan Nagamany 2009 Desalination at low temperatures and low pressures *Desalination* **244** 239–47
[16] Pingniang Shen, Zhiyuan Liu, Kaizhong Huang, Qiong Xiao 2010 An Engineering Study and Analysis of the Factors Affecting Evaporation Rate Under the Vacuum Drying Process *Modernization of Traditional Chinese Medicine and Materia Medica* **12(1)** 118-21
[17] Pal G 2012 Vacuum systems of the Large Cyclotron at VECC being presented in this *International Symposium on Vacuum Science &Technology and their Applications for Accelerators IVS-2012* (Kolkata India)
[18] Marshall T, Girard C 2001 Modelling of ice formation and condensation on a cryogenic surface *Fusion Engineering and Design* **54** 473–483

Journal of Physics: Conference Series **390** (2012) 012046 doi:10.1088/1742-6596/390/1/012046

K-130 Cyclotron vacuum system

R C Yadav, S Bhattacharya, R B Bhole, Anindya Roy, Sarbajit Pal, C Mallik, R K Bhandari

Variable Energy Cyclotron Centre, 1/AF, Bidhan Nagar, Kolkata-700064

E-mail: yadav@vecc.gov.in

Abstract. The vacuum system for K-130 cyclotron has been operational since 1977. It consists of two sub-systems, main vacuum system and beam line vacuum system. The main vacuum system is designed to achieve and maintain vacuum of about 1 x 10^{-6} mbar inside the 23 m^3 volume of acceleration chamber comprising the Resonator tank and the Dee tank. The beam line vacuum system is required for transporting the extracted beam with minimum loss. These vacuum systems consist of diffusion pumps backed by mechanical pumps like roots and rotary pumps. The large vacuum pumps and valves of the cyclotron vacuum system were operational for more than twenty five years. In recent times, problems of frequent failures and maintenance were occurring due to aging and lack of appropriate spares. Hence, modernisation of the vacuum systems was taken up in order to ensure a stable high voltage for radio frequency system and the extraction system. This is required for efficient acceleration and transportation of high intensity ion beam. The vacuum systems have been upgraded by replacing several pumps, valves, gauges and freon units. The relay based control system for main vacuum system has also been replaced by PLC based state of the art control system. The upgraded control system enables inclusion of additional operational logics and safety interlocks into the system. The paper presents the details of the vacuum system and describes the modifications carried out for improving the performance and reliability of the vacuum system.

1. Introduction

A large room temperature K-130 cyclotron was commissioned at Variable Energy Cyclotron Centre (VECC), Kolkata in 1977. High vacuum is one of the basic requirements for preventing high RF voltage breakdown and loss of charged particles acceleration due to molecular collisions with residual gases. The vacuum system is thus one of the main subsystems of the cyclotron. The vacuum system has been operating since the cyclotron was commissioned. Initially, the cyclotron was used to accelerate light ions, viz., proton, deuteron and alpha particles using internal PIG ion source. Later, after development of an ECR ion source, heavy ions were accelerated during 1997-98. From 2007, the PIG ion source with modified central region was again used to accelerate the light ions for ultimate production of Radio-active Ion Beam. The requirement of high vacuum was stringent for heavy ion acceleration, where beam current were in the order of nano-amperes, compared to light ion acceleration where beam current were in the order of micro-amperes.

2. Vacuum system

The K-130 cyclotron has two major vacuum spaces to serve different functions.

i. A large lumped pumping system for cyclotron chamber (23 m^3) i.e. cyclotron vacuum system. The vacuum enclosure of the Cyclotron comprises the stainless steel Dee tank and the Resonator Tank made out of mild steel and copper cladded steel for acceleration.

ii. A distributed pumping system for beam transport line, i.e., beam line vacuum system. The beam lines carry the high energy beams after their extraction from the cyclotron to the designated destinations in the caves for experiments.

2.1 Cyclotron vacuum system

The cyclotron vacuum system [1] is further subdivided into two part i.e. Main vacuum system & Additional vacuum system. Both systems are shown schematically in Figure 1.

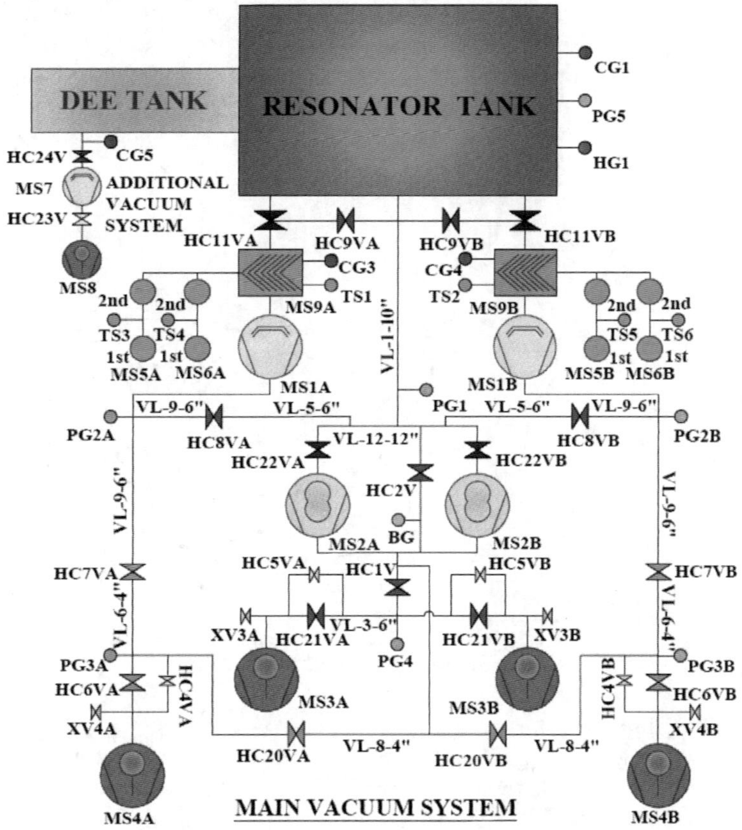

Fig. 1. Cyclotron Vacuum System

2.1.1. *Main vacuum system*

The main vacuum system has three major components of the high vacuum pumping system namely the 2.5 ton electro-pneumatically operated main gate valve, refrigerated Chevron baffle and fractionating oil diffusion pump having pumping capacity of 42000 l/s each for air. A pair of above three assemblies is suspended from the two large opening in the resonator tank. Chevron baffles over the diffusion pumps are cooled to -60°C with the help of Freon refrigeration units. There are two roots pumps (1000 l/s) and four mechanical pumps (50 l/s) for roughing of entire system as well as backing to diffusion pump. A set of roots and rotary mechanical pump is used for roughing the entire system from the atmospheric to few microns before 889 mm diffusion pumps are used for high vacuum. Several HV valves and pirani gauges have been installed in fore line to provide maximum flexibility to change the operation sequence of system whenever found necessary.

2.1.2 *Additional vacuum system*

The Addition vacuum system includes a 25 cm oil diffusion pump (2000 l/s) with its associated gate valve and baffle backed by a rotary pump (40 m^3/h). This is located on the Dee Tank at extreme end to provide extra pumping close to the accelerating region. This was found necessary as region where the ion beam circulates is a far away from the main pumping ports, in addition, it has several impedances in its pumping path by way of the moveable RF panels and the Dee aperture etc.

2.2 Beam line vacuum system

Most part of the beam line is of cylindrical shaped. The beam transport lines running to 65 meters in length in four different channels are basically made of Aluminium tubes, 100 mm in diameter. Various beam optics and diagnostic elements are mounted on it. There are nineteen pumping units. Pumping units consisting of a combination of 10 cm diffusion pump (280 l/s); rotary pumps (5 m^3/h) and gate valve have been installed at short distances apart along the length of the beam line to achieve a uniform vacuum throughout the length. Vacuum in the range of 10^{-6} Torr is found to be good enough to keep beam transport loss to a minimum. The locations of all pumping units are shown in Figure 2.

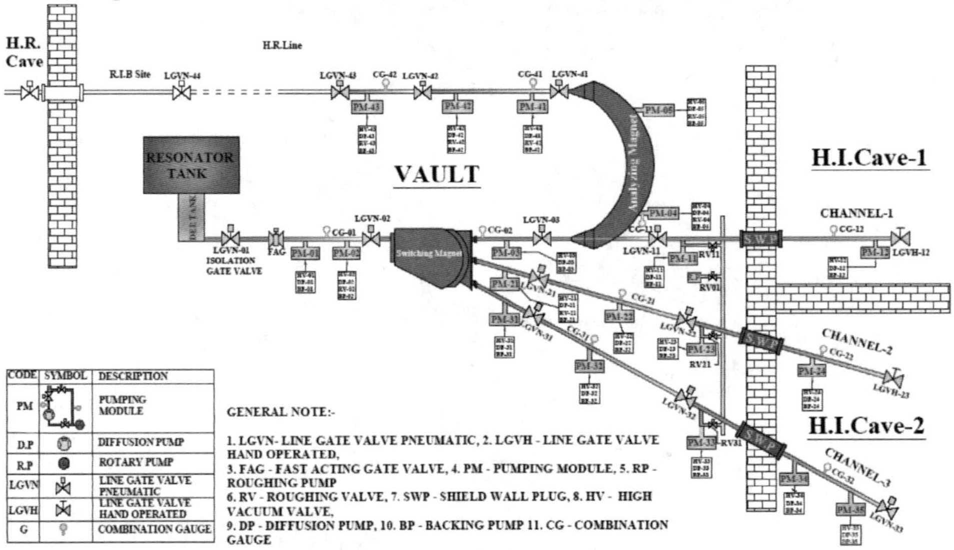

Fig. 2. Beam Line Vacuum System

3. Reason for modification

After many years of operation, the cyclotron vacuum system components and its control system started malfunctioning. Moreover due to temperature cycling, a number of cracks were observed in the backing line. The maintenance became impossible especially due to non-availability of spares. This was disturbing the smooth operation of cyclotron. In order to increase the reliability and to achieve stable vacuum in the cyclotron chamber and the beam lines, some modifications were necessary. To reduce the down time, replacement with improved version of vacuum pumps, gate valves, vacuum gauges and PLC based control system were necessitated.

4. Modification scheme

Under the modernization program, the major task was to incorporate suitable modifications to accommodate specially the huge outer casing of 889 mm diffusion pump and similar size of high vacuum gate valve in vacuum system. In addition to these, many other modifications have been done in plumb line for installation of other components. Some of these activities associated with both vacuum systems are mentioned below.

4.1 Cyclotron vacuum system

In the process of improvement in vacuum system, we took a number of steps as given below.

A. The outer casing of one 889 mm diffusion pump has been replaced using the existing jet assembly. This was done because of multiple leaks on pump body.
B. Two new 889 mm high vacuum gate valves have been installed at inlet of main diffusion pumps. The old gate valves were not isolating cyclotron chamber from vacuum system to carry out maintenance in chamber at short notice.
C. Two 250 mm gate valves used at roughing port have also been replaced for proper isolation of system while chamber is at atmosphere.
D. Two 300 mm high vacuum gate valves used at inlet of roots pump were replaced because of severe defect in carriage assembly and recurring leak in old type of limit switch arrangement.
E. All manual gate valves at inlet of rotary pumps have replaced with electro pneumatically operated gate valves to enable the complete remote operation as well as incorporate auto operation of system.
F. One Electro-pneumatic, 25 mm right angle valve at inlet of each rotary pump as bypass valve for high conductance inlet valve to facilitate the controlled pumping especially at high pressure to prevent oil loss through exhaust port.
G. Four 100 mm and three 150 mm gate valves used in backing line have also been replaced to improve isolation to more option in the system operation.
H. One 889 mm chevron baffle has been replaced because of leak in its refrigeration line with a refurbished baffle. This baffle has two refrigeration lines. One line is being used keeping other at redundancy.
I. Temperature Sensor has put in boiler of both diffusion pumps to monitor oil temperature.
J. One pair of Freon units has been upgraded to improve better cooling of one chevron baffle that operate around -60^0 C regularly.
K. To monitor vacuum level at different locations of systems, several Pirani and Pirani-cold cathode combination gauges have been used in places of old gauges.
L. A residual gas analyzer has been installed to measure the gas composition, interpretation of gas data and for trouble –shooting.
M. The common operator control panel for both systems has been designed ergonomically. Relay based control system has been up-graded to PLC based control [2] of vacuum system. The PLC based system, hooked to LAN, with features like audio-visual alarm for various failures like system interlocks, field component, operational interlock, control hardware, communication etc is implemented to assist in both manual and automatic control of the vacuum system.
N. Many safety features have been incorporated into the systems, both at vacuum level and operation level to prevent operator's errors such as turning on/off pumps in wrong sequence and opening the valve at the wrong time during process.

4.2 Beam line vacuum system

The following steps were taken to improve the vacuum:

A. Remote operation of all line gate valve located at different location have been implemented.
B. A 10 cm Isolation gate valve i.e. interface valve between Dee tank and beam line is operated from control console and it has been interlocked with pressure of both side.
C. Roughing line has been added in two pumping units i.e. one pumping unit in common line between Dee tank & switching magnet chamber and second unit located at switching magnet chamber.
D. Thermal switches have fitted at each diffusion pump to prevent overheating incase of cooling system failure.
E. A common roughing system for three high intensity beam line has been commissioned using a rotary pump, a main valve and another one valve for each beam line.

F. Remote display of vacuum level through the PC has been implemented even during cyclotron operation. For this several Pirani - Cold cathode combination gauges have been used at least one in between two line gate valves to monitor vacuum level as well as for trouble-shooting.
G. Two 150 mm pumping units for new vacuum chamber for the 159.5^0 analyzing magnet and three 100 mm pumping for 8 meter long beam line have been installed to facilitate the transport of beam to radio-active ion beam facility. Roughing line has been included in each of these newly installed pumping units to utilize the same rotary pump as backing and roughing pump. Three full range gauges have been used in this new part.

After implementing above changes in respective system, each system was operated separately and vacuum was achieved ~ 5 x 10^{-6} Torr in the beginning. Helium leak detector was connected for leak testing with fore line of diffusion pump. Cyclotron and beam line were leak tested with 2x 10^{-8} mbar l/s helium background and no leak was found. Now we are achieving regularly vacuum ~ 2 x 10^{-6} mbar in cyclotron chamber as well as beam line.

5. Performance after modification

Consequent to this refurbish, the performance of entire vacuum system has remarkably improved. Presently, the vacuum system is functioning smoothly. All interlocks have also been tested. All Vacuum components installed recently including the newly upgraded Freon units are performing well. Baffle temperature is operated at a temperature of about -60^0 C regularly. Vacuum level of the order of 10^{-6} mbar has also been achieved in resonator tank and beam line within reasonable time. The cyclotron chamber pump down curve is shown in Figure 3.

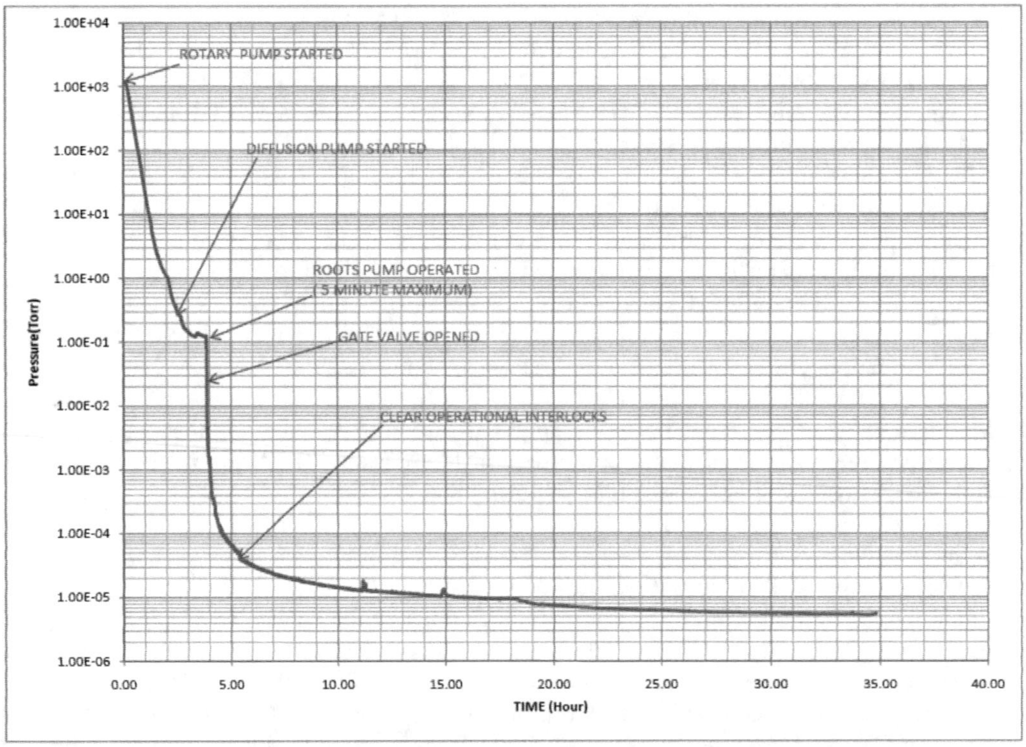

Fig. 3. Cyclotron Chamber Pump down (With Two Diffusion Pumps)

After implementation of changes in vacuum system and other sub-system, the cyclotron has been operated and beam of good intensity could be accelerated in the cyclotron. The achieved transmission of beam along the radius inside accelerator chamber is shown in Figure-4. Residual gas in the cyclotron is regularly monitored by using a RGA. Fig. 5 shows a typical bar chart for normal operation of the cyclotron.

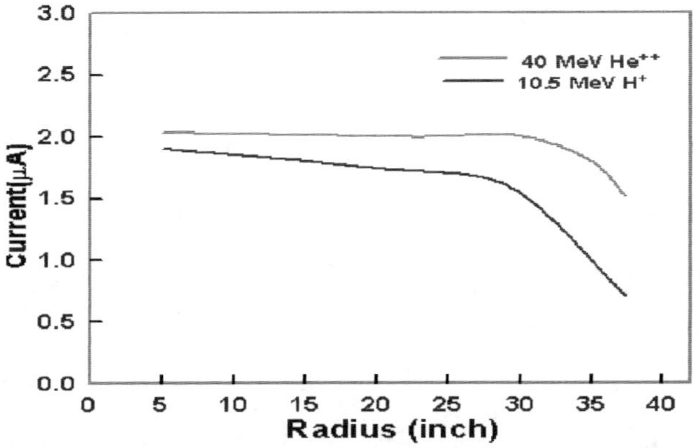

Fig. 4. Transmission curve for alpha (40 MeV) and Proton (10.5 MeV) beam

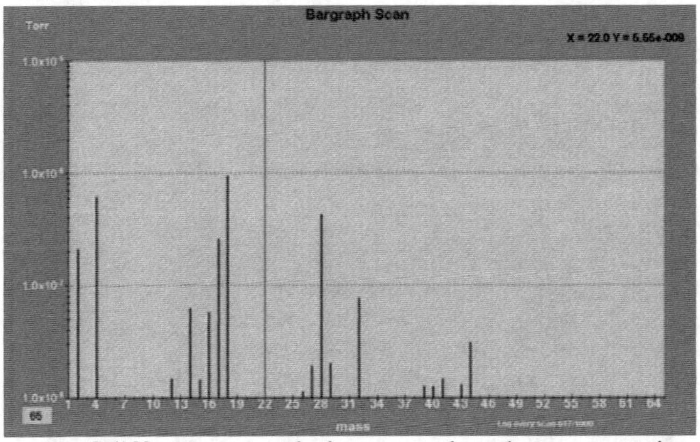

Fig. 5. Partial pressure of different gases during normal cyclotron operation at 4.7 x 10^{-6} Torr

6. Conclusions

New PLC based control system of cyclotron vacuum system has made the systematic operation of vacuum system more reliable. Auto start and shutdown of cyclotron vacuum system has been implemented and working satisfactorily. A PLC control system along with new MMI has been implemented. Online monitoring of vacuum level in different part of vacuum system as well as an accelerator chamber and beam line is helping diagnosis of system failure. The vacuum level in beam line at different locations is monitored through PC. Its vacuum system is operated with relay based control system.

Further improvement in vacuum system is planned by reducing gas loads in the beam line, reducing water leaks in the resonator tank, providing chilled water for cooling diffusion pump. Reduction of power fluctuations and a new PLC based control system for the beam line vacuum will improve the reliability of the vacuum system.

References

[1] D.L. Bandypadhyay and S. Bhattacharya, Indian Particle Accelerators Conference 2005 (InPAC-2005), March 1 – 5, 2005, VECC, Kolkata. India

[2] Anindya Roy, R. B. Bhole, J. Akhtar, R. C. Yadav, Sarbajit Pal, D. Sarkar, R.K. Bhandari, International Vacuum Electronic Conference (IVEC-2011), February 21 – 24, 2011, Bangalore, India

International Symposium on Vacuum Science & Technology and its Application for Accelerators IOP Publishing
Journal of Physics: Conference Series **390** (2012) 012047 doi:10.1088/1742-6596/390/1/012047

Achieving ultra high vacuum conditions in SMARTEX-C: control of instabilities and improved confinement

Lavkesh Lachhvani[1], Sambaran Pahari[2] and Y C Saxena[1]

[1] Institute for Plasma Research, Near Indira Bridge, Gandhinagar-382428, India
[2] Bhabha Atomic Research Centre, Visakhapatnam-530012, India

E-mail: `lavkesh@ipr.res.in`

Abstract. SMall Aspect Ratio Toroidal Experiments in a C-shaped trap (SMARTEX-C) attempts to confine electrons in a toroidal trap using static electric and magnetic field. Confinement of these electrons is largely limited by charge losses due to the presence of background neutrals and by short pulsed magnetic field. Ions formed due to electron impact ionization of background neutrals lead to instabilities that are ultimately observed to limit the confinement. Lowering the amount of neutrals in the trap is therefore of paramount importance. In SMARTEX-C, prior to electron injection, the trap is pumped down to base pressures of the order of $2 \pm 1 \times 10^{-8}$ mbar. However, as one turns on the electron source (tungsten filament emitting thermionically) pressure in the system increases to $\sim 1 \times 10^{-7}$ mbar. Partial pressure analysis of the vacuum system indicates predominant presence of H_2 during filament operation. Replacing the copper current-leads for electron source with stainless steel-304 led to reduction of H_2 outgassing. Additionally, viton seals have been replaced with Aluminum wire-seals that allow baking at elevated temperatures and further reduce out-gassing. All these measures have led to an improvement in the base pressure to $7 \pm 1 \times 10^{-9}$ mbar even during filament operation. Confinement of electron plasmas in the trap has thus significantly improved.

1. Introduction

Electron plasmas confined in a uniform magnetic field (in cylindrical chamber) have become an important part of fluid dynamics research [1] and its distinct feature of confinement [2], equilibria [3], transport and Kelvin-Helmoholtz instability makes it exciting branch of plasma physics. Such traps have demonstrated confinement for hours and even days [4] when operated at pressures $\sim 1.33 \times 10^{-10}$ mbar. The successful confinement under under ultra-high vacuum (UHV) conditions follows from conservation of angular momentum of a collection of like charges; presence of neutrals in the trap leads to an external torque and hence causes rapid collisional losses [2].

In contrast, poor confinement ($\sim 100\,\mu$s) of electron plasmas observed in in non-uniform magnetic field, such as toroidal geometries, [5, 6, 7] were attributed primarily to toroidal effects namely, curvature and and ∇B drifts. However, with efficient injection of electrons along field lines especially in C-shaped traps [8, 9], confinement of toroidal electron plasma has significantly improved; strong $E \times B$ drifts are seen to compete and overcome the toroidal drifts. Lifetimes of the order of few milliseconds were successfully reported in SMARTEX-C. At this stage the lifetime in SMARTEX-C was largely thought to be limited due to a short pulse magnetic field. However, in recent times, even in the presence of a steady state B field of 10 ms, SMARTEX-C demonstrated only a limited improvement in confinement [10].

The limited confinement of plasma in SMARTEX-C is now determined by losses that are dominated by instabilities, primarily driven by ions. Ion resonance instability [11, 12, 13] occurs when a small number of ions trapped in the electron potential well oscillate with a frequency close to that of the diocotron frequency. This drives the diocotron mode unstable. The presence of ions results from electron impact ionization of the residual gas in the trap. Growth rates of the instability are accordingly observed to get effected by the energy of injected electrons and the ionization cross-section of the residual gas [11].

Any further improvement in confinement in SMARTEX-C is therefore crucially dependent on controlling ion driven instabilities. In other words, it underlines the requirement of UHV in SMARTEX-C. In this paper we present our efforts towards up- gradation of vacuum conditions that allow operations in SMARTEX-C at base pressures of the order of $\sim 1 \times 10^{-9}$ mbar. Section 2 explains the trap and its limitations including residual gas analysis of the trap under vacuum. Several measures taken to improve the vacuum are discussed in section 3. The improved vacuum condition in SMARTEX-C along with the significant improvement obtained in confinement are reported in the concluding section.

2. SMARTEX-C Trap and its operation

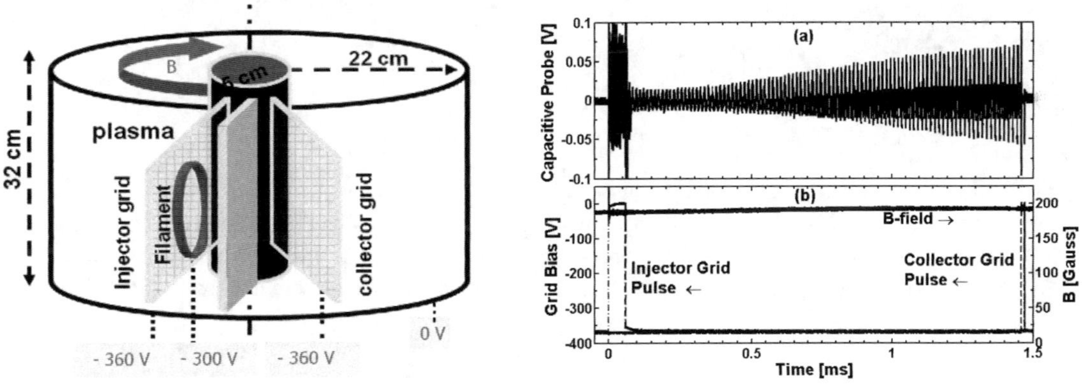

Figure 1. Left: Trap geometry and arrangement of electrodes, Right: Typical pulses of (a) diagnostics signal from midplane capacitive probe across $1 \times 10^3\,\Omega$(b) magnetic field (solid), injector grid(dash-dot) bias, collector grid(dash) bias.

SMall **A**spect **R**atio **T**oroidal **EX**periment - **C (SMARTEX-C)** attempts to confine electrons in a C-shaped toroidal trap using static axial electric and toroidal magnetic field. The device, geometry of electrodes and diagnostic are shown in figure 1 (left). The plasma chamber is a torus with rectangular cross-section. Trap is pumped down to 1×10^{-8} mbar. Injector grid and collector grid are kept $270°$ apart from each other. Vessel wall and inner wall are held at ground potential. A single circular filament of diameter 10 cm is placed on a poloidal section, which emits electrons thermionically parallel to B field. A 100-200 Gauss toroidal pulsed magnetic field is generated for $\sim$ 10 ms. As it reaches a steady state, electrons are injected by switching the injector grid to ground for a brief period of 60 µs (*inject mode*). This is followed by the *hold mode* when the injector grid is turned "off" (reverts to negative bias) and the electrons are confined between two negatively biased grids. Capacitive probe diagnostics kept flush with inner wall responds to electrostatic activity in the plasma and therefore provides an order of magnitude estimate of the confinement time [14]. The trap is operated in an"inject-hold-dump" mechanism akin to a cylindrical penning trap [1] and is described in detail in [13].

2.1. Vacuum-System

The trap volume is $\sim 50\,\mathrm{l}$. Prior to electron injection the trap is evacuated to base pressures of 1×10^{-8} mbar by a 500 $\mathrm{L\,s^{-1}}$ Turbo Molecular Pumping (TMP) backed by rotary pump (30 $\mathrm{m^3/h}$). The conduction of the pumping port limits the effective pumping speed to $259\,\mathrm{l\,s^{-1}}$. All the ports and flanges of the vessel are vacuum-sealed. The pump down is followed by a bake-out of chamber up to 150 °C for 48 - 60 h. Pressure is monitored by Bayard-Alpert ionization gauge. After a cycle of pump down and bake out ultimate base pressure obtained in the device is of the order of $2 \pm 1 \times 10^{-8}$ mbar. However, pressure rises during filament operation up to 1×10^{-7} mbar.

2.2. Poor vacuum conditions: Reasons, analysis and effect on plasma confinement

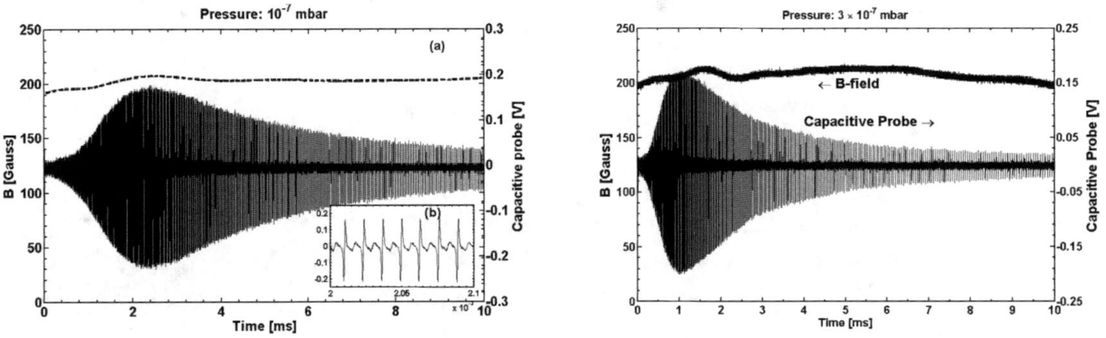

Figure 2. Left: (a) Diocotron modes reaching peak amplitude at 1 ms and decay in oscillation amplitude driven by collisional losses and ion-resonance instability limiting the confinement of electron plasma.(b) Inset graph shows zoomed diocotron oscillations for the period of 2.0 - 2.1 ms. Right: Diagnostics signal at 3×10^{-7} mbar pressure, faster growth rate of instability due to higher pressure

The high operating pressures in SMARTEX-C restricts the lifetime of the plasma. At 1×10^{-7} mbar the trapped electron plasma shows coherent periodic oscillations as shown in figure 2 left (a). These are the familiar diocotron oscillations [1] albeit with toroidal signatures (figure 2 left (b)) [8]. The amplitude of oscillations grow and then saturates before decaying. The unstable non linear mode is primarily responsible for the loss of charges. As discussed earlier, the mode is driven unstable by ions whose oscillating frequencies are close to the diocotron frequency ($\sim 100\,\mathrm{kHz}$ for $n_e = 5 \times 10^6\,\mathrm{cm^{-3}}$). The presence of ions can be traced to the existing high operating pressures in the trap at 1×10^{-7} mbar fractional ionization due to electron (300 eV) impact ionization of H_2 can be estimated to be 0.69×10^{-2}. This fractional ionization goes down by one order if, one reduces the operating pressure from 1×10^{-7} mbar to 1×10^{-8} mbar. The effect on confinement can be seen from figure 2 (right), where operating pressure is 3×10^{-7} mbar. 1/e decay of peak amplitude can be used to estimate the confinement time of the plasma. Higher pressures evidently lead to rapid onset of the instability and shorter confinement time.

Several factors that effect the gas load in SMARTEX-C and eventually the confinement are discussed below.

2.2.1. Gas Load due to real and virtual leaks: All the ports of SMARTEX-C are vacuum sealed using aluminum wire seals or copper gaskets. Top and bottom flange of the vacuum vessel due to their large sizes were however sealed with high vacuum Viton O-rings and had leak rates of the order of $1 \times 10^{-9}\,\mathrm{mbar\,l\,s^{-1}}$. Outgassing rate and permeation rate of these Viton O-rings contributed to the virtual leak and increased the total gas load by an order of magnitude. These O-rings also limited the baking of the vacuum vessel to $150 - 175°\mathrm{C}$. Theroretical estimation of total gas load due to real and virtual leaks is $3.5 \times 10^{-7}\,\mathrm{mbar\,L\,s^{-1}}$. After two-three baking cycles, these O-rings were vulnerable to leaks further increasing the gas load.

2.2.2. Higher neutral pressure during filament operation: Once the heating of the filament is turned on, the pressure in the trap slowly increases from 2×10^{-8} mbarto $3 \pm 1 \times 10^{-7}$ mbar. Thereafter it reduces but steadies at 1×10^{-7} mbar. The electron injection and confinement is carried out at these pressures. The unwanted gas load evidently results from higher outgassing from vessel walls and other in-vacuum components at high temperature. Efforts were therefore initiated to diagnose and identify the source of this gas load at elevated temperatures.

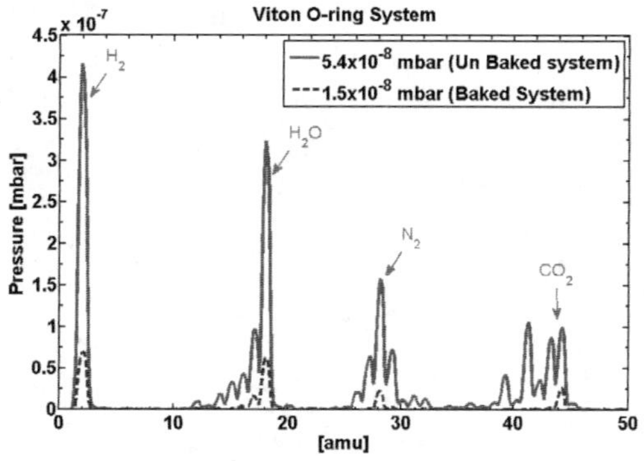

Figure 3. Left: Partial pressure analysis at base pressure before baking system (in green) and after baking at 175 °C for 48 - 72 h condition (in blue) in *Viton* sealed vacuum system, Right: Partial pressure analysis at base pressure (in blue) and filament operation condition (in red) in *Viton* sealed vacuum system.

A Residual Gas Analyser (RGA) was used and mass spectrum of background neutrals in the trap was acquired during various stages of pump down. When system is exposed to atmosphere for prolonged time and injector assembly is newly assembled consisting of new tungsten filament and alumina bushes, it takes 24 to 48 hours of pump down for pressures to reach 1×10^{-7} mbar. Mass spectrum shows higher amount of water vapor pressure at this stage. A complete baking cycle at 150 °C for 72 h and simultaneous operation of the filament source reduces water vapor from the system by an order of magnitude as evident from RGA spectrum in figure 3 (left) . Total pressure of the system is reduced to 1.5×10^{-8} mbar.

Figure 4. Copper strips used as injection leads for tungsten filaments

When the filament is heated to 2800 K with a dc current of 20 A, it expectedly raises the

temperature of injector assembly and surrounding
vessel wall through radiation. RGA spectrum (figure 3 right) now reveals the dominant presence of Hydrogen.

There are two likely sources; high outgassing of H_2 from copper strips that act as current leads to the filament (figure 4) and/or outgassing of the Stainless Steel vessel at elevated temperature. The latter indicated occurrence of an in-sufficient baking of the vacuum vessel. Increase in N_2, CO and CO_2 in RGA spectrum during filament operation could also be due to high permeation rate of Viton O-rings, used to seal the top and bottom flanges of the vacuum vessel.

3. Measures to attain UHV

In view of the analysis presented in the previous section several measures were taken to reduce the neutral pressures: (1) Viton O-rings were replaced with Aluminum wire seal (2) ETP copper strips were replaced with SS 304 strips and (3) Baking was carried out at higher temperatures of 250 - 275 °C for prolonged duration.

Smooth, parallel flanges with three-delta finish were fabricated to obtain sealing with wire seals. To obtain a leak rate better than 1×10^{-10} mbar L s^{-1}, 3 mm thick circular Al wire seals of diameter ~ 512 mm were prepared. Viton O-rings were replaced with Aluminum wire seal. Necessary torque T, required to tighten the bolts (having diameter D) of wire-seal flange (having radius R) for a specific leak rate can be given by, $T = \frac{2\pi RFD}{KN}$ where, K is the constant dependant on accuracy of threads, smoothness and lubrication, F is the sealing force required per unit length and N is the number of bolts. All quantities are in SI. As reported in [15], to attain specific leak rate of 1×10^{-10} mbar L s^{-1} a sealing force of 250–270 kN m^{-1} is required. Theoretically, 40 N m should be enough to tighten the wire seal flange of 526 mm diameter having 36 bolts of $M10$ standard.

The wire seal flanges were leak tested up to 1×10^{-9} mbar L s^{-1}. Wire seals also made it possible to bake the system at 250 – 275°C for 48 – 72h. Permeation through Viton O-rings was also reduced. Theroretical estimation of total gas load resulting from vessel, trap components and sealants reduced from 2.5×10^{-6} to 3.5×10^{-7} mbar L s^{-1} after taking measures.

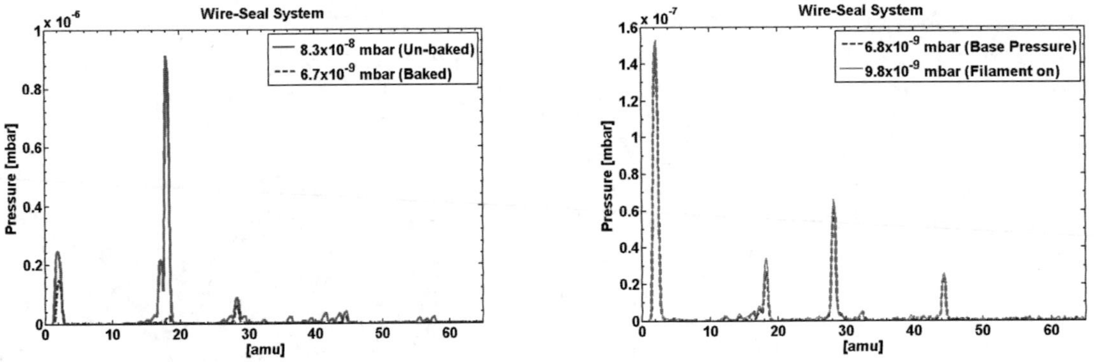

Figure 5. Left: Partial pressure analysis at base pressure before baking system (in green) and after baking at 250 °C for 48 - 72 h condition (in blue) in Wire-sealed vacuum system, Right: Partial pressure analysis at base pressure (in blue) and filament operation condition (in red) in Wire-sealed vacuum system.

Additionally presence of hydrogen significantly reduced on replacing copper current leads with SS leads. RGA spectrum under improved condition reveals that outgassing of Hydrogen is absent during filament operation (figure 5 right). High temperature baking also resulted in very low partial pressure of water vapor as seen in RGA spectrum of baked system (figure 5 left). Replacement of Viton O-rings with Al wire-seals has removed the CO and CO_2 from the vacuum system.

4. Results

Measures reported in the previous section result in lower base pressure of $4 \pm 0.2 \times 10^{-9}$mbar. Significantly the pressures during filament operation also remained as low as $6 \pm 1 \times 10^{-9}$mbar.

Controlled and reproducible experiments are possible at pressures as low as $8 \pm 1 \times 10^{-9}$mbar. The improved vacuum conditions eventually leads to successful control of the diocotron instability and considerable increase in the confinement time as described below.

Electrons were injected at a background pressure of 2×10^{-8} mbar. Diocotron oscillations as seen from wall probe is shown in figure 6. The oscillations are linear (small amplitude) and growth is largely arrested till 3 ms. Beyond this period, the oscillations grow slowly and are seen to last longer than the duration of the steady-state magnetic field. The confinement time is well beyond 20 ms . At 6×10^{-9} mbar and a B field of 30 ms a further improvement upto 70 ms has been obtained . Controlled ultra-high vacuum conditions resulting in slow growth-rate of ion-resonance instability therefore marks a significant improvement in lifetime of toroidal electron plasmas in SMARTEX-C and lifetimes well beyond 100 ms can be expected with an extended B field.

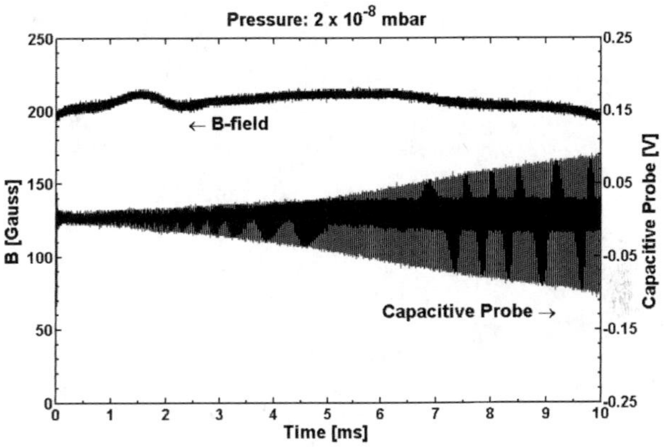

Figure 6. Capacitive probe signal showing diocotron modes, increase in amplitude is observed till 10 ms.

Acknowledgement

Authors wish to acknowledge, Siju George, Dharmendra Rathi and Firoz Pathan for helping us in helium leak-testing of vacuum system and and in constructing wire-seals. In addition to this, authors thank Dr. B. Ganguly for lending us Residual gas analyzer, whenever needed.

References

[1] Malmberg J H and deGrassie J S 1975 *Physical Review Letters* **35** 577–580
[2] O'Neil T M 1980 *Physics of Fluids* **23** 2216–2218
[3] Driscoll C F, Malmberg J H and Fine K S 1988 *Phys. Rev. Lett.* **60**(13) 1290–1293
[4] Malmberg J H and Driscoll C F 1980 *Phys. Rev. Lett.* **44**(10) 654–657
[5] Daugherty J D, Eninger J E and Janes G S 1969 *Physics of Fluids* **12** 2677–2693
[6] Khirwadkar S, John P, Avinash K, Agarwal A and Kaw P 1993 *Physical Review Letters* **71** 4334–4337
[7] Zaveri P, John P I, Avinash K and Kaw P K 1992 *Physical Review Letters* **68** 3295–3298
[8] Pahari S, Ramachandran H S and John P I 2006 *Physics of Plasmas* **13** 092111
[9] Marler J P and Stoneking M R 2007 *Journal of Physics: Conference Series* **71** 012003
[10] Lachhvani L 2009 *Towards long time confinement in SMARTEX-C* Master's thesis University of Pune
[11] Peurrung A J, Notte J and Fajans J 1993 *Physical Review Letters* **70** 295
[12] Marksteiner Q R, Pedersen T S, Berkery J W, Hahn M S, Mendez J M, Durand de Gevigney B and Himura H 2008 *Phys. Rev. Lett.* **100**(6) 065002
[13] Pahari S 2004 *Electron plasma studies in a toroidal penning trap* Ph.D. thesis Instittute for Plasma Research
[14] Trivelpiece A W and Gould R W 1959 *Journal of Applied Physics* **30** 1784–1793
[15] Kahn Z, Kotwal S, Patel R J and SST V 2000 Performance study of wire-seal Tech. Rep. IPR/TR-66/2000

In-service helium leak testing of vacuum furnace

Anis Ahmad, S. K. Tripathi, P. S. Sawant, D. Mukharjee and B. K. Shah

Quality Assurance Division, Bhabha Atomic Research Centre,Trombay, Mumbai, India, Pin-400085

E-mail: ahmad@barc.gov.in, anis1958@gmail.com

Abstract. Helium leak detection of vacuum furnaces and equipments used for processing of nuclear material is generally carried out by utilizing vacuum spray technique. In this technique helium leak detector is connected to the furnace, back ground reading is noted and helium gas is sprayed on all the suspected joints. Any increase in back ground is noted as leak signal. Processing of Zirconium alloy cladded fuel pins is carried out in vacuum furnace of about 3 meter length and 500 mm inside diameter. Furnace is connected with two numbers of rotary vacuum pump and one number of diffusion pump for creating vacuum (1 x 10^{-6} torr) inside the furnace. It is desirable that furnace should have good vacuum and best possible leak tightness during dynamic and static vacuum. During dynamic vacuum at higher temperature although required vacuum is achieved the furnace may have fine leakage through which air may enter and cause oxidation of clad tube leading to change in its coloration. This change in coloration will cause rejection of fuel element. Such fine leakages may not be reflected in the dynamic vacuum of the system at high temperature. During trial run change in coloration of outside surface of clad tube was observed although dynamic vacuum of the furnace was in the range of $1x10^{-6}$ torr range. To eliminate such possibilities of oxidation due to fine leakages in the system, it was decided to carry out in-service leak testing of the furnace. Helium leak testing of the furnace was carried out by using vacuum spray method and leaks observed were repaired and furnace was retested to ensure the leak tightness. The in-service helium leak testing of the furnace helped in maintaining its leak tightness during service under dynamic vacuum and prevent oxidation of fuel element. This paper describes the techniques of in- service helium leak testing, it's importance for detection of fine leak under dynamic vacuum and discusses details of the testing method and result obtained.

1. Introduction

Leak detection forms an important part of quality control step in chemical, petrochemical, fabrication and nuclear industry. Designer's have been specifying leak tightness as per different codes like ASME, ASTM, BIS etc. as required by their intended use. In nuclear industry because of radioactive hazard considerations leak tightness in many furnaces and equipments are specified up to$1x10^{-9}$std.cc/sec. Helium leak testing of enclosed chambers are carried out to ensure the high degree of leak tightness in nuclear industry. There are generally four ways of carrying out helium leak detection namely vacuum spray method, pressure sniffer method, pressure vacuum method and back pressure filling or bombing method. Helium leak testing of vacuum furnaces and equipments are carried out by using vacuum spray method.

2. Leak Detector

In the present work Console model Helium Leak Detector (HLD) has been used. The basic principle of helium leak detector is based on mass-spectrometer, which is kept under high vacuum of the order of 10^{-5} torr. Helium atoms entering in the mass-spectrometer from the leak are first ionized by bombardment of electrons and the positive ions so created are accelerated by an electric field and the ions are separated from each other depending on their mass, by application of magnetic field over it. The helium ions follow a definite path. A slit is used to allow selectively only helium ions. These ions are collected at an electrometer tube, which generate a signal in mili-volt range in proportion to leak rate. The signal is then suitably amplified and displayed on a digital LED display or can be coupled to data acquisition computer through RS-232 port provided in the helium leak detector.

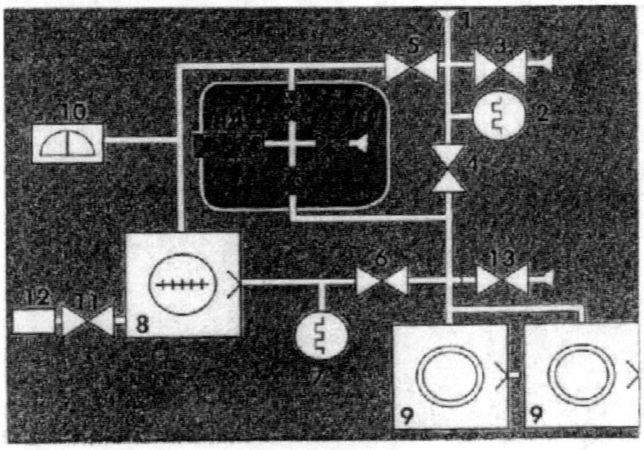

Fig. 1. Schematic of Helium Leak Detector

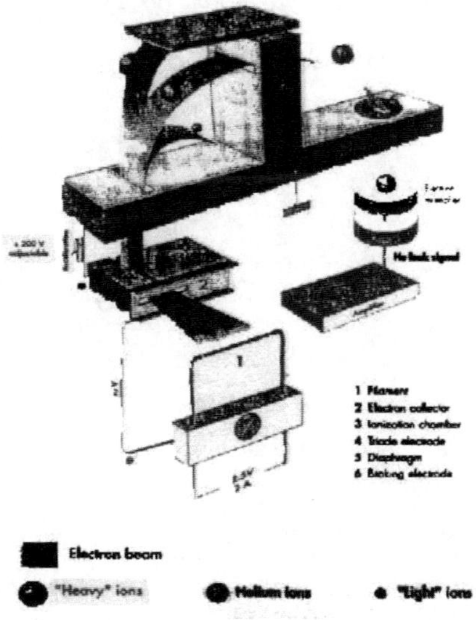

Fig. 2. Schematic Diagram of Mass Spectrometer

Figure 1. gives the layout of typical helium leak detector and figure 2. gives the schematic of mass-spectrometer tube. It consists of 2 nos. of roughing pump each of 20 m^3/hr. pumping capacity and a 130 litre per second hybrid-turbo molecular pump. The detector has measurement range of $2x10^{-11}$ to $1x10^{-1}$ mbar l/sec. of helium. Two standard permeation type calibrator namely internal and external calibrators do calibration of helium leak detector. In auto mode the detector calibrates itself using internal calibrator of $1x10^{-7}$ mbar l/sec. and only if it is satisfactory gives green signal for further operations. For external calibration of helium leak detector a reference leak standard with leak rate in the range of $1x10^{-8}$ to $1x10^{-9}$ mbar l/sec. may be used.

3. Vacuum Furnace

The vacuum furnace has a large size diameter of about 500 mm x 3 meter height. It is used for processing of fuel elements made of Zirconium clad tube. The furnace is designed to achieve around 1000^0C temperature and $1x10^{-5}$ torr vacuum. The vacuum system consist of two numbers of rotary pumps of pumping speeds of 90 CFM and 15 CFM respectively and one diffusion pump of 250 LPM pumping speed. The layout of the furnace is shown in the figure 3. The heating in the furnace is carried out by using seven numbers of circular heaters mounted around the furnace and one number plate type of heater below the furnace.

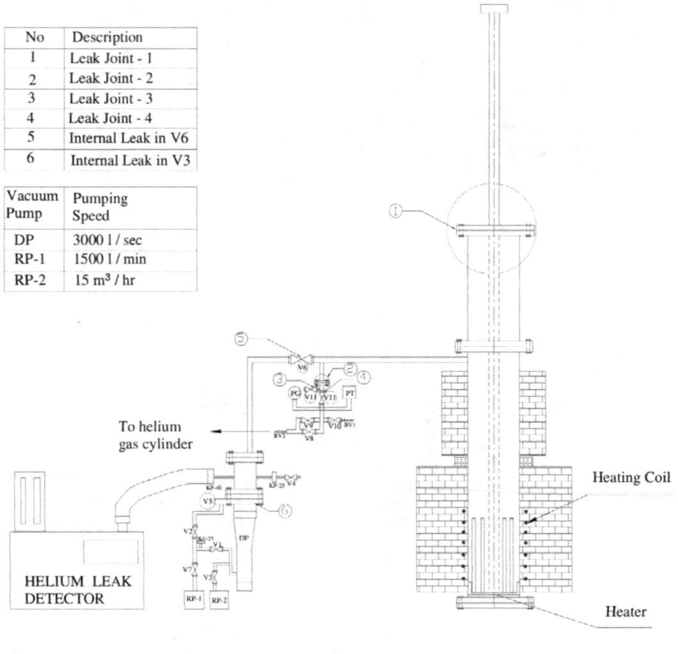

Fig. 3. Helium Leak Testing of Vacuum Furnace

4. Leak Detection Procedure

The helium leak detector was connected to vacuum furnace as shown in figure 3. The leak detection was carried out by using vacuum spray method. The dynamic vacuum of the furnace achieved was about 1 x 10^{-6} torr. Initially helium leak testing of the furnace was carried out in gross mode and helium gas was sprayed over all the joints like- flanges, tube fitting connectors, gauge head joints etc. more than fifty joints were sprayed with helium gas throughout the furnace one by one by keeping sufficient time to stabilize the reading, leak was observed in only one joint identified as joint no-1, which was taken up for repair. After repairing of the leaky joint re-testing of joint-1 was carried out in gross test mode and no leak signal was observed and then fine leak testing of the furnace was planned and carried out by using vacuum spray method. Helium gas was sprayed over all the joints mentioned above numbering more than fifty throughout the furnace one by one by keeping sufficient time to stabilize the reading. All the detected leaky Joints were covered by the cello tape to avoid false leak signal during testing of further joints. Six joints showed clear leak indication beyond the back ground reading, these joints were repaired and again helium leak tested after the repair. The results have been provided in the table-1. The helium leak detector was calibrated with a standard calibrator of standard leak 1.2 x 10^{-9} mbar l/sec. of helium before and after the test cycle.

Table 1

Sr. no.	Leak Location	Back ground reading (m bar l / sec)	Reading after helium gas spray (m bar l / sec)	Reading after repair (m bar l / sec)
1	Joint no. – 1	3 x 10^{-7}	2 x 10^{-4}	3 x 10^{-9}
2	Joint no. – 2	3 x 10^{-7}	2 x 10^{-6}	5 x 10^{-9}
3	Joint no. – 3	9 x 10^{-7}	1 x 10^{-5}	7 x 10^{-9}
4	Joint no. – 4	9 x 10^{-7}	4 x 10^{-7}	8 x 10^{-9}
5	Valve - 6	3 x 10^{-7}	1 x 10^{-4}	Not Repaired
6	Valve- 3	3 x 10^{-7}	1x10^{-3}	Not Repaired

5. Conclusion

For good and unambiguous helium leak testing the back ground signal (BG) should ideally be kept at least ten times less than the specified acceptance limit. Normally it is desirable to maintain background of the helium leak detector as less as possible (1x10^{-9} mbar l/sec.) for better performance of the helium leak detector. However in case of maintenance of furnaces and vacuum equipment etc. it is difficult to get such a low back ground because of associated leakage from leaky joints. In modern leak detector like one used here has facility of both the gross and fine leak testing. Gross leak testing can be done just at 6 mbar vacuum and fine leak testing is possible if vacuum achieved is better than 2x10^{-2} mbar. It has been our experience that after spraying the helium gas, if leak is present meter reading will shoot up immediately which indicates clear leak signal.

Before in service helium leak testing of the furnace, it was observed that coloration of outside surface of clad tube was getting changed during trial run indicating oxidization of the surface, although dynamic vacuum of the furnace was in the range of 1x10^{-6} torr. After leak detection and corrective actions were taken significant improvement in coloration of clad tubes of fuel elements was observed, which indicates that oxidization of the tube surface has reduced considerably due to absence of air the furnace. Hence it is concluded that in service helium leak testing of vacuum furnaces helps in production of quality nuclear products as per requirement.

Acknowledgment

Authors are extremely thankful to Shri S. Anantharaman, Head, Quality Assurance Division & PIED for permitting publication and presentation of this technical paper in " International Symposium on Vacuum Science & Technology and its Application for Accelerators" organized by Indian Vacuum Society at VECC, Kolkata. We gratefully acknowledge the encouragement provided by Dr. G. J. Prasad, Director, NFG in carrying out this work. We are also thankful to Shri R. P. Singh former Director, NFG for his guidance and encouragement.

Generation of vapor and concomitant plasma production in an electron-beam evaporator

A. Majumder, G. K. Sahu, S. Barauh, B. Jana, V. K. Mago, A. K. Das and K. B. Thakur

Bhabha atomic Research Centre, Mumbai, India

abhi@barc.gov.in

Abstract. Electron beam heating is a technique to generate vapor of refractory and high melting point metals. Vapor production finds application in thin film deposition and laser-based purification of materials. A strip electron-gun whose filament is heated by AC current is generally used because of larger molten pool formation and quiet evaporation. Electron-gun thus generates vapor. The incident beam of electrons is backscattered with large angular distribution. Both the electron groups, namely the primary and the backscattered electrons participate in production of plasma by electron-impact ionization. The plasma is weakly ionized (~ 0.1% degree of ionization) with ion density ~ 10^8 cm^{-3} and has low electron temperature (~ 0.3 eV). The vapor and the simultaneously produced plasma expand in the space above the target. Plasma expands by ambipolar diffusion in a transverse magnetic field while the vapor expands as a collision-less atomic beam. In this paper we study vapor and plasma formation of copper and zirconium. Details shall be discussed.

1. Introduction

The technique of electron beam (e-beam) heating has become a favored method of atomic beam generation of refractory and high melting point materials [1, 2]. This finds application in laser-atom interaction studies of refractory materials such as charge-transfer collision experiments [3], production of isotopically pure elements [4], purification of alloy [5] etc. In such investigations, high-energy electrons from continuous electron gun (e-gun) impinge on a metal target. This creates a molten pool which becomes the source of evaporating atoms / atomic beam. By shining lasers on the atomic beam at a distance away from the evaporating target, the atoms are ionized. A photoplasma [6] is produced at the laser-atom interaction region, which becomes the focus of investigation. Weakly ionized, low-temperature plasma gets formed in the atomic beam near the target due to the presence of continuous e-beam. This plasma, sometimes called background plasma, expands with the atomic beam and interferes with the process parameters of the photoplasma experiments. There is a need to efficiently remove or deflect out the background plasma from the atomic beam so that the experiments become free from such interferences. It is therefore essential to understand the characteristics and dynamics of vapor and the concomitant plasma produced by continuous e-beam.

Formations of vapor and plasma generation are inherent characteristics of e-beam melting. This paper discusses few aspects of the studies carried out in our laboratory with copper (Cu) and zirconium (Zr) with e-gun. Techniques to characterize the vapor and plasma are mentioned. Typical numbers are provided. The mechanisms of vapor and plasma formation and propagation are highlighted.

2. Experimental details

The experimental system consisted of an e-beam evaporator having a strip-type e-gun (rating: 50 kV, 2 A). It had segmented hot tantalum cathode that produced a thin sheet of electrons by thermionic emission. The electron trajectories were bend by 270^0 by a pair of magnetic coils in Helmholtz configuration that produced ~ 50 G magnetic field along the direction of the length of the chamber. This configuration was adopted to protect the anode-cathode space of e-gun from vapor entry causing discharge. It also prevented the back streaming of ions to the filament causing reduction in filament life. The evaporator was a double-walled water-cooled stainless steel cylindrical vacuum chamber (90 cm diameter, 80 cm length) operated typically at 5x10-5 mbar. The ingots, either copper or zirconium were placed in a water-cooled copper crucible.

The molten pool was photographed with an internal standard and analyzed by image-processing software. Optical filters were used to avoid saturation of the photograph pixel. The dimension of the strip was measured to be of 12 cm length and 0.6 cm width

The temperature of the evaporating liquid surface was measured by two-color pyrometer (M/s Keller, PZ40) using a periscopic arrangement. The two monitoring wavelengths were 950 nm and 1050 nm. Process generated thin film acted as the mirror. The pyrometer read the average temperature over a 0.6 cm diameter zone focused on the central region of the evaporating source and was kept at a distance of ~ 100 cm from the source.

A water-cooled quartz crystal thickness monitor (QCM) (M/s Sigma SQM-160) was used to measure the atom fluxes. It had a shutter that was controlled electronically. Measurements were carried out for various e-beam powers.

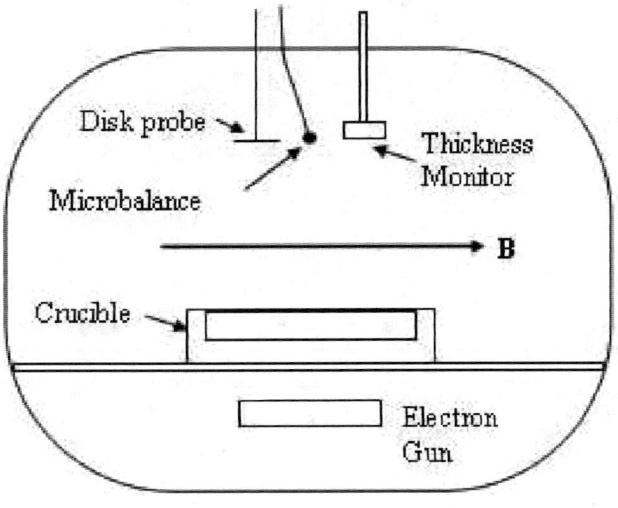

Fig. 1. Schematic setup (length view) of the electron beam evaporator along with diagnostics.

Velocity of the atomic beam was measured by a microbalance (M/s Setaram, model B-24). It was placed at the same height as that of the thickness monitor. Its mass resolution was 0.03 microgram. Appropriate precautions were taken to protect it from unwanted vapor.

For plasma diagnostics disk Langmuir probe was used. It was made of molybdenum (1 cm in radius) and was 0.2 cm thick. It had alumina insulation at the back so that down-streaming ions were not collected on it. Figure 1 shows the schematic of the experimental chamber along with the diagnostics.

3. Results and discussions

When high-energy (~ 45 kV) electrons from the e-gun impinge the target, they transfers concentrated heat flux. As a result the target melts and becomes the source of evaporating atoms. Photographs of melt pools of Cu and Zr are shown in Fig. 2. For Cu, the highest temperature measured was ~ 1950 K (at ~ 90 kW e-beam power) whereas the corresponding value for Zr was 3250 K at 100 kW. The temperature of the evaporating area depends on the incident power density of the e-gun and thermo-physical properties of the target material.

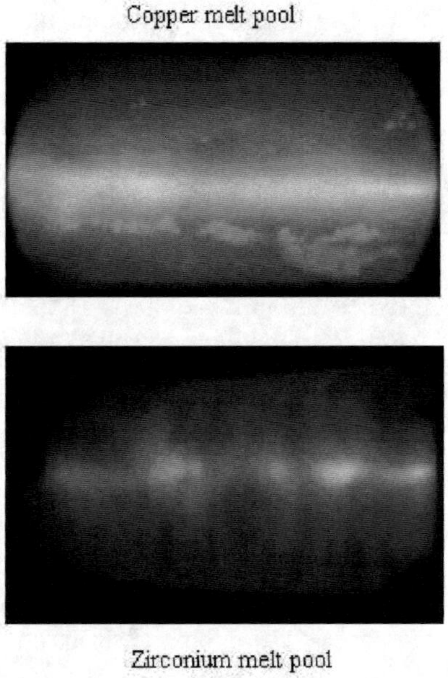

Fig. 2 Photographs of melt pools of Cu and Zr

Figure 3 shows the measured deposition rate as a function of e-beam power for Zr at a height of ~ 40 cm from the crucible. Typically, deposition rates vary in the range of tens of angstrom. But for Cu they are hundreds of angstrom. The measured velocities of Cu and Zr are similar, ~ $1x10^5$ cm/s. The derived atom densities from the deposition rate are ~ $1x10^{13}$ cm^{-3} and ~ $1x10^{11}$ cm^{-3} for Cu and Zr respectively. Such wide difference (two orders of magnitude) in the atom densities is due the difference in vapor pressure of the two metals. The vapor undergoes expansion in the evaporator. Near the target the atom-atom mean free path is smaller than the dimension of the vapor source. So the atoms collide among themselves and undergo adiabatic free expansion. For a point source, the spatial distribution of atoms follows r^{-2} dependence, where r is vertical distance (height) from the evaporating source. For an infinite linear source, the atoms follow r^{-1} dependence. So the atom distribution for a finite linear source (our case) will range between $r^{-(1\ to\ 2)}$.

The expanding vapor intercepts electron flux and gets ionized. Typically measured plasma (ion) densities of Cu and Zr at a height of ~ 40 cm are ~ $1x10^9$ cm^{-3} and $1x10^8$ cm^{-3} respectively. The plasma formation mechanism can occur by two processes: electron-impact ionization of neutral atoms of the beam and thermal ionization. Their relative magnitudes depend upon various parameters like temperature of the evaporating source, e-beam currents, energies and path lengths, ionization cross sections etc. Electron-impact ionization is the predominant process in most cases. The primary electrons (e-beam electrons) and backscattered electrons from the target contribute to ionization and hence in plasma generation. The plasma is weakly ionized and has low electron temperature (~ 0.1 eV).

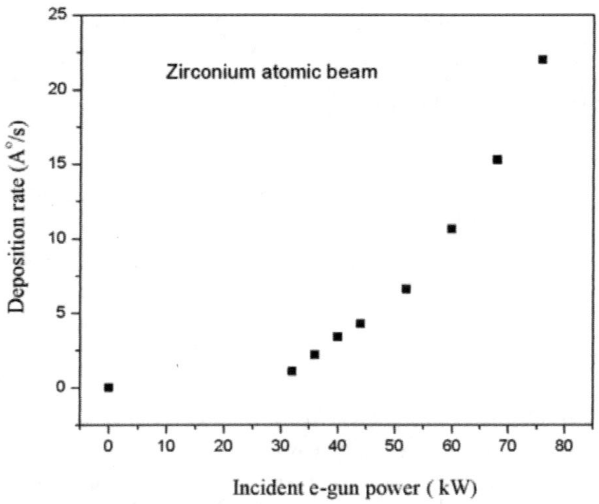

Fig.3 Deposition rate of Zr as a function of e-beam power.

The plasma co-expands with vapor thereby changing the spatial characteristics of both the vapor and itself. Ambipolar diffusion governs the plasma motion. The plasma potential decreases linearly in the propagation direction away from the source. Electric field, which is the negative gradient of potential, remains constant along the streamline and is directed upwards. As a result the electrons are decelerated while the ions are accelerated and gain drift energy. The plasma density decreases due to volume expansion and ion acceleration from ambipolar diffusion. When there exists a global transverse magnetic field, the Lorentz force creates a spontaneous polarizing electric field within the plasma and as a result cross-field propagation occurs in the vertical direction.

Acknowledgement
The authors thank. Dr. L M. Gantayet, Director, Beam Technology Development Group, BARC for his constant support and encouragement. They appreciate the technical help offered by Ms. R. A. Patankar and Mr. M. Sunkanna during the experiments.

References

[1] Schiller S, Heisig U and Panzer S 1982 *Electron Beam Technology* (New York: J. Wiley)
[2] Thakur K B, Sahu G K, Gaur S J, Das R C, Tak A K, Patankar R A, Bhowmick G K, Manohar K G, Jagatap B N and Venkatramani N 2005 *Vacuum* **77** 443
[3] Tamura K, Adachi H, Ogura K, Ohba H and Shibata T 1999 *Jpn. J. Appl. Phys.* **38** 6512
[4] Letokhov V S, Mishin V I and Puretzky A A 1977 *Prog. Quant. Elect.* **5** 139
[5] Hiroshi M 1998 *Appl. Phys. Lett.* **72** 1948
[6] Majumder A, Mago V K, Ray A. K, Kathar P T and Das A K 2005 *Appl. Phys B* **81** 669

Simulation of electron beam from two strip electron guns and control of power density by rotation of gun

G.K.Sahu, S.Baruah and K.B.Thakur

Laser & Plasma technology Division, Bhabha Atomic Research Centre, Trombay, Mumbai-400085, India

Email: gksahu@barc.gov.in

Abstract: Electron beam is preferably used for large scale evaporation of refractory materials. Material evaporation from a long and narrow source providing a well collimated wedge shaped atomic beam has applications in isotopic purification of metals relevant to nuclear industry. The electron beam from an electron gun with strip type filament provides a linear heating source. However, the high power density of the electron beam can lead to turbulence of the melt pool and undesirable splashing of molten metal. For obtaining quiet surface evaporation, the linear electron beam is generally scanned along its length. To further reduce the power density to maintain quiet evaporation the width of the vapour source can be controlled by rotating the electron gun on its plane, thereby scanning an inclined beam over the molten pool. The rotation of gun has further advantages. When multiple strip type electron guns are used for scaling up evaporation length, a dark zone appears between two beams due to physical separation of adjacent guns. This dark zone can be reduced by rotating the gun and thereby bringing two adjacent beams closer. The paper presented here provides the simulation results of the electron beam trajectory and incident power density originating from two strip electron guns by using in-house developed code. The effect of electron gun rotation on the electron beam trajectory and power density is studied. The simulation result is experimentally verified with the image of molten pool and heat affected zone taken after experiment. This technique can be gainfully utilized in controlling the time averaged power density of the electron beam and obtaining quiet evaporation from the metal molten pool.

1. Introduction

Electron beam is preferably used for evaporation of refractory material in metallurgical industry owing to its highest achievable power density, crucible free operation and clean environment. Generally pencil type electron gun is used for applications requiring low evaporation rate (few grams/hr) like optical coating industry. In order to increase evaporation area for application requiring higher evaporation rate, either multiple electron guns are used or single high power pencil beam is scanned over a limited region (scanning amplitude proportional to the spot diameter) [1-3]. In specialized applications like isotopic purification required in nuclear industry, a very high evaporation rate (few kgs/hour) along with stringent conditions of quiet evaporation from a linear vapor source generating a wedge shaped atomic beam is essential. The linear vapor source can be conventionally generated by placing multiple pencil electron guns in a row or by scanning a high power density pencil beam over a line. But, the requirement of quiet evaporation puts an upper limit to the power density used to avoid splashing of liquid metal. As an alternate to multiple pencil type beams or scanning pencil type beam, a true linear electron beam is preferred to obtain high evaporation rate with low incident power density. An indigenously developed strip filament type electron gun [4] is used for this purpose. Such type of electron beam is instrumental in generating high evaporation rate as well as avoiding splashing.

To further scaling up of linear vapor source (increasing length) and to generate a larger molten pool to feed the vapor source, the electron beam is scanned along its length by gainfully utilizing the AC magnetic field generated by the filament heating current [5]. The scanning amplitude is typically of the

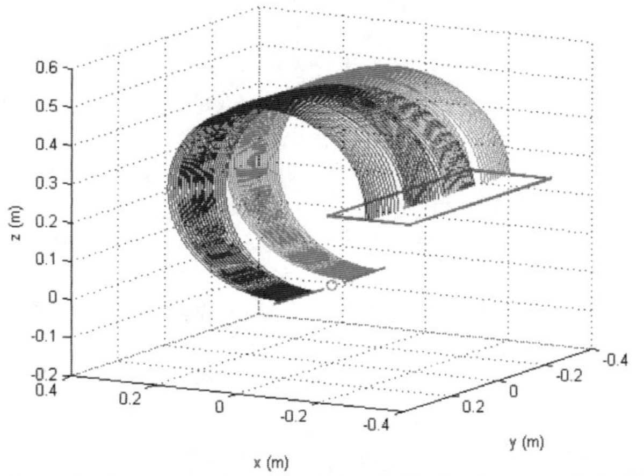

Fig. 1. Electron beam trajectory from two strip type guns deflected in circular path due to axial magnetic field. Each individual beam is scanned along the target length.

order of electron beam length. Still further scaling up of the evaporation length requires use of multiple strip electron guns placed adjacent to each other. When multiple strip electron guns are used, due to finite geometrical spacing between two guns, a dark zone is inevitable on the target surface. To overcome the dark zone, the individual electron gun can be rotated suitably to bring the electron beam from two separate guns closer to each other. A typical configuration of strip electron beam from two strip type filaments traversing through 270^0 bending before falling on the target surface is shown in Figure-1. The rotation of the gun serves another purpose of controlling the power density of the electron beam. When the gun is rotated, an inclined beam is getting scanned over the length thereby increasing the overall width of the beam and hence reducing time averaged power density.

The concept of how the rotation of the gun leads to reducing average power density and also reducing dark zone is illustrated in Figure-2. The filament length of the electron gun in its normal configuration is parallel to the axial magnetic field. The electron beam is deflected in circular fashion

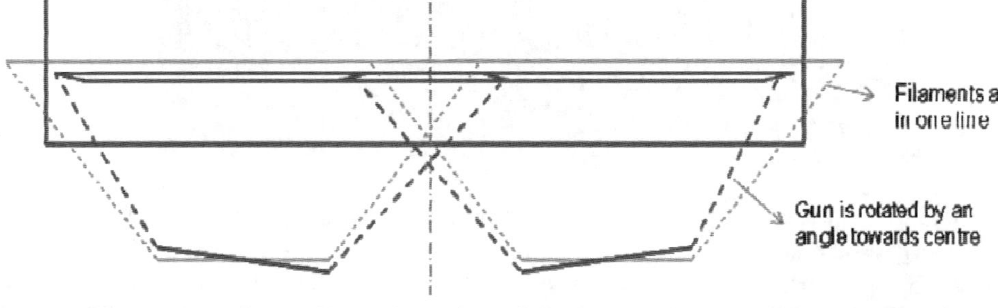

Fig. 2. Illustration of the effect of rotation of the gun on the overall length and width of two strip beams and hence the average power density of the incident electron beam.

(Figure-1) and fall on the target in a line parallel to axis of magnetic field. When such a beam is scanned along the axial direction the overall beam width is the minimum. On the other hand when the electron gun is rotated by a small angle making an angle with the axial magnetic field, the electron beam traverses a spiral path and falls on the target at a location closer to the centre. Moreover, the electron beam on target is also inclined with respect to the axial direction. When such a beam is scanned along the axial direction, the overall width of the beam increases. Although the scanning length is reduced, the time averaged power density over one scanning cycle is also reduced due to increase in overall width of the scanned electron beam.

The paper describes the simulation carried out for the power density distribution and change in beam dimension due to rotation of the electron gun.

2. Experimental Setup

The experimental system consists of a large cylindrical vacuum vessel, inside which a copper block (target of 600 mm x 200 mm surface area) is kept inside water cooled copper crucible. The target is heated by electron beam from two electron guns kept below the crucible. Each electron gun consists of a strip type tantalum filament of length 130mm and width 3mm generating thermionic electrons by I^2R heating. The sheet of electron beam accelerated up to 60 keV, focused and extracted by properly shaped grid electrode and anode. The extracted beam is deflected by 270^0 in a circular path of 260 mm radius by an axial magnetic field (~32 Gauss) generated by a pair of Helmholtz coils. The magnetic field is generated in the vicinity of the filament due to filament heating current. Due to tapering of the magnetic field towards end of the filament due to its finite size, electron trajectories are not only permanently shifted but bunch at one end and rarified at other end. On either side of the electron emitting filament, 30 mm of dummy filament is inserted to avoid non-uniform magnetic field due to filament heating current [5].

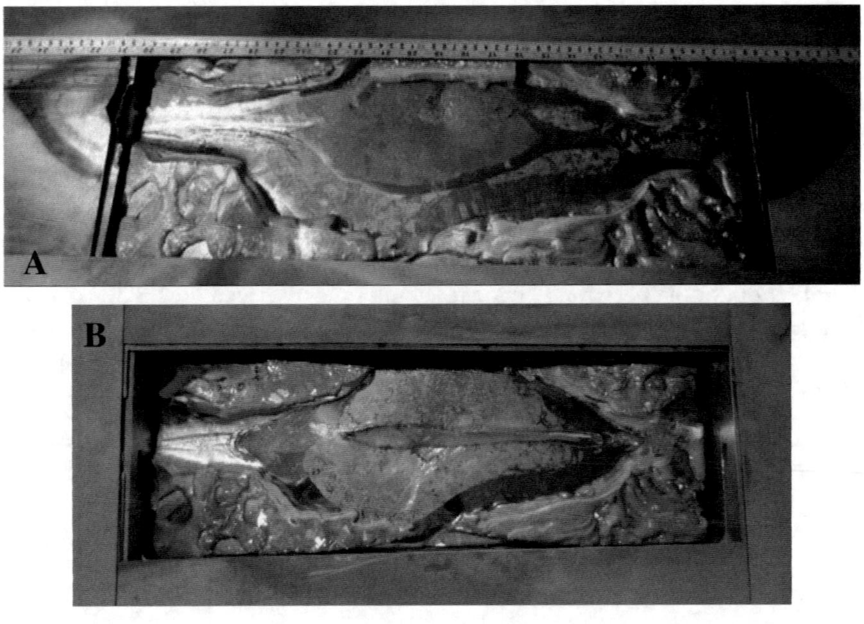

Fig. 3. The Image of the copper target (600 mm long) after impinged with electron beam from two strip type electron guns. A] When the two guns were co-linear. B] When two guns are rotated by 3^0 towards centre. It can be seen that the overall length of the electron beam is > 600 mm in first case (heat effects are seen on the moly plate kept on the crucible) and less than 600 mm in second case as predicted by simulation.

The dummy filament length, grid electrode and physical separation between two guns add to the spacing between the active filaments. In the present configuration a minimum 150 mm separation between active filaments of two adjacent guns is required. The electron beam is scanned along its length by the AC magnetic field generated by the filament heating current. The scanning of the two electron beams are in phase so that a constant separation is maintained between two electron beams while scanning. The amplitude of the scanning depends on the filament heating amplitude. In this experiment an AC filament heating current of 90 Amp rms is used to provide oscillation amplitude of 370 mm (±120mm oscillation and 130mm electron beam length).

A molybdenum sheet is kept on the boundary of the crucible to observe if the electron beam is falling beyond the crucible length of 600 mm. As the electron beam falls on it a heat mark is observed (Figure-3). As in-situ measurement of the electron beam length is not feasible in the present experimental system, the heat mark on molybdenum sheet will be an indication of the length of the scanned electron beam.

Two electron guns are initially placed with the filament length co-linear with each other. In this case, the electron beams from both the guns move parallel to each other. The beam impression on the target is observed with the heat mark on molybdenum sheet and molten pool size of copper target. The same experiment is repeated by rotating the guns by 3^0 on its plane in such a fashion that the filament makes an angle with magnetic field axis. The electron beam from both guns in this case traverses towards the centre of the crucible and the gap between the two beams is reduced.

3. Simulation of Electron Beam and results

Simulation of electron beam from the filament up to the target is carried out by solving Lorentz equation numerically in time step of 1 x 10^{-12} sec. The electrostatic field in the region between anode and cathode, the magnetic field due to filament heating current and the bending magnetic field are incorporated into the Lorentz equation for solving the electron beam trajectory. As filament heating current of 50 Hz frequency is used, the simulation is carried out for 20 msec (one cycle) so that the time averaged beam characteristics can be obtained.

Electron beam is assumed to be emitting uniformly from the surface of active filament (130mm x 3mm). Emission current of 2.5 amp from each gun accelerating to 60 kV is considered. Total power incident on the target is 300 kW (150 kW per gun). The pervience of each gun is ($J/V^{3/2}$), where J is emission current density (0.64 amp/cm^2) and V is electron beam energy (60 kV). Thus the pervience of each gun is 0.044 μ Pervience. The space charge repulsion within the electron beam is neglected as

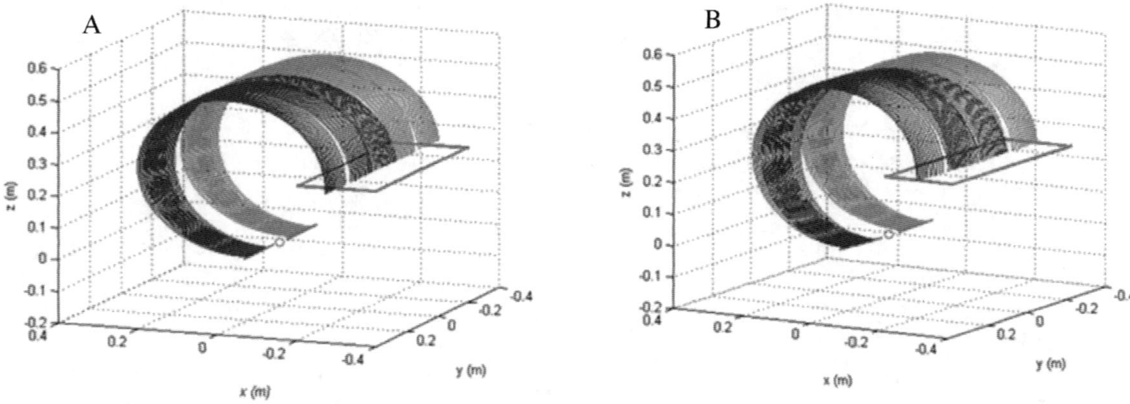

Fig. 4. The simulation result of electron beam trajectory from two strip guns. Trajectory is plotted for 0, +ve peak and –ve peak of filament current. [A] the filaments of two guns are co-linear and the scanned beam length is more than the target length of 600 mm. [B] the filaments are at 3^0 angle and the electron beam is now contained within target length.

the pervience of the electron beam under consideration is much less than the pervience required for onset of charge repulsion within the electron beam.

The simulation is carried out using MatLab program to visualize the trajectory when the gun is rotated by an angle. The simulation is carried out for two cases (Figure-4). One for the filaments from both guns positioned in one line and in second case filament is making an angle 3^0 with respect to the co-linear position. The filament heating current is 90 amp AC is used for both the cases for oscillating the beam along its length. In first case the scanned electron beam length is higher than the second case. Initially at the beginning of electron beam trajectory, the separation between two beams is 150 mm. In first case as both the electron beams travel parallel to each other, the gap of 150 mm is maintained till the target surface. When the electron beams are made to oscillated in phase, the gap between the two beams remains throughout the oscillation period.

In case of rotated beam the gap between two beams is reduced along the beam trajectory from filament to the target. For 3^0 rotation, it is found that the gap 0f 150 mm has reduced to 20 mm. The reduction in gap has advantageous in avoiding the dark zone in evaporation source.

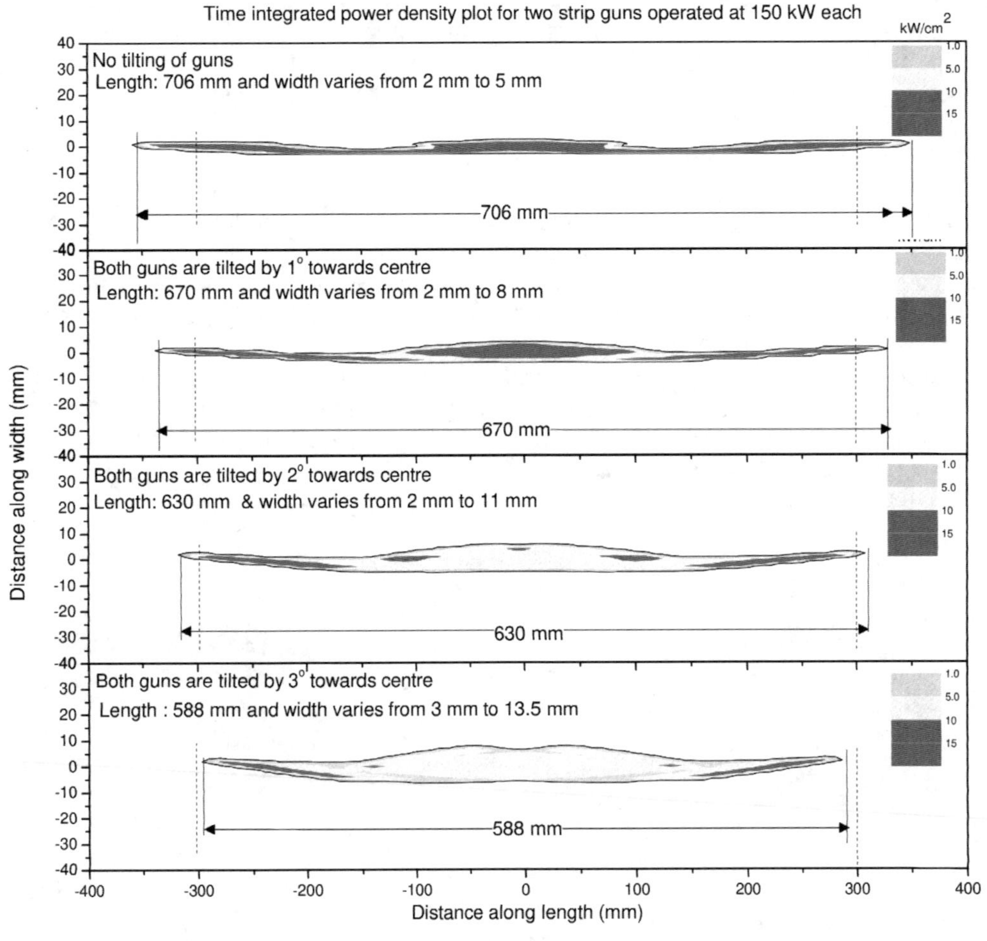

Fig. 5. The time averaged power density and location of the electron beam on a target from two strip type electron guns. The guns are rotated by various angles (0^0, 1^0, 2^0 and 3^0) towards each other. It can be seen that as the angle of rotation is increased, the length of the electron beam reduced from 706 mm to 588 mm where as the average width has increased from 3.5 mm to 8.25 mm. This has resulted in reduction of power density from >15 kW/cm^2 to <10 kW/cm^2.

The evaporation rate form the linear vapor source depends on the source temperature which in-turn is dependent on incident power density, material property of target material and cooling. The rotation of the gun has effect on the overall size of the beam and the time averaged power density. The simulation using in-house developed electron beam trajectory program is carried out for 0^0, 1^0, 2^0 and 3^0 rotation. When the gun is rotated the electron beam image on target is also rotated with respect to the scanning direction. When such inclined beam is scanned, the time averaged width of the beam increases with angle of rotation and correspondingly the time averaged power density is decreased. The incident power density is plotted for two 150 kW electron beams as described earlier. As can be seen from Figure 5, as the angle of rotation is increased, the length of the electron beam reduced from 706 mm to 588 mm where as the average width has increased from 3.5 mm to 8.25 mm. This has resulted in reduction of power density from >15 kW/cm^2 to <10 kW/cm^2.

4. Conclusions

The results of simulation of electron beam from two strip electron guns shows the overall scanned length, width and time averaged power density of the incident electron beam can be changed by rotating the electron guns. This method can thus be gainfully utilized to suit the process requirement like containing the electron beam within the limit of the crucible length, reducing the dark zone between two strip electron beams and also to reduce the time averaged power density for obtaining quiet evaporation.

Acknowledgement

Authors are grateful to V.K.Mago, D.Das, A.K.Das and L.M.Gantayet for several fruitful scientific and technical discussions.

References

[1] Schiller S, Heisig U, and Panzer S, *Electron Beam Technology* (Wiley New York, 1982).
[2] N. Rykalin, A. Uglov, I. Zuev, and A. Kokara, *Laser and Electron BeamMaterial Processing Handbook* ~MIR, Moscow, 1988
[3] P. H. Ron, Bull. Indian Vac. Soc. **1**, 3 ~1998
[4] Thakur K B, Sahu G K, Tamhankar R V and Patel K 2001 *Rev. Sci. Instrum.* **72** 207–15
[5] Sahu G K, Baruah S, Thakur K B 2010 *Vacuum* **85** 236-244

Issues related to nanoparticles generation by exploding wire method

Rashmita Das, Basanta Kumar Das, Rohit Shukla and Anurag Shyam

Energetics and Electromagnetics Division, Bhabha Atomic Research Centre, Autonagar, Visakhapatnam, Pin-530012, India

E-mail: rashmita78@gmail.com

Abstract. Nano science and nanotechnology continue to grow as fields of scientific research and commercial development. At this time, focus is to develop the particles of desired size with significant change of properties than bulk material. But the sample preparation methods involved in each process, surface functionlization and existing characterization techniques remains an important and complicated issue of nanoscience for the development of new materials till the present date. Thus for a science ,that is all about size, these important size issues need to be fully understood as it impacts not only the electronic and optical properties but also environmental and biological interactions. We are producing metal nanoparticles by exploding wire method. Vacuum plays an important role in the production of pure nanopowder as well as to avoid contamination and storage of the material to produce stable nanoparticles. In this paper we will discuss about the complications and intermediate process involved in the production of nano crystalline by exploding wire method.

1. Introduction

Nano technology is the modern incarnation of age old existing synthesis of nano materials like carbon black or beautiful texture in the paintings. With the advancement of science and detection techniques, it became a leading research in all basic cognitive sciences such as physical, earth science, biology and medicines with hope that, by reducing more and more the scale at which matter is manipulated, unprecedented control over material properties can be achieved[1-3] . The existing definition of nano materials shows that, nano material is a material, which is intentionally produced in the nano scale i.e approx 1-100nm, to have a specific property or a composition. This size boundary refers to the nano scale around which the properties of materials are likely to change significantly from bulk one. This impacts many collective properties, i.e colour, electrical conductivity, ability to absorb light, magnetic properties.,which characterize bulk solids? Effort towards the functionalization, formulation, of nano materials is required to make them applicable in industry.

2. Issues related to nano particles generation

Many alternative methods have been proposed to realize production of particles with a controllable morphology: mechanical milling, precipitation, freeze-drying, spray –draying, emulsion based methods, chemical synthesis, and vapour deposition methods. Achievement of desired properties of nano materials depends upon various factors.

2.1 Size

Many fundamental properties are size dependant. Experimental setup for the production of nanoparticles is shown in figure 1and collected sample is shown in figure 2.

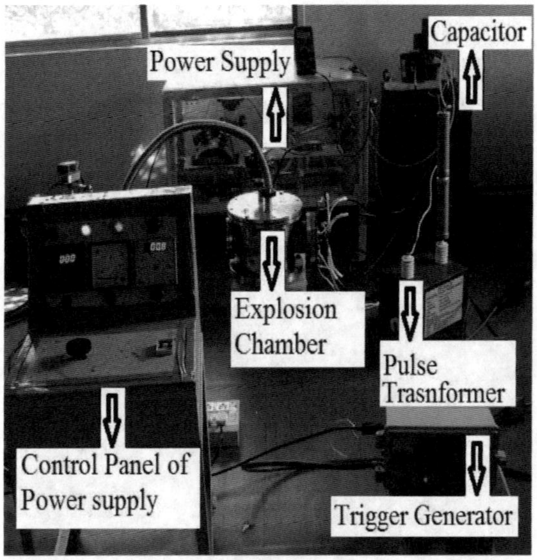

Fig. 1. Experimental setup for nano particles production

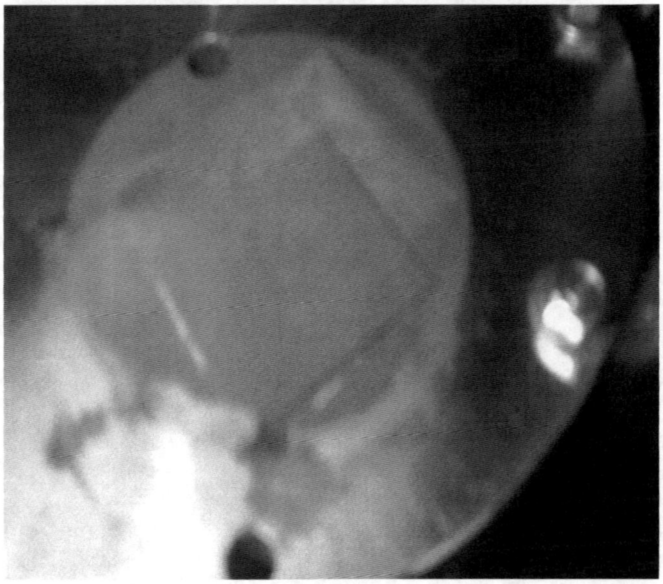

Fig. 2. Photograph of collected sample

On the nano scale, the total free energy is a sum of free energy of the free energy of the bulk and surface of the nanoparticles [4] . Surface free energy contributes largely with the decrease of size. Size affects the particle motion. In aqueous environments and as aerosol, there is a tendency of nanoparticles to form aggregates that are much larger than the primary size particles initially produced. So integral properties needs to be specified by number concentration, surface concentration and mass concentration.

2.2 Shape

Recently, there are number of research papers emphasizing the shape dependant properties variation on the nano scale [5]. Scanning electron microscope SEM) and transmission electron (TEM) and atomic force microscope are useful tools to see shapes and measure the distribution of nanoparticles. SEM and TEM images of collected sample are shown in figure 3 and figure 4.

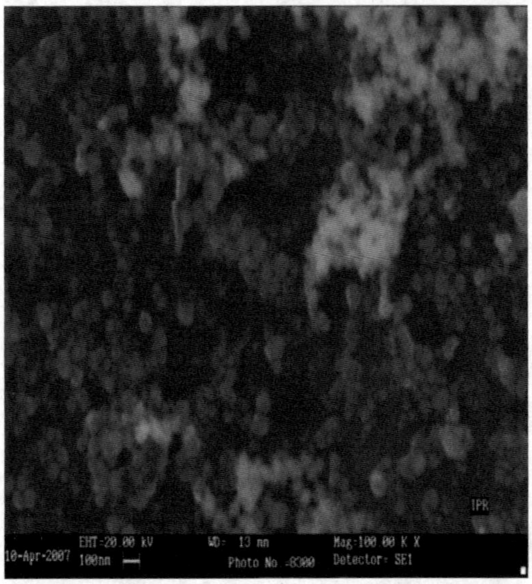

Fig. 3. Image of Scanning Electron Microscope of the nano particles sample

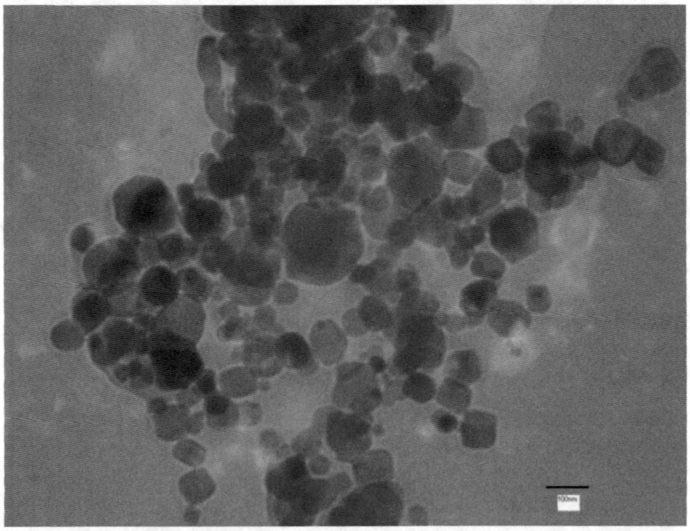

Fig. 4. Image of Transmission Electron Microscope of the nano particles sample

For spherical particles, the measured diameter from different in struments can be related because no corrections need to be made for shapes, but for non spherical particles or agglomerates and aggregates that are irregularly shaped, volume equivalent diameter need to be formulated

2.3 Surface contamination
Surface contamination is the most important part of the processing the materials which has the direct impact on the functionalization. Vacuum plays the important role in every stage of processing like production, storage, characterization. So in situ characterization techniques become important to avoid the surface contamination in larger extent and to avoid the loss of chemical activity of the material..

2.4 Excessive heat production problems
There are many processes where the generation of nanoparticles involves generation of heat as in exothermic reaction as well high temperature in laser ablation, exploding wire method, and plasma assisted nanoparticles generations process etc [6]. The rapid cooling process and sample collection process becomes crucial to get the desired shape size of the nanoparticles and of desired range.

2.5 Yield of the product
In all the above process, though the aim is to produce samples of mono dispersed, single-sized particles, more often than not there is a range of nanoparticles sizes each slightly different than the other than that produced. At this time it is often difficult to make the sample of nanoparticles containing exactly same number of atoms as like any molecular formula.

2.6 Characterization Techniques
To fully understand the properties of functional nano structures , it is necessary to know the portion of all the atoms in the nanostructures..It is difficult to quantify surface properties such as surface coverage and the number of functionalized groups at the interface of nanoparticles. Because of these difficulties, the development of suitable characterization techniques remains an important priority.

2.7 Environmental effect
nanoparticles are also difficult to handle in production process. Their impact on health and the environment has also become a serious concern. The process to recapture nanoparticles when they are released into the environment is difficult.

3. Conclusion
Critical issues involved in the generation of nanoparticles have been discussed. The whole process is a multi component dependant. New characterization techniques are required for the study of nanoparticles surface. The role of aggregation, reactivity of nano material surfaces and the functionality of the material after certain time period remains important challenge in nano material research.

References
[1] P.K Dash and Y. Balto, Research Journal of Nanoscience and Nanotechnology 1 (1) : 25:33, 2011
[2] A.A.Ogwu, T.H.Darma and E.Bouquerel, Electrical resistivity of copper oxide thin films prepared by reactive magnetron sputtering, 24, 172-177, 2007
[3] Rita John and S.Sasi Florence, Structural and optical properties of ZnS nanoparticles synthesized by solid state reaction method, 6(10) : 535-539, 2009.
[4] Vicki H. Grassian, When size really matters: Size dependant properties and surface chemistry of metal and metal oxide nanoparticles in gas and liquid phas environments, J.Phys.Chem.C 2008,112, 18303-18313
[5] Asep Bayu Dani Nandiyanto, Kikuo Okuyama, Progress in developing spray-drying methods for the production of controlled morphology particles: From the nanometer to submicrometer size ranges, Advanced powder technology, 22, 1-19, (2011).
[6] Richard C. Flagan, Melissa M.Lunden, particle structure control in nanoparticle sysnthesis from the vapor phase, Materials Science and Engineering A 204,113-124, (1995)

Journal of Physics: Conference Series **390** (2012) 012052 doi:10.1088/1742-6596/390/1/012052

Experience with helium leak and thermal shocks test of SST-1 cryo components

Rajiv Sharma, Hiren Nimavat, G L N Srikanth, Nitin Bairagi, Pankil Shah, V L Tanna and S Pradhan

Institute for Plasma Research, Bhat, Nr. Indira Bridge, Gandhinagar-382 428, Gujarat

Email: rajivs@ipr.res.in

Abstract. A steady state superconducting Tokamak SST-1 is presently under its assembly stage at the Institute for Plasma Research. The SST-1 machine is a family of Superconducting SC coils for both Toroidal field and Poloidal Field. An ultra high vacuum compatible vacuum vessel, placed in the bore of the TF coils, houses the plasma facing components. A high vacuum cryostat encloses all the SC coils and the vacuum vessel. Liquid Nitrogen (LN_2) cooled thermal shield between the vacuum vessel & SC coils as well as between cryostat and the SC coils. There are number of crucial cryogenic components as Electrical isolators, 80 K thermal shield, Cryogenic flexible hose etc., which have to be passed the performance validation tests as part of fulfillment of the stringent QA/QC before incorporated in the main assembly. The individual leak tests of components at RT as well as after thermal cycle from 300 K to 77 K ensure us to make final overall leak proof system. These components include, Large numbers of Electrical Isolators for Helium as well as LN_2 services, Flexible Bellows and Hoses for Helium as well as LN_2 services, Thermal shock tests of large numbers of 80 K Bubble shields In order to validate the helium leak tightness of these components, we have used the calibrated mass spectrometer leak detector (MSLD) at 300 K, 77 K and 4.2. Since it is very difficult to locate the leaks, which are appearing at rather lower temperatures e.g. less than 20 K, We have invented different approaches to resolve the issue of such leaks. This paper, in general describes the design of cryogenic flexible hose, assembly, couplings for leak testing, test method and techniques of thermal cycles test at 77 K inflow conditions and leak testing aspects of different cryogenic components. The test results, the problems encountered and its solutions techniques are discussed.

1. Introduction

In SST-1 Tokamak, hundreds of electrical isolators have been used in the SST-1 machine to isolate the LHe feed nozzles, which are at ground potential from the magnet system, expected to at high voltage ~ 2 kV during fault condition. Similarly, LN_2 Electrical Breaks connects the whole LN_2 distribution to the 80 K thermal shield hydraulically but at the same time it provides electrical isolation between sub-systems up to the desired voltage level (~ 1 kV). High voltage is expected during any transient operation in SST-1 e.g. eddy current flow during the magnet quench, plasma disruption; ramp-up / down of PF coils etc. As a part of quality control and acceptance criteria, these insulation breaks has been undergone the performance tests at IPR to meet our design criteria and operation needs. In SST-1 Tokamak, the thermal shield '[2]' will be maintained at 80 K with Helium leak tightness of $< 1.0 \times 10^{-8}$ mbar-l/s at service conditions in order to reduce the heat load on superconducting magnet system at 4.5 K.

2. Design considerations and requirements

The major considerations are the helium leak tightness < 1.0 x 10^{-8} mbar l/s at cryogenic temperature. In the 'Table 1', description of the different types of cryogenic systems and component's technical requirements, acceptance criteria at our site will be discussed in this section.

Table 1. Technical requirement and acceptance criteria of different cryogenic components

Cryogenic components	Technical specification
Helium and Nitrogen Electrical Insulation Breaks	▪ Compatibility with temperature at 4.5 K and 77 K ▪ Working condition (inside): 4 bar SHe flow at 4.5 K and 2.5 bar LN_2 flow at 77 K, Compatibility with vacuum (outside): 10^{-5} mbar ▪ Helium leak rate better than 10^{-8} mbar l/s @ 4.5 K and 77 K with Super-critical helium flow, LN_2 flow with to vacuum respectively. ▪ Acceptable Helium leak tightness after minimum 5 thermal shocks using LN_2: < 1.0 x10^{-8} m bar-l/s ▪ Electrical Isolation: ~ 2 kV in Helium breaks and ~ 1 kV (maximum) due to eddy current in LN_2 breaks
80 K LN_2 thermal shield	▪ Type of the shield and material : Bubble type, SS304L ▪ Working temperature range : 80 K – 300 K ▪ Cooling loop : Inlet/supply at bottom and outlet/return at top. ▪ Allowable temp on the shield : ≤ 85K (max.) ▪ Allowable He leak rate : < 10^{-8} mbar-l/s at service conditions.
Cryogenic Flexible hose	▪ Material and structure: S S 316 L, Corrugated Stainless steel structure with a stainless steel braided hose cover of SS 304 wires ▪ Dimension as per DN standard: ϕ 25.5 ID/35 mm OD, length 500-600 mm with both end welded ASME B1.3 Section IX of 1" NB Sch 40 S pipe ▪ Working Operating Pressure Environment: (a) Internal pressure Maximum 12 bar at 300 K and 4 bar at 4.5 K inside hose (b) External vacuum of order of 10^{-5} mbar over hose ▪ Working Temperature range**:** 4.5 K – 300 K, Flow medium: Helium gas/Liquid ▪ Acceptable Helium leak rate: ≤ 1x 10^{-8} mbar- l/s at 300 K and after five thermal cycles from 300 to 77 K and reverse ▪ Bending radius/requirement (Minimum): ~ 0-250 mm (In static & dynamic conditions)
Dummy Heat Exchanger Assembly	▪ Material: Copper, tube dimensions: ϕ 11 mm, 1 mm thick, Length: ~ 80 m ▪ LN_2 fluid inflow condition in the Cu tube and Helium leak rate of heat exchanger at 77 K: <1 x 10^{-8} mbar l/s

3. Thermal shock test at 77 K

As acceptance criterion, we have carried out the thermal cycles of the components from 300 K–77 K - 300 K for 5 cycles considering the European standard ANSI/EIA-364-32C-200. The thermal cycles tests have been performed in two ways:

- Inflow of LN_2 fluid in the components, as shown in Fig. 1(a).
- Dipping the LN_2 fluid into the components chamber, keep submerged in liquid until boiling has subsided to its original rate (approx. minimum 30 minutes) till components is returned to thermal stability at standard ambient conditions and then repeat this cycle up to 5 times as shown in Fig. 1(b).

The liquid nitrogen for the test has been used from the LN_2 tanks of 35-m^3 storage capacity at

2.75 bar (g) maximum operating pressure. We have introduced a metal cryogenic flexible hose with high-pressure vacuum sealing Viton 'O' ring assembly for thermal cycle test connection assembly. In past, we have faced cracking and leaking problem of PVC hose during the thermal shock test and it is vital considering the safety aspect of cryogen hazards.

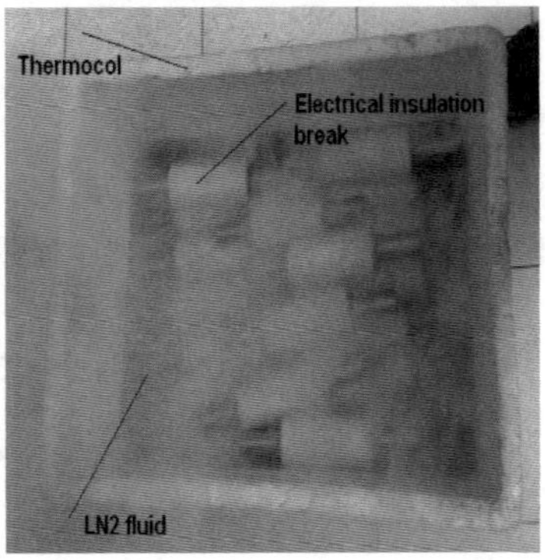

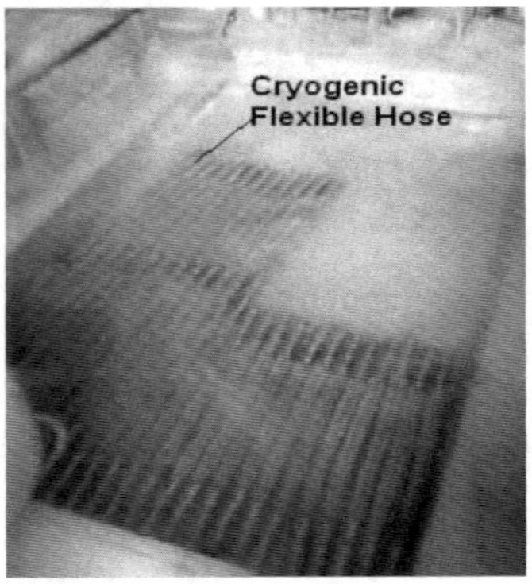

Fig. 1(a). Thermal shock inflow condition Fig. 1.(b). In bath condition

4. Helium leak testing of cryogenic components at 300, 77 and 4.2 K and test results

To ensure the complete leak tightness at cryogenic temperature, leak testing of different cryogenic components are carried out at following three stages:

- Helium leak testing at 300 K of individual insulation breaks, helium flexible hoses, LN_2 panels and dummy heat exchanger assembly to assure elimination of any leak components before it is welded to the respective assembly.
- Helium leak testing of components along with weld joints.
- Leak test of whole assembly at 77 K inflow condition and after passing this temperature, test is to be performed at 4.2 K as per operational needs.

4.1 Electrical Insulation Breaks helium leak test at 4.5 K

The insulation breaks isolator assembly integrated with superconducting toroidal field coil accommodated as shown in Fig. 2. Helium leak observation monitored with the vacuum gauge reading deterioration at different temperature level from 300 K to 4.5 K and RGA instrument was used to detect the helium atomic mass with varying partial pressure of helium with respect to time. The test results are:

- Vacuum level in test chamber at 5.0 K (magnet average temp): 1.7 x 10^{-6} m bar
- Gross helium leak rate from RGA helium acceptable partial pressure: 1.7 x 10^{-7} m bar l/s

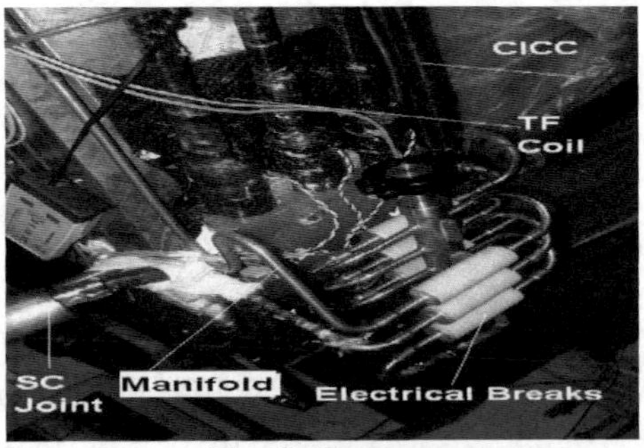

Fig. 2. Test set-up

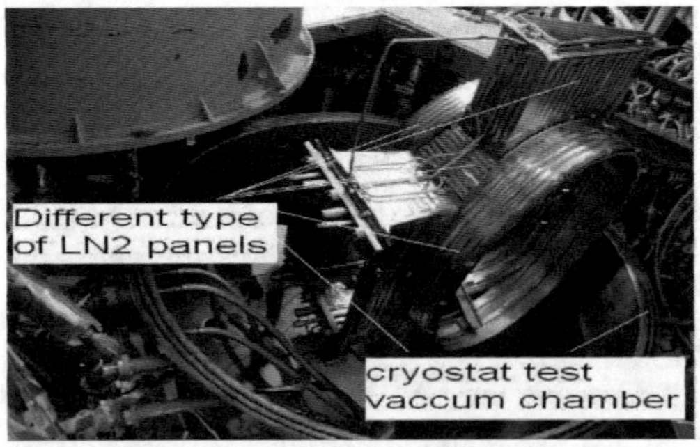

Fig. 3. Test set up of helium leak at 77 K

4.2 LN_2 panel helium leak test at 77 K

Helium leak testing of different panels integrated assembly has been performed in the test chamber as shown in 'figure 3' at 77 K, by pressuring the total hydraulic assembly with the helium gas @0-8 bar, 300 K. The vacuum level and RGA partial pressure readings continuous monitored. If any deterioration in the vacuum and RGA pressure readings beyond the acceptable limit tend to helium leak at 77 K, that assembly after warm upto 300 K and then check the helium leak rate to locate and identify the sections of leak.

4.3 Cryogenic flexible helium hose leak test at 77 K and 4.5 K

The helium leak test was also carried out inflow condition of LN_2, hose welded with supply pipe and test chamber as shown in 'figure 4'. When the temp maintained 77 K then pressurizes the loop with helium gas and any leak can be monitored on helium leak detector online that is mounted on the test chamber, the test result shown below:

- Helium leak rate observed on flexible hose at 300 K: 5.0 x 10^{-9} mbar l/s
- Helium leak rate after thermal shock observed: 3.0 x 10^{-9} mbar l/s
- Helium leak rate LN_2 inflow condition: 1.2 x 10^{-9} mbar l/s

The helium leak tightness, minimum bend radius test as per BS 6501: Part 1:1991 standard of hose assembly in static condition at @4 bar, 4.5 K is to planned to ensure the flexibly criteria for uses, the test set up assembly is shown in 'figure 5'.

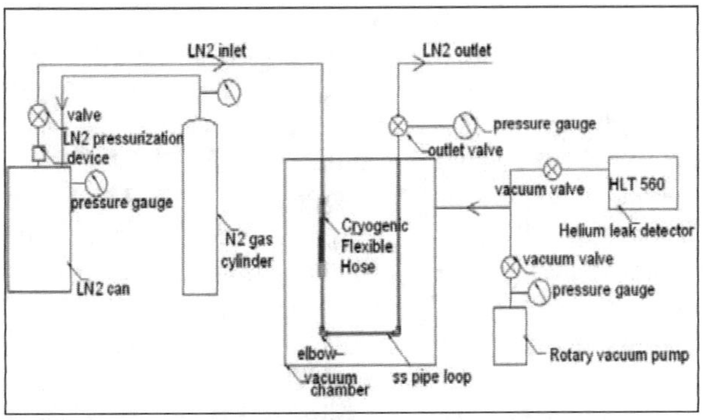

Fig. 4. Schematic of helium leak test of cryogenic flexible assembly hose at 77 K

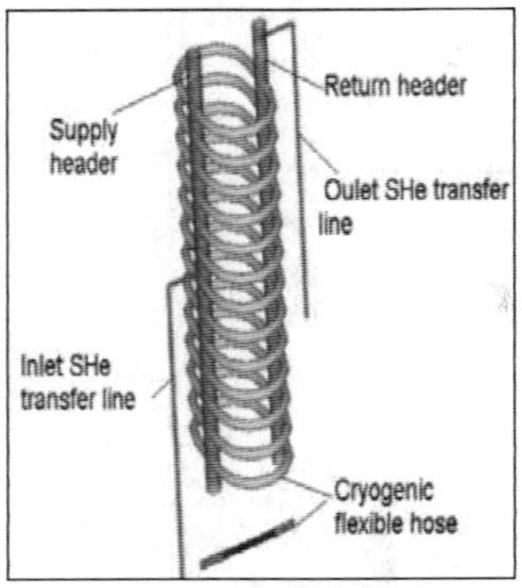

Fig. 5. Helium leak test at 4.5 K

4.4 Dummy heat exchanger helium leak test at 77 K

Helium leak testing at 77 K has been done inflow condition of LN_2 (approx. 100 liter LN_2 consumed) as shown in 'figure 6'. The test results and parameters are listed below:

- Helium leak rate observed at all brazing joints, over tube surface: <1 x 10^{-8} mbar l/s

Pressurized by Nitrogen gas through pressure device: 1-10 psi

5. Problem encountered and solutions techniques:

We have faced problems in the different components described above during thermal shock; leak test at cryogenic temperature, experience of cold leak, investigation and solutions will be discussed in this section.

5.1 Observation of cold leak in helium insulation breaks

When one finds the leaks at cold temperature, it is essential that the test set up assembly warmed up to 300 K and carried out the helium leak test in vacuum mode to localize and confirm the leak. Usually the high-pressure helium gas based sniffer method is adopted '[1]'. The HLT 560 helium leak detector was used in sniffer mode, and soap bubble was observed at one end of the interface of G-10/SS of insulation breaks as shown in 'figure 7 and figure 8'

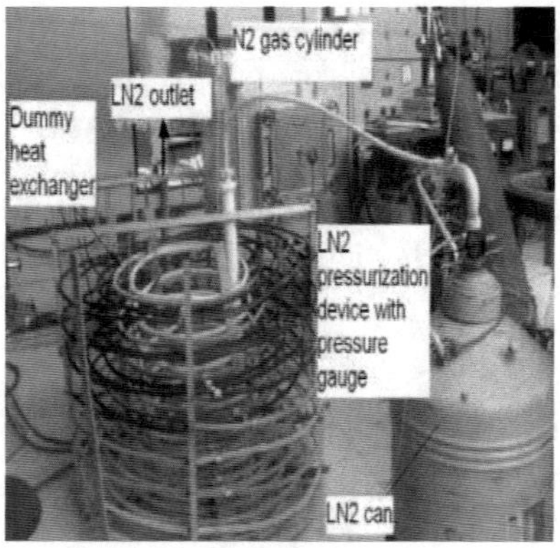

Fig. 6. Test set up of helium leak test inflow condition of LN_2

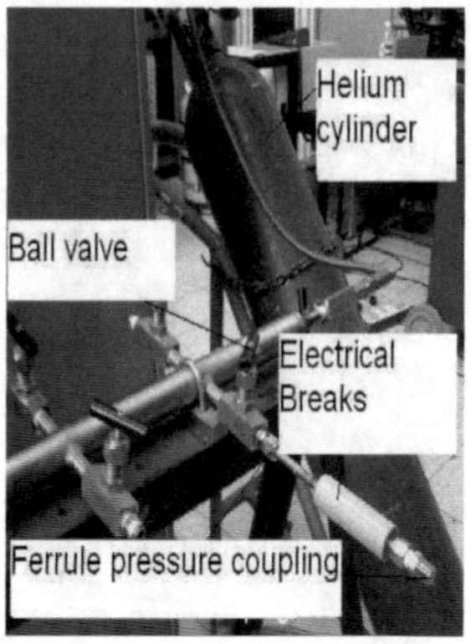

Fig. 7. Electrical break test with high-pressure test set up

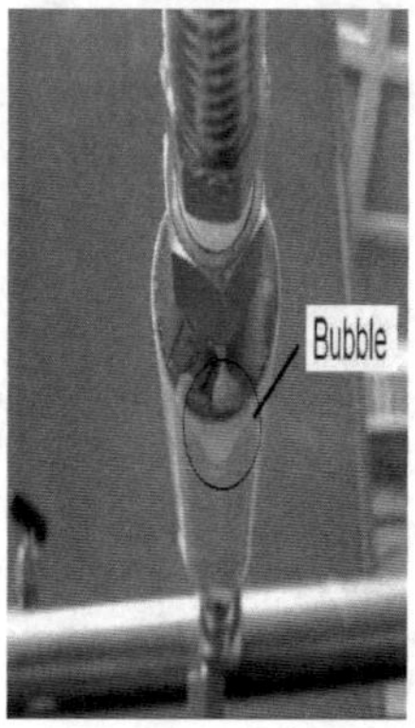

Fig. 8. Bubble leak view (encircled)

5.2 Failure of cryogenic flexible hose at cryogenic temperature

We observed the helium leak in 3 numbers of cryogenic flexible hoses in after thermal cycles test at 77 K, the helium leak rate at 300 K detected of order of ($2x10^{-6}$ to $1x10^{-7}$) mbar l/s at one end of flexible hose, Helium leak rate was detected $7x10^{-5}$ mbar l/s at 300 K the interface weld joint as shown in black circle marked as shown in the 'figure 9'.

In other cryogenic flexible hose of helium leak rate $9.0x10^{-3}$ mbar l/s, the helium leak detector was tripping due to over range of its sensitivity and pressure range < 10^{-3} mbar in leak detector after spraying the helium on suspected weld joint area. In order to locate the leak, masking technique was developed i.e. covered with sealed polythene bag and rest area is sealed by adhesive tape. The initial helium background was $3.0x10^{-8}$ mbar l/s; changed to $5.2x10^{-9}$ mbar l/s after the suspected marked weld area covered with PU bag with adhesive tape. Removing the adhesive tape at one location helium background was increased to $2.5x10^{-8}$ mbar l/s indication of leak in that particular location as shown in 'figure 9'. We have repeated this on all location by masking and unmasking of the entire weld to find and locate the leak. Thus the leak was located and cryogenic flexible hoses were sent for re-welding or replacement.

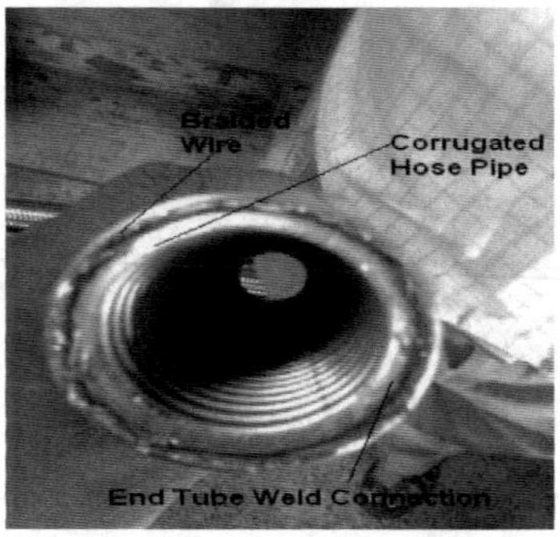

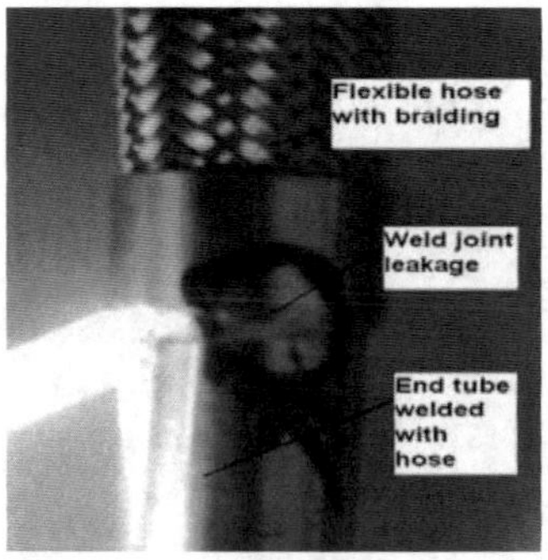

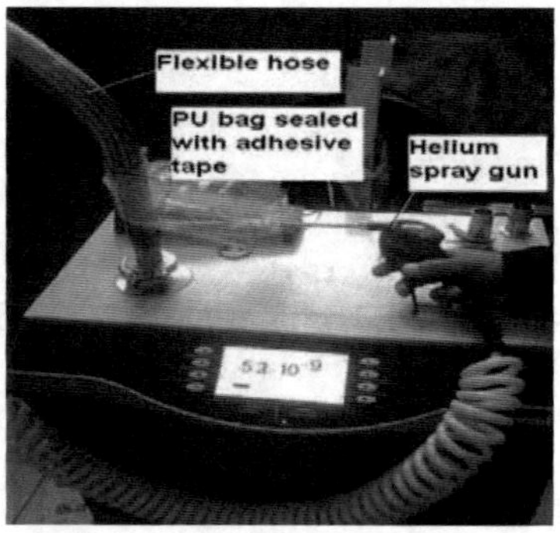

Fig. 9. Cryogenic flexible hose showing internal interfaces of welding with marked enlarged view of leak zone and detected by masking-unmasking technique

5.3 Problems during thermal cycles test of LN_2 panels

In order to save the LN_2 fluid consumption, we have performed the test 1-4 numbers of in batch. During the test of LN_2 shield top port panel section type, which consists of 2 panels, both panels were not able to cooled to LN_2 temp during helium leak at 77 K experiment, the same set up of panels had not also cooled in thermal cycle test. After that we cut the U pipe inter connection of both panel, single panel has been cooled entirely as shown in 'figure 10'. We analyses that possible factors/reasons could be for this phenomenon are listed in the following paragraph:

Journal of Physics: Conference Series **390** (2012) 012052 doi:10.1088/1742-6596/390/1/012052

(i) More pressure, velocity and flow requirement of LN_2 fluid through panels due to its not straight and plane geometry construction as hydro-forming bubble type or some portion may be blocked during fabrication (ii) high heat flux rate requirement (iii) thermal resistance between LN_2 fluid and SS panel surface (iv) parallel cooling hydraulic is recommended for application

5.4 Problems and solution during helium leak test at 300 K with silicon flexible tube

For helium leak testing at 300 K, silicon tube was used in past, during test we experiences that due to flexibility, a small components can not be stand in vertical position as in case of Electrical breaks, application of sealing material as vacuum grade tacky tape at the interfaces of tube to component, holding the electrical breaks by hand during spraying helium for test, replacement of tacky tape after number of times of its uses. As large quantities of cryogenic components had to test, vacuum grade especially grade high-strength silicone rubber bush sealing was designed and used for the leak test, all problems specified above were eliminated. The helium background was also improved from $1.0x10^{-8}$ mbar l/s to 10^{-9} to 10^{-10} mbar l/s range and much time reduction of leak test. The photographs of silicon tube and silicon bush arrangement for helium leak test are shown in Fig. 11 and Fig. 12.

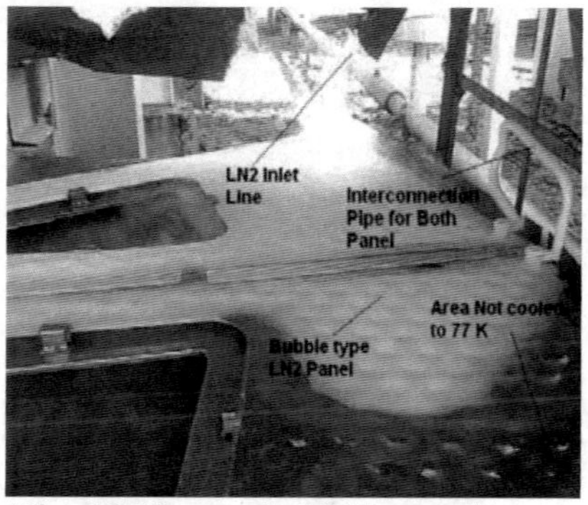

Fig. 10 Noncooled LN_2 panel

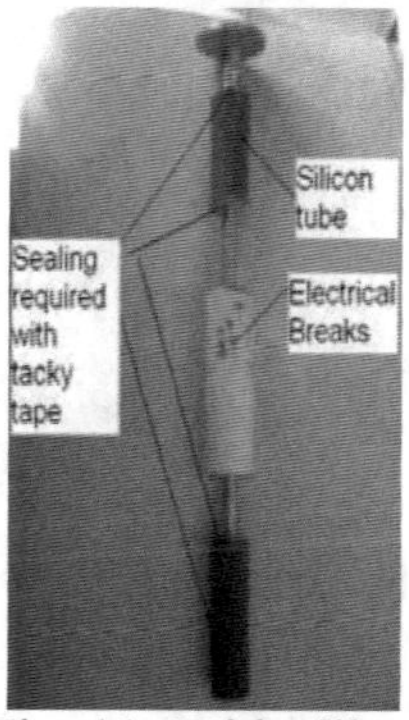

Fig. 11. With silicon tube

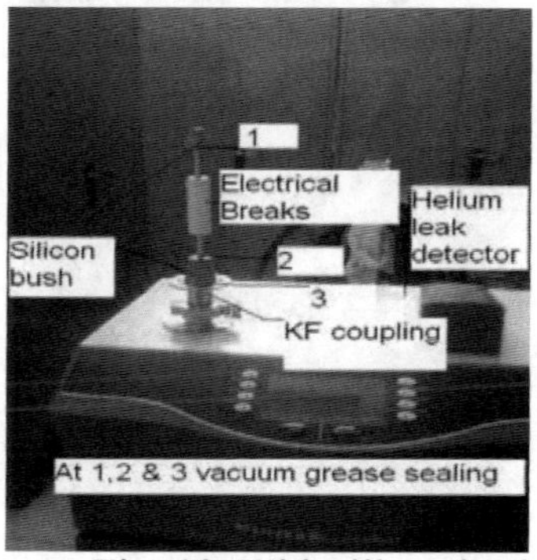

Fig. 12. With silicon bush

6. Conclusion

For the quality control, performance validation and to ensure the helium leak tightness at cryogenic temperature, SST-1 cryogenic components viz. LN_2 panels, Electrical Breaks, Cryogenic flexible hoses were subjected to thermal cycles test at 77 K in actual operating scenario prior to the integrated assembly in the SST-1 machine. We have in-house developed the test facility and successfully carried out the testing of large numbers as 184 numbers of LN_2 panels, 400 numbers of electrical Insulation breaks of helium and nitrogen service, 50 numbers of cryogenic flexible hoses. Cold leaks which were developed at cryogenic temperature and after thermal cycle at 77 K, a masking/unmasking and thermal cycle-pressure helium test techniques were developed to locate the leaks and make cold leaks accessible to 300 K temperature environment. The cryogenic flexible leakage at the weld interface joint, we analyses that due to low temperature, inner bellow pipe contracts and exert force on the outer pipe since end pipe, inner pipe and braded wire are welded together. The electrical breaks, LN_2 panels have been installed, tested and integrity with the SST-1 machine found satisfactory and acceptable.

Acknowledgement

The authors are grateful to acknowledge the help and supports provided by the Cryo crews, Magnet division, Vacuum Division and Device Integration Team (DIT) members during the cold testing of SST-1 machine.

References

[1] K. Zapfe *Leak Detection Technical Note* 2007-08 Feb2007
[2] Manoj Kumar Gupta, Vijay Bedakihale, Ziauddin Khan, V.L. Tanna and D. Chenna Reddy Liquid Nitrogen Thermal Shield for SST-1 *NSC* ***22*** 513-118, 2008

International Symposium on Vacuum Science & Technology and its Application for Accelerators IOP Publishing
Journal of Physics: Conference Series **390** (2012) 012053 doi:10.1088/1742-6596/390/1/012053

Brazing of photocathode RF gun structures in Hydrogen atmosphere: Process qualification, effect of brazing on RF properties and vacuum compatibility

Ajay Kak[1], P. Kulshreshtha[2], Shankar lal[3], Rakesh Kaul[4], P. Ganesh[4], K.K. Pant[3] and Lala Abhinandan[1]

[1] Mechanical & Optical Support Section, RRCAT, Indore, India
[2] Accelerator Magnet Technology Division, RRCAT, Indore, India
[3] Free Electron Laser Laboratory, MAASD, RRCAT, Indore, India
[4] Laser Material Processing Division, RRCAT, Indore, India

Email: lala@rrcat.gov.in, kakajay@rrcat.gov.in, shankar@rrcat.gov.in

Abstract. In this paper, we report on the development of a brazing process for an ultra-high vacuum (UHV) compatible photocathode RF gun structure developed at our Centre. The choice of brazing alloy and its form, brazing clearance between parts to be joined and the brazing cycle adopted have been qualified through metallographic examination of identical joints on an OFE copper prototype that was cut open after brazing. The quality of brazed joint not only affects the UHV compatibility of the gun, but also influences the RF parameters finally achieved. A 2-D electromagnetic code, SUPERFISH, was used to predict the variation in RF parameters before and after brazing considering actual brazing clearances provided between the parts to be joined. Results obtained from low power RF measurements on the brazed gun structure confirm the integrity of the brazed joints and show good agreement with those predicted by electromagnetic simulations. The brazed gun structure has been leak-tested and pumped down to a vacuum level limited by the vacuum compatibility of the flange-fittings employed in the setup.

1. Introduction

Laser driven photocathode RF guns are used to generate ultra short pulses of electrons with high peak currents and low emittances required for advanced accelerator applications such as Free Electron Lasers (FELs), colliders, ERLs, Inverse Compton Scattering, Plasma Wake field accelerators, etc. [1]. In a Laser photocathode RF gun, electrons generated by shining a laser beam on a cathode are accelerated to high energies by an electromagnetic field setup in the gun by feeding high radiofrequency (RF) power to the structure. A typical design of photocathode RF gun is the 1.6 cell BNL/SLAC/UCLA type S-band gun, consisting of two cells- a full-cell and a half cell, each of which has ports for different use. The two cells and the ports are machined from Oxygen Free Electronic (OFE) grade copper and AISI Stainless Steel (SS) 316L respectively, and brazed together after tuning of RF properties of the independent cells to form a photocathode RF gun structure. The requirement of a good quality electron beam from a photocathode gun imposes stringent conditions upon the RF parameters, and consequently on the geometrical features of the gun, and a small variation in RF parameters due to different possible reasons can seriously affect the quality of electron beam from the gun[2,3]. Variation in geometrical dimensions of gun cells due to brazing is one of the most important contributors to the deterioration in RF parameters of a tuned gun structure. Therefore, adequate care has to be taken while designing the brazed joints and finalizing the brazing cycle to minimize such deleterious effects of distortion due to brazing.

Hydrogen atmosphere is preferred for brazing copper since the oxides of copper can easily be removed by reactive hydrogen at brazing temperatures. The issue of hydrogen diffusion in copper during the brazing process can be resolved by subsequent baking of the brazed component in a vacuum baking furnace. Brazing of a photocathode RF gun structure involves joining of two RF cavities (full-cell and half-cell) and one rectangular waveguide made of Class 1 OFE Copper (Cu), six cylindrical ports and two seal plates made of AISI SS 316L. Since brazing of a photocathode RF gun involves joining of copper to SS, gold based alloys are usually employed worldwide at brazing temperatures of ~ 1000° C. In order to avoid such high brazing temperatures, all SS components of the photocathode RF gun were electroplated with copper and Eutectic Copper-Silver (72%-28%) filler alloy (BVAg8) was used in the brazing. Before brazing of the actual gun structures, similar joints were qualified by brazing of multiple dummy samples with SS to copper joints.

Brazing of a photocathode RF gun structure is complex as the profile, cross sectional area and angle of various joints are different. The requirement of minimization of variation in RF parameters of gun due to distortion caused by brazing makes the task further challenging. To minimize these distortions and to achieve UHV compatibility, proper fixtures were designed and brazing cycles were finalized after testing over multiple full-scale gun prototypes made of Electrolytic Tough Pitch (ETP) copper before brazing the final gun structures [4]. The volume of brazing filler required was determined by calculating the volume of joints to be filled and care was taken to minimize the flow of filler inside the gun cells, since this can seriously compromise the high electric field gradient supporting capability of the gun. This has been confirmed by metallographic test of brazed joints of a gun prototype which was brazed by employing the same brazing cycle with the same amount of filler alloy as was used in the final gun structures. The brazing procedure adopted and calculations of brazing joint and filler volumes for the photocathode RF gun is given in section 2. Section 3 discusses the metallographic evaluation of the brazed joints, which is followed by a discussion of predicted and measured variation in RF parameters of the gun structure due to brazing in section 4. UHV compatibility of the brazed gun structure is discussed in section 5 followed by conclusion and discussions in section 6.

2. Brazing procedure and filler requirement estimation

Brazing of photocathode gun was carried out in two brazing cycles in a hydrogen furnace (HT 1800 G, Linn, Germany) having hot zone of 30cmx28cmx30cm. The same filler alloy (BVAg8) was employed for both cycles. In the first cycle, the two cells were brazed together and 6-ports were simultaneously brazed on the two cells. The second cycle involved brazing of the waveguide at the RF port on the full cell. All joints brazed together in the first cycle were tubular butt-lap joints. The second cycle involved brazing of a rectangular butt-lap joint. While achieving the required brazing clearance in the butt joint was relatively easy, care had to be taken during machining of the port openings on the gun cell and the ports themselves to ensure brazing clearance of the order of 30-50 μm in the tubular lap joint region. This was ensured by inspection of each component after machining and after electroplating. After each brazing cycle, leak test of all brazed joints were performed using helium leak detector [4].

The different butt and lap joints involved in brazing of the photocathode gun structure are shown in Fig.1 and 2. The formula used to calculate volume of brazing joints and respective filler volumes required for each joint are given in table I. Copper silver eutectic (Cusil, BVAg8) was used as brazing alloy, in form of a ring of 0.1 mm thick foil for the butt joints and a ring of wire of 1mm diameter for lap joints. From practical considerations 30-50 % extra volume of filler alloy was taken.

The brazing cycle and fixtures used during the brazing are important to ensure UHV compatibility of the joints. Since the first cycle involves brazing of six orthogonal joints and two joints at 22.5 degree, proper fixturing is a crucial element of the brazing process. Fixtures used were machined out of AISI SS 304 and the tuner ports were held in position using tungsten wire. More details of the fixtures and the brazing cycle employed are discussed in Ref. 4.

Table 1: Calculations of volume of brazing joints.

Joint	Formula used to calculate joint volume	Volume of joint (mm^3)
Half cell - seal plate	$\pi\left(\frac{OD_{lip}^2-OD_{half\,cell}^2}{4}\right)t+\pi\left(\frac{OD_{lip}^2-ID_{seal\,plate}^2}{4}\right)\times h_{seal\,plate}$	449.76
Full cell – half cell	$\pi\left(\frac{OD_f^2-ID_f^2}{4}\right)t-\left(\frac{OD_f-ID_f}{2}\right)b_{wo}t+d*b_{wo}t+\pi\left(\frac{ID_f^2-OD_{iris}^2}{4}\right)L_{iris},.$	401
Full cell – tuner port	$\pi\left(\frac{ID_h^2-ID_p^2}{4}\right)t+\pi\left(\frac{ID_h^2-OD_p^2}{4}\right)L,$	25.30
Full cell - vacuum port		62.47
Full cell – beam exit port		27.52
Full cell - half cell		401
Half cell – laser port		35.63
Gun - Waveguide	$2(B_f-B_{wi})\times(A_f-A_{wi})\times t+2(B_f-B_{wo})\times(A_f-A_{wo})\times h$	401.45
Waveguide – RF port flange	$2(B_f-B_{wi})\times(A_f-A_{wi})\times t+2(B_{Rf}-B_{wo})\times(A_{Rf}-A_{wo})\times h$	365.58

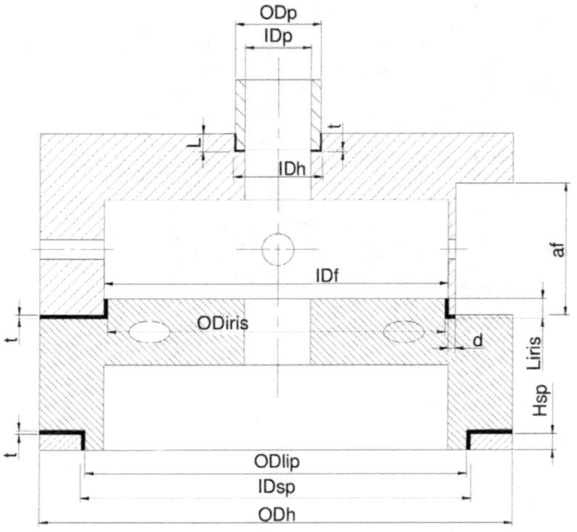

Fig.1: Details of different brazing joints in photocathode gun brazed in first cycle.

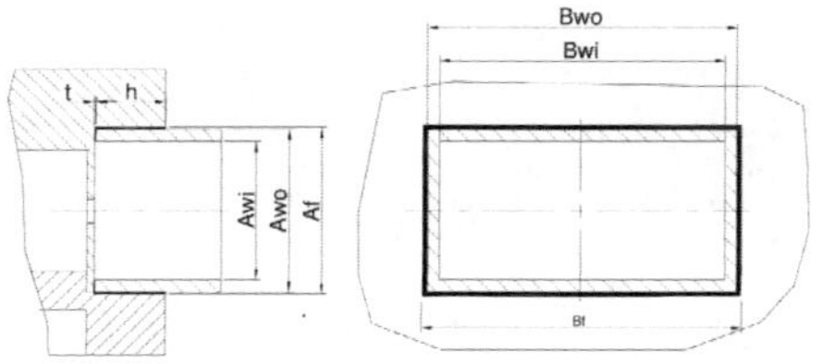

Fig. 2: Details of photocathode gun to waveguide brazing.

3. Evaluation of brazed joints

The photographs of gun assembly after first and second cycles are shown in Fig. 3 (a) and (b), while a cross-section of a prototype gun cut open for metallographic evaluation is shown in Fig. 3 (c). This section includes Cu-Cu as well as Cu-Cu plated SS joints which were examined under a microscope at 20X and 50 X to evaluate the depth of penetration of brazing alloy and the micro-structure of the joint interface. Micrographs of Cu-Cu and Cu-SS brazed cross-sections are shown in the Fig. 4. Full penetration of the brazing alloy, confined to joint interface is seen in Cu-Cu joint, while in the case of Cu-SS joint penetration is observed along the grain boundaries into the copper matrix. All the joints were found to be micro-structurally sound and free of defects like porosity and lack of wetting. Thickness of braze joint in the case of Cu-Cu is in the range of 20 μm with some excursions up to 60 μm, while in the case of Cu-SS joint it is in the range of 25 μm with some excursions up to 40 μm. Figure 4B clearly shows penetration of the filler alloy into the copper matrix along the grain boundaries for Cu-SS joint.

Fig. 3: Gun assembly after: (a) first brazing cycle, (b) second brazing cycle and (c) cross section of a brazed gun prototype

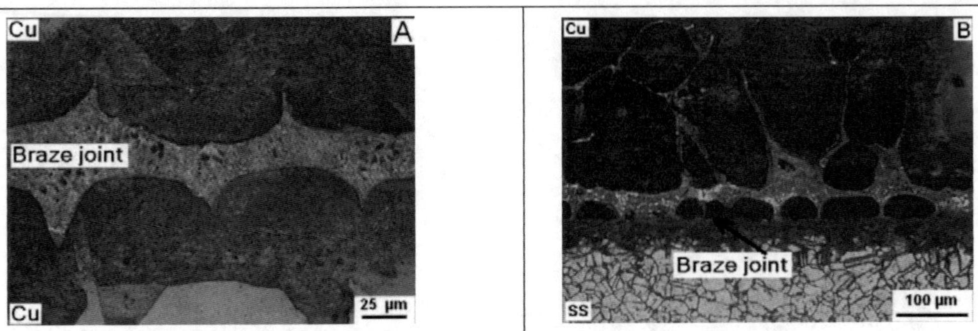

Fig. 4: Micrographs of: (A) copper to copper and (B) copper to SS brazed joint

4. Effect of brazing on RF properties

From brazing considerations, a clearance of ~ 25-50 μm is usually maintained at the location of the joint between the full and half-cells of a photocathode gun as shown in Fig. 5. This clearance is present during tuning, but gets filled up with the brazing filler after brazing, which causes a change in independent full cell frequency (f_f) and waveguide to independent full cell coupling coefficient (β_f) before and after brazing. SUPERFISH simulations show that f_f decreases linearly with brazing clearance with a gradient of -0.127 MHz/10 μm, while β_f after brazing becomes 1.17 times its value before brazing [3]. This change is like a step and is independent of variations in the value of the brazing clearance. Low power RF measurements on the brazed gun structure confirm that experimentally measured variation in RF parameters is very close to the predictions as given in table II, which further confirms that the brazed joints show good RF continuity, distortion due to brazing is nominal, and calculation of filler volumes is appropriate.

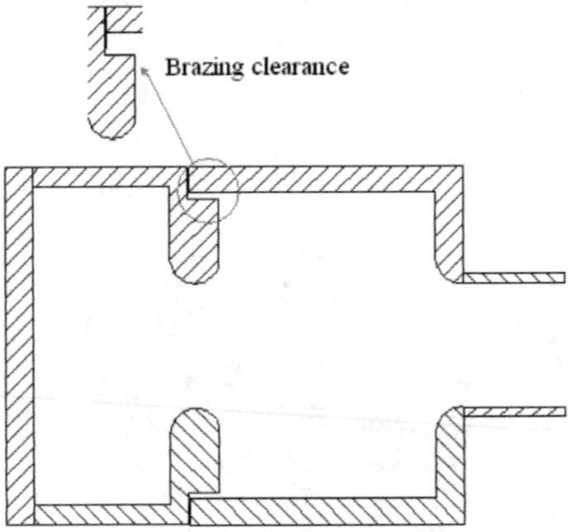

Fig. 5: Brazing clearance between the full-cell and half-cell

Table II: Comparison of variation in RF properties of photocathode gun due to brazing.

Variation in Parameters	Prediction		Experimentally measured	
	Gun 1	Gun 2	Gun1	Gun 2
$\Delta\beta_f$	1.17	1.17	1.26	1.27
Δf_f (MHz)	0.457	0.457	0.54	0.44

5. Testing of UHV compatibility

Before brazing of final gun structures, all brazed prototypes were subjected to leak testing using a Mass Spectrometer Lead Detector (MSLD) (Alcatel, ASM 142) using Helium. The measured leak rate at all joints was $\leq 5\text{x}10^{-10}$ mbar l/s, which was acceptable. Figure 6 shows the photocathode gun being leak-tested and a test setup where a Sputter Ion Pump of 140 l/s is employed backed by a turbo-molecular pump [Pfeiffer, TSU 260D] of 210 l/s to ultimately achieve a vacuum level of $3\text{x}10^{-8}$ mbar within 48 hours.

6. Conclusion

A technique to braze a photocathode RF gun in hydrogen furnace with predicted variation in RF parameters along with UHV capability has been developed, qualified and successfully used to braze multiple photocathode RF gun structures. Metallographic examination confirms that all joints are micro-structurally sound and free of defects. Further study to qualify the brazing joints is presently underway.

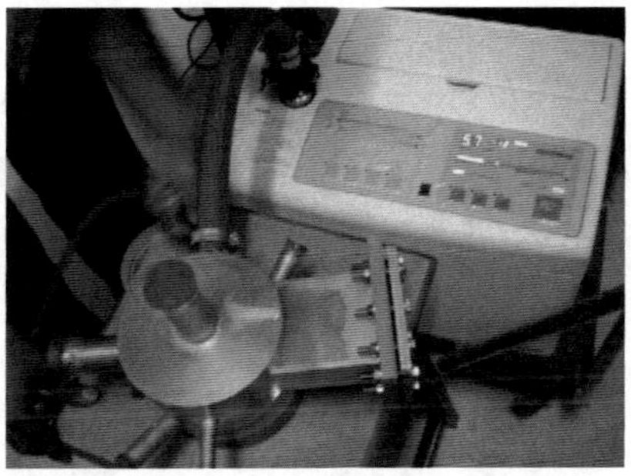

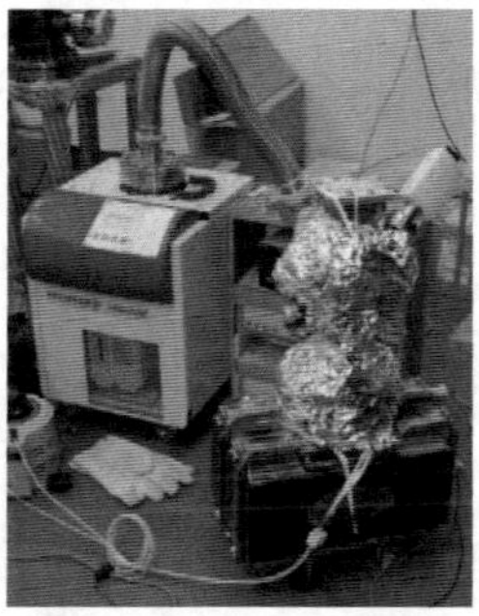

Fig.6: Photocathode RF gun being leak tested, and on the vacuum test setup

Acknowledgement

Technical assistance of Mr. Rajkumar B. Bhowate, FEL Lab and Chemical Treatment Facility, RRCAT is gratefully acknowledged

Reference

[1] Palmer D T, Miller R H, Winick H, Wang X J, Batchelor K, Woodle M H, and Ben-Zvi I, *Proc. SPIE* **2522**, 514 (1995).

[2] Shankar Lal, Pant K K and Krishnagopal S,"A new two-step tuning procedure for a photocathode gun", *Nucl. Instrum. Methods Phys. Res.* A **592**,180 (2008).

[3] Shankar Lal, Pant K K and Krishnagopal S,"A novel scaling law relating the geometrical dimensions of a photocathode radio frequency gun to its radio frequency properties", *Rev.Sci. Instrum.* **82**, 123304 (2011).

[4] Kak Ajay, Kulshrestha P, Shankar Lal, "Development of brazing technique for brazing a photocathode gun", *Proceeding of InPAC* 2009, paper 233.

Journal of Physics: Conference Series **390** (2012) 012054 doi:10.1088/1742-6596/390/1/012054

Vacuum Control Systems of the Cyclotrons in VECC, Kolkata

Anindya Roy[1], Javed Akhtar, R. C. Yadav, R. B. Bhole, Sarbajit Pal, D. Sarkar, R. K. Bhandari

Variable Energy Cyclotron Centre, Department of Atomic Energy, India
1/AF, Bidhan Nagar, Kolkata – 700064, India

E-mail: r_ani@vecc.gov.in

Abstract. VECC has undertaken the modernization of the K-130 Room Temperature Cyclotron (RTC) (operational since 1978) and commissioning of K-500 Superconducting Cyclotron (SCC) at present. The control system of RTC vacuum system has been upgraded to Programmable Logic Controller (PLC) based automated system from relay based manual system. A distributed PLC based system is under installation for SCC vacuum system. The requirement of high vacuum in both the cyclotrons ($1x10^{-6}$ mbar for RTC and 5×10^{-8} mbar SCC) imposes the reliable local and remote operation of all vacuum components and instrumentation. The design and development of the vacuum control system of two cyclotrons using the Experimental Physics and Industrial Control System (EPICS) distributed real-time software tools are presented.

1. Introduction

At VECC, Kolkata the K-130 Room Temperature Cyclotron has been delivering light heavy ion beams up to 15 MeV/nucleon since 1978. The K500 Superconducting Cyclotron, under commissioning activity at the centre, is expected accelerate heavy ion beams to energy up to 80 MeV/A for fully stripped light heavy ions and about 10 MeV/A for heavy ions [1]. The vacuum systems of the two cyclotrons are entirely different from the point of view of requirement, construction and operational philosophy.

1.1. RTC Vacuum system

The main vacuum system of RTC (as shown in Figure 1) which maintains a vacuum level of $1x10^{-6}$ mbar inside 23 m^3 volume of Dee tank and resonator tank, was designed as two redundant sets of pumping system using oil diffusion pumps backed by rotary pumps and roots pumps. The oil (DC-704) diffusion pumps, having a pumping speed of 41000 litre/sec of air, are being operated with a fore pressure tolerance of $4x10^{-1}$ mbar. There are Freon cooled (approx. -60ºC) chevron baffles at the inlet of the diffusion pumps to arrest oil migration into the high vacuum chamber. The backing pumps i.e. two rotary pumps and a roots pump, of individual pumping systems are interconnected through twenty two electro-pneumatic Ultra High Vacuum (UHV) gate valves to ensure maximum availability of the system. Also, there are various vacuum gauges e.g. hot cathode gauge, cold cathode gauge and pirani gauge, to measure vacuum level at various points of the system. The individual pumping systems are capable to operate from 'Roughing' to 'Backing' mode. The beam line vacuum system of

[1] To whom any correspondence should be addressed.

RTC comprises of several pumping modules. Each pumping module consists of diffusion pump backed rotary pump and associated isolation valves and gauges. The control system of main vacuum system is upgraded to PLC based system from initial relay based system to incorporate auto-operation and PC based remote monitoring and control of the system using Supervisory Control Software. A similar control system is under installation for beam line vacuum system.

1.2. SCC Vacuum system

The SCC vacuum system is divided into four sub-systems i.e. beam chamber, liner, cryostat outer vacuum chamber (OVC) and beam lines. In order to maintain 5 x 10^{-8} mbar inside the SCC beam chamber, several technically dry (oil-less) pumping modules and cryo-panels are used. Each pumping module of the system consists of a Turbo-pump, backed by a Scroll Pump integral with a VPI (Vacuum Pump Isolation) Valve. There are liquid Nitrogen and liquid Helium cooled cryo-panels installed within the median plane valley of the beam chamber. The liquid Nitrogen Cooled panels remove moisture and gas species with higher molecular weight. The liquid Helium cooled panels, protected by liquid nitrogen cooled chevron baffles, remove mainly Nitrogen, Oxygen and Argon. The bottom surfaces of the liquid cooled panels are bonded with activated charcoal for removing lightest gases, Hydrogen, Helium, and Neon, mainly by cryo-adsorption. The liquid Helium cooled panel is by direct heat source from room temperature surfaces.

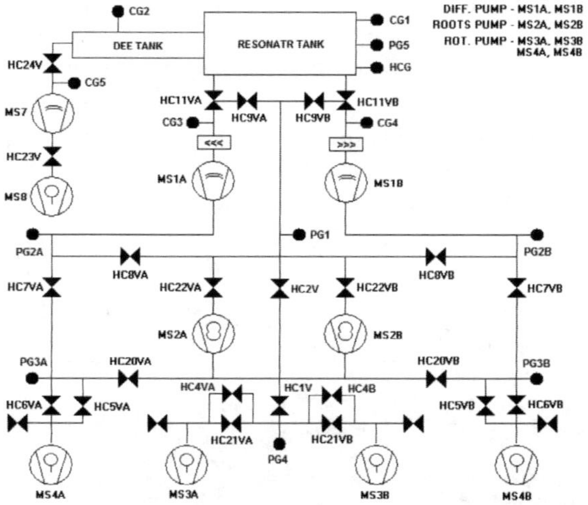

Figure 1. The schematic of Vacuum system of RTC

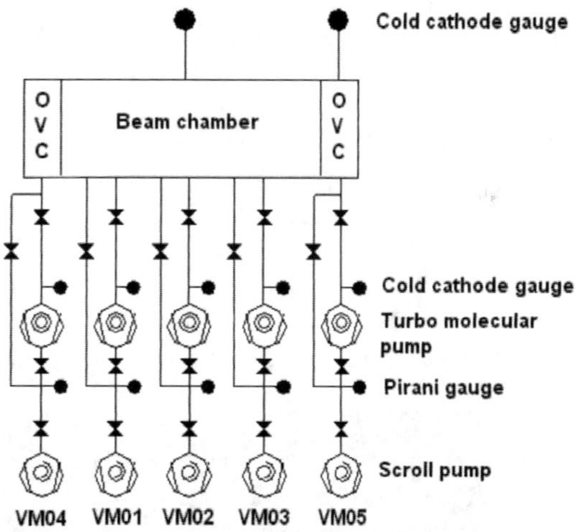

Figure 2. The schematic of Beam chamber and Cryostat OVC Vacuum subsystem of SCC

The liner vacuum system is comprised of two rotary pumps, four valves and four Pirani gauges to maintain a pressure in the order of 10^{-2} mbar inside liners. The OVC and beam line vacuum system consist of 23 pumping modules. All subsystems, except liner subsystem, are designed to have redundancy at pumping module level. The schematic of the beam chamber and cryostat OVC subsystem is shown in Figure 2.

2. Control System

The control systems of the vacuum system of two cyclotrons are comprised of several standalone PLC nodes. At present main vacuum system of RTC is controlled by a PLC node situated at the Pit area of the cyclotron. The control system for SCC vacuum which is also comprised of three standalone PLC nodes for main machine, vault beam lines and cave beam lines, is under installation.

2.1. Control hardware

The control systems are implemented using Schneider make PLC with integrated Modbus-TCP connectivity. The field components are distributed judiciously among the PLC input/output modules to overcome the single field component or PLC hardware failure. The systems are designed to incorporate 'auto', 'manual' and 'maintenance' mode of operations. Each PLC node contains a hardware local panel, consisting of indications and push buttons for in-situ operation in manual mode and maintenance mode. The Graphical User Interfaces (GUI) running on several PCs facilitate remote monitoring and operation of the system in manual and auto mode.

2.2. Supervisory Control System

The Experimental Physics & Industrial Control System (EPICS), a standard open-source dual layer software tool for designing distributed control system, is adopted to implement the supervisory control software [3]. The indigenously developed MODBUS-TCP based Input Output Controller (IOC), which communicates with PLC, is in the lowest layer. The Operator Interface (OPI), sitting on top of IOC, communicates with the IOC for monitoring and supervisory control. The OPIs (shown in Figure 3) are also developed in house using VB6 on Windows XP to include various common features of Windows and connectivity to Access/Excel used as local database. Several reusable ActiveX components e.g. text display, image display, alarm window, set-point input, button etc. are developed using the E-Z (Easy) Channel Access library to reduce OPI development time [2]. The OPIs incorporate the features e.g. system 'mimic' for ease of operation, on-line trending of selected parameters, audio-visual alarm, user authenticated secure mode for control of various components and modification of set-points etc [2]. The OPIs facilitate the user to operate the systems in manual and auto mode. The auto mode of operation is designed with a facility for user to select field components for particular run. This design philosophy reduces the operational complicacy and also includes the advantages of redundant field components. This operational philosophy has been tested satisfactorily in RTC vacuum system. In RTC vacuum system, the user selection is restricted to the choice of diffusion pump, roots pump and rotary pumps. The similar method is also being adopted in SCC vacuum system, though the user selection is restricted to the choice of pumping module instead of pumps. The OPIs also provide diagnostic views of control hardware. An indigenously developed MySQL based EPICS tool is used to log selected vacuum parameters for historic analysis.

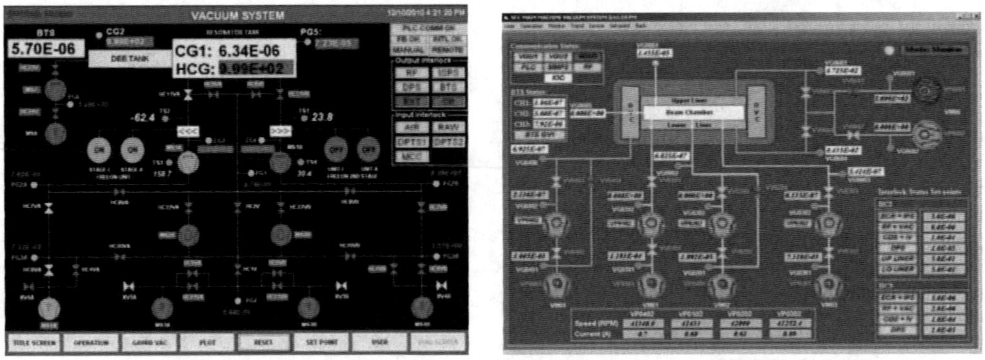

Figure 3. The operator interfaces of Vacuum system of RTC & SCC

3. References

[1] A. Sur, R. K. Bhandari, "Superconducting Cyclotron and its Vacuum System", Proceedings of the International Symposium on "Vacuum Science and Technology", TIFR, India, 2007.

[2] Anindya Roy, et. al., "Vacuum Control System of VEC", Proceedings of the Indian Particle Accelerator Conference, RRCAT, India, 2009.

[3] Anindya Roy, et. al., "Vacuum System of the Cyclotrons in VECC, Kolkata", Proceedings of XIIth IEEE IVEC, MTRDC, India, 2011.

Three stage vacuum system for ultralow temperature installation

N. K. Das, J. Pradhan, Md. Z. A. Naser, B. Ch. Mandal, A. Roy, P. Kumar, C. Mallik and R.K. Bhandari

Variable Energy Cyclotron Centre, Department of Atomic Energy, 1 / AF, Bidhan Nagar, Kolkata- 700 064, India;

E-mail: nkdas@vecc.gov.in

Abstract. We use a three stage vacuum system for developing a dilution fridge at VECC, Kolkata. We aim at achieving a cooling power of 20μW at 100mK for various experiments especially in the field of condensed matter and nuclear physics. The system is essentially composed of four segments-bath cryostat, vacuum system, dilution insert and ^{3}He circulation circuit. Requirement of vacuum system at different stages are different. The vacuum system for cryostat and for internal vacuum chamber located within the helium bath is a common turbo molecular pump backed by scroll pump as to maintain a vacuum ~10^{-6}mbar. For bringing down the temperature of the helium evaporator, we use a high throughput Roots pump backed by a dry pump. The pumping system for ^{3}He distillation chamber (still) requires a high pumping speed, so a turbo drag pump backed by a scroll pump has been installed. As the fridge use precious ^{3}He gas for operation, the entire system has been made to be absolutely leak proof with respect to the ^{3}He gas.

1. Introduction

With a mission to build up a dilution refrigerator to obtain a cooling power of 20μW at a temperature of 100mK, we are using a bath cryostat with 25 litres of available LHe capacity with a hold time of 30hrs. The cryostat incorporates an internal vacuum chamber (IVC) which contain the dilution insert. The dilution insert is consisted of a helium evaporator, heat exchanger, ^{3}He distillation chamber and a mixing chamber. The helium evaporator has a volume of 99cc and is made of OFE copper. Capacities of distillation chamber and mixing chamber are of 40cc and 10cc respectively. The schematic view of the system is shown in figure1. It operates at sub-atmospheric pressure and necessitates all subsystems and network of capillary interconnections truly leak proof. Vacuum of the system relies on three distinct stages-a common turbo-pumping station for cryostat and IVC, a root pump backed by dry pump is used for helium evaporator and a turbo molecular drag pump backed by scroll pump for ^{3}He circulation.

IVC is housed within the helium bath of the cryostat with its top flange immersed in LHe. As such the leak tightness of the flange was ensured with the help of indium o-ring corner seal. The IVC is made of OFE copper with a height and diameter of 350mm and 130mm respectively. It is evacuated to about 10^{-6}torr using turbo molecular drag pump. IVC plays a significant role by way of thermally isolating the very low temperature system from the LHe bath. The evaporator which is used for precooling and condenses the circulating gas is again placed inside the IVC. The vapour pressure of the evaporator is reduced separately by a root pumping system (250m^3/hr). ^{3}He circulation unit pumps out the primarily ^3He vapour from the ^3He distillation chamber which contains a mixture of super fluid ^4He and ^3He. Since the distillation chamber maintains a temperature of 0.7K, the vapour pressure of ^3He is about 1000 times higher than that of ^4He. Therefore, the vapour above the super fluid will be essentially ^3He.

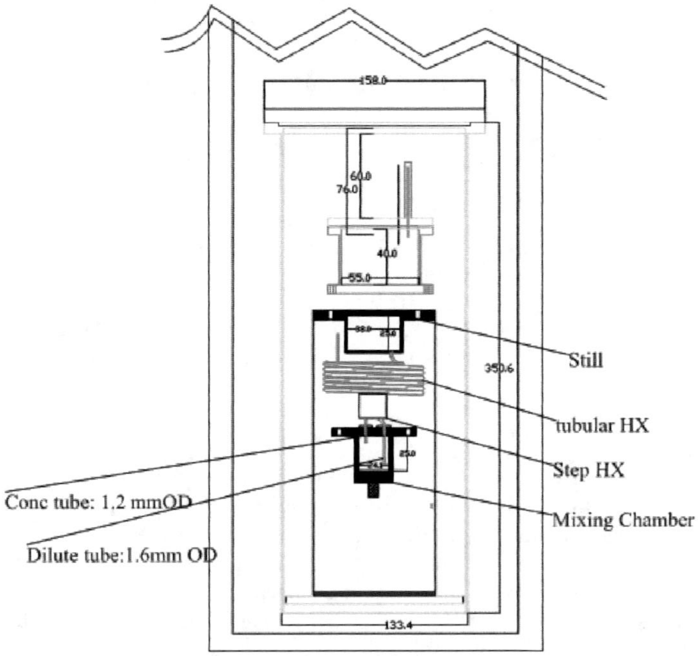

Fig. 1. Schematic view of the dilution fridge

2. Superfluid suppression

A common method of reducing film flow in dilution refrigerators is a knife-edge device. The principle behind the knife-edge film killer is that a superfluid film is reduced when it flows around a sharp bend. If the bend is sharp enough, the film does not flow past it. Unfortunately, there is a catch: the film evaporates where it stops, cooling the material of the knife-edge. As the knife-edge cools, some of the helium vapor condenses on the evaporator. This newly condensed, somewhat reduced, film then continues to flow downstream.

3. Evaporator pumping system

This is the continuously filled evaporator (1K pot) design and tested in the laboratory and found to offer a temperature less than 1.0K by withdrawing vapour from the evaporator. In order to minimize the heat load and to prevent film creep across the pumping tube, size optimization of the pumping line and pump-out port has been performed. Five nos thermal Baffles-are used along the pumping line as thermal barriers to radiative heat load.

3.1. Basis to choose the pump

To achieve the available cooling power of 10 mW@1.0K, the required steady state molar flow rate that has to be pumped out by the pump is calculated to be 250 μmols/sec. For the minimum ultimate evaporator temperature of 1K the vapour pressure at the pot has to be maintained 0.1 torr. Initially if we consider the pressure drop along the pumping line be negligible, then from the ideal gas equation

$P\dot{V} = \dot{n}RT$, where P is the pressure at the pumping line at the top of the cryostat, T is the room temperature (300K), $\dot{n}$ and $\dot{V}$ are the molar flow rate and the pumping speed at the top of the cryostat. For P = 0.1 Torr, The pumping speed is calculated as 168 m^3/hr. To minimize the vibration arises due to the pump the pump is placed few meter away from the cryostat. Considering this the pump of pumping speed 250 m3/hr is chosen.

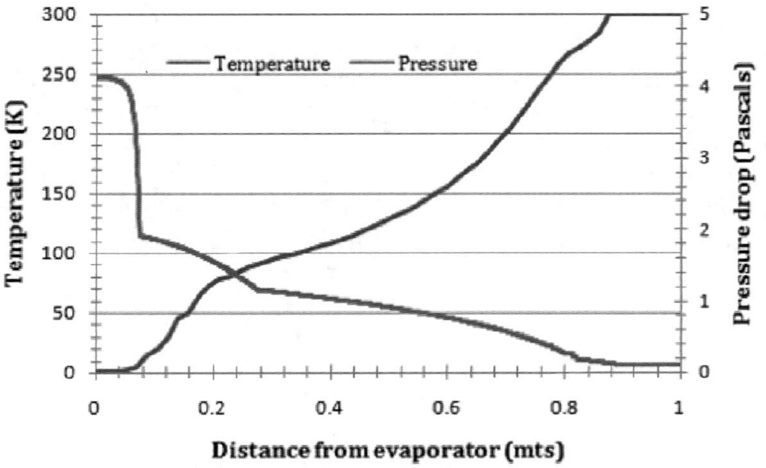

Fig. 2. Temperature and pressure distribution along the pumping line of the evaporator (1K pot).

4. Optimization of the pumping line inside the cryostat

The geometry of pumping lines are chosen such that the conduction heat load to the 1K pot as well as to the LHe bath are minimum and the pressure drop along the pumping line is as minimum as possible. The wall thickness of the pumping section connected to the evaporator must thin, small diameter to keep the cross sectional area for heat conduction as low as possible[1]. The superfluid that climbs up the pumping line should evaporate before reaching the pumping section that have larger diameter to pump out the more vapour emerging out of the superfluid evaporation as well as normal fluid evaporation. With a view to make an optimal balance between thermal conduction load ($Q \propto r$) from and viscous flow conductance ($C \propto D^4$) of the pumping line, it has been designed in three sections having different diameters.

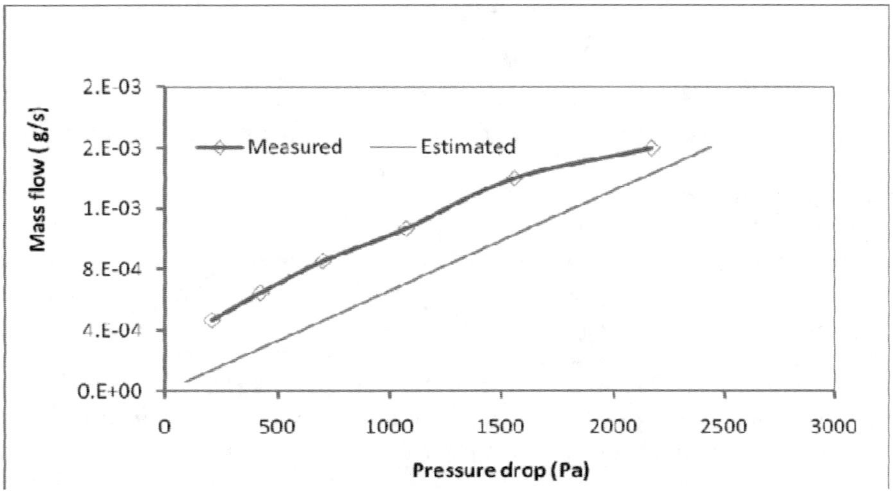

Figure 3: Mass flow rate through evaporator pumping lines for different heater settings.

The pumping line consists of mainly three sections with different length and diameter in order to make balance between the thermal load and pumping speed. The conductance of the piping system is primarily depends on the sizing of orifice diameter at the entry of 1K pot. The ANSYS FLOTRAN™ CFD module was used to obtain pressure and temperature distribution for sizing of the piping lines. We have obtained ultimate temperature of less than 1K and cooling power of ~17milliwatt at 1.2K continuously. We could control the temperature by fiddling with needle valve from outside. The absolute pressure was measured at room temperature near to the cryostat. The standard chart provides the vapour pressure inside the pot from its temperature. The above figure shows the variation of mass flow rate obtained from the heater power with the pressure drop along the lines. The slope of curve which is proportional to the conductance of pipe system is same as the estimated value of 30m3 /hr. The vertical shift of the measured curve from the calculated value is possibly due to error in our meter reading, which is not accurate enough to monitor the small pressure variations

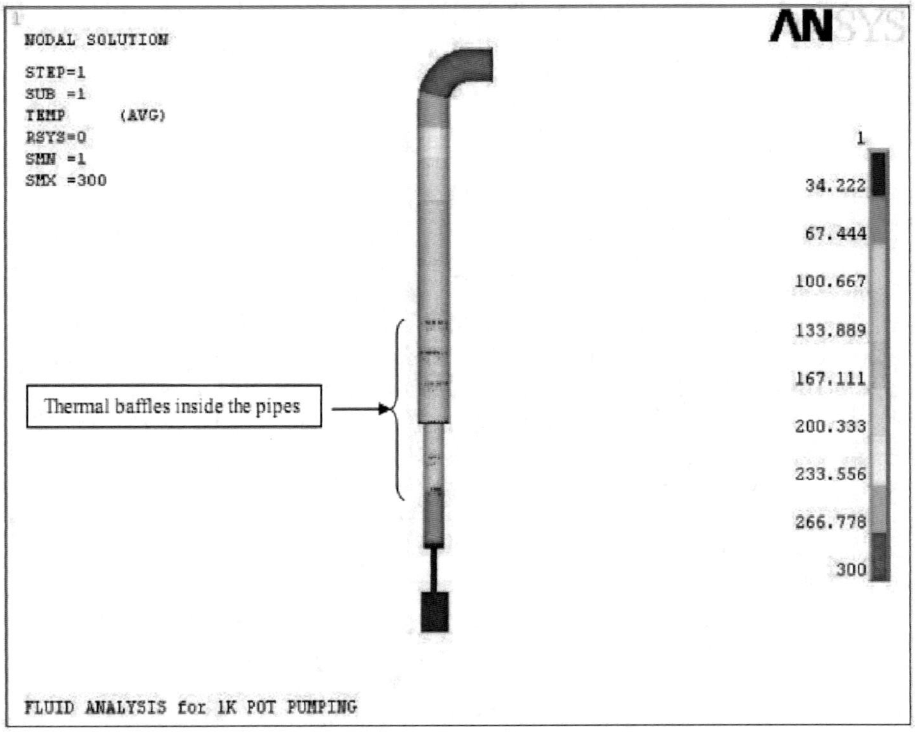

Fig. 4. Temperature contour of the different sections of the evaporator (1 K pot) pumping system.

5. ^{3}He Circulation Circuit

An important feature of the refrigerator is the ^{3}He circulation system. The cooling power of dilution fridge is dependent on the ^{3}He circulation rate. As only modest pressures are required to maintain the cold of the Still, much higher circulation rates are available by using a turbo-drag pump to circulate ^{3}He rather than a diffusion pump. The turbo-drag pump also requires a comparatively small backing pump than that of a diffusion pump. The pumping system chosen for this application is a turbo-drag pump (Edwards; model No: EXD556H) backed by helium sealed scroll pump (Edwards; model No: XDS35i). The maximum ^{3}He circulation rate available with this combination of pumps was 30µmoles/s. As a way of minimizing the heat conduction through the pumping line, the diameter of the line has been incrementally increased towards the 300K. Length of 150 mm(OD:9.5 mm) from still followed by 1 inch pipe of length 250 mm and 200 mm of 2" diameter pipe have been used to make the line. Even though the evaporating ^{3}He in the Still is pumped through 1mtr flexible line after 300K, the Still pressure has been estimated to be about 0.1 torr. The total conductance of the pumping line is estimated of 15 m^3/hr. This high circulation rate minimises the cool-down time of the refrigerator.

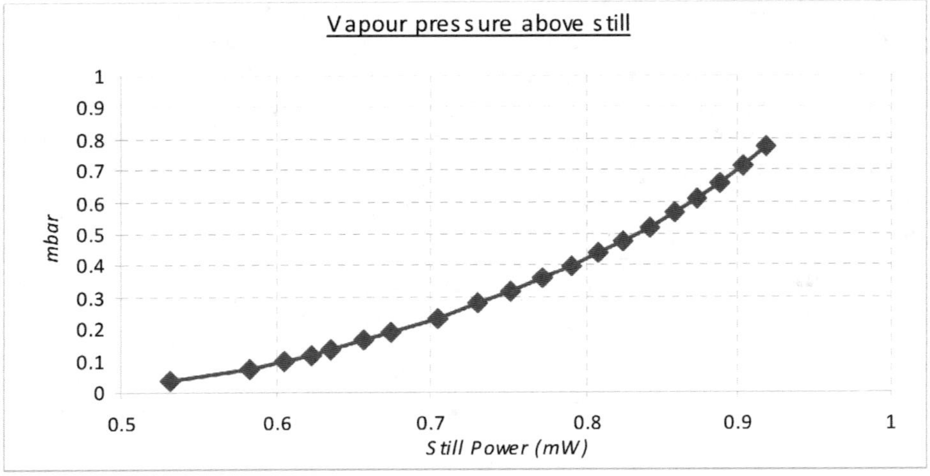

Fig. 5. Variation of vapour pressure with still power

In order to circulate and get the continuous cooling in the mixing chamber of the dilution fridge, the vapour above the STILL is required to pump out continuously. The vapour constitute mainly of ^{3}He, which at the temperature of STILL ~ 0.7K is about 1000 times more vapor pressure than ^{4}He. The heater is provided around the STILL to maintain the required flow and its vapour pressure variation with heater power is depicted in figure 5 above. The pumping lines for the STILL are designed taking into account the compromise between reducing the thermal load and increasing pumping speed[2]. The orifice of diameter 2mm is provided at the entry to restrict the super fluid film flow, which basically limits the conductance of the pumping lines.

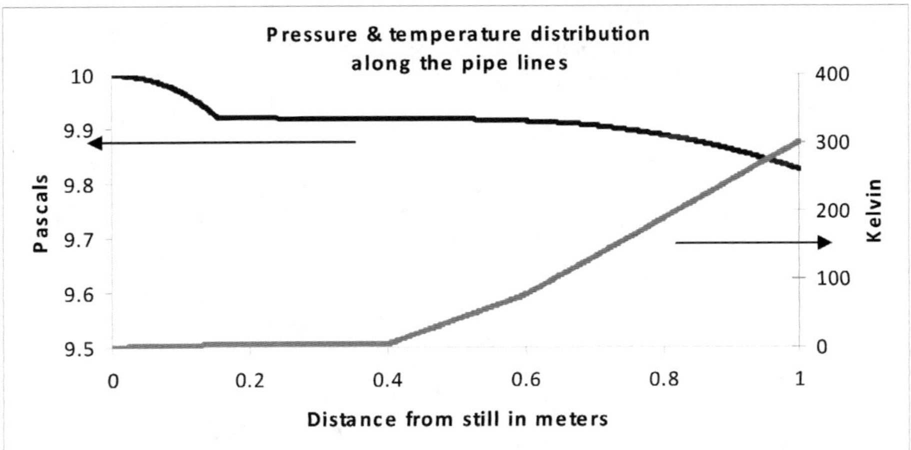

Fig. 6. Pressure and temperature distribution along the still pumping lines.

The pressure and temperature distribution along the three sections of pumping lines is shown. The lines are anchored at suitable positions at different temperatures for reducing the heat load at lower temperature.

Journal of Physics: Conference Series **390** (2012) 012055 doi:10.1088/1742-6596/390/1/012055

6. Leak testing of the vacuum system

Vacuum and pressure leak test were carried out for all joints, seals and welding points. Convenient and reliable sealing in the IVC and helium evaporator and 300K flange were made from indium o-ring to avoid the spillage of liquid helium . Indium material used is ultra-pure (99.9 minimum purity) to prevent hardening of the material at sub-Kelvin temperatures, as well as to restrict impurities of elements with low vapour-pressure. The indium seal joints have undergone for the helium leak test under following conditions-room temperature leak test of the indium seal joints, test after repeated thermal shock at liquid nitrogen temperature, helium leak test of indium seal at liquid nitrogen temperature in flow condition. The schematic view of the leak-testing set up is shown in figure 7. The average measured leak rate of all seals was ~4.0×10^{-10}mbar-l/s by helium leak detector.

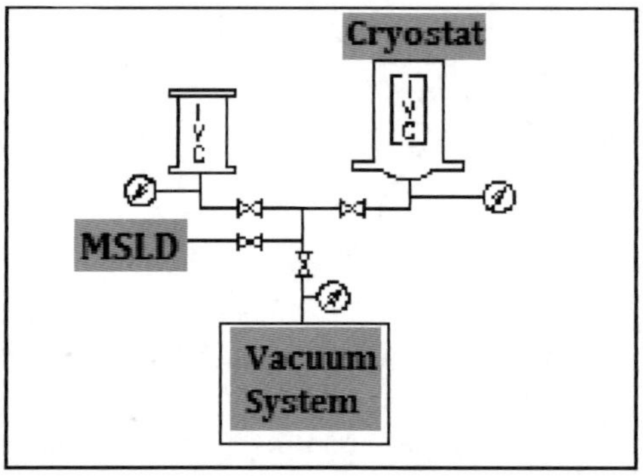

Fig. 7. Schematic diagram for leak testing set up.

7. Conclusion

Three sets of pumping system have been optimally selected to run the dilution refrigerator. Sizing of the pumping line & capacities have been taken in to account to satisfy the requirement of the system. Leak testing of all the joints seals and the pumping line have been thoroughly carried out.

Reference

[1] DeLong L E et.al. 1971 *Rev. Sci. Instr.* **42**, Number 1, pp 147-150
[2] Richardson L J et.al. 1978 *Cryogenics* February pp 109-111

Development of a high vacuum sample preparation system for helium mass spectrometer

P Kumar, N K Das, C Mallik and R K Bhandari

Variable Energy Cyclotron Centre, 1/AF, Bidhan Nagar, Kolkata – 700064

Email: pradeep_vecc@yahoo.co.in

Abstract. A high vacuum sample preparation system for the $^3He/^4He$ ratio mass spectrometer (Helix SFT) has been developed to remove all the gaseous constituents excluding helium from the field gases. The sample preparation system comprises of turbo molecular pump, ion pump, zirconium getter, pipettes and vacuum gauges with controller. All these are fitted with cylindrical SS chamber using all metal valves. The field samples are initially treated with activated charcoal trap immersed in liquid nitrogen to cutoff major impurities and moisture present in the sample gas. A sample of 5 ml is collected out of this stage at a pressure of 10^{-2} mbar. This sample is subsequently purified at a reduced pressure of 10^{-7} mbar before it is injected into the ion source of the mass spectrometer. The sample pressure was maintained below 10^{-7} mbar with turbo molecular vacuum pumps and ion pumps. The sample gas passes through several getter elements and a cold finger with the help of manual high vacuum valves before it is fed to the mass spectrometer. Thus the high vacuum sample preparation system introduces completely clean, dry and refined helium sample to the mass spectrometer for best possible analysis of isotopic ratio of helium.

1. Introduction

The $^3He/^4He$ ratio provides useful information on the origin of helium (primordial or radiogenic). It is also an important indicator to the study of magmatic volatiles [1]. The most vital part of any mass spectrometric analysis is the sample preparation. Several sample extraction and cleaning techniques [2][3] have been used for helium isotopic analysis using high vacuum sampling systems. The mass spectrometer requires completely clean and dry samples at a pressure lower than 10^{-7} mbar for precise measurements. Helium bearing gas samples collected at atmospheric pressure from different geothermal field stations contains impurities like N_2, O_2, Ar, CH_4 etc. This paper describes the fabrication, design and operational details of cleaning of field samples at a vacuum of the order of 10^{-7} mbar for helium isotope analysis.

2. Design, Fabrication and Operation

The sample preparation system is fabricated with all stainless steel components (316L) and cleaned prior to use on the system. The cleaning is done by boiling the fabricated parts in 5% HNO_3 followed by cleaning with acetone and alcohol followed by drying. Conflate flanges with OFHC copper gasket are used to assemble various components of the sample preparation system.

2.1 The Pretreatment stage

The samples are collected from geothermal field stations in pre evacuated stainless steel sampling cylinders (500 cc, double ended) for analysis. These sample gases were first treated with liquid nitrogen cooled activated charcoal trap for removal of most of the impurities like moisture, nitrogen, oxygen, hydrocarbons and argon. The schematic of the assembly of the charcoal trap is shown in figure – 1.

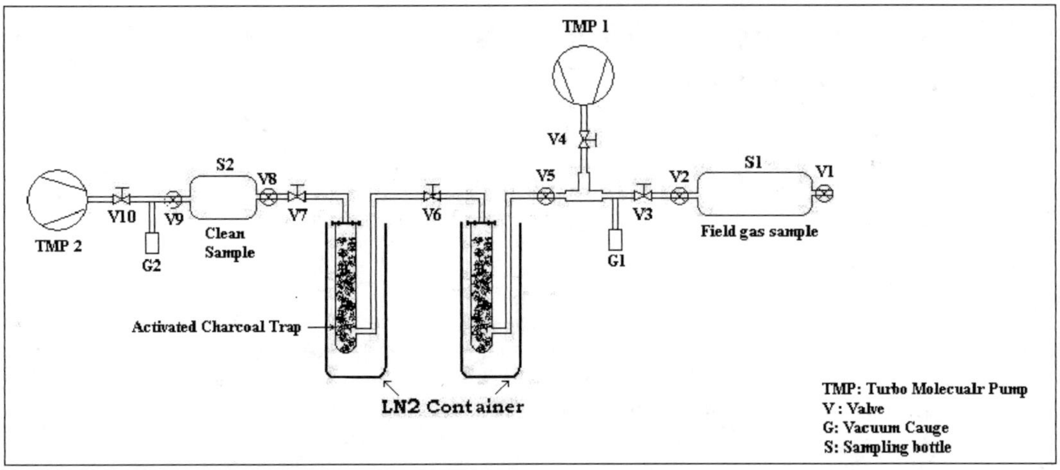

Fig. 1. Schematic of pretreatment stage charcoal trap assembly.

The pre treatment stage consists of two turbo molecular vacuum pumps along with two charcoal traps with interconnected tubings, valves and vacuum gauges. All the interconnected tubings are ¼" SS tubes and the valves are Swagelok SS 4L valves. The vacuum gauges used are compact full range gauges (Pfeiffer).

Sample collected from the pretreatment stage in a cylinder that was evacuated to a vacuum level of 10^{-6} mbar using TMP 2 with V8, V9 and V10 valves open. The charcoal traps were first evacuated to a vacuum level of 10^{-5} mbar with use of TMP 1 and V4, V5 and V6 open. While evacuating the charcoal traps V3 and V7 are kept closed. When the vacuum level in the charcoal traps reached to the desired level, liquid nitrogen was poured to the containers followed by closing of V4 and V6. Valve V5 was kept close and sampling bottle (S2) is evacuated to 10^{-6} mbar. After a time of about 10 minutes allowed for cooling the charcoals, valve V2 and V3 and V5 are opened and the field sample was allowed to pass through the first trap and vacuum decreases. After few minutes valve V6 was opened to allow the sample gas to pass the second charcoal trap. With liquid nitrogen cooled charcoal, most of the impurities like moisture, nitrogen, oxygen, hydrocarbons present in the sample get trapped up on the charcoal leaving relatively pure helium sample. At this stage vacuum inside the charcoal trap was of the order of few mille-bar. After done this valve V7 and V8 was opened while V9 and V10 were kept closed and the helium sample was collected in a 5 ml SS sample cylinder for further cleaning. 5 ml of the relatively pure helium sample was collected at a vacuum pressure of 10^{-2} mbar from the pretreatment stage.

2.2 The Sample Preparation system

Figure 2 shows the schematics of the high vacuum sample preparation system. The interconnecting vacuum tubes are fabricated from ¾" SS pipes. The whole system is assembled on a table top for easy access and operation.

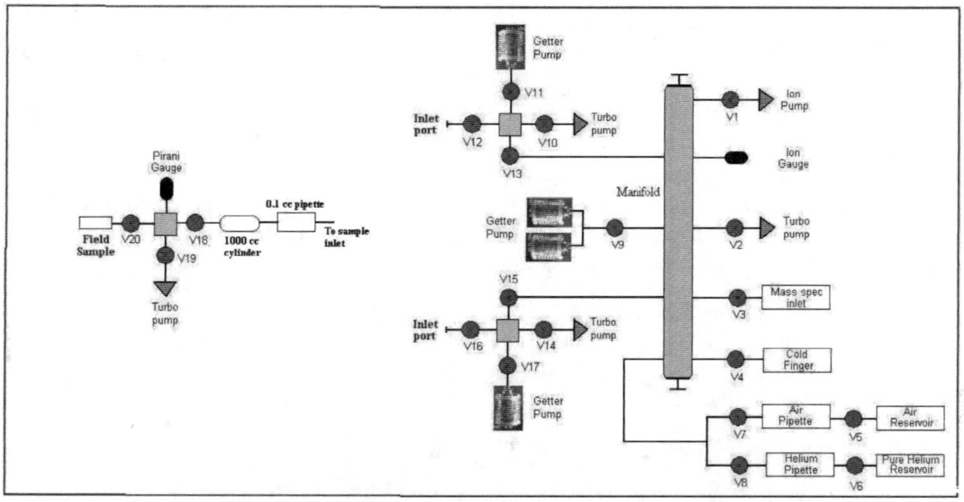

Fig. 2. Schematic of sample preparation system

The sample preparation system is an all metal high vacuum system consists of the following components: 1: Turbo-molecular drag vacuum pump (Pfeiffer, 60 l/s) with two stage diaphragm backing pump 2: StarCell Ion pump (Varian, 20l/s) for clean vacuum 3: SAES NP10 zirconium getter pumps (Zr 84 wt%, Al 16 wt%) in jackets that can be water cooled 4) Ultra low volume micro ion gauge (Granville Phillips) 5: Liquid nitrogen cooled cold finger (0.5" dia, 4" long) 6: All metal UHV valves (Varian) 7: SS manifold (volume 355 cc). The turbo molecular drag pump has a DN63CF inlet port and the ion pump has a DN40CF inlet port. The all metal valves have DN16CF connections. The pumps, gauges, cold finger were connected to the SS manifold using the all metal valves. The total volume of the gas preparation line including the getters is 355 cc. There are two inlets lines to enable admission of the sample gas to final clean up line. Each of the inlets are pumped by turbo molecular pump backed by diaphragm pump and are fitted with NP10 getter pump.

Sample from the pretreatment stage collected at a pressure of $\sim 10^{-2}$ mbar and expanded in a 1000 cc SS cylinder. The cylinder was first evacuated using a turbo molecular pump to a vacuum pressure $\sim 10^{-7}$ mbar and isolated from the pump by closing valve V19. The vacuum pressure was monitored using a pirani vacuum gauge. The sample gas from the 5 ml cylinder transferred to the 1000 cc cylinder by opening valve V18 and allowed to expand and the sample gas pressure decreases from 10^{-2} mbar to $\sim 0.5 \text{ X } 10^{-4}$ mbar.

Helium sample from the 1000cc cylinder at a pressure $\sim 10^{-4}$ bar was introduced through one of the inlet port using a 0.1 cc pipette with valve V16 open. The sample gas exposed to the four getters, the ion pump and the cold finger at a vacuum pressure of $\sim 10^{-8}$ mbar via opened valves V1, V4, V11 and V17. Since the internal volume of the sampling system including the getters is 355 cc the pressure will reach to $1 \text{ X } 10^{-6}$ mbar. At this time the turbo pump was isolated from the manifold via closed valves V2, V10 and V14. Expansion of 0.1 cc of the sample in the manifold occurs and pressure of the sample gas drops to $\sim 1 \text{ X } 10^{-6}$ mbar. Sample gas at a pressure of $\sim 10^{-6}$ mbar is not suitable to inject into the mass spectrometer. The valve V9 was closed to trap the sample in two getters (volume 103 cc) and the remaining manifold was evacuated by opening valve V2, V10 and V17 by turbo pump. Again on improvement of vacuum pressure to an order of 10^{-8} mbar, the turbo pumps are then isolated from the manifold and V9 was opened to expand the trapped sample gas further. By this time the sample gas pressure further drops to $\sim 2 \text{ X } 10^{-7}$

Journal of Physics: Conference Series **390** (2012) 012056 doi:10.1088/1742-6596/390/1/012056

mbar. As the sample gas exposed to the four getter pumps remaining active gases are caught in to it. With the cold finger the remaining moisture etc. are cleaned up. At this stage the sample gas in the gas preparation system is ready for inject into the mass spectrometer for analysis. The mass spectrometers flight tube with inlet pipe has a volume of 1275 cc. and kept at a vacuum of ~10^{-9} mbar. With opening of the inlet valve V3 sample get injected to the mass spectrometer. Prior to injection of the sample; turbo pump, ion pump, cold finger and the getters are isolated from the manifold with closing respective valves. The injected sample expands into the mass spectrometer and final vacuum pressure during analysis was of the order 5 X 10^{-8} mbar. Figure 3 shows the enlarged view of the sample preparation system along with the assembly with the mass spectrometer. Figure 4 illustrates the relative intensities of helium - 3 and helium - 4 of air sample analyzed with the mass spectrometer.

Fig. 3. The sample preparation system

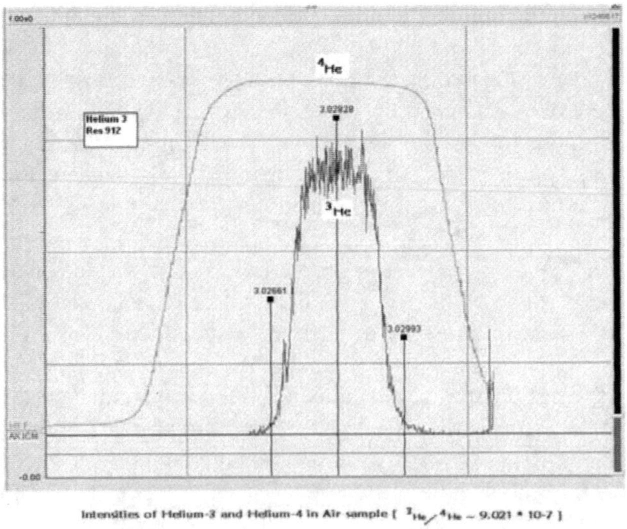

Fig. 4. Intensity of helium -3 and Helium -4 in air sample

3. Conclusion

Several helium samples collected from various geothermal fields in West Bengal, Jharkhand, Bihar and Odisha were analyzed successfully for helium isotope ratio. The developed sample preparation system is efficient to deliver clean samples of helium at a suitable vacuum pressure for representative isotopic ratio analysis.

Acknowledgement

We are thankful to Department of Atomic Energy, Government of India, for supporting this research work.

References

[1] Das N K, Ghose D, Bhandari R K, Sen P and Sinha B 2009 *Current Science* **96** pp 1031- 1032
[2] Beyerle U, Hertig W A, Imboden D M, Baur H, Graf T and Kipfer R 2000 *Environ. Sci. Technol.* **34** pp 2042 - 2050
[3] Jenkins W J, Lott D E, Cahill K, Curtice J and Landry P 2010 *The Go-Ship Repeat Hydrology Manual: A Collection of Expert Reports and Guidelines* **134** 14

Ion pump using cylindrically symmetric spindle magnetic field

M H Rashid

Variable Energy Cyclotron Centre, 1/AF- Salt Lake City, Kolkata-700064, India

E-mail: haroon@vecc.gov.in

Abstract. For all accelerators and many research and industries, excellent vacuum conditions are required and the highest possible pumping rates are necessary. For most applications the standard ion sputtering pump (ISP) meets these requirements and is optimal for financial point of view also. The physical principle of the ISP is well known and many companies manufacture variety of ISP. Most of them use dipole magnetic field produced by permanent magnet and electric dipole field between the electrodes in which tenuous plasma is created because of interaction of between the relatively fast electrons slow residual gas atoms. Performance of an ISP depends basically on the electron cloud density in between the titanium electrodes but in the available present configurations no consideration has been given to electron confinement which needs a mirror magnetic field. If this is incorporated it will make a robust ISP surely; furthermore, the requirement of constant feeding of high voltage to electrodes for supplying sufficient number of electrons will be reduced too. A study has been performed to create sufficient rotationally symmetric spindle magnetic field (SMF) with inherent presence of magnetic mirror effect to electron motion to confine them for longer time for enhancing the density of electron cloud between the electrodes. It will lessen the electric power feeding the electrodes and lengthen their life-time. Construction of further compact and robust ISP is envisaged herein. The field simulation using the commercially available permanent magnet together with simulation of electron motion in such field will be presented and discussed in the paper.

1. Introduction

The first major use of vacuum technology in industry occurred in about 1900 in manufacturing of electric light bulbs. The advent of nuclear energy in the 1950's provided impetus for development of vacuum equipment on a large scale. Increasing applications for vacuum processes were steadily discovered, as in space simulation and microelectronics. Any accelerator person cannot think about an accelerator, a beam-line, scattering or reaction chamber, and an ion or electron source without vacuum creation. The high energy beam accelerators like synchrotrons, storage rings and colliders need very good vacuum conditions because the beam of ions or electrons circulate a large number of times and unwanted collisions with residual gas atoms reduces the beam and its life-time. Vacuum is very important for transmission or scanning electron microscopes and they are needed to meet usual or unusual configurations economically. A focused ion beam system assist in achieving very small semiconductor structure and micron scale mechanical structures, which also need very clean vacuum environment. A wide variety of gases are used in such system and puts extra requirement of multiple level of vacuum at different positions. In many experiments, it is not only necessary to reach the pressures of of the order of 10^{-9} torr but to reduce the hydrocarbons in the residual gases to an absolute minimum. To achieve a vacuum of this order the vacuum vessel and the equipment inside must be cleared of residual gas (degassed) to the greatest extent possible, by means of a roughing pump. To eliminate hydrocarbons, the unit is pumped down to about 10^{-3} torr using sorption pumps and from there, ion sputtering pumps (ISP's) and titanium sublimation pumps complete the task down to 10^{-9} torr or below. Wide application of ISP is also attributed to its operation in any position and without water or liquid nitrogen supplies. They have a long life and can provide very clean, ultrahigh vacuum, free of organic contamination, moving part and vibration. Due to achieving clean vacuum, bakeability, low power consumption and high pumping speed, an ISP is the ideal cost-effective companion.

The working of an ISP was reported first by J. Plucker [1] and he found that with time the voltage requirement to maintain a current in a gas discharge tube was more. Later, it was explained that the current decreased because of pressure decrease in the tube and the relationship was found to be directly proportional. This property was used by F. Penning [2] to measure the gas pressure inside the tube and extended measurement in the range 10^{-3} to 10^{-5} torr in a vacuum vessel. Later some industries exploited the invention in developing ISP's and ion gauges for creating vacuum and measuring pressure respectively. Apart from an ISP, various other types of pumps like diffusion pump, cryo pump, turbo-molecular pump etc. are commercially available in the market for various applications.

The residual atoms/ molecules inside a vessel spread out in any direction through diffusion and under pressure gradient to another vessel, if they are connected to provide gas conductance. Both electric field and magnetic field are used in an ISP. The neutral atoms and molecules when enter the pump chamber consisting of invisible fields, there is predominant ionization process due to presence of energetic electrons. We know that electron is the best ionizing agent as an electron with sufficient energy can impart maximum energy to the bound outer electrons in an atom. These electrons get excited so much that they break away from the atom and mostly single charge ions are continuously created. The charged particles get energy from the electrostatic field, which is created due to application of appropriate voltage between the cathode and anode electrodes. As soon as ionization starts, more and more electrons get sufficient energy for further ionization and a chain reaction is established. Ions also get sufficient energy and strike the cathode made of titanium material, which is highly reactive with hydrogen, oxygen, nitrogen, etc., while being lost on the cathode surface. Furthermore, the titanium atoms are vaporized or sputtered from the cathode and they entrap (engrave) those reactant atoms and product molecules on the cathode surface at the vicinity of the interior surface of the pump chamber. The active gasses are pumped by chemical combination with the sputtered titanium, the inert gasses by ionization and burial in the cathode and the light gasses by diffusion into the cathode. Magnetic field is also present in the vacuum chamber. The most important function of magnetic field is to increase the path length (mean free path) of the ionizing energetic electrons in the vacuum chamber. Another effect of the magnetic field is to reflect most of the electrons before being lost on the chamber surface. Thus, the average life time of the electrons increases in the chamber causing more and more ionization. The created ions also gain energy and always move in a curved path under the influence of the EM fields and ultimately get lost on the cathode surface incessantly.

A typical ISP consists of two flat rectangular cathodes with a stainless steel anode between them made up of many open-ended boxes. This assembly, mounted inside a narrow box attached to the vacuum system, is surrounded by a permanent magnet (PM). The anode is operated at a potential of about 7 kV, whereas the cathodes are at ground potential. They use PMs to generate sufficient field (about 2 kG) with little mirror effect. But there is ample scope of improving the performance significantly of an ISP by reconfiguring the magnetic field, which will reduce the electric load on electrodes too. Here, we take advantage of the anti-solenoid field created by either a pair of coaxial coil energized oppositely or a radially magnetized permanent magnet ring. It creates magnetic field with convexity of field lines towards the centre of the chamber. The field has great inherent benefit of confining the electrons inside the pump chamber for longer time to achieve maximum number of ionization per energetic electron. Application of the field will enhance further its pumping speed and reduce the electrical power consumption to make it more cost-effective.

2. Spindle magnetic field

First, we decide the geometry of the ISP as well as the magnet system with yoke. The geometry should be such that the pieces of the PM's are arranged outside the chamber. The highly permeable iron yoke is enhances the field and reduces the volume of the PM material and it is properly placed around the PM ring and chamber ends. The yoke also protects the PM and keep them tightly bound around the curved surface of the cylindrical metal chamber. The geometry of the magnet and chamber is fixed after field optimization. The optimized geometry is shown in figure 1 below. It seems to be a big ISP to evacuate large vessel. Here, just an example of a big ISP is depicted for the purpose of following the procedure to design appropriate ISP's suitable for applications.

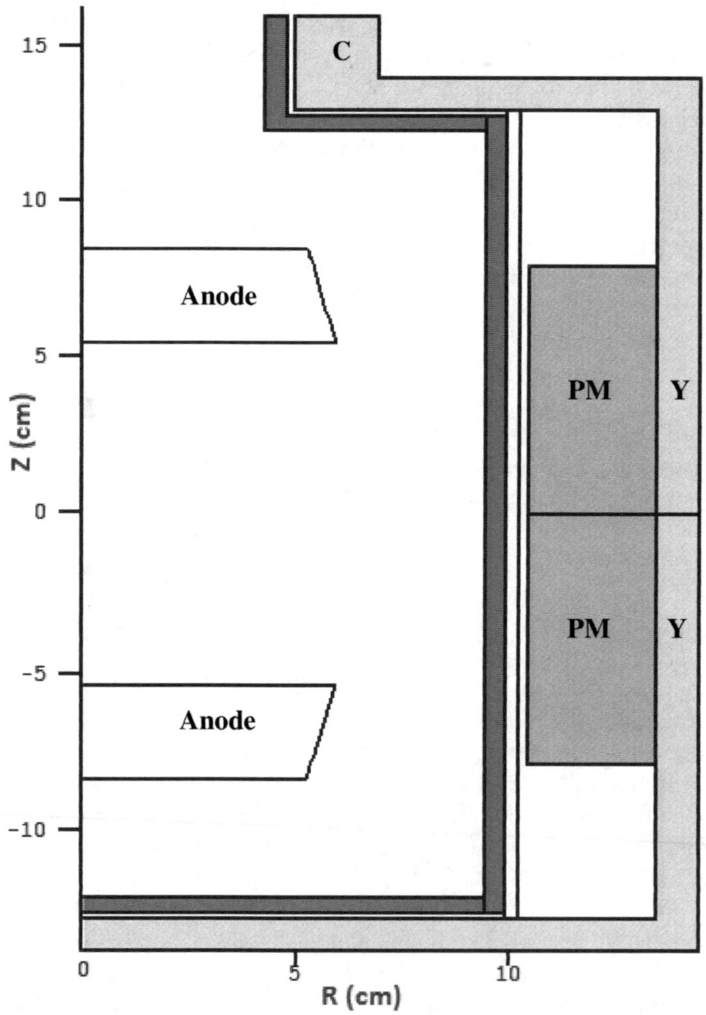

Fig. 1. The pump is rotationally symmetric about the z-axis. The innermost shaded region is the cathode at the vicinity of the chamber surface. The anodes are cylindrically and radially gridded and several kV positive potential is applied. The rare-earth PM pieces are densely packed and are radially magnetized. The inner diameter of the PM ring is 22 cm. The outer shaded area levelled as 'Y' is the iron yoke of 1 cm thickness. The pump is connected to the vessel to be evacuated at C. Pumping speed will depend on the cathode surface area, applied potential, the produced magnetic field, ionization process and burial (capture) process of ions and atoms on the cathode surface.

2.1. Magnetic field computation

In the above geometry the rotationally symmetric magnetic field is generated due to the combination of the arranged PM's and the highly permeable magnetic material (low carbon magnetic steel), which provide ruggedness and strength in structure of the pump. The magnetic lines of force (MLF) emanating radially from the PM surface, facing the chamber cylinder, are bent and meet the end yokes axially such that their convexity is towards the centre of the chamber. The field was calculated using the PANDIRA code [3]. The field configuration is shown in the following figure 2. The PM used for calculation has remnant field and coercive force 8.5 kG and 8.5 kOe respectively. The radial and axial dimensions of the PM ring are 3 cm and 14 cm respectively. The yoke works as a shield also for external field and the pump will continue to work unaffected in an external magnetic field.

The computed field has been plotted in figure 3 in 3D. In any r-z plane, field plot depicted in the figure is same. It is seen that the magnetic field at the vicinity of the centre characterised by r=z=0 is zero. And it rises from the centre in radial and axial direction both. This may be termed as zero-B field configuration. The magnetic field achieved at the chamber surface in z=0 radial plane and at r=0, end plate surfaces are almost same value ~1.86 kG. So, a symmetric spindle magnetic field (SMF) [4] is created to achieve regular magnetic mirror ratio for achieving good quality confinement of electrons for longer time. The legend of the shaded plot region is shown in the top region of the figure and the field values are in Gauss. The maximum field achieved, which is more than 3 kG, is near the ends at the highest radius, 9.5 cm. The maximum axial width of the vacuum chamber is 24 cm.

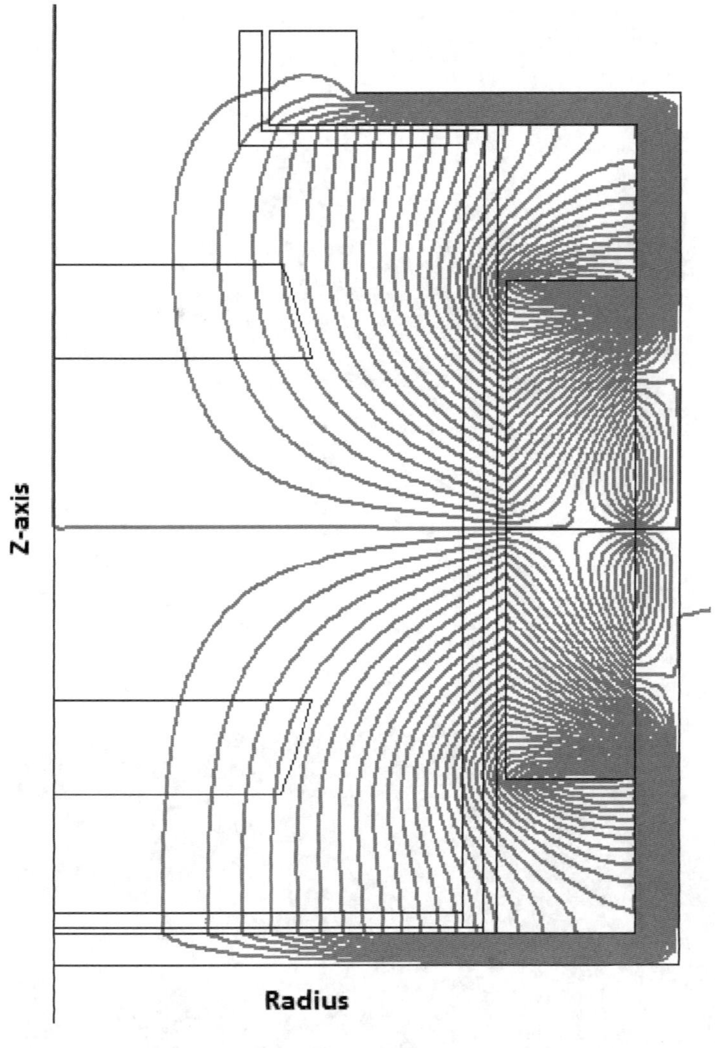

Fig. 2. Plot of the MLF's due to the combined effect of the PM magnetization and induced magnetization of iron yoke. The vacuum chamber with electrodes is immersed in the created magnetic field. This clearly shows the configuration of the SMF. The maximum field inside the yoke is about 10.3 kG. These MLF's coerce the electrons to gyrate around them and follow their path. The electrons as well as ions have high probability of being lost at the cylindrical surface of cathode and on the cathodes at the ends.

2.2. Motion of charged particles

Motion of the charged particle in the computed field has been achieved by TrapCAD code [5]. The SMF in the r-z plane is given as input. Here, simulation of motion of single charge oxygen ion has been performed. One after another, 11 ions were launched around the centre of the chamber with components of energy parallel and perpendicular to the MLF 1 keV and 2 keV respectively at different launching positions. The LHS and RHS figures in figure 4 show the projections of the same plot in x-y and x-z plane respectively. It is clear that 7 ions are lost on the cylindrical surface and the rest on the end electrodes. The trajectory of an electron has been depicted in figure 5. The electron was projected at z=5 cm and r=4.0 cm with energy components 2 keV and 3.5 keV respectively in parallel and perpendicular direction with respect to the MLF at the launching position. The LHS and the RHS figures, in figure 5, are depicted in the tilted x-y-z coordinate system and the projection in the x-y plane respectively of the same trajectory. It is seen that the electron is reflected a large number of times. Number of reflection of the electrons depends on their energy components and total energy, launching position, magnetic field components and its gradient and curvature. Presence of electric field and scattering with other similar or different particles also affect their behaviour. When the charged particles pass through the zero field area at the centre, they experience Coulomb- like scattering and probability of losing it increases. So, their density at the centre decreases.

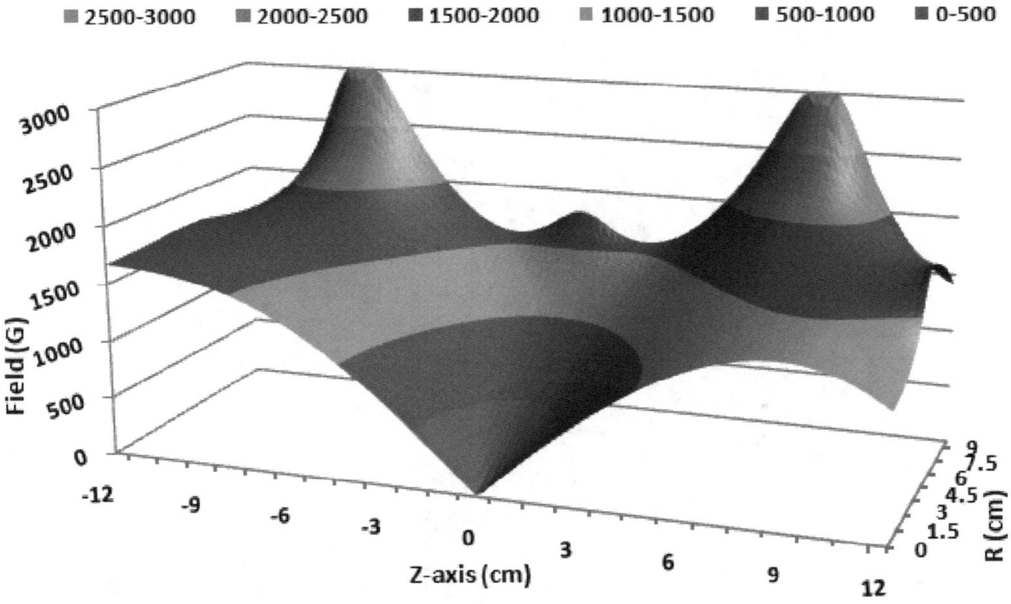

Fig. 3. The 3D plot of the computed magnetic field in the r-z plane.

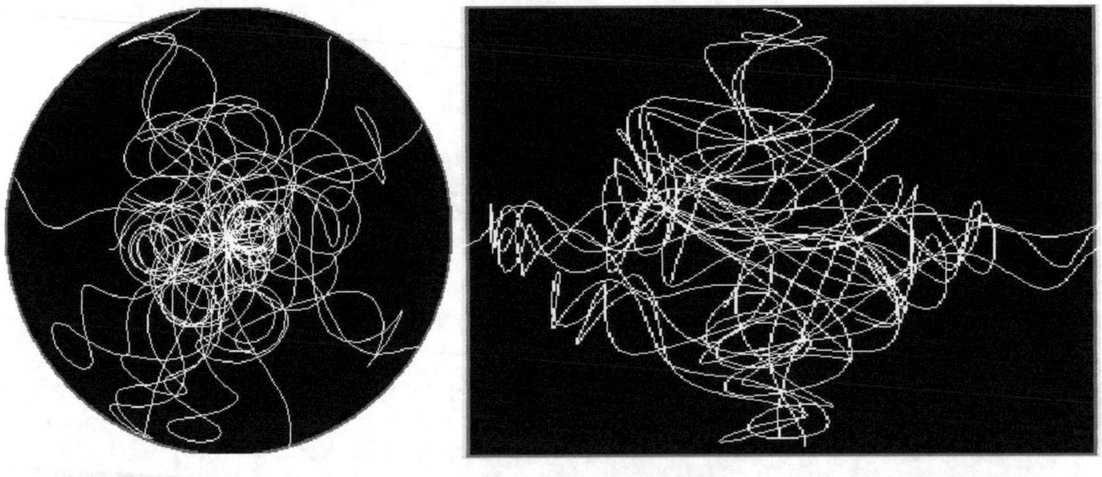

Fig. 4. Trajectories of 11 oxygen ions in the chamber immersed in the computed SMF.

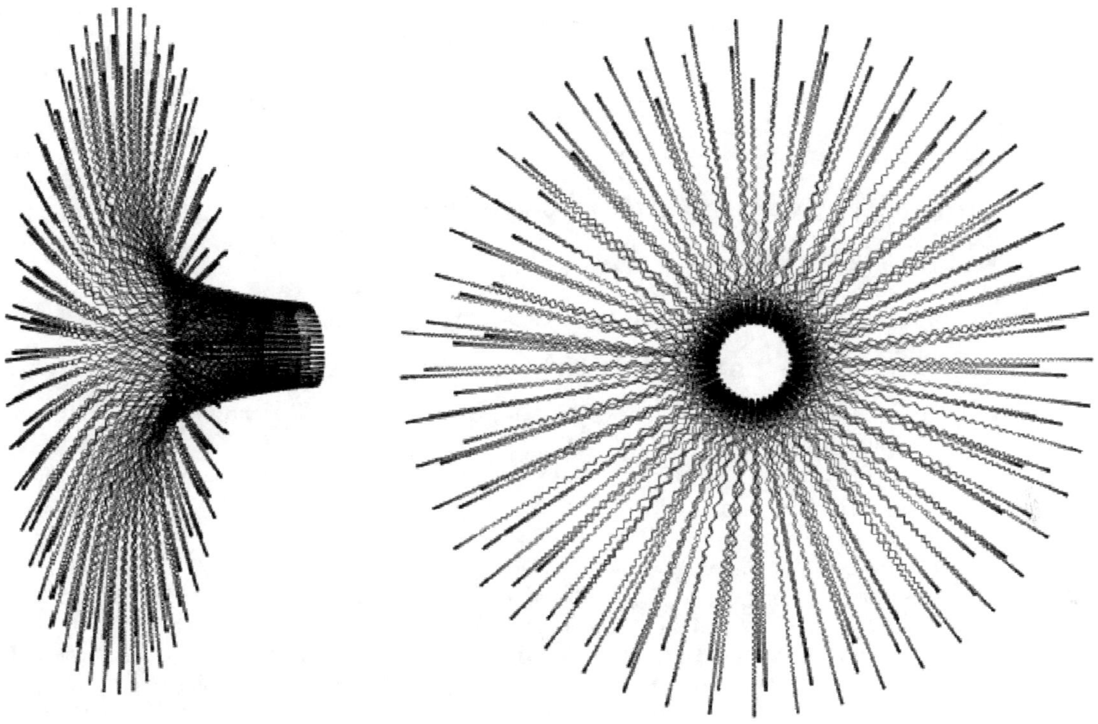

Fig. 5. Electron trajectory in the computed SMF inside the chamber.

When one hundred electrons were launched distributed uniformly in the vacuum chamber several times with different total energy and its components, it was found that about 42 electrons were left on the average in the vacuum chamber after 200 ns. So, it proves that the average containment time of the electrons is more than 200 ns. Though electrons are almost tied with the MLF's, they drift due to various reasons also. But the heavier particles are not so because of high momentum and they are bound to be lost on cathodes once they are ionized and reach its surface through curved path.

3. Conclusion

The described structure was taken as an example and shown that an ISP can be constructed using a SMF produced by a PM ring to suit the evacuation requirement. The magnetic mirror ratio on a curved MLF's terminating on the electrode surface increases appreciably as one move towards the centre from the chamber surface. It causes more electron confinement because of multiple mirror reflections, so more ionization. Thus, a compact and more powerful ISP can be constructed due to rapid ion loss.

The most common parameter used to express the capability of a pump used for evacuation of a given vessel is pumping speed. The ionization rate, titanium sputtering process, gas burial and re-emission on the titanium anode, which are related to magnitude and structure of electric and magnetic field, determine the pumping speed of the ISP. Experiments only will corroborate the proposed idea.

References

[1] Dahl Per F 1997 *Flash of the Cathod Rays: A History of J. J. Thompsons Electron* (CRC Press).
[2] Druvesteyn M J and Penning F M 1940 (1941) *Review of Mod. Phys.* **12 (13)** 87 (72).
[3] Menzel M T and Stokes H K 1987 Poisson *Superfish Group of Codes* **LANL** LA-UR-87-115.
[4] Rashid M H and Bhandari R K 2004 *Indian J. Phys* **78** p 995.
[5] Feket E, Derzi A, Biri S and Ivan I 2006 *The Upgraded TrapCAD code* ATOMKI Annual Rep. ref. code P20638.

International Symposium on Vacuum Science & Technology and its Application for Accelerators IOP Publishing
Journal of Physics: Conference Series **390** (2012) 012058 doi:10.1088/1742-6596/390/1/012058

Design and development of pulsed electron beam accelerator *'AMBICA – 600'*

Rishi Verma[1], Pankaj Deb, Rohit Shukla, Surender Sharma and Anurag Shyam

Energetics & Electromagnetics Division, Bhabha Atomic Research Centre, Autonagar, Visakhapatnam (A.P.), India 530012

E-mail: rishiv@barc.gov.in

Abstract. Short duration, high power pulses with fast rise time and good flat-top are essentially required for driving pulsed electron beam diodes. To attain this objective, a dual resonant Tesla transformer based pulsed power accelerator 'AMBICA-600' has been developed. In this newly developed system, a coaxial water line is charged through single turn Tesla transformer that operates in the dual resonant mode. For making the accelerator compact, in the high power pulse forming line, water has been used as dielectric medium because of its high dielectric constant, high dielectric strength and high energy density. The coaxial waterline can be pulsed charged up to 600kV, has impedance of ~5Ω and generates pulse width of ~60ns. The integrated system is capable of producing intense electron beam of 300keV, 60kA when connected to impedance matched vacuum diode. In this paper, system hardware details and experimental results of gigawatt electron beam generation have been presented.

1. Introduction

In the last few years, pulsed beam technology has reached to very sophisticated level. The growth of interest in the intrinsic behavior of intense relativistic electron beams and its utilization in various fields has also been grown very rapidly. Intense electron beams can be used for variety of applications like – high power microwave generation, heating of plasmas to high temperatures, plasma confinement, collective acceleration of ions, short pulse gas lasers, controlled thermo-nuclear fusion, flash X-Ray generation, material modification studies etc.

The basic structure of an electron beam accelerator for all mentioned applications consists of prime power source, high voltage generator, pulse forming line and a field emission (FE) diode. The sequential integration of these major constituents is shown in Fig. 1.

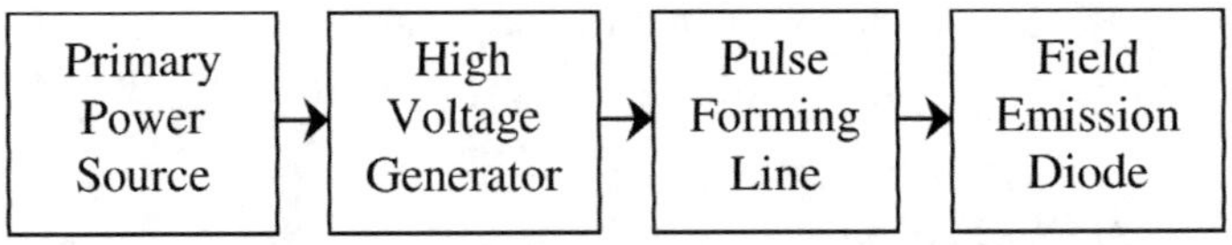

Fig. 1. Basic structure of a pulsed electron beam accelerator.

[1] To whom any correspondence should be addressed.

The DC charging of a high voltage energy storage capacitor provides primary power source to the system. Then a high voltage generating system is needed to transform the few ten's of kV in the primary power source to values in the range of several hundreds of kilo-Volts. Such high voltages can be produced either by using Marx generator or Pulse transformer. While utilizing Pulse transformer technique for voltage amplification, by tuning the parameters (LC) of primary (L_1C_1) and secondary (L_2C_2) circuits such that frequencies of free oscillations in the primary and secondary circuits become identical; the system operates in Tesla transformer mode. The main advantages of the Tesla transformer generator as compared to the Marx generator are [1]: (i) Possibility to work with rather high pulse-recurrence frequency (ii) Lower cost due to a small number of used high voltage capacitors that reduces weight and system dimensions and (iii) More reliability in the performance because of the absence of large number of sparkgap switches. The transient behavior of Tesla transformer is analyzed using simple equivalent circuit as shown in Fig. 2.

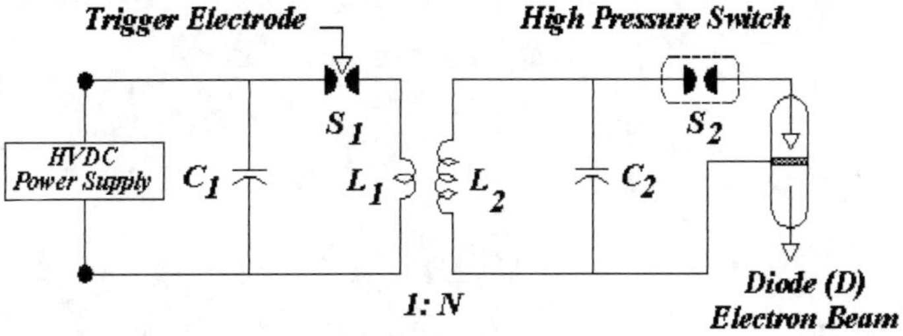

Fig. 2. Equivalent circuit of Tesla transformer based accelerator.

In brief, the principle of the Tesla transformer is based on the energy transfer between the two magnetically air-coupled and perfectly tuned resonating circuits: R_1-C_1-L_1 and R_2-C_2-L_2. R_1 and R_2 are the winding resistances of L_1 and L_2 respectively (not included in the figure). Suppose that, with open switch S_1, V_{DC} is charging C_1. When C_1 is sufficiently charged, S_1 is closed resulting in the discharge of C_1 in L_1. Finally, the primary capacitor C_1 acts together with the primary coil L_1 as a resonating circuit resulting in a damped oscillating wave of a certain frequency f_1. Because of the coupling between the two circuits, the oscillation frequency of the primary circuit appears in the secondary circuit as well. The condition required for the dual tuned circuits to operate in resonance mode is given as [1,2]:

$$f_1 = \frac{1}{2\pi\sqrt{L_1C_1}} = \frac{1}{2\pi\sqrt{L_2C_2}} = f_2 \quad (1)$$

$$L_1C_1 = L_2C_2 \quad (2)$$

The magnitude of amplified output voltage V_{C2} across C_2 depends on the pre-charging voltage V_{C1} at C_1 and the efficiency η of the energy transfer i.e. [2]:

$$V_{C2} = \eta\, V_{C1}\sqrt{C_1/C_2} \quad (3)$$

For the maximum energy transfer, it is mandatory to have a coupling coefficient between windings of the transformer accordingly as [1,2]:

$$k = M\sqrt{L_1L_2} = 1;\ 0.6;\ 0.385 \quad (4)$$

Pulse forming line is required for compressing the duration of the high voltage pulse (which is typically in the range of few μs, as an output from high voltage generator) in the range of 30 to 100ns. Another important role of the pulse forming line is to match the driver/accelerator impedance with vacuum diode. The typical impedance of vacuum diode used for various applications is in the range of

5 to 20Ω. Pulse forming/compression along with impedance matching is normally done either by using water or oil as dielectric in the coaxial line of defined length. In our system a high voltage capacitor has been used for primary energy storage, Tesla transformer for high voltage generation and water coax for pulse compression and impedance matching.

2. Experimental Setup

The dual resonant Tesla transformer [2] based AMBICA-600 pulsed power system comprises of primary energy storage capacitor, single turn pulse transformer, coaxial waterline, high voltage holding switch and field emission diode with voltage and current diagnostics. The overall view of the integrated system has been shown below in Fig. 3.

Fig. 3. Overall view of the integrated AMBICA – 600 system.

2.1. Primary energy storage

To energize the primary of the Tesla transformer, a 7.1μF low inductance capacitor that can be charged up to maximum limit of 40kV was used. The capacitor has self- mounted triggerable spark gap on top of it that facilitates direct connection between capacitor and energy transfer switch. The switch is triggerable in the range of 8kV to 40kV by adjusting appropriate Nitrogen (N_2) gas pressure in the gap. Delay and jitter in the main spark gap is minimized by utilizing a trigger voltage of magnitude double the maximum voltage to be switched i.e. ~80kV. The peak current limit of this capacitor for pulsed operation is ~150kA. For making the connection between primary energy storage capacitor and Tesla transformer while keeping the bank inductance low, eight numbers of custom made double braided coaxial cables (equivalent to RG 217 or YK198) have been used in parallel. These indigenously developed cables have impedance of ~19Ω and inductance of ~100nH/m. The material of double layer braid is silver plated copper and Teflon has been used as dielectric in between the layers. These cables have passed high voltage standoff test up to 40kV DC.

2.2. High voltage generator

Efficient design of Tesla transformer depends upon parameters like coefficient of coupling, tuning ratio, voltage step up ratio and dielectric strength. The design for a given secondary output voltage and coupling coefficient is an iterative process and was started by assuming values for the step up ratio and the primary input voltage which in turn fixes the ratio L_1/L_2. For attaining high reliability and repeatability, trade off has been made between good coupling coefficient and high voltage stand off. The construction of an indigenously designed single turn pulse transformer has been shown in Fig. 4.

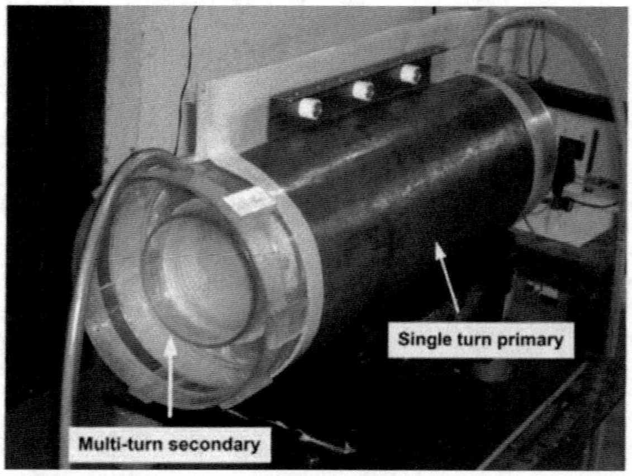

Fig. 4. Constructional view of the indigenously designed pulse transformer.

The single turn primary (outside) is made up of 1mm thick Copper sheet having width ~600mm. It is fixed on insulated Acrylic cylindrical support of mean diameter 250mm. The measured inductance of the single turn primary is ~90nH and the total inductance (L_1) of primary side (including contribution of all connections) is ~220nH. The secondary winding has been concentrically supported inside the primary. It has 70 turns made of polyethylene insulated RG213 inner conductor. The inductance of the secondary winding (L_2) is ~200μH. The mutual inductance (M) and coupling coefficient (*k*) obtained for the Tesla mode operation are 2.42μH and 0.54 respectively. The secondary output of the transformer is connected to the input of coaxial waterline.

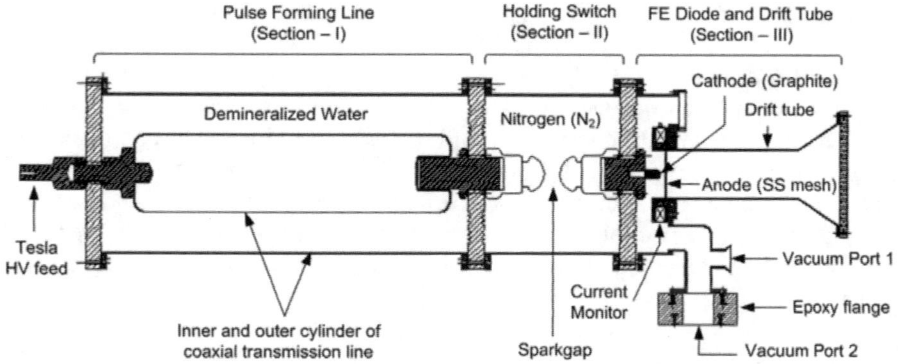

Fig. 5. Constructional schematic of the integrated Pulse Forming Line assembly.

2.3. Pulse forming line and High voltage holding switch

A water line that acts as the secondary capacitor during the resonance charging has been used for pulse compression. Since the dielectric constant (ε_r) of the deionised water is ~80, therefore it facilitates compact construction with high energy density. The constructional layout of the used coaxial waterline along with high voltage holding switching and the vacuum diode assembly has been shown in Fig. 5. The most important consideration in the design of pulse forming line is to choose appropriate physical dimensions (mainly diameter and length) of inner and outer conductors since choice of these parameters decide [3-5] (i) driver impedance (ii) pulse duration of the compressed pulse (iii) electric field stress (across the inner and outer conductors) and lastly but not the least (iv) tuning of the secondary circuit. In the utilized water coax, the inner conductor diameter, outer conductor diameter and their common lengths are 250mm, 500mm and 1000mm respectively. These physical parameters

Journal of Physics: Conference Series **390** (2012) 012058 doi:10.1088/1742-6596/390/1/012058

define pulse forming line characteristics as (i) impedance ~5Ω (ii) duration of compressed pulse ~60ns and (iii) capacitance (C_2) of the pulse forming line (PFL) as ~7nF.
In coaxial configuration, the electric field stress E_r at a radius r is given by the relation:

$$E_r = \frac{V/r}{\ln(b/a)} \tag{5}$$

Here a and b are the diameters of the inner and outer conductors respectively, and V is the voltage between two conductors. For the maximum charging voltage of 600kV, the estimated electric field stress is ~70kV/cm close to the surface of inner conductor and ~35kV/cm near the surface of outer conductor. The resistance of the water line is estimated using expression:

$$R_w = \frac{\rho}{2\pi l}\ln\left(\frac{b}{a}\right) \tag{6}$$

For the utilized de-ionised water resistivity ρ of 1MΩcm, the total resistance of the coaxial line corresponds to ~1.1kΩ. For maintaining resistivity >1MΩcm, water is continuously kept in closed loop circulation though a large capacity monobed deioniser (Cole-Parmer, model # WW-1503-05).

The high voltage holding switch adjoined to the PFL output, discharges the pulse forming line at peak charging voltage, into the field emission diode that generates intense relativistic electron beam (REB). The sparkgap is made up of brass electrodes and mounted on Acrylic flange having equidistant circular grooves along the radial surface to avoid breakdown. The clearance between the facing tips of the electrodes is kept as 5cm. The breakdown voltage is further adjusted by a controlling the Nitrogen filling gas pressure inside the chamber. The sparkgap is designed to operate in the self-breakdown mode and it can be adjusted to switch voltage in the range of 150kV – 600kV.

2.4. Field Emission Diode

Conversion of high voltage nanosecond pulses into electron beam requires the use of low impedance vacuum diode. The diode assembly consists of cold cathode and closely spaced from it is the anode. When a fast rising high voltage pulse is applied then due to field enhancement plasma is formed over cathode surface and emission of electron starts [6,7]. The most commonly used materials as cold cathode for 'field enhanced' electron emission are Aluminium, Graphite and Copper. The current density J emitted from the cathode is space charge limited and in the one dimensional case, obeys the Child-Langmuir Law given as [6]:

$$J = 2.33\times 10^{-6}\left(\frac{V^{3/2}}{d^2}\right) Amp/m^2 \tag{7}$$

Here, V is the accelerating voltage and d is the anode cathode gap spacing in MKS units. The effective gap spacing d tends to decrease with time as the cathode plasma expands towards the anode with velocity, typically of the order of 3cm/μs. This phenomenon is called 'plasma gap closure' and causes the current to grow with respect to time until the diode completely shorts. For planar diode consisting of cylindrical cathode of radius r_c and distance between the anode-cathode (AK) spacing d, impedance of the diode Z_d is given by relation [6-8]:

$$Z_d = 136\times\frac{(d/r_c)^2}{\sqrt{V}} \tag{8}$$

Here, V is the voltage applied across field emission diode in MV. In our case, the diode consists of a planar graphite cathode (of 40mm diameter) and stainless steel anode mesh (of 240mm diameter). The typical impedance of diode for cathode-anode spacing of 3mm, at 200kV is ~6.84 ohms. It may be noted that, impedance of the pulse forming line is ~5Ω and the actual voltage that appears across the diode depends on the division ratio of impedance of the pulse forming line and the dynamic impedance of diode.

3. Diagnostics and Experimental results

A capacitive voltage divider and self-integrating Rogowski coil have been used to measure the peak charging voltage of PFL and diode current pulse respectively. The most important parameter of the generator is output current. The bandwidth response of self-integrating Rogowski coil was initially confirmed by 60ns duration, 60A square pulse. It was found that the high frequency response was remarkably good while the calibrated factor was 0.01V/A. To investigate the performance of the integrated system, numbers of experiments were performed. The coupling between primary and secondary circuit and charging of the PFL was non-intrusively measured by an indigenously developed capacitive voltage divider installed at the HV feed point of waterline.

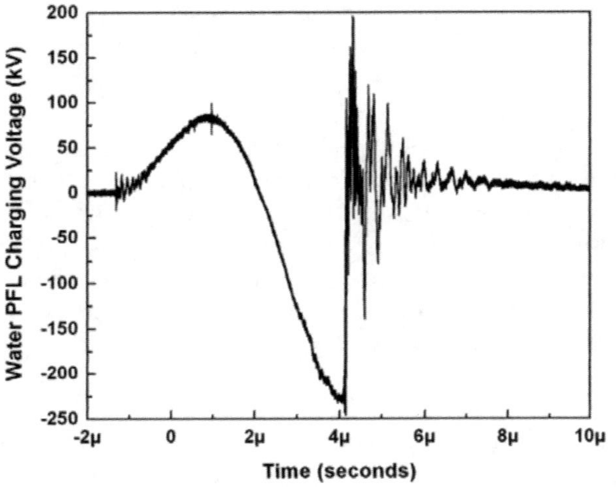

Fig. 6. Oscillogram of the PFL charging voltage.

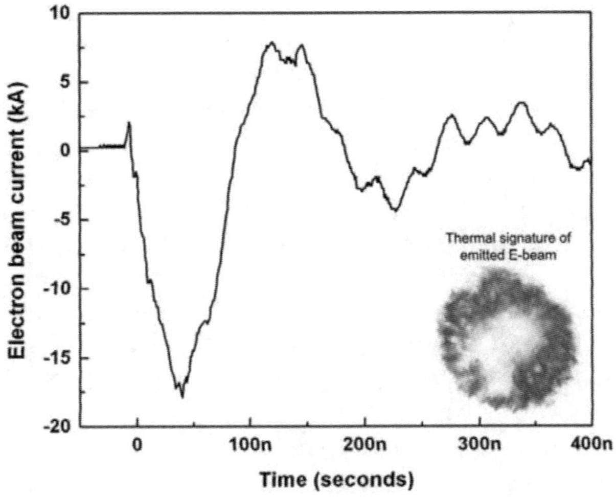

Fig. 7. Field emission diode current waveform.

The typical oscilloscope trace of dual resonant Tesla output voltage charging the waterline, along with the firing of hold-off switch at the negative peak (at ~230kV), has been shown in Fig. 6. The N_2 gas pressure and spacing between the switching electrodes of high voltage holding switch is so adjusted that the switch breaks down at the second peak of the pulsed high voltage that charges pulse forming line. The experimentally measured diode current at cathode-anode spacing of 3mm was ~17kA while the duration (FWHM) of the pulse was ~60ns. This value of diode current is in agreement with Child-Langmuir relation. The oscillogram of diode current waveform is shown in Fig. 7. The peak power of emitted electron beam was ~2.4GW. The background pressure in the diode as well as in the drift region was maintained < 5×10^{-4} Torr and no external magnetic field was applied. The signature/impression of emitted electron beam obtained on thermal paper (in a single shot) is also shown in the inset.

From the thermal impression of emitted electron beam it is inferred that in cathodes having flat profile, radial distribution of energy flux is mostly concentrated at the edge because of the high electric field and hence the beam is emitted in annular fashion.

4. Conclusion

A dual resonant Tesla transformer based pulsed electron beam accelerator 'AMBICA–600' has been fabricated and tested successfully. It is remarkable to note that in this system, the length of water based pulse forming line (ε_r ~80) is ~1/4 the length of conventionally used pulse forming line using oil dielectric (ε_r ~4.7), hence new accelerator is relatively much compact. In wide range of experiments electron beams of 100–200keV range with peak current up to 20kA lasting for 60ns have been obtained. The system is presently being used for conducting short pulse high power microwave (HPM) generation experiments using Virtual Cathode Oscillator device as load. Along with the utilization of this newly developed system for various applications, we also plan to study the effect of other candidate cathode materials and their geometries on extracted beam current. Some experiments with various anode mesh transparencies will also be preformed since the extent of transparency may affect the space charge cloud formation considerably.

Acknowledgment

Authors are highly thankful to Shri G. P. Shrivastava, Director, E&I Group and Shri S. Bhattacharya, Associate Director (T), E&I Group, BARC, Mumbai for giving us consistent encouragement and support to continue this project.

References

[1] Kolchanova V. A., Nosov G. V., Petrov A. V., Usov Yu. P., 2001 *Issues of Nuclear Science and Techniques* **5**, 43.
[2] Jain K. K., Chennareddy D., Jhon P. I. and Saxena Y. C., 1986 *Sadhana* **9**, 19.
[3] Jain K. K and John P. I., 1981 *Proc. Indian Acad. Sci.* **4,** 75.
[4] John P. I., 1982 *Proc. Techn. of Electrical Insulation and High Voltage Pulse Techniques*, BARC, Mumbai, India.
[5] Katsuki S., Takanao D., Namihira T., Akiyama H. and Majima T., 2001 *Rev. Sci. Instrum.* **72**, 2759.
[6] Nation J. A., 1979 *High Power Electron and Ion Beam Generation Particle Accelerators* **10**, 1.
[7] Boscolo I., Leo M., Luches A. and Provenzano L., 1977 *Rev. Sci. Instrum.* **48**, 747.
[8] Suzuki T., Murakami H., Saito Y., Yamagishi A. and Inaba H., 1980 *Rev. Sci. Instrum.* **51**, 1485.
[9] Benford J. and Swegle J., 1992 *High Power Microwaves*, Boston: Artech House.
[10] Buttram M. T., 1979 *IEEE Trans. on Nuclear Sciences* **NS-26,** 4183.

Influence of substrate bias voltage on structural and optical properties of RF reactive magnetron sputtered WO_3 thin films

V. Madhavi, P. Kondaiah and S. Uthanna

Department of Physics, Sri Venkateswara University, Tirupati – 517 502, India.

E-mail: madhuvphysics@gmail.com

Abstract. In this investigation, tungsten oxide (WO_3) thin films have been deposited on unheated glass substrates by RF magnetron sputtering of tungsten target at constant oxygen partial pressure of $6x10^{-2}$ Pa, sputtering power of 150 W and at different substrate bias voltages in the range from 0 to -150 V. The structural and optical properties of the films were investigated by using X-ray diffraction, Fourier transform infrared spectroscopy and UV-Vis-NIR spectrophotometers. The films formed up to substrate bias voltage of -50 V were found to be amorphous in nature, while those films formed at -100 V showed a weak (200) reflection indicates the presence of monoclinic phase of WO_3 in amorphous matrix. The optical transmittance of the films was in the range 78 - 95 % in the wavelength range of 500 – 1200 nm. The absorption band edge was shifted towards higher wavelength side with increasing substrate bias voltage. The optical band gap of the films decrease from 2.81 to 2.69 eV with increase of substrate bias voltage from 0 to -150 V respectively. The refractive index of the films increased from 2.19 to 2.27 with increase of substrate bias voltage from 0 to -150 V respectively.

1. Introduction:

In recent past, tungsten oxide (WO_3) has been extensively studied for its distinctive physical and chemical properties especially the electrochromic, photochromic, gas sensors properties. Electro chromism in amorphous tungsten oxide films has been extensively studied since it was discovered in 1961. The improvement of material properties requires a closer inspection of preparation conditions. Various techniques such as thermal evaporation [1], pulsed-laser evaporation [2], RF sputtering [3], spray pyrolysis [4] and sol-gel processing [5] etc., have been employed for deposition of WO_3 films. The chemical composition, microstructure and physical behaviour of the deposited films relies more on the control of deposition parameters such as substrate temperature, gas pressures, sputtering power and substrate bias voltage. The primary purpose of biasing is to prevent the deposited film from becoming contaminated with gas atoms and other gas/metal reaction products. A negative potential applied to the substrate vary the ion energy, and further influence the nucleation and growth kinetics of the films. Subsequently, the microstructure and physical properties of the films are greatly modified. The present study is aimed to investigate the influence of substrate bias voltage on the growth characteristics of RF reactive magnetron sputtered WO_3 films.

2. Experimental

The tungsten oxide thin films were deposited onto glass and Si substrates from tungsten metal target (99.99%) at room temperature by RF reactive magnetron sputtering technique. Pure tungsten target of 50 mm diameter and 3 mm thick was used as a sputter target. Pure oxygen and argon were used as reactive and sputtering gases respectively. The required quantities of oxygen and argon gases were admitted into the sputter chamber through the fine controlled needle valves and their flow rates were monitored individually employing Aalborg mass flow controllers. Tungsten target was pre-sputtered in pure argon atmosphere for 15 minutes before deposition of each film in order to remove the oxide layer formed if any on the target. A shutter was provided in between the sputter target and the substrates to isolate the substrates during the pre-sputter process. The WO_3 films were deposited under various substrate bias voltages (V_b) in the range 0 to – 150 V by fixing the other process parameters constant as shown in table 1.

Table.1. Deposition conditions maintained during the preparation of RF magnetron sputtered WO_3 films.

Sputter target	: Tungsten (50 mm dia, 3 mm thick)
Target to substrate distance	: 50 mm
Base pressure	: $2x10^{-4}$ Pa
Oxygen partial pressure	: $6x10^{-2}$ Pa
Sputtering pressure	: 4 Pa
Sputter power	: 150 W
Substrate bias voltage	: 0 to -150 V
Sputter time	: 120 min

The thickness of the films was determined using Dektek depth profilometer. The crystallographic structural properties were determined by X-ray diffractrometer. The chemical binding configuration was analysed with Fourier transform infrared spectrophotometer. The surface morphology of the films was analysed with scanning electron microscope. Optical transmittance spectra of the films were recorded with UV-Vis-NIR double beam spectrophotometer in the wavelength range 300 – 1200 nm. The refractive index of the deposited films in the visible region 400 - 800 nm was measured using Filmetrics F-20.

3. Results and Discussion

The measured thickness of the films was in the range of 370 – 390 nm. Fig 1.Illustrates the dependence of deposition rate on the substrate bias voltage for the WO_3 films.

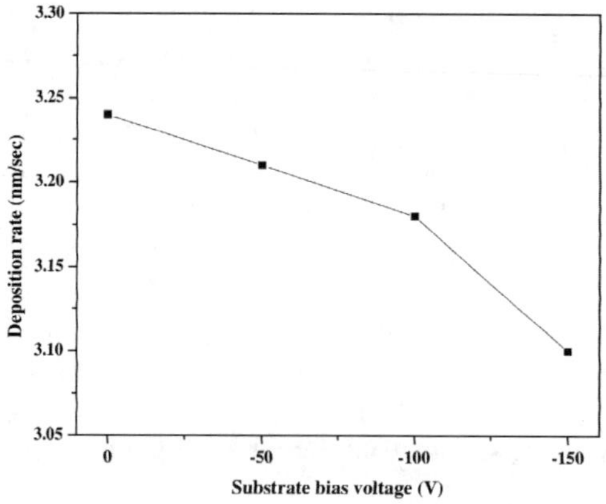

Fig. 1. Variation of deposition rate with substrate bias voltage of WO_3 films.

The deposition rate was decreased from 3.24 to 3.10 nm/sec with increase of substrate bias voltage from 0 to – 150 V respectively. As increase of substrate bias voltage, the molecules get re-sputtered from the substrates and hence the deposition rate was decreased. Such a decrease in deposition rate with substrate bias voltage was also noted [6]

Fig 2. shows the XRD patterns of WO_3 thin films formed at different substrate bias voltages. It is evident from the figure that no peaks attributed to the WO_3 can be found when films sputtered at unbiased and biased at -50 V, indicating that the films are in x- ray amorphous. The amorphous diffraction pattern may result from the amorphous property and internal stress of tungsten oxide films, which is usually a property of the films prepared by sputter deposition at room temperature [7]. When the substrate bias voltage was increased from -100 to -150 V a weak diffraction peak is present at 2□ = 24.36^{o} which corresponds to the (200) plane of monoclinic structured WO_3 phase [JCPDS 43-1035].

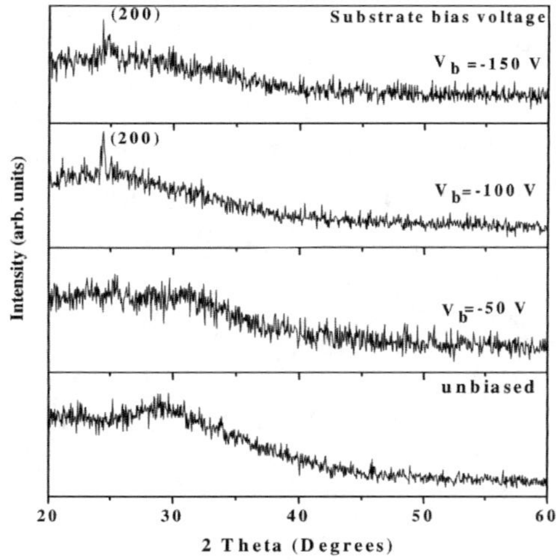

Fig. 2. XRD profiles of WO_3 films deposited at different substrate bias voltages.

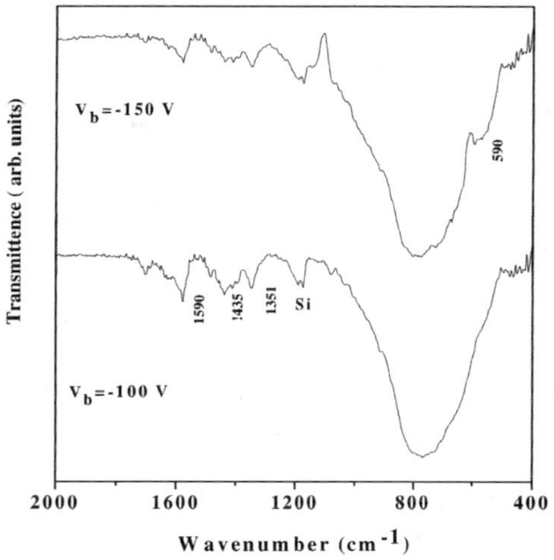

Fig. 3. FTIR spectra of WO_3 films formed at -100 and -150 V.

Before giving the experimental results lets recall briefly the tungsten oxide structure and infrared features WO_3 consists of packed corner sharing WO_6 octahedral. The corresponding tungsten oxide vibrations are characterized by three infrared regions at 900 - 600 cm^{-1}, 400 - 200 cm^{-1} and < 200 cm^{-1} corresponds to (O – W - O) stretching, bending and lattice modes respectively [8]. Fig 3. Shows the broad hump around 1000-500 cm^{-1} region arise from the □ (W - O_{inter} - O) and □ (W - O_{inter} - W) stretching frequencies, respectively [9]. At substrate bias voltage of -150 V a small absorption peak located at 590 cm^{-1} is related to W - O corner deformation mode of WO_3. The general broadness of the IR peaks conformed the films to amorphous nature as observed by XRD. The weak bands situated at 1351 and 1590 cm^{-1} resulting from weakly bound water contaminants due to exposure of the sample to the ambient atmosphere. Such a presence of water contamination was also reported by Nguyen et al. [10]. Another weak band at 1435 cm^{-1} have been assigned to the stretching vibration of NH_4 due to contamination with ambient atmosphere [11]

Fig 4. shows the optical transmittance spectra of WO_3 films deposited at room temperature with different substrate bias voltages. The optical transmittance of all the films is above 80% in the visible region. The transmittance slightly decreases as the substrate bias voltage increase. The optical absorption edge was shifted to higher wavelength with increase of substrate bias voltage from 0 to -150 V. The fundamental absorption edge of WO_3 films is formed by non-direct permitted transition of the 2p electron from the valance band of the oxygen to the 5d condition bands of tungsten [11]. The optical absorption coefficient (□) of the films was calculated from the optical transmittance (T) data where the reflection losses were taken into account. The optical band gap (E_g) of WO_3 films was determined by extrapolating the linear portion of the $(□h□)^2$ versus photon energy (h□) to $(□h□)^2 = 0$. The optical band gap of the unbiased films was 2.81 and it is found to decreased to 2.69 eV with increase of substrate bias voltage to -150 V as shown in Fig 5.

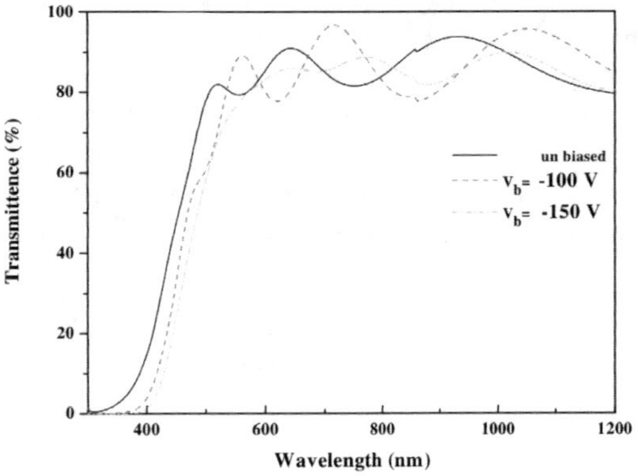

Fig. 4. Transmittance spectra of WO_3 films.

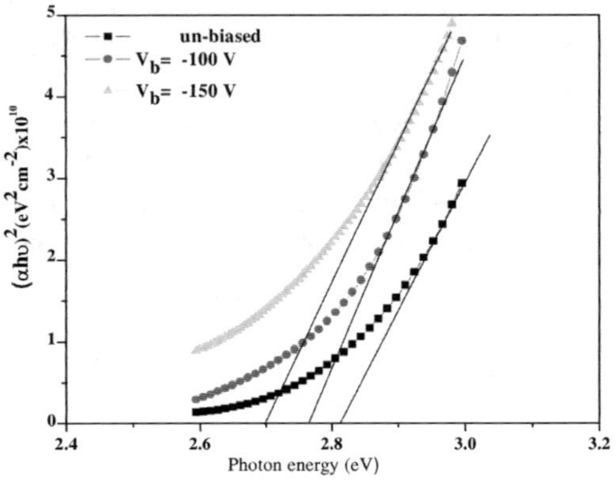

Fig. 5. (□h□)^2 versus photon energy (h□) of WO_3 films.

The optical band gap of bulk WO_3 is reported to be 2.62 eV, the larger band gaps observed are possibly due to amorphous nature of the WO_3 films. The decrease of band gap is due to the creation of oxygen ion vacancies in the films. Following the earlier report that the denser films shows lower band gap than the porous one [12], the present observation may be attributed to the difference in the density of the films. The refractive index of the films decreased with increase of wavelength as shown in Fig.6. The refractive index of the films at wavelength of 550 nm increased

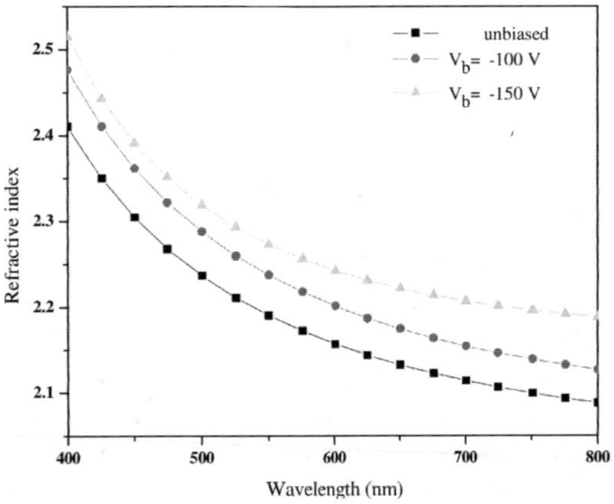

Fig. 6. Wavelength dependence of refractive index of WO_3 films with wavelength.

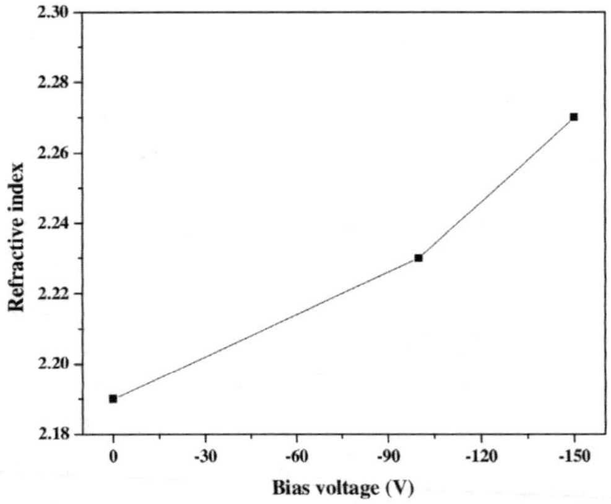

Fig. 7. Substrate bias voltage dependence of refractive index of WO 3 films.

from 2.19 to 2.27with the increase of substrate bias voltage from 0 to – 150 V as shown in Fig 7. The un biased films showed the lower refractive index due to the low packing density because of the amorphous nature. The increase in the refractive index with the substrate bias voltage was due to the improvement in the crystallinity and packing density of the films. Such an increase in the refractive index with the substrate bias voltage was also observed in Dc magnetron sputtered Ta_2O_5 films [13].

Fig 8. Shows the scanning electron microscopic images of WO_3 films deposited at different substrate bias voltages. From Fig 8 it is clear that agglomeration of grains takes place with increase of substrate bias voltage. Applying negative bias voltage to the substrates increases ion bombardment and reducing the nucleation energy, giving rise to the variations in the surface morphology and microstructures of the films.

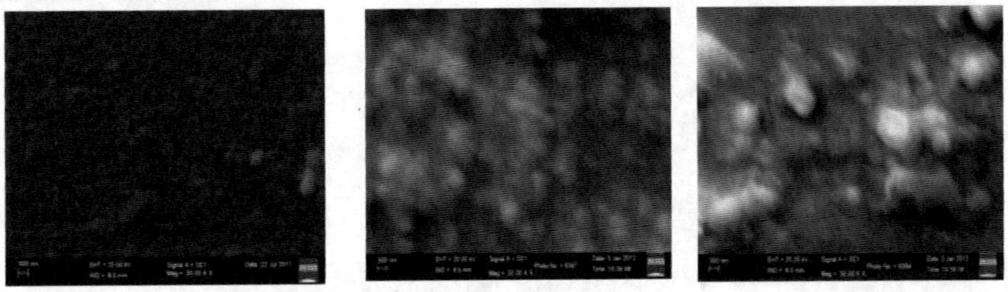

(a) (b) (c)

Fig. 8. SEM images of WO_3 films formed at substrate bias voltages (a) unbiased, (b) -100 and (c) -150 V.

4. Conclusions

WO_3 thin films were deposited by RF magnetron sputtering technique at different substrate bias voltages. The effects of substrate bias voltage on the structural, optical and morphological properties of the films were investigated. The deposition rate was decreased from 3.24 to 3.10 nm/sec with increase of substrate bias voltage from 0 to –150 V. The amorphous nature of WO_3 films were observed at unbiased and biased at -50 V. The predominant peak of (200) reflection was observed at higher bias voltages of -100 and -150 V. From the FTIR studies it was observed that the broad hump around 1000-500 cm^{-1} region arise from the □ ($W - O_{inter} - O$) and □ ($W - O_{inter} - W$) stretching frequencies. The optical band gap was decreased from 2.81 to 2.69 and the refractive index increased from 2.19 to 2.27 with increase of substrate bias voltage from 0 to -150 V respectively.

Acknowledgement

One of the authors V. Madhavi, thankful to UGC for awarder of UGC-RFSMS Junior Research fellowship to carry out the present work.

5. References

[1] Feng M, Pan A L, Zhang H R, Li Z A, Liu H W, Shi D X, Zon B S and Gao H 2005 *Appl. Phys. Lett.* **86** 141901
[2] Lu Y F and Qiu H 2000 *J. Appl. Phys.* **88** 1082
[3] Li W J and Fu Z W 2010 *Appl. Surf. Sci.* **256** 2447
[4] Leon J M, Acosta D R, Pal U and Castaneda L 2010 *Electrochim. Acta* doi:10.1016/j.electacta.2010.11.038
[5] Deepa M, Srivastava A K, Kar M and Agnihotry S A 2006 *J. Phy. D: Appl. Phys.* **39** 1885
[6] Wang Q M and Kim K H 2008 *J. Vac. Sci. Technol.* A **26** 1258
[7] Wang X G, Jiang Y S, Yang N H, Yuan L and Pang S J 1999 *Appl. Surf. Sci.* **143** 135
[8] Daniel M F, Desbat B, Lassegues J C, Gerand B and Figlarz M 1987 *J. Solid. State. Chem.* **67** 235
[9] Srivastava A K, Deepa M, Singh S, Kishore R and Agnihotry S A 2005 *Solid State Ionics* **176** 1161
[10] Nguyen T A, Park S, Kim J B, Kim T K, Seong G H, Choo J and Kim Y S 2011 *Sensors and Actuators B.* doi:10.1016/j.snb.2011.08.028
[11] Rougier A, Portemer F, Quede A and El Marssi M 1999 *Appl. Surf. Sci.* **153** 1
[12] Sun X, Liu Z and Cao H 2010 *J. Alloys and Compounds* **5045** 5418
[13] Chandra S V J, Chandra Sekhar M, Mohan Rao G and Uthanna S 2009 *J Mater Sci: Mater Electron.* **20** 295

Conceptual Design of Vacuum Chamber for testing of high heat flux components using electron beam as a source

M S Khan, Rajamannar Swamy, S S Khirwadkar and Prototype Divertors Division

Institute for Plasma Research, Bhat, Gandhinagar-382428, Gujarat.

E-mail: khan@ipr.res.in

Abstract. A conceptual design of vacuum chamber is proposed to study the thermal response of high heat flux components under energy depositions of the magnitude and durations expected in plasma fusion devices. It is equipped with high power electron beam with maximum beam power of 200 KW mounted in a stationary horizontal position from back side of the chamber. The electron beam is used as a heat source to evaluate the heat removal capacity, material performance under thermal loads & stresses, thermal fatigue etc on actively cooled mock – ups which are mounted on a flange system which is the front side door of the chamber. The tests mock – ups are connected to a high pressure high temperature water circulation system (HPHT-WCS) operated over a wide range of conditions. The vacuum chamber consists of different ports at different angles to view the mock –up surface available for mock -up diagnostics. The vacuum chamber is pumped with different pumps mounted on side ports of the chamber. The chamber is shielded from X – rays which are generated inside the chamber when high-energy electrons are incident on the mock-up. The design includes development of a conceptual design with theoretical calculations and CAD modelling of the system using CATIA V5. These CAD models give an outline on the complete geometry of HHF test chamber, fabrication challenges and safety issues. FEA analysis of the system has been performed to check the structural integrity when the system is subjected to structural & thermal loads.

1. Introduction

The Divertor is one of the key components of the TOKAMAK. Its function is to extract heat and helium ash (the products of the fusion reaction) and other impurities from the plasma. It comprises of mainly two parts viz. a supporting structure made primarily from stainless steel and the plasma-facing component. A cross-sectional view of ITER Tokamak is shown in figure 1. The Plasma Facing Components (PFC) will be made of tungsten and Carbon Fiber Composites (CFC). Plasma facing materials have to withstand particle and heat loads from the plasma and neutron loads during reactor operation [1-2]. A broad spectrum of high heat flux test facilities are being used worldwide to investigate the thermal response of plasma facing materials and components to fusion relevant thermal loads [3-4]. These tests cover both normal operation scenarios with cyclic thermal loads and power densities in the range of several MW/ m^2 and transient heat load tests to simulate short events such as edge-localized modes, plasma disruptions and vertical displacement events.

Journal of Physics: Conference Series **390** (2012) 012060 doi:10.1088/1742-6596/390/1/012060

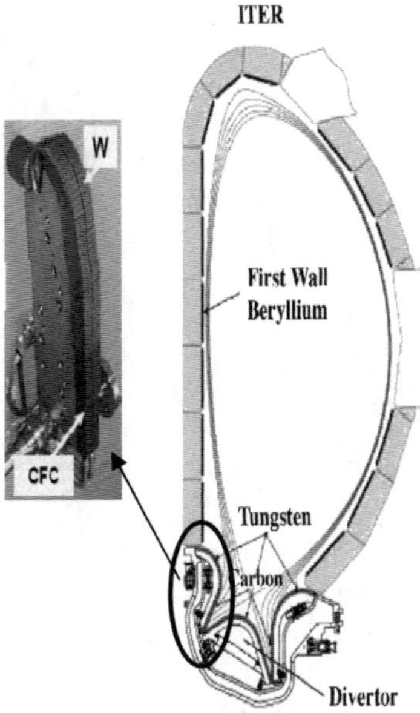

Fig. 1. Cross-sectional view of ITER Tokamak showing divertor and First wall [1]

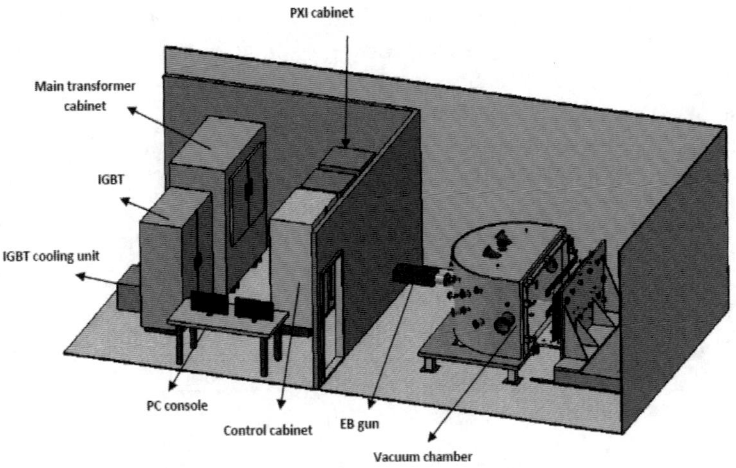

Fig. 2. Schematic of proposed High Heat Flux Test Facility at IPR

2. Objectives of High Heat Flux (HHF) test facility

The primary objectives of HHF test facility is to provide the condition needed to evaluate the performance of PFMs and plasma facing components under ITER specific thermal loads i.e. quasistationary heat fluxes up to approximately 10MW/m^2 (< 20MW/m^2 during slow transients for a duration of <10s) and sufficiently large number of cycles. To verify heat transfer coefficients/correlations currently in use by comparing finite-element calculations with measured temperatures. A secondary objective will be to collect enough data on temperature distribution in the test module so that future calculation of heat transfer coefficients by inverse conduction analysis is possible. A schematic of the proposed high heat flux test facility at IPR (HHF test facility) is shown in figure 2 and different types of HHF components with testing parameters are shown in Table 1.

Table 1: Different types of HHF components and testing parameters [1-2]

High heat flux components		**ITER Divertor**
Design	Flat tile Brush Monoblock	
Heat flux	Average MW/m^2 Maximum MW/m^2	3-5 10-20
Maximum heat load MJ/m^2 Life time Years No. of full load cycles Neutron damage dpa Structure material		10 3 3000 0.2 CuCrZr & CFC/W
Coolant Pressure MPa Temperature ^{0}C Velocity m/s Leak rate g/s		Water 4 100 – 150 9 – 11 < 10^{-7}

3. Worldwide HHF Test Facility

In order to study the effect of these high heat loads, several electron beam-testing facilities in different countries have been utilized for testing of high heat flux components of next step fusion devices. In principle all the testing facilities are comparable. They consist of an electron gun, a beam sweeping system, a vacuum chamber in which samples can be inserted, and a number of diagnostic devices. However, some machine parameters are quite different. Table 2 represents the comparison study on various electron beam facilities worldwide based on different parameters listed in table 2.

Table 2: Machine parameters for the electron beam facilities all over the World [3-4]

FACILITIES	JUDITH 2	JEBIS	TSEFEY	FE 200	EBTS	EB 1200	IPR
Max voltage (KV)	30-60	100	30	200	30	40	45
Max current (mA)	4000	4000	2000	1000	1000	30000	4444
Max power (KW)	200	400	60	200	30	1200	200
Max heated area (m^2)	0.25	0.18	0.25	1.0	0.10	0.27	0.16
Beam generation	Thermal emission	Plasma discharge	Thermal emission	Thermal emission	Thermal emission	Thermal emission	Thermal emission
Particle type	e^-	e^-	e^-	e^-	e^-	e^-	e^-
Power density (GWm^{-2})	10	2	0.2	60	-	10	1.2

4. Different subsystems

(a) Vacuum Chamber
(b) Electron beam equipment
(c) Different types of diagnostic

Here, we will discuss only about conceptual design of Vacuum chamber. We have chosen a D-shape vertical chamber as it reduce the volume, surface area as compare to horizontal cylindrical chamber. Electron gun is projected horizontally from backside and will be supported separately with a separate arrangement. The dimensions of the chamber are assumed to be of diameter 2.4 m, height 1.5 m with double wall cooling system to accommodate full-scale mock-up for testing. Conceptual design of D – shape double wall vertical chamber is shown in figure 3. The mounting arrangement for mock – up is designed in such away that it is supported from outside such that weight of the mock – up is not coming on to the chamber. A front door flange with cantilever support plates is designed and shown in figure 4. The supporting structure for the chamber is designed to be a flat surface with leg type support. The position of diagnostic ports are designed in such away that all the ports can view the total area of the mock – up such that the data at any point on the mock – up can be easily done through the diagnostic devices.

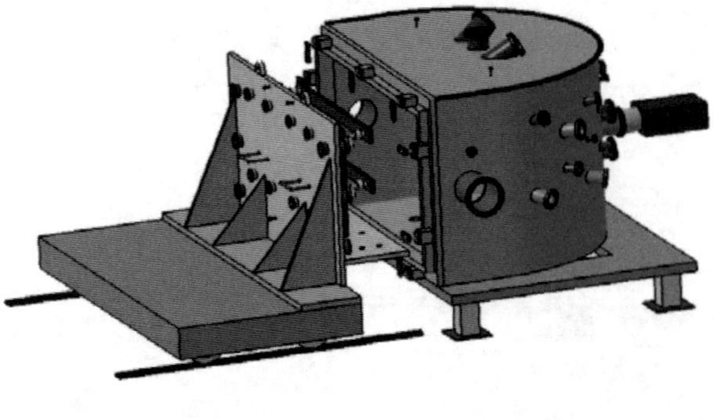

Fig. 3. Conceptual design of D-shape double wall vertical chamber

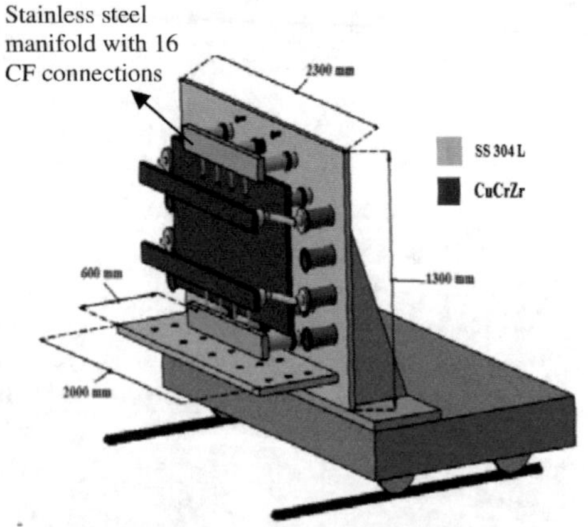

Fig. 4. Front door with struts and bracket arrangement

For the positioning the different ports, the semicircular area of the chamber is divided into 15^0 angles and is projected vertically. On each 15^0 axis, the angular position of ports are done based on viewing the port at a particular point, such that all the ports through out the circumference are projected at particular points at the front end which makes easy observation during experiment. The front door flange is used for mounting PFC with the help of cantilever support plate and the PFCs are masked with copper blocks. The distance between mock – up and electron beam point is calculated to be of 1.7 m such that the beam can raster the total surface area of the mock – ups of different sizes ranging from small scale to large scale.

5. Design procedure for cylindrical shell

The required thickness of cylindrical shell or tube exclusive of corrosion allowance under external pressure either seamless or with longitudinal butt-welded joint as per *ASME Boiler and Pressure Vessel Code* Section VIII Division [5] is determined and shown in table 3.

Table 3: Design parameter for Shell thickness and Flat head

S.No	Parameter	Unit	Value
Design of shell thickness for external pressure as per UG-28 (C)			
1.	External pressure	Pe (Bar/PSI)	1.013/14.7
2.	Temperature	T(°C)	27
3.	Material for construction • Shell plates, flanges, blanks: ASTM/ASME SA 240 Grade SS 304L • Collar for the rectangular front plate: forged plate F304L • Reinforcing bars (channel) made from plates of ASTM/ASME SA 240 Grade SS 304L or SA 182 F304L • Column: MS channels of IS2062		
4.	1st assumed thickness (due to external pressure)	ta (mm)	8
5.	Unsupported length	L (mm)	1500
6.	Outside diameter	Do (mm)	2400
7.	Allowable pressure Pa = 4B/(3(Do/t)), where t = ta	Pa(PSI)	52.44, Safe
8.	Provided thickness	tp (mm)	12
Design of Flat head (Non circular cross section)			
9.	Thickness $t = d\sqrt{\frac{ZCP}{S \cdot e}}$	t (mm)	Top plate: 20 Bottom plate: 20 Front plate: 20

5.1 Static FE analysis

Results of finite element analysis consist of nodal displacements and element stresses. The structural analysis has been made by means of CATIA V5 R18 "Generative Structural Analysis tool". The aim of this analysis is to determine the structural stress distribution in the system. Data used for structural analysis is shown in table 4 and the results of structural analysis are shown in figure 5a and figure 5b.

Figure 5(a) VonMises stress in Pa

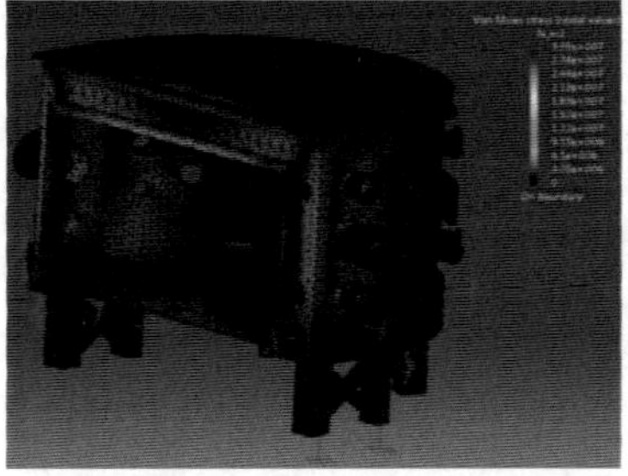

Figure 5(b) Total deflection in mm

Table 4: Data used for structural analysis

Structural analysis: CATIA V5 R12 Generative Structural Analysis tool	
Material	SS 304L for chamber and mild steel for the chamber
Element type	Solid 185 having three (3) degrees of freedom
Boundary conditions	• External surface load: pressure $1x10^5$ Pa • Gravity load • Lower key points for support columns as a fix point
Vector sum deflection	1.3 mm
Von Mises stress Max	30.05 MPa, within the yield strength of the material and is safe
Radiation analysis: ANSYS 12	
Method	AUX 12 radiation matrix generator method
Element type	Solid Tet 10 node 87, Shell 57 with temperature as a one degree of freedom
Radiation surface	PFC and VV
Boundary conditions	• Temperature on the surface of PFC 1773 K • Space temperature: 300 K
Maximum surface temperature	643.7 K

5.2 Thermal Radiation analysis

Thermal responses of the PFC and VV during electron beam operation have been calculated analytically and 3-D finite element analysis using ANSYS. All the ports, which are attached to the chamber, are ignored and heat flux is calculated. In the FE model the radiation is included through the use of AUX 12 radiation matrix generator method. The radiation matrix is calculated on the basis of the Stefan – Boltzman law of radiation. The maximum surface temperature of the chamber is found to be 643.7 K. Results are shown in figure 6(a) and figure 6(b).

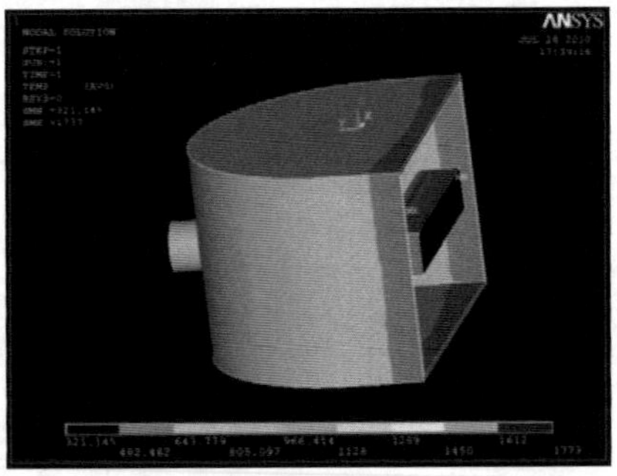

Fig. 6(a). Contour plot of nodal temperature

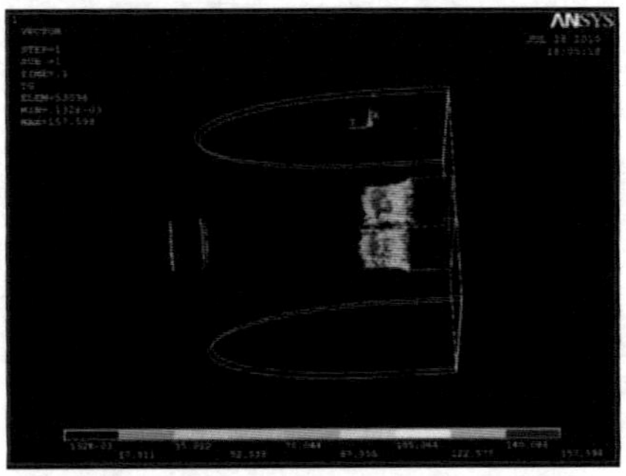

Fig. 6(b). Vector plot representing thermal gradient

5.3 Vacuum Pump calculations

The pump down time (τ) from the initial (atmospheric) pressure to the working pressure of the system is to be calculated for a given volume of chamber (litre) with a given pumping speed (S_p) of a particular pump. The ultimate pressure (P_u) in the chamber can be expressed as

$$P_u = \frac{Q_g}{S} = \frac{\left(\frac{Q_g}{V}\right)}{\left(\frac{V}{S}\right)} = \left(\frac{Q_g}{V}\right)\tau = \frac{\Delta P}{\Delta t}\tau \;\;\ldots\ldots\ldots (1)$$

Where Q_g is the total gas load, V is the volume of chamber, S is the pumping speed in chamber and C is the conductance of an elbow

$$\frac{1}{S} = \frac{1}{S_P} + \frac{1}{C} \;\ldots (2)$$

$$C = 3.81\sqrt{\frac{T}{M}}\frac{D^3}{(L_1 + L_2)} \;\ldots\ldots\ldots\ldots\ldots\ldots (3)$$

Time required for lowering the pressure in the chamber from P_i (initial) to P is given by

$$t = \left(\frac{V}{S_t}\right)\left[1+\left(\frac{S_t}{C}\right)\right]\ln\frac{P_i - \left[1+\left(\frac{S_P}{C}\right)\right]P_0 - \left[1+\left(\frac{P_0}{P_P}\right)\right]P_u}{P - \left[1+\left(\frac{S_P}{C}\right)\right]P_0 - \left[1+\left(\frac{P_0}{P_P}\right)\right]P_u} \quad \text{.................................(4)}$$

- The total effective specific speed S_{eff} = 2S = 3178.2 lt/sec
- The ultimate pressure in the chamber Pu = 6.884 x 10^{-7} Torr
- Total out gassing rate Q_D = 1.094 x10^{-3} Torr litre/sec
- Time required is t = 68.04 min

6. Conclusions:

A conceptual design approach of setting up HHF test facility at IPR has been discussed and high lighted the features of the vacuum chamber with the design calculations. Theoretical calculations were done to support the conceptual design and FEA structural analysis was carried out to identify the deflections and observed to be in the safe limit. Through radiation analysis, the surface temperature inside the chamber is determined which gives an outline for the cooling of the chamber.

7. References:

[1] H. Bolt et al 2002, *J. Nucl. Mater.* **43** 307.
[2] R. Tivey et al 1999, *Fusion Eng. Design.* **46**, 207.
[3] M. Rodig, M. Akiba, P. Chappuis, R. Duwe, M. Febvre, A. Gervash, J. Linke, N. Litounovsky, S. Suzuki, B. Wiechers, D.L. Youchison 2000, *Fusion. Eng. Design.* **51** 715.
[4] J. Linke, I.V.Mazul, R.Nygren et.al 2007, *Journal of Nuclear materials.* **367** 1422.
[5] Dennis Moss, "Pressure Vessel Design Manual", 3rd Edition, Elseiver Science Publishers.
[6] ASME Boiler and Pressure Vessel Code (BPVC): BPVC Section VIII – Rules for construction of pressure vessels Division 1.

International Symposium on Vacuum Science & Technology and its Application for Accelerators IOP Publishing
Journal of Physics: Conference Series **390** (2012) 012061 doi:10.1088/1742-6596/390/1/012061

A Low Impedance Marx Generator as a Test bed for Vacuum Diodes

Biswajit Adhikary, P Deb, R.Verma, R. Shukla, S.K.Sharma P.Banerjee, R Das, T Prabaharan, BK Das and Anurag Shyam

Energetics and Electromagnetics Division, Bhabha Atomic Research Centre, Visakhapatnam, India

E-mail: biswajitadhikary@yahoo.com

Abstract. A low impedance Marx generator was developed, which will serve as a test bed for Vacuum diodes of various electrode materials and geometries. The vacuum diodes will be used for high power microwave generation. The generator is capable to supply ~3GW of pulsed power to the vacuum diodes which is sufficient enough to produce plasma within the diode for electron beam generation. A vacuum of 10^{-5}Torr is required for virtual cathode formation within the diode, when the beam current exceeds the space charge limiting current. A vacuum diode of reflex triode geometry has been designed and vacuum of 10^{-5}Torr has been achieved. The repetitive operation of the vacuum diode depends upon the recovery of the diode, the importance of the vacuum system on the recovery of the diode will be explained. A vacuum system with high voltage isolator has been installed for getting the desired vacuum within the diode. The design criterion of the vacuum system will be discussed. The 300kV/1.8kJ Marx generator which will power the vacuum diode has six stages with stage capacitance and voltage of 240nF and 50kV respectively. It has an impedance of ~7 ohm and can deliver 200kV voltage across the diode in critically damped load condition. The generator has a very fast rise time of 200ns.The operational characteristics of the Marx generator are determined experimentally. The results have been analyzed and compared to an equivalent circuit model of the system.

1. Introduction

Intense pulsed relativistic electron beams (REB) is generated in a vacuum diode when a high voltage pulse is applied across the electrodes of the vacuum diode. The relativistic electron beams has applications in wide variety of fields, from high power microwave (HPM) generation, flash X-ray generation, pumping of lasers, surface modification of materials etc. Vacuum diodes require high voltage pulse. A pulsed power driver has been developed to provide the high voltage pulse to the vacuum diode. A typical pulsed power system that may deliver the pulsed power to the vacuum diode consists of a power supply to charge the Marx bank, the Marx bank itself, a pulse forming line (PFL) or conditioning circuit a peaking switch and at last Vacuum diode (Load). The PFL is required to sharpen the pulse to enhance the power, as a typical Marx bank has slow rise times. The PFL makes the system bulky and complex. It also reduces the voltage at the load to half (if the impedance is matched). A Marx bank with a fast rise time can directly drive a vacuum diode (without using PFL). Such a system is more compact, and more voltage can be delivered to the vacuum diode. The vacuum diode chosen for this experiment was of reflex triode geometry.

2. Experimental set up

The charging power supply, the Marx bank and the vacuum diode constitute the whole system. The system is compact enough to be truck loaded (Fig:1).

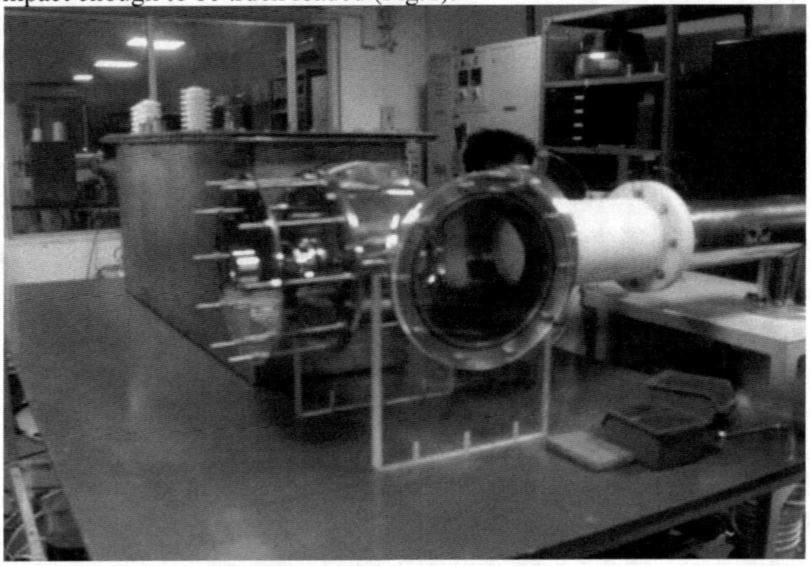

Fig 1:Marx bank with reflex triode

2.1 The Fast Marx bank

To achieve a multi hundred kilo volt output a Marx bank is used due to it's high energy storage capability. The energy rating of the Marx bank is taken to be as 1.8kJ. Six stages are used each with stage capacitance of 240nF.The capacitors are chosen for their availability, physical dimensions and voltage rating of 50kV. For the Marx bank to directly drive a vacuum diode a pulsed voltage of $\geq$ 200kV [1] is desirable. Thus taking V_{out} = 200kV, and choosing critically damped condition of Marx bank loading if n = no. of stages, V = stage charging voltage, the output voltage is ~70% of nxV [2]. So, $V_{out} = 0.7(nxV)$.Thus, $n = (200 \times 10^3)/ 0.7 \times 50 \times 10^3 \approx 6$. Thus with six stages the maximum erected open circuit voltage is 300kV and the stored energy rating is 1.8kJ.The fast rise time of 200ns the system helped in uniform plasma formation on the cathode surface. The fast rise time is achieved by reducing the inductance of the Marx circuit.

2.2 The Vacuum diode

The vacuum diode geometry chosen for these set of experiments is of 'reflex triode' for Virtual cathode oscillator (Vircator) mode of operation. In the Vircator mode the high power electron beam from the diode is injected into an open propagation region in a vacuum of 10^{-5} mbar, through a mesh. When the beam current exceeds the space charge limiting current a space charge cloud called as 'Virtual Cathode' is formed. The beam propagates for a certain distance and then reflects from the virtual cathode. The oscillating electrons between the cathode and the virtual cathode generate electromagnetic radiation (HPM). A good vacuum system with fast pumping will remove the ablated material from the cathode and anode making the diode a fast recovery to be ready for the next high voltage pulse.

3. Operation of The Marx bank

The developed Marx generator has a design open-circuit voltage of 300 kV, with an estimated FWHM pulse width close to 700 ns. Energy rating of the Marx generator is 1.8 kJ, so an estimated peak power of ~ 3 GW is expected from the device.

Journal of Physics: Conference Series **390** (2012) 012061 doi:10.1088/1742-6596/390/1/012061

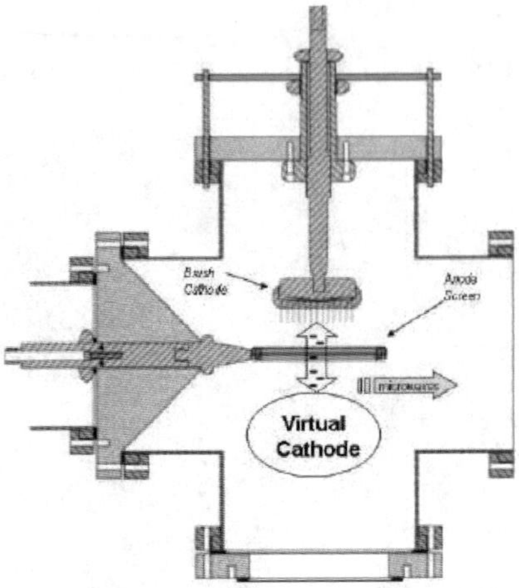

Fig 2: Vacuum diode of reflex triode geometry

With a charging voltage of 50 kV DC, an open-circuit voltage of 300 kV is achieved. The output voltage is measured by using spherical gaps with diameters of 150 mm. The spherical gaps are used in place of the load to measure the open-circuit voltage of the Marx generator. The 300 kV open-circuit output voltage of the Marx generator is confirmed by using a gap spacing of ~ 300 mm.

A photograph of a shot taken by using the open-shutter digital camera with spherical gaps in place of the load is shown in Fig. 3. Each spark gap at atmospheric pressure has a self-breakdown voltage of 12 kV approximately, so the maximum output voltage, which we can expect without pressurizing the spark gap, is 72 kV. To increase the charging voltage, the gaps must be pressurized.

Fig.3. Marx generator (in open air) in operation is framed by a digital camera in the open shutter mode in which the shutter of the camera is left open and the room is kept dark. The first light that it gets (spark gap arcing light) is captured by it.

3.1. Experimental results

Short circuit mode of Marx bank operation: After confirmation of the open-circuit voltage, the Marx generator is loaded. The voltage measurements are done by using a 20-MHz, RC-compensated voltage probe with an attenuation factor of 10,000:1. The waveforms are recorded on a 1-Ghz, 2-Gs/s digital storage oscilloscope. A negative pulse is obtained as it is to be connected to the cathode of the diode.

Journal of Physics: Conference Series **390** (2012) 012061 doi:10.1088/1742-6596/390/1/012061

The Marx inductance is determined from the experimental results. For finding it, the Marx circuit is shorted, without any load and charged to voltages of about 12 kV and the current waveform of 12 kA in the first peak is recorded using a current monitor. The period, T, of the 1600 ns pulse, is obtained from the current waveform (Fig. 4). $T = 2\pi\sqrt{L_M C_M}$,

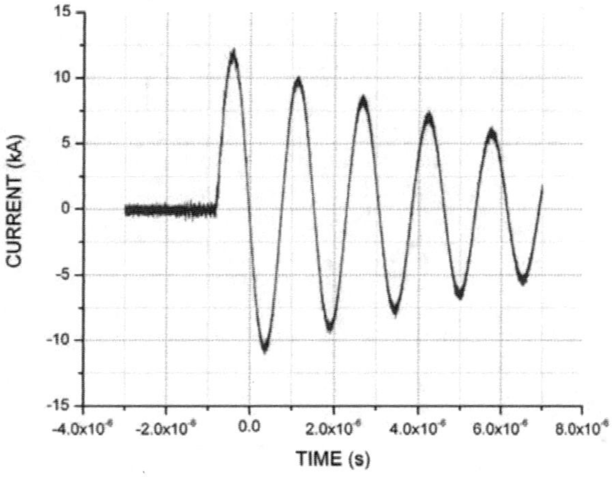

Fig. 4. Current waveform of the Marx generator operated in the short-circuit mode.

where L_M = 1500 nH & C_M = 40 nF. Thus, Z_M = 6.25 Ω.

The estimated pulse width can be calculated as 80% of half the period of the short-circuit mode [2]. The Marx impedance thus calculated helped in preparing the load, which is ~ 1.5 times the Marx impedance, to damp the Marx circuit. The dummy load is made from a copper-sulphate solution.

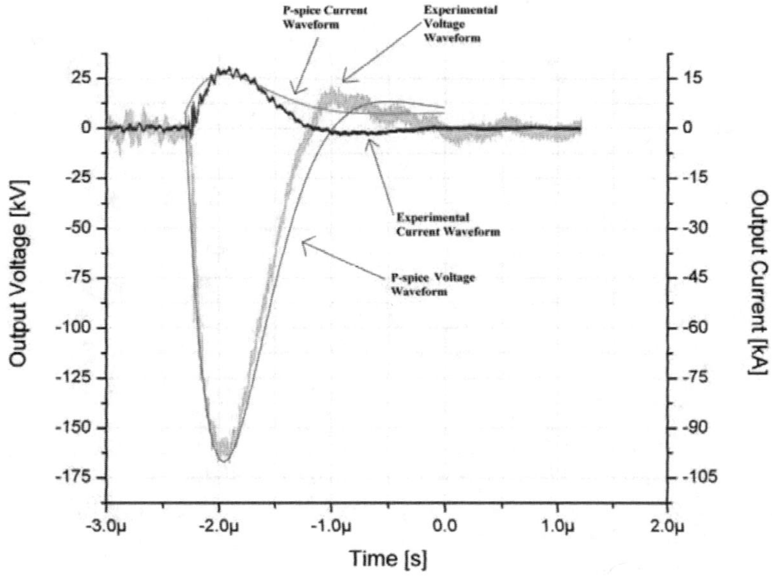

Fig. 5. Marx generator experimental and simulated output voltage and output current 9 □ load at a charging voltage of 40 kV.

Marx bank characterization on resistive load:

The Marx output when measured with a load of 9 Ω and charging voltage of 40kV, is 163 kV, which is ~ 68% of the open-circuit output voltage of 240 kV and the voltage reversal is ~ 10%. The Marx output current is 18 kA. The 9 Ω load represents the damped load condition. The rise time is 200 ns and FWHM pulse width is 640 ns. So a pulsed power of ~ 3 GW is generated (see Fig: 5).

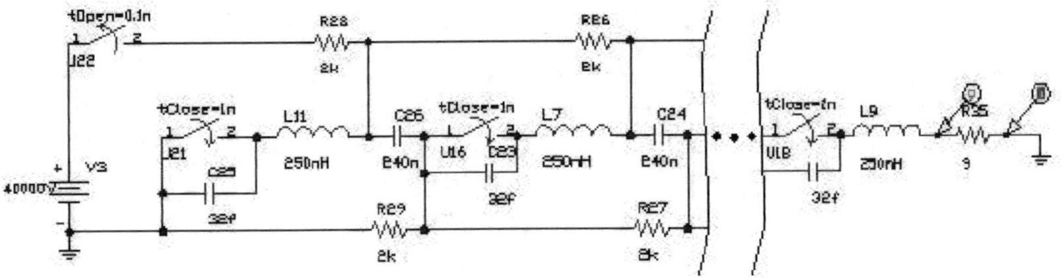

Fig. 6. Simulation circuit with a 9 Ω load

4. Simulation

The six stage Marx generator circuit is simulated in P-spice [3]. The stage capacitance is 240 nF, and the total inductance of the circuit is divided into 6 stages to give a stage inductance of 250 nH. The spark gap capacitance is 32 femto F. Because it is operating in open air, the return conductor is kept a large distance away from the stage capacitor, so the stray capacitance is taken to be negligible. The simulated Marx circuit (Fig. 6) delivers 157 kV, 18.75 kA to a 9-ohm load for a charging voltage of 40 kV . Rise time is 200 ns. Table 1 shows the results.

Table 1. Results

Charging Voltage	Experimental Output Voltage/Current/$t_{r(\text{rise time})}$	Simulated OutputVoltage/ Current/$t_{r(\text{rise time})}$	Load Resistive
40 kV	163 kV/18 kA/200 ns	157 kV/18.75 kA/200 ns	9 Ω

5. Summary

A fast Marx bank has been developed which gives rise time of 200ns and pulsed power 3GW to the vacuum diode. The vacuum diode of the reflex triode geometry is directly connected to the Marx bank with critically damped condition.

References

[1] Y. J. Chen, M.S. Thesis, Texas Tech University, USA, (2005).
[2] W.J.Carey, J.R. Mayes, "Marx generator design and performance," Power Modulator Symposium, 2002 and 2002 High Voltage workshop. Conference Record of the Twenty-fifth International, 625 (2002)
[3] www.orcad.com (student version).

Vacuum system design for twenty-cell PWT Linac structure

S K Gupta, S Chouksey

Materials & Advanced Accelerator Sciences Division, Raja Ramanna Centre for Advanced Technology, Indore (MP), India

Abstract. This paper describes the vacuum system design for a 20-cell Plane Wave Transformer (PWT) type LINear ACcelerator (LINAC) structure to obtain a vacuum better than 5×10^{-8} mbar everywhere inside the linac structure. Specific material out-gassing rate for a 20-cell linac is determined on the basis of vacuum achieved in a similar type of 8-cell linac structure, which is smaller in length, and presently under operation in the Compact Ultrafast TEra-hertz Free Electron Laser (CUTE-FEL) beam transport line at RRCAT.

1. Introduction

Twenty-cell PWT linac structures are being developed at RRCAT to serve as an injector for an Infra-Red Free Electron Laser (IR-FEL) and also for the Indus booster synchrotron. The required vacuum level inside the linac has to be better than $5X10^{-8}$ mbar to support high RF field gradient and also to minimize beam loss due to charge exchange with the residual gas molecules. This vacuum level is designed to be achieved by pumping of the linac structure through its vacuum port using a pumping chamber and a combination of Turbo Molecular Pump (TMP) and Sputter Ion Pump (SIP). For the pumping port, a number of small slots and holes have been machined on the surface of linac tank (vacuum envelope), as a large single slot cannot be made on the tank surface to prevent the leakage of magnetic field lines, which are along the circumference of the tank inner surface. 20-cell PWT linac can also be used as an Integrated photo-injector.

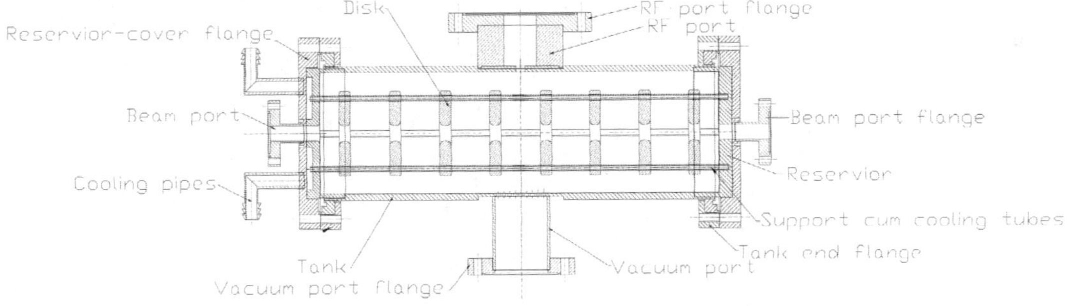

Fig. 1. 8-cell PWT Linac schematic

2. PWT Linac structure

Typical 8 and 20-cell PWT linac structures consist of 8 and 20 Oxygen Free Electrolytic (OFE) copper disks respectively (see figure 1). Four hollow stainless steel (SS316L) tubes support the disk array and also carry flowing water to cool the structure. Tubes are brazed to an OFE copper reservoir at one end and have propped support at the other end reservoir. The reservoir to which tubes are brazed serves as water storage unit. Both reservoirs are brazed to cover flanges made of SS316L, which connect with the end flanges of a tank made of SS316L (copper coated inside). The tank serves as the outer boundary and vacuum envelope for the structure. RF and vacuum ports are connected to the tank and beam ports are welded to the cover flanges at both ends of the linac. Two cooling pipes are also connected to reservoir through the cover flanges to carry the water from the chiller to reservoir.

3. Vacuum system design for 20-cell PWT Linac

Pumping gas: Air (Molecular Weight, M= 29)
Operating temperature, T: 313 K (40 C)

3.1 Symbols used
A: Out-gassing surface area of linac, cm^2
a: Arithmetic average of a_1 and a_2, cm
a_1, a_2: Length of rectangular cross-sections of tapered pipe at top and bottom respectively, cm
A_a: Area of apertures, cm^2
b: Width of rectangular cross-section, cm
C_1: Conductance of linac tank apertures (Slots and Holes), l/s
C_2: Conductance of linac pumping port, l/s
C_3: Conductance of pumping chamber, l/s
C_{eq}: Equivalent conductance of C_1, C_2 and C_3, l/s
C_L: Conductance of linac, l/s
D: Inner diameter of pipe, cm
D_1, D2: Inner diameters of tapered pipe at top and bottom respectively, cm
ID: Inner diameter, cm
K: Correction factor for molecular flow
L: Length, cm
l/s: Litre per second
OD: Outer diameter, cm
P_L: Pressure at farthest end of linac from outlet, mbar
P_o: Pressure at the outlet of linac, mbar
q: Specific out-gassing rate, mbar-l/sec-cm^2
S_p: Pumping speed at SIP inlet, l/s
S_o: Pumping speed at linac outlet, l/s

3.2 Formulae used
Conductance of apertures

$$C = 3.64 \times \left(\frac{T}{M}\right)^{\frac{1}{2}} \times A$$

Conductance of pipe

$$C = 3.81 \times \left(\frac{T}{M}\right)^{\frac{1}{2}} \times \frac{D^3}{L}$$

Conductance of tapered pipe with circular cross-section

$$C = 7.62 \times \left(\frac{T}{M}\right)^{\frac{1}{2}} \times \frac{D_1^2 \times D_2^2}{(D_1 + D_2) \times L}$$

Conductance of tapered pipe with rectangular cross section

$$C = 19.4 \times \left(\frac{T}{M}\right)^{\frac{1}{2}} \times \frac{\left(b/a\right)^2}{1+\left(b/a\right)} \times \frac{a_1^2 \times a_2^2}{(a_1+a_2)\times L} \times K$$

Where, K= 1.2258 (for b/a= 0.2955)

$$C_{eq} = \frac{C_1 \times C_2 \times C_3}{C_1C_2 + C_2C_3 + C_3C_1}$$

$$S_o = \frac{S_p \times C_{eq}}{S_p + C_{eq}}$$

$$P_o = \frac{q \times A}{S_o}$$

$$P_L = P_o + \frac{q \times A}{4C_L}$$

3.3. Specific out-gassing rate determination

Determination of the specific out-gassing rate is an important step in vacuum design calculations. Instead of using data from literature for out-gassing rates for copper and copper coated SS316L having undergone similar surface treatment as in our case, for the present calculations it has been determined from the operating vacuum conditions of an 8-cell PWT linac that is presently employed in the CUTE-FEL beam-line. Since it's configuration and construction is similar to the 20-cell linac structure, both are expected to have similar vacuum properties. A single port pumping (Lumped pumping) scheme was adopted in 8-cell linac (see figure 2). Dimensions of linac, vacuum port and pumping chamber are given in Figure 2. Twenty slots and nine holes have made in the linac tank to pump the linac structure (see figure 2). On the basis of vacuum achieved at the outlet of 8-cell linac (5×10^{-8} mbar), the specific out-gassing rate has been determined to be 5.2×10^{-10} mbar-l/sec-cm^2 (see table 1) using the relevant expressions described above.

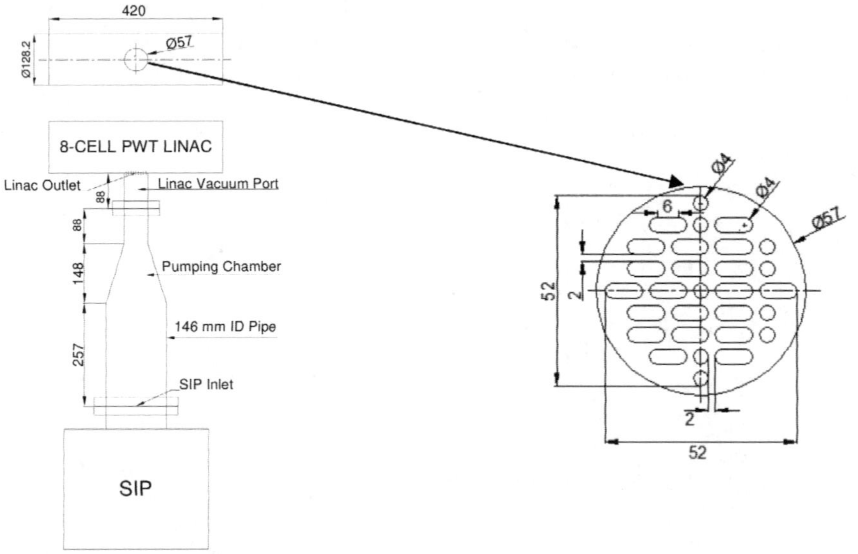

Fig. 2. Existing vacuum pumping scheme for 8-cell PWT Linac (Dimensions are in mm)

Summary of calculations (table 1):

Table 1

A	C_1	C_2	C_3	C_{eq}	S_p	S_o	q
3533	100.9	261.9	161	50.1	140	36.9	5.2×10^{-10}

3.4. Vacuum calculations and analyses for different pumping schemes

3.4.1.
Case-1**:** Considering the **s**ame pumping scheme as adopted for the existing 8-cell linac.
Summary of calculations (table 2):

Table 2

q	A	C_1	C_2	C_3	C_{eq}	S_p	S_o	P_o	P_L
5.2×10^{-10}	8452	100.9	261.9	161	50.1	140	36.9	1.2×10^{-7}	1.24×10^{-7}

As the vacuum level of $P_L= 1.24\times10^{-7}$ mbar (see table 2) inside the linac is poorer than our required vacuum level of 5×10^{-8}mbar, it is needed to improve the design of vacuum system. From above results (see table 2) it is clear that pumping speed at linac outlet (S_o) is less due to low equivalent conductance (C_{eq}), which needs to be increased.

3.4.2.
Case-2**:** Conductance of the linac tank aperture is the lowest of the three conductances (see table 2). This can be increased by increasing the number of slots machined on the linac tank. The limit on the number of slots comes from the number that can be accommodated inside the vacuum port whose OD is limited by the linac tank OD (141.3mm in our case). A seamless SS316L pipe of 127mm (5" Gauge) O.D. has selected for the vacuum port which is the maximum OD pipe commercially available below 141.3mm OD. The number of slots can therefore be increased to 72 from 20 (see figure 3), resulting in an increase in C_1 to 328.3 l/s from 100.9 l/s (see table 3). The conductance of vacuum port has also increased consecutively to 2410.2 l/s from 261.9 l/s, as ID of vacuum port pipe is increased to 121 mm from 57 mm (see table 3).

The conductance of pumping chamber can be increased by using a 146 mm ID pipe throughout the length of the pumping chamber in place of a narrow pipe (57 mm ID) and conical section (see figure 3). As a result of these changes, the conductance of pumping chamber is increased to 801.5 l/s from 161 l/s (see table 3).

The equivalent conductance is also increased substantially to 212.3 l/s from 50.1 l/s (see table 3).

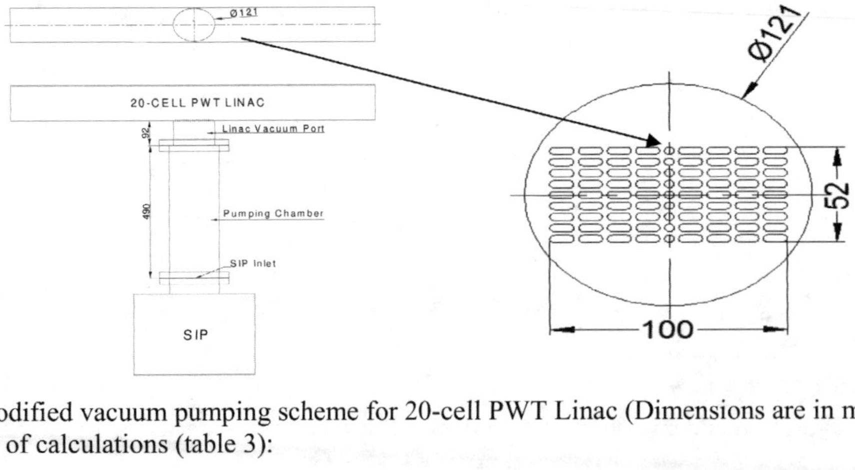

Fig. 3. Modified vacuum pumping scheme for 20-cell PWT Linac (Dimensions are in mm)
Summary of calculations (table 3):

Table 3

q	A	C_1	C_2	C_3	C_{eq}	S_p	S_o	P_o	P_L
5.2×10^{-10}	8452	328.3	2410.2	801.5	212.3	140	84.3	5.2×10^{-8}	5.63×10^{-8}

The pressure in 20-cell linac is still higher than the required value. From above results (see table 3) it is clear that equivalent conductance (212.3 l/s) is higher than the pumping speed at SIP inlet (140 l/s), which now needs to be increased. Using the commercially available next higher capacity pump of 270 l/s pumping capacity, the ultimate pressure is reduced to 4.12×10^{-8} mbar (see table 4), which is acceptable.
Summary of calculations (table 4):

Table 4

S_p	S_o	P_o	P_L
270	118.8	3.69×10^{-8}	4.12×10^{-8}

The required vacuum (better than 5×10^{-8} mbar) has now been achieved inside the 20-cell linac. It is because of increased pumping speed at linac outlet to 118.8 l/s from 84.3 l/s (see table 4).

The pressure difference in 20-cell linac (which is 2.5 times longer than 8-cell linac) is not much (see table 4). Therefore, distributed pumping is not required.

Though, the required vacuum level has been achieved, we attempt to improve vacuum further, as it can possibly open other alternate uses of the structure.

3.4.3.
Case-3**:** The conductance of the linac tank apertures needs to be increased to a value close to the next higher conductance value (C_3= 801.5 l/s, see table 3) to increase the equivalent conductance effectively. To achieve this, the number of slots is increased to 180 from 72 (see figure 4) resulting in an increase in C_1 to 800.4 l/s from 328.3 l/s (see table 5).

Design of the linac vacuum port is modified to accommodate the additional slots. A tapered vacuum port with rectangular section is proposed to be used (see figure 4). This change further increases the conductance to 3738 l/s from 2410.2l/s (see table 5), while conductance of pumping chamber is same as in case2.

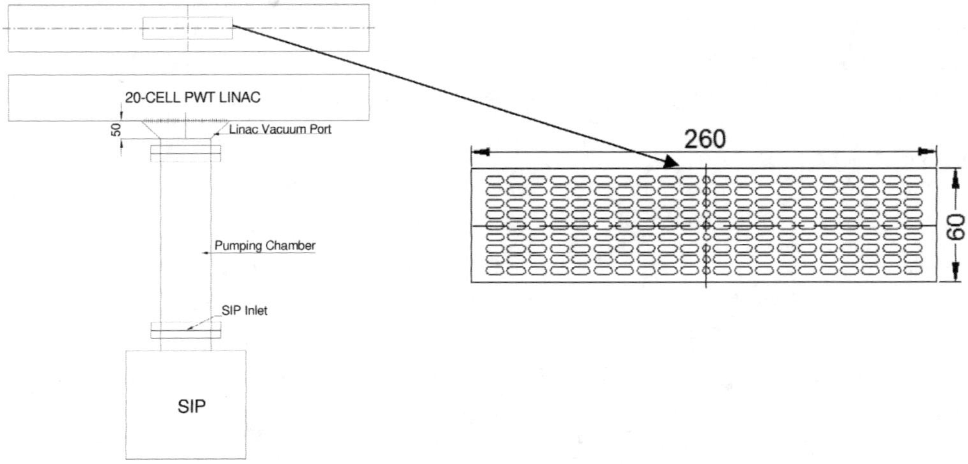

Fig. 4. Optional vacuum pumping scheme for 20-cell PWT Linac (Dimensions are in mm)

Summary of calculations (table 5):

Table 5

q	A	C_1	C_2	C_3	C_{eq}	S_p	S_o	P_o	P_L
5.2×10^{-10}	8452	800.4	3738	801.5	361.7	270	154.6	2.83×10^{-8}	3.26×10^{-8}

As equivalent conductance is higher than pumping speed at SIP inlet (see table 5), SIP of 500 l/s capacity can be used.
Summary of calculations (table 6):

Table 6

S_p	S_o	P_o	P_L
500	209.8	2×10^{-8}	2.43×10^{-8}

From above results (see table 6) it is clear that in case-3 vacuum has not increased much (from 4.12×10^{-8} mbar to 2.43×10^{-8} mbar), while pumping speed at linac outlet increased substantially high from 118.8 l/s to 209.8 l/s. It means that specific out-gassing rate is limiting the vacuum inside the 20-cell PWT linac. Vacuum can be improved effectively only by reducing specific out-gassing rate of the linac structure. Case-2 is hence the optimized design of the 20-cell PWT linac vacuum system, as simpler than case-3.

4. Future plan

A twenty-cell PWT linac can also be used as an Integrated photo-injector in which vacuum better than 10^{-9} mbar is required to support higher RF field gradients and special photocathode materials.
From conventional 20-cell PWT linac vacuum system design, it is seen that specific out-gassing rate should be approximately two orders of magnitude higher than required operational vacuum level.
Specific out-gassing rate of 20-cell PWT linac can be decreased to 10^{-11} mbar-l/sec-cm^2 by baking at higher temperature for longer duration [1], [2]. The 8-cell PWT linac was baked at 150 C for 24 hours resulting in a relatively high specific out-gassing rate.

5.Conclusions

The vacuum system design for 20-cell PWT linac, which shall be used as an injector for IR-FEL and for Indus booster synchrotron, has been finalized for a vacuum better than 5×10^{-8} mbar. Case-2 design has been found optimum and finalized, while case-1 design could not fulfill the vacuum requirement. It is also concluded that the improvement in the vacuum of 20-cell PWT linac (better than 10^{-9} mbar) for Integrated photo-injector application is possible by reducing the specific out-gassing rate (less than 10^{-11} mbar-l/sec-cm^2) of the linac structure by baking at higher temperature for longer duration.

Acknowledgements

Authors thank K K Pant for his help and support. Authors also thank D P Yadav, V Sathe, R Sridhar and K C Ratnakala for their advice and many useful discussions.

References

[1] Ratnakala K C, Tiwari S K, Shukla S K, Kotaiah S *Outgassing rate measurement of copper Plated stainless steel,* IVS2007, TIFR, Mumbai, Nov., 2007
[2] Arnaud G, Di Bartolo G, Robeiri A, Roch M *Tesla HF Couplers: Outgassing of copper plated AISI 304L samples,*LAL/RT 02-01, TESLA Report 2003-14, October 2002.

Thick SS316 materials TIG welding development activities towards advanced fusion reactor vacuum vessel applications

B Ramesh Kumar [1], **R Gangradey**

Institute for Plasma Research, Bhat, Gandhinagar-382428.

[1]E-mail: buddu@ipr.res.in, brkumar75@gmail.com

Abstract. Advanced fusion reactors like ITER and up coming Indian DEMO devices are having challenges in terms of their materials design and fabrication procedures. The operation of these devices is having various loads like structural, thermo-mechanical and neutron irradiation effects on major systems like vacuum vessel, divertor, magnets and blanket modules. The concept of double wall vacuum vessel (VV) is proposed in view of protecting of major reactor subsystems like super conducting magnets, diagnostic systems and other critical components from high energy 14 MeV neutrons generated from fusion plasma produced by D-T reactions. The double walled vacuum vessel is used in combination with pressurized water circulation and some special grade borated steel blocks to shield these high energy neutrons effectively. The fabrication of sub components in VV are mainly used with high thickness SS materials in range of 20 mm- 60 mm of various grades based on the required protocols. The structural components of double wall vacuum vessel uses various parts like shields, ribs, shells and diagnostic vacuum ports. These components are to be developed with various welding techniques like TIG welding, Narrow gap TIG welding, Laser welding, Hybrid TIG laser welding, Electron beam welding based on requirement. In the present paper the samples of 20 mm and 40 mm thick SS 316 materials are developed with TIG welding process and their mechanical properties characterization with Tensile, Bend tests and Impact tests are carried out. In addition Vickers hardness tests and microstructural properties of Base metal, Heat Affected Zone (HAZ) and Weld Zone are done. TIG welding application with high thick SS materials in connection with vacuum vessel requirements and involved criticalities towards welding process are highlighted.

1. Introduction

The development of fusion reactors like ITER and upcoming Indian DEMO are having material challenges in terms of their compatability with operational conditions[1,2,3]. The stringent conditions like structural, thermo mechanical and neutron irradiation are the major challenging issues towards the development of the reactor sub system components. Fusion reactor construction utilizes different stainless steel materials like SS 304L, 316 and 316 L(N) for the fabrication of various components like vacuum vessel, plasma facing first wall components, mechanical support structures and other subsystems. The SS 316 grade materials are primary structural material candidates in fusion reactor components due to their superior mechanical and corrosion resistant properties[4]. The fusion reactors plasma are proposed by using D-T plasma operation and produces high energy 14 MeV neutrons

[1] To whom any correspondence should be addressed.

which can damage the other structural components in longer. The new design concepts by employing double walled vacuum vessel with using shields blocks of Borated steel materials in combination with water circulation are planned [5]. The steel blocks and water in combination shield effectively the energitic neutron generated from the fusion plasma reaction. In this regard, the fabrication of vacuum vessel and its associated sub systems materials selection is a challenging task in terms of their mechanical and irradiation properties compatability. The fabrication of vacuum vessel sector components will be carried out with different type of welding processes for joining of SS materials in various forms of shells, ribs and key components to make final sector modules [6,7]. TIG welding is one of the major technique used for joining of the stainless steel plates for fabrication of different vacuum vessel components[8,9]. Some technical codes and standards are followed for the fabrication with welding procedures in ITER vacuum vessel components[10]. Thick materials welding is challenging as per the demands like 20 mm – 60 mm range in different type of steels . The present paper reports charatcerization of TIG welded samples with 20 mm and 40 mm thick SS 316 material. The characterization of these samples for Tensile tests, Bend tests, Impact tests and Hardness measurements have been carried out to analyze weld joint strength. Microstructural analysis for weld zone and HAZ are carried out.

2. Samples preparation with TIG welding process

SS 316 materials with 20 mm thick and 40 mm thick are obtained and prepared for the TIG weld process as per standard procedure. The elemental analysis was carried out for 20 mm and 40 mm thick SS 316 base materials. The filler wires used are ER 316 grade and 2.4 mm diameter. The chemical composition data is shown as per Table 1. Rest other elements are in balanced range as per required. The 20 mm base metal properties are UTS 589MPa, Yield strength ~310 MPa and Elongation 55%. For 40 mm thick plate, base metal properties UTS~ 596 MPa, Yield strength ~389 MPa and Elongation 59%. These measured with test pieces in laboratory.

Table 1. Chemical composition of SS 316 L 10 mm material.

Element (in %)	C	S	P	Si	Mn	Cr	Ni	Mo
SS 316- 20 mm	0.02	0.012	.020	0.4	1.5	17.7	13.10	2.85
SS 316- 40 mm	0.024	0.015	.037	0.43	1.25	16.55	10.39	2.05
Filler wire ER 316	0.019	0.003	0.024	0.470	1.570	18.5	12.20	2.30

2.1 TIG welding 20 mm thick plate

The SS 316 sample welded plates of 20 mm thickness are fabricated by TIG welding process. The sample plates of dimensions 150 mm X 300 mm X 20 mm are made cut from SS 316 single plate sheets. The final welded sample size is 300 mm X 300 mm X 20 mm. The samples are fabricated with single V groove edge preparation and joined with multipass TIG welding process by keeping the root gap of 2 mm. The developed 20 mm thick sample was shown in figure 1(a) for reference. The process parameters used for TIG welding for both the samples are mentioned in Table 2.

Fig. 1(a). 20 mm welded sample

Fig. 1(b). 40 mm welded sample

Table 2. Weld parameters used for the preparation of sample

Weld Samples	Root gap	Groove Angle	Current (Amps)	Ar Shielding Gas Flow rate	Distortions	No. of passes
20 mm sample	2 mm	60	70	7 lpm	$< 3.5^0$	14
40 mm samples	2 mm	60,60	80-120	9 lpm	$< 2.7^0$	13,14 (27)

2.2. TIG welding 40 mm thick plate

The SS 316 sample welded plates of 40 mm thickness are fabricated by TIG welding process. The sample plates of dimensions 150 mm X 300 mm X 40 mm are made cut from SS 316 single plate sheets and final welded sample size is 300 mm X 300 mm X40 mm. The samples are fabricated with double V groove (60^0) design edge preparation with 2.5 mm root gap and joined with multipass TIG welding process. In initial 3 passes, the current of 80A was used and after 4th pass onwards 100A current was maintained with constant voltage of 12V. The process parameters used for TIG welding are shown in Table 2. The developed 40 mm thick weld sample was shown in figure 1(b). The developed samples are subjected to non destructive examination by X ray radiography and Ultrasonic tests. Both the samples are reported with no significant detectable defects like pores, voids or incomplete penetration.

3. Mechanical properties characterization

The TIG welded SS 316 samples are subjected to Tensile properties, standard Bending tests (180^0) side bend, hardness tests, Impact fracture V notch tests. In addition optical macro and microstructures are carried out. The details of the tests and results are described in the following sections.

3.1. Tensile tests

The SS 316 TIG weld samples are prepared as per the standard tensile test specimens. FIE make Universal testing Machine (model UTES-40) is used for the testing of the samples. All the tested specimens were broken out of the weld region, in base metal indicating of the higher strength of the welded joints compared with base metal.

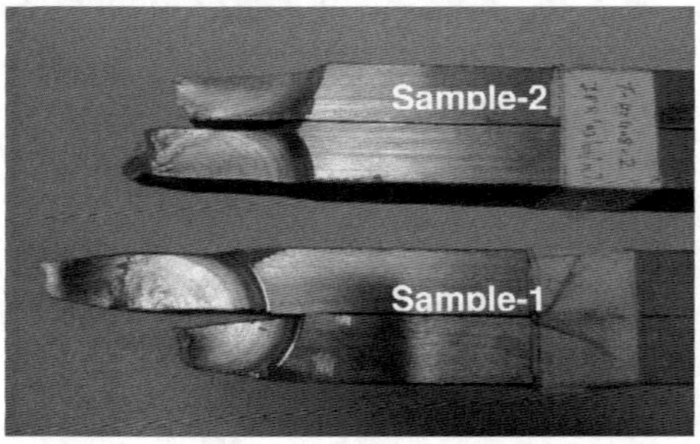

Fig. 2 (b). 20 mm Tensile samples

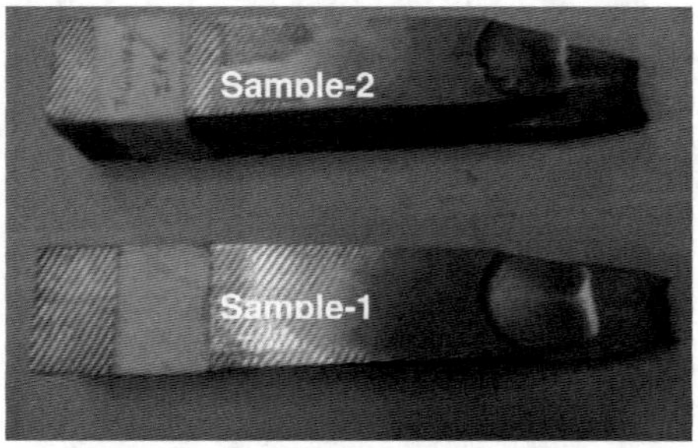

Fig. 2 (b). 40 mm Tensile samples

Table 3. Tensile Test parameters of samples

Sample type	Base Metal (UTS)	Weld sample (UTS)
20 mm thick	589 MPa	609 MPa, 609 MPa
40 mm thick	595 MPa	598 MPa , 608 MPa

The tensile properties data obtained with this weld process is used for the data base generation purpose and comparison to the optimum conditions of fusion reactor structural material data [5,10]. The welded samples tested for tensile test results are shown in figure 2(a) for 20 mm sample and figure 2(b) for 40 mm sample. The samples indicate the broken region which is away from the weld region indicating of good weld quality joint strength. The obtained tensile parameters are given in the Table 3.

3.2. Bend tests

The TIG welded SS 316 samples are further tested for bend tests in order to check the weld joints mechanical strength and weld quality. The fabricated samples are subjected to the standard 180^0 bend tests. The 20 mm thick weld samples are machined as per the required standard size and are subjected to side bend tests. Some examples with bend test samples are described in figure. 3(a) and 3(b). The samples as shown in figure. 3(a) have no apparent defects on welded joint surface after bending test for 20 mm thick sample. This in addition represents complete penetration of filler material and no pores or voids inside the weld joint. Figure 3(b) was shown with 40 mm thick weld samples subjected to side bend tests. These samples exhibited good weld joint strength and without any opening on the weld line or shown any defects like cracks or openings. The samples are successfully passed for the subjected bend test requirements.

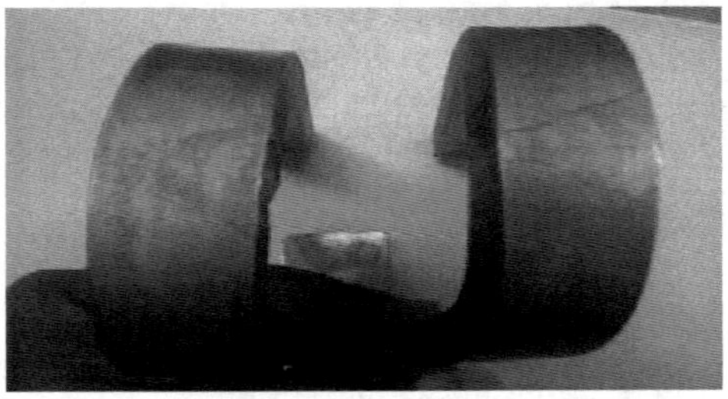

Fig. 3 (a). Bend tests of 20 mm sample

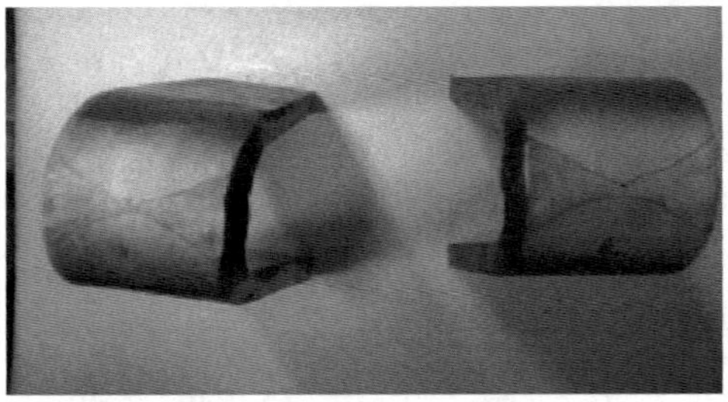

Fig. 3 (b). Bend tests of 40 mm sample

3.3. Vickers Hardness tests

The developed SS 316 TIG weld samples are tested for Vickers Hardness values for base metal, HAZ and weld zone. A standard load of 10 Kg is applied onto the all welded samples for the hardness measurements. The values of Micro Vickers hardness measurements for each Base metal, HAZ (Heat affected zone) and WZ (weld zone) are shown in Table 4. The measurements indicate low hardness values at base metal region in comparison with higher hardness values at weld zone and heat affected zone region due the work hardening during heating and cooling of weld process.

Table 4. Hardness measurements for welded samples.(10Kg Load)

Weld sample	BM	HAZ	WZ
20 mm Thick	190	226	210
40 mm Thick	205	233	240

3.4. Impact tests

The TIG welded SS 316 samples of 20 mm and 40 mm are subjected to the Impact fracture tests for Base metal, HAZ and Weld zone. This impact energy value which indicates the material resistance limit towards the applied sudden stress intensity or fracture energy which it can absorbs. In nuclear reactor grade components the ductile to brittle transition temperature range for the SS grade materials is very important parameter to qualify them for the structural integrity. The standard ASTM A 370 specification with 2 mm was prepared with standard Charpy V notch with 45^0 angle with all the samples before subjecting to the impact energy tests.. The SS 316 weld samples were tested for Impact tests at 0^0 C temperature. The specimens are prepared with standard size 10 mm X 7.5 mm X 55 mm from Base metal, HAZ and Weld Zone. In case of 20 mm sample the weld zone has exhibited higher impact energy (187J)compared to base metal(160J) which is indicative of good quality of welded region. Where as in case of 40 mm sample, the welded zone has shown less impact energy (71J) in comparison with base metal(119J) and HAZ(71J). values.

Table 5. Impact Energy tests

Sample	BM	HAZ	WZ
20 mm	160 J	139 J	187 J
40 mm	119 J	110 J	71 J

3.5. Macro & Micro structures
The TIG welded SS 316 samples are subjected to optical macro and microstructural analysis. The weld samples are chemically etched for the HAZ and weld zone clear identification.

3.5.1. 20 mm weld sample.
The 20mm sample was analyzed for optical macrostructures which was in shown in figure 4(a) clearly indicated with multipass weld zone. In figure 4(b) the weld zone microstructures are found with fine grain boundaries with austenitic structures. Figure 4(c) indicates the HAZ and weld zone region. In this Base metal structures are also seen in away regions.

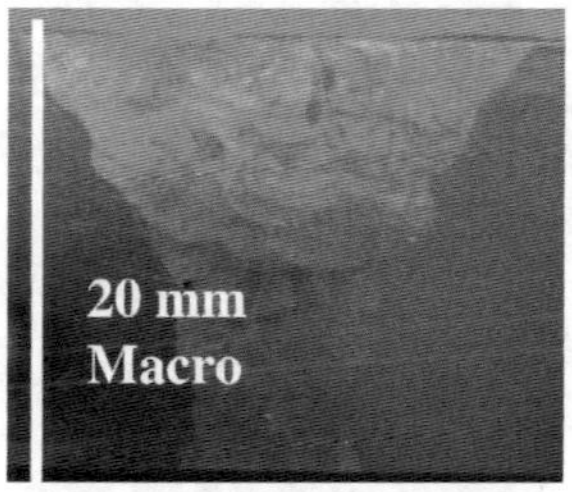

Fig. 4 (a). Macro

Fig. 4 (b) .Weld Zone

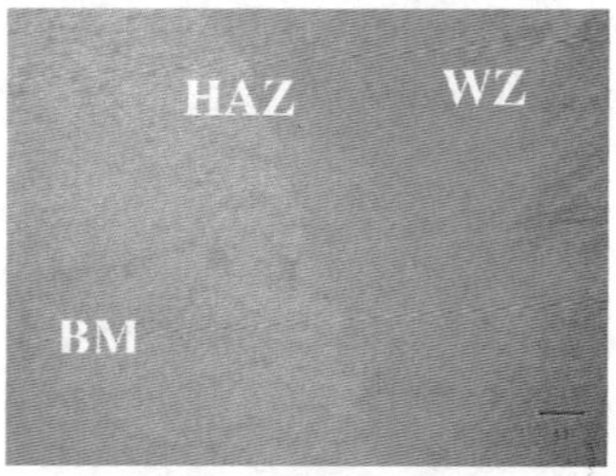

Fig. 4 (c) .HAZ

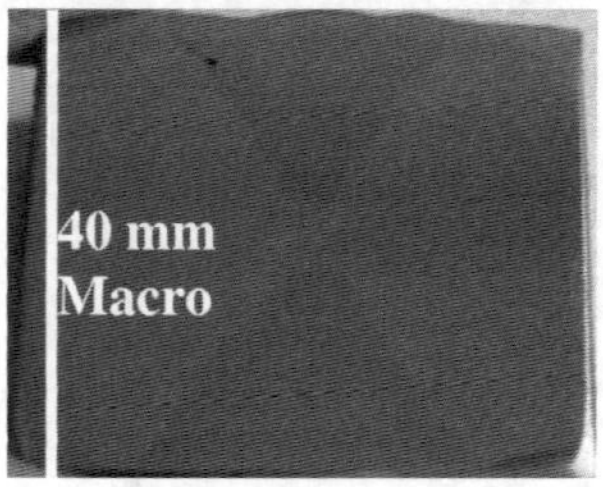

Fig. 5 (a). Macro

Fig. 5 (b). Weld Zone

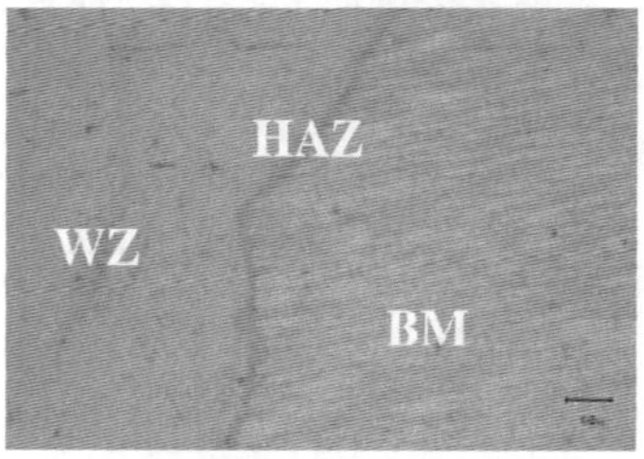

Fig. 5 (c). Macro

3.5.2. 40 mm weld sample.
The macrostructures of 40 mm thick SS 316 welded samples are shown in figure 5(a). Weld zone microstructures are shown in figure 5(b). The grain boundaries with austenite nature are observed in these samples also. The HAZ structures are shown in figure 5(c). In the same figure base metal microstructures can be seen in away regions. The weld zone structures in both cases were observed with a coarsened fine grain boundaries with austenitic structures. In addition the weld zone structures are free from any kind weld defects like cracks or inclusions.

5. Critical challenges with high thickness plate welds in Vacuum vessel
Advanced upcoming fusion reactors like ITER, Indian DEMO etc. are being planned with double walled vacuum vessel geometry for the safe operation and protection of the reactor subsystems from the 14MeV neutrons. In this regard the fabrication of the vacuum vessel components has to be done

by taking the considerations like welded joints mechanical stability, integrity, corrosion resistant materials and long life. In this regard, the fabrication of majority of the components are to be developed with different weld processes in both thin and thick plates SS materials of various grades. The double wall supporting structures, thick plates, joining ribs, shells and key components are fabricated with variety of joint configurations like plane plates, curved plates like D shape or near to D shapes and T type butt welds [5-7]. VV components are directly in contact with high pressure water circulation to remove heat loads on vessel and in addition to thermalization of high energy neutrons and shielding them to protect other subsystems. In this regard the components of the weld joints should be of leak proof and free from any kind of weld defects with higher mechanical properties. In addition, other advanced weld techniques like Laser beam welding , Electron Beam welding and Hybrid Laser TIG processes also will be used to develop the vessel structural components. In case of Laser welding process high thick materials welding is a challenging as these are limited due to the limited penetration depths and higher thick SS welds are still challenging. Hence Hybrid Laser TIG process is under consideration for the development of some critical components in VV modules. In case EBW, the vacuum based welding process, higher penetration depths (60 mm) are possible but are very expensive and limited with dimensional requirements. In addition, this is very expensive and industrially needs to be established to cater the upcoming fabrication requirements for large scale jobs.

6. Summary and Outlook

The present work reports the TIG welding process characterized 20 mm and 40 mm thick SS 316 materials. The 20 mm and 40 mm samples are characterized for mechanical Tensile properties, Bend tests, Impact tests. Further Hardness measurements nd microstructural studies are carried out. Microstructures reveal the good weld process and produced no significant defects. The thick SS material welding processes have to be established for the requirements of VV components towards upcoming fusion reactor components fabrication and development.

References :

[1] V. Barabash , The ITER International Team, A. Peacock, S. Fabritsiev, G. Kalinin, S. Zinkle, A. Rowcliffe, J.-W. Rensman, A.A. Tavassoli, P. Marmy, P.J. Karditsas, F. Gillemot, M. Akiba, Materials challenges for ITER – Current status and future activities, Journal of Nuclear Materials 367–370 (2007) 21–32.

[2] ITER Materials Properties Handbook, Austenitic Steel 316L(N)-IG, ITER Document No. S 74 MA 2 AA04, File Code: ITER-AA04-2105.

[3] R. Srinivasan, S.P. Deshpande &The Indian DEMO team, Strategy for the Indian DEMO design, Fusion Engineering and Design, 83(2008) 889-892.

[4] A.-A.F. Tavassoli, 16-8-2 weld metal design data for 316L(N) steel , Fusion Engineering and Design, 83(2008)1467-1470.

[5] L. P. Jones, P. Aubert, V. Avilov, F. Coste, W. Daenner, T. Jokinen, K. R. Nightingale, M. Wykes, "Towards advanced welding methods for the ITER vacuum vessel sectors", Fusion Engineering and Design 6 9 (2003) 215-220.

[6] M. Oznozuka et.al, "Manufacturing and maintenance technologies developed for a thick wall structure of ITER vacuum Vessel", Fusion Engineering and Design 55 (2001) 397-410.

[7] K. Ioki et. al., "Design Improvement and R&D achievements for VV and in Vessel components towards ITER construction", Nuclear Fusion 43 (2003) 268-273.

[8] G. Kalinin et. al. "ITER R&D: "Vacuum Vessel and In-Vessel Components: Materials Development and Test", Fusion Engineering and Design 55 (2001) 231-246.

[9] K. Ioki ,C. Bachmann, P. Chappuis, et.al, ITER vacuum vessel: Design review and start of procurement process, Fusion Engineering and Design 84 (2009) 229–235.

[10] M. Nakahira, Technical Code issues of ITER Vacuum Vessel and their resolutions, Journal of Nuclear Science and Technology, 40(2003) 687-694.

International Symposium on Vacuum Science & Technology and its Application for Accelerators IOP Publishing
Journal of Physics: Conference Series **390** (2012) 012064 doi:10.1088/1742-6596/390/1/012064

Characterization of titanium-zirconium-vanadium non evaporable getter coated vacuum chambers

K V A N P S Kumar[1], T Bansod [1], C Mukherjee[2], G Singh[3], Pragya Tiwari[4], B K Sindal[1] and S K Shukla[1]

[1]Ultra High Vacuum Technology Division, RRCAT, Indore - 452 013
[2]Mechnanical & Optical Support Section, RRCAT, Indore-452 013
[3]Laser Material Design & Development Section, RRCAT, Indore- 452 013
[4]Indus Synchrotron Utilization Division, RRCAT, Indore- 452 013

E-mail: kkumar@rrcat.gov.in

Abstract. Non evaporable getter (NEG) coating technology development started at UHVTD, Raja Ramanna Centre for Advanced Technology (RRCAT). An UHV compatible, cylindrical magnetron sputtering system for Ti-Zr-V (Titanium-Zirconium-Vanadium) NEG coating was designed, developed and successfully installed at UHVTD, RRCAT. Surface studies play important role in this technology development. Inside surfaces of many vacuum chambers were coated with varying thickness and compositions using this setup. Preliminary studies of adhesion, crystallographic structure & size of crystallites, surface morphology, trapping of argon gas during coating and its release after getter activation were carried out and some results are reported in this paper.

1. Introduction

Getter coatings are chemical pumps which are able to chemisorb the active gases on their surfaces. The alloys of titanium, zirconium, vanadium and Iron are the most commonly used getter materials used in accelerators.

To coat the inner surfaces of vacuum chamber, magnetron sputtering is the most efficient method thus ensures better adhesion and microstructure formation. An UHV compatible, cylindrical magnetron sputtering system for NEG coating was designed and developed in house at UHVTD, RRCAT, India [1]. Using this system, many chambers were coated with varying thicknesses of NEG films [2]. The characteristics of NEG thin films in terms of microstructure, adhesion, surface roughness, grain formation was studied. The barrier effect of NEG coating in controlling hydrogen (H_2) outgassing from the system was also observed. The phase formation and crystallite size was calculated using X-ray diffraction. For surface roughness and to see the structure of individual grains, Atomic Force Microscope (AFM) was used. Composition analysis of coated film was carried out by Energy Dispersive X-ray Spectroscopy (EDX) and the formation of grains in the coated film was obtained by Scanning Electron Microscope (SEM).

[1] Corresponding author – K.V.A.N.P.S. Kumar

Journal of Physics: Conference Series **390** (2012) 012064 doi:10.1088/1742-6596/390/1/012064

2. Substrate Preparation Methods

Ti-Zr-V films were deposited on stainless steel, aluminium and glass substrates. The typical dimensions were 10X10X2 mm for metal substrates and 50X25X2 mm for glass substrates. The metal substrates were used for X-ray ray diffraction (XRD), microstructure, and compositional studies while glass substrate was used for surface roughness measurement by AFM. Prior to loading in the coating chamber, all the substrates were chemically and ultrasonically cleaned. Titanium (Ti), Zirconium (Zr), Vanadium (V) wires were twisted and placed coaxially in cylindrical vacuum chamber connected to a DC Power Supply (Parameters:- 1kV and 1A). To study wall thickness uniformity, substrates were placed at different positions of the chamber. The detail of the coating procedure is explained elsewhere [1].The details of Ti-Zr-V target preparation and other parameters are given in table 1.

Table-1: Details of Ti-Zr-V target preparation, deposition parameters and its composition.

Target Wire ratio (Ti:Zr:V)	Substrate belongs to	Deposition Parameters (Dishcarge Voltage, Discharge Current, Argon gas Pressure, Magnetic Field, Temperature of substrate)	Ti-Zr-V Composition	Thickness of coated film on substrate (μm)
1:2:2	Chamber_D	-600V, 150 mA, $1.3X10^{-2}$ mbar, 250 Gauss, 150^0C	Ti-25%,Zr-27%, V-48%	1.65
2:2:1	Chamber_LA	-600V, 200 mA, $2.5X10^{-2}$ mbar, 220 Gauss, 150^0C	Ti-59%,Zr-16%, V-25%	3.6
2:2:1	Chamber_LB	-600V, 250 mA, $2.5X10^{-2}$ mbar, 220 Gauss, 150^0C	Ti-59%,Zr-18%, V-23%	4

3. Characterization Studies

The analysis of surface roughness and grain formation was carried out using an AFM (AFM: SOLVER PRO, NT-MDT) on glass substrate. The compositional studies and image capture were done using SEM (PHILIPS XL30) on metal substrates. X-ray diffraction of all the substrate was carried out using Rigaku Diffractometer (GEIGERFLEX) with curved crystal graphite 224R monochromator. The XRD data was recorded at Cu-K_α in the 2θ range of 15^0-70^0 at a step of 0.02^0.

3.1. Adhesion

In order to perform coatings satisfactorily, they must adhere to the substrates on which they are applied. The scotch tape test was used to test the adhesion of the coating on the chamber wall. One coated chamber (which was activated several times at higher temperatures up to 350-450^{0}C) was subjected for scotch tape test. The adhesion was tested as per the US Military Standard (M-13508-C). The coated film passed the test.

3.2. Morphological, structural & chemical composition analysis

Using AFM and SEM, the surface morphology of NEG films were studied. It was clear from the comparison with AFM data of National Synchrotron Light Laboratory (LNLS), Brazil [3] that the desired characteristics related to the nanometre structure of the films were met for NEG coating carried out at RRCAT.

3.2.1. Surface Roughness Studies of thin films.

The important parameter in evaluation of Ti-Zr-V NEG film is its surface roughness. The rougher the surface, larger the actual area available for pumping. With increased roughness more sites for pumping becomes available, as a result pumping speed also enhances. It also provides more multiple surface reflections and the probability that the molecule is trapped and therefore leads to gettering action. The 2D and 3D AFM images of the films are given in the figure 1 to figure 6 for all the three substrates prepared.

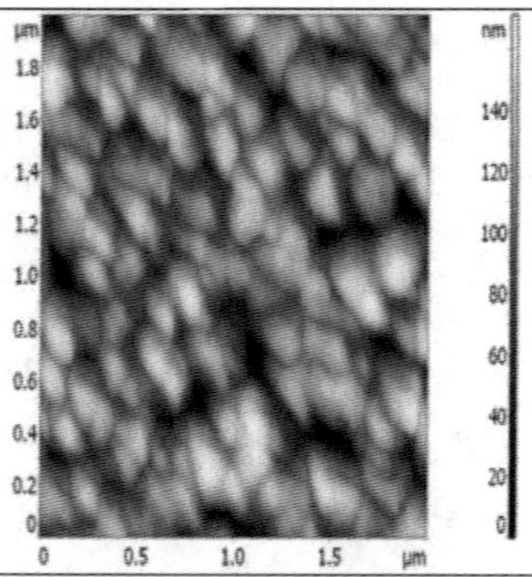

Fig. 1. AFM 2D image of Chamber_LB showing grain information

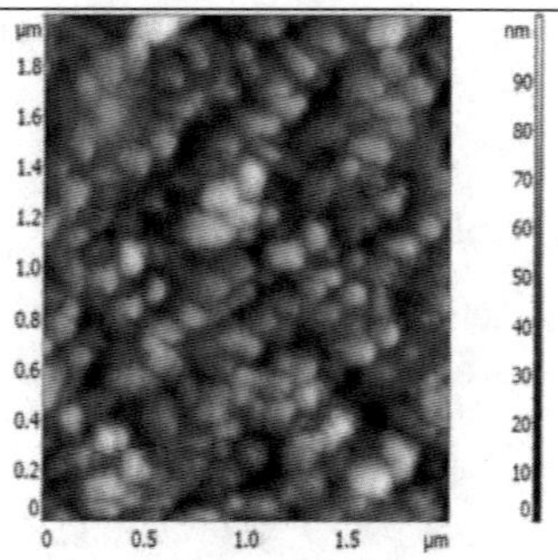

Fig. 2. AFM 2D image of Chamber_LA showing grain information.

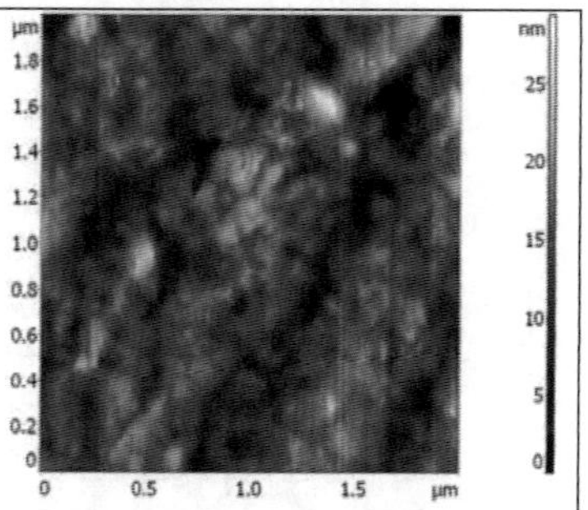

Fig. 3. AFM 2D image of Chamber_D showing grain information.

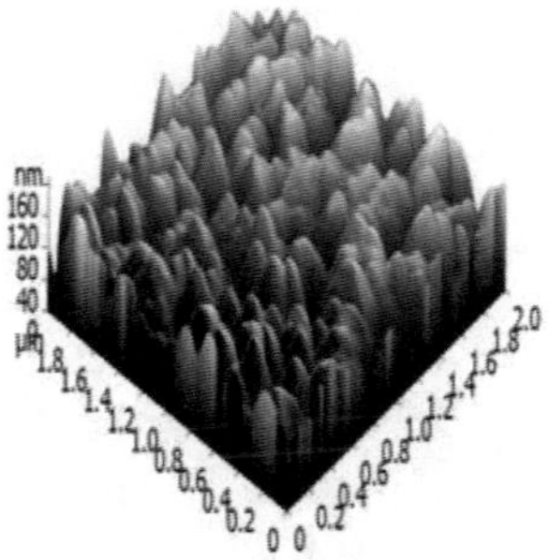

Fig. 4. AFM 3D image of Chamber_LB showing surface roughness

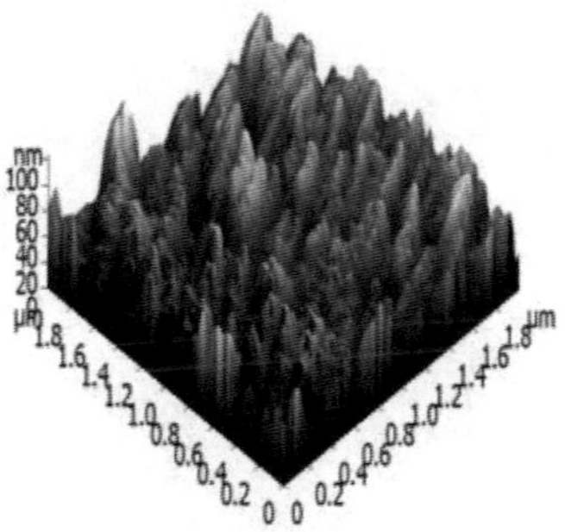

Fig. 5. AFM 3D image of Chamber_LA showing surface roughness

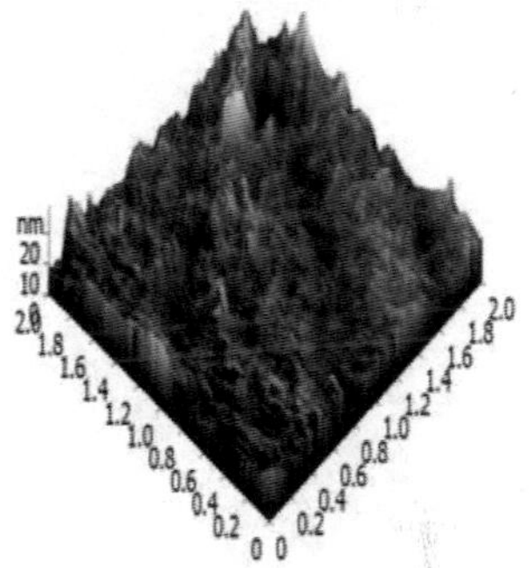

Fig. 6. AFM 3D image of Chamber_D showing surface roughness

The observation of surface roughness found from Ti-Zr-V coated glass substrates using AFM is given in table 2.

Table 2. Details of Surface roughness of TiZrV coated substrates

Substrate Belongs to	Average roughness(nm)	R.M.S. roughness (nm)	Peak to Peak roughness (nm)
Chamber_D	2.2	2.93	29
Chamber_LA	10.9	13.6	103
Chamber_LB	21.0	26.0	168

The AFM results of Ti-Zr-V film from UHVTD are comparable to that of LNLS, Brazil [3]. The observed peak roughness of the films deposited at LNLS is of 80 nm. Also, the LNLS film had less uniform grains indicating larger roughness, smaller grain size, and larger surface area. These characteristics are very interesting for NEG application. The observations obtained from films coated of RRCAT are in agreement with the observations given by LNLS group [3]. The measured pumping speed of 0.7 $l/s/cm^2$ for H_2 found with the Ti-Zr-V film deposited on Chamber_D using a pumping speed setup [2].

3.2.2. X-ray Diffraction Analysis.
The X-ray diffraction pattern of Ti-Zr-V film coated over stainless steel (SS) substrate and glass are shown in figure 7 and figure 8. A diffraction peak due to film was observed near $2\theta \approx 38.30^0$ for both the substrates. Two substrate related peaks were also observed for the film coated over SS. The observed peak can be indexed with (002) peak of titanium (JCPDS 05-682). The crystallite size (d) was calculated using Debye Scherer relation:

$$d = \frac{0.9\lambda}{\Delta \cos\theta}$$

Here λ is the wavelength used i.e. 0.15418nm for Cu-K_α , Δ is FWHM of diffraction peak, θ is the angular position of recorded peak in radians. The instrumental broadening was calculated by carrying out X-ray diffraction on silicon powder. The data was corrected for instrumental broadening before calculating the crystallite size. The calculated crystallite size for films deposited over SS substrate is 14 nm.

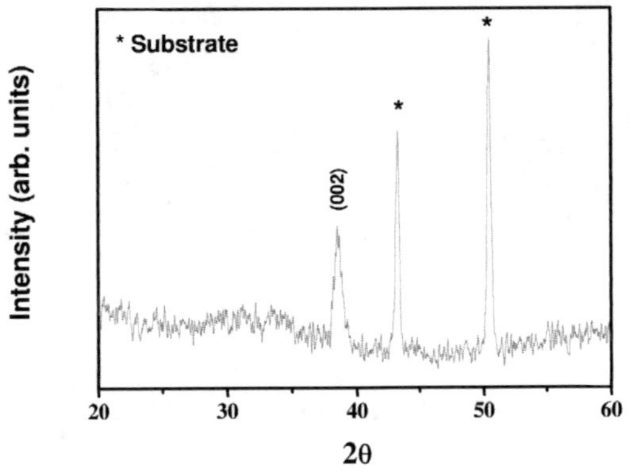

Fig. 7. XRD pattern for TiZrV coated film on Stainless Steel substrate of Chamber_LB

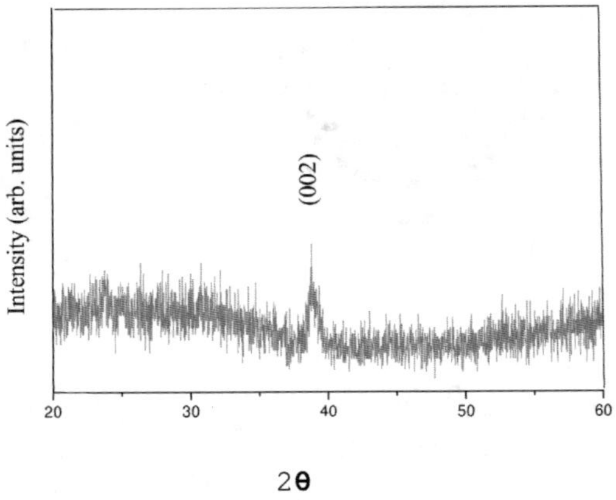

Fig. 8. XRD pattern for TiZrV coated film on Glass Substrate of Chamber_LB

3.2.3.SEM/EDX Studies.
Images of the surface of the films were obtained in a XL 30 Philips microscope. The grain information (in air exposed NEG sample) obtained through SEM can be seen from the figure 9. The grains are clearly visible and they are in conical form. This implies that the grain formation found to be uniform throughout the region of film and uniform in shape. The arrangement of grains of chamber_LB was in agreement with 2D AFM image observations as shown in figure1 to figure 3. The composition details are given as ternary diagram in figure 10.

Journal of Physics: Conference Series **390** (2012) 012064 doi:10.1088/1742-6596/390/1/012064

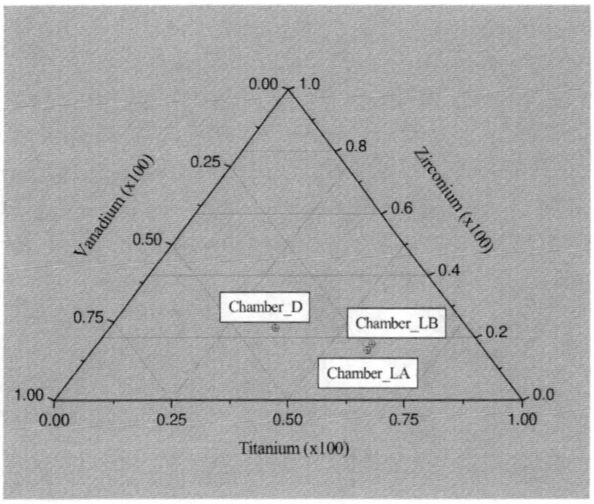

Fig. 9. SEM Micrograph of Chamber_LB

Fig. 10. Composition in Ternary diagram

3.2.4. Vacuum characterization of NEG coated chambers.

Ti-Zr-V non evaporable getter coatings help in achieving ultra high vacuum of the order of 10^{-11} mbar and below. Their presence acts as barrier for hydrogen even in unbaked systems. Reduction in hydrogen partial pressure by three orders was observed in the present study. The figure 11 shows the details of residual gas present in vacuum chamber under different conditions. Trapped argon gas content was found during the activation studies of chamber at an elevated temperature of 300^0C. The evolution of gases corresponding at 300^0C was shown in figure 11. The order of gas release observed was as follows Hydrogen (H_2), Methane (CH_4), Argon (Ar) followed by CO, H_2O and CO_2. This behaviour was as expected in an ultra high vacuum environment.

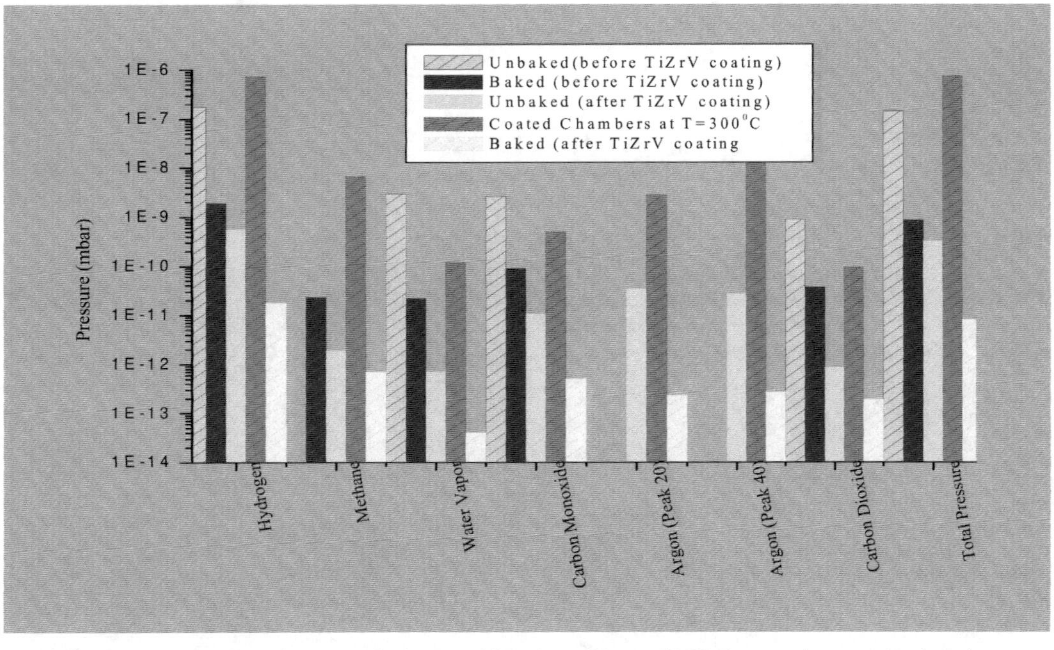

Fig. 11. Details of gas evolution and barrier effect of NEG coated vacuum chambers.

4. Conclusion

The characterization of Ti-Zr-V non evaporable getter coated chambers provided very encouraging results. The adhesion property of the film was good even after repeated activation of the chamber. The AFM images gave an excellent information about the surface roughness and grain information. The data obtained by SEM also confirmed that the grain formation was uniform and grains are arranged in regular manner. XRD measurements also gave an indication about the nano-crystalline formation of the film and estimation of the grain size. The trapping of the argon gas during the coating can be avoided by mild baking and prolonged pumping of the coated chamber before removing from the coating system. The ultimate vacuum of the order of $2.6X10^{-12}$ mbar was achieved in both chamber_LA, Chamber_LB and $1X10^{-11}$ mbar in the chamber_D.

Acknowledgement

The encouragement of Director, RRCAT is sincerely acknowledged. The authors would also like to acknowledge the support from all the members of UHVTD, RRCAT.

Reference

[1] Shukla S K, Sindal B K, Tripti Bansod and Kumar K V A N P S Conference Proceedings by IVS Symposium on Thin Films:Science & Technology, November 09-12, 2011, BARC, Mumbai.

[2] Tripti Bansod, Sindal B K, Kumar K V A N P S and Shukla S K Conference Proceedings by IVS Symposium on Thin Films: Science & Technology, November 09-12, 2011, BARC, Mumbai.

[3] Marcelo J. Ferreira, Denise A, Tallarico, Pedro A.P.Nascente AIP conference Proceedings 1092, 168 (2009) (Synchrotron Radiation in Materials Science:6th International Conference, edited by R. Magalhaes-Panigo 168-172 2009).

International Symposium on Vacuum Science & Technology and its Application for Accelerators IOP Publishing
Journal of Physics: Conference Series **390** (2012) 012065 doi:10.1088/1742-6596/390/1/012065

ZnO/Ti Thin Film: Synthesis, Characterization and Methane Gas Sensing Property

D Chakraborty, R Gayen, S Hussain[1], R Bhar ,A K Ghoshal and A K Pal

Department of Instrumentation Science, Jadavpur University, Kolkata-700032
[1]UGC-DAE CSR, Kalpakkam Node, Kokilamedu-603104, India

E-mail : dcphy123@gmail.com

Abstract. Thin films of ZnO were synthesised on glass substrate by dc sputtering technique at a system pressure of 10^{-4} Torr. A layer of TiO_2 was deposited onto the above ZnO films by high pressure sputtering technique. The composite structure was then subjected to rapid thermal annealing at a pressure of 10^{-2} Torr in argon ambience. Microstructural and compositional analyses were carried out by SEM, EADX and XRD studies. The optical band gap pure ZnO (~3.45 eV) was found to shift to higher energy (~3.54 eV) for the ZnO/Ti structures. FTIR measurement indicated the presence of a prominent absorption peak at ~490 cm^{-1} due to ZnO stretching mode. Methane gas sensing measurement was carried out at 353K, indicated that the highest sensitivity was nearly 60% with a response time ~52 min when the film were exposed to 10 % vol. of methane gas.

1. Introduction

The detection of different kinds of explosives and hazardous gases in the atmosphere is generally made by using metal oxides and porous silicon based sensors. These sensors are quite expensive and are not readily available in the market. Hence, it is of prime interest to the researches to synthesize new materials which would allow developing sensors of low cost, high sensitivity, selectivity, short response and recovery times as well as of small dimensions to enhance their portability and use.

The current day gas sensors, generally made of thin film structures of different metal oxides (SnO_2, TiO_2, ZnO, etc.), are being studied widely [1-6]. Most of them have the ability to transform the change in gas concentration to electrical signal immediately (i.e. electrical resistance or change in conductivity of these materials), which allow using rationally simple scheme for implementing the final equipment. Besides, thin film sensors using less electrical input have also been manufactured. Among these, zinc oxide (ZnO) is one such material which has been studied in various forms for the detection of different gases (Table 1). However, as can be seen from Table 1, the sensors made of ZnO have high working temperatures.

ZnO is also a multifunctional material. Because of its high chemical stability, low dielectric constant, large electrochemical coupling coefficient and high luminous transmittance, ZnO materials have been widely used as dielectric ceramic, pigment, catalyst and sensing material. It is sensitive to many sorts of gases, and has satisfactory stability. Its gas selectivity can be improved by doping additives and catalysts. However, the working temperature of these sensors is rather high, normally 400-500°C and their gaseous

Journal of Physics: Conference Series **390** (2012) 012065 doi:10.1088/1742-6596/390/1/012065

selectivity, comparatively poor. Studies on gas sensing materials by improving their preparation method and decreasing their working temperature are currently one of the major research topics [13].

Table 1 Application of ZnO for detection of different gases

Sensor	Gas	^{0}C	K	Ref
ZnO nanorods	ethanol,H_2	350	623	[7]
ZnO nanowire	ethanol	300	573	[8]
Al-doped	ZnONOx	100-300	373-573	[9]
ZnO thin films	CH_4	150-350	423-623	[10]
SnO_2–ZnO	ethanol acetone	300	573	[11]
Fe_2O_3 -ZnO	NH_3	350	623	[12]

In this connection methane sensor made of nanocrystalline Ti/ZnO composite films has been reported. The films show good sensing behaviour at a temperature ~353K, quite low as compared to the other methane sensors reported before.

2. Experimental

ZnO /Ti composite thin films were deposited on cleaned glass substrates. ZnO films were deposited by d.c. sputtering technique. The substrate was maintained at room temperature and the argon gas pressure during deposition was kept constant at ~0.3 mbar. The deposition was carried out for 3 hours at a constant power (~ 60 W). Subsequently, titanium was sputtered at room temperature at an argon pressure of ~ 0.3 mbar at a constant power of ~75 W. The deposition was carried out for 10 minutes and a very thin layer of Ti was deposited. The base pressure was maintained better than ~10^{-5} mbar. The ZnO/Ti composite structures so obtained were subjected to rapid thermal annealing inside a quartz chamber. Annealing was carried out for 3 minutes at temperature ~ 573K in partial argon gas pressure (~10^{-4} Torr). Gold contacts were deposited on the ZnO/Ti thin films for sensitivity measurements.

Optical studies were performed by measuring transmittance in the wavelength range λ=200-2500 nm using a spectrophotometer (Hitachi-U3410) at room temperature. The spectra were recorded with a resolution of λ ~ 0.07 nm along with a photometric accuracy of ± 0.3% for transmittance measurements. Compositional information was obtained by EDAX (Oxford Instruments' INCA Energy 250 Microanalysis System), which indicated that the films were nearly stoichiometric. The surface morphology of the films was studied by Field Emission Scanning Electron Microscope (FESEM) (Carl Zeiss SUPRA® 55 with GEMINI® Technology with resolution ~0.8 nm @ 15 kV). AFM pictures were recorded by using a Nanosurf Easy Scan 2 in contact mode while X-Ray Diffraction (XRD) studies were carried out by Rigaku MiniFlex using Cu Kα line (0.154 nm). FTIR spectra were recorded in the range of 400-4000 cm^{-1} by using a Nicolet™-380 FTIR.

3. Results and Discussion

3.1. Microstructural Studies.

3.2.1. SEM studies

The SEM micrograph of pure ZnO on the glass substrate sputtered for 3 hrs is shown in Fig. 1(a). It is clearly identified that the ZnO particles form a cluster over the surface of the glass substrate. As seen from the micrograph, the film is compact and quite homogeneous in nature. Fig 1(b) shows the micrograph of a ZnO/Ti film subjected to RTA 3minutes at ~573 K. Cracks have developed on the surface of the film.

Journal of Physics: Conference Series **390** (2012) 012065 doi:10.1088/1742-6596/390/1/012065

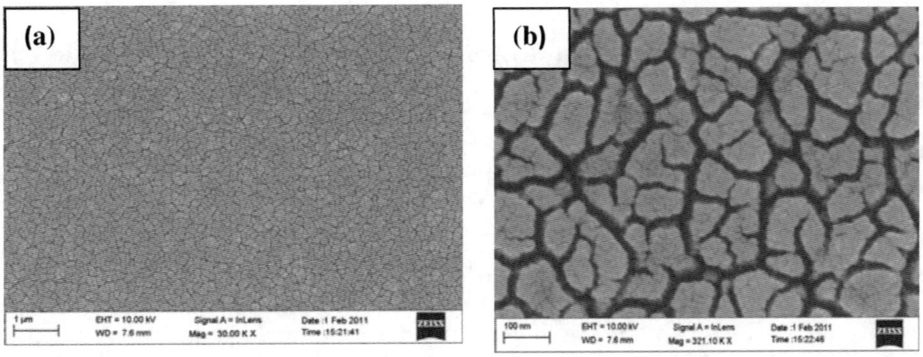

Fig. 1: SEM micrographs of: a) pure ZnO on glass substrate and b) ZnO/Ti after Rapid Thermal Annealing

3.1.2. EDAX Studies.

Energy Dispersive X-Ray method was used to determine the qualitative and quantitative composition of the surface of the film. Fig. 2 shows the EDX spectrum of a ZnO/Ti film. The spectrum is dominated by the peaks of zinc (Zn), oxygen (O_2), gold (Au) and with suppressive peaks of titanium (Ti).The percentage weight of the elements present as obtained from the analysis of the spectra are oxygen (18.22%), titanium (0.76%) and zinc (81.03%). The gold peaks have been excluded from the analysis.

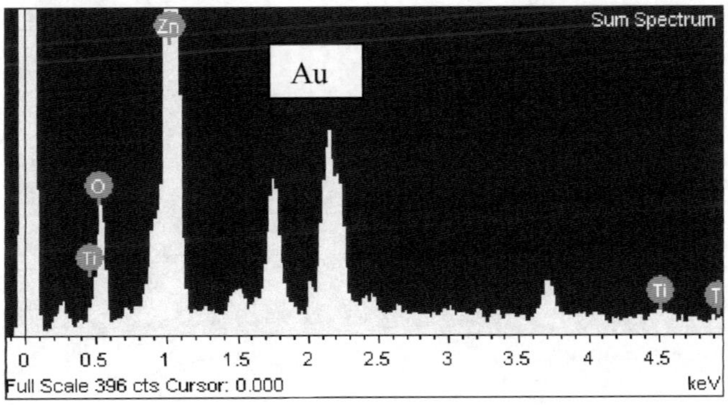

Fig.2. EDAX spectrum of a representative ZnO/Ti film

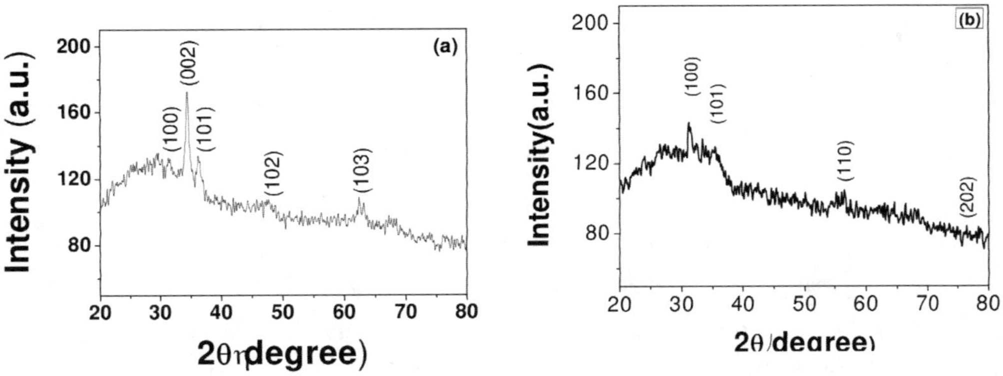

Fig. 3. XRD traces of: a) pure ZnO sample and b) ZnO/Ti sample

3.1.3. XRD Studies.

X-Ray diffraction pattern was obtained from Rigaku Miniflex.Diffraction pattern of pure ZnO is shown in Fig.3a. XRD peaks at 2θ ~ 31^0, 34^0, 36^0, 47^0 and 62^0 are due to the reflections from the (100), (002), (101), (102) and (103) planes of ZnO, respectively and the XRD pattern of ZnO/Ti (Fig.4b) depicts that peaks at 2θ ~ 31°,35°, 56°and 76°corresponds to the reflections from (100), (101), (110) and (202) planes of ZnO, respectively. There was no observable peak for reflection from Ti.

3.2. Optical Studies.

The band gap of ZnO was found by the optical procedure, in which the transmittance vs. wavelength was obtained by UV-VIS-NIR Spectrometer. The band gap was determined by extrapolating the linear portion of the plot of $(\alpha h\nu)^2$ versus hν (Fig.4 c, d). The band-gap of pure ZnO was ~3.45 eV and ZnO/Ti was shifted to ~3.54 eV due to Burnsten Moss Shift

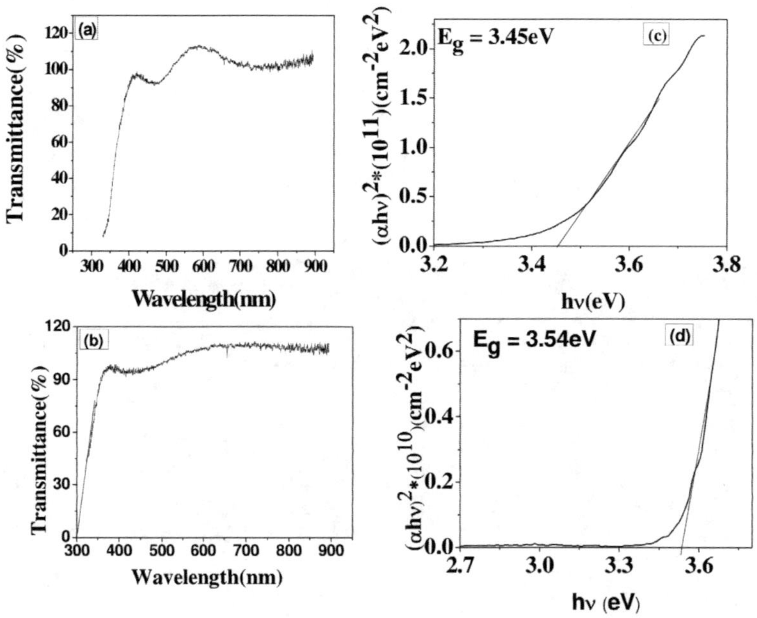

Fig. 4. Transmittance vs. wavelength plot of a) pure ZnO thin film, b) ZnO/Ti thin film. $(\alpha h\nu)^2$ vs. hν plot; c) pure ZnO thin film, b) ZnO/Ti thin film

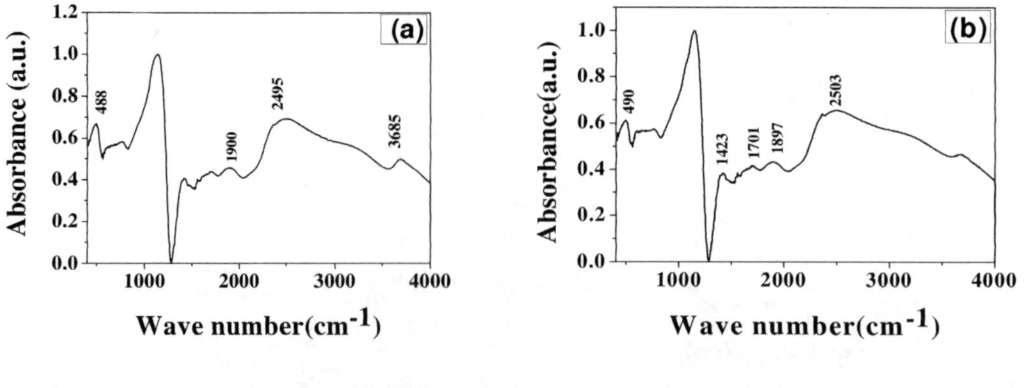

Fig.5. FTIR spectra of: a) ZnO and b) ZnO/Ti film

3.3. FTIR Studies.

FTIR absorbance spectrum was recorded in the range of 400 to 4000 cm $^{-1}$ of the pure ZnO and ZnO/Ti is shown in the Fig.5. The spectral region encompasses every stretching and bending mode. It may be observed that FTIR spectra for both pure ZnO and ZnO/Ti film (Fig.6) show a prominent peak at ~488 cm^{-1} due to Zn-O stretching mode. Peak at ~1701cm^{-1} (Fig. 5(b)) is due to OH bending of water.

4. Sensitivity Studies.

After characterising the material, the sensitivity tests of the material towards methane gas was carried out. For this, the sample was mounted on a platform. Electrical contacts were made to an ammeter and a voltmeter through the gold contacts which was deposited by the evaporation technique on the corners of the sample. The chamber was isolated using a glass bell jar.

Before injecting the gas into the chamber, the sensor was heated to ~60° C. It was observed that as the temperature was increased, the resistance of the sensor material decreased with increasing temperature. The temperature was maintained at 60° C for gas sensing experiment. Resistance remained all most steady [Fig (6)] at the controlled temperature of 60° C. Then at this temperature, at a steady value of resistance, methane gas (10% volume concentration) was injected into the chamber. The change in the electrical resistance of the sensor was calculated by studying the change of current. The sensitivity(S) of the material was calculated based on the following formula:

$$S = [\{R_{air} - (R_{gas})\, eq/R_{air}] \times 100 \qquad (1)$$

Where, the sensor's resistance in presence of the target gas (i.e. R_{gas}) and that in air (i.e. R_{air}) [14]

Fig (7) shows the plot of the sensitivity versus time interval. It is found that S is less than unity for reducing gases. The percent sensitivity i.e. the percent reduction of sensor resistance is given by [14]:

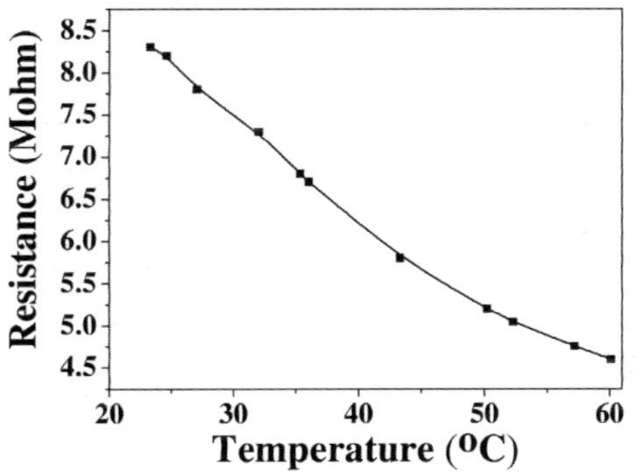

Fig. 6. Resistance vs. Temperature

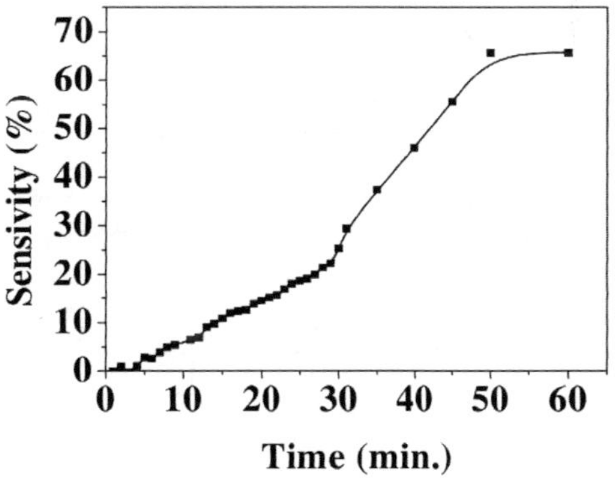

Fig. 7. Sensitivity vs. time

5. Conclusion.

In this work deposition and characterization of dc-sputtered ZnO/Ti thin film and its use for detecting methane gas was carried out. The sensitivity measurement of the film was carried out at ~330K in presence of 10%vol. concentration of the methane gas. Its sensitivity may be increased by increasing uniformly deposition of titanium (Ti) over ZnO layer or by using other catalyst like palladium or by using multiple layers. Experiments in these directions are in progress. Also the selectivity and stability of the sensors are to be examined. The above result may be utilized to design a methane gas sensor for environment health and safety.

References

[1] Adamyan A.Z., Adamyan Z.N., Aroutiounian V.M.,2007, *Physica E*, **38**, N 1-2, pp. 164–167.

[2] Arakelyan V.M., Martirosyan Kh.S., Galstyan V.E., Shahnazaryan G.E., Aroutiounian V.M ,2007, *Phys. Stat. Sol. (c)*, **4**, N 6, pp. 2059-2062.

[3] Mor G.K., O.K, . Paulose M, Ong K.G., Grimes C.A.,2006, *Thin Solid Films*, **496**, pp. 42-48.

[4] Aygun S, Cann D,2005, *Sensors and Actuators B*, **106**, pp. 837-842.

[5] Jun Youn-Ki, Kim Hyun-Su, Lee Jong-Heun, Hong Seong-Hyeon,2005, *Sensors and Actuators B*, **107**, pp. 264–270.

[6] Choi U.S., Sakai G., Shimanoe K., and Yamazoe N.,2005, *Sensors and Actuators B*, **107**, pp. 397-401.

[7] Bie Li-Jian, Yan Xiao-Na, Yin Jing, Duan Yue-Qin, Yuan Zhi-Hao,2007, *Sensors and Actuators B*, **126**, pp. 604–608.

[8] Hsueh Ting-Jen, Hsu Cheng-Liang, Chang Shoou-Jinn, and Chen I-Cherng,2007, *Sensors and Actuators B*, **126**, pp. 473-477.

[9] Navale S.C., Ravia V., Mulla I.S., Gosavi S.W., Kulkarni S.K.,2007, *Sensors and Actuators B*, **126**, pp. 382–386.

[10] Bhattacharyya P., Basu P.K., Saha H., Basu S.,2007, *Sensors and Actuators B*, **124**, pp. 62–67.

[11] Kim Ki-Won, Cho Pyeong-Seok, Kim Sun-Jung, Lee Jong-Heun, Kang Chong-Yun, Kim Jin-Sang, Yoon Seok-Jin,2007, *Sensors and Actuators B*, **123**, pp. 318-324,.

[12] Patil D.R., Patil L.A,2007, *IEEE Sensors Journal*, **7**, N 3, pp. 434-439.

[13] JiaqiangXu, Qingyi Pan, Yu'anShun, ZhizhuangTian, Sensors and Actuators B: Chemical ,Volume **66**,Issues 1-3,25 July 2000,Pages277-279

[14] .Mitra P and Mukhopadhaya A.K.,2007, Bulletin of The Polish Academy Of Sciences Technical Sciences; Vol **55**, No. **3**.

International Symposium on Vacuum Science & Technology and its Application for Accelerators IOP Publishing
Journal of Physics: Conference Series **390** (2012) 012066 doi:10.1088/1742-6596/390/1/012066

Vacuum System of the Ion Source and Injection Line of a High Current Compact Cyclotron

V. S. Pandit, R. C. Yadav, G. Pal, C. Nandi, A. Goswami, P. Sing Babu, S. Srivastava, A Misra

Variable Energy Cyclotron Centre, 1/AF, Bidhannagar, Kolkata-700064, India

E-mail: pandit@vecc.gov.in

Abstract. A 2.45 GHz microwave ion source and the transport system of high current proton beam for injection study in a compact cyclotron have been designed, fabricated indigenously, installed and commissioned. Presently it is under testing for beam characterization and inflection study. The vacuum system consists of a plasma chamber with 8mm slit in the plasma electrode, two-segment ceramic insulators (Al2O3) column, which supports the beam extraction electrodes and isolates the high voltage deck from the beam line and a 3m long low energy beam transport line (LEBT) equipped with several diagnostic elements such as faraday cup, slits, beam viewer etc. There is a provision to connect a needle valve to feed gas into the chamber for beam space charge neutralization studies. An online vacuum data logging software has been developed for continuous monitoring of the vacuum near the extraction zone. Monitoring system being very sensitive to the high voltage breakdown, is designed with a spark protection circuit which provides an isolation of 10kV using an optical isolation technique. In this paper we present design philosophy, operating experience and some experimental results. We have studied the behaviour of vacuum system near the extraction zone under various operating conditions of the high current ion source.

1. Description of Ion source

The 2.45GHz microwave ion sources are now a days widely used to produce several mA of proton beam for various applications. These sources are based on the principle of off resonace microwave discharge. The main features of such sources are high yield, high proton fraction and long term stability and reliability which meet the requirement of a driver for an ADS system. At the Variable Energy Cyclotron Centre, Kolkata a 2.45 GHz microwave ion source and the injection line to transport high current proton beam have been designed, fabricated indigenously, installed and commissioned (shown in Fig. 1) which produces protons beam at 80 keV. The main objective of this development is to carry out detailed injection study of high beam current in a compact cyclotron. The ion source consists of a plasma chamber, two solenoids and a triode ion extraction system. The diameter of the apertures in the plasma electrode, accelerating electrode and de-accelerating electrode are 6mm, 8mm and 8mm (new one) respectively. The plasma chamber is a double walled water-cooled cylindrical stainless steel chamber of 100 mm length and 90 mm diameter. The microwave power from the 2.45 GHz, 1.2 kW magnetron is coupled to the chamber through a three stubs tuning unit, an auto tuner and water cooled ridged wave-guide. Ion source with adjustable solenoid, its power supplies, microwave generator (2.45 GHz, 1.2 kW), a high precision gas flow system etc., all are kept at a high voltage deck ~100 kV. High voltage deck is separated from the ground through polypropylene insulators. A two-segment ceramic insulators (Al_2O_3) column, which supports the beam extraction electrodes, separates the high voltage deck and the beam line at the ground potential. Power to the various subsystems on the deck is supplied using a 150kV, 30kW isolation transformer. Since all subsystems of the ion source are placed on the high voltage deck, an isolated control system is used for remote monitoring and operation of various sub-systems e.g. microwave generator, microwave tuner, solenoid magnet power supplies and gas feeding system. The system architecture is modular with dedicated control nodes for individual sub system. The PC based supervisory console is connected with the control nodes in RS485 multi-drop fashion through an indigenously developed optical fibre based serial link for electrical isolation. A protection circuit is also implemented to reduce communication failure due to frequent spark.

The injection beam line consists of two magnetic solenoids (SOL1: 40 cm, 3.6 kG) (SOL2: 40 cm, 3.3 kG), with some diagnostic elements slits, faraday cup and emittance monitoring box. The beam from the ion source is expected to contain a substantial fraction (~10 to 20 %) of molecular hydrogen ion. Two motor controlled independent slits, one set for the *x*-plane and other set for the *y*-plane is used to control the size of the beam and to reject the molecular hydrogen beam. We have also provided another water cooled fix slit of 4 cm dia after the first solenoid and before the waist position of proton. Beam current measuring equipments used in the beam line are; a water-cooled faraday cup (up to 10 mA only) with secondary electron suppresser and a DCCT. Three turbo pumps having pumping speed of 520 l/s are used to evacuate the entire system. We have achieved vacuum of the order of 1.5×10^{-7} Torr near the extraction zone and in the beam line. Control units for adjusting current in the solenoids, movement of solenoids, tuning of microwave power, adjustment of gas flow etc. is placed on the high voltage deck and control and monitoring of the various voltages and current is done with a PC at ground potential through optic fibre.

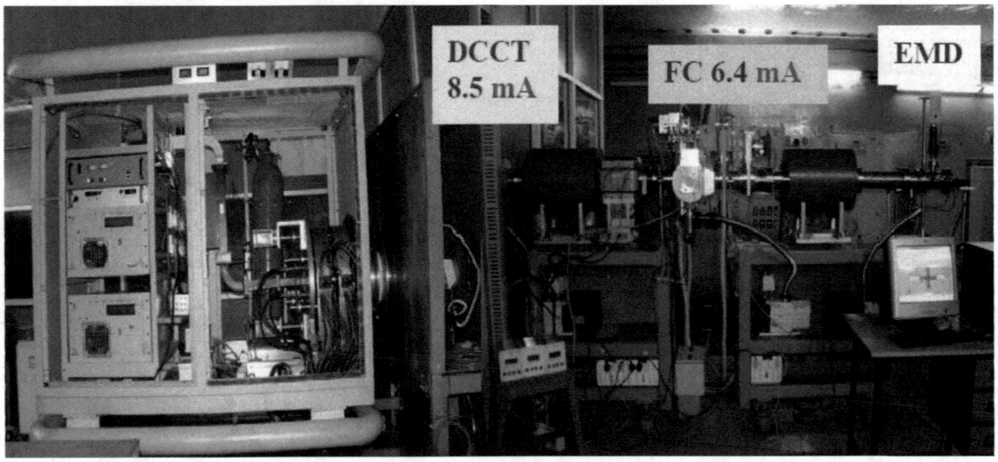

Fig. 1. Ion source on high voltage deck and LEBT with two solenoid magnets. At the end of beam line we see locally made water-cooled beam dump cum faraday cup.

We have obtained 6.4 mA beam current on FC and 8.5 mA on DCCT at 400 watt of microwave power and extraction voltage of 80 kV. We transported this beam up to the last beam dump near the diagnostic chamber. We observed increase in the beam current (I > 10 mA) at the DCCT with increase in microwave power. Beam spot of 80 keV, 5 mA proton beam on water cooled alumina plate is shown in Fig. 2. Most of the ring type shadow around the hot spot is due to other beams and neutrals generated in the extraction region. At present we are testing the source for long term stability and beam quality improvement.

A spiral inflector will be used to inflect the beam in the median plane of the 10 MeV cyclotron and to place the beam on the proper orbit. We have designed and studied the various optical properties of the inflector in the presence of space charge effects. We observed that the influence of space charge is drastic on emittance growth with upright ellipse as initial conditions. However, with optimized tilted ellipse the emittance growth is minimal. Thus the beam optimization for matching is very much important and necessary to avoid the beam loss. This inflector has already been fabricated and tested with high voltage upto 25kV. In order to study the inflection and transmission of high beam current through the spiral inflector, we have designed and fabricated a small magnet (shown in Fig. 3) having a similar characteristics as the central region of 10 MeV cyclotron. Magnet has already been assembled with a vacuum chamber with inflector inside and is connected in the beam line. The preliminary testing of beam transmission at low beam current ~ 1 mA is going on. A beam transmission of 50% has been achieved and efforts are on to improve the transmission efficiency. We would like to point out here that we need first to optimize the beam transmission at low current to avoid any damage to the inflector. The fabrication of a sinusoidal beam buncher together with a fast faraday cup is near completion. These components will be installed very soon after the second solenoid magnet to study the behavior of beam bunching at high current.

Fig. 2. Beam spot of 80 keV, 5 mA on water cooled alumina plate. Most of the ring type shadow around the hot spot is due to other beams and neutrals generated in the extraction region.

Fig. 3. Magnet assembly with vacuum chamber to house the spiral inflector and the spiral inflector on based plate of the chamber.

Journal of Physics: Conference Series **390** (2012) 012066 doi:10.1088/1742-6596/390/1/012066

2. Estimation of vacuum requirement

Low conductance has been utilized to achieve steady state pressure differential among inter-connected spaces and also to control the pressure differential as per the process requirement. Ion source is one of such application where the plasma chamber pressure needs to be controlled at different set values as per the ion production requirement while extraction chamber pressure is to be maintained regularly at pressure lower than 1×10^{-5} mbar to avoid sparks and discharge against the high extraction voltage. Schematic diagram of ion source and beam line components are shown in Fig. 4 highlighting the main vacuum parts. Plasma and extraction chambers are connected through 8 mm aperture for the extraction of ion beam. Upstream pressure controller has been connected to the plasma chamber to control hydrogen gas flow rate in the plasma chamber to reach the set pressure. Pumping modules have been designed for ion source and installed at extraction chamber. The pumping modules have been designed in view of acheiving plasma chamber vacuum in the range of 1×10^{-3} mbar for gas flow rate settable in the range of 2-5 sccm, keeping the extraction chamber vacuum better than 1×10^{-5} mbar. After extraction the ions move through a beam transport line of 3.2 meter long and diameter 100 mm. The beam line consists of two magnetic solenoids with some diagnostic elements slits, faraday cup and emittance monitoring box. The Vacuum system has been divided in two parts to estimate the pumping requirement, the ion source and extraction part and beam transport line.

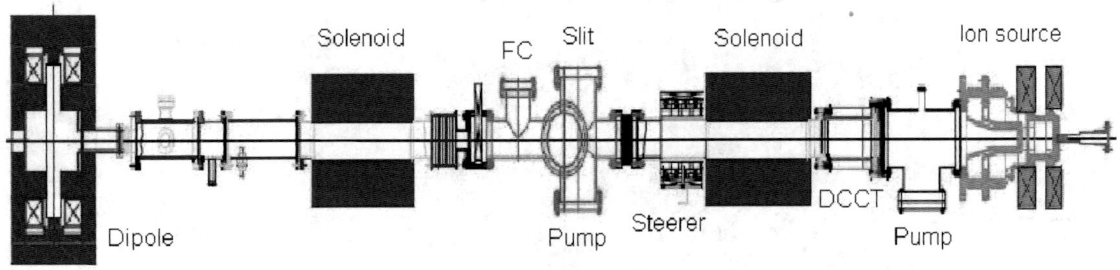

Fig. 4 Schematic of vacuum system of ion source and beam transport line.

The pumping system has been designed to get pressure in the plasma chamber in range of ~ 10^{-3} mbar with a load of gas flow rate ~ 2 sccm of hydrogen and maintaining the pressure better than 1.5 x 10^{-5} mbar in the extraction chamber. The total estimated outgassing load from plasma chamber and extraction system is ~2 x 10^{-5} mbar l/s. The plasma and extraction chambers are pumped by two 500 l/s turbo molecular pump having high compression ratio for hydrogen. These are electrically isolated by ceramic of low out-gassing rate. The base pressure of 1×10^{-6} mbar in the plasma chamber and ~2 $\times10^{-7}$ mbar in extraction zone have been routinely obtained without any gas load.

A uniform vacuum of 1 x 10^{-6} mbar is needed to be maintained from the extraction chamber to end of beam line which is 3.2 m long. Major portions of the beam line consist of tubes of diameter 100 mm except at two locations where 152 mm crosses are connected for assembly of pump and diagnostic components. This offers an overall volume of approximately 37.4 liter and a surface area of ~1.45 m^2. The total estimated gas load from line is 1.5 x 10^{-4} mbar l/s. We have used metal gaskets at demountable joints and same have been considered at the time of estimating the load. A 500 l/s turbo molecular pump backed by scroll pump has been used to evacuate the beam line at distance of 1.5 meter from extraction chamber. At the end of the beam line we have installed a vacuum chamber to in-house the spiral inflector. The total estimated gas load from this chamber is approximately ~7 x 10^{-6} mbar l/s. This chamber is pumped by a turbo pump of speed 60 l/s and partly pumped by the beam line turbo pump.

3. Online Vacuum Monitoring

The online vacuum data logging software have been developed with a spark protection hardware circuit for continuous monitoring of the vacuum near the extraction electrode and in the backing line of the pump for 2.45 GHz microwave ion source. The monitoring system which we put initially was found very sensitive to voltage breakdown at high voltage deck and used to hang even with a minor spark. Once, the entire circuit as well as the serial port of the pc was damaged by the induced voltages due to sparks. In order to make the system reliable in such kind of environment we have carried out detailed study and some modifications in the circuit. Now we have added a spark protection circuit which provides an isolation of 10kV using an optical isolation technique. It uses 6N137 digital opto isolator IC, along with max232 which is basically used for isolating the low voltage electronics with the high voltage system without affecting the communication speed. The spark protection system uses two symmetrical circuits. One circuit convert max232 voltages coming from PC to TTL level, so that opto-isolator can work in TTL voltages and provide an isolation of 10kV. The other circuit again transforms the TTL voltage level into max232 voltage level. In this way the data are transmitted through a RS232 cable to the vacuum gauge with isolation. This circuit has been tested for continuous operation and found to be very satisfactory. The peak in the Fig. 5 shows the increase in the vacuum due to spark produced near the extraction electrode where as second plot in Fig. 5 indicates the reduction in the pressure when the ion source was operating. Pressure resumed the almost same value after switching off the source. We have also provided facility to log the reading in the Microsoft excel sheet for further analysis of the data. The temperature of the vacuum gauge is also monitored to detect any malfunctioning of the vacuum gauge.

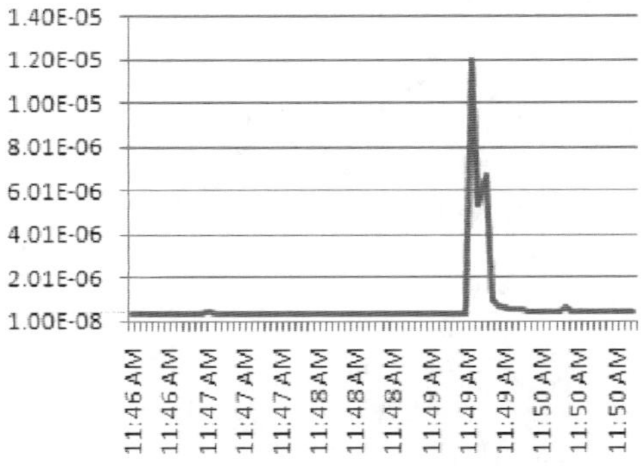

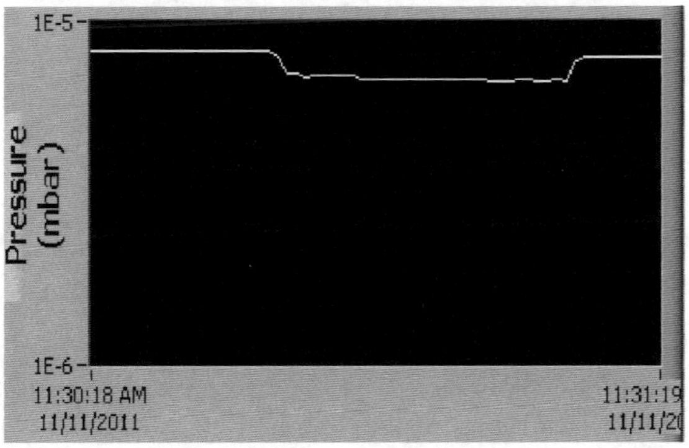

Fig. 5. Online display of the pressure near the extraction zone. Bump indicates the pressure rise due to spark of high voltage. The second plot shows the duration of reduced pressure zone in which ion source was operating.

4. Operational experience

The key issue during the designing and commissioning of the ion source was to get a satisfactory working regime. In this kind of source two positions of ECR points are very important. One ECR point should be located very near to the microwave window which is very vital for microwave power coupling into the source plasma. The other ECR point is required to be present at the extraction hole of the plasma electrode to create dense plasma for better ion extraction. The magnetic field at the middle of chamber helps the plasma in radial confinement and axial diffusion. The optimization of the required field values were set up by adjusting the locations of the solenoid coils and currents.

During the initial operations of the source, we faced few problems and taken appropriate actions:

- Initial promising operations were halted several times due to failure of the electronic components at the high voltage deck induced by HV sparking. To solve these problems we have grounded most of the floating conductors inside the deck and also done some modification by connecting a high resistance between isolation transformer and the deck. We placed appropriate filters and tested the system up to 90 kV with negligible leakage current (200 μA).

- There was a serious problem in microwave coupling to the plasma. We observed very high reflected power and lots of heating in the ridged wave-guide. Locally made control of manual tuner did not work properly. We inserted an auto tuner in the system. We designed and fabricated a new water-cooled ridged wave-guide and replaced it with the existing one.
- The thermal fracture of microwave window from the back streaming electrons and source plasma heating was another problem. We have now placed a 5 mm thick boron nitride plate behind the water-cooled plasma chamber. A RF quartz window is now placed for vacuum sealing just before the 90 deg bent in the waveguide. This will save the system in case of fracture of boron nitride microwave window.
- We faced lots of glow discharge and internal arcing during initial operation in the extraction region. Earlier our extraction slits were supported by cylindrical electrodes without any holes for the pumping. The main cause of the above problem was suspected to be the back streaming electrons that are produced by the proton impact ionization of neutrals in the extraction system. The electrons thus produced are then accelerated by electric field back towards the plasma chamber where they are focused on the axis by the magnetic field cause the discharge. In order to improve the vacuum in this region we machined several holes in both biased and ground electrodes. This process reduced the frequency of discharge by a considerable amount and improved the performance of the source.

The outlined strategy and improvement in the subsystems improved the reliability and performance of the source in terms of the long term stability. We have operated the source continuously for 8 hours with current more than 5mA on the Faraday cup. There were only 3-4 sparks which stopped the beam which was recovered within 2 minutes after the tripping of the high voltage.

References

[1] V. S. Pandit *et al*, 2007 *Symposium on Ion Beam Technology and its Applications*, p 109.
[2] V. S. Pandit *et al.*, 2011 *Indian Particle Accelerator Conference* (InPAC-2011, IUAC, NewDelhi).
[3] P. Sing Babu, A. Goswami, V. S. Pandit, 2011 Physics of Plasmas **18** 103117.
[4] P. Sing Babu, A. Goswami, V. S. Pandit, 2011 Nucl. Inst. And Meth. A **642** 1.
[5] A. Goswami, P. Sing Babu, V. S. Pandit, 2010 *19th International Conference on Cyclotrons and their Applications*, (Lanzhou).
[6] A. Goswami, P. Sing Babu, V. S. Pandit, 2011 *Indian Particle Accelerator Conference* (InPAC-2011, IUAC, NewDelhi).
[7] Labview User Guide, version7.1.

Vacuum Technology in the study of Graphene

A. K. Ghoshal [a], S. N. Banerjee [b]D. Chakraborty[c]

[a] Guru Nanak Institute of Technology, Kolkata
[b] S N Bose National Centre for Basic Sciences, Kolkata
[c]Department of Instrumentation Science,Jadavpur University,Kolkata:700032

e-mail: ghoshal2416@gmail.com

Abstract. Graphene, an allotrope of carbon is a two-dimensional sheet of covalently bonded carbon atoms that has been attracting great attention in the field of electronics. In a recent review graphene is defined as a flat monolayer of carbon atoms tightly packed into a 2-D honeycomb lattice. A survey has been made of the production processes and instrumentation for characterization of graphene. In the production of graphene, the methods mainly used are Epitaxial growth, oxide reduction, growth from metal-carbon melts, growth from sugar. In the characterization of graphene, the instruments that are mainly used to study the atomic properties, electronic properties, optical properties, spin properties are Scanning Electron Microscopy, Transmission Electron Microscopy, Raman Spectroscopy. In all these instruments high or ultra-high vacuum is required. This paper attempts to correlate vacuum technology in the production and characterization of graphene.

1.Introduction.

Graphene, a carbon allotrope with a two-dimensional honeycomb structure, has become an important player at the forefront of condensed matter physics, drawing the attention of theorists and experimentalists alike due to its challenging nature as a many-body problem, its unusual electronic properties and possible technological applications (see Refs. [1] and references therein). Graphene also belongs to a large class of planar condensed-mattersystems, which includes other graphite-related materials as well as high-Tc superconductors. The mobility of carriers in graphene has been reported as high as 15,000 cm2V-1s-1 with the potential to reach 100,000 cm2V-1s-1 with the reduction of impurities [1]. The increase of charge carriers does not significantly reduce the mobility,unlike 3-dimensional semiconductors where dopants and phonons act as scattering centers and reduce mobility. From Lattice Monte Carlo simulations of the phase diagram of graphene as a function of the Coulomb coupling between quasiparticles, it was shown that the graphene in vacuum is likely to be an insulator([1]).

An important property of graphene is optical transparency.This somewhat limits observation and detection of the material, however Raman spectroscopy and electron microscopy are viable techniques to identify single-layer graphene [1]. The optical transparency of graphene attracts attention because the conductivity is not adversely impacted by the thinness of the material. The single-atom layer is not depleted of charge carriers (in contrast to conventional semiconductors) and may offer solutions to the opto-electronics industry.

Graphene can be fabricated in many forms: few-layer sheets vs. free-standing sheet, nano-scale vs. micro-scale.However, to be industrially viable the method needs to be scalable and relatively low cost. The application or end-use of the product usually dictates the method of fabrication. In particular, a method to fabricate large-area free-standing sheets of single-layer graphene needs to be obtained. This requires an approach that is not limited in size to the starting material. For graphene to be competitive with current or other next-generation products it needs to be affordable. This requires a low-to-moderate fabrication expense. The next three sections will describe some of the patented and patentpending methods to fabricate grapheme

The inherent properties of graphene appeal to many industries, including the electronics industry. The high mobility of charge carriers and the high thermal conductivity offer the industry a potential alternative to silicon and diamond. The nature of the carriers has the potential to create next generation solid state devices (ballistic transistors, spin transistors, etc.). The optical transparency of grapheme coupled with the electronic properties may provide the optoelectronics industry with photovoltaic solutions such as transparent electrodes. Other industries also have the opportunity to use graphene's inherent properties. The carbon base of graphene offers compatibility with organic-based materials.High strength, lightweight materials may be formed with the addition of graphene to conventional plastics.

This paper attempts to correlate vacuum technology in the production and characterization of graphene.

2.Production Processes.

Three primary synthesis methods have emerged for the production of graphene: (i) micromechanical cleavage (Giem 2007, Park 2009), (ii) chemical vapor deposition (CVD) (O'biren, 2010), and (iii) chemical exfoliation (Srinivason 2007, Park 2009).[1]

Micromechanical exfoliation of graphite or the 'scotch tape' method is relatively easy and can yield high quality, large area (~ 1 mm) sheets; however, this technique is low yield and only useful for experimental purposes. In the CVD or epitaxial approach, graphene is grown on a metal, generally Cu, substrate. Single layer and bi-layer graphene sheets, measuring up to approximately 1 cm square have been produced successfully by this method. However, neither method is currently capable of producing bulk quantities of graphene. The chemical methods of synthesizing graphene (Srinivason 2007, Park 2009) have emerged as the main approach for large quantity production. This can be done through the chemical exfoliation of graphite oxide (GO). The chemical exfoliation starts with natural graphite and reacts with a mixture of strong acids, such as nitric and sulfuric, and an oxygen donating compound. This reaction induces stress within the graphite layers by intercalating the graphite sheets with oxygen atoms and expanding their separation distance (Srinivason, 2007). The end result is graphitic oxide.

The level of oxidation varies based on the method and initial type of graphite used (Park, 2009). Each method leaves different functional groups at the edge, or sometimes also on the surface of the graphene sheet that influence the chemical and physical properties of the GO. Prior experimental results show that often, chemical exfoliation method can create irreversible defects in the graphene sheets, which can decrease the conductance by 3 orders magnitude compared to a defect-free graphene-sheet (Park, 2009). Therefore, our efforts have been focused on optimizing the oxidation and exfoliation processes to gain finer control on surface defects and edge/surface functional groups.[1]

3.Characterization Instruments

3.1Scanning Electron Microscope(SEM.).

ll electron beam instruments are built around an electron column, which produces a stable electron beam, controls beam current, beam size and beam shape, and rasters the beam. Electron optics are a very close analog to light optics, and most of the principles of an electron beam column can be understood by thinking of the electrons as rays of light and the electron optical components as simply their optical counterparts.

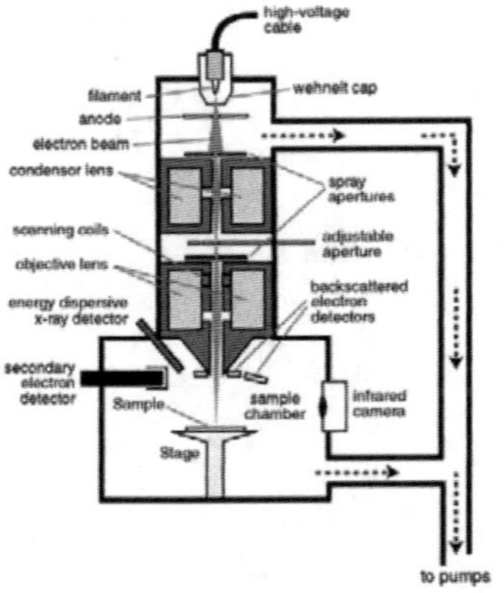

Fig. 1. Block diagram of SEM

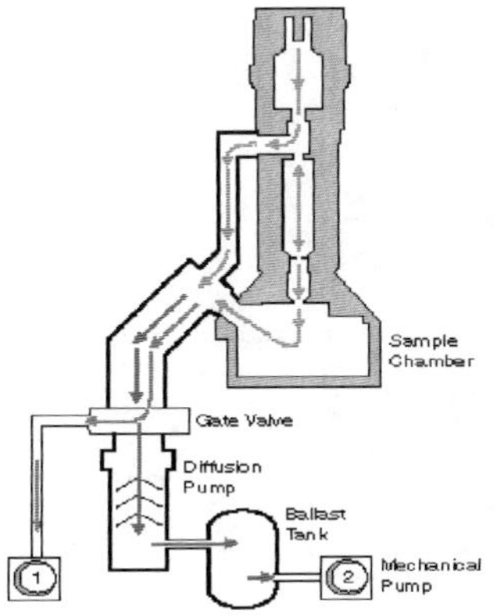

Fig. 2. Schematic diagram of SEM vacuum system

3.1.1. Vacuum system of a SEM .

The pump-down sequences for a microprobe and scanning electron microscope are very similar. Initially, a mechanical pump use used to evacuate the column and sample chamber. Pressure is monitored using a thermocouple gauge. Thermocouple gauges work down to a pressure of about 10^{-3} torr. Once the vacuum is sufficiently good, the "gate" valve that isolates an oil diffusion pump from the column is opened. This allows the diffusion pump to evacuate the column further. There is a water-cooled baffle located above the pump to prevent oil vapor from entering the electron column. A cold-cathode (ion) vacuum gauge, which operates at pressures down to 10^{-6} torr, is used to monitor the pressure at higher vacuums. A cold-cathode gauge gets dirty with time and reads a better vacuum than actually exists.

During initial pumping (green and blue) the column is evacuated using mechanical pump 1. Once a vacuum of about 1 torr is achieved the gate valve opens to access the oil diffusion pump (green and red), isolating mechanical pump 1. The diffusion pump is kept at operating vacuum by mechanical pump 2. The ballast tank serves as a vacuum reservoir and allows the diffusion pump to function for short intervals with pumps 1 and 2 turned off. This minimizes mechanical vibration and permits high magnification operation

3.2.Transmission Electron Microscope(TEM.)

Due to strong interactions of electron with matter, gas particles must be absent in the column. The required high vacuum is maintained by a vacuum system typically comprising a rotary pump (pre-vacuum pump), a diffusion pump and one or more ion getter pumps.

There are three main types of vacuum pump: rough pumps, high vacuum pumps which need backing, and high vacuum pumps which don't need backing.

A *rough pump* can pump down from atmospheric pressure, but can only reach a rather modest vacuum, at best 0.01 mbar (0.00001 of an atmosphere).

High vacuum pumps can reach good or ultra-high vacuum (UHV), between log(-6) - log(-10-11) mbar. A high vacuum pump that needs backing means that it needs another pump (a rough pump) to work on the 'back end of it' (the high-pressure end), to take away from it air that it has pumped out of the vacuum chamber. Examples of such pumps are *diffusion pumps* and some types of high-vacuum *turbo pumps*.

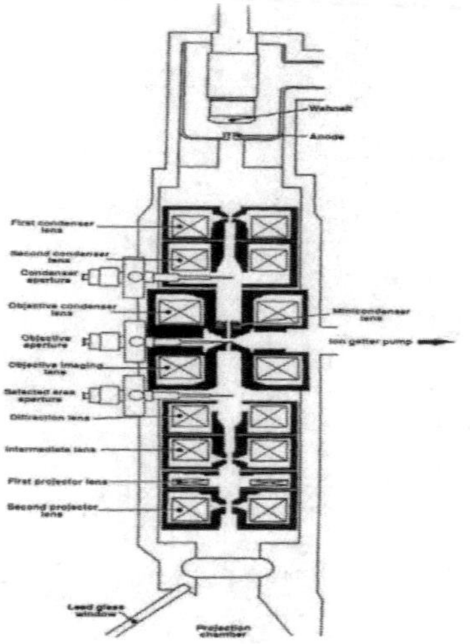

Fig.3. Block diagram of TEM

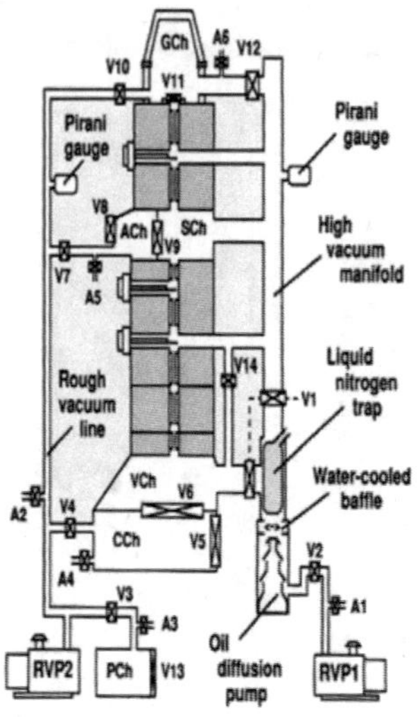

Fig.4 Block diagram of vacuum system of TEM

Some UHV pumps, like *ion pumps*, do not need backing: the vacuum they deal with has such a low pressure that the pump can dispose of residual gas atoms simply by accelerating them into the solid walls of the pumping chamber (this seems improbable, but it does work).

3.3.Raman Spectroscopy

Raman spectroscopy is a useful technique for the identification of a wide range of substances - solids, liquids, and gases. It is a straightforward, non-destructive technique requiring no sample preparation. Raman spectroscopy involves illuminating a sample with monochromatic light and using a spectrometer to examine light scattered by the sample.

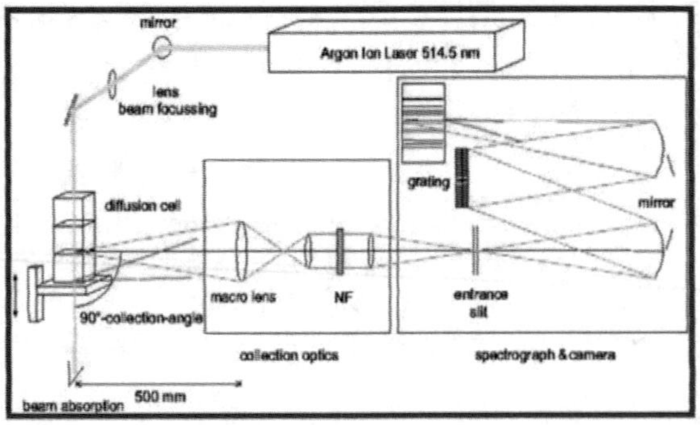

Fig. 5.Block diagram of Raman Spectrometer

Fig. 6. SEM micrograph of graphene

At the molecular level photons can interact with matter by absorption or scattering processes. Scattering may occur either elastically, or inelastically. The elastic process is termed Rayleigh scattering, whilst the inelastic process is termed Raman scattering. The electric field component of the scattering photon perturbs the electron cloud of the molecule and may be regarded as exciting the system to a 'virtual' state. Raman scattering occurs when the system exchanges energy with the photon, and the system subsequently decays to vibrational energy levels above or below that of the initial state. The frequency shift corresponding to the energy difference between the incident and scattered photon is termed the Raman shift. Depending on whether the system has lost or gained vibrational energy, the Raman shift occurs either as an up- or down-shift of the scattered photon frequency relative to that of the incident photon. The down-shifted and up-shifted components are called respectively the Stokes and anti-Stokes lines. A plot of detected number of photons versus Raman shift from the incident laser energy gives a Raman spectrum. Different materials have different vibrational modes, and therefore characteristic Raman spectra. This makes Raman spectroscopy a useful technique for material identification. There is one important distinction to make between the Raman spectra of gases and liquids, and those taken from solids - in particular, crystals. For gases and liquids it is meaningful to speak of the vibrational energy levels of the individual molecules which make up the material. Crystals do not behave as if composed of molecules with specific vibrational energy levels, instead the crystal lattice undergoes vibration. These macroscopic vibrational modes are called *phonons*.

Raman spectroscopy is a purely optical experiment. The sample is cooled to cryogenic temperatures under a vacuum and exposed to radiation, usually in the visible spectrum. This radiation causes the atoms in the sample to enter an excited state. When the atoms return to a lower energy state, they emit radiation, which is measured by a detector. This luminescence yields a great amount of information about the electronic structure of the sample.

4.Study of graphene by Scanning Electron Microscopy, Transmission Electron Microscopy, Raman Spectroscopy.

The SEM micrograph of graphene samples can examine the nature of structure of growth of the on the substrate. at 80 kV acceleration voltage.Example:

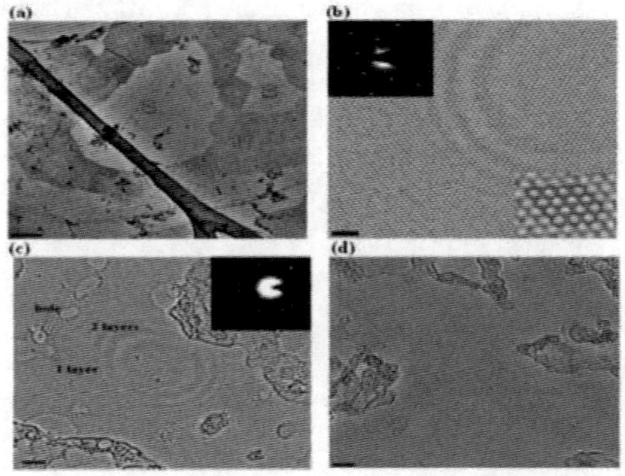

Fig. 7: TEM images of few-layer graphene (a) single layer graphene (S) and bilayer graphene (B) supported by multi-layer graphene (M) (scale bar : 250 nm), (b)Atomic resolution image of the single layer in (a) (scale bar : 1 nm), inset: electron diffraction pattern corresponding to a single layer graphene (left) and higher magnification image (right) (c) bilayer graphene having a mismatch of crystal orientation (scale bar : 2 nm), inset : electron diffraction pattern corresponding to a turbostratically arranged bilayer (right) and (d) chemically bonded few-layer graphene having different orientations (scale bar : 2 nm).[Growth and properties of few-layer graphene prepared by chemical vapour deposition[2].

TEM micrograph can detect the types of layer in a graphene i.e. single layer,multilayer.,bi-layer. Atomic resolution imaging of graphene has been achieved in TEM using an aberration-correction technique in combination with a monochromator;point defects, formation and annealing of Stone-Wales defects, and multiple pentagons/heptagons combinations were observed *in situ* .The stability and dynamics of the edge of a hole in a suspended graphene platelet has been studied; TEM images with a stated sub-Ångstrom resolution showed that both 'armchair' and 'zigzag' configurations could be formed during the edge reconstruction and 'zigzag' edges were observed to be particularly stable under electron irradiation.

4.1 Raman Spectroscopy.

Raman spectroscopy has been used to characterize grapheme and several review articles have been published discussing the optical phonon spectrum and Raman spectrum of graphene. The Raman spectra of graphene includes the G peak located at ~ 1580 cm − 1 and 2D peak at ~ 2700 cm − 1 ,caused by the in-plane optical vibration (degenerate zone center *E* 2g mode) and second-order zone boundary phonons, respectively.The D peak, located at ~ 1350 cm − 1 due to first-order zone boundary phonons, is absent from defect-free graphene, but exists in defected graphene. It was proposed that Raman could be used to distinguish the 'quality' of graphene and to determine the number of layers for n-layer graphene (for n up to 5) by the shape, width, and position of the 2D peak. As shown in Figure 5 , the 2D peak shifts to higher wavenumber values and becomes broader for an increasing number of layers. The shifting and splitting of Raman modes can be used to analyze mechanical strain in graphene. For example, Raman spectra of epitaxial graphene grown on SiC show a signifi cant phonon 'hardening' (blue shift of the G and 2D peaks), mainly due to the compressive strain that occurs when the sample is cooled down after growth.It has been stated that the substrates play a negligible role in the Raman spectrum of micromechanically cleaved graphene transferred onto them, indicating the weak interactions between the transferred graphene and such substrates. The frequency of the G and 2D peaks can also be tuned by charge doping through electron-phonon coupling changes.The Raman spectral signatures of epitaxial graphene grown on SiC,especially the width of the 2D peak, have been correlated to the carrier mobility of the graphene.] The intensity ratio of the D and G peak has been used as a metric of disorder in graphene,such as arising from ripples, edges, charged impurities, presence of domain boundaries, and others. For edges, the intensity of the D peak depends on the edge structure; it is weak at the zigzag edge and strong at the armchair edge. [3]

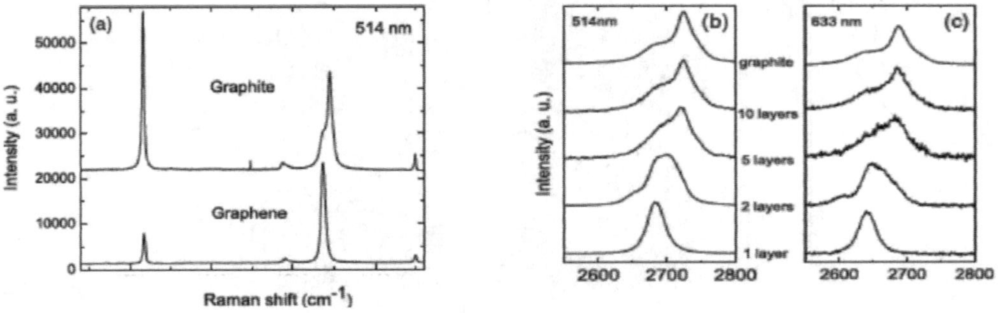

Fig. 8.(a) Comparison of Raman spectra at 514 nm for bulk graphite and graphene. They are scaled to have similar height of the 2D peak at 2700 cm − 1 ; (b) Evolution of the spectra at 514 nm with the number of layers; (c) Evolution of the Raman spectra at 633 nm with the number of layers.

4.Conclusion.

This paper attempts to correlate vacuum technology in the production and characterization of graphene.

References.

[1] Drut Joaqu´ın E. and L¨ahde Timo A.; 2010 ;Is graphene in vacuum an insulator?; arXiv:0807.0834v2.

[2] Johnson Patricia 1, Pesce-Rodriguez Rose, Griep Mark H., Behler Kris, Mallick Govind, Sarney Wendy, and Karna,,Bulk Shashi P; 2010;Synthesis And Characterization Of High-Purity Graphene

[3] Yanwu Zhu , Shanthi Murali , Weiwei Cai , Xuesong Li , Ji Won Suk , Jeffrey R. Potts ,and Rodney S. Ruoff; 2010;Graphene and Graphene Oxide: Synthesis, Properties,and Applications;Adv. Mater. **XX**, 1–1.

Journal of Physics: Conference Series **390** (2012) 012068 doi:10.1088/1742-6596/390/1/012068

Studies of cryocooler based cryosorption pump with activated carbon panels operating at 11K

S. Kasthurirengan[1], Upendra Behera[1], Ranjana Gangradey[2], Swarup Udgata[3] and V. Krishnamoorthy[3]

[1] Centre for Cryogenic Technology, IISc, Bangalore 560012, India
[2] Cryopump Group, Institute of Plasma Research, Gandhinagar, Gujarat 382428, India
[3] I-Design Engineering Solutions Ltd., Ubale Nagar, Wagholi, Pune 412207, India

E-mail: fantasrini@gmail.com

Abstract. Cryosorption pump is the only solution for pumping helium and hydrogen in fusion reactors. It is chosen because it offers highest pumping speed as well as the only suitable pump for the harsh environments in a tokamak. Towards the development of such cryosorption pumps, the optimal choice of the right activated carbon panels is essential. In order to characterize the performance of the panels with indigenously developed activated carbon, a cryocooler based cryosorption pump with scaled down sizes of panels is experimented. The results are compared with the commercial cryopanel used in a CTI cryosorption (model: Cryotorr 7) pump. The cryopanel is mounted on the cold head of the second stage GM cryocooler which cools the cryopanel down to 11K with first stage reaching about ~50K. With no heat load, cryopump gives the ultimate vacuum of 2.1E-7 mbar. The pumping speed of different gases such as nitrogen, argon, hydrogen, helium are tested both on indigenous and commercial cryopanel. These studies serve as a bench mark towards the development of better cryopanels to be cooled by liquid helium for use with tokamak.

1. Introduction

Cryopumps belong to the class of entrapment or capture vacuum pumps and they retain the gas molecules by sorption and / or by condensation on its internal surfaces. Thus the performance of cryopump is governed by the interplay of these two pumping mechanisms. The equilibrium pressure of adsorbed gas particles is significantly lower than the corresponding saturation pressure for cryo-condensation. This is due to the fact that the dispersion forces between the gas molecules and the surface are greater than between the gas molecules themselves in the condensed state. The adsorption takes place at the porous surfaces. Hence, porous materials are of great importance towards this technology.

Commercially available cryo-pumps are intended for industrial application and operate in the temperature range of 20-25K, since they do not require the pumping of lighter gas molecules like hydrogen, deuterium and helium etc. However, for fusion devices which involve extremely high temperatures and magnetic fields, the only possible pump that can be used is the cryosorption pumps to pump these gases. Many investigators [1-3] have carried out experimental studies towards the development of cryopanels with increased pump efficiency for use with the next generation tokamak fusion devices. These studies also indicate that activated carbon is the right material for these applications.

Both adhesion and proper cooling of the charcoal at panel surface are critical aspects in the development of cryopanels. Also, the poor thermal conductivity of the charcoal degrades the performances of the panels used in the cryopump. Literature survey indicates that gases such as hydrogen, helium and their isotopes can be pumped around 20 K, but helium cannot be pumped at this temperature. Very few literatures are available which demonstrate the pumping of helium at ~ 4.5 K. Hence, there is a need for studies of measurements of pumping speeds of cryosorption panels (with the specific activated charcoal with a suitable binder) for different gases inclusive of Helium in the temperature range from 4.5 K to 10 K. Such a study will also enable to benchmark the performances against a standard commercial panel.

Towards the above goal, the present work has been undertaken using a commercially available two stage GM cryocooler based cryopump (Cryotorr 7) reaching the lowest temperature of 11 K. Various indigenous activated carbon panels are mounted on the cold head and tested with gases such as nitrogen, argon, hydrogen and helium. These experimental results are discussed in this work.

2. Experimental Setup

Figure 1 shows the schematic of the cryopump and the experimental setup. This is based on the commercially available CTI cryopump of Model Cryotorr 7. The housing of a cryopump is made of stainless steel with an 'O' ring sealed high vacuum flange of 300 mm diameter. The top flange has provisions to connect the pressure gauges and also for gas into the system. The two stage GM Cryocooler is driven by a 3 kW helium compressor along with a rotary valve. The system reaches no-load temperatures of ~ 11 K in the second stage and ~ 50K in the first stage. Temperatures of the first stage and second stage of GM Cryocooler are measured by silicon diodes (SI-410). The cryopanel is mounted on second stage cold head of the GM cryocooler. Indium is used to improve thermal conductivity. Heat switch is mounted in between cold head and cryopanels, which helps to heat the sample without affecting the performance of cryocooler.

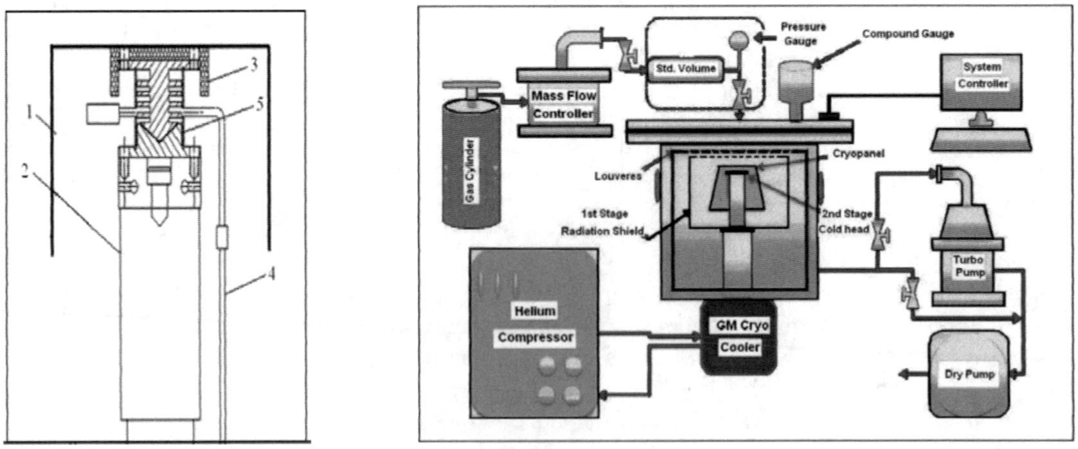

Fig. 1 Schematic of the cryocooler with activated carbon cryopanels and actual experimental setup. 1. Cryopanel coated with activated carbon, 2. Two stage GM cryocooler, 3. Heater, 4. Gas line for Heat Switch, 5.Heat Switch

Pfeiffer Vacuum gauges are used to measure the pressures of the system. Aalborg mass flow controller GFC 17 (0-50 ml/min) is used to control the mass flow during the gas inlet for measurement. Test gas mixing facility is used to mix the gas in different percentages. All the gases used for measuring pumping speeds have ~ 99.999% purity. Rotary vane pump is used to create initial vacuum of the system.

3. Experimental Results

The experimental results of cool-down behaviour of the GM Cryocooler, ultimate pressures obtained under no load conditions and also the pumping speeds for different gases are presented below

3.1 Cool-down Behavior

Figure 2 plots the cool-down characteristic of two stage GM cryocooler, with Cryopanel mounted on the second stage cold head. The cryocooler reaches the no load temperature of ~ 11 K and hence the cryopanel mounted on the second stage can only reach the lowest temperature of this value. The typical cool down time is of the order of 200 minutes

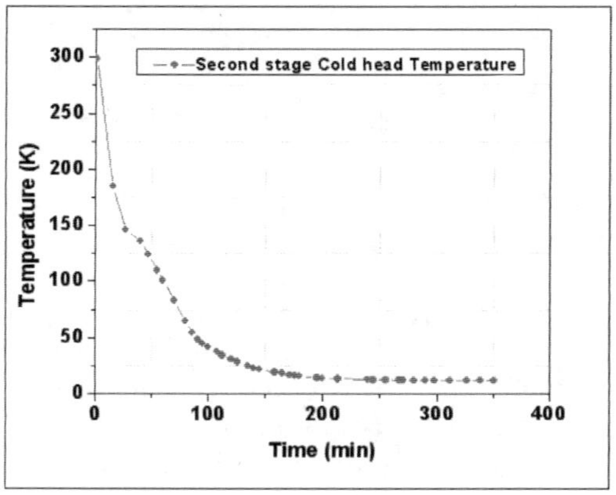

Fig. 2. Cool down curve of two-stage GM cryocooler

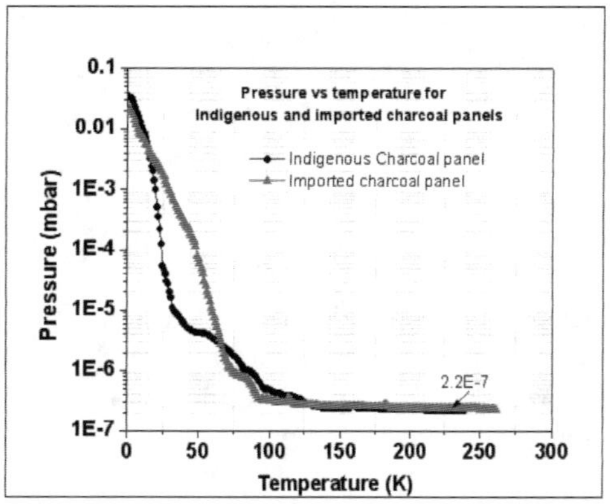

Fig. 3. Ultimate pressures reached with no gas load on the panels

3.2 Ultimate pressures reached on no gas load

Rotary vane pump is used initially to develop the base pressure levels of ~ 2.2E-2 mbar. Figure 2 shows the pressure versus temperature wherein the ultimate pressure of ~ 2.2 E-07 at ~11 K has been achieved for both imported and indigenous charcoal panels. This indicates that without gas load both the panels perform similar.

3.3 Pumping Speed studies of different gases on different types of charcoal panels

Experiments have been conducted for different gases such as Nitrogen, Argon, Hydrogen and Helium as adsorbate. The imported charcoal panel (original panel used in the Cryotorr 7 Cryopump) has been used as benchmark for comparison of indigenous activated carbon panels. The imported activated carbon consists of granules of sizes ~ 2 to 3 mm. On the other hand, three types of indigenous activated carbons have been studied namely granules with sizes ranging from 0.5mm to 1 mm. The prepared panels had approximately the same physical surface areas as that of the imported charcoal panel

3.4 Pumping Speeds for Nitrogen and Argon

Figure 4(a) shows the pumping speed versus pressure of different cryopanel for nitrogen gas (of purity 99.999%). The pumping speed in general decreases with increasing pressure for all the cryopanels. However, the commercial cryopanel with the largest granular size has the highest pumping speed.

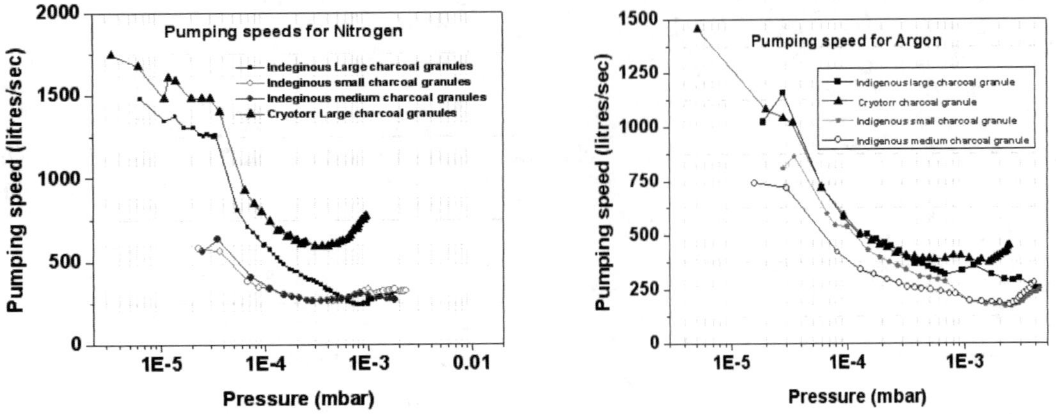

Fig. 4. Pumping speeds for (a) nitrogen and (b) argon gas with different activated carbon cryopanels

The indigenous cryopanel with largest granular size activated carbon has slightly lesser pumping speed compared to the commercial one. The activated carbon panels having medium and smaller granular sizes have still lesser pumping speeds. These results can be understood as follows. Assuming that the physical areas of the panels to be the same and the adhesive closes almost the same number of pores in each panel, the granules of activated carbon of larger size has increased internal surface area. Hence, this leads to increased pumping speeds for the panels with larger activated carbon granules. It is also observed that when the experiments are repeated with the same cryopanel, the pumping speeds improve with each run.

This is due the better cleanup of the pores with the subsequent runs. Figure 4(b) shows the pumping speed versus pressure for different cryopanels for argon gas (of purity 99.999%). Here again, the pumping speed decreases with increasing pressure. Also, the commercial cryopanel shows the highest pumping speed. The cryopanel with indigenous activated carbon with larger granular size has pumping speeds similar to the commercial one. The panels with medium and smaller granules have slightly lesser pumping speeds.

3.4 Pumping Speeds for hydrogen and helium

Figure 5(a) plots the pumping speed versus pressure for indigenous and imported activated carbon cryopanels with the adsorbate of Hydrogen (3%) in nitrogen. Since the temperature of the panel rises rapidly, due to the increased thermal conductivity of hydrogen, experiments could not be conducted with pure hydrogen and so the above gas mixture has been chosen. Even in this case, it is observed that the imported cryopanel performs better compared to the indigenous one. Also, the pumping speeds for Hydrogen are found to be relatively lower compared to those of Nitrogen and Argon.

When the experiment was conducted using helium gas (99.999%), there was sudden increase in temperature when helium was introduced in the system and the temperature kept on increasing due to which the pressure of the system also increased. Similar to the case of hydrogen, helium – nitrogen gas mixture was chosen as adsorbate in the subsequent experiments. The experimental results for small granular charcoal panels with helium gas mixtures in nitrogen as adsorbate are shown in Figure 5(b). The results indicate the extremely low pumping speeds of this system, that too in the limited pressure range.

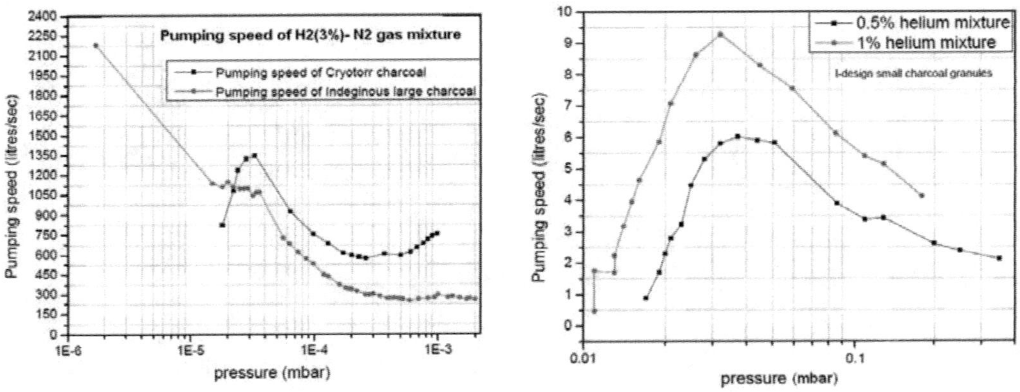

Fig. 5. Pumping speeds for (a) hydrogen (3%) in nitrogen gas mixture and (b) helium (0.5% and 1%) in nitrogen gas mixtures as adsorbate.

It is seen from the above table that the system performance is better for the imported cryopanel with larger granular activated carbon for all gases. With hydrogen as adsorbate, the pumping speeds are lower compared to those for nitrogen and argon. In the case of helium, the pumping speeds are extremely low. The above results clearly indicate that for pumping helium gas we need to have cryopanels operating at temperatures much lower than 11 K.

It is seen from the above table that the system performance is better for the imported cryopanel with larger granular activated carbon for all gases. With hydrogen as adsorbate, the pumping speeds are lower compared to those for nitrogen and argon. In the case of helium, the pumping speeds are extremely low. The above results clearly indicate that for pumping helium gas we need to have cryopanels operating at temperatures much lower than 11 K.

Table 1. Comparison of pumping speeds for different gases for different cryopanels at 3.0E-5 mbar.

Panel	N_2	Ar	H_2^*	He
Without charcoal panel	192	127	50	0.2
Cryotorr 7 charcoal panel	1450	1135	250.2	0.15
small charcoal granule panel	600	825	100.8	0.14
Medium Charcoal granule panel	575	720	155.3	0.23
Large Charcoal granule panel	1370	1129	215.3	0.22

3.4 Effect of Degassing on Cryopanels

Figure 7 indicates that degassing of the cryopanels are very much needed to improve the performance of cryopump. The pumping speed of cryopump increased after it degassed upto 350K. This degassing temperature of the panels is limited upto 350K due to the presence of the adhesive. Heat switch is mounted in between GM cryocooler cold head and cryopanels to heat up the panels without affecting the performance of Cryocooler.

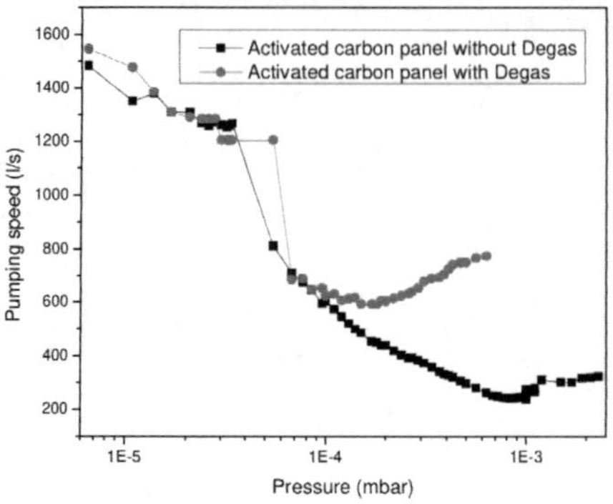

Fig 6. Pumping speeds of indigenous charcoal panel with nitrogen with and without degassing the sample

4. Conclusion

Towards the development of cryosorption pumps and to benchmark the performances of indigenous activated carbon based cryopanels, pumping speed measurements have been made using a CTI Cryosorption pump (model Cryotorr 7). The indigenously fabricated panels are mounted at the second stage cold head of the GM Cryocooler which reaches 11 K on its second stage cold head. The experimental studies show that the performance of the indigenous activated carbon panels is better compared to the original panel for all gases. For hydrogen, the pumping speed is much lower compared to nitrogen and Argon. For pumping helium, the studies clearly indicate the need for having temperatures much lower than 11 K. Hence attempts are now in progress to develop an experimental setup based on a Cryocooler which reaches 4.2 K.

Acknowledgements

The authors acknowledge the financial support of BRFST, IPR Ahmadabad for this research work.

References

[1] C. Day, et al. Pumping Performance of cryopanels coated with activated carbon. Adv. Cryog. Eng., 43 (1998), pp. 1327–1334.

[2] Hauer V, Day C. Cryosorbent Characterization of Activated Charcoal in the COOLSORP Facility, FZK Report, FZKA 6745, 2002..

[3] C. Day, et al., R&D progress for the ITER vacuum pumping system, Proceedings of the 20th Symposium on Fusion Engineering, San Diego, IEEE Catalog Number 03CH37469, 2003, pp. 458–461..

[4] M. Dremel, et al., Design and Manufacturing of cryosorption pumps for TEST BEDS of ITER relevant neutral beam injectors, Proceedings of the International Congress of Cryogenic Engineering, Beijing, in press..

[5] Design Description Document, DDD 5.3, The "ITER Neutral Beam System" and DRG1 of ITER documentation.

International Symposium on Vacuum Science & Technology and its Application for Accelerators IOP Publishing
Journal of Physics: Conference Series **390** (2012) 012069 doi:10.1088/1742-6596/390/1/012069

Fabrication of an Ultra High Vacuum Compatible Faraday Cup for Qualification of Electron Gun for 10 kW Industrial LINAC

Ajay Kak, A. Kher, S. C. Vishwakarma, Abhay Kumar, Manoj Gandhi, Pramod Radheshyam, Ajay Kumar and Lala Abhinandan

Raja Ramanna Centre for Advanced Technology, Indore 452013, India

Email: lala@rrcat.gov.in

Abstract. In this paper, we report fabrication and testing of a unique Faraday Cup designed for qualification of Electron Gun for a 10 kW industrial LINAC. This is a compact device consisting of a tapered copper cavity electrically isolated by a ceramic cylinder metalized at both ends. Kovar tubes matching with the diameter of ceramic isolator were used at both ends to facilitate high quality UHV compatible joints with copper cavity at one end and standard knife edge stainless steel flange at the other end. The Kovar tube was flared at both ends to form a collar matching with the outer diameter of the ceramic isolator. The joint between Kovar collar and ceramic isolator was done by hydrogen brazing using copper silver eutectic alloy. All the joints were tested with helium leak detector and leak rates were found to be below 1 x 10^{-10}mbar.litre/second.

1. Introduction

A Faraday cup is a device used for measuring current in a beam of charged particles. The device intercepts the electron beam and conducts the current to a measuring instrument through an electrical lead. Secondary emission and back scattering of the striking electron beam are the main reasons of measurement inaccuracies. The accuracy of measurement can be improved by giving proper geometrical shape to the cup and by applying bias voltage to the cup.

A Faraday cup has been developed for qualifying the electron gun being developed for a 10 kW industrial RF linear accelerator. The electron gun has been designed to give mono-energetic electrons of 70 keV energy and 1 ampere pulsed beam current with 15 microsecond pulse width at 1Hz - 300 Hz pulse repetition rate. This configuration demands the Faraday cup to dissipate average power of 315W.

The Faraday cup measures the collected current by measuring the voltage drop across a resistor as the current passes to the ground under the potential developed by the same collected electrons. The collector of the device has to be isolated from the electron gun's anode which is also at the ground potential. This isolation has been achieved through the use of a ceramic isolator(99.5% Alumina) between the collector and the knife edge flange of the device attached to the gun's anode. Kovar transition tubes must be used between the knife edge flange on the side of the electron gun's anode and the ceramic isolator and also between the ceramic isolator and the collector cup. The joint between the Kovar tube and collector cup is a brazed joint and experiences the shear stresses due to thermal expansion of the collector cup. The physical dimensions of the faraday cup have been selected to keep the ceramic and Kovar tubes away from beam path. This does not allow a temperature difference between Kovar and ceramic. However, the brazed joints between Kovar and ceramic and Kovar and collector cup do see the effect of thermal expansion of collector cup. Hence, in order to limit these thermal stresses for retaining the vacuum leak tightness of the brazed joints, it is important to limit the temperature rise of the collector cup.

As the energetic electrons strike the walls of the collector cup, they undergo secondary emission, elastic scattering and inelastic scattering. Elastic scattering is a relatively low energy phenomenon and the secondary emission coefficients are also very small at higher energies (>20 keV). At higher energies, re-diffusion or inelastic scattering of primaries dominates. These phenomena can lead to some electrons leaving the collector and escaping to the knife edge flange side beyond the ceramic isolator. Extra electrons would flow from earth to make up for these electrons and thus would produce inaccuracy in the measurement. These phenomena can be controlled by having geometry of the collector that invites multiple surface interactions of secondary and backscattered electrons, thus increasing the probability of retaining them inside the cup. Another method to do the same thing is by using a positive bias voltage between the collector cup and the ground so that these secondary and backscattered electrons are forced to go to the ground. Looking at all the requirements of the design, OFE copper appears to be appropriate material for the collector cup due to its high thermal conductivity and brazability. Although, all good thermal conductors have large secondary emission coefficients (SEC), its value decreases sharply with increasing impact energy of primary electrons and at impact energies larger than 20 keV, the value approaches 1% and goes down even further beyond this impact energy. A conical taper has been provided in the cup to distribute the heat and decrease the heat flux. This also helps to increase the probability of the surface interactions of the secondary and backscattered electrons with the collector cup surface. It has four circumferential fins on the external surface for dissipation of heat load of 315 W. A small fan with a 12V DC motor is mounted near the fin surfaces to allow a forced air flow convection conditions. The gap between the fins has been kept more than twice the thickness of boundary layer to allow optimum flow conditions. Figure-1 explains the engineering design features of the gun.

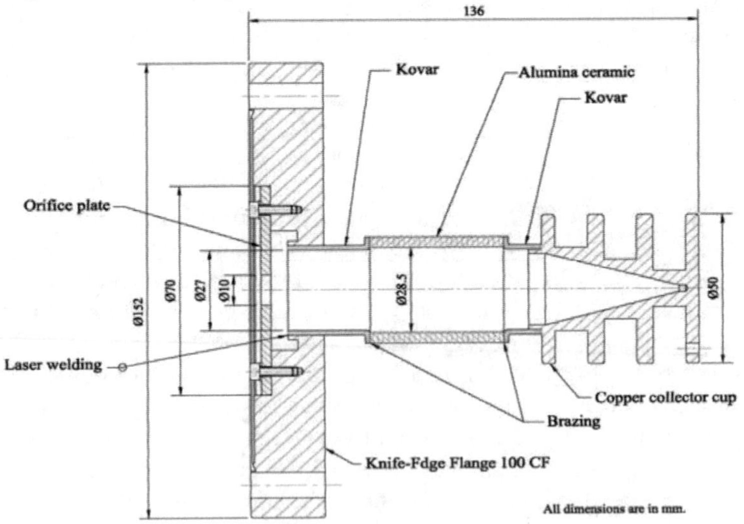

Fig.1

2. Design for Manufacturing (DFM)

Vacuum leak tightness of the order of 1 x 10^{-9}mbar.litre/sec was required from the brazed and the welded joints. The brazed joints shall also have enough strength to take the thermal stresses due to differential thermal expansions of the material at both sides of the joint. The forces of handling during the cleaning and assembly have to be taken by these joints.

The brazed joint between Kovar pipe and ceramic pipe uses a flared construction of Kovar described below. Copper has a coefficient of expansion that is approximately three times that of Kovar. This allowed a large clearance at room temperature for putting the brazing filler foils between them. Due to small clearance at brazing temperature, the joint strength is expected to be on its theoretical maximum (Figure-2). Although, the joint between Kovar and copper sees a residual tensile stress but since copper is completely annealed at brazing temperature, the yield strength of the material becomes very small and thus it yields under the residual stresses at brazed joint. This provides a high strength brazed joint with negligible residual stresses.

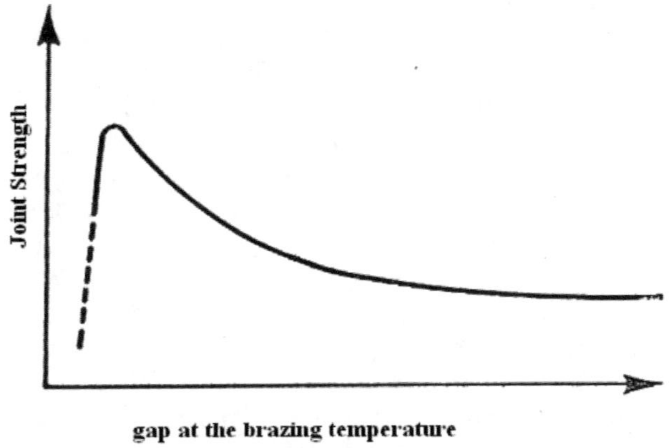

Fig. 2. variation of brazed joint strength with the gap

The TIG (GTAW) welding between stainless steel and Kovar leads to excessive thermal stresses on the brazed joint between Kovar and ceramic due to high heat input of the process. The design has been done to mitigate the problem in two ways. The length of the Kovar pipe between the knife edge flange and the alumina ceramic pipe was kept on the higher side to allow flexibility resulting in lower thermal stresses on the brazed joint during the welding operation. Higher length of Kovar pipe also increased the thermal conduction length and thus decreased the temperature of the brazed joint. The usual TIG welding process was replaced with indigenously developed Nd-YAG laser welding to have high heat intensity and low total heat input.

The orifice plate has been kept removable to retain the possibility of changing the same with the ones with higher or lower orifice diameters.

3. Fabrication

As explained above the design features were realized by careful fabrication of the device. The main problem in fabrication of this assembly was to prepare Kovar tube with collar to suit outer diameter of metalized ceramic isolator. One alternative was to fabricate the collar by welding a Kovar flange on the tube followed by brazing with ceramic isolator. However, it would involve more number of components and also increase the number of joints. Further, Kovar is difficult to machine as it is soft and gummy. The tools tend to plough the alloy instead of cutting into it, and do not easily form chips. Surface scale oxide tightly adheres to and penetrates the surface of the tool [1]. Cold working requires a temperature of 450^0 - 500°C. Finally, it was decided to go for flaring of the Kovar tube by making an outer and inner guide to hold Kovar tube and then applying pressure manually on the tube through tailstock of lathe machine at room temperature. After a few trials, a proper flared tube was made. The flatness of the flared end was achieved by careful machining of the face.

The joint between Kovar flared tube and ceramic isolator was done by hydrogen brazing using copper silver eutectic having melting point of 779^0C [2]. Hydrogen brazing ensures removal of oxides of copper by reduction with hydrogen at brazing temperatures. The Kovar interface is initially plated with a thin layer of electroless copper to prevent intergranular penetration of silver bearing brazing alloy which may lead to stress corrosion cracking. Proper heating and cooling cycle lasting for more than 12 hours in a Linn (model HT 1600 G) hydrogen atmosphere furnace was employed to minimize the stress on the interface of ceramic isolator and Kovar tube [3]. The moly-manganese metallized layer consisting of a glassy phase and modified refractory metal on the face of the ceramic isolator helps its proper bonding with Kovar flared tube. Finally, the joint between Kovar tube and knife edge flange was done by laser welding method using a 250 watt average power fiber coupled pulsed Nd:YAG laser having pulse duration in the range of 2-20 ms and repetition rate from 1-100 Hz.

4. Measurement Circuit

After the assembly of the Faraday cup with the electron gun, high voltage pulses of negative polarity of 2 μs pulse-width duration, repeating at 1Hz and having rise and fall time of 350 ns were applied to the electron gun's cathode with respect to its anode. The pulse amplitude was gradually raised from 15 kV to 40 kV while keeping filament heater current at estimated RMS value of about 35 A. The application of high voltage pulse was synchronized at the zero crossing of the line voltage.

As explained earlier, the collector cup is isolated from the gun's anode. With this arrangement, we could separate out the displacement current flowing between the cathode and anode from the beam current. A pulse beam current measurement circuit was developed earlier during a dedicated experiment using a pulsed 40 kV electron gun. The same measurement circuit was employed with this Faraday cup. Several improvements were made in the circuit to reduce the noise during the qualification of the Faraday cup. The connection between the Faraday Cup and the ground was established through a 50 Ω resistor and a 50 Ω coaxial cable. The inner conductor of the coaxial cable was connected to the copper collector cup. The outer conductor was connected to the nearest point of the grounded anode flange, for keeping the unshielded wire length to a minimum. The 50 Ω resistor was placed between the inner conductor and the outer conductor. A common-mode and differential filter was used in the beam current measurement circuit to remove noise from the measurement signal. The voltage drop across the 50 Ω resistor was measured. The beam current was also monitored with the help of Tektronix made current probe, TCP-303, and amplifier, TCPA- 300, on digital storage oscilloscope, TPS-2024, to cross-check the measurements made with the 50 Ω resistor. Figure-3 explains the measurement circuit. The effect of a positive bias was also studied but no change in the collected beam current was observed when the bias voltage was varied from 0 to 50 V. This indicates that secondary and backscattered electrons are being retained in the collector cup effectively only by the geometry of the cup.

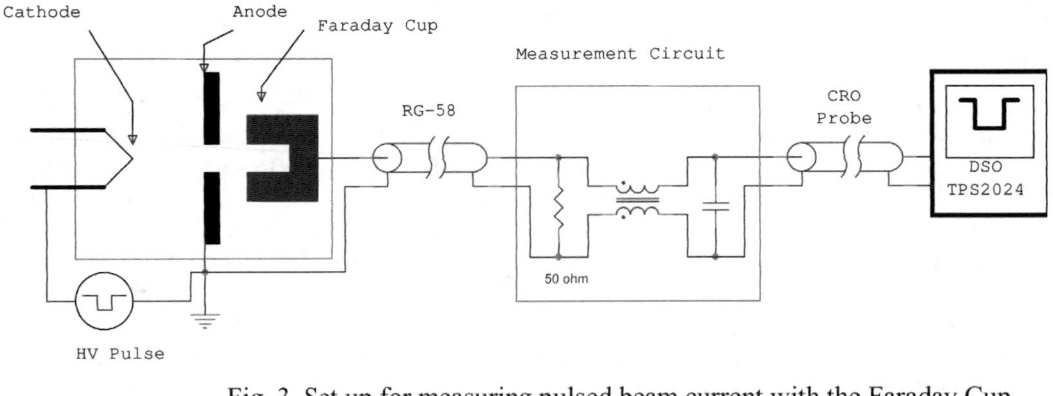

Fig. 3. Set up for measuring pulsed beam current with the Faraday Cup

5. Measurement Results and conclusion:

The Faraday cup measurement has been used in the low power tests on a vertical test bench for initial qualifications of the electron gun being developed. The Faraday cup and the vertical test bench is shown in Figure 4 and Figure 5 respectively. Figure-6 gives a picture of current and voltage pulse on digital storage oscilloscope. The Faraday cup has measured up to 400 mA current at 40 kV pulse voltage. This finding is extremely useful and confidence building as it confirms that the gun will deliver 1 ampere beam current at 70 kV pulse voltage. Development of a horizontal test bench is in progress for measurement of other important beam parameters like transverse beam emittance and beam size. 70 kV pulse power supply matching with gun's requirements is also under development.

The fabrication technique used in the development of Faraday cup has shown that faithful devices can be made with joints made by furnace brazing and laser welding and they are superior to the conventional joining techniques due to their unique ability to join dissimilar materials. The experience gained in the design and manufacturing of Faraday cup would pave way for development of beam diagnostics devices where such fabrication techniques can be employed with success.

Fig. 4. Faraday cup-ready for installation

Fig. 5. Vertical qualification bench with electron gun and Faraday cup

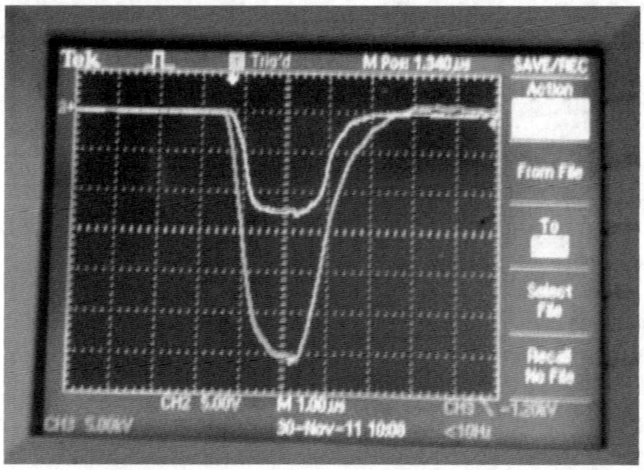

Fig. 6. Faraday cup's current and high voltage pulse across the electron gun shown in blue and pink coloured waveforms respectively on digital storage oscilloscope (Scale: Current - 100 mA/div; Voltage - 5 kV/ div)

References

[1] www.kovaralloy.com
[2] Kak A. et al 2009 InPAC Section 19 - Abstract 233
[3] Kumar R et al 2009 InPAC section 19 Abstract 233

Optimization of Electron Beam Transport for a 3-MeV DC Accelerator

S. Baruah[1], D. Bhattacharjee[2], R. Tiwari[3], G.K. Sahu[1], K. B. Thakur[1], K.C. Mittal[2] and L.M. Gantayet

[1] Laser & Plasma Technology Division
[2]Accelerator & Pulse Power Division
Beam Technology Development Group, BARC, Mumbai

E-mail: sbaruah@barc.gov.in

Abstract. Transport of a low-current-density electron beam is simulated for an electrostatic accelerator system. Representative charged particles are uniformly assigned for emission from a circular indirectly-heated cathode of an axial electron gun. The beam is accelerated stepwise up to energy of 1 MeV electrostatically in a length-span of ~3 m using multiple accelerating electrodes in a column of ten tubes. The simulation is done under relativistic condition and the effect of the magnetic field induced by the cathode-heating filament current is taken into account. The beam diameter is tracked at different axial locations for various settings of the electrode potentials. Attempts have been made to examine and explain data on beam transport efficiency obtained from experimental observations.

1. Introduction

A 3-MeV, 30-kW DC Accelerator is under development at Electron Beam Centre, Kharghar, Navi Mumbai for industrial applications. Initial beam trials were conducted for up to 1 MeV beam energy and the beam transmission efficiency was found to decrease with the increase of either the anode-cathode potential difference or the total beam current (at a constant anode-cathode potential difference). In order to find a tangible explanation to this, beam trajectory simulation was carried out considering all the ion-optical elements from the cathode up to the end of the accelerating column containing ten accelerating tubes. The simulation is done without considering space-charge of the beam for this low perveance (~10^{-9} A/$V^{3/2}$) gun following Schiller *et al.* [1].

2. Electron gun and the accelerating column

The DC accelerator setup is explained elsewhere in details [2]. The accelerator primarily consists of an electron gun, an accelerating column, a magnetic steering, a focussing arrangement and a magnetic beam scanner for product irradiation as shown in Figure 1. The electron gun [3] uses a LaB_6 pellet of diameter 8 mm and thickness 1 mm for thermionic emission, which is indirectly heated by a tungsten filament. The filament power is maintained between 100 W and 150 W. The anode-cathode potential difference can be varied from 1 - 5 kV. The accelerating column consists of 10 accelerating tubes. Each tube is about 30 cm long and has 20 titanium electrodes (dynodes). It can sustain a voltage of 335 kV with evacuation inside (10^{-7} mbar) and pressurized SF_6 environment outside (6 atm.) [4]. After the accelerating column, a beam line (ID ~ 95 mm) is used for propagation of the beam towards a 2D steering magnet (inside the pressure vessel) located 400 mm away from the exit of the accelerating column. This is followed by a vacuum pumping port, a focusing coil, a scanning magnet and a scan-horn. The focusing coil can provide a maximum of 5000 ampere-turns for reducing the beam size. The scanning magnet is at a distance of about 3 m from the exit of the accelerator column. It is used to scan the electron beam on the product requiring irradiation. The scan-horn has an exit window of size 1000 mm x 50 mm with a 50-micrometer thick titanium foil. The beam current is collected by a 3-mm thick aluminium collector kept below the scan-horn. The total beam current is obtained from the power supply circuit, using the return currents from the dynodes and the collector.

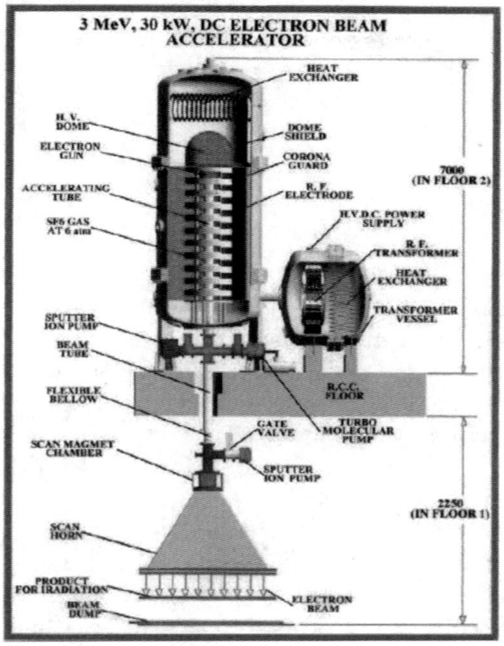

Fig. 1. Schematic of the 3-MeV DC Accelerator

3. Beam trials and gun alignment

Beam trials were conducted at 1 MeV beam energy [5]. The beam current transmission efficiency is shown in Figure 2 (Curve A) as a function of the anode-cathode potential difference, $V_{A\text{-}C}$, of the electron gun maintaining the beam power at 1 kW. The steering and the focusing coils were optimized to obtain the highest possible beam current. The curve shows the maximum transmission efficiency at $V_{A\text{-}C}$ = 1.5 kV, which falls gradually with the increase of $V_{A\text{-}C}$ from 1.5 kV to 4.5 kV.

The beam transmission was measured as a function of the beam current from 1 mA to 4 mA keeping $V_{A\text{-}C}$ constant at 1.5 kV. The transmission efficiency was found to decrease gradually from ~78% to 66% with the increase of the electron beam current. [5].

It was later found that both the anode and the cathode were tilted from their positions by about 1 mm and 0.5 mm, respectively with respect to the beam axis. After correcting the alignment of the anode and the cathode, beam transmission was measured again. The transmission efficiency was found to be slightly lower than before, while the general trend of the curve remained same (Curve B, Figure 2). To understand the behaviour of the electron beam propagation through the ion-optics, simulations were carried out using charged particle simulation software SIMION [6]. The results from the simulation are presented in the next section.

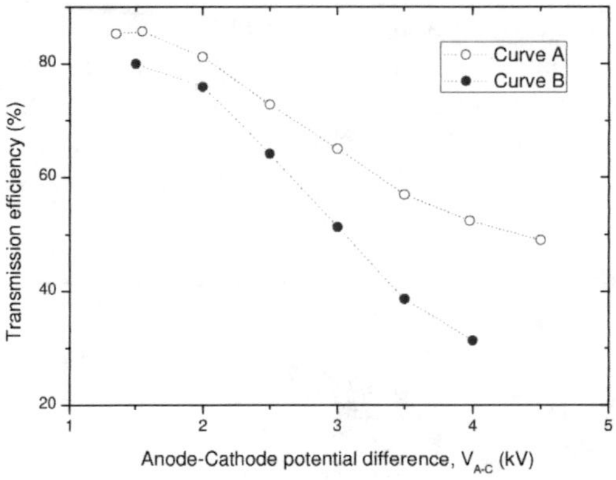

Fig. 2. Transmission efficiency as a function of the anode-cathode potential difference [5].

4. Simulation of the beam trajectories

The electron beam simulations are carried out for a beam energy of 1 MeV. An exact replica of the electron gun along with the accelerating column is modelled in 3D taking advantage of their cylindrical symmetry. The direction of beam propagation is taken as the X-axis of the coordinate system and hence, the beam cross-section lies in the YZ plane. A sectional view of the electron gun with the first accelerating tube is shown in Figure 3. Circular distribution is used for assigning the initial locations of the electrons (1000 – 5000) on the surface of the disc-type cathode (diameter 8 mm) of the electron gun. The cathode and the grid electrode (separation ~ 8 mm) are kept at the same potential (-1 MV) and a potential difference of V_{A-C} = 4.5 kV is assigned between the anode and the cathode (separation ~ 25 mm). The gun is followed by an accelerating column consisting of 10 tubes (each of length ~30 cm). Each tube has 20 electrodes, and each consecutive electrode of the ten accelerating tubes is kept at a higher potential with respect to the preceding one so that the final energy of the electron at the exit of the accelerating column is 1 MeV. The computational volume is 500x16x16 cm^3, and 10 grid points per 1 mm is considered for computing the electric field.

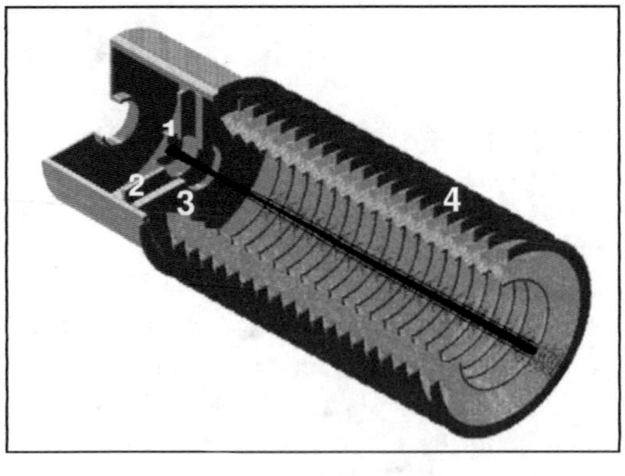

Fig. 3. A 3D sectional view of the geometry of the electron gun with the first accelerating tube.
(1) Cathode, (2) Grid, (3) Anode, (4) First acceleration tube.

Simulation is performed with V_{A-C} = 4.5 kV and for a beam energy of 1 MeV (after full acceleration). The trajectories are monitored and the beam-spots are recorded at different axial locations. With the prescribed setting of voltages for the electrodes (~200) of the accelerating tubes, the beam-spots at the cathode, at the anode exit and at the end of the accelerating column are shown in Figure 4. Even at the exit of the accelerating column, the beam does not diverge more than 18 mm in diameter. As expected, the beam-spot is symmetric on the YZ plane with its center coinciding with the X-axis. This indicates that the beam transmission efficiency should not deteriorate provided the beam is not influenced by (i) space charge, (ii) initial thermal transverse velocity, (iii) filament current-induced magnetic field and, (iv) misalignment of electrode(s) or mis-assignment of voltage to any electrode(s).

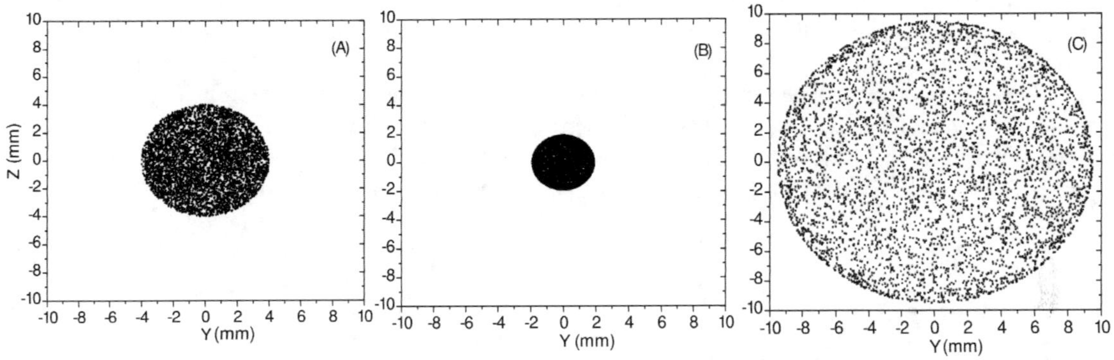

Fig. 4. Beam-spots at different axial locations of the accelerating column.
(A) On the cathode surface, (B) At the exit of the anode, and, (C) At the exit of the accelerating column. The diameter of the beam at the exit of the accelerating column is found

For a 1 mA electron beam with $V_{A\text{-}C}$ = 4.5 kV, the electron gun has a low perveance (of the order of 10^{-9} A/$V^{3/2}$). Therefore, intrinsic space charge of the beam is negligible [1]. Furthermore, the initial thermal transverse velocity is also not expected to play a role as it dominates beam propagation only when the beam perveance is too low [1].

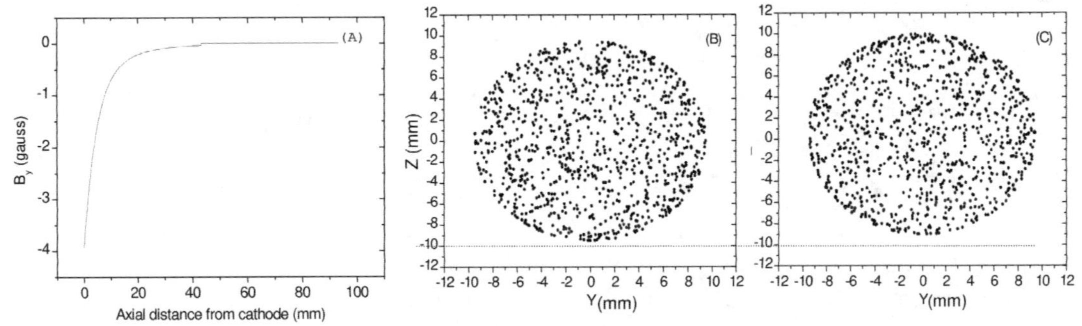

Fig. 5. Beam-spots for showing effect of filament current-generated magnetic field.
(A) Filament current-induced magnetic field at different axial locations, (B) Beam-spot at the exit of the accelerating column without considering the magnetic field, (C) Beam-spot at the exit of the column considering the effect of the magnetic field. The beam-spot is found to shift vertically by < 1 mm.

The aforesaid electron gun uses a linear coil-type tungsten filament (wire diameter = 0.5 mm) of length 14 mm and diameter 6 mm for cathode heating. A typical current of 18 A through the coil produces a magnetic field of about 4 gauss (Figure 5(A)) at the location of electron emitting surface of the cathode in a direction opposite to the net current in the coil. Incorporating this magnetic field, the beam spot at the exit of the accelerating tube is found to exhibit a vertical shift of <1 mm as shown in

Figure 5(C). Increasing the filament current would increase the shift and the beam-spot would start to appear distorted. However, for the typical current of 18 A, these effects are negligible.

Mechanical misalignment, as earlier noticed in the experiment, could have serious repercussions, especially when it comes to the cathode or the anode of the electron gun. Similarly, voltage mis-assignment to the electrode(s) could also lead to straying of the beam off the axis.

A test of beam transmission can be made with a beam very tightly focussed towards the exit of the accelerating column to eliminate dependencies on the steering and the focussing coils. Such a beam can be achieved by slightly modifying the cathode-grid-anode arrangement. Figure 6(C) shows such an arrangement. By bringing the electron emitting cathode surface to the same plane as that of the grid and also by placing the cathode holder in such a way that it does not protrude ahead of the cathode surface, beam cross-over as seen in the simulated trajectory for the original design (Figure 6(A)) can be avoided. The modification thus leads to a continuously converging beam towards the exit of the accelerating column. Near the exit of the accelerating column, the beam diameter reduces from 18 mm (Figure 6(B)) for the original design to about 1 mm (Figure 6(D)) with the modified arrangement.

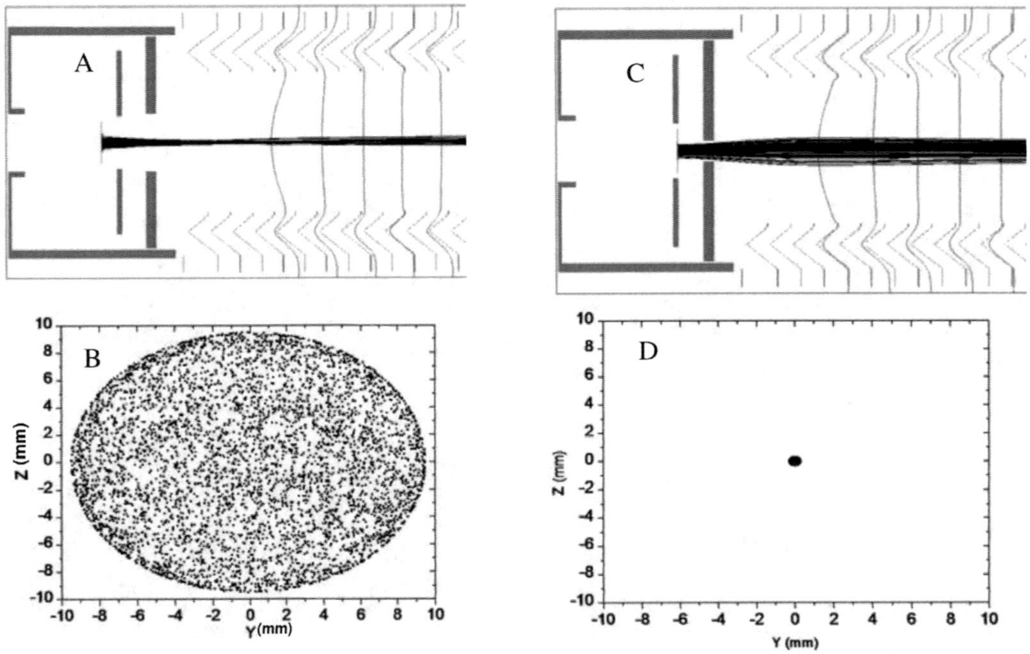

Fig. 6. Existing configuration (A) of the gun electrodes can be modified (C) to reduce beam diameter at the exit of the accelerating column from about 18 mm (B) to 1 mm (D).

5. Conclusion

Transport of a low-density electron beam for a DC accelerator system has been simulated. Beam-spots at different axial locations of the accelerator are recorded at the prescribed voltage settings and it has been found that the beam efficiency should not deteriorate till the exit of the accelerating column. It is also pointed out that the intrinsic space-charge or the initial thermal velocity spread of the electron beam is not expected to influence the beam propagation. Mechanical misalignment or mis-assignment of voltage to any electrode(s) can play a role in the experimentally observed deterioration of the beam transport efficiency. A modified configuration of the electron gun is proposed for checking beam transmission eliminating dependencies on the steering as well as the focussing coils.

Acknowledgement
Authors are grateful to A. K. Das and D. P. Chakravarthy for fruitful scientific discussions and helpful support.

7. References

[1] Schiller S, Heisig U, and Panzer S 1982 *Electron Beam Technology* (New York : Wiley)
[2] Majumder R, Mittal K C, Nanu K et al. 2003 *Proc. of Indian Particle Accelerator Conference (InPAC)*, Raja Ramanna Centre for Advance Technology, Indore, India 241
[3] Jain A, Chindarkar A R and Mittal K C 2007 *Proc. of Asian Particle Acclerator Conference (APAC),* Raja Ramanna Centre for Advance Technology, Indore, India 348
[4] Jayaprakash D, Mishra R L, Ghodke S R, Mukeshkumar, Mahendrakumar, Nanu K and Mittal K C 2009 *Bulletin of Indian Vacuum Society* **12(3)** 3
[5] Acharya S, Nanu K, Mittal K C et al. 2011 *Proc. of Indian Particle Accelerator Conference (InPAC)*, Inter-University Accelerator Centre, New Delhi, India
[6] Dahl D 2000 *Int. J. Mass Spec.* **200** 3

Development of Electron Guns for Linacs and DC Accelerator

D Bhattacharjee, R Tiwari, D Jayaprakash, R L Mishra, A R Tillu, B Nayak, R B Chavan, Shiv Chandan, V Yadav, S R Ghodke, Mahendra Kumar, K C Mittal, D P Chakravarthy and L M Gantayet

BARC, APPD, EBC Kharghar, Navi Mumbai, India.

dhruva.bhattacharjee@gmail.com

Abstract. Electrons guns for RF linacs and DC Accelerators are designed and developed at Electron Beam Centre (EBC)/APPD/BARC. Planar geometry grid and Pierce geometry grid configuration diode and triode guns with LaB6 cathode are developed. The cathode assembly consists of cups and heat shields made out of Tantalum and Rhenium sheets. The cathode assembly and the electron guns are tested on a test bench for beam characterization. The paper presents the development of the electron guns.

1. Introduction

The Accelerator and Pulse Power Division of BARC has developed three industrial electron accelerators for radiation applications - a) a 500 keV Cockcroft-Walton accelerator (at BRIT, Vashi), b) a 10 MeV RF linac at EBC, c) a 3 MeV DC Accelerator at EBC; and one 9 MeV RF Electron Accelerator at ECIL, Hyderabad, for cargo scanning. A 6 MeV RF electron accelerator is under development at EBC and will also be used for cargo scanning applications.

These accelerators require electron guns with special requirements. The two DC machines require an almost parallel beam with less than 1.0 degree divergence from the gun to enable the beam to drift through the accelerating modules. A 5 kV thermionic DC electron gun is used as injector for the two accelerators. The RF linac structure is similar for the 10 MeV, 9 MeV and 6 MeV linacs and so a 50 keV thermionic pulsed electron gun is used as injector for them with the beam current between 500 mA to 1 A.

For irradiation of the products, the accelerated beam (using accelerating tubes or linac structures), passes through the magnetic sweep scanner, scan horn and is brought into air through a titanium foil of 25 ▯ - 50 μm thickness, for irradiating the products. The beam size is typically 20 mm diameter just outside the titanium foil. For cargo scanning application, the accelerated beam (using linac structures) passes through a focusing solenoid coil and is focused to a size of < 2 mm diameter on a Tungsten target to generate the x-rays.

2. Gun Design

2.1. Gun for DC Accelerator

A triode gun, having a flat cathode and parallel discs with apertures for the grid and anode has been designed. The electron trajectories have been obtained using the SLAC gun design code EGUN [1] and also with CST Particle Studio 'figure 1'. A Pierce geometry gives a beam current of 26 mA at 5 kV and with a beam size of about 3.5 mm diameter 'figure 2'.

Journal of Physics: Conference Series **390** (2012) 012071 doi:10.1088/1742-6596/390/1/012071

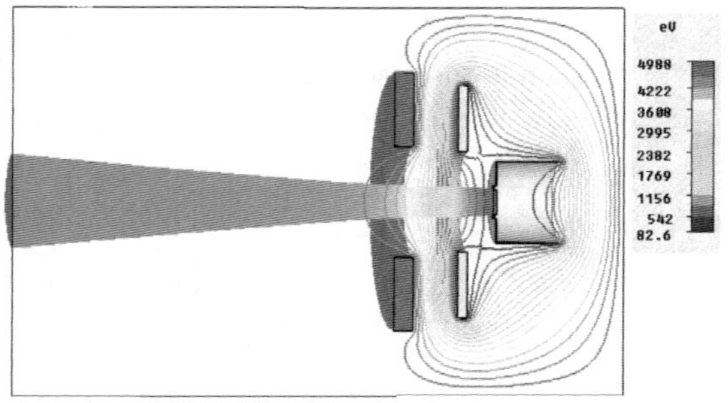

Fig. 1. A CST Particle Studio simulation of Planar Diode Gun at 5 kV and 48 mA with perveance 1.4 x 10^{-7} $A/V^{3/2}$. A beam size of 12.7 mm diameter is obtained at 19 mm from anode.

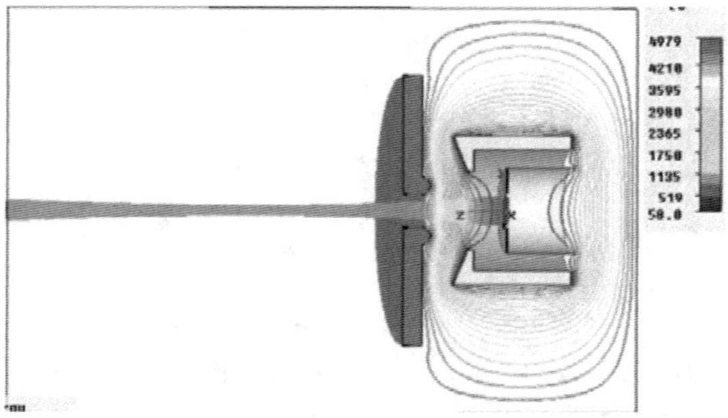

Fig. 2. A CST Particle Studio simulation of Pierce Diode Gun at 5 kV and 26 mA with perveance 7.54 x 10^{-8} $A/V^{3/2}$. A beam size of 3.3 mm diameter is obtained at 19 mm from anode.

2.2. Guns for RF Accelerators

A triode electron gun operating in diode mode (cathode-grid shorted) was designed. Two different configurations were tried; (a) planar configuration with the cathode as flat disc and parallel discs with apertures for the grid and anode 'figure 3', (b) Pierce grid configuration with cathode as flat disc, and grid as a shaped electrode and a shaped anode 'figure 4'.

The 10 MeV accelerator at EBC, and the 9 MeV accelerator at ECIL, Hyderabad, used planar electron guns with 8 mm and 6 mm cathode diameter beam emission areas respectively. The cathode for these guns is a LaB6 pellet of dimensions 10 mm dia. x 1 mm thk.

It is known that a converging or collimating beam of smaller beam diameter at the linac input will improve the capture efficiency and hence the transmission of the beam. For the 9 MeV ECIL Linac, the output beam should be of diameter $\leq$ 2 mm at the Tantalum target. To achieve this, a Pierce gun design approach was attempted.

A design goal of beam current of 500 mA at 40 kV with a beam size less than 3.5 mm and using an 8 mm cathode diameter for beam emission was decided. To design the electron gun, simulations were conducted using the CST Particle Studio. It is a Diode Type thermionic Gun. It incorporates a Pierce geometry Grid, also known as Focusing electrode. Studies were conducted to obtain good beam quality, beam size and beam current by varying (a) Planar anode inner diameter, (b) Pierce angle of the Grid, (c) cathode-anode gap, (d) cathode-grid gap, (e) anode shapes.

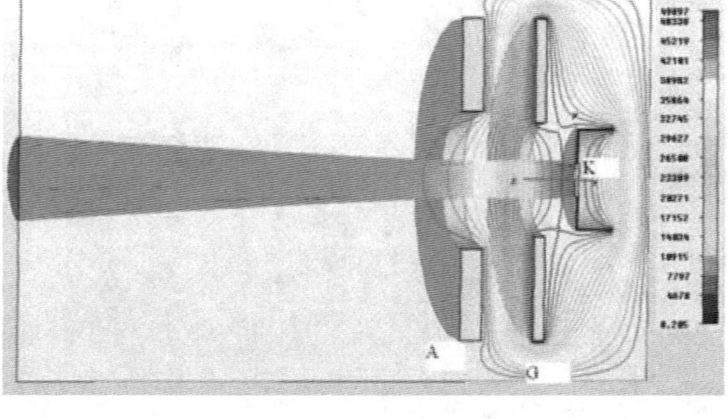

Fig. 3. A CST Particle Studio simulation of Planar Diode Gun at 50 kV and 1.154 A with perveance 1.03 x 10^{-7} $A/V^{3/2}$. A beam size of 13.7 mm diameter is obtained at 49 mm from anode.

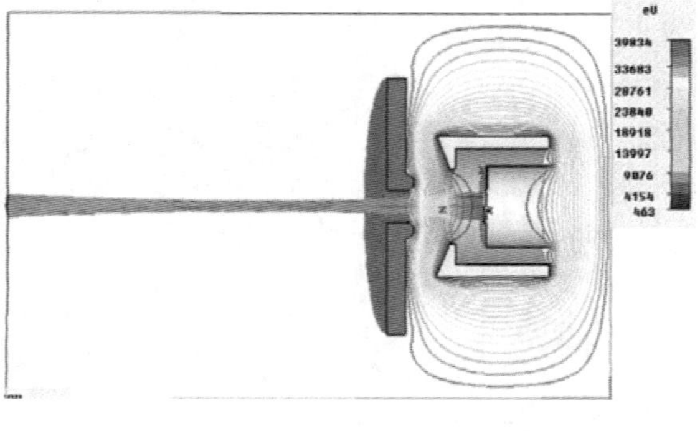

Fig. 4. A CST Particle Studio simulation of Pierce Diode Gun at 40 kV and 598 mA with perveance 7.53 x 10^{-8} $A/V^{3/2}$. A beam size of 3.0 mm diameter is obtained at 49 mm from anode.

Journal of Physics: Conference Series **390** (2012) 012071 doi:10.1088/1742-6596/390/1/012071

The simulations resulted in the required 40 kV Diode electron gun with Pierce Focusing electrode (grid) of ID 23 mm and angle ~ 67.5 °, to give ~ 600 mA of beam current and having a diameter of less than 4 mm at a distance of 100 mm from the cathode. The anode is also shaped with aperture of 10 mm. The cathode-grid distance is 9 mm and the anode-cathode distance is 25 mm. The gun is operated in the space charge region.

3. Choice of cathode

The cathode is a LaB6 Pellet of dia. 10 mm and thickness 1 mm. The Lab6 pellet is indirectly heated by a coil filament made out of Tungsten wire. A filament power of 270 W raises the temperature of the LaB6 pellet enough to emit the required current of 500 mA at 40 kV.

LaB6 cathode is chosen because it has low thermionic work function 2.66 eV, high melting point 2715 °C, high current density, it can work in vacuum of the order of 10^{-6} – 10^{-5} mbar. LaB6 offers the capability of long life and orders of magnitude less sensitivity to air exposure than conventional dispenser cathodes. In the case of mild cathode poisoning resulting in lower beam emission, it is observed that after conditioning the cathode to higher filament power a few times, there is a an improvement in the output beam current.

4. Cathode assembly

The cathode pellet is housed in a 0.05 mm thk. rhenium cup which is then housed in an outer cylindrical tantalum cup of 25 mm diameter and 0.15 mm thickness. The pellet is held in place by two thin strips of rhenium which are spot welded to the tantalum cup. The opening of the rhenium and tantalum cups decides the beam emission area of the cathode to be of diameter 8 mm or 6 mm.

The Lab6 pellet is indirectly heated by a coil filament made out of tungsten wire of diameter 0.5 mm and consisting of 9 turns with coil diameter of 5 mm 'figure 5'. The gap between the tungsten wire and the pellet is about 1.5 mm. The radial heat shields consist of two cylindrical heat shields of tantalum (0.15 mm thk.) and having diameters 20 mm and 25 mm. The inner cylindrical Ta heat shield also has two Tantalum discs as heat shields in the axial direction. The inner and outer cylindrical heat shields are connected by Ta strips (2-3 mm wide, 0.15 mm thk.). The outer cylindrical heat shield is connected to the ceramic base by 4 Ta strips (2-3 mm wide, 0.25 mm thk.) and a SS 304 cup. The filament is spot welded to two tantalum rods (diameter 2.5 mm) which are rigidly supported on a ceramic base. These Ta rods have ceramic sleeves on them to electrically isolate them from the heat shields. The ceramic base sits on the SS 304 cup. One of the Ta rods is shorted to the outer cylindrical heat shield to maintain same potential of the filament and the cathode.

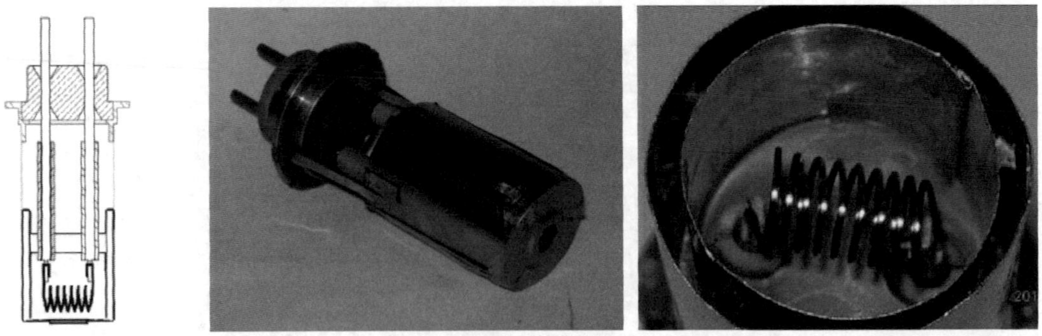

Fig. 5. Schematic of cathode assembly, fabricated cathode assembly and filament with heat shields.

5. Gun fabrication and testing

The grid and anode electrodes are made of SS 304. The grid is supported on 4 rods connected to the top metallic flange. This flange also has 4 no. of 14 kV, 56 A capacity HV ceramic feedthroughs. Two feedthroughs are connected to the filament and remaining two is spare. The anode is isolated from the

HV top flange by a ceramic tube of OD 160 mm and length 200 mm. The grid is isolated from the cathode assembly by using MACOR discs. The sealing for vacuum is obtained using Indium wire gaskets or rectangular Viton gaskets, which are placed between the ceramic tube and the corresponding metallic flanges. Initially the anode flange was placed below the ceramic insulator tube. Later a gun chamber of height ~ 180 mm mm consists of two DN 100 ports and one DN 63 port was added. The anode flange is connected below the gun chamber and the sealing is with an aluminium wire gasket (0.5 mm diameter). Recently, the size of the gun is redesigned to a smaller one with the ceramic insulator of size OD 120 mm and length 150 mm and the gun vacuum chamber of height 140 mm. Also, the anode is connected to a pipe connected to an intermediate flange and not at the end of the gun vacuum chamber. The grid is supported on a pipe 'figure 6'. All SS 304 components of the electron gun were electropolished before assembly. Cathode and grid are shorted externally. The cathode, grid and anode are separated by insulators and can take independent potentials. The cathode, grid and anode are aligned using Aluminium and SS jigs.

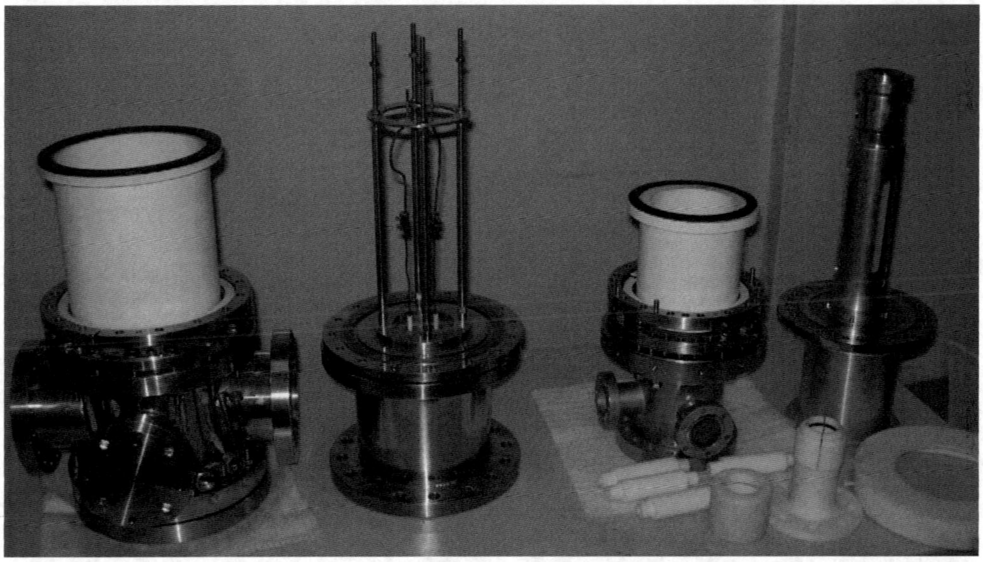

Fig. 6. Comparison of electron gun sizes. The right one is the gun for the 6 MeV linac.

The high voltage testing of the guns were carried out on a test bench, using a 50 kV, 10 μs pulsed solid state gun modulator, a 70 kV HV Isolation Transformer for filament heating and a diffusion pumping system 'figure 7'. The vacuum levels were of the order of 2 x 10^{-5} mbar. With cathode conditioning, the beam current improved in the 2nd and 3rd runs. The effective capacitance between cathode and anode with cathode grid shorted was ~37 pF. For the planar configuration, the beam sizes were measured on graphite plate and also with segmented graphite faraday cup placed at various distances from the anode. For the Pierce configuration (8 mm cathode), the beam spot sizes were measured on graphite plate and as puncture impressions on four 12 μm aluminium foils followed by an aluminium collector; kept at various distances (48 mm, 55 mm, 62 mm, 69 mm and 74 mm) from the anode. At the filament power of 264 W and extraction voltage of 40 kV, a beam current of 500 mA was extracted and a beam size of ≤ 3.5 mm diameter was obtained 'figure 8','figure 9' and 'figure 10'. With the 6 mm cathode, a beam current of 350 mA was achieved at 50 kV extraction voltage, 270 W filament power, 2 x 10^{-5} mbar vacuum and beam spot size ≤ 3.5 mm. The test bench based on diffusion pumping system is replaced by a TMP based system 'figure 7'.

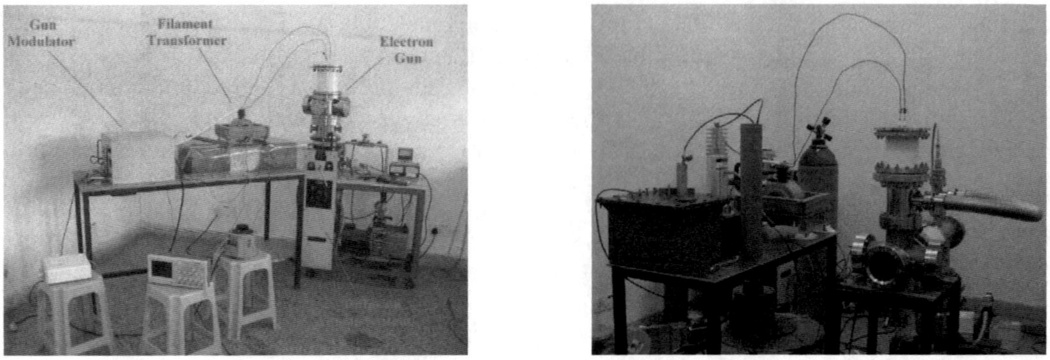

Fig. 7. Electron gun on DP based test bench (left) and on the TMP based test bench (right).

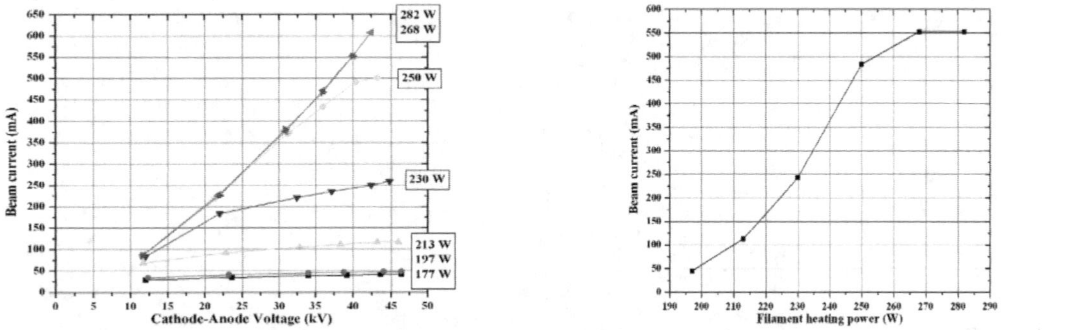

Fig. 8. Beam characteristics (left) and variation of beam current at 40 kV (right) of Pierce Gun with, 8 mm diameter cathode.

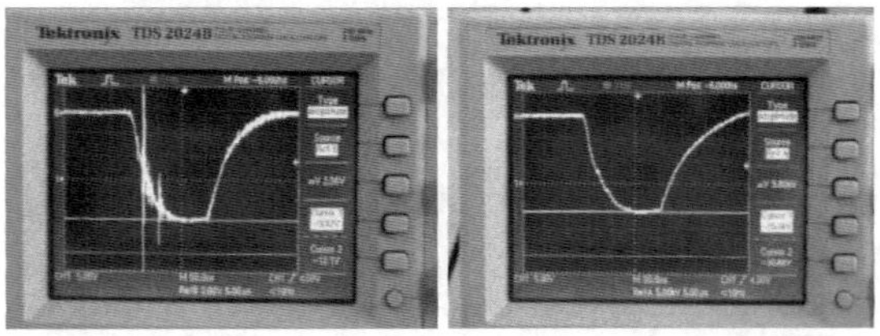

Fig. 9. Beam current waveform 542 mA (left) and Extraction voltage 41.6 kV (right).

Fig. 10. Beam spot sizes on Al foils and Al collector disk.

The results of tests on various gun configurations are summarised in table 1 below.

Table 1. Results of various electron guns.

Type of Gun	Cathode diameter (mm)	Extraction voltage (kV)	Beam current (mA)	Filament power (W)	Beam diameter (mm)
Planar	8	50	850	295	≤ 13
Planar	6	50	250	231	≤ 5
Pierce	8	40	500	264	≤ 3.5
Pierce	8	50	520	285	≤ 3.5
Pierce	6	50	350	270	≤ 3.5
Pierce	8	66	800	300	----

6. Problems and Solutions:

With the cathode assembly within the ceramic isolator tube of the gun, it was observed that the ceramic tube was getting heated up to 80 ºC, when the filament power was 280 W. Frequent leaks occurred on the locally developed gun ceramic tube. Leaks were temporarily sealed by a vacuum sealant, which opened up with time. The indium wire gasket for the ceramic-metal seal was replaced by VITON gasket. To improve vacuum in the gun and avoid heating of the ceramic, a 3-port gun vacuum chamber was incorporated. The cathode heating is within this chamber and the ceramic is less heated. There was a case of reduction in beam current due to cathode poisoning, which was then cleaned, activated and put back. Once, filament broke while replacing the contaminated cathode pellet.

While cathode conditioning at 280 W on test bench, due to accidental slippage of the viton gasket, the complete cathode assembly got heavily contaminated and had to be discarded 'figure 11'. In one case, the filament got shorted to the Rhenium strips holding the LaB6 pellet. Due to this there was a large flow of current and the Rhenuium strip got removed 'figure 12'. We have incorporated Aluminium wire gasket (0.5 mm diameter) for the vacuum sealing between the collar of ceramic isolator and the SS. A leak rate of 5 x 10^{-10} mbar l/s is achieved.

Fig. 11. Heavy contamination of cathode assembly after accidental exposure to air, during cathode conditioning.

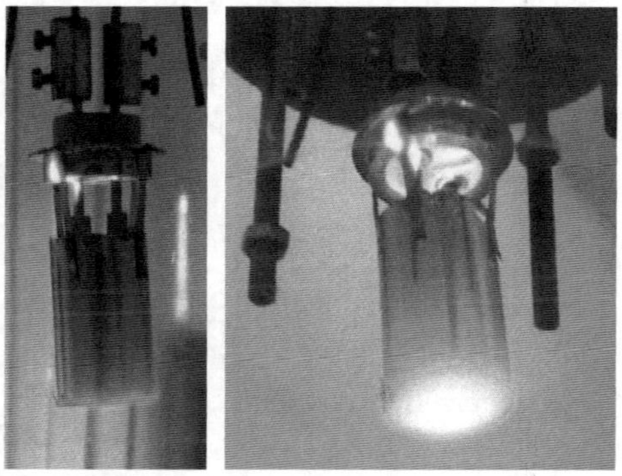

Fig. 12. Large filament current flow (30 A) through the Ta rod and Ta strip due to internal shorting of the filament with the body.

7. Conclusions:

Simple triode gun designs have been indigenously designed, developed, tested and successfully operated both for the planar and Pierce configurations. These guns use planar LaB6 pellet as cathode, which can operate well in vacuum of the order of 5 x 10^{-6} mbar. Efforts are being made to optimize the filament heating and studies are being done to understand the effect of magnetic field of the filament on beam optics of the gun.

References

[1] Jain A et al. 2007 Proc. APAC 2007 WEPMA011.
[2] D. Bhattacharjee et al. 2011 Proc. INPAC 2011 163.
[3] D. Bhattacharjee et al. 2011 Proc. INPAC 2011 166.
[4] CST Particle Studio Software.
[5] Humphries S *Charged Particle Beams*.
[6] Septier A *Focusing of Charged Particles* vol II.

Pulsed and RF glow discharge in Helium atmosphere

Pooja Gulati[1], U. N. Pal[1], N. Kumar[1], V Srivastava[1], Ram Parkash[1] and Vimal Vyas[2]

[1]Microwave Tubes Area, CSIR -Central Electronics Engineering Research Institute (CSIR-CEERI), Pilani, Rajasthan-333031, India
[2]Department of Physics, Banasthali University, Rajasthan, India

E-mail: pgulati1585@rediffmail.com

Abstract This paper reports the optical and electrical characterization of sinusoidal and pulse glow-discharge plasma in helium. A homogeneous type of discharge has been observed for different operating conditions in helium DBD. The image of discharges makes sure that the diffuse discharge covers the entire surface of the electrodes. Optical emission spectroscopy has been used to determine the main emission lines of the helium glow discharge plasma. The internal plasma parameters of DBDs have been investigated at different operating conditions. It has been observed that the DBDs are becoming more intense with increase in pressure and applied power. The effect of the excitation mechanism on the emission properties and discharge parameters for the helium DBD is studied experimentally and further analysed by using plasma simulation tool OOPIC-Pro.

1. Introduction

Dielectric barrier discharges (DBDs), also referred to as silent discharges, are generated in discharge configuration with at least one dielectric barrier between two planar or cylindrical electrodes connected to an ac or pulse power supply. It is an easy way to generate low temperature plasma at atmospheric pressure and have very attractive features for various industrial applications [1-3] because it can provide nonequilibrium plasma conditions about atmospheric pressure. Typical dielectric materials are glass, quartz, ceramics or polymers that can prevent the transition from glow to arc. The most interesting property of DBD is that the breakdown is initiated in a large number of independent current filaments with nanosecond duration. In DBDs, recent studies have shown that the use of fast rising voltage waveforms can dramatically increase the electrical-VUV conversion efficiency compared with conventional sinusoidal voltage waveforms [4]. In this paper, we have studied the excitation mechanism of primary and secondary discharges in helium DBD by using unipolar pulse waveform and sinusoidal waveform with emphasis on time-resolved spectral and photographic diagnostics of the discharges. For further investigation of the discharge parameters in helium DBDs, we have performed experiments at different operating conditions. We have also studied the effect of the excitation conditions on the discharge parameters for the helium DBD by using plasma simulation tool OOPIC-Pro.

2. Experimental set-up

Figure 2.1 shows the schematic diagram of experimental setup. The source frequency is kept at 30 kHz for both power supplies. The DBD cell has been mounted inside the ultra high vacuum chamber. An Ocean Optics spectrometer (HR4000) with the resolution of 0.75nm having grating 300 lines/mm has been mounted with the vacuum chamber for the spectroscopic observations of the discharge. The vacuum chamber has been mounted on an ultra-high vacuum pump system using rotary and turbo molecular pumps. The base pressure is kept at ~$1x10^{-6}$ mbar. At room temperature, the helium gas of 99.9% purity (BOC Gases) has been filled in the DBD cell. The pressure of the gas has been measured by pressure gauges (Pfeiffer: APR262, Pfeiffer: PKR251) and the pressure has been maintained by vacuum valves (Matheson: 316L, Varian: 9515091).

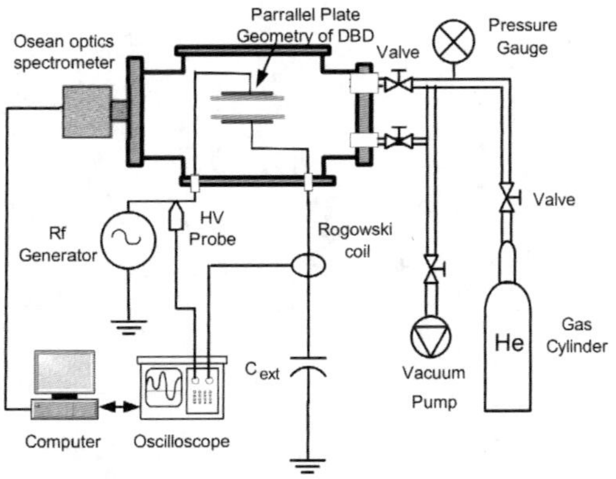

Figure 2.1: Schematic view of Experimental set-up

The upper copper electrode acts as anode which is connected to the high-voltage (HV) power supply (Huttinger HF Generator TIG 10/100 PSC) for sinusoidal waveform and with the high voltage pulse power supply up to 5 kV maximum for the pulse waveform while the lower electrode is grounded. The electrodes are insulated by the Teflon cover to avoid long path discharges in the geometry and whole assembly is covered in the Perspex box to avoid the discharge with the chamber. A 1: 1000 high voltage probe (Tektronix P6015A) measures the voltage across the DBD and the Rogowski-type Pearson current monitor, Model 110 (0.1VA−1, 1 Hz–20 MHz, 20 ns usable rise time) measures the total current flowing through the DBD. The total current and applied voltage waveforms are visualized by means of a four-channel Tektronics DPO 4054 digital oscilloscope.

3. Results and discussions

In this work, we have illustrated the dependence of different discharge parameters on the operating conditions. Figure 3.1 shows the average image of discharges taken with a digital camera (SONY DSC-P100, exposure time: 25 ms). The image indicates that the diffuse discharge covers the entire surface of the electrodes. Figure 3.2 and 3.3 shows the total current trace together with the applied voltage using pulse and sinusoidal waveforms respectively. Both the image and the total current confirm the diffused type discharge in helium DBD. The total current waveform for sinusoidal and pulse excitation shows that even for a same applied potential more current have been observed in case of pulse waveform as shown in figure 3.1 and 3.2 [5].

Figure 3.1: Diffused discharge in between the gap

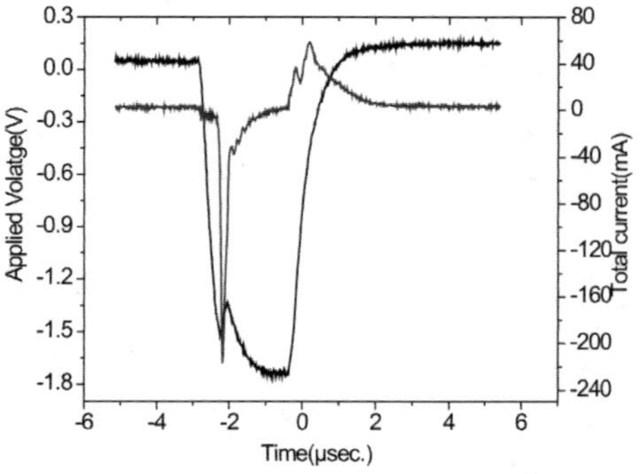

Figure 3.2: V-I waveforms for Pulse at 600mbar and 30kHz

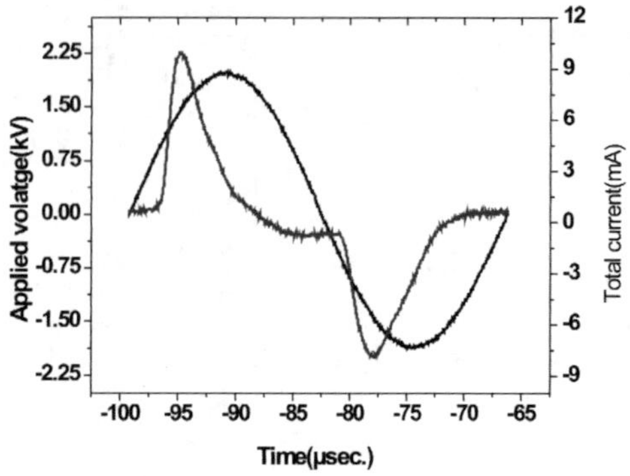

Figure 3.3: V-I waveforms using sine wave at 600mbar and 30kHz

The average deposited power for ac excitation can be determined from the Lissajous ($V-Q$) diagram, where the area enclosed within the $V-Q$ curve is the energy deposited into the plasma per discharge cycle [6]. Figure 3.4 shows the Lissajous diagram for fixed applied voltage with different pressure at a 30-kHz frequency using sinusoidal waveform. The total amount of power consumed by the DBDs is required to estimate the total efficiency of the system and the required power of the HV supply (pulse or sine wave). The energy consumed by the plasma for one cycle has been calculated from the area of parallelograms for different operating conditions and is shown in figure 2.4. This figure depicts the increase in consumed energy as the operating pressure increases. The same pattern can also be seen in case of pulse excitation.

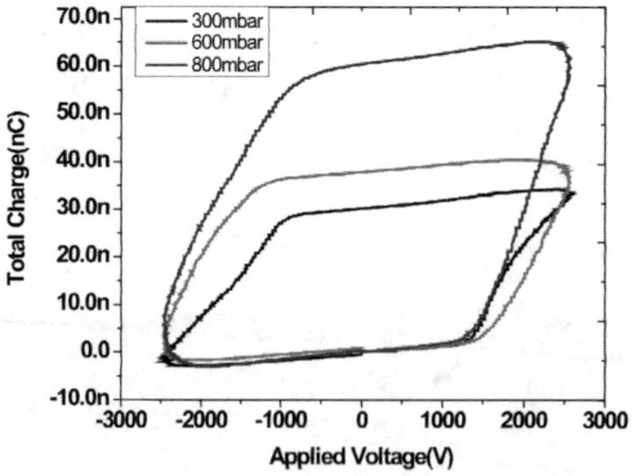

Figure 3.4: Lissajous $V-Q$ diagram of the Helium DBD in the case of three different pressures at 30 kHz for sine wave

3.1 Study of emission spectrum: To study the emission of the helium spectrum, we have used an Ocean Optics spectrometer and the spectrum is shown in figure 3.5 which clearly depicts almost all the helium neutral lines. From figure 3.5, it is observed that with the increase in applied power the spectral line intensities of all helium neutral lines is increasing for both excitation methods however the pulse excitation deposits more power.

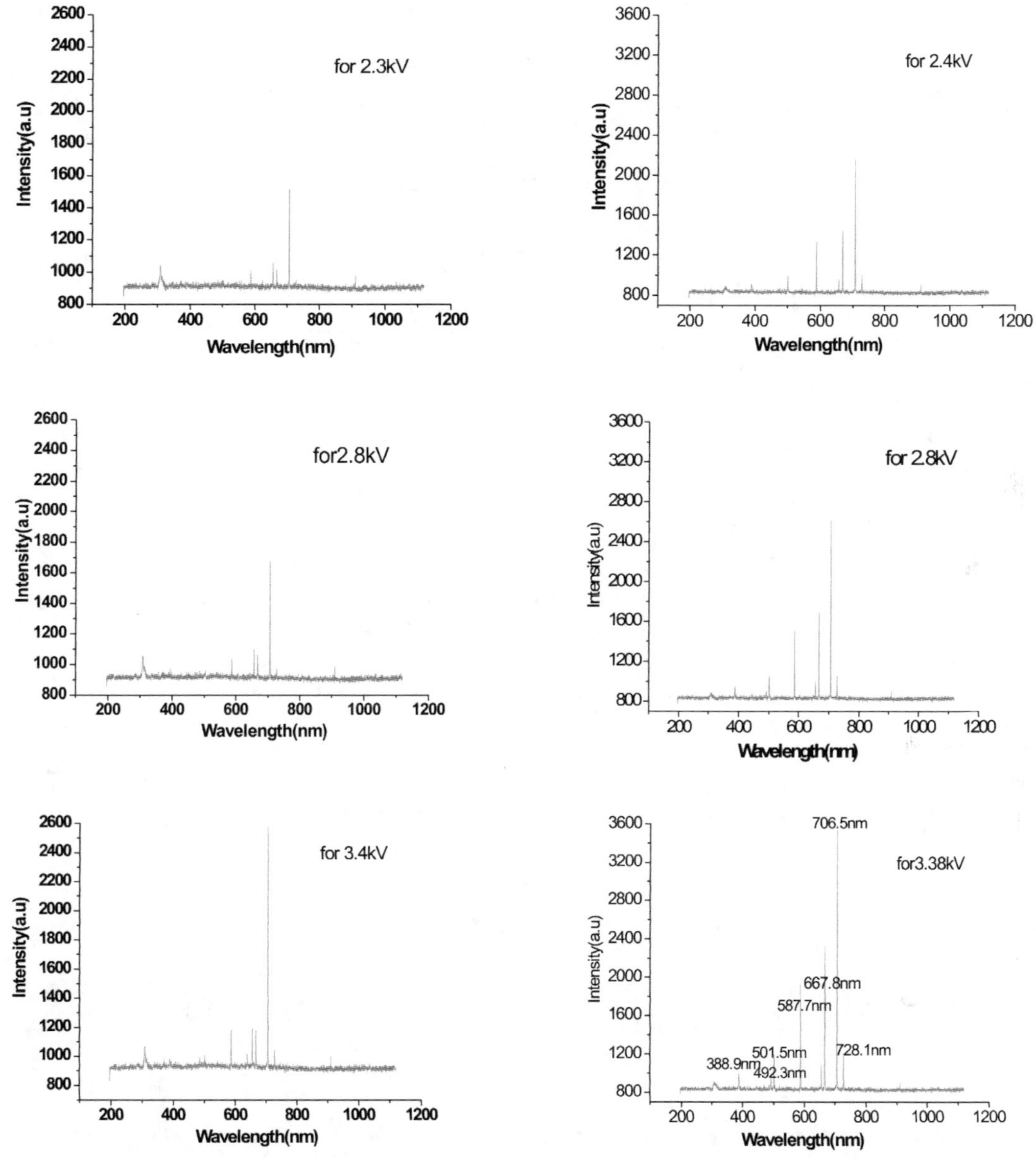

(a) (b)

Figure 3.5 (a) Emission spectra for helium using sinusoidal waveform for 300 mbar, (b) Emission spectra for helium using Pulse for 300mbar

Figure 3.6 represents increase in intensities of the spectral lines when the operating pressure is in the range of 300-600mbar while it decreases for pressure greater than 600 mbar. This behavior can be easily seen in the 706.5nm line. This line is dominant to all other He I (helium neutrals) lines. Figure 3.7 demonstrates that several emissions spectral lines for various applied potential using sinusoidal waveform that indicates the increase in intensities with the increase in applied power. This same behaviour has also been observed in case of unipolar pulse.

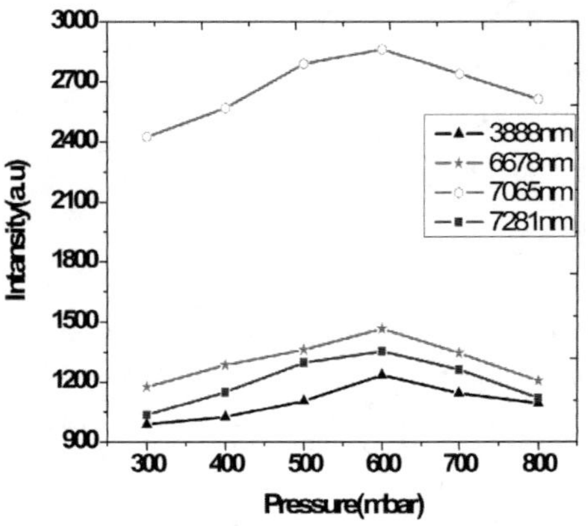

Figure 3.6: Intensities of different emission lines Vs. Pressure using pulse waveform

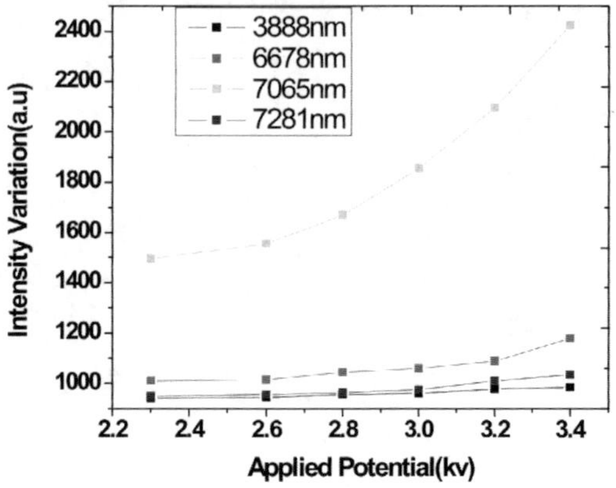

Figure 3.7: Intensities variation of emission lines Vs. applied potential sinusoidal Waveform

3.2 Kinetic Simulation using PIC Code:

The 2D object-oriented particle in cell code OOPIC Pro has been used to study the electrical as well as the kinetic behaviour of the DBDs [7]. We have analysed the effect of operating parameters over the discharge pattern and concentration of the electrons and ions, to analyze the discharge kinetics in more details. Figure 3.8(a) shows the simulation model of the helium DBD configuration without discharge and 3.8 (b,c,d) show phase space of the electrons at different operating pressure. The electrons of the order 10^3 /cm^3 are loaded uniformly in the gas gap along with the neutral He gas at different gas pressure to initiate the discharge at room temperature. The unipolar-pulsed voltage is applied across equipotential boundary condition to excite the neutral He gas atom. The following figures confirm that electron concentration is continuously increasing with the increasing gas pressure for the same simulation time as observed in the experiments.

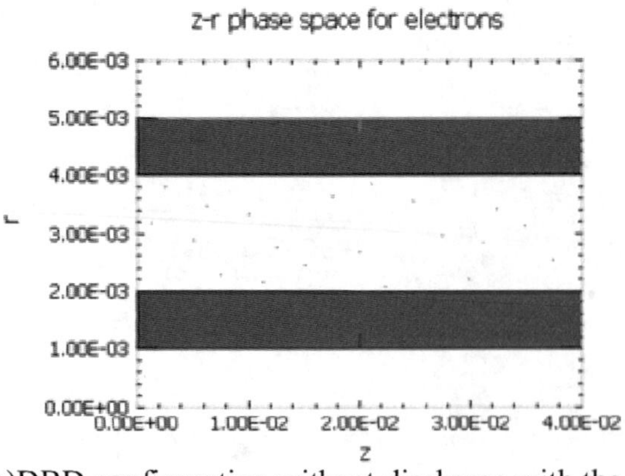

(a)DBD configuration without discharge with the initial electrons

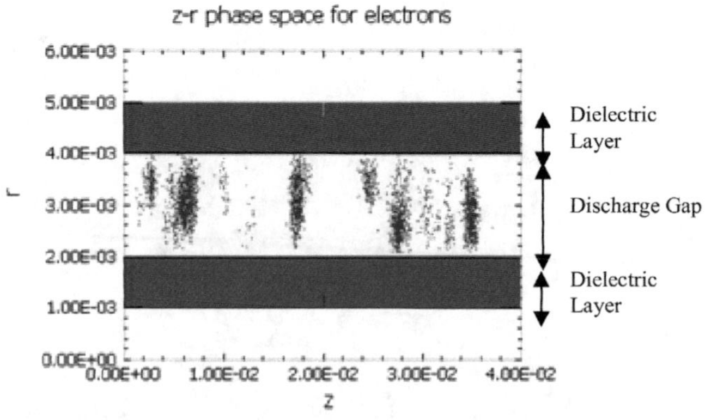

(b)Discharge at 300 mbar for 123nsec. Simulation time

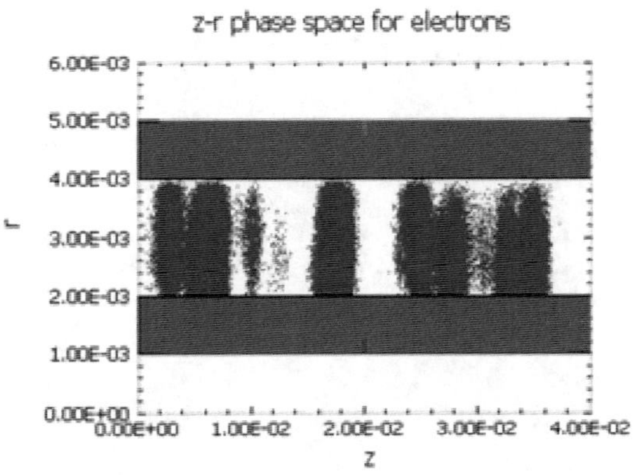

(c)Discharge at 600mbar at 123nsec. Simulation time

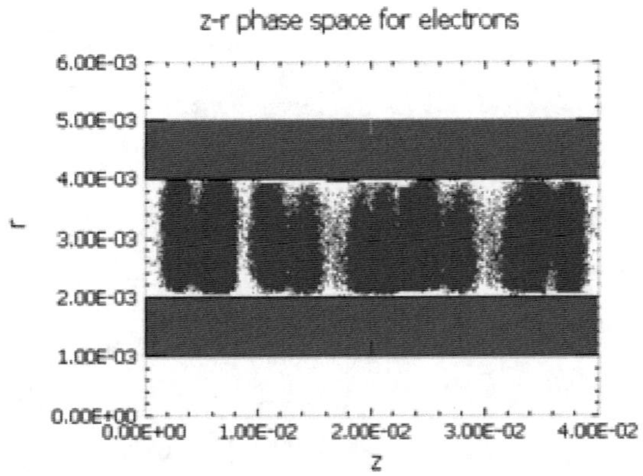

(d)Discharge at 800mabr at 123nsec. Simulation time

Figure 3.8

4. Conclusion

The homogeneous type of discharges has been observed in helium DBD at different operating conditions. The simulated electron concentration is in close agreement with the reported values in the literature. Kinetic simulation confirms that electron concentration is continuously increasing with the increased gas pressure for the same simulation time. Pulse excitation deposited more power so the more current and highly intense spectral lines of helium discharge for the same repetition rate as for the sinusoidal waveform. The pulse excitation is found to be more efficient than sinusoidal waveform for the same operating conditions.

5. Acknowledgement

The work has been carried out under CSIR Network project (NWP0024). The authors wish to acknowledge valuable discussions with Mr. Niraj Kumar, Mr. M.S Tyagi and Mr. B.L Meena.

6. References:

[1] U Kogelschatz, B Eliasson and W Egli, 1999 *Pure Appl. Chem.* **71,** 1819–28.
[2] P Rehn, A Wolkenhauer, M Bente, R Forster and W Viol, 2003 *Surf. Coat. Technol.* **174–5,** 515–8.
[3] E E Kunhardt ,2000 *IEEE Trans. Plasma Sci.* **28,** 189–200.
[4] R. P Mildren and. R. J Carman, 2001 *J. Phys. D, Appl.Phys.*, **34,** L1–L6
[5] K. Takaki and T. Fujiwara, 2001 *IEEE Trans. Plasma Sci.*, **29**, 518–523.
[6] James M Williamson, Darryl D Trump, Peter Bletzinger, and Biswa N Ganguly, 2006 J. Phys. D: *Appl. Phys.* **39** 4400–4406.
[7] OOPIC-Pro, Tech X Corporation, http://www.techxcorp.com.

Experimental Investigation of Pseudospark generated electron beam

Niraj Kumar, D K Verma, J Prajapati, M Kumar, B L Meena, M S Tyagi, V Srivastava and U N Pal

Microwave Tubes Area, Central Electronics Engineering Research Institute (CEERI) (Council of Scientific and Industrial Research (CSIR)), Pilani, Rajasthan-333031, India

E-mail: niraj.kr@rediffmail.com

Abstract The pseudospark (PS) discharge is, however, more recently recognized as a different type of discharge which is capable of generating electron beams with the highest combined current density and brightness of any known type of electron source. PS discharge is a specific type of gas discharge, which operates on the left-hand side of the hollow cathode analogy to the Paschen curve with axially symmetric parallel electrodes and central holes on the electrodes. The PS discharge generated electron beam has tremendous applications in plasma filled microwave sources where normal material cathode cannot be used. Analysis of the electron beam profile has been carried out experimentally for different applied voltages. The investigation has been done at different axial and radial location inside the drift tube in argon atmosphere. This paper represents experimentally derived axial and radial variation of the beam current inside the plasma filled drift tube of PS discharge based plasma cathode electron (PCE) gun. With the help of current density estimation the focusing and defocusing point of electron beam in axial direction can be analyzed. It has been further confirmed the successful propagation of electron beam in confined manner without any assistance of external magnetic field.

1. Introduction

The Pseudospark discharge can produce the electron beam of high current density (>10^4A cm^{-2}), high brightness (up to 10^{12}Am^{-2}rad^{-2}) and narrow beam diameter. Interest in this high-quality, high-current electron beams has proved their potential applications in diverse areas such as electron-beam lithography and plasma processing techniques [1]. One of the major advantages for this type of gun is elimination of back-ion bombardment problem. It would add great advantage for high-power microwave tube community. Furthermore it can enhance the overall efficiency of microwave tubes.

Investigation has been done on the effect of seed electrons and physical dimension of hollow cathode on Pseudospark discharge [2,3]. Plasma electron gun is electron beam source where the extraction of electron is based on gas discharge. This kind of e-gun has been described in detail elsewhere [4,5]. To create the plasma one needs a discharge, such as, for example that used in ion sources. Thus there are two main components of the plasma cathode electron gun: a plasma generator based on hollow cathode pseudospark discharge and a beam extraction system. The discharge system for plasma cathode electron (PCE) guns does not have a hot filament or similar kind of thermionic solid cathode, and so it is ''filament less'' or ''cold cathode'' which is a primary distinction between a plasma cathode e-gun and a thermionic e-gun. Due to the absence of a hot electrode, the plasma cathode system is basically more reliable, with a longer lifetime, and can generate electron beams at much higher background gas pressure, even in the fore-vacuum pressure range. A characteristic feature of this type of discharge is the generation of distinctive electron beams, before and at the time of voltage collapse [6]. H.Yin and colleague [6] had developed Pseudospark plasma electron gun with a single hole channel for microwave applications. However, little work has been done on the analysis of electron beam profile inside the drift space in gaseous atmosphere. Our focus is to develop high energy, intense electron beams for high power microwave applications. In this paper, we present some results for analysis of electron beam profile generated by Pseudospark experiment using a single hole channel.

2. Experimental set-up

The initial study of electron beam production has been carried out on a single-gap Pseudospark system for a wide range of parameters, including applied voltage and gas pressures. The experiments has been performed for the single-gap Pseudospark sourced PCE-gun for electron beam generation. Figure 1 shows the schematic view of the experimental setup. The hollow cathode geometry has been used to design PCE gun. The hollow cathode is cylindrical in shape having a total height of 59.7 mm. The inner and outer diameter is 59.4 mm and 65.4 mm respectively. The thickness of the cylinder and the aperture size is 3 mm each. The anode dimensions are same as that of the cathode. The anode and cathode are assembled in a ceramic casing with 3 mm gap between them. A Circular disc (copper) of dia 30mm is kept at ground. With the help of current transformers (Model 110, Pearson Current Monitor) beam current at collector can be measured. The hollow cathode is connected to the dc power supply 25 kV, 1 mA through a 5 MΩ charging resistor. The anode is kept at ground. The PCE Gun has been evacuated using TMP (Varian Turvo V-301) and then refilled with argon gas in a controlled manner using mass flow meter (Matheson:8272-0453). There was no external guiding magnetic field applied to the drift space. The charging voltage was measured using a voltage probe (Tektronix P6015A). Measurements of the beam current were realized using circular rings where each ring are connected to wire which are passing through current transformer and further these transformers are connected to the digital oscilloscope (Tektronix DPO 4054) which synchronously displays the voltage and current waveform. The actual developed PCE gun is shown in Fig. 2.

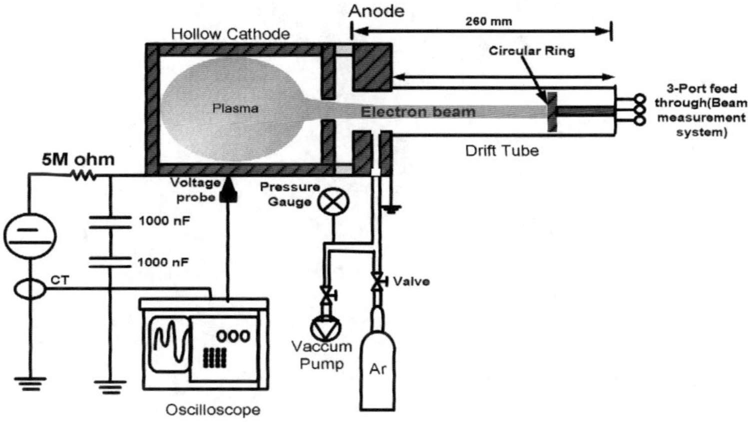

Fig. 1. Schematic view of Experimental set-up

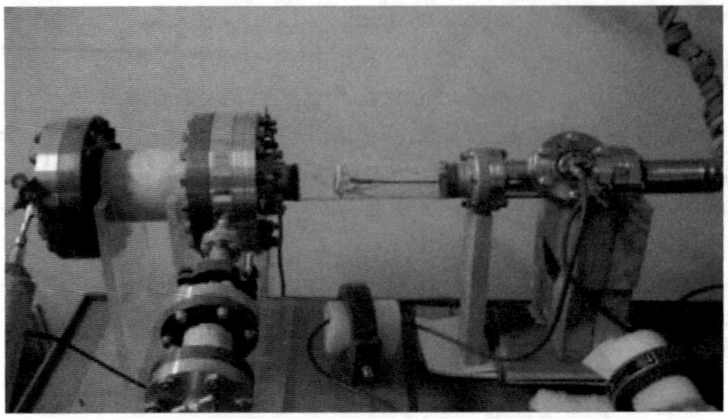

Fig. 2. Developed single gap PCE Gun

3. Results and discussions

Experiments have been performed for the analysis of beam current distribution inside the drift space of PCE gun. Initially the PCE gun has been evacuated up to ~10^{-6} mbar. Then argon gas is filled inside the pseudospark chamber in a controlled manner. The high voltage applied across the pseudospark chamber was then increased slowly until breakdown occurred. All the discharge took place nearly around 10^{-1}mbar. A graph has been plotted between different breakdown voltages and pressure which is shown in Fig. 3 at fixed gap between anode and cathode of 3mm. This curve further confirms the pseudospark nature of discharge. Real time waveform for beam current and discharge current has been plotted for 16kV in Fig. 4 which clearly demonstrate the pseudospark characteristics. A typical experimental trace for different breakdown voltages vs. beam current &discharge current has been reported in Fig. 5 at fixed location z=14.5cm inside the drift tube. Beam current and discharge current is found to be increasing with respect to applied voltage. Figure 6 represents the variation of beam current inside the drift space at two distinct locations which are start and end of drift space which is 15cm long. During propagation of beam inside the drift space, there is beam loss mainly due to recombination with ions, collision with walls etc. Analysis for axial and radial variation of electron beam profile has been carried out. The electron Beam is found to be focused inside the drift space [7]. Radial variation of beam current density at two different distinct axial locations (z=0 & 107 mm) for 20kV applied voltage is shown in Fig. 7. Due to the beam loss during propagation, reduction in current density has been observed.

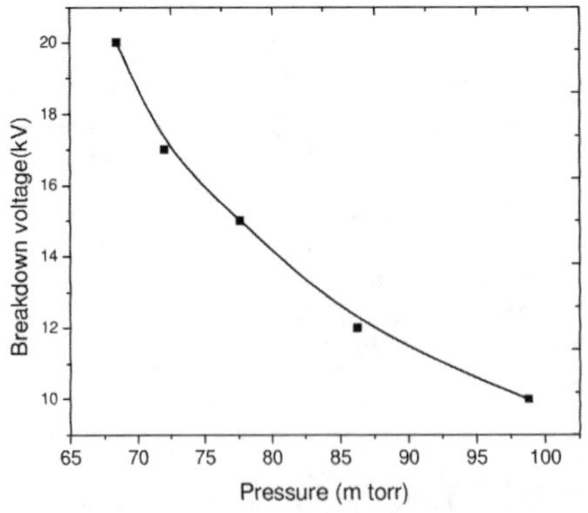

Fig. 3. Experimentally deduced breakdown voltage vs. applied pressure at (gap=3 mm, cathode aperture=3 mm) *Ref.* [7]

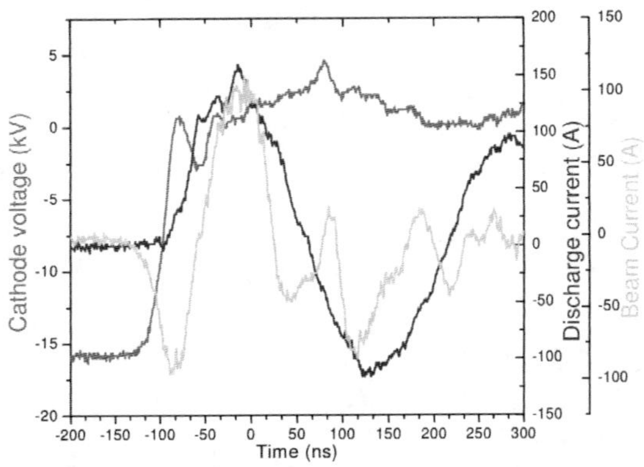

Fig. 4. Typical trace for beam current and discharge current @ 16kV

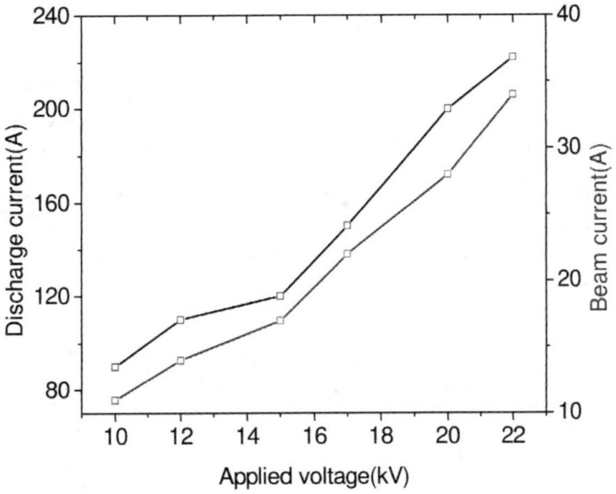

Fig. 5. Beam current and discharge current variation vs. applied voltage at z=14.5cm

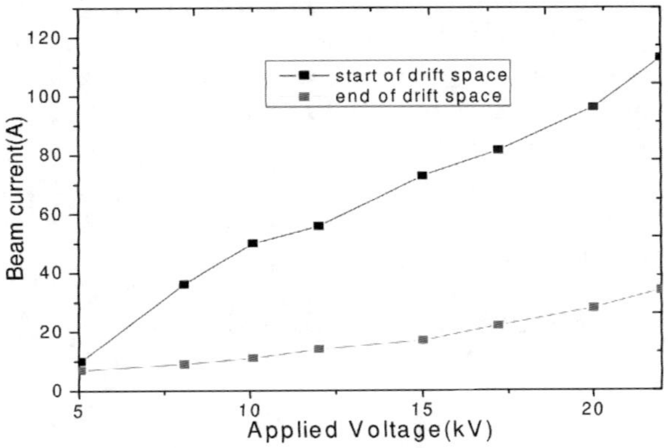

Fig. 6. Beam current vs. applied voltage

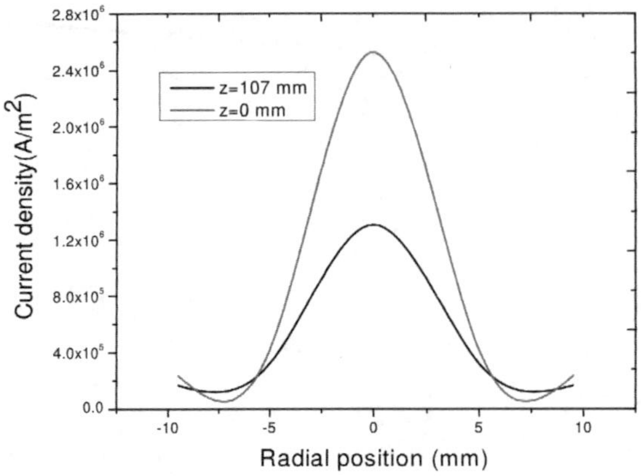

Fig. 7. Radial beam current density at 20kV

4. Conclusion

We have successfully designed and developed pseudospark based electron gun. The pseudospark discharge has been found to be a promising source of high brightness and high intensity electron beam pulses. The experimental investigation has been carried out for axial and radial variation of the beam current inside the drift space of PCE-gun. Analysis of the electron beam profile has been done experimentally for different applied voltages. It has been proved the successful propagation of electron beam in confined manner without any assistance of external magnetic field.

Acknowledgement

The work has been carried out under CSIR Network project (NWP0024). The authors wish to acknowledge valuable discussions with Dr. H K Dwivedi and Dr. Ram Prakash.

References:

[1] M. A. Gundersen and G. Schaefer 1990 *NATO ASI Series B* **219.**

[2] G. L. Vyas , B. L. Meena, U. N. Pal, M. Kumar, M. S. Tyagi, and A. K. Sharma,2010 *8th International Vacuum Electronics Sources Conference IVESC.*

[3] U. N. Pal, V. P. Dubey, M. K. Barik, V. Lamba, D. K. Verma, N. Kumar, M. Kumar, B. L. Meena, M.S. Tyagi and A.K. Sharma 2011 *XII IEEE International Vacuum Electronic Conference* .

[4] D. M. Goebel, J. M. Butler, R. W. Schumacher, J. Santoru, and R. L. Eisenhart,1994 *IEEE Trans. Plasma Sci.* **22**, 547–53.

[5] D. M. Goebel, R. W. Schumacher, and R. L. Eisenhart,1998, *IEEE Trans.Plasma Sci.* **26** 354-365.

[6] H. Yin, W. He, G. R. M. Robb, A. D. R. Phelps, K. Ronald, and A. W.Cross,1999, *Phys.Rev. ST Accel.Beams,* **2** 2 020701.

[7] Niraj Kumar, U. N. Pal, D. K. Verma, J. Prajapati, M. Kumar, B. L. Meena, M. S. Tyagi & V.Srivastava,2011 *J Infrared Milli Terahz Waves*,**32** 12.

Design and Characterization of X-Band Sheet Beam Klystron Cavity

A S Nirmala Devi*, A K Bandyopadhyay, L M Joshi, B Kumar and R Meena

Microwave Tubes Division, CSIR-Central Electronics Engineering Research Institute, Pilani, Rajasthan, India

*E-mail : nirmala@ceeri.ernet.in

Abstract. Sheet Beam Klystron (SBK) is a device employing rectangular/elliptical cross section beam. The barbell cavity for SBK has been designed providing uniform flat electric field profile across the width of the cavity. The loop coupling arrangement has also been designed using CST Microwave studio software tool and optimized for maximum return loss. The designed barbell cavity has been re-optimized after the insertion of the loop coupling to provide flat electric field profile across the width of the cavity. The cavity has also been fabricated and initial characterization has been carried out. The simulated results and the experimental results matches closely.

1. Introduction

Attaining high power at higher frequency regimes is still an active area of research in microwave tubes. The major limitation is that the device size scales down as wavelength. So devices which overcome these shortcomings using various techniques are been continuously studied. Sheet Beam Klystron (SBK) is one such device which is expected to give good results in both high and low frequency regimes which employs a rectangular cross section beam instead of circular. It is similar to arranging several cylindrical beams in linear fashion. Thus for the given beam voltage and beam current density since beam area is increased, the beam current is increased proportional to the beam width. Thus higher power is achieved at reduced input voltage. This futuristic device has wide range of applications like directed energy weapons, linear accelerators and muon accelerators [1-3]. Despite many advantage, there are several critical issues in this device such as high aspect ratio sheet electron beam generation and need for design with high transmission efficiency. Also because of diocotron effect the breakdown of the beam might occur [4,5]. Also the drift tube can propagate all the TE modes and hence there is a serious problem of feedback and oscillations. Due to the large lateral dimension of the beam SBK requires large drift tubes and overmoded cavities. R/Q of the cavity is also much less than that of the conventional Klystron [6]. Major objectives in the cavity design of sheet beam tubes are a flat field across the whole beam width, a large (unloaded) shunt impedance and reduction of transit time effect [7].

2. Concept of the barbell cavity

Due to the usage of high aspect sheet beam, the cavity used for such a beam should be specially designed to provide uniform field across the overall width of the beam. The barbell cavity is one such cavity to provide the desired uniform flat field profile. The barbell cavity comprises of three sections. The centre one is a simple rectangular waveguide operated closed to cut off with a quarter wave section on either side of the waveguide as shown in figure 1. As the central waveguide is operated close to the cut off, the group velocity of the wave is nearly zero which maintains the electric field profile to be uniform along the overall width of the cavity. To terminate the open waveguide, the quarter wave sections are used. The quarter wave section transforms the short at the wall to open at the waveguide ends [8].

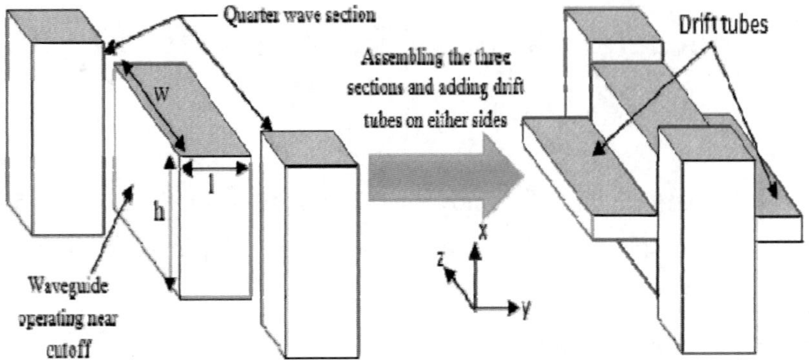

Fig. 1. Barbell cavity showing the middle waveguide section and the two quarter wave section along with the drift tubes

The height of the quarter wave sections are greater than that of the middle waveguide section but has the same length as the waveguide section. The width of the quarter wave sections are not exactly λ/4. The greater the height difference the greater is the width required to get the flat field profile.

3. Design of the X-band SBK cavity

The operating frequency is chosen in the X-Band at 11.4 GHz. As we know the middle section has to be near cut off. So

$$\lambda = \frac{c}{f} = \frac{3 \times 10^{8}}{11.4 \times 10^{9}} = 26.32\,mm$$

For a wave to propagate in a wave the broader dimension 'h' should be,

$$h \geq \frac{\lambda}{2}$$

Hence for 11.4 GHz, $h \geq 13.16$

for the waveguide to operate close to cut off the broader dimension 'a' is chosen as 14 mm. In this way we have h=14 mm.

The width of the waveguide depends on the beam width along with the needed clearance. The length of the middle section is equivalent to the gap in the conventional re-entrant cavity. So the length depends on the wavelength of the RF to be amplified. To get better coupling coefficient we need the length in which the beam is exposed to the RF field should be less than □/2. The cavity length 'l' has been chosen accordingly.

The design of the quarter wave sections are done by using CST MWS [9]. The length of the quarter wave section should be same as that of the middle waveguide section length 'l'. Since the middle section height h = 14 mm, the quarter wave section height should be greater than 14mm. As already mentioned the higher the height of the quarter wave section larger is the width needed to get the flat field. The height and width of the quarter wave sections has also been optimized with the 3D simulation tool as shown in figure 2.

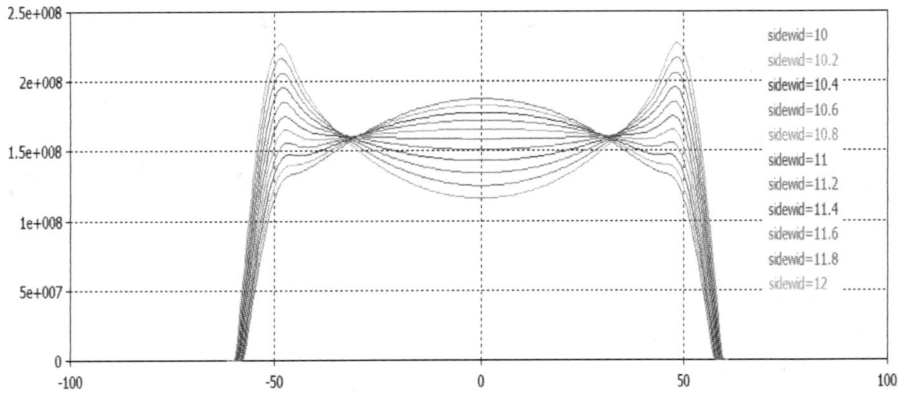

Fig. 2. Field profile along the cavity width for varying widths of the quarter wave sections

The parameter 'sidewid' represents the width of the side quarter wave section. Thus for a width of 11 mm the desired uniform field profile has been achieved.

4. Design of loop coupling for the X-Band barbell cavity

The major parameters of the loop coupling are

- Outer diameter (OD) of the inner conductor forming the loop
- Inner diameter (ID) of the outer conductor radius
- The radius of the loop
- The length projection of the inner conductor into the cavity before forming the loop

They are given in the following figure 3.

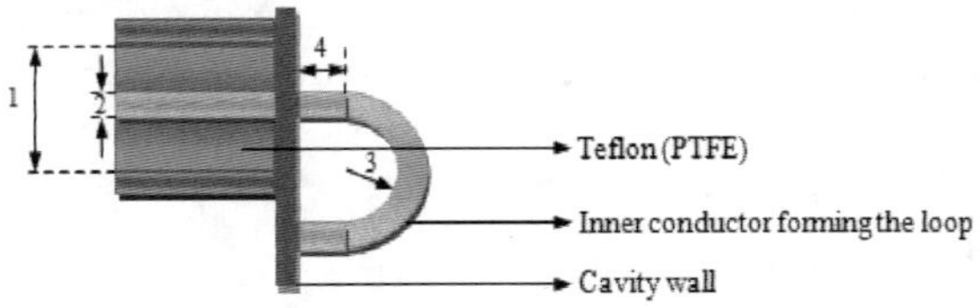

Fig. 3. Coupling loop

All the loop dimensions are optimized for the best return loss at the operating frequency using CST MWS. The final optimized dimensions are given in table 1.

Table 1. Optimized loop parameters

Loop conductor radius	0.62 mm
Loop radius	2 mm
Loop inner length	2.5 mm
Dielectric used	Teflon
Dielectric constant	2.08
ID of the outer conductor	4.13mm

5. Electric field profile across the cavity

The next step is to insert the loop in the cavity designed. So the cavity is redesign with the loop. Then the eigenmode analysis of the cavity is carried out and the field profile across the cavity width is found to be no longer uniform. The field profile in the presence and absence of the loop is given in figure 4. So it is first necessary to recover the distorted field profile.

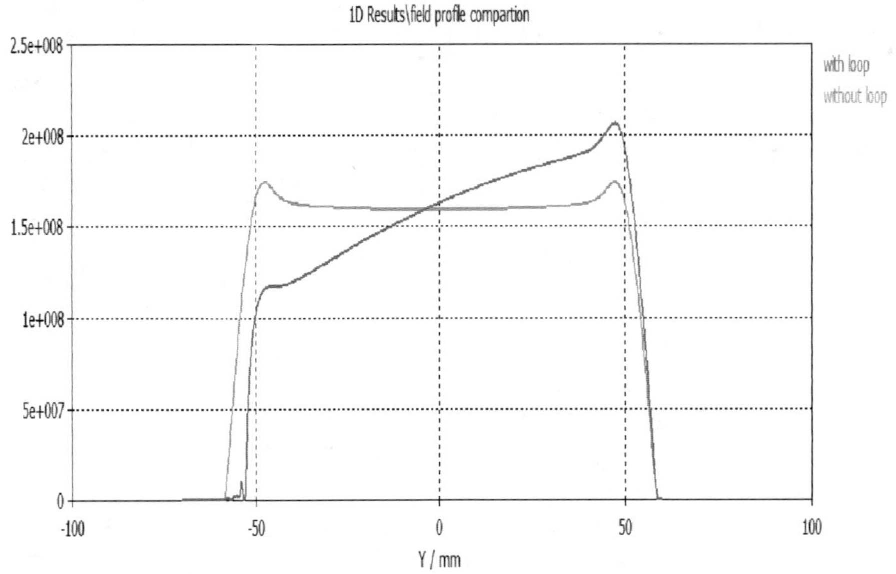

Fig. 4. Field profile across the cavity width with and without the loop

By slightly altering the width of the quarter wave section in which the loop is placed, the flat uniform electric field profile can be regained back. After many iterations in the CST MWS, the final dimensions of the barbell cavity including the loop and losses of the dielectric in the loop is given in the table 2

Table 2. The cavity parameters for various cases

Parameters	Dimensions
Beam height	6 mm
Beam width	95 mm
Cavity height	14 mm
Cavity length	10 mm
Left quarter wave section height	16.5 mm
Right quarter wave section height	16.5 mm
Left quarter wave section length	10 mm
Right quarter wave section length	10 mm
Left quarter wave section width	11.54 mm
Right quarter wave section width	11.5 mm

6. Fabrication of the designed cavity

The central barbell structure is fabricated and is shown in the figure 5(a).

Fig. 5(a). Fabricated barbell structure

The fabricated drift tubes along with the barbell structure is shown in the figure 5(b).

Fig. 5(b). Fabricated Barbell cavity with drift tubes on either sides

7. Characterization of the fabricated barbell cavity

The cavity is excited using the copper antenna and is inserted through one of the drift tubes. The experimental set up is shown in the figure 6. The probe is inserted in a hole offset from the centre. This is to excite all the modes in the given frequency range. The depth of insertion also affects the resonant frequencies of the cavity mode. The lighter the insertion better is the accuracy of the observed results. The deeper the insertion the modes will appear prominently but the frequency of the corresponding modes will be shifted. This is due to the fact that the exciting probe acts also as a perturbing antenna which affects the field pattern considerably.

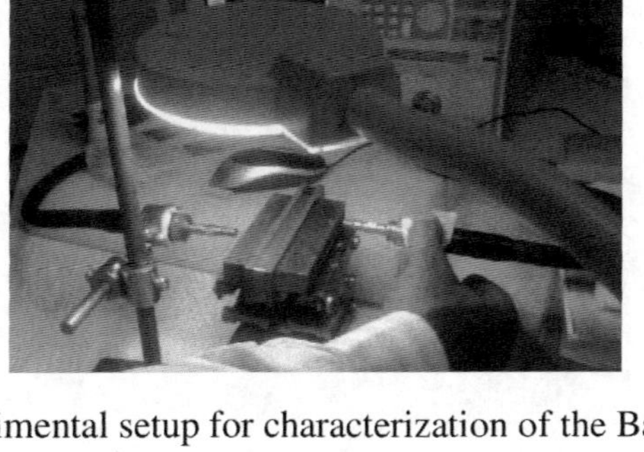

Fig. 6. Experimental setup for characterization of the Barbell cavity

The Vector Network Analyzer is used for characterization of the barbell cavity. The measured and simulated results are tabulated in table 3

Table 3. The measured and simulated results

MODE NO:	CST MWS simulated results (GHz)	Experimentally measured results (GHz)
1	11.4257	11.4250
2	11.4853	11.47
3	11.712	11.7388
4	12.1412	12.128
5	12.684	12.685
6	13.3503	13.351
7	14.1738	14.1738

To identify the modes one more copper wire as perturbing object has been used, and is inserted in various holes of the cavity one by one. If the wire is at the field maximum of the excited mode, then there will be severe disturbance or change in the corresponding resonant frequency. But if the wire is at the field minimum, the change or shift in the resonant frequency of the corresponding mode is very less.

8. Conclusion and future work

The barbell cavity for SBK has been designed and fabricated. The initial experimental verification of the modes inside the cavity was done and the results quite matches with the simulation. The future work includes the insertion of the loop followed by experimentally measuring the R/Q of the cavity along the overall width of the cavity.

Acknowledgement

The authors thank the Director, CEERI, Pilani for his constant encouragement and permission to publish this paper.

References

[1] Benford J 2008 IEEE Trans. Plasma Sci. **36(3)** 569
[2] Barker R J and Schamiloglu E 2001 *High Power Microwave Sources and Technologies* (III series, Inc., New York: IEEE press)
[3] Booske J H 2008 *Phys. Plasmas* **15** 055502-1
[4] Booske J H, McVey B D and Antonsen T M, Jr. 1993 *J. Appl. Phys.* **73(9)** 4140
[5] Panda P C, Srivastava V and Vohra A 2011 *Indian J. Pure Appl. Phys.* **49(6)** 410
[6] Solyga and H.Henke April 1998 Two-Dimensional design of a low voltage mm-wave sheet beam klystron *ITG Conference: Displays and Vacuum Electronics*
[7] G. Caryotakis 2001 A Sheet Beam Klystron Paper Design *5th MDK Workshop* CERN, Geneva.
[8] D. Yu and P.B. Wilson May 1993 Sheet-beam klystron rf cavities *Proceedings of 1993 IEEE Particle Accelerator Conference* Washington D.C p.2681-3
[9] CST Particle Studio, CST Studio Suite^{TM} 2010, Darmstad, Germany

Journal of Physics: Conference Series **390** (2012) 012075 doi:10.1088/1742-6596/390/1/012075

A Large High Vacuum Reaction Chamber for Nuclear Physics Research at VECC, Kolkata

S Kundu, S Bhattacharya[1], J K Meena, T K Ghosh, T Bhattacharjee, P Mukhopadhyay, C Bhattacharya, T K Rana, K Banerjee, G Mukherjee, S R Banerjee, D L Bandyopadhyay, M Ahammed and P Bhattacharya

Variable Energy Cyclotron Centre, 1/AF, Bidhan Nagar, Kolkata - 700 064, INDIA

Abstract. A large, segmented, horizontal axis, reaction chamber (SHARC) has recently been fabricated, installed and integrated with the beam line in the VECC superconducting cyclotron (SCC) experimental area. It is a cylindrical, three segment, stainless steel chamber of length 2.2 m, diameter 1 m. Two pairs of parallel rails have been provided internally for placement of the target assembly and detector systems within the chamber. The whole target assembly can be placed anywhere on the rail to facilitate optimum flight path. The nominal vacuum of ~$1X10^{-7}$ mbar has been obtained in ~8 hrs by means of two turbo molecular (1000 l/s) and two cryo pumps (2500 l/s) backed by mechanical pumps. The whole vacuum system as well as the target positioning (vertical and rotational movements) operations are fully automated with manual override option; both are monitored and controlled locally as well as remotely through the local and remote control units providing real time status display.

1. Introduction

The K500 superconducting cyclotron (SCC) at the Variable Energy Cyclotron Centre (VECC), is likely to open up new areas of intermediate energy experimental nuclear physics research and vigorous activities are on to develop new, state-of-the-art experimental setups to facilitate high quality research. Vacuum reaction chamber is an essential component of any accelerator based experimental setup in nuclear physics and the design of the chamber depends upon the experimental programme [1]. It may either be a very specialised chamber for dedicated experimental setup [2, 3, 4], or it may be a versatile general purpose reaction chamber [5] which has sufficient flexibility to incorporate various types of detector systems for different types of experiments. The present chamber belongs to the second category, and the aim here was to design a multipurpose reaction chamber which will cater to the needs of different types of experiments involving large number of charged particle detectors of various kind forming arrays inside the chamber, along with the provision to place gamma and neutron detectors at suitable locations outside the chamber. With this aim, a large, Segmented, Horizontal Axis, Reaction Chamber (SHARC) has recently been fabricated, and then installed and integrated with the beam line in the SCC experimental area. SHARC has been designed in-house at VECC and fabricated by M/s. Vacuum Techniques, Bengaluru. Here we report salient features of SHARC, which will be used mainly for charged particle based experimental studies of intermediate energy heavy-ion collisions, using the accelerated ion beams (~5-80 MeV/A) from SCC.

[1] Email for correspondence: saila@vecc.gov.in

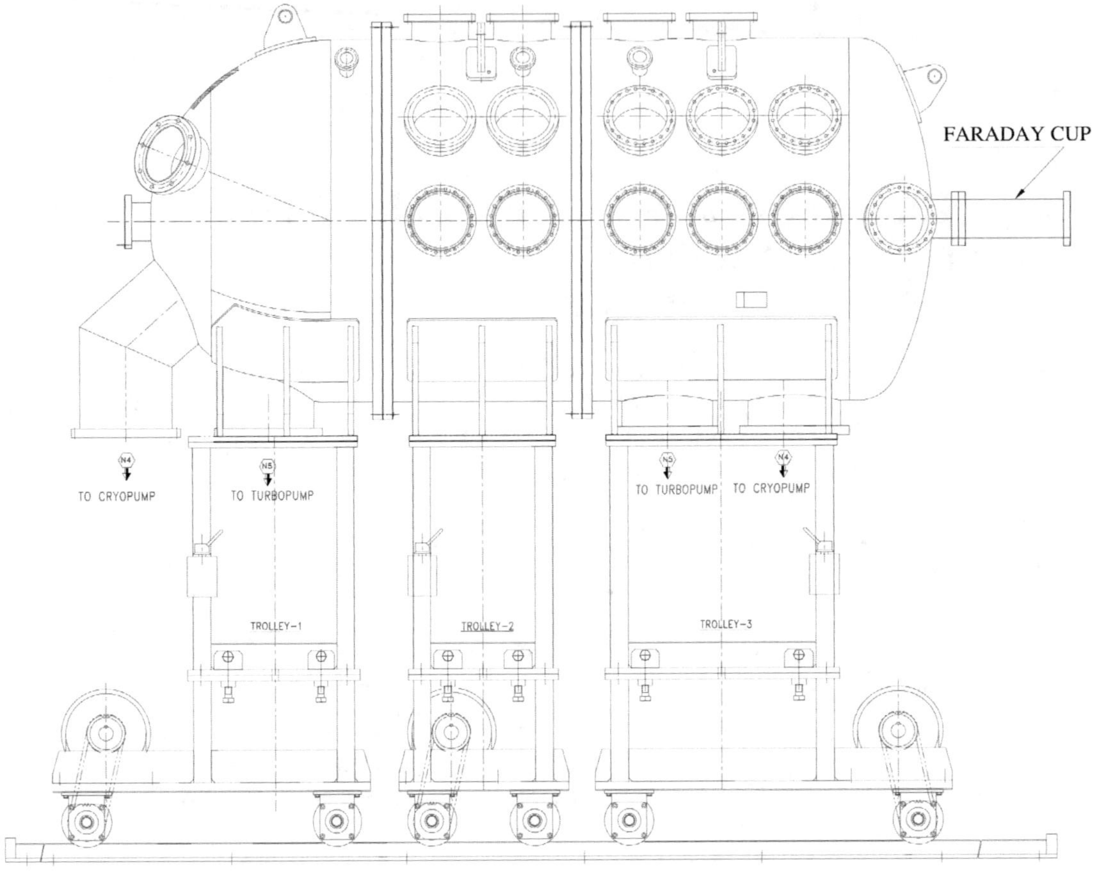

Fig. 1. Schematic diagram of SHARC without vacuum pumping system.

2. Basic design parameters

SHARC was basically designed as the primary facility to be used for experiments in SCC beam line I. The dimensions of SHARC were optimised to have maximum flexibility for versatile experimental setups; in particular, optimum shape was decided to be cylindrical with its axis coinciding with the beam axis to avail maximum possible flight path for time of flight measurements. Other salient design considerations were,

i. The chamber should be suitably segmented for easy access inside the chamber. Each segment should have independent movement arrangement.
ii. The chamber vacuum control system should have maximum flexibility. The vacuum system should have both auto and manual modes of operation with dynamic display of complete status. There should be facility for remote monitoring of vacuum status with emergency shutdown option.
iii. Target ladder should have the provision to be placed any position inside the chamber along the beam axis. The vertical and rotational movements of the ladder have to be controlled and monitored in auto / manual mode.
iv. Both vertical position and rotational angle of the ladder should be accurately measured. The complete target ladder control and monitoring should also be available at the remote station.
v. The chamber will have a Faraday cup of sufficiently large aperture to stop the beam and measure the beam current .

3. Details of SHARC

A schematic drawing of SHARC is displayed in Fig. 1 and the photograph of the chamber installed in its present position (SCC beam hall – 1) in Fig. 2.

3.1. Mechanical details

SHARC is cylindrical, three segment, stainless steel (SS 304L) chamber of length 2.2 m, diameter 1 m and wall thickness ~10 mm; the front (beam-entry) end is hemispherical in shape of radius 500 mm and the rear end is elliptical dish (2:1) shaped. The segments are mounted on separate support structures which rest on external rails such that each segment can move independently on the rails by automatic gear-motor control mechanism having built-in limit switch locking facility with manual override option. All three segments (and the rear part in particular) may be rolled back on rail to open up the chamber to give accessibility for the installation of users' equipments inside.

Fig. 2. SHARC installed in SCC beam hall-1

There is arrangement to manually align the chamber with the beam line axis precisely in horizontal/ vertical/ rotational degrees of freedom. Two pairs of rails are provided within the chamber for mounting and placement of the target ladder system and the user designed detector assemblies at any position within the chamber. A generalized detector mounting table (made of Aluminum alloy 6061-T6) with precision alignment mechanism on manually movable stands with locking arrangement on rails is also provided as default arrangement. There is no rotating arm inside; users are encouraged to fabricate their own detector stands as per requirement. Two glass windows are kept for viewing the target and detector positions inside.

3.2. Target assembly

The target ladder system has been so designed to take the maximum advantage of the full length of the chamber. The whole assembly is mounted on one pair of internal rails and may be placed at any position within the chamber to optimize the flight path. The target assembly includes a ladder that can hold six targets at a time in a column (see Fig. 3). Target ladder can be rotated with angular precision 0.1^0 and it can be moved vertically with accuracy of 0.1 mm. Ladder movement mechanism is constituted of Programmable Logic Controller (PLC) driven, remote / locally operated, vacuum compatible stepper motors (2 nos.), which are used to execute (i) up/down and (ii) rotational (360^0, both clockwise and anti-clockwise, along the vertical target ladder axis) movements of the target ladder. Target ladder movements (rotational/ vertical) can be measured through PLC.

Fig. 3. SHARC target assembly.

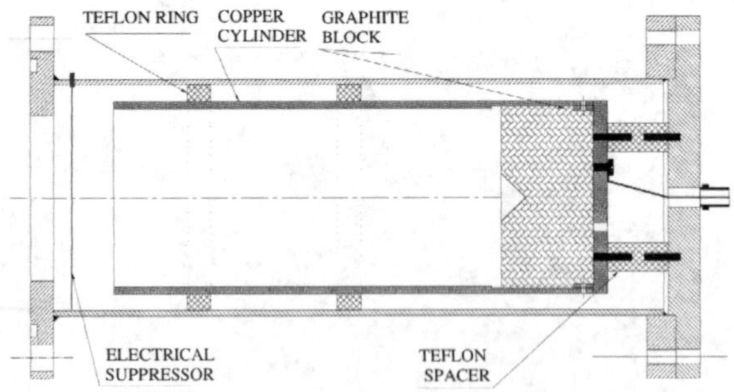

Fig. 4. Schematic drawing of Faraday cup.

3.3. Input / Output ports

A number of ports (24 nos., size – 25 cm dia) are provided on different locations of the chamber body (see Fig. 1 & 2) to take the detector and other readout signals out of the chamber. Flanges with standard LEMO connectors as well as indigenously designed and fabricated flanges with Flat Ribbon Connector (FRC) connectors are used for this purpose.

3.4. Faraday cup

A Faraday cup has been fitted with the chamber (not shown in Fig. 1) for the measurement of incident flux. The schematic drawing of the cup is shown in Fig. 4. The Faraday cup has large aperture of diameter 75 mm to collect the beam transmitted through the target. A Graphite block has been used to stop the beam. An electron suppressor ring has also been provided to repel back the electrons emitted from target as well as the secondary electrons from Faraday cup.

4. Vacuum pumping system

To achieve optimum vacuum performance, all inside surfaces were given smooth granular finish and were finally electro-polished with bright finish. Clean vacuum ~1×10^{-7}mbar is achieved by means of two turbo-molecular (1000 l/s, Varian 1001 Navigator) and two cryo pumps (2500 l/s, CTI-Cryogenics) backed by mechanical pumps (Adixen ACP 40, $37m^3$/hr.). The complete vacuum pumping system is shown schematically in Fig. 5. Typically, nominal vacuum ~$1X10^{-7}$ mbar is obtained in 8 hrs.

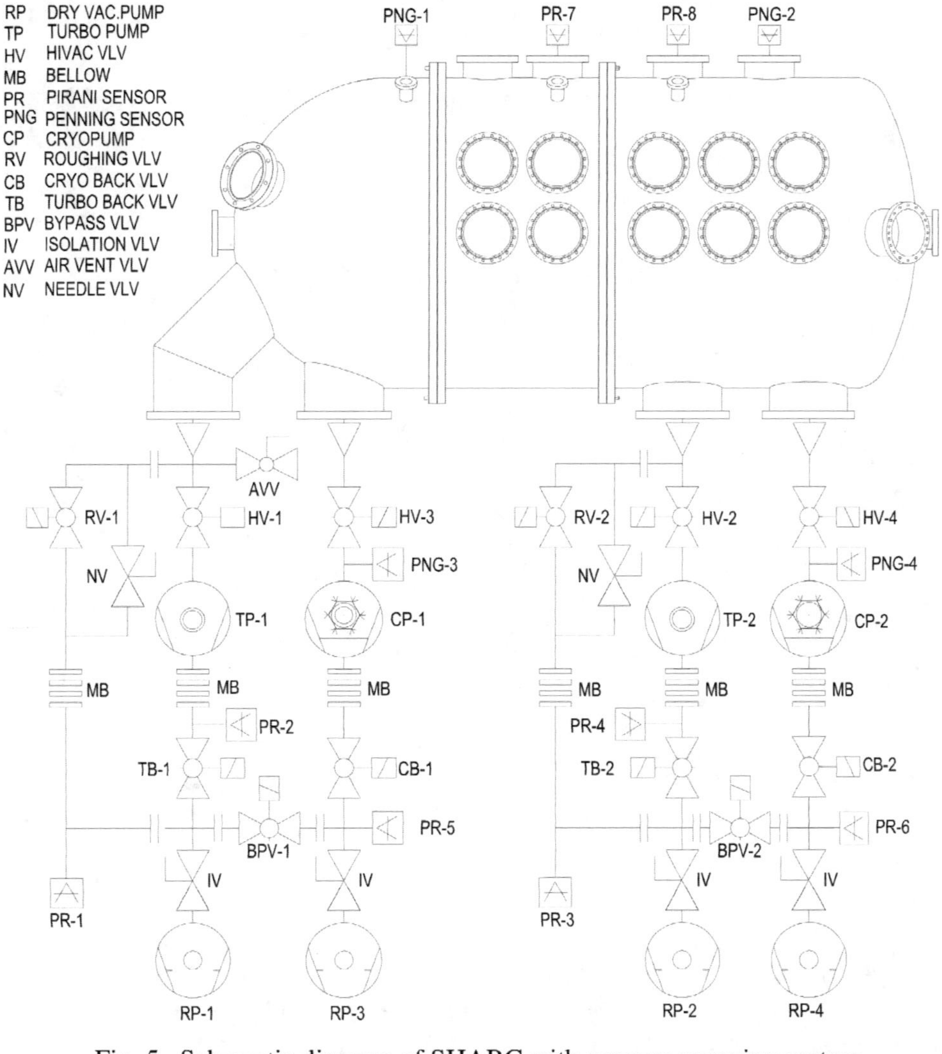

Fig. 5. Schematic diagram of SHARC with vacuum pumping system .

The whole vacuum operation sequence is pre-programmed and fully automated by using a PLC. A typical vacuum pumping sequence is displayed in Fig. 6. The vacuum operation is controlled locally as well as remotely through compact control units, which provide continuous display of the status of the full pumping sequence for monitoring purpose. There is also a provision for slow manual pumping under demanding experimental conditions.

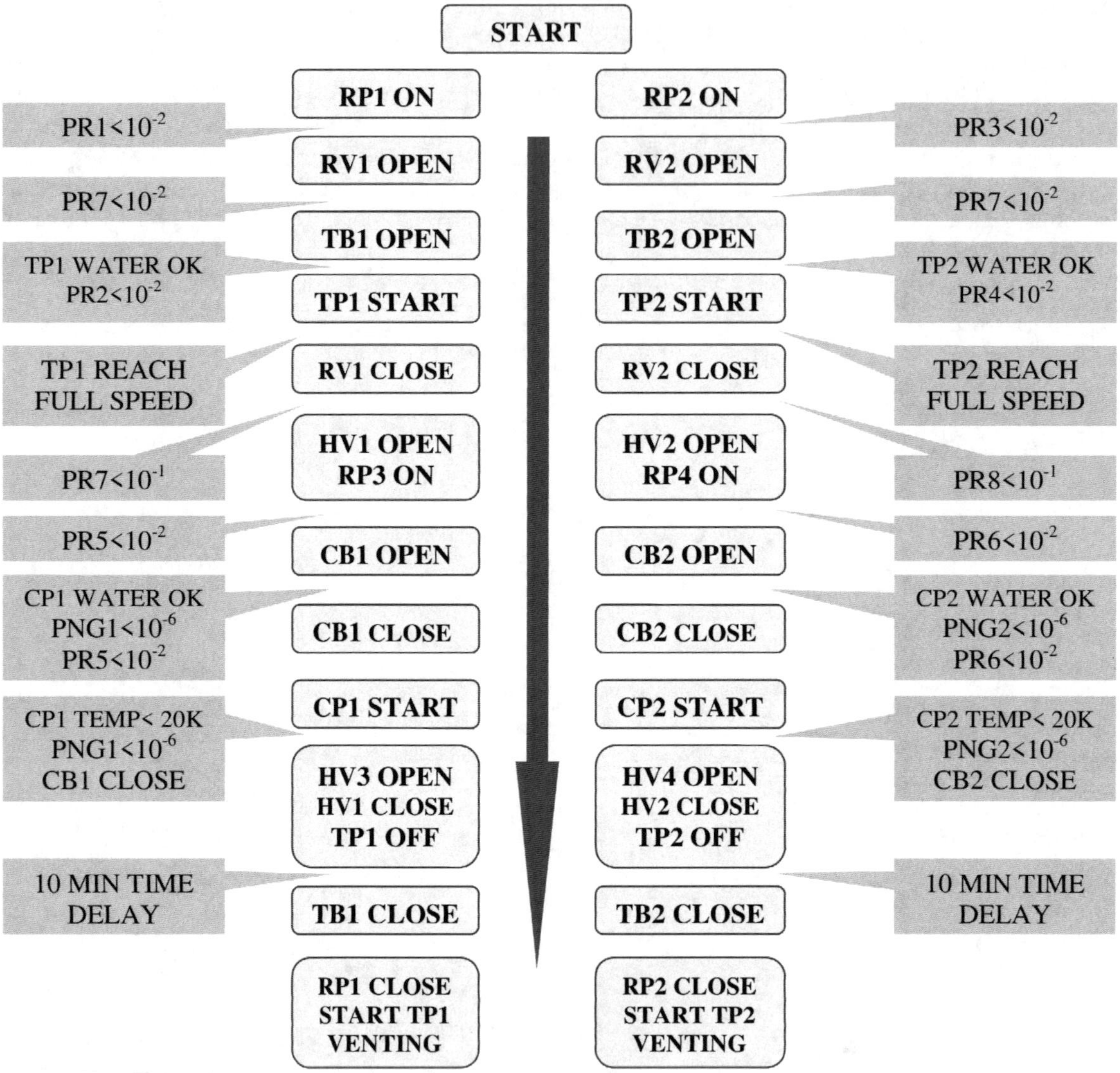

Fig. 5. A Typical pumping sequence. Pressure is in mbar.

5. Summary

The large multipurpose reaction chamber SHARC, planned for nuclear physics experiments with SCC, has been successfully commissioned and integrated with the 0^0 beam line in SCC beam hall-1. The fully automated vacuum system and the movable target assembly, which are unique, result in ease of operation and offer more experimental flexibility for the users.

References

[1] S. Bhattacharya, Pramana, 75 (2010) 305.
[2] http://www.iuac.res.in/research/nuclear_physics/GDA/GDA_main.html
[3] http://www.physics.wayne.edu/~emiller/Positron/experimentrt.html
[4] http://brahma.iuc.res.in/~iuc_cc/veccinga3.html
[5] J. L. Yntema and H. W. Ostrander, Nucl. Instr. And Meth. 16 (1962) 69.

Design and development of collector for C-band 250 kW CW Klystron

Suman Baloda, O S Lamba, Meenu Kaushik, Richa, Prachi Bansal, Kumud, Pradeep, D Kant, LM Joshi

Microwave Tubes Area, CSIR - Central Electronics Engineering Research Institute, Pilani, Rajasthan-333031

Email: osl@ceeri.ernet.in

Abstract. The paper presents the design and development of collector for C-band 250 kW high power klystron. The design criteria for the collector assembly is selection of material, vacuum and high temperature compatibility, proper electron beam dispersion, minimum back scattering of electrons and thermal design for proper cooling at high power dissipation. All these aspects have been discussed for collector development in details. The collector has been designed in TRAK and then beam propagation has been analyzed in MAGIC 2D software. The thermal simulation has been done using ANSYS 11.0 (multi-physics). The outer surface of the collector has been grooved to facilitate its proper cooling. Design results are presented for water cooling with different flow rates and channel dimensions. OFHC copper material is chosen for collector which is suitable for vacuum and hydrogen brazing operations and good thermal properties for efficient cooling.

1. Introduction

Klystron is a high power microwave amplifier, which operates under ultra high vacuum of order 10^{-8} torr or better. Klystrons are widely used in radar communication, material processing and particle accelerator for medical, industrial and scientific applications. Klystron mainly consists of electron gun, RF cavity stack, RF window, collector and electromagnet. Each of these sub assemblies is separately designed, fabricated and evaluated. The paper mainly focussed on the design and development of collector. The function of the collector is to collect spent electron beam coming out of the tube and effectively dissipate the heat generated in it. The electron beam that acquires kinetic energy through the acceleration in the gun delivers part of its energy to RF field in the process of interaction in the RF section. The final energy exchange takes place in the output cavity. The collector is placed next to output cavity. The energy still left the beam which depends on the efficiency of RF interaction is allowed to be dissipated on the collector resulting in heating of latter. The collector is designed to remove the heat efficiently thus produced. Also electrons impinging the collector surface may produce secondary electrons which under favourable conditions may travel towards RF interaction region resulting in instabilities & other undesirable effects. It therefore is necessary to design the collector so as to trap the secondary electrons in the collector itself.

The shape of the collector surface generally has a short tube at the entry, a more strongly divergent conical part near the entry into the collector & more extended conical section at the end. The following considerations should be taken into account to design a collector for klystron:

- The thermal conductivity of the collecting cylinder should be high.
- Diameter of the collecting cylinder should be small compared to its length so that the probability of secondary electrons finding their way out is low.
- The material of which collector is made should be a poor secondary emitter.
- The beam should be diverged by shielding the collector from magnetic field used to focus the beam in the interaction region

The aspect ratio of the collector is chosen so as to have largest surface area to dissipate maximum heat by increasing its surface area. Sometimes the surface is made ducted to increase the effective surface area. A forced liquid cooled collector has been designed and developed to dissipate 500 kW power. In an ideal collector, cooling technique should be implemented such that heat generated inside the collector would be removed as fast as possible to prevent temperature rise of collector beyond a safe limit. Also, it should have low weight, small size, and proper internal shape so that secondary electrons do not travel back in the drift region.

1.1. Specifications of this particular klystron

- Operating Frequency 5 GHz
- Output Power 250 kW CW
- Beam Voltage 60 kV
- Beam Current 10 Amp
- Focusing Electromagnet
- No. of cavities 6
- Efficiency 40 % (min.)
- Gain 45 dB (min.)
- Cooling Water

2. Electrostatic Design Approach

The electrostatic design optimized through electron beam dispersion in the collector. The electron beam is launched from cathode with the help of all the optimized beam parameters. Various effects like beam scattering, scalloping and beam overlapping are observed. The electron beam profile observed for 10 KeV to 70 KeV beam energy and beam was uniformly distributed over the surface. The collected power in the collector is calculated. The collector is again simulated in MAGIC2D for validation and beam profile is further analysed. When beam propagates into the collector, secondary emission and back transmission of electrons into the RF section are the major issues of concern. The results of beam dispersion are shown in fig 2.

2.1. Derived Beam Parameters

- Beam Radius 3.00 mm
- Beam current 10 A
- Beam voltage 60 kV

Fig. 1. Beam profile in gun and collector in TRAK

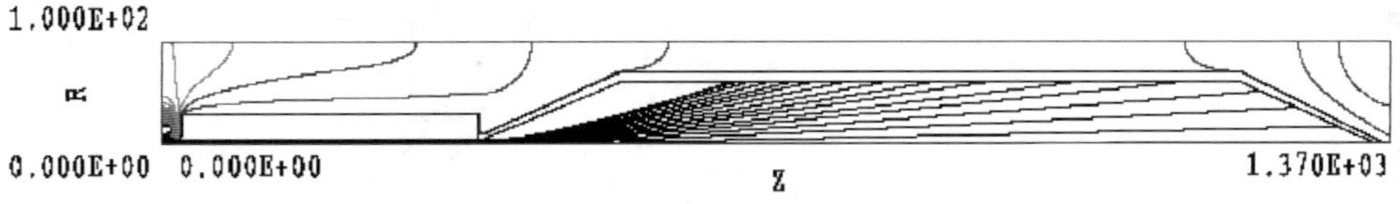

Fig. 2. Electron beam profile in collector

3. Thermal Analysis of Collector

A forced water cooled collector has been designed to dissipate 500kW power. Physical dimensions have been determined after analyzing the temperature profile of the collector using ANSYS code. The temperature rise of the surface calculated for 100 l/min. to 400 l/min. flow rates. The desired flow rate is optimised for safe limit of temperature rise of collector. The temperature analysis is carried out for plain as well as grooved surface.

Heat transfer occurs between a moving fluid and solid surface. The rate of heat transfer between a surface and a fluid is given by: $h = \{K_w/D_H\}.N_u$
where K_w is Thermal Conductivity ($Wm^{-1}K^{-1}$),
D_H is Hydraulic diameter (m) and N_u is Nusselt number.
$N_u = 0.023\ R_e^{0.8}\ .\ P_r^{n}$
$P_r =$ Prandtl number
$R_e =$ Reynolds number
n = 0.4 for heating (wall hotter than the bulk fluid) & 0.33 for cooling (wall cooler than the bulk fluid)
$R_e = D\rho v / \mu$
Typical value of heat transfer Coefficient h in $Wm^{-2}K^{-1}$
For different condition of the flow different values of h are obtained and are given as input to the ANSYS and temperature distributed on geometry surface is obtained as shown in Table1.

3.1. Dimensions of Collector

• No. of Grooves	110
• Output Diameter	14 cm
• Input Diameter	12 cm
• Collector length	85 cm
• Total Surface Area	9021.27 cm^2
• Collector Dissipation	500 kW

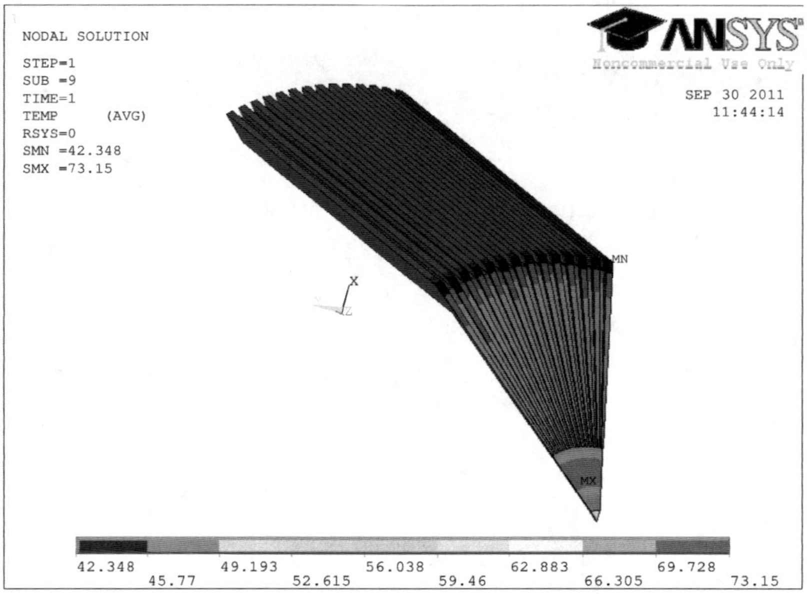

Fig. 3. Temperature profile of collector

Table 1 Simulated temperatures with varying flow rates.

Water Flow (lpm)	Applied heat flux (w/m^2)	Cooling coefficient (w/m^{2}K)	Simulated temp.(^{0}C) in ANSYS
100	554245.68	3502.34	67.31-116.63
200	554245.68	6098.37	48.84-85.56
300	554245.68	8435.27	42.34-73.15
400	554245.68	10617.89	39.03-66.20

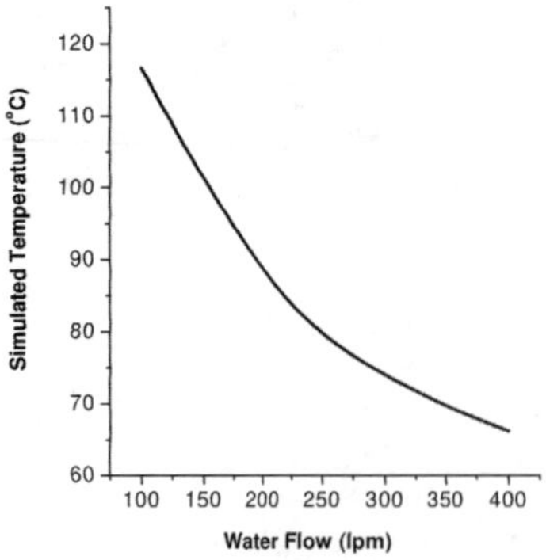

Fig. 4. Plot of water flow vs simulated temperature

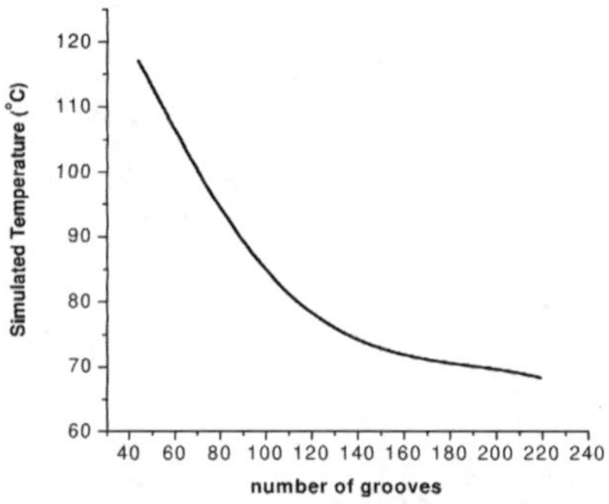

Fig. 5. Plot of numbers of grooves vs simulated temperature at 500 kw power dissipation

4. Temperature variation with grooves

The outer surface of collector grooved to increase the surface area. The number of grooves increased from 44 to 219. It is observed that temperature fast decreased with increase number of grooves but after 110 grove minor changes observed in temperature as shown in fig. 5.

5. Engineering design of collector

The simulated dimensions of collector converted to the engineering drawing of collector parts for workshop fabrication. The cylindrical collector both ends are made conical to avoid back scattering of electrons after hitting the collector surface. All parts are fabricated separately and brazed in hydrogen furnaces to make vacuum tight joints. The collector assembly leak checked for leak rate of 10^{-11} torr litre/sec.

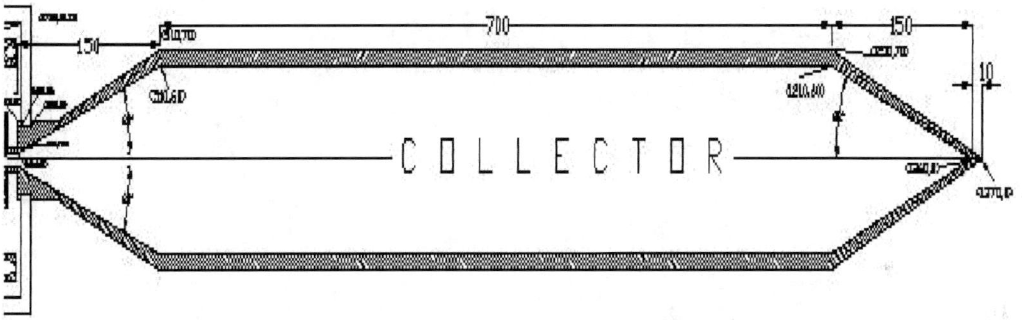

Fig. 6. Drawing of collector

Fig. 7. Fabricated Plain collector assembly

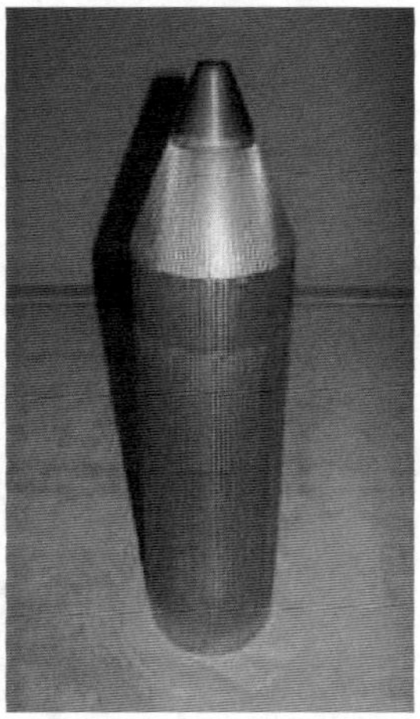

Fig. 8. Fabricated 110 Grooved Collector

6. Conclusion

The electrostatic, thermal and mechanical design of collector assembly with cooling system for 250 KW C band klystron has been done successfully. Thermal analysis of collector has been done using ANSYS code and we have got the surface temperature of the collector to be 42.34 ^{0}C -73.15 ^{0}C which is quite easy to maintain. Thermal simulation results obtained from ANSYS clearly indicates that the grooved surface collector of diameter 14 cm and length 85 cm and water flow rate of 300 lpm kept the temperature in safe operating limit. The fabrication of collector is under progress.

Acknowledgement

The authors are thankful to Dr. Chandra Shekhar, Director CEERI and Dr SN Joshi, project advisor, Microwave Tube Area for permission to present this work and their encouragement.

References

[1] "*Design Aspects of a 1 KW Tunable D/E Band Klystron*", *Journal of the IETE*, Vol.39, Nov.-Dec.1993, pp345-350.

[2] Allan W. Scott, "Cooling of Electronic Equipments", John Willey and sons.

[3] Frank P Incropera, *"Introduction to Heat Transfer"*, John Willey and sons.

[4] ANSYS user manual, supplied with software.

Journal of Physics: Conference Series **390** (2012) 012077 doi:10.1088/1742-6596/390/1/012077

Studies of Adsorption Characteristics of Activated Carbons in between 4.5 to 10 K for Cryopump Applications

V. Krishnamoorthy[1], Satya Swarup Udgata[1], Vijai Shankar Tripathi[2], Ranjana Gangradey[3], Srinivasan Kasthurirengan[4] and Upendra Behera[4]

[1] I-Design Engineering Solutions Ltd., Ubale Nagar, Wagholi, Pune 412207, India
[2]HEG R&D Centre, Mandideep, Raisen Dt. 462 046, India.
[3] Cryopump Group, Institute of Plasma Research, Gandhinagar, Gujarat 382428, India
[4] Centre for Cryogenic Technology, IISc, Bangalore 560012, India.

E-mail: svkrishnamurthi@gmail.com

Abstract. Cryosorption pump is the only solution to pump helium, hydrogen and its isotopes in fusion devices. To design such pumps, knowledge of adsorption characteristics of activated carbons in the temperature range from 4.5 to 77 K is needed, but is very scarce in the open literature. Hence an experimental setup is designed and developed to measure adsorption characteristics of activated carbons down to 4.5 K. For this purpose, a commercial micropore analyzer operating down to 77K is coupled to a two-stage GM cryocooler, to enable cooling the sample temperature down to 4.5 K. A heat switch is mounted in between the second stage cold head and the sample chamber helps to vary the sample temperature from 4.5 K to 77K without affecting the performance of the cryocooler. The details of the experimental setup is presented elsewhere. We present here the experimental results of adsorption of different types of activated carbons in the temperature range 4.5K to 10 K using Helium gas as adsorbate. These results are evaluated in terms of surface area, pore sizes and their distributions. Also the effect of epoxy based adhesive used in bonding the activated carbons to the panels is evaluated. These results will be useful towards the selection of the right activated carbons for the development of cryosorption pumps.

1.Introduction

The pumping of helium, hydrogen and its isotopes from fusion exhaust is possible only by the use of Cryosorption pumps, since they offer high pumping speeds and are capable of operating in the tokomak environment of radiation and high magnetic field. Normally, the cryo-condensation pumps operating at 4.2K, helps to condense all the gases except helium and hydrogen. To condense these gases, the porous sorbent is coated on the cryopanels. Among the various sorbent materials used for such applications, activated carbon provides the higher pumping speed and enables pumping of hydrogen and its isotopes in the presence of helium [1]. Also, activated carbon used requires moderate regeneration temperature (400K). Activated carbon adsorbent is suitable for use when handling large volume flow rates and to pump impure gases encountered in industrial processes particularly for fusion applications [3]. To have the sorbent adhered to the cryosurface, it needs to be bonded to the surface with an adhesive. Thus the fabrication of the cryosorption panel involves the selection of the right sorbent (with the proper isotherms and pore sizes) and the adhesive (high thermal conductivity and stability).

Journal of Physics: Conference Series **390** (2012) 012077 doi:10.1088/1742-6596/390/1/012077

To have high pumping speeds for cryosorption pumps for helium, the knowledge of the adsorption characteristics of the activated carbons (mounted on the panel by special epoxy resins) is necessary. The sticking coefficient for helium for such cryopanels is also reduced because some of the pores of such activated carbons are also blocked by adhesive. So there is a need of choosing the best activated carbons with high sticking coefficient for helium. However, such data is not available in open literature in the temperature range below 77K [1]. Hence there is a need to characterize and analyze the adsorption characteristics of various activated carbons especially in the temperature range from 4.5 K to 10 K using a specially built experimental setup.

2. Experimental setup

There are no commercial instruments available to measure the characteristics of porous materials in the above temperature range. Hence the above experimental setup has been setup with the commercially available micropore analyzer suitable for use upto 77K along with the two stage GM Cryocooler and this is shown in Figure 1.

Fig. 1. Experimental setup to measure adsorption characteristic at 4.2K

A commercial micropore analyzer (Autosorb IQ) is procured from Quantachrome Instruments Inc, USA. This instrument is suitable for measurement of the adsorption characteristics of porous materials at 77 K using liquid nitrogen. This instrument has provision to degas the sample in which sample under study is heated and pumped out. After degassing, the sample under study is moved analyzer port, wherein the measurements are performed. The sample is maintained at 77 K with the help of a liquid nitrogen bath. The warm and cold zone volumes are measured automatically and used in the measurements.

For measurements down to 4.5 K, the sample needs to be cooled to this temperature. This is done by using a Cryocooler (Model SRDK-415 from Janis Research Systems Inc. USA). This two stage GM Cryocooler provides a refrigeration power of ~ 1.5W at 4.2 K at its second stage and ~ 35W at 50 K in its first stage respectively. The sample under study is now mounted on the second stage cold head of the Cryocooler replacing the standard LN2 dewar. The thermal load of second stage to the second stage is minimized using the cooling power of the first stage as the thermal radiation shield.

Several layers of superinsulation are wrapped around first and second stage cold head to reduce radiation losses. Turbo molecular pumping system is used to evacuate the interspace of the vacuum jacket. Silicon diode sensors (DT670 and SI410, from M/s SI, USA) are used to monitor the temperatures of second stage cold head. The cold head temperature can also be increased above 4.2 K with the help of two cartridge heaters (each of 50 ohms). The sample temperature can be maintained at any specific temperature in the range from 4.5 K to 77 K using a temperature controller (Model 332 from M/s Lakeshore).

For 4.2 K measurements, the special sample chamber used is made of stainless steel mounted on the 2nd stage cold head of the Cryocooler and replaces the standard glass sample chamber used normally for 77 K measurements. Also, for experiments at temperatures other than that of the cold head, the sample chamber needs to be raised in its temperature, without affecting the performance of the Cryocooler. For this purpose, a heat switch is interposed between the sample chamber and the cold head. The stainless steel sample chamber is of conflate type (5 cm diameter) sealed by copper gasket. A heat switch is mounted in between sample chamber and the cold head of the Cryocooler, which helps to carry out experiments at temperatures from 4.5 K to 100 K without affecting the performance of the Cryocooler.

This heat switch consists of two interpenetrating copper parts welded into a thin walled stainless steel cylinder such that they are separated by a narrow gap. Normally, the upper and lower copper blocks do not touch each other. Hence, when this gap is evacuated, the heat switch is open. The top part of the switch can be maintained at a temperature different from that of its bottom due to the poor heat transfer of the thin walled stainless steel cylinder connecting the upper and lower parts of the heat switch. However, when helium is filled in the gap, the heat switch is closed since the top part of the switch will be at the same temperature as that of its lower part i.e. that of the cold head.

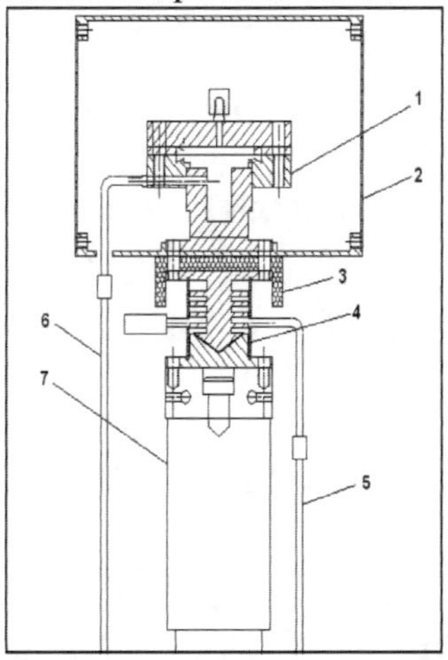

Fig. 2. (a) Drawing of sample chamber mounting on the cold head. (b) Photograph of the sample mount on the cold heat exchanger of the GM cryocooler. 1. Sample chamber, 2. Radiation shield, 3. Heater, 4. Heat Switch, 5. Gas line for Heat Switch 6. Gas line for adsorption, 7. 2nd stage cold head.

The sample port has a volume of ~ 1 cm^3 and is connected to the adsorption analyzer with the help of 0.125 in diameter stainless steel tube with appropriate end coupling. This smaller cross section pipe used to reduce void volume. Also similar pipe connections are used for gas entry into or evacuation of the heat switch. The sample needs to be degassed outside since the cold head of GM Cryocooler cannot be overheated. After GM Cryocooler reaches the set cold end temperature, the adsorption measurements are carried out by the analyzer by a dosing routine run by a predefined program.

3. Experimental Results

The adsorption characteristics of several types of activated carbons are studied using different adsorbates. The preliminary experiments were carried out with nitrogen gas as adsorbate at 77K and later these samples are evaluated at 4.2K-10K using helium gas as a adsorbate. In order to determine surface areas and the pore size distributions of the sample, the adsorption isotherms are needed.

3.1 Experimental studies at 77K using nitrogen gas

The adsorption isotherms plot the specific volume adsorbed by the sample as a function of relative pressure (pressure/saturation vapor pressure) at a constant temperature. Figure 3(a) shows the adsorption isotherms for different activated carbon samples. Also included in the same figure is the adsorption isotherm for standard alumina sample (SARM-2005) for comparison, which has a surface area of ~ 105 m^2 per gram. It is seen that activated carbon has higher volume of adsorption compared with the standard sample. Most of the carbon samples exhibit IUPAC classification type-I isotherms with an almost vertical rise at lower relative pressure range above 0.1,The alumina sample exhibit type-V isotherms, It almost rises to vertical at higher relative pressure range around 0.7

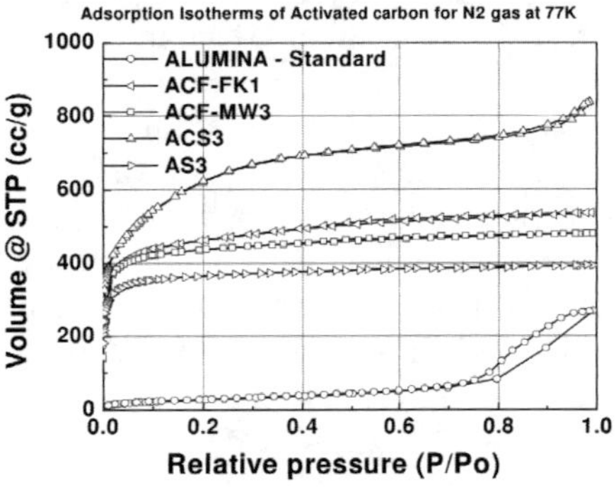

(a)

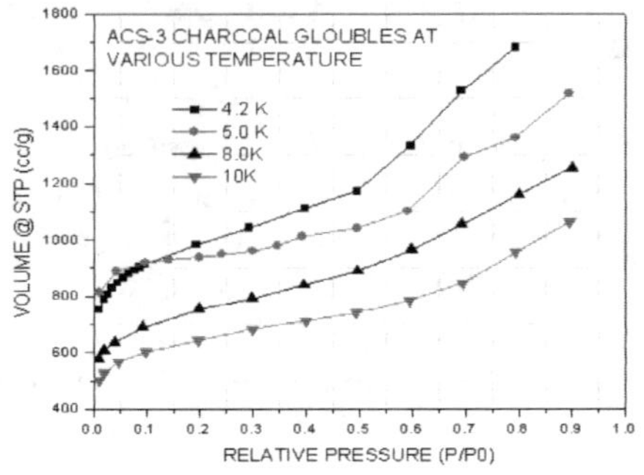

(b)

Fig. 3. (a) Adsorption isotherms of Standard Alumina sample and the activated carbon samples for nitrogen gas adsorption at 77 K. (b) Adsorption isotherms of ACS-3 activated carbon samples with helium gas at 4.2 to 10K

3.2 Experimental studies at 4.5 to 10 K using helium gas

The experimental procedure of measurements at 4.5 K is different when compared to 77 K measurement. At 77K, the dead volume measurement is done using helium gas because adsorption of sample at 77K with helium is nearly zero. However, at 4.5 K, helium gas itself is chosen as the adsorbate. Hence, the blank (adsorption of the sample chamber without the sample) measurements at these temperatures are carried out separately. This data is used for the measurements with the sample mounted in the sample chamber. The typical adsorption isotherms of a typical activated carbon globule sample namely ACS 3 is shown in Figure 3(b). It is observed that the adsorbed volume gradually increases with lowering of temperature. The adsorption volumes are also higher at lower temperatures for helium as adsorbate.

Table 1 Total surface areas of the activated carbon samples by BET method at different temperatures.

SAMPLES	SURFACE AREA MEASUREMENT –(m^2/g)				
	77K (Nitrogen as adsorbate)	4.2K (Helium as adsorbate)	5K (Helium as adsorbate)	8K (Helium as adsorbate)	10K (Helium as adsorbate)
Blank	0.00	0.12	0.00	0.00	0.00
Carbon pellets	1003.1	1465.1	1031.5	894	887.2
ACS-3	2038.2	2700.3	2639.45	1963.4	1736.8
ACFNW3	1773.2	1951.01	1924.1	1589.3	1335.2
Srilanka Charcoal Granules	1364.5	1801.2	1797.6	1188.5	1069.2
ACF- FK3	2230	2920.2	2697.2	2582.6	2014.9
ACS3+SS plate+ adhesive	1927	2600.7	2424.4	1831.2	1516.44
Charcoal granule Coarse	1126.74	1668.84	1576.49	1374.52	1235.61

4. Surface areas and pore size distributions

The adsorption surface areas of the samples can be determined by assuming that multilayer adsorption occurs over the sample surface starting from the initial monolayer. Thus, the standard BET technique is used in calculation of the surface areas of the activated carbons starting from the adsorption isotherms. The software provided along with the Autosorb IQ is used for the above calculations. The pore sizes are arrived at using the Density Functional Theory (DFT) method, using the experimentally measured isotherm data. The Autosorb IQ software enables the calculation of the pore size distributions.

Table 1 shows the surface areas of different activated carbon samples as determined by BET method with nitrogen as adsorbate at 77 K and helium as adsorbate in the temperature range from 4.5-10 K. We note that the surface areas are in general larger at lower temperature and the surface area with nitrogen at 77K nearly matches with that of helium in the temperature range between 5-8 K.

Pore size distribution is one of the important characteristic of a porous material. The volume adsorbed by the sample depends on the pore size and pore type. The overall pore sizes for the activated carbon is in the range from 10 to 50 Å. The pore sizes above 20 Å contribute less to the adsorption. Activated carbon samples ACS-3, ACF- NW3, ACF-FK2 have pore sizes below 20 Å and they are highly microporous [4]. On the other hand, SARM - Alumina sample has pore diameter of ~ 25 Å and it is mesoporous. Table 1 shows the average pore sizes measured for different activated carbon samples. To understand the influence of the pore size distribution on the adsorption, we chosen activated carbon globules ACS-3 for the analysis. Figure 4 shows the pore size distributions of activated carbon ACS3 at 77K and at 4.5 K.

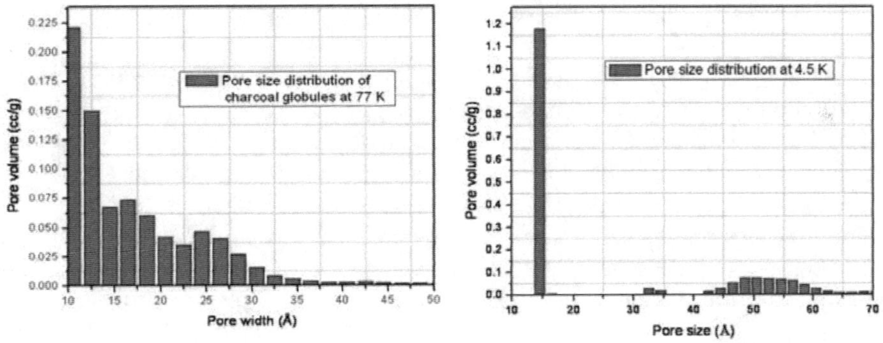

(a) (b)

Fig. 4. (a) Pore size distribution of ACS-3 at 77K using nitrogen gas as adsorbate.(b) Pore size distribution of ACS-3 at 4.5 K

From the above figure it is seen that at 77K, the pore sizes have a distribution over the range from 10 A to 35 A. On the other hand, the 4.5K studies show that the major pore sizes are concentrated at about 15 A. The pore size distributions at higher temperatures indicate a similar behaviour except that the peak around 15 A gradually decreases with increasing temperatures.

5. Studies on Activated Carbon Coated on Adhesive Panel

To understand the actual performance of cryopanels, experiments have been conducted on small size cryopanels wherein the activated carbons are mounted using an epoxy based adhesive. ACS-3 charcoal globule is chosen because it offers higher volumes of adsorption compared to other samples. The base stainless steel cryopanel needs an adhesive coating onto which ACS-3 sample is adhered. It is observed that the epoxy based adhesive performs better than other types and hence this is chosen.

The adsorption characteristics of the cryopanels coated with activated carbon ACS-3 samples were studied as follows. Initially the plain ACS-3 samples were characterized. Next, a stainless steel sheet of size 20 mm x 10 mm x 1 mm is finely coated with adhesive on its surface and experimented for its adsorption characteristics. Subsequently, the same size stainless steel plate coated with the adhesive and filled with activated carbon ACS-3 is studied for its adsorption characteristics. The performance characteristics of the various samples are shown in Table 2.

The results indicate that there is a reduction in the surface area of the ACS-3 coated onto the cryopanel and this value is ~ 10 % of that of the bare activated carbon ACS-3. This is easily understood since some of the pores are closed by the adhesive. Also, the average pore size of ACS-3 coated on the panel is found to be greater than that of the bare ACS-3. This is perhaps due to the interference of adhesives into the porous structure of the sample. Even though there is a decrease in the surface area and rearrangement of pore distribution, ACS-3 still exhibits micropore structure.

Table 2. Comparison of surface areas and average pore sizes of ACS-3 under different conditions.

S.No	Sample under study	Surface area in 77K with N_2 (m^2/g)	Average pore size in 77K with N_2 (Å)	Surface area in 4.5K with He(m^2/g)	Average pore size in 4.5K with He (Å)
1	Activated carbon ACS-3	2031.6	7.60	2700.1	6.21
2	Stainless steel sheet	0.00	0.00	0.00	0.00
3	Stainless steel coated with Adhesive	0.00	0.00	0.00	0.00
4	Stainless steel coated with Adhesive and ACS-3	1978.00	12.3	2600.7	11.81

6. Conclusion

An experimental setup has been designed and developed to measure adsorption characteristics of activated carbons down to 4.5 K. For this purpose, a commercial micropore analyzer operating down to 77K is coupled to a two-stage GM cryocooler, to enable cooling the sample temperature down to 4.5 K. A heat switch is mounted in between the second stage cold head and the sample chamber helps to vary the sample temperature from 4.5 K to 77K without affecting the performance of the cryocooler. The experimental results of adsorption of different types of activated carbons at 77K with nitrogen as adsorbate and in the temperature range 4.5K to 10 K with helium as adsorbate are presented. These results are evaluated in terms of surface area, pore sizes and their distributions. Also the performances of the small size cryopanels wherein the activated carbon ACS-3 is adhesively bonded is studied. It is observed that ACS-3 activated carbon is one of the best samples that can used for the development of cryopanels.

Acknowledgements

The authors wish to acknowledge the financial support given by NFP, IPR, Ahmedabad, which enabled setting up of this experimental setup. They also acknowledge the valuable support given by M/s I-Design Pvt. Ltd., Pune and CCT staff towards the fabrication of this experimental setup.

References

[1] Hauer.V, and Chr. Day., *Cryosorbent characterization of activated charcoal in the Coolsorb Facility*, Forschungszentrum Karlsruhe, Karlsruhe, 2002.

[2] Autosorb-*1 Gas sorption analyzer system manual*, Quantachrome Corporation, Boynton Beach, Florida. P/N 05098, 2009.

[3] Blacher.S, B. Sahouli.B, Heinrichs, P. Lodewyckx. R. PirardSmith, H. L. and Jones, *micro pore size distributions of activated carbon*, Langmuir 16 (2000) 16, 6754-6756.

[4] Chr.Day, *Colloids and surfaces* A 187-188 (2001) 187

[5] Chr.Day, *use of porous materials for cryopumping*, hand book of surface and interference of materials, Academic press, 2001, Vol 5, p 265

Journal of Physics: Conference Series **390** (2012) 012078 doi:10.1088/1742-6596/390/1/012078

Development of Magneto-static Solver Module for the design of compact and light-weight Traveling Wave Tubes

A Mercy Latha, V Srivastav, R K Gupta and S K Ghosh

Microwave Tubes Division, CSIR-Central Electronics Engineering Research Institute, Pilani-333031, Rajasthan, India

Email: amercylatha@gmail.com

Abstract- Traveling Wave Tubes are the unique class of microwave amplifiers with broad bandwidth, which makes it preferred for space applications. The constraints for a space application are very stringent demanding compactness and less weight. Entire size of the tube is mainly decided by the size of the collector, as it is the bulkiest component in a tube. Hence, if the size of the collector is reduced then the entire tube size will reduce. Researches reveal that in order to reduce the size of the collector, application of the magnetic field is necessary. Therefore, to study the effect of magnetic field on the collector performance, a magneto-static solver module has been developed. This paper describes the development of the magneto-static solver.

1. Introduction

Traveling Wave Tubes (TWTs) are the most preferred microwave amplifiers used for space application. The reason behind this is that only TWTs are able to satisfy almost all of the space requirements. Major constraints imposed to qualify for space include high efficiency, compactness, less weight, high reliability and long life. These requirements are critical to achieve and many of these are even contradictory. Hence, achieving all, at the same time, becomes a big challenge.

A TWT consists of mainly electron gun, slow wave structure, couplers (both input and output), magnetic focusing system and collector. Overall efficiency of TWT is decided by its electronic, circuit and collector efficiency [1]. Among these efficiencies, attaining high collector efficiency is of prime importance because most of the power is wasted as heat in collector region only. High efficiency in collector is achieved by various techniques like multi-stage depression [2], modification of collector geometry [3– 6], using secondary electron suppression coatings etc [7, 8].

Further, size and weight of a TWT largely depends on the collector, as collector is comparatively larger and heavier than other components of TWT. Hence, reducing the collector dimensions will yield compactness and light weight to the tube. But, generally the overall diameter of the collector depends on the universal beam spreading nature of the beam due to space charge effect. Hence, any attempt to reduce the size of the collector can be fruitful only if magnetic field is applied on the collector region [9]. In order to study the effect of magnetic field on collector performance, a magneto-static solver module is needed. Therefore, the authors were motivated to develop the magneto-static solver module.

The objective of a magneto-static solver module is to compute the static magnetic field profile around the magnet. The magnetic field configurations generally used in any TWT, may be, free permanent magnet, solenoid or periodic permanent magnet. The magneto-static solver module in its present form has the feature to simulate any of these magnetic configurations. The theory for the development of the magneto-static solver module is described in section 2. All simulation results along with their corresponding validation, done in CST STUDIO SUITE [10], are included in section 3. Final concluding remarks are mentioned in section 4.

2. Theory

Determination of magnetic field components in vicinity of a magnet can be done by any of the following methods, as mentioned in [11],

- Method based on determining the distribution of microscopic ampere's current
- Method based on the Poisson and Laplace equations, determining the magnetic scalar potential
- Method based on a system of equivalent magnetic dipoles (superposition of elementary results obtained for elementary magnetic dipoles)

2.1 Free Permanent Magnet

Permanent magnets are basically ferromagnetic materials. Free permanent magnets can be used at the entrance of the collector region so that the axial magnetic field produced by these magnet aid in better collection of the electrons.

For magnetic material with straight line normal demagnetization curves, it is possible to calculate with reasonable accuracy the flux density at a distance X from the pole surface, on the magnet's centerline [11].

- Rectangular magnet

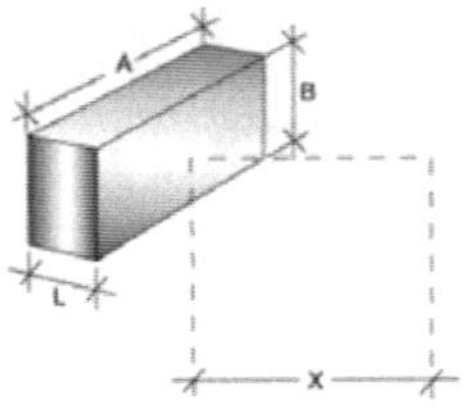

Fig. 1. Geometry of free rectangular permanent magnet

$$B_X = \frac{B_r}{\pi}\left[\tan^{-1}\frac{AB}{2X\sqrt{4X^2+A^2+B^2}} - \tan^{-1}\frac{AB}{2(L+X)\sqrt{4(L+X)^2+A^2+B^2}}\right]$$

- Cylindrical magnet

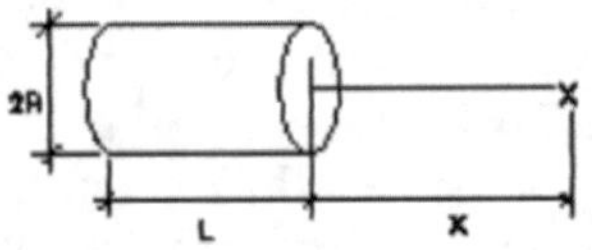

Fig. 2. Geometry of free cylindrical permanent magnet

$$B_x = \frac{B_r}{2}\left[\frac{(L+X)}{\sqrt{R^2+(L+X)^2}} - \frac{X}{\sqrt{R^2+X^2}}\right]$$

- Ring shaped magnet

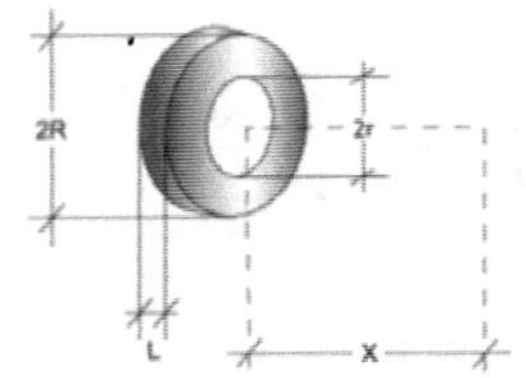

Fig. 3. Geometry of free ring-shaped permanent magnet

$$B_x = \frac{B_r}{2}\left[\left(\frac{(L+X)}{\sqrt{R^2+(L+X)^2}}-\frac{(L+X)}{\sqrt{r^2+(L+X)^2}}\right)-\left(\frac{X}{\sqrt{R^2+X^2}}-\frac{X}{\sqrt{r^2+X^2}}\right)\right]$$

2.2 Solenoid

The magnetic field around a solenoid can be derived by computing the magnetic vector potential [12]. Magnetic fields are produced by currents, which have associated directions. Hence, it is obvious that magnetic potential is a vector quantity. The magnetic vector potential is high in the vicinity of currents and tends to have the same direction as the currents which produce it.

The radial and axial magnetic flux density expressions are derived to be,

$$B_z = B_{z0} - \frac{B_{z0}''}{2^2}r^2 + \frac{B_{z0}^{iv}}{2^2 4^2}r^4 + \cdots$$

$$B_r = -\frac{B_{z0}'}{2}r + \frac{B_{z0}'''}{2^2 4}r^3 - \frac{B_{z0}^{v}}{2^2 4^2 6}r^5 + \cdots$$

where B_{z0} is the magnetic flux density at the axis.

The magnetic flux density along the axis of a circular coil is found experimentally to be approximately,

$$B_{z0} = \frac{\mu_0 a^2}{2}\frac{NI}{(a^2+z^2)^{3/2}}$$

where, a – radius of the coil, N – number of turns and I –current flowing through the coil. The degree of approximation is fairly good, if the dimensions of the winding cross-section are small compared with coil radius.

2.3 Periodic Permanent Magnet

Periodic permanent magnet (PPM) structure is most commonly used for focusing of the electron beam in slow wave structure of a TWT. It generally consists of an assembly of ring-shaped magnet separated by soft iron pole pieces, as shown in figure 4. A practical PPM configuration used in a TWT consists of more than 25 such assemblies of ring magnet and pole piece.

The analysis of the PPM structure can be done in a simplified way by analyzing only a half-period PPM cell (next half period will only change in polarity), neglecting the end effects and taking advantage of periodicity and symmetry [13].

- Calculate total permeance of all flux paths
- Calculate the load line slope
- Determine the operating point of the magnet
- Finally, compute axial magnetic field which is given by,

$$B_z(z,r) = \sum_{n=1,3,5,\ldots}^{\infty} \frac{4B_g \sin(n\pi g/L)}{I_0(2n\pi r_{f1}/L)n\pi} \times I_0(2\pi nr/L)\cos(2\pi nz/L)$$

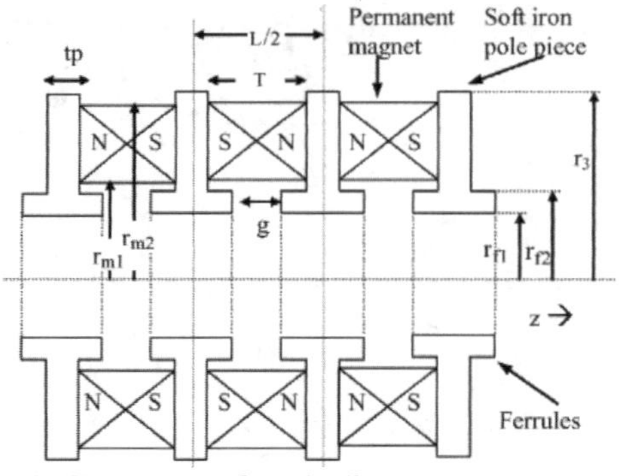

Fig. 4. Geometry of periodic permanent magnet

3. Validation of Results

Above mentioned theory has been implemented in MATLAB and a magneto-static solver module has been developed. The module is developed in a user friendly manner with good graphical user interfaces. The results of each of the magnetic configurations with their validations are as follows.

3.1 Free Permanent Magnet

Permanent magnets are basically ferromagnetic materials. Free permanent magnets can be used at the entrance of the collector region so that the axial magnetic field produced by these magnet aid in better collection of the electrons. For magnetic material with straight line normal demagnetization curves, it is possible to calculate with reasonable accuracy the flux density at a distance X from the pole surface, on the magnet's centre line [11].

- Rectangular magnet

A rectangular free permanent magnet with all dimensions - length, breadth and width, taken as 1mm (similar to a unit cube) has been simulated. The remanant magnetic flux density has been set to 1T. The comparison of the magnetic field profile along axial direction, obtained from the magneto-static solver and CST STUDIO SUITE is shown below in figure 5.

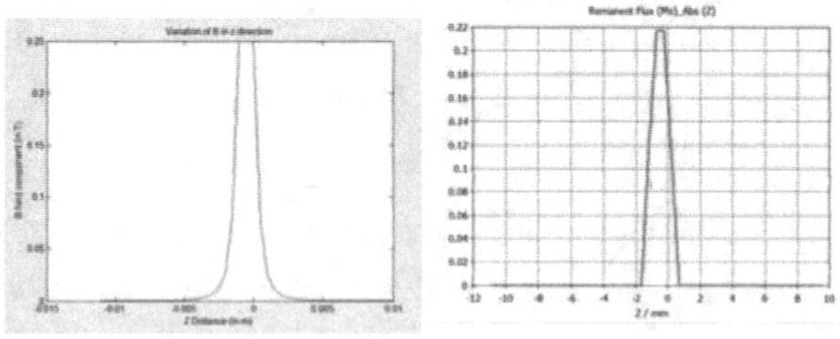

(i) (ii)

Fig. 5. Comparison of the B field along Z direction obtained from (i) MATLAB Magneto-static solver, (ii) CST STUDIO SUITE

From figure 5, it is evident that both the profiles match approximately but not accurately. The reason for this deviation is that several approximations and assumptions were taken during the analytical derivation of the expression for magnetic field. Flux linkage effects were not taken into account while the derivation of magnetic field expression but CST STUDIO SUITE considers all the flux linkages. This is the obvious reason for this deviation.

- Cylindrical magnet

Cylindrical free magnet of radius 1mm and length 1 mm has been simulated with remanant magnetic flux density, 1T. The validation of magnetic field profile along axial direction obtained from the magneto-static solver by the ones obtained by CST STUDIO SUITE, are depicted below in fig. 6.

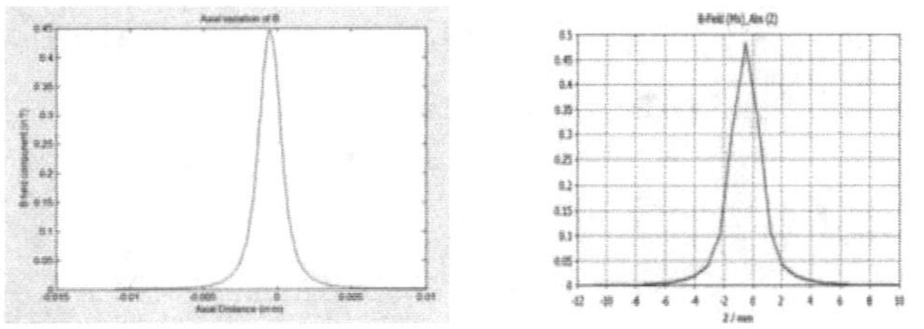

(i) (ii)

Fig. 6. Comparison of the B field along axial direction obtained from (i) MATLAB Magneto-static solver, (ii) CST STUDIO SUITE;

It could be observed from the above results that both the profiles agree closely, except for quiet smoothness at the tip of magnetic field, obtained by this magneto-static solver module. Again this is due to the negligence of flux linkage effect in the analytical expression.

- Ring shaped magnet

A ring-shaped free permanent magnet of outer radius 2mm and inner radius 1 mm with unit thickness has been simulated with remanant magnetic flux density of 1T. The comparison of axial magnetic field profile obtained by this magneto-static solver and by CST STUDIO SUITE is shown in figure 7, side by side, and could be observed to have very good agreement between them, thus validating the magneto-static solver module for free permanent magnets.

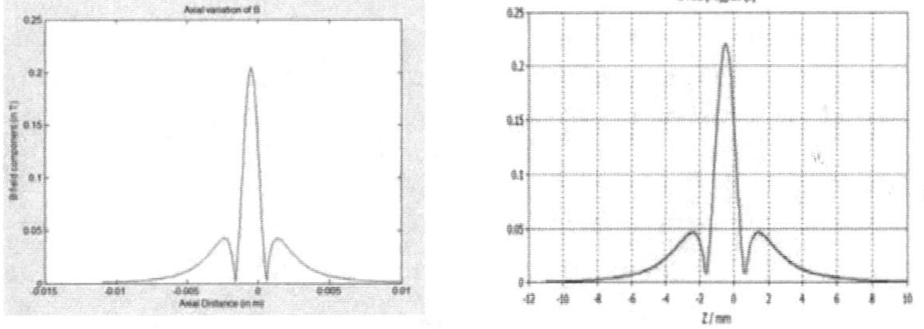

(i) (ii)

Fig. 7. Comparison of the B field along axial direction obtained from (i) MATLAB Magneto-static solver, (ii) CST STUDIO SUITE

3.2 Solenoid

Solenoid of diameter 30mm, length 15mm and 100 turns has been simulated with remanant magnetic flux density of 1T. The 2D results for the B field obtained from the magneto-static solver and CST STUDIO SUITE are displayed below in figure 8.

The figure 8 depicts the 2D patterns of the radial magnetic field. From the above comparisons, it is evident that magnetic field patterns agree though having slight deviations. The deviations are due to the fact that, in the field expression of Br and Bz, only finite number of terms (in particular only four terms) are taken into account. Accurate field patterns could be obtained only if an infinite series is taken into account. But, in the implementation of the magneto-static solver, it was assumed that later terms will be of very small value to affect the accuracy and hence was avoided, which led to the deviation in the magnetic field patterns.

3.3 Periodic Permanent Magnet

A periodic permanent magnet configuration has been simulated whose dimensions and demagnetization properties have been taken from [13]. The axial magnetic field profile has been obtained by the magneto-static solver as shown in figure 9. This agrees very closely with the published result [10], which is also shown for comparison below.

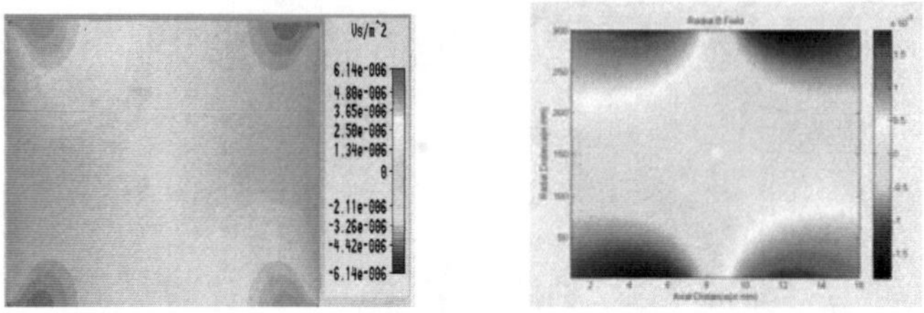

(i) (ii)

Fig. 8. Comparison of the 2D results of radial B field obtained from (i) MATLAB Magneto-static solver, (ii) CST STUDIO SUITE;

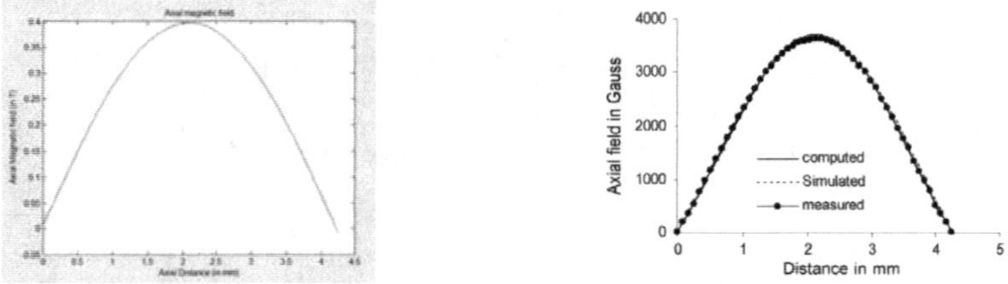

(i) (ii)

Fig. 9. Comparison of the B field along axis (i) MATLAB Magneto-static (ii) Published work

4. Conclusions

The development of the magneto-static solver has been elaborated for various magnetic configurations. From the results obtained, it can be concluded that magneto-static solver has very good agreement with CST PARTICLE STUDIO and thereby validating it.

Acknowledgement

Authors are thankful to Director, CEERI, Pilani for his kind permission to publish this work. Authors are also grateful to other team members for their co-operation for the improvement of the work.

References

[1] D.S. Komm, R.T. Benton, H.C. Limburg, W.L. Menninger, X. Zhai, IEEE Trans on Electron Devices, Jan 2001, vol 48, pg. 174
[2] Henry G. Kosmahl, Proc. of IEEE, vol. 70, no. 11, Nov. 1982
[3] F. Sterzer, IRE Trans. On Electron Devices, vol. ED-5, pp. 300, Oct. 1958
[4] Takanori Okoshi, IEEE Trans on Electron Devices, vol. ED-19, no. 1, Jan. 1972
[5] D. A. Dunn, W. R. Luebke, G. Wada, IRE Trans on Electron Devices, Jul 1959.
[6] D. A. Dunn, R. P. Borghi and G. Wada, IRE Trans. Electron Devices, pp. 262, 1960.
[7] P Ramins and B T. Ebihara, IEEE Trans. Electron Devices, vol. 33, no. 11, pp. 1915, 1986.
[8] Arthur N. Curren, Peter Ramins, , IEDM 85, pg. 361
[9] Pagonakis I.G, Hogge J.P, Alberti S, Avramides K.A, Vomvoridis J.L, IEEE Trans. on Plasma Science, vol. 36, no. 2, pp. 469 – 480, 2008.
[10] CST STUDIO SUITE 2010 operating Manual, Darmstadt, Germany, Licensed 2010
[11] Ana Mladenovic Vuckovic and Slavoljub Aleksic, Facta Universitatis Ser.: Elec. Energ. vol. 23, no.3, pp 259, 2010.
[12] Willis W.Harman., "Fundamentals of Electronic Motion" , McGraw-Hill Book Company, 1953
[13] Mrityunjay Santra, Lalit Kumar, Jyothi Balakrishnan, IEEE Trans. On Electron Devices, vol. 56, no.5, pp. 974, 2009.

Accelerator Vacuum Protection System

Pradip Barua, Ashok Kothari , M. Archunan , Rajan Joshi

Inter-University Accelerator Centre, Aruna Asaf Ali Marg, Post Box - 10502, New Delhi – 110067

Email: barua123@gmail.com

Abstract. A new and elaborate automatic vacuum protection system using fast acting valve has been installed to avoid accidental venting of accelerator from experimental chamber side. To cover all the beam lines and to reduce the system cost, it has been installed at a common point from where all the seven beam lines originate. The signals are obtained by placing fast response pressure sensing gauges (HV SENSOR) near all the experimental stations. The closing time of the fast valve is 10 milli-second. The fast closing system protects only one vacuum line at a time. At IUAC, we have seven beam lines so one sensor was placed in each of the beam lines near experimental chamber and a multiplexer was incorporated into the fast closing system. At the time of experiment, the sensor of the active beam line is selected through the multiplexer and the Fast closing valve is interlocked with the selected sensor. As soon as the pressure sensor senses the pressure rise beyond a selected pressure, the signal is transferred and the fast valve closes within 10 to 12 millisecond.

1. Introduction

We have a 15UD PELLETRON Accelerator at IUAC. The ultra high vacuum (1.0×10^{-10} Torr) condition is maintained throughout the accelerator and seven beam lines. A combination of Ion pump & Getter pump is used to create UHV and the pumping stations are installed at various locations, keeping in mind the requirements of the distributed pumping, taking care of the conductance limitation of such a large vacuum system. The whole vacuum system is divided into individual sections using all metal pneumatic straight through valves. The closing time of these valves is approximately 100 milliseconds.

The operation of the valve is controlled by a valve controller which takes 'vacuum ok' interlock signal available from Ion pump controller from both side of the valve. When vacuum ok interlock signal from either side of the valve is failed, the controller closes the isolation valve immediately. The operation is shown in Fig. 1.

In addition to the all metal pneumatic straight through valves there are two high speed, spring actuated, all metal fast acting straight through valves. These valves are installed at the top and bottom side of accelerator for emergency closing, in case of catastrophic failure of ultra high vacuum system. The fast valve gets actuated by a coaxial presser sensor installed at critical locations. The closing time of these valves is approximately 35 milliseconds.

With the interlocks provided at all critical points, whole system is protected against any vacuum failure or accidental venting of the experimental chambers.

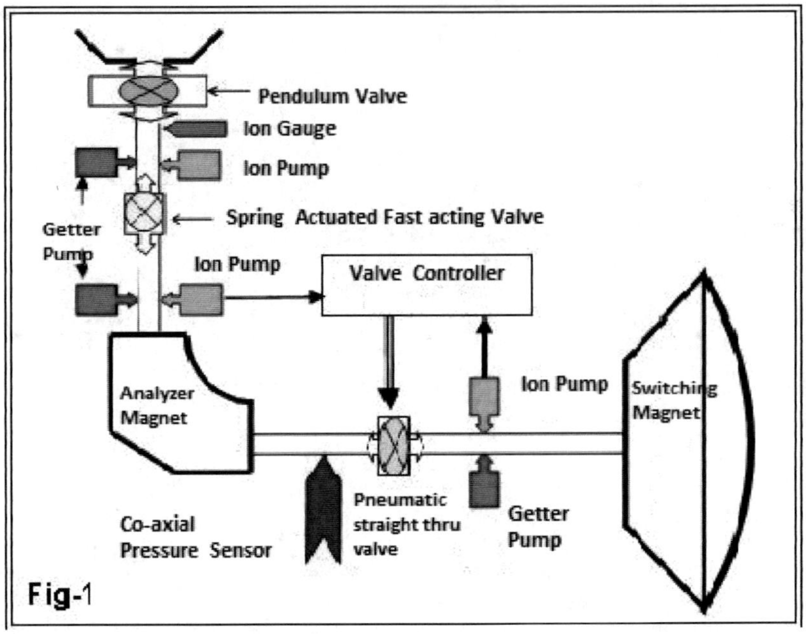

Fig. 1. Schematic of Valve operation

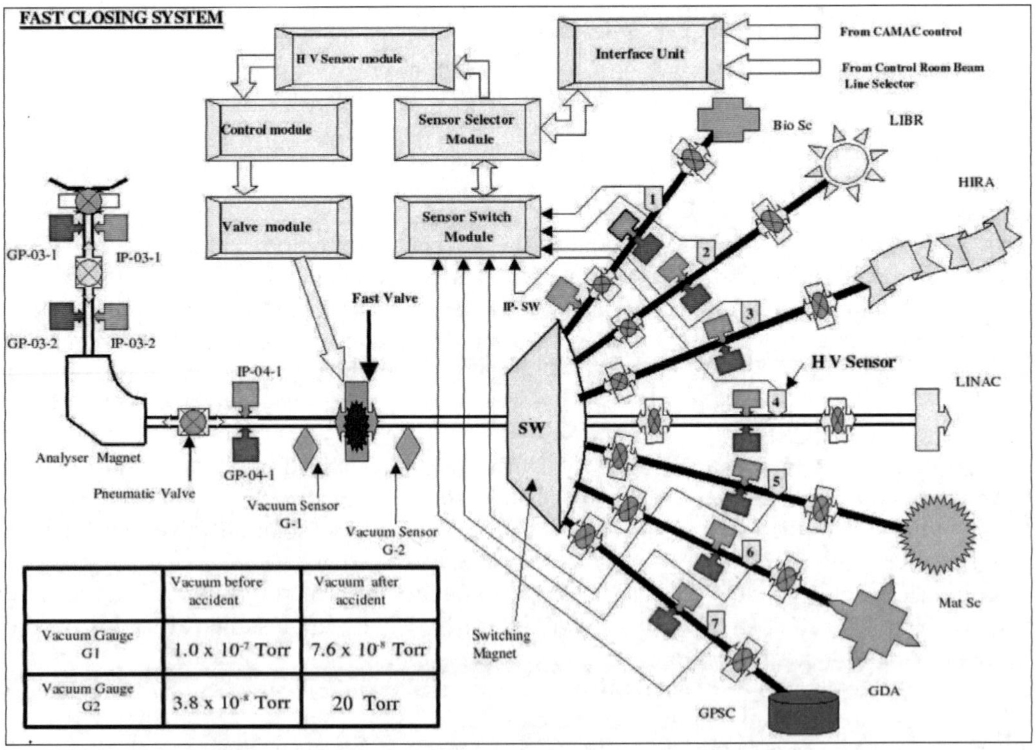

	Vacuum before accident	Vacuum after accident
Vacuum Gauge G1	1.0×10^{-7} Torr	7.6×10^{-8} Torr
Vacuum Gauge G2	3.8×10^{-8} Torr	20 Torr

Fig. 2. Fast closing valve Installation and Test results

2. Fast acting vacuum protection system

With the available interlocks and fast acting valves, having response time of 35 ms, we have had few vacuum accidents resulting in vacuum disturbance inside the accelerator system. Most of the accidents happen from experimental chamber sides. So, to avoid accidental venting of the accelerator vacuum system from the experimental chamber end, a new and better vacuum protection system was designed & commissioned. The heart of the vacuum protection system is a fast acting valve installed at a common point from where all the seven beam lines originate. The system consists of a valve, control electronics & fast response pressure sensing gauges near all the experimental stations. The fast closing system couples with one sensor at a time [protects only one vacuum line at a time]. As soon as the pressure sensor senses the pressure rise beyond a selected pressure, the signal is transferred and the fast valve closes within 10 to 12 millisecond. The valve control electronics module and pressure sensors were procured from VAT. The standard components work with one pressure sensor only. So it was suggested to VAT to provide a multiplexing unit so as to enable same control electronics to couple with pressure sensors placed at different locations, as per requirement.

At IUAC, we have seven beam lines so one sensor was placed in each of the beam lines near experimental chamber and a multiplexer was incorporated into the fast closing system. At the time of experiment, the sensor of the active beam line is selected through the multiplexer and the Fast closing valve is interlocked with the selected sensor. The schematic of the system is shown in Fig. 2

2.1 Design of Fast Closing vacuum protection system

Fast closing system has three electronic modules (valve module, control module and HV sensor module), seven high vacuum sensor and one fast closing valve, one multiplexer unit which is the combination of sensor selector and switch box. One interface unit has been made which works between sensor selector box, CAMAC and control room beam line selector, for remote operation. Interface unit is connected with CAMAC, which sends signal to the valve module for open / close & read the status of valve and modules. If vacuum is OK on both the sides of the valve, the valve can be opened.

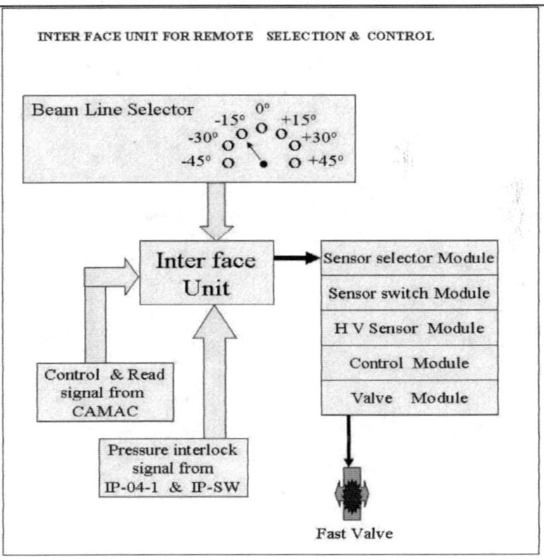

2.1.1 *Test results*

Actual test to test the efficacy of the system was done after commissioning the fast valve. In this test, accident conditions were simulated and two full range (atmosphere to 10^{-9} Torr) vacuum sensors G1 and G2 (one gauge at each side of the valve) were monitored before and after the vacuum accident, refer Fig 2. The valve functioned perfectly well and vacuum was protected in the desired line. The test result is as shown below.

	Vacuum before accident	Vacuum after accident
Vacuum Gauge G1	1.0×10^{-7} Torr	7.6×10^{-8} Torr
Vacuum Gauge G2	3.8×10^{-8} Torr	20 Torr

Journal of Physics: Conference Series **390** (2012) 012079 doi:10.1088/1742-6596/390/1/012079

3. Off line testing of Fast Closing System

Complete setup for testing the closing time of valve is shown in Fig 3 below. All seven fast valve sensors were mounted on a small vacuum chamber. Initially valve was opened by pressing open switch and resetting the system. Then vacuum was created with the help of a turbo pumping system till vacuum reached 10^{-7} Torr. Then suddenly the system was vented. This drastic change of vacuum was sensed by HV sensor module and triggered control & valve module for closing the valve. One TTL high signal is available from HV sensor when it senses vacuum disturbance and it starts counter. One optical transmitter & a receiver is used to generate stop signal for the counter. When valve closes no light is received at the optical receiver and it sends low signal to stop counter. The counter shows 100509 count for 1 second. The valve test was done 10 times and the average closing time found is 11.6 ms.

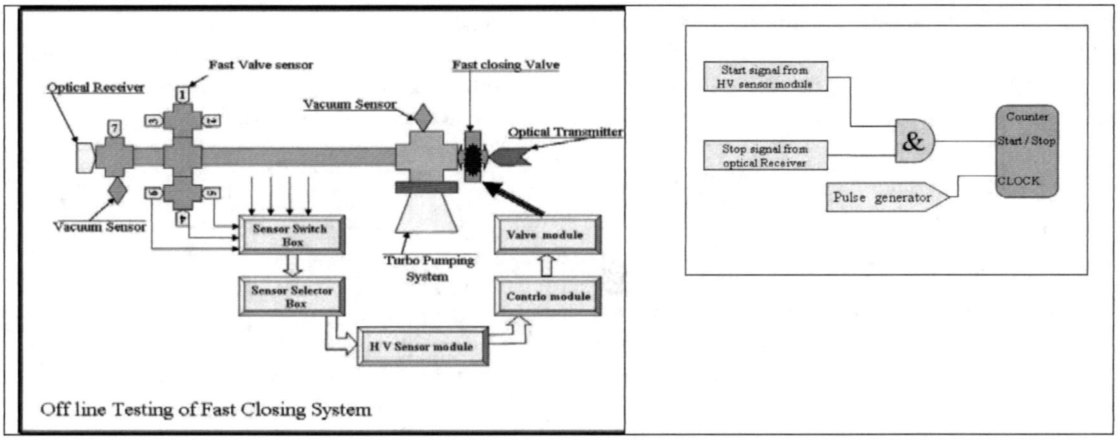

Fig 3. Valve timing test set up

References

[1] VAT Vakuumventile AG , Series 75 fast closing valve .

International Symposium on Vacuum Science & Technology and its Application for Accelerators IOP Publishing
Journal of Physics: Conference Series **390** (2012) 012080 doi:10.1088/1742-6596/390/1/012080

Helium Leak Detection of Vessels in Fuel Transfer Cell (FTC) of Prototype Fast Breeder Reactor (PFBR)

N. G. Dutta

Kay Bouvet Engineering Pvt. Ltd N-3, Addl. Midc Area Satara 415004.

E-Mail: ngdutta@kaybouvet.com

Abstract. Bharatiya Nabhikiya Vidyut Nigam (BHAVINI) is engaged in construction of 500MW Prototype Fast Breeder Reactor (PFBR) at Kalpak am, Chennai. In this very important and prestigious national programme Special Product Division (SPD) of M/s Kay Bouvet Engg.pvt. ltd. (M/s KBEPL) Satara is contributing in a major way by supplying many important sub-assemblies like- Under Water trolley (UWT), Airlocks (PAL, EAL) Container and Storage Rack (CSR) Vessels in Fuel Transfer Cell (FTC) etc for PFBR. SPD of KBEPL caters to the requirements of Government departments like - Dept. of Atomic Energy (DAE), BARC, Defense, and Government undertakings like NPCIL, BHAVINI, BHEL etc. and other precision Heavy Engg. Industries. SPD is equipped with large size Horizontal Boring Machines, Vertical Boring Machines, Planno milling, Vertical Turret Lathe (VTL) & Radial drilling Machine, different types of welding machines etc. PFBR is 500 MWE sodium cooled pool type reactor in which energy is produced by fissions of mixed oxides of Uranium and Plutonium pellets by fast neutrons and it also breeds uranium by conversion of thorium, put along with fuel rod in the reactor. In the long run, the breeder reactor produces more fuel then it consumes. India has taken the lead to go ahead with Fast Breeder Reactor Programme to produce electricity primarily because India has large reserve of Thorium. To use Thorium as further fuel in future, thorium has to be converted in Uranium by PFBR Technology.

1. Introduction:

Leak for various reasons is an undesirable feature in a component or system. Still leaks do occur in practice and can some time cause significant loss in terms of both money and human life. Hence it is essential that leaks are effectively detected and remedial measures to plug the leaks are taken sufficiently in advance. Over the years more than twenty different leak test methods have been developed with different detection sensitivity. Helium leak detection is widely used in industry for detection of fine leaks in components /systems.

By far the most sensitive and versatile leak test today is Helium Mass spectrometer Leak test. It gives reliable, repeatable results and is normally specified for critical area applications like Nuclear, Space etc. The principle of detection is that of a "Mass spectrometer" Developed by Dr. A.O. Nair as far back as in 1937. In a mass spectrometer which is kept under high vacuum (of the order of 10^{-5} torr) helium atoms entering the spectrometer from the leak, are first ionized by bombardment of electrons over it and positive ions so created are accelerated by an electric field and the ions are separated from each other depending on their mass by charge ratio, by application of a magnetic field over it in a perpendicular direction. The helium ions follow a definite path. A slit is used to allow only helium ions to cross this slit and these ions are collected at an electrometer tube, which generates a signal in millivolt range, which is then suitably amplified and displayed on a meter.

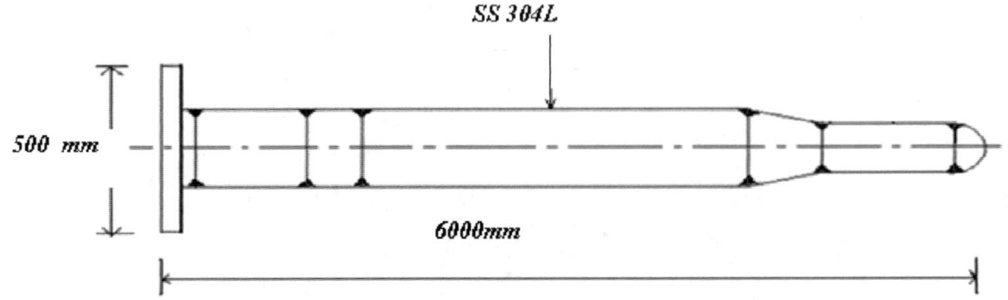

Fig.1. FTC Vessel

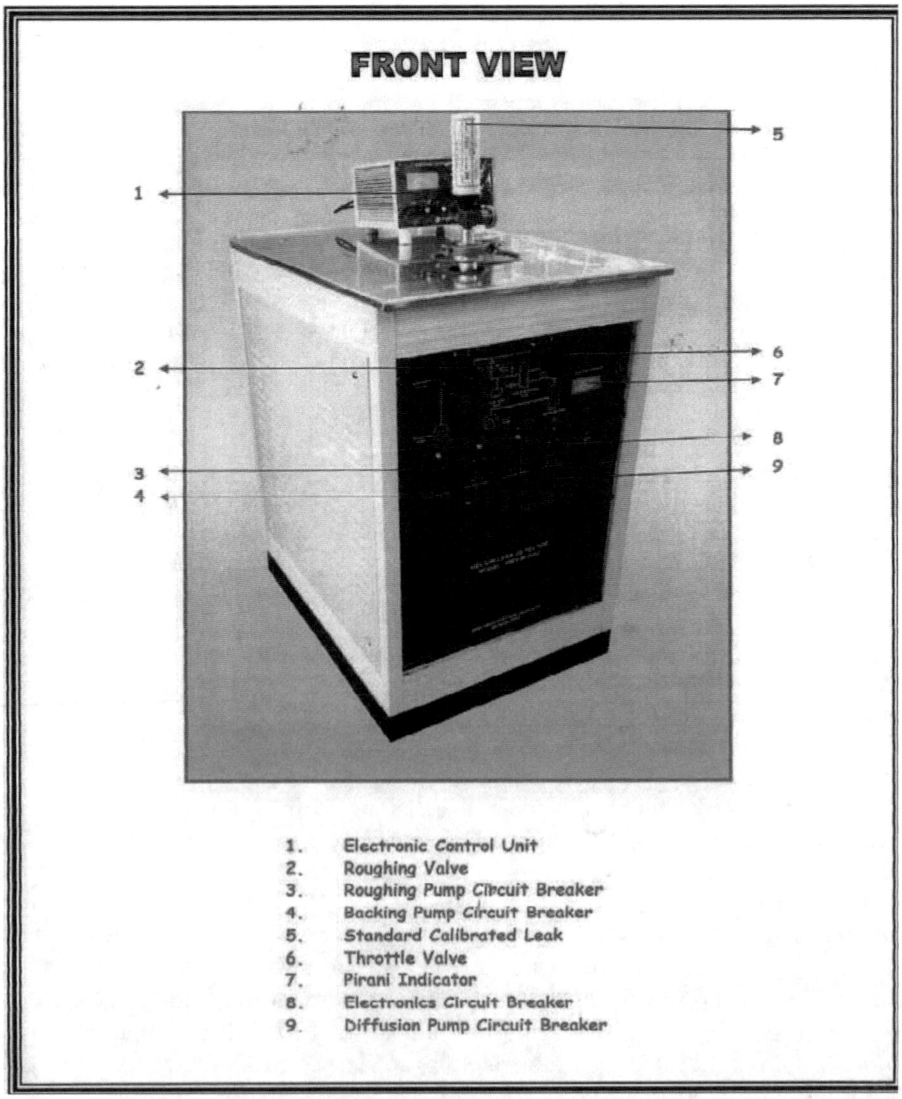

Photo 1. HLD Model HMS-M-300(M/s HHV, Bangalore)

2. FTC Vessels

In PFBR Reactor there are 4 types of vessels in Fuel Transfer Cell (FTC), Fresh SA Preheating Facility (FSPF)-3Nos, Spent SA washing facility (SSWF)-4Nos, Absorber SA Storage Facility (ASSF)-8Nos, Canned SA Storage (CSAS)-3Nos These vessels are made of SS304L 150 to 250 NB Sch 40-80 pipes (welded & seamless) & are around 6 to 7 meters long. There are around 5 to 6 welds in the full length of the assemblies along with flange weld at the top and closed dish end at bottom. Helium leak tightness of $1X10^{-8}$mbar.lit/sec is specified on these welds. 11 no of such vessels are fabricated and supplied to BHAVINI. The vessels have undergone several quality control checks like – Dimension, RT and LPT of welds, Material Test Control (Chemical, YS, UTS, % El etc.), Helium Leak Test and final Performance Test.

3. Helium Leak Detector (HLD)

Helium Leak Detector (HLD) model HMS-M-300F supplied by Hind High Vacuum Company Pvt. Ltd., Bangalore has been used to test these FTC vessels. **Photo-1** shows the photograph of the Detector used. The sensitivity of the detector is $1x10^{-10}$ mbar.lit/sec.

4. Helium Leak Test (HLT) of FTC Vessel

Fig-2 shows the typical arrangement used for carrying out HLT of FTC vessels. The acceptance limit for local leak is $1x10^{-8}$mbar.lit/sec.

4.1 Calibration of Leak-Detector:

The Leak-Detector is started as per Instruction-Manual. When the Spectrometer Vacuum reads 10^{-5} milibar, put ON the filament & connect the standard leak ($4x10^{-8}$mbar.lit/sec.) to detector. The leak indicator-meter should read $4x10^{-8}$mbar.lit/sec. The HLD is thus calibrated with std leak before start of testing. HLD is calibrated at start & end of shift & should show the reading of standard leak.

Std. Leak= $4 x10^{-8}$ mbar.lit/sec.
Observed Leak= $4 x10^{-8}$ mbar.lit/sec.
Positive detection of Helium =OK

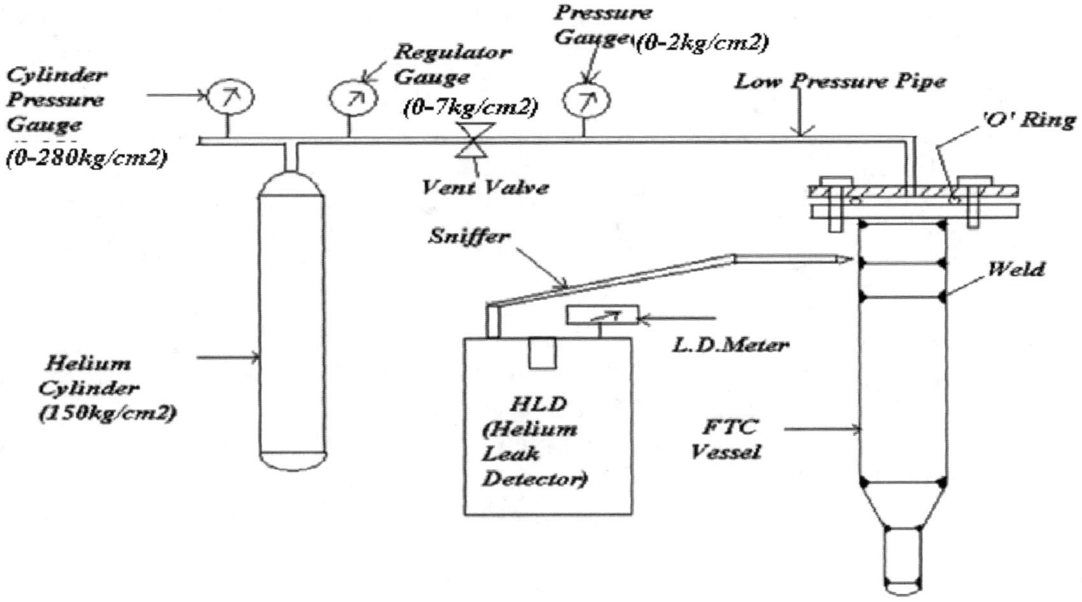

Fig. 2. HLT of FTC Vessel

4.2 Procedure for Sniffer Test:
Pressurise FTC vessel with mixture of Helium & Ar gas to 1.1 bar(g) with Helium concentration of 20 to 25% by closing all openings with appropriate fixtures. Scan all weld joints, flanges etc. with sniffer as per std practice (ASME-V).
Before HLT, air-pressure drop test 1.1kg(g) for 1 hr & soap solution test were also carried out to eliminate gross leaks.The testing of each vessel was witnessed by BHAVINI's third party QS Engineer.

5. Observations:

Table-1: Test Result of FTC vessel

Weld No.	Background Reading (with sniffer) mbar.lit/sec	Observed Reading mbar.lit/sec	Remark
4C1	1×10^{-8}	1×10^{-8}	OK
4C2	1×10^{-8}	1×10^{-8}	OK
4C3	1×10^{-8}	1×10^{-8}	OK
4C4	1×10^{-8}	1×10^{-8}	OK
4C5	1×10^{-8}	1×10^{-8}	OK
4C6	1×10^{-8}	1×10^{-8}	OK

6. Conclusion

Helium Leak Test using needle valve type sniffer is capable of detecting fine leaks of order of 1×10^{-8} mbar.lit/sec. This has been successfully carried out on 11 nos. of FTC vessels supplied to BHAVINI by M/s Kay Bouvet Engineering Pvt. Ltd. Satara. The tests were witnessed by BHAVINI third party Qs engineers & were satisfied with the test & have issued shipping release (SR) for shipment of these vessels.

Acknowledgement
The author would like to acknowledge with thanks the encouragement & guidance given in preparation of this paper by Joint MD Shri. Niraj Chandra. He would also like to express sincere thanks to QA Engineers-S/Shri Jayanand Kadam, Suryakant Patil, Rahul Nipane & Jivan Chavan for carrying out HLT on FTC vessels & also in preparation of this technical paper.

References:

[1] Dutta N.G., SS More, 2010, Fabrication & Quality Assurance of some Important Components & Sub-assemblies for Prototype Fast Breeder Reactor (PFBR) project. Asian Nuclear Prospects (ANUP-2010), Chennai, IGCAR, Indian Nuclear Society (INS) & IAEA.

[2] DUTTA N.G.,2002 , Overview of Leak-Testing technology, VAC-NEWS, June2002, Indian Vacuum Society (IVS).

Indigenous development of ultra high vacuum (UHV) magnetron sputtering system for the preparation of Permalloy magnetic thin films

Jakeer Khan, N Selvakumar, Prasanta Chowdhury [a)] and Harish C Barshilia
Surface Engineering Division, National Aerospace Laboratories (CSIR), Bangalore–560 017, India

[a]email: pchowdhury@nal.res.in

Abstract We have designed and developed an indigenous ultra high vacuum (UHV) sputtering system which can deposit magnetic thin films with high purity and good uniformity. The equipment consists of state-of the-art technologies and sophistication. Turbo-molecular pump combined with sputter ion pump is used to pump down the vacuum chamber up to 10^{-10} to 10^{-11} mbar of pressure. With this system it is possible to deposit coatings of various materials on a sample size of diameter 3". The $Ni_{81}Fe_{19}$ ferromagnetic thin films, with Tantalum (Ta) as a buffer and cap layers have been deposited on silicon substrates using this ultra high vacuum (UHV) sputtering system. The magneto transport measurement study indicated a significant variation in the AMR values of the films for varying thicknesses of tantalum and NiFe layers.

1. Introduction

To achieve the desired properties of magnetic materials, one of the most important parameter is the purity of the films without any oxide contamination. It is proven that high vacuum chamber with base pressure up to 10^{-6} mbar, any metallic films are prone to absorb the gas impurities during deposition. Hence for the deposition of these coatings, a special ultra high vacuum sputtering system has been designed and developed and in this system, the maximum attainable vacuum was 10^{-10} to 10^{-11} mbar. Fig. 1 shows the photograph of the UHV sputtering chamber. Deposition of the Permalloy thin films on silicon substrates by using the UHV sputtering system provided us high purity samples. This UHV sputtering system is a balanced magnetron sputtering system, facilitated with four DC power supplies. The UHV sputtering system consists of several parts which are assembled together as follows

2. Sputtering system details

2.1. Vacuum chamber

The vacuum chamber is cylindrical in shape and is equipped with two circular glass windows on either sides through which we can observe the experimental process. Inside the chamber there is a substrate holder which can be rotated using a DC stepper motor and it's axis of rotation was controlled by PCI based electronic device using a Labview software program. A feed through facilitates the substrate to be heated up to 800°C using a heater which is fixed below the base of the sample holder. The chamber has the provision for four sputtering guns. The sputter guns are installed with planar circular magnetrons that are water cooled.

Fig.1. Photograph of the UHV sputtering system

2.2. Rotary Pump

The first and foremost step in pumping the chamber to Ultra High Vacuum is to create a rough vacuum. That is, the chamber must be pumped from the atmospheric pressure to a pressure of approximately 10^{-3} mbar. This process is known as Roughing and is done by using a two stage rotary pump.

2.3. Turbo Molecular Pump

Once the roughing vacuum is achieved, then the further pumping is carried out by Turbo Molecular pump (TP). The turbo pump can pump down the vacuum chamber up to a vacuum of 10^{-6} to 10^{-8} mbar. The UHV sputtering system is equipped by a PFEIFFER VACUUM LTD., made turbo molecular pump having a maximum attainable speed of 49200 rotations per minute. The Turbo molecular pump can be isolated from the vacuum chamber using a manually operated separation gate valve. The extent of evacuation of the carrier gases inside the vacuum chamber can be controlled by partially closing the turbo molecular pump gate valve while doing the deposition. In this way we can maintain the desired deposition pressure of the gasses, which are introduced inside the vacuum chamber.

2.4. Sputter Ion Pump

To achieve the ultra high vacuum inside the vacuum chamber, it is further facilitated with a sputtering ion pump. After achieving the vacuum of 10^{-8} mbar, sputter ion pump is used to pump down the vacuum chamber up to 10^{-10} to 10^{-11} mbar of pressure. The sputter ion pump also can be isolated from the vacuum chamber by means of a gate valve. The process of maintaining the ultimate ultra high vacuum (i.e., 10^{-10} to 10^{-11} mbar) inside the chamber is done by the help of Sputter ion pump. Inside the sputtering ion pump, a swirling cloud of electrons produced in hollow Penning cells ionizes incoming gas atoms and molecules and a strong electrical potential, typically 3 kV to 7 kV, is used to trap these gas ion and further to accelerate them into a solid electrode. The swirling ions strike the chemically active cathode inducing sputter and finally pumped down by chemisorptions which effectively remove them from the vacuum chamber and the result is a net pumping action.

2.5. Pressure Gauge

The measurement of vacuum pressure inside the chamber is done by a pressure gauge which is directly connected to the vacuum chamber. The UHV system is equipped with a four mode digital Pirani-Penning pressure gauge. Up to the roughing vacuum of 10^{-3} mbar, the pressure is read by the Pirani Gauge. A Pirani gauge is a Wheatstone-bridge circuit with the sensing resistor exposed to the gases pressure in the system. The measuring range is of the order of 10^{-1} to 10^{-3} mbar. However lower pressure readings below 10^{-3} mbar (i.e., from 10^{-4} to 10^{-11} mbar) are measured by the Penning gauge which can be pneumatically controlled by a switch provided on the control panel. The penning gauge, however, is an ionization gauge. It uses the idea of measuring the current of a glow discharge as an indicator of the gas pressure. A four mode digital Pirani- Penning gauge was used here.

2.6. Mass Flow Controller

Mass Flow controller is used to control flow rate of the gases inside the vacuum chamber. Three mass flow controllers for controlling the flow rates of different gases e.g., Argon, Nitrogen and Oxygen gases are connected through a common gas pipe to the vacuum chamber and all the three mass flow controllers are controlled by three different digital electronic controllers.

2.7. Load Lock Chamber

A load lock chamber is used as a part of the UHV chamber for performing the sample loading at a pressure of 10^{-6} mbar. So without losing the vacuum inside the vacuum chamber, samples can easily be taken out or put inside the chamber easily by the load lock technique. The load lock chamber is a small spherical chamber with a loading and locking mechanism for the sample holder. The load lock chamber is separately pumped to a vacuum of 10^{-6} using a rotary and turbo molecular pump. For this, the turbo molecular pump is having a maximum speed of 75000 rotations per minute. After pumping the load lock chamber to high vacuum the sample holder mounted with the sample is transferred to the chamber. The load lock chamber also can be isolate from the vacuum chamber by means of a gate valve.

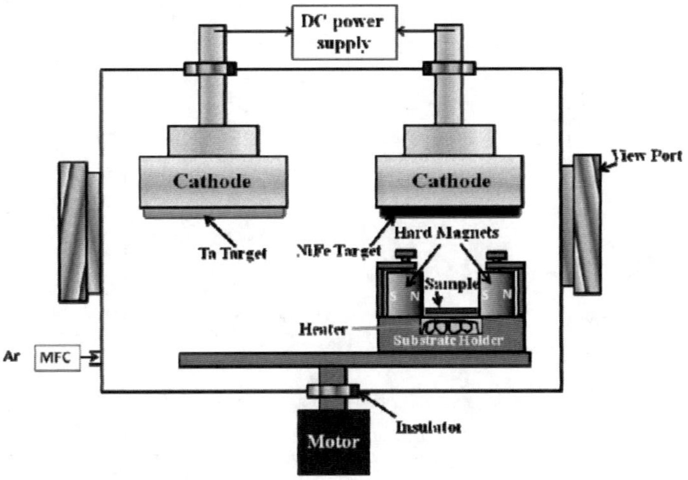

Fig.2. Block diagram of the UHV system chamber showing the arrangement of the various parts.

3. Experimental details

Highly pure ferromagnetic thin films of $Ni_{81}Fe_{19}$ (i.e., Permalloy) have been deposited on the silicon substrates (dimensions: 20 mm×20 mm), while the tantalum (Ta) was used as a buffer layer and cap layer for the Permalloy magnetic thin films on silicon substrate. Thin films were deposited for the varying thickness of the Ta and $Ni_{81}Fe_{19}$ thin films, respectively. To get the high purity Permalloy thin films, the deposition of these films was carried out in a controlled environment of Ar gas using a special ultra high vacuum (UHV) sputtering system at a base pressure of 10^{-9} to 10^{-10} mbar. Prior to the deposition of Permalloy thin films on the silicon substrates, the vacuum system was pumped down to a base pressure of 10^{-9} to 10^{-10} mbar. After this, the deposition pressure (Ar gas pressure) was kept constant at particular value of $1.35x10^{-3}$ mbar and all the experiments were performed at this constant pressure of Ar gas.

First a 10 nm thick buffer layer of was deposited on silicon substrates. For this a highly pure (purity = 99.999%) Ta-sputtering target was used and the DC power supply used for the buffer layer was 10 W. After this for the deposition of the $Ni_{81}Fe_{19}$ thin film, a highly pure (purity = 99.999%) compound sputtering target of $Ni_{81}Fe_{19}$ was used and the DC power supply used for this target was 14 W. After this, for the same deposition parameters as for buffer layer, a capping layer of Ta (10nm) was deposited over the top of the Permalloy thin film. Both the thin films were deposited at 8 SCCM of Ar-gas flow rate. During the in suit deposition of $Ni_{81}Fe_{19}$ thin films two hard magnets having a magnetic field of 250 gauss were placed parallel to thin film plane which helped us to aligned the spin of individual domains in the applied direction of external magnetic field as given in the Fig.2. The magnetic characterization of these coatings was carried out by the magneto transport system and vibrato sample measurement (VSM) system.

4. Results and Discussion

4.1. Effect of varying thickness of NiFe thin films on its AMR value

$Ni_{81}Fe_{19}$ thin films of various thicknesses were deposited on silicon substrates and the variations in the AMR ratio with respect to the thickness were measured in the magneto transport system. Fig .3 shows the dependence of the AMR value of the NiFe thin film on its varying thickness. On increasing the thickness of the NiFe, increase in the AMR value were observed, indicating a size effect in the resistivity which is strongly influenced by the orientation of the magnetization with respect to the current direction [1].

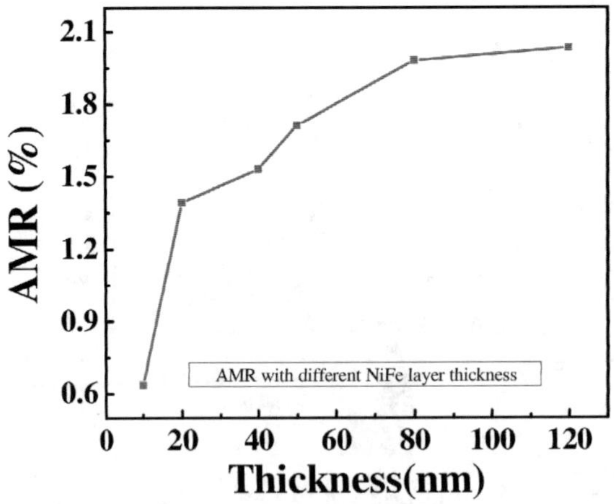

Fig.3. Variation of AMR ratio with respect to varying thickness of NiFe thin films.

4.2 Effect of varying thickness of Ta layer on NiFe thin film's AMR value

Ta/NiFe/Ta thin films were deposited on silicon substrates for varying Ta layer thickness and the effect of Ta layer thickness on the AMR values of the thin films was studied. Fig.4 shows the variation of the AMR values with respect to the Ta layer thickness. Up to 3 nm thickness of Ta layer increase in the AMR value was observed, however for the 4-5 nm thickness of the Ta layer, decrease in AMR value was observed. The reduction in AMR values is expected due the fact that the Ta interfaces, primarily, form magnetically dead layers with NiFe films and the result is the decrease in the total magnetic moment of NiFe film, further, the alloy phases also cause the degradation of the magnetocrystalline anisotropy and consequently leads to lower value of AMR in Permalloy thin films [2,3].

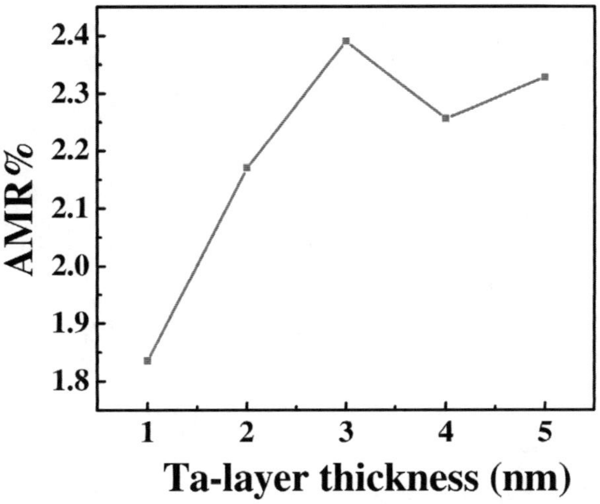

Fig.4 Variation of AMR ratio with respect to varying thickness of Ta thin films.

5. Conclusions

An indigenous UHV sputtering system was designed and developed for deposition of magnetic thin films. Permalloy ($Ni_{81}Fe_{19}$) magnetic thin films having Ta as a buffer and cap layer were deposited on the silicon substrates using this UHV sputtering system. The magneto transport studies confirmed that the AMR ratio is strongly depends on the thickness of the Ta and NiFe layers.

Acknowledgements

The authors thank the Director, NAL (CSIR) for giving permission to publish these results.

References

[1] Rijks G S M and Lenczowski S K J 1997 Phys. Rev. B. **56** 1
[2] Ding L, Teng J, Zhang Q, Feng C, Wang M L, Yu G H and Wang S 2009 Appl. Phys. Lett. **94**, 162506
[3] Kowalewski M, Butler W H, Moghadam N, Stocks G M and Schulthess T C, 2000 J. Appl. Phys. **87** 9

AUTHOR INDEX

AUTHOR INDEX

AUTHOR INDEX

Institute of Physics
Dirac House, Temple Back
Bristol BS1 6BE UK